OTHER VOLUMES IN THIS SERIES

OTHER VOLUMES IN THIS SERIES

Selected Tables in Mathematical Statistics

Volume 10

Selected Tables in Mathematical Statistics

Volume 10

Edited by the Institute of Mathematical Statistics

Coeditors
R. E. Odeh
University of Victoria

J. M. Davenport
Texas Tech University

Managing Editor
N. S. Pearson
Bell Communications Research

AMERICAN MATHEMATICAL SOCIETY
PROVIDENCE, RHODE ISLAND

1980 *Mathematics Subject Classification* (1985 *Revision*)
Primary 62Q05; Secondary 62E15, 62E99, 62G15, 62J99

International Standard Serial Number 0094-8837
International Standard Book Number 0-8218-1910-0
Library of Congress Card Number 74-6283

Contents

Preface

This volume of mathematical tables has been prepared under the aegis of the Institute of Mathematical Statistics. The Institute of Mathematical Statistics is a professional society for mathematically oriented statisticians. The purpose of the Institute is to encourage the development, dissemination, and application of mathematical statistics. The Committee on Mathematical Tables of the Institute of Mathematical Statistics is responsible for preparing and editing this series of tables. The Institute of Mathematical Statistics has entered into an agreement with the American Mathematical Society to jointly publish this series of volumes. At the time of this writing, submissions for future volumes are being solicited. No set number has been established for this series. The editors will consider publishing as many volumes as are necessary to disseminate meritorious material.

Potential authors should consider the following rules when submitting material.

1. The manuscript must be prepared by the author in a form acceptable for photo-offset. The author should assume that nothing will be set in type although the editors reserve the right to make editorial changes. This includes both the introductory material and the tables. A computer tape of the tables will be required for the checking process *and* the final printing (assuming it is accepted). The authors should contact the editors prior to submission concerning the requirements for the computer tape.

2. While there are no fixed upper and lower limits on the length of tables, authors should be aware that the purpose of this series is to provide an outlet for tables of high quality and utility which are too long to be accepted by a technical journal but too short for separate publication in book form.

3. The author must, whenever applicable, include in his introduction the following:

   (a) He should give the formula used in the calculation, and the computational procedure (or algorithm) used to generate his tables. Generally speaking FORTRAN or ALGOL programs will not be included but the description of the algorithm used should be complete enough that such programs can be easily prepared.

(b) A recommendation for interpolation in the tables should be given. The author should give the number of figures of accuracy which can be obtained with linear (and higher degree) interpolation.

(c) Adequate references must be given.

(d) The author should give the accuracy of the table and his method of rounding.

(e) In considering possible formats for his tables, the author should attempt to give as much information as possible in as little space as possible. Generally speaking, critical values of a distribution convey more information than the distribution itself, but each case must be judged on its own merits. The text portion of the tables (including column headings, titles, etc.) must be proportional to the size 5–1/4″ by 8–1/4″. Tables may be printed proportional to the size 8–1/4″ by 5–1/4″ (i.e., turned sideways on the page) when absolutely necessary; but this should be avoided and every attempt made to orient the tables in a vertical manner.

(f) The table should adequately cover the entire function. Asymptotic results should be given and tabulated if informative.

(g) Examples of the use of the tables must be included.

4. The author should submit as accurate a tabulation as he/she can. The table will be checked before publication, and any excess of errors will be considered grounds for rejection. The manuscript introduction will be subjected to refereeing. Since an inadequate introduction may lead to rejection, the author should strive for an informative manuscript, which not only establishes a need for the tables, but also explains in detail how to use the tables.

5. Authors having tables they wish to submit should send two copies to:

> Dr. Robert E. Odeh, Coeditor
> Department of Mathematics
> Univerity of Victoria
> Victoria, B. C., Canada V8W 2Y2

At the same time, a third copy should be sent to:

> James M. Davenport, Coeditor
> Department of Mathematics
> Texas Tech University
> P. O. Box 4319
> Lubbock, Texas 79409

Additional copies may be required, as needed for the editorial process. After the editorial process is complete, a camera-ready copy must be prepared for the publisher.

Authors should check several current issues of the *The Institute of Mathematical Statistics Bulletin* and *The AMSTAT News* for any up-to-date announcements about submissions to this series.

Acknowledgments

The tables included in the present volume were checked at the University of Victoria by Dr. R. E. Odeh. The editors and the Institute of Mathematical Statistics wish to express their great appreciation for this invaluable assistance. So many other people have contributed to the instigation and preparation of this volume that it would be impossible to record their names here. To all these people, who will remain anonymous, the editors and the Institute also wish to express their thanks.

Selected Tables in Mathematical Statistics
Volume 10, 1986

THE DISTRIBUTION OF POSITIVE
DEFINITE QUADRATIC FORMS

B.K.Shah*
N.Y.S. Department of Health
Helen Hayes Hospital
W. Haverstraw, N.Y.10993

ABSTRACT

Convergent expressions for the distribution function of positive definite quadratic forms of samples from the normal distribution with mean zero and variance one have been obtained in Laguerre form, power series and Ruben type respectively by Gurland (7), Pachares (12) and Ruben (13). A recurrence relation to calculate the coefficients in a Laguerre series representation was obtained by Shah (14). Since the Laguerre series representation is orthogonal to a chi-square density function, the Laguerre series will be used to tabulate the percentile points of the distribution of positive definite quadratic forms for n = 2 to 10.

INTRODUCTION

Let $y_1, y_2, \ldots, y_n$ be independently and normally distributed random variables with mean zero and variance one. Let A be a positive definite matrix of order nxn, and let $\underline{y}'(1xn) = (y_1, y_2, \ldots, y_n)$. The positive definite quadratic form $\underline{y}'A\underline{y}$ can be written as $Q = \lambda_1 z_1^2 + \lambda_2 z_2^2 + \ldots + \lambda_n z_n^2$, where the z_i's are linear functions of the y_i's and are also independent and normally distributed with mean zero and variance one, and

---

Received by the editors September 1982 and in revised form January 1984.
1980 Mathematics Subject Classification. Primary 62Q05; Secondary 62E99.
*This work was carried out when the author was at the Nathan Kline Institute for Psychiatric Research. Orangeburg, N.Y. 10962.

1

the λ_i's (i = 1,2,...,n) are the positive characteristic roots of the matrix A.

Gurland (7), Pachares (12) and Ruben (13) obtained various representations of the distribution of Q. Shah (14), however, obtained computationally convenient and manageable expression for the distribution of Q using a Laguerre series:

$$Pr\{ Q \leq x \} = \{\Gamma(m)\}^{-1} \int_0^{x/2\bar{\lambda}} e^{-y} y^{m-1} dy \tag{A}$$

$$+ \sum_{k=1}^{\infty} a_k \Gamma(k)\{\Gamma(k+m)\}^{-1} (\bar{\lambda})^{-k} e^{-x/2\bar{\lambda}}(x/2\bar{\lambda})^m L_{k-1}^{(m)}(x/2\bar{\lambda}),$$

where $m = n/2, \Gamma(m) = \int_0^{\infty} \exp(-x)x^{m-1}dx$, $\bar{\lambda} = (\max\lambda_j + \min\lambda_j)/2$,

$$L_r^{(m)}(x) = \sum_{j=0}^{r} (-x)^j \Gamma(m+r+1)\{\Gamma(m+j+1)\Gamma(j+1)\Gamma(r-j+1)\}^{-1}, \quad r = 0,1, \ldots,$$

$$a_0 = 1, \quad a_i = (1/i) \sum_{j=o}^{i-1} a_j b_{i-j}, \quad i = 1,2,\ldots, \text{ and}$$

$$b_i = (1/2)(\bar{\lambda} - \lambda_i)^{-1}.$$

It can be seen from equation (A) that if the λ_1's are all equal, then $a_i = 0$ (i = 1,2,...) and the equation (A) reduces to a chi-square distribution with n degrees of freedom except for a scalar factor.

During a computer experiment on an IBM 360/50 conducted early in 1965, we found that the infinite series (A) with $\bar{\lambda}$ = (max λ_j + min λ_j)/2 converges faster than that with any other value of $\bar{\lambda}$. Kotz et al.(10) arrived at the same conclusion mathematically in 1967.

TABLES

A computer program has been prepared utilizing equation (A) in the Fortran IV language in double precision. Without loss of generality, one can normalize λ's by setting $\Sigma\lambda_i = 1$. For a given n and λ's, the computer program obtains $p = \Pr(Q \leq x)$ at $x = 0.01(0.01)10.0$. Using the familiar Newton-Raphson iterative method, we then calculated the critical values corresponding to $p = 0.001, 0.005, 0.01(0.01)0.19, 0.20(0.05)0.80, 0.81(0.01)0.99, 0.995, 0.999$. Critical values are tabulated for n = 2(1)10 and are given for λ's in multiples of 0.05. For example, for n = 2, $\lambda_1 = 0.45$, $\lambda_2 = 0.55$ and p = 0.03, the critical value from Table 2.1 on page 18 is 0.03031. For convenience of printing, the λ's are multiplied by 100. In the last column of each n, the λ's are all equal to 100/n. If n =3, p = 0.8 and $\lambda_1 = \lambda_2 = \lambda_3 = 100/3$, the critical value from table 2.4 is 1.54721.

ACCURACY

The degree of accuracy in calculating probability integral depends upon the number of terms of equation (A) included in the computation. During a computer experiment, the following rule was discovered : In order to obtain an answer correct to 6+m decimal places, one needs to calculate at most $(m+n+1)(\max\lambda_j/\min\lambda_j)$ terms, m = 0,1,.... Therefore in order to obtain the value of the probability integral correct to six decimal places, one requires at most $(n+1)(\max\lambda_j/\min\lambda_j)$ terms. The following table orivides some results of the experiments.

TABLE A

| Values of n & λ's | x | No. of terms required to obtain accuracy to | | | | | |
| | | Six decimal places | | Seven decimal places | | Nine decimal places | |
		Actual	Using the rule	Actual	Using the rule	Actual	Using the rule
$n = 4$	0.5	24	25	30	30	37	40
$\lambda_1=0.5, \lambda_2=0.3$	2.5	23	25	30	30	33	40
$\lambda_3=0.1, \lambda_4=0.1$	4.5	21	25	23	30	34	40
$n = 4$ $\lambda_1=0.865, \lambda_2=0.045$ $\lambda_3=0.045, \lambda_4=0.045$	2.0	82	95	99	114	145	152
$n = 6$	0.5	21	28	26	32	38	40
$\lambda_1=0.4, \lambda_2=0.2$	2.5	23	28	28	32	36	40
$\lambda_3=\lambda_4=\lambda_5=\lambda_6=0.1$	4.5	17	28	21	32	32	40

We used the above rule to truncate the infinite series (A) and to calculate its value to six decimal places. The following table provides a comparison of the accuracy of the result with the values found by Imhof (8) and Davies (4) using the method of numerical integration.

TABLE B

| n | λ's | x | Pr($Q \leq x$) | | |
			Imhof $\varepsilon=0.001$	Davies	Eq.(A)
3	0.6, 0.3, 0.1	0.1	0.0543	0.0542	0.054214
3	0.6, 0.3, 0.1	0.7	0.4935	0.4936	0.493562
3	0.6, 0.3, 0.1	2.0	0.8760	0.8760	0.876041
6	0.3,0.3,0.15,0.15,0.05,0.05	0.1	0.0070	0.0064	0.006453
6	0.3,0.3,0.15,0.15,0.05,0.05	1.0	0.6003	0.6002	0.600205
6	0.3,0.3,0.15,0.15,0.05,0.05	3.0	0.9839	0.9838	0.983897

It can be seen from the above table that Imhof's approximation for $n = 6$ and $x = 0.1$ differs from Eq. (A) by five units in
the fourth decimal place.

The printed critical values were then obtained using the
Newton-Raphson iterative method and were rounded to the fifth
decimal place.

When the λ's are equal, the distribution of Q reduces to
that of a fixed scalar multiple of a chi-square distribution
with n degrees of freedom. The following table compares the
accuracy of percentile points of the tabled value for $n = 2$ and
$n = 10$ with that of the chi-square distribution. We have divided
the critical values of chi-square by $n = 1/\overline{\lambda}$ in order to compare
it with the tabled values. The critical values of chi-square
divided by n are correct to eight decimal places.

TABLE C

| | | Pr(Q $\leq$ x) = p | |
| | | x^2 percentage points divided by n | Tabled values |
n	p		
2	0.01	0.0100 5033	0.0100 5
	0.50	0.6931 4718	0.6931 5
	0.99	4.6051 7025	4.6051 7
10	0.01	0.2558 2122	0.2558 2
	0.50	0.9341 8178	0.9341 8
	0.99	2.3209 2512	2.3209 3

It can be seen from Table C that all the critical values
given in the fourth column are correct to five decimal places.

INTERPOLATION

For intermediate values of a given vector, $\underline{\lambda}'$ say, an approximate value can be obtained by selecting a tabulated vector $\underline{\lambda}$ that satisfies the following ordered set of rules:

(a) $\sum\limits_{1}^{n} (\lambda_i - \lambda_i')^2$ = minimum.

(b) $\sum\limits_{1}^{[n/2]} (\lambda_{n-i+1} + \lambda_i - \lambda_{n-i+1}' - \lambda_i')$ = maximum ,

where $[n/2]$ is the integer part of $n/2$.

(c)   Compare $|\lambda_{n-i+1} - \lambda_i|$ for each set of λ's until a minimum is found, i.e., one only proceeds to the next rule if the previous one fails to yield a unique tabulated vector $\underline{\lambda}$.

Approximate critical values of some selected vectors $\underline{\lambda}'$ were obtained by the above procedure, and these values were then compared with the exact values. The percent errors in approximating the upper percentile points were less than 1%. The following examples illustrate the procedure of selecting the vector $\underline{\lambda}$.

Example 1. We wish to find x such that $Pr(0.43Z_1^2 + 0.29Z_2^2 + 0.28Z_3^2 \leq x) = 0.95$. From the table of quadratic forms for $n = 3$, we need to find vectors satisfying rule (a). It is simple to verify that the vectors $(0.45, 0.30, 0.25)$ and $(0.40, 0.30, 0.30)$ yield the same minimum value. Table D compares the critical value obtained by the above procedure with that obtained through the use of Eq. (A).

TABLE D

i	λ_i'	Choices of λ_i	
		(1)	(2)
1	0.43	0.45	0.40
2	0.29	0.30	0.30
3	0.28	0.25	0.30
Rule (a)		0.0014	0.0014
Rule (b)		0.0100	0.0100
Rule (c)		0.2000	0.1000
x	2.63459	2.65132	2.61908

It can be seen from the above table that rule (a) and rule (b) provide identical values but rule (c) is satisfied, so that $\underline{\lambda}$ = (0.40,0.30,0.30) is used for approximating the critical value. The last row gives the exact critical values of Q for each $\underline{\lambda}$ and $\underline{\lambda}'$ and it can be seen that the relative error in selecting this vector is less than 0.70%.

Example 2. We wish to obtain x such that
$$Pr(0.425z_1^2 + 0.375z_2^2 + 0.175z_3^2 + 0.025z_4^2 \le x) = 0.95 .$$
The following table is self explanatory:

TABLE E

i	λ_i'	Choices of λ_i		
		(1)	(2)	(3)
1	0.425	0.40	0.45	0.40
2	0.375	0.40	0.35	0.35
3	0.175	0.15	0.15	0.20
4	0.025	0.05	0.05	0.05
Rule(a)		0.0025	0.0025	0.0025
Rule(b)		0.0000	0.1000	0.0000
x	2.65358	2.63511	2.64686	2.58374

It can be seen from the above table that rule (b) is satisfied, indicating that there is no need to satisfy rule (c). The second vector namely, $(0.45, 0.35, 0.15, 0.05)$ is the best candidate to approximate the percentile point. The relative percent error is less than 0.3%.

Example 3: We wish to obtain $\Pr(0.425Z_1^2 + 0.375Z_2^2 + 0.125Z_3^2 + 0.075Z_4^2 \le x) = 0.95$. The following table shows that there are four likely candidates:

TABLE F

i	λ_i'	\multicolumn{4}{c}{Choices of λ_i}
		(1)	(2)	(3)	(4)
1	0.425	0.40	0.40	0.45	0.45
2	0.375	0.40	0.40	0.35	0.35
3	0.125	0.10	0.15	0.10	0.10
4	0.075	0.10	0.05	0.10	0.05
Rule(a)		0.0025	0.0025	0.0025	0.0025
Rule(b)		0.0000	0.1000	0.1000	0.0000
Rule(c) $i=1$			0.3500	0.3500	
$i=2$			0.2500	0.2500	
x	2.63175	2.62672	2.63511	2.63594	2.64686

It can be seen that vectors 1 and 4 are eliminated by the application of rule (b). Vectors 2 and 3 are indistinguishable by rule (c) for $i = 1$ and $i = 2$. Thus we can select either of these two vectors, and the maximum percent error is less than 0.16%.

Example 4: We wish to calculate x such that
$$\Pr(0.75Z_1^2 + 0.125Z_2^2 + 0.075Z_3^2 + 0.05Z_4^2 \le x) = 0.95.$$
There are two candidates, which are given in the following table, for further selection:

TABLE G

i	λ_i'	Choices of λ_i	
		(1)	(2)
1	0.75	0.75	0.75
2	0.125	0.15	0.10
3	0.075	0.05	0.10
4	0.05	0.05	0.05
Rule (a)		0.00125	0.00125
Rule (b)		0.00000	0.00000
Rule (c) $_{i=1}$		0.70000	0.70000
$i=2$		0.10000	0.00000
x	3.15428	3.15927	3.15264

It can be seen that rule (a) and rule (b) are not able to discriminate between the two candidates and for i = 1, rule (c) also does not discriminate between them. However, at i = 2, we find that $|\lambda_3 - \lambda_2|$ is a minimum. We, therefore, select vector 2 for approximation. The relative percent error is less than 0.1%.

APPLICATIONS

Application 1. A general class of problems arises in missile firing operations when the hit probability of a weapon depends on the combination of several random errors. For simplicity consider errors in two dimensions. Let T be the true position of a target and I be the point of impact of a weapon aimed at A. Let (x_1,y_1) and (x_2,y_2) be the components of the vectors TA and AT respectively. If R is the radius of effectiveness of the weapon, then the probability of a hit p is the probability that the resultant vector TI has length no greater than R,

$$p = \Pr\{ x_3^2 + y_3^2 \le R^2 \} \tag{B}$$

where $(x_3, y_3) = (x_1 + x_2, y_1 + y_2)$.

Assume that each (x_k, y_k), $k = 1, 2$, is distributed according to a bivariate normal distribution with mean zero and variance covariance matrix $\| _k\sigma_{ij} \|$. Then (x_3, y_3) will have a bivariate normal distribution with zero means and variance covariance matrix $\| _1\sigma_{ij} + _2\sigma_{ij} \| = \| \lambda_{ij} \|$. For this illustration assume that the components of each error are independent, i.e., $\| _k\sigma_{ij} \|$, $k = 1, 2$, is diagonal. If $x_3 = x\sqrt{\lambda_{11}}$ and $y_3 = y\sqrt{\lambda_{22}}$. then x^2 and y^2 each have a chi-square distribution with one degree of freedom. We may write (B) as

$$p = \Pr\{ a_1 x^2 + a_2 y^2 \le t \},$$

where $a_i = \lambda_{ii}/\sigma^2$, $\sigma^2 = \lambda_{11} + \lambda_{22}$ and $t = R^2/\sigma^2$.

Assume that $_1\sigma_{11} = 100$, $_1\sigma_{22} = 400$, $_2\sigma_{11} = 100$ and $_2\sigma_{22} = 1400$. The interest is to obtain the radius of effectiveness of the weapon so that the probability of hit is 95%.

We first calculate the a_{ii}'s. Since $\lambda_{11} = _1\sigma_{11} + _2\sigma_{11} = 200$, $\lambda_{22} = _1\sigma_{22} + _2\sigma_{22} = 1800$ and $\sigma^2 = \lambda_{11} + \lambda_{22} = 2000$, we have $a_1 = \lambda_{11}/\sigma^2 = 0.1$ and $a_2 = \lambda_{22}/\sigma^2 = 0.9$. For n = 2, $a_1 = 0.10$, $a_2 = 0.90$ and p = 0.95, the tabulated value is 3.56514. Thus

$$\Pr\{ a_1 x^2 + a_2 y^2 \le 3.56514 \} = 0.95.$$

Hence $R^2/\sigma^2 = 3.56514$ giving R = 84.44099.

Application 2. Consider $_1\sigma_{11} = 100$, $_1\sigma_{22} = 400$, $_2\sigma_{11} = 100$, $_2\sigma_{22} = 1400$ and R = 40 and the framework of Application 1. The interest is to find the probability of hit.

We find $a_1 = 0.9$, $a_2 = 0.1$ and $t = R^2/\sigma^2 = 40^2/2000 = 0.8$. The value of $\Pr\{ 0.9x^2 + 0.1y^2 \le 0.8 \}$ can be obtained from the table using an inverse interpolation formula given in Aitken (1).

Applying Lagrange's interpolation formula, we found this value to be 0.615887, which agrees with Grad and Solomon (5) who gave values to four decimal places.

Application 3. Chernoff and Lehman (2) have shown that when the maximum likelihood estimates of parameters in the chi-square test of goodness of fit are based on the individual observations rather than on the cell frequencies, then the asymptotic distribution of $\sum_{i=1}^{k} (O_i - E_i)^2/E_i$ is that of $\sum_{i=1}^{k-s-1} x_i^2 + \sum_{i=k-s}^{k-1} \theta_i x_i^2$, where O_i and E_i are observed and expected frequencies in the ith cell, k is the number of cells, x_i's are independently and normally distributed random variables with mean zero and variance one, s is the number of parameters to be estimated and the θ_i are the roots of a determinantal equation $(0<\theta_i<1)$.

Chernoff and Lehman considered the fitting of a normal distribution with mean μ and variance σ^2 to grouped data. In the case of four cells $(-\infty,-1),(-1,0),(0,1),(1,\infty)$ and $\mu = 0$ and $\sigma = 2.5$, the following values for the two roots were obtained:

$$\theta_1 = 0.80 \quad \text{and} \quad \theta_2 = 0.20.$$

The probability α is then given by

$$\Pr \{x_1^2 + 0.8x_2^2 + 0.2x_3^2 \geq x\} = \alpha.$$

This can be easily standardized so that the coefficients sum to one and its value can be obtained from the table. We find

$$\Pr \{0.5x_1^2 + 0.4x_2^2 + 0.1x_3^2 \geq x/2\} = \alpha.$$

For n = 3 and $\alpha = 0.05$ (i.e. p = 0.95 in the table), we find x/2 = 2.81754 or x = 5.63508. The critical value for the chi-square distribution with one degree of freedom at $\alpha = 0.05$ is 3.84146, indicating that in the normal case the use of maximum likelihood estimates in forming χ^2 may lead to a serious

underestimate of the critical value. For more than four cells,
which is usually the case in practical situations, the table for
the distribution of positive definite quadratic forms can be very
helpful in finding critical values.

Application 4. The exponential distribution as a failure model
plays an important role in reliability studies. Consider the
exponential life distribution with location parameter $\mu \geq 0$ and
scale parameter $\theta > 0$. Suppose that a sample of size c from a
population with this life distribution has been subjected to life
testing and that the test is terminated when the rth failure
occurs, $r \leq c$. This is known as Type II censoring in which the
number of failures is considered fixed, and the only random
variates are the failure times. Let $x_{(1)}, x_{(2)}, \ldots, x_{(r)}$ be the
time of failures arranged in an ascending order. We wish to
determine a lower confidence bound for $R(t) = \exp\{-(t-\mu)/\theta\}$, the
population survival proportion at time t. The maximum-likelihood
estimators of μ and θ are :

$$\hat{\mu} = x_{(1)} \quad \text{and} \quad \hat{\theta} = \{ \sum_{i=1}^{r} x_{(i)} + (c - r)x_{(r)} - cx_{(1)} \}/r,$$

respectively. Grubbs (6) has shown that

$$\Pr\{ \ln(1/R(t)) > \eta\} = \Pr\{ (x_{(1)} - \mu + Z(r - 1)\theta*)/\theta > \eta \},$$

where $Z = (t - x_{(1)})/[(r-1)\theta*]$ and $\theta* = r\hat{\theta}/(r-1)$, and that
$Q = \ln(1/R(t)) = (\chi^2(2) + Zc\chi^2(2r-2))/2c$, since $2c(x_{(1)} - \mu)/\theta$
and $2(r-1)\theta*/\theta$ are independently distributed as chi-square
random variables with 2 and (2r-2) degrees of freedom, respectively. If Z is considered fixed, then the distribution of Q is a
weighted sum of chi-square variates. When Z is positive, the
tables of the distribution of positive definite quadratic forms
can be used to find the percentile points of Q and subsequently
that of R(t).

As an illustration, let five components be subjected to a life test, where test is to be terminated at the second failure. And let $x_{(1)} = 36$ and $x_{(2)} = 57.5$ be the observed failure times in hours (say), and the interest is to determine the 90th percentile point for $R(48)$. Using this information, we find that $Z = 0.13953$. Noting Cochran's theorem (3), we write

$$Pr\{(\chi^2(2) + Zc\chi^2(2))/2c > \eta\}$$
$$= Pr\{(\chi^2(1) + \chi^2(1)/2c + Z(\chi^2(1) + \chi^2(1))/2 > \eta\}$$
$$= Pr\{0.1(\chi^2(1) + \chi^2(1)) + 0.06977(\chi^2(1) + \chi^2(1)) > \eta\}$$
$$= Pr\{0.29452(\chi^2(1) + \chi^2(1)) + 0.20548(\chi^2(1) + \chi^2(1)) > x\},$$

where $x = 2.94521\eta$. It can be seen from the above that Q is a weighted sum of four chi-square variates with one degree of freedom. Therefore, we need to find a table of positive definite quadratic form for $n = 4$. From the table, the closest vector to the coefficients of the four chi-square variates is $(0.30, 0.30, 0.20, 0.20)$. For $\lambda_1 = \lambda_2 = 0.30$, $\lambda_3 = \lambda_4 = 0.20$ and $p = 0.10$, we find $x = 0.26149$. Hence $\eta = 0.26149/2.94521 = 0.08878$. Thus

$$Pr\{ Q > 0.08878\} = Pr\{ R(48) > 0.91504\}.$$

Using the normality approximation given by Mann et al. (11), the percentile point is 0.9185, while the exact value is 0.91522.

Application 5. Let $X_1, \ldots, X_n$ be independently and normally distributed variables with mean μ and variance σ^2. When a trend underlies a sequence of observed values $X_1, X_2, \ldots, X_n$, the usual mean square estimator of variance $\sum_1^n (X_i - \bar{X})^2/(n-1)$ can be replaced by the mean square successive difference δ^2 (Von Neumann et al. (16)), where $\delta^2 = \sum_1^{n-1} (X_{i+1} - X_i)^2/2(n-1)$. The δ^2 does not eliminate the effect of trend, but it reduces the effect considerably more than the usual mean square estimator.

Using an orthogonal transformation on the X's, the statistic δ^2 can be reduced to a linear function of chi-square variates giving

$$\delta^2 = \sum_1^{n-1} \lambda_i y_i^2 ,$$

where y_i (i = 1,2,...,n-1) are normally and independently distributed with mean zero and variance σ^2 , and λ_j = {1 - cos(jπ/n)}/(n-1), j = 1,2,...,n-1. It can be seen that the statistic δ^2 is a quadratic form in normal samples, so that the tables of positive definite quadratic forms can be used to set confidence limits on δ^2 when the data are not a random sample from a population with mean μ.

Application 6.   In comparisons among class means, any linear combination

$$L = d_1\bar{y}_1 + \ldots + d_k\bar{y}_k,$$

is called a comparison of treatment means if $\sum_1^k d_i = 0$,where each d_i is fixed, $\bar{y}_i = \sum_{j=1}^{n_i} y_{ij}/n_i$, n_i is the number of replications in the ith class and the observations y_{ij} are assumed to be independently and normally distributed with mean μ_i and variance one. For any comparison $\sum d_i\bar{y}_i$ among the class means,an unbiased estimate of its variance is

$$V = \sum_1^k d_i s_i^2/n_i = \sum_1^k d_i^2 S_i^2/n_i\nu_i$$

(see Snedecor and Cochran (15) p.269,324), where $\nu_i = n_i - 1$, s_i^2 is the mean square within the ith class and $S_i^2 = \nu_i s_i^2$. Since $d_i^2/n_i\nu_i$ is a positive quantity and S_i^2 is distributed as a chi-square random variable with ν_i degrees of freedom, the distribution of the V-statistic is the same as that of Q given in (A).

There are situations in which it becomes necessary to test

whether a sample value V is consistent with the postulated population value of σ_V^2. This problem arises in applications where σ_V^2 has been obtained from a very large sample and may be assumed known. Let the null hypothesis value be $\sigma_V^2 = \sigma_{V_o}^2$. Usually, the tests wanted are one-tailed tests. When the alternative is $\sigma_V^2 > \sigma_{V_o}^2$, compute

$$Q = \Sigma \, d_i^2 \, s_i^2 / n_i \nu_i \sigma_{V_o}^2 = \Sigma \lambda_i s_i^2 \, ,$$

where $\lambda_i = d_i^2 / n_i \nu_i \sigma_{V_o}^2$. If Q exceeds the tabulated value of $Q_{0.05}$, then the test is significant at the 5% level.

For more applications, readers are encouraged to refer to Johnson and Kotz (9).

ACKNOWLEDGEMENTS

The author is indebted to Professors F.J.Anscombe, L.J. Savage and C.G. Khatri for their encouragements. I am also thankful to Professors R.E. Odeh and James Davenport for their valuable assistance in preparing these tables. Thanks also goes to Professor James Dickey, Dr. Margarett Hoff, Dr. N. Shirlene Pearson, Mr. Joseph Wanderling and Mr. Neil Shah for their assistance in proof reading and programming.

The work was supported in part by the Office of Naval Research and by the General Research Support Grant from the Department of Human Health Services.

REFERENCES

1. Aitken, A.C. (1932). On interpolation by iteration of proportional parts, without the use of differences. Proc. Edinburgh Math. Soc.3, 56-76.

2. Chernoff,H., and Lehmann, E.L.(1954). The use of maximum likelihood estimates in χ^2 tests for goodness of fit. Ann. Math. Stat. 25, 579-586.

3. Cochran, W.G. (1934). The distribution of quadratic forms in
    a normal system, with applications to the analysis of
    variance. Proc. Camb. Phil. Soc. 30,178-198.

4. Davies, R.B. (1980). The distribution of a linear combination
    of χ^2 random variables. Applied Statistics 29, 323-333.

5. Grad, A., and Solomon, H. (1955). Distribution of quadratic
    forms and some applications. Ann. Math. Stat. 26, 464-477.

6. Grubbs,F.E. (1971).Fiducial Bounds on Reliability for the
    Two Parameter Negative Exponential Distribution. Technome-
    trics,13, 873-876.

7. Gurland, J. (1955). Distribution of definite and of
    indefinite quadratic forms. Ann. Math. Stat. 26, 122-127,
    Corrections in Ann. Math. Stat. 33, 813 (1962).

8. Imhof, J.P. (1961). Computing the distribution of quadratic
    forms in normal variables. Biometrika 48,419-426.

9. Johnson, N.L., and Kotz, S. (1972). Distributions in
    Statistics. Vol.2. John Wiley & Sons Inc. New York.

10. Kotz, S., Johnson, N.L., and Boyd, D.W. (1967). Series
    representations of distributions of quadratic forms in
    normal variables.I. Central case. Ann. Math. Stat. 38,823-
    837.

11. Mann,N.R., Schafer, R.E., and Singpurwalla N.D. (1974).
    Methods for Statistical Analysis of Reliability and life
    Data. John Wiley & Sons Inc. New York.

12. Pachares, J. (1955). Note on the distribution of a definite
    quadratic form. Ann. Math. Stat. 26, 128-131.

13. Ruben, H. (1960). Probability content of regions under
    spherical normal distributions, I. Ann. Math. Stat. 31,
    598-618.

14. Shah, B.K. (1968). Distribution theory of quadratic forms.
    Ph.D. dissertation presented to the Faculty of the Graduate
    School of Yale University.

15. Snedecor, G.W., and Cochran, W.G. (1967). Statistical
    Methods. The Iowa State University Press, Ames, Iowa.

16. Von Neumann, J., Kent, R.H., Bellinson, H.R., and Hart, B.I.
    (1941). The mean square successive difference. Ann. Math.
    Stat. 12, 153-162.

The tables calculate the critical values of a positive definite quadratic form $Q = \Sigma \, \lambda_i z_i^2$, where $Z_1, Z_2, \ldots, Z_n$ are independently and normally distributed random variables with mean zero and variance one, and λ_i's ($i = 1, 2, \ldots, n$) are positive values. For a given n,p and a set of λ's, the tables give the value of x (say x_p) such that $\Pr\{ \, Q \leq x_p \, \} = p$.

TABLE INDEX

Sample size (n)	Table no.	Page nos.
2	1	18-19
3	2	20-26
4	3	27-39
5	4	40-56
6	5	57-75
7	6	76-92
8	7	93-107
9	8	108-118
10	9	119-127

TABLE 1.1 N= 2

P ↓	λ: 5 95	10 90	15 85	20 80	25 75
0.001	0.00044	0.00060	0.00071	0.00080	0.00087
0.005	0.00219	0.00301	0.00358	0.00401	0.00434
0.010	0.00441	0.00605	0.00719	0.00805	0.00871
0.020	0.00892	0.01220	0.01449	0.01620	0.01752
0.030	0.01354	0.01846	0.02189	0.02446	0.02644
0.040	0.01828	0.02483	0.02939	0.03283	0.03546
0.050	0.02313	0.03131	0.03701	0.04130	0.04460
0.060	0.02809	0.03791	0.04474	0.04989	0.05384
0.070	0.03319	0.04462	0.05259	0.05859	0.06320
0.080	0.03841	0.05146	0.06056	0.06741	0.07268
0.090	0.04377	0.05842	0.06865	0.07635	0.08228
0.100	0.04928	0.06551	0.07686	0.08541	0.09199
0.110	0.05492	0.07274	0.08521	0.09460	0.10184
0.120	0.06072	0.08011	0.09369	0.10393	0.11181
0.130	0.06668	0.08762	0.10230	0.11338	0.12192
0.140	0.07280	0.09527	0.11106	0.12297	0.13216
0.150	0.07910	0.10308	0.11996	0.13271	0.14254
0.160	0.08557	0.11104	0.12901	0.14259	0.15306
0.170	0.09224	0.11917	0.13821	0.15261	0.16372
0.180	0.09909	0.12747	0.14757	0.16279	0.17454
0.190	0.10615	0.13593	0.15709	0.17313	0.18551
0.200	0.11342	0.14458	0.16678	0.18363	0.19664
0.250	0.15324	0.19073	0.21793	0.23870	0.25480
0.300	0.19982	0.24239	0.27413	0.29859	0.31763
0.350	0.25466	0.30069	0.33630	0.36410	0.38588
0.400	0.31955	0.36699	0.40558	0.43625	0.46048
0.450	0.39653	0.44308	0.48346	0.51635	0.54264
0.500	0.48800	0.53122	0.57187	0.60610	0.63392
0.550	0.59684	0.63438	0.67337	0.70778	0.73639
0.600	0.72681	0.75648	0.79150	0.82451	0.85288
0.650	0.88308	0.90291	0.93121	0.96072	0.98744
0.700	1.07335	1.08136	1.09978	1.12301	1.14603
0.750	1.30968	1.30358	1.30848	1.32168	1.33805
0.800	1.61255	1.58912	1.57610	1.57416	1.57943
0.810	1.68399	1.65655	1.63932	1.63356	1.63588
0.820	1.75995	1.72829	1.70659	1.69670	1.69576
0.830	1.84096	1.80483	1.77837	1.76402	1.75947
0.840	1.92764	1.88674	1.85524	1.83605	1.82751
0.850	2.02073	1.97473	1.93784	1.91341	1.90046
0.860	2.12111	2.06965	2.02699	1.99686	1.97900
0.870	2.22989	2.17253	2.12367	2.08732	2.06402
0.880	2.34843	2.28467	2.22909	2.18595	2.15657
0.890	2.47845	2.40768	2.34481	2.29420	2.25801
0.900	2.62218	2.54369	2.47280	2.41396	2.37008
0.910	2.78253	2.69545	2.61569	2.54768	2.49510
0.920	2.96350	2.86675	2.77705	2.69874	2.63619
0.930	3.17068	3.06289	2.96188	2.87185	2.79778
0.940	3.41229	3.29164	3.17753	3.07394	2.98632
0.950	3.70113	3.56514	3.43544	3.31578	3.21191
0.960	4.05870	3.90375	3.75486	3.61548	3.49151
0.970	4.52550	4.34583	4.17200	4.00713	3.85703
0.980	5.19293	4.97797	4.76862	4.56767	4.38054
0.990	6.35473	6.07843	5.80747	5.54433	5.29369
0.995	7.53701	7.19835	6.86486	6.53883	6.22444
0.999	10.33768	9.85143	9.37011	8.89582	8.43212

TABLE 1.2 N= 2 (CONT.)

P ↓	λ: 30 70	35 65	40 60	45 55	All λ's = 100/2
0.001	0.00092	0.00095	0.00098	0.00100	0.00100
0.005	0.00460	0.00478	0.00491	0.00499	0.00501
0.010	0.00922	0.00959	0.00985	0.01000	0.01005
0.020	0.01853	0.01928	0.01980	0.02010	0.02020
0.030	0.02796	0.02908	0.02985	0.03031	0.03046
0.040	0.03748	0.03898	0.04001	0.04062	0.04082
0.050	0.04712	0.04899	0.05028	0.05104	0.05129
0.060	0.05687	0.05911	0.06066	0.06157	0.06188
0.070	0.06673	0.06935	0.07116	0.07222	0.07257
0.080	0.07671	0.07970	0.08177	0.08298	0.08338
0.090	0.08681	0.09017	0.09250	0.09386	0.09431
0.100	0.09703	0.10076	0.10334	0.10486	0.10536
0.110	0.10737	0.11148	0.11432	0.11598	0.11653
0.120	0.11785	0.12232	0.12542	0.12723	0.12783
0.130	0.12845	0.13330	0.13664	0.13861	0.13926
0.140	0.13919	0.14440	0.14801	0.15012	0.15082
0.150	0.15006	0.15564	0.15950	0.16177	0.16252
0.160	0.16108	0.16703	0.17114	0.17356	0.17435
0.170	0.17223	0.17855	0.18292	0.18548	0.18633
0.180	0.18354	0.19022	0.19484	0.19755	0.19845
0.190	0.19500	0.20204	0.20691	0.20978	0.21072
0.200	0.20661	0.21402	0.21914	0.22215	0.22314
0.250	0.26716	0.27635	0.28271	0.28645	0.28768
0.300	0.33229	0.34320	0.35076	0.35521	0.35667
0.350	0.40270	0.41525	0.42396	0.42909	0.43078
0.400	0.47930	0.49337	0.50315	0.50892	0.51083
0.450	0.56320	0.57863	0.58939	0.59574	0.59784
0.500	0.65586	0.67244	0.68403	0.69088	0.69315
0.550	0.75924	0.77664	0.78886	0.79611	0.79851
0.600	0.87598	0.89375	0.90632	0.91381	0.91629
0.650	1.00981	1.02732	1.03983	1.04733	1.04982
0.700	1.16628	1.18258	1.19442	1.20158	1.20397
0.750	1.35404	1.36766	1.37788	1.38417	1.38629
0.800	1.58778	1.59632	1.60335	1.60788	1.60944
0.810	1.64211	1.64922	1.65532	1.65934	1.66073
0.820	1.69962	1.70510	1.71016	1.71359	1.71480
0.830	1.76069	1.76433	1.76819	1.77096	1.77196
0.840	1.82575	1.82732	1.82983	1.83183	1.83258
0.850	1.89535	1.89456	1.89552	1.89665	1.89712
0.860	1.97013	1.96665	1.96584	1.96597	1.96611
0.870	2.05088	2.04432	2.04148	2.04045	2.04022
0.880	2.13859	2.12850	2.12330	2.12093	2.12026
0.890	2.23450	2.22034	2.21239	2.20845	2.20727
0.900	2.34023	2.32132	2.31015	2.30436	2.30259
0.910	2.45791	2.43341	2.41843	2.41044	2.40795
0.920	2.59042	2.55930	2.53974	2.52909	2.52573
0.930	2.74184	2.70274	2.67759	2.66369	2.65926
0.940	2.91816	2.86926	2.83717	2.81919	2.81341
0.950	3.12872	3.06748	3.02651	3.00325	2.99573
0.960	3.38923	3.31190	3.25911	3.22875	3.21888
0.970	3.72932	3.62984	3.56037	3.51983	3.50656
0.980	4.21594	4.08306	3.98755	3.93078	3.91202
0.990	5.06458	4.87044	4.72455	4.63514	4.60517
0.995	5.93018	5.67165	5.46954	5.34181	5.29832
0.999	7.98611	7.57356	7.22655	6.99122	6.90776

TABLE 2.1 N= 3

P ↓	λ: 5	5 90	10	5 85	15	5 80	20	5 75	25	5 70
0.001		0.00320		0.00395		0.00443		0.00477		0.00503
0.005		0.00951		0.01173		0.01316		0.01418		0.01494
0.010		0.01533		0.01890		0.02120		0.02285		0.02407
0.020		0.02496		0.03072		0.03446		0.03716		0.03915
0.030		0.03345		0.04109		0.04609		0.04970		0.05238
0.040		0.04135		0.05072		0.05688		0.06135		0.06468
0.050		0.04892		0.05990		0.06717		0.07245		0.07639
0.060		0.05627		0.06879		0.07713		0.08320		0.08774
0.070		0.06349		0.07749		0.08686		0.09371		0.09883
0.080		0.07063		0.08607		0.09645		0.10404		0.10974
0.090		0.07772		0.09456		0.10593		0.11427		0.12054
0.100		0.08481		0.10300		0.11535		0.12443		0.13126
0.110		0.09190		0.11142		0.12474		0.13455		0.14194
0.120		0.09903		0.11985		0.13412		0.14466		0.15261
0.130		0.10620		0.12830		0.14351		0.15478		0.16328
0.140		0.11344		0.13680		0.15294		0.16492		0.17398
0.150		0.12076		0.14534		0.16242		0.17511		0.18472
0.160		0.12817		0.15396		0.17195		0.18536		0.19552
0.170		0.13568		0.16265		0.18156		0.19567		0.20638
0.180		0.14330		0.17144		0.19125		0.20607		0.21733
0.190		0.15105		0.18032		0.20103		0.21656		0.22836
0.200		0.15894		0.18932		0.21092		0.22715		0.23950
0.250		0.20078		0.23633		0.26226		0.28193		0.29699
0.300		0.24768		0.28758		0.31754		0.34052		0.35822
0.350		0.30120		0.34434		0.37788		0.40395		0.42415
0.400		0.36318		0.40808		0.44452		0.47332		0.49583
0.450		0.43580		0.48056		0.51896		0.54997		0.57446
0.500		0.52163		0.56403		0.60307		0.63555		0.66157
0.550		0.62374		0.66137		0.69932		0.73222		0.75913
0.600		0.74592		0.77635		0.81100		0.84291		0.86978
0.650		0.89322		0.91413		0.94279		0.97175		0.99730
0.700		1.07292		1.08208		1.10152		1.12487		1.14725
0.750		1.29640		1.29135		1.29775		1.31187		1.32833
0.800		1.58301		1.56047		1.54921		1.54901		1.55534
0.810		1.65064		1.62406		1.60860		1.60474		1.60834
0.820		1.72255		1.69172		1.67179		1.66396		1.66453
0.830		1.79925		1.76391		1.73922		1.72708		1.72429
0.840		1.88132		1.84118		1.81143		1.79459		1.78807
0.850		1.96946		1.92420		1.88904		1.86707		1.85640
0.860		2.06451		2.01376		1.97280		1.94524		1.92994
0.870		2.16752		2.11085		2.06365		2.02997		2.00950
0.880		2.27978		2.21668		2.16272		2.12233		2.09605
0.890		2.40292		2.33279		2.27148		2.22368		2.19087
0.900		2.53904		2.46117		2.39179		2.33579		2.29558
0.910		2.69091		2.60444		2.52613		2.46097		2.41233
0.920		2.86232		2.76616		2.67784		2.60238		2.54402
0.930		3.05855		2.95133		2.85165		2.76443		2.69478
0.940		3.28741		3.16731		3.05447		2.95362		2.87062
0.950		3.56101		3.42555		3.29706		3.18004		3.08095
0.960		3.89972		3.74528		3.59755		3.46069		3.34157
0.970		4.34190		4.16274		3.99001		3.82752		3.68225
0.980		4.97416		4.75969		4.55138		4.35264		4.17019
0.990		6.07475		5.79892		5.52895		5.26781		5.02147
0.995		7.19476		6.85656		6.52402		6.19989		5.88947
0.999		9.84797		9.36216		8.88175		8.40921		7.94904

TABLE 2.2 N= 3 (CONT.)

P ↓	λ:	30	5 65	35	5 60	40	5 55	45	5 50	10 10	10 80
0.001		0.00521		0.00534		0.00543		0.00547		0.00487	
0.005		0.01549		0.01588		0.01614		0.01626		0.01443	
0.010		0.02497		0.02560		0.02601		0.02621		0.02319	
0.020		0.04062		0.04167		0.04234		0.04266		0.03755	
0.030		0.05437		0.05577		0.05667		0.05712		0.05006	
0.040		0.06713		0.06888		0.07000		0.07055		0.06160	
0.050		0.07931		0.08139		0.08272		0.08338		0.07254	
0.060		0.09110		0.09350		0.09504		0.09579		0.08307	
0.070		0.10263		0.10534		0.10708		0.10794		0.09332	
0.080		0.11398		0.11699		0.11894		0.11989		0.10337	
0.090		0.12520		0.12852		0.13067		0.13172		0.11327	
0.100		0.13635		0.13998		0.14232		0.14347		0.12306	
0.110		0.14745		0.15138		0.15392		0.15517		0.13278	
0.120		0.15854		0.16277		0.16551		0.16685		0.14246	
0.130		0.16963		0.17417		0.17710		0.17854		0.15211	
0.140		0.18075		0.18559		0.18871		0.19025		0.16177	
0.150		0.19191		0.19705		0.20037		0.20200		0.17144	
0.160		0.20312		0.20856		0.21208		0.21381		0.18114	
0.170		0.21440		0.22015		0.22386		0.22569		0.19088	
0.180		0.22577		0.23181		0.23572		0.23765		0.20067	
0.190		0.23722		0.24357		0.24768		0.24969		0.21053	
0.200		0.24877		0.25542		0.25973		0.26184		0.22047	
0.250		0.30833		0.31648		0.32176		0.32436		0.27165	
0.300		0.37159		0.38122		0.38746		0.39054		0.32616	
0.350		0.43947		0.45053		0.45772		0.46126		0.38514	
0.400		0.51297		0.52538		0.53346		0.53744		0.44992	
0.450		0.59323		0.60687		0.61576		0.62015		0.52206	
0.500		0.68168		0.69635		0.70595		0.71070		0.60357	
0.550		0.78015		0.79560		0.80575		0.81078		0.69708	
0.600		0.89114		0.90699		0.91745		0.92266		0.80611	
0.650		1.01814		1.03383		1.04429		1.04951		0.93561	
0.700		1.16631		1.18103		1.19098		1.19598		1.09266	
0.750		1.34370		1.35617		1.36483		1.36926		1.28808	
0.800		1.56394		1.57207		1.57816		1.58139		1.53965	
0.810		1.61505		1.62194		1.62729		1.63016		1.59916	
0.820		1.66912		1.67460		1.67910		1.68157		1.66251	
0.830		1.72649		1.73038		1.73391		1.73593		1.73014	
0.840		1.78758		1.78967		1.79209		1.79359		1.80256	
0.850		1.85288		1.85291		1.85407		1.85497		1.88040	
0.860		1.92300		1.92068		1.92039		1.92060		1.96441	
0.870		1.99865		1.99364		1.99169		1.99110		2.05551	
0.880		2.08077		2.07266		2.06877		2.06725		2.15485	
0.890		2.17050		2.15879		2.15264		2.15005		2.26389	
0.900		2.26933		2.25342		2.24462		2.24075		2.38448	
0.910		2.37923		2.35838		2.34643		2.34103		2.51909	
0.920		2.50290		2.47614		2.46039		2.45315		2.67108	
0.930		2.64408		2.61017		2.58978		2.58028		2.84515	
0.940		2.80834		2.76560		2.73942		2.72708		3.04823	
0.950		3.00431		2.95038		2.91676		2.90074		3.29108	
0.960		3.24655		3.17791		3.13433		3.11335		3.59180	
0.970		3.56248		3.47341		3.41566		3.38754		3.98449	
0.980		4.01415		3.89385		3.81376		3.77417		4.54610	
0.990		4.80125		4.62257		4.49851		4.43563		5.52393	
0.995		5.60389		5.36259		5.18834		5.09770		6.51917	
0.999		7.51076		7.11647		6.80823		6.63739		8.87712	

TABLE 2.3 N= 3 (CONT.)

P ↓	λ: 15	10 75	20	10 70	25	10 65	30	10 60	35	10 55
0.001		0.00545		0.00587		0.00616		0.00638		0.00652
0.005		0.01615		0.01737		0.01825		0.01888		0.01931
0.010		0.02594		0.02788		0.02930		0.03031		0.03100
0.020		0.04195		0.04508		0.04736		0.04900		0.05011
0.030		0.05587		0.06001		0.06304		0.06522		0.06669
0.040		0.06869		0.07376		0.07747		0.08014		0.08195
0.050		0.08082		0.08676		0.09110		0.09424		0.09636
0.060		0.09248		0.09924		0.10420		0.10778		0.11020
0.070		0.10381		0.11135		0.11690		0.12091		0.12363
0.080		0.11488		0.12320		0.12931		0.13373		0.13674
0.090		0.12578		0.13483		0.14151		0.14633		0.14962
0.100		0.13654		0.14632		0.15353		0.15876		0.16232
0.110		0.14721		0.15769		0.16544		0.17106		0.17488
0.120		0.15781		0.16899		0.17726		0.18326		0.18735
0.130		0.16836		0.18023		0.18901		0.19540		0.19975
0.140		0.17890		0.19143		0.20073		0.20749		0.21210
0.150		0.18943		0.20263		0.21243		0.21955		0.22442
0.160		0.19998		0.21382		0.22412		0.23162		0.23673
0.170		0.21055		0.22504		0.23583		0.24369		0.24906
0.180		0.22116		0.23628		0.24756		0.25578		0.26140
0.190		0.23182		0.24757		0.25933		0.26791		0.27378
0.200		0.24254		0.25891		0.27115		0.28009		0.28620
0.250		0.29743		0.31678		0.33135		0.34202		0.34935
0.300		0.35525		0.37739		0.39417		0.40653		0.41502
0.350		0.41710		0.44179		0.46065		0.47461		0.48423
0.400		0.48415		0.51109		0.53186		0.54731		0.55800
0.450		0.55781		0.58658		0.60901		0.62580		0.63746
0.500		0.63983		0.66985		0.69358		0.71150		0.72400
0.550		0.73250		0.76293		0.78747		0.80619		0.81933
0.600		0.83891		0.86861		0.89321		0.91226		0.92575
0.650		0.96342		0.99075		1.01434		1.03302		1.04643
0.700		1.11245		1.13507		1.15605		1.17331		1.18597
0.750		1.29598		1.31051		1.32644		1.34064		1.35151
0.800		1.53069		1.53221		1.53924		1.54767		1.55499
0.810		1.58610		1.58422		1.58882		1.59562		1.60191
0.820		1.64506		1.63947		1.64134		1.64630		1.65142
0.830		1.70799		1.69832		1.69716		1.70004		1.70384
0.840		1.77538		1.76125		1.75670		1.75722		1.75952
0.850		1.84781		1.82879		1.82044		1.81830		1.81888
0.860		1.92601		1.90161		1.88900		1.88383		1.88244
0.870		2.01085		1.98052		1.96312		1.95449		1.95082
0.880		2.10339		2.06651		2.04370		2.03111		2.02482
0.890		2.20501		2.16087		2.13192		2.11477		2.10543
0.900		2.31745		2.26523		2.22928		2.20683		2.19391
0.910		2.44305		2.38175		2.33777		2.30911		2.29195
0.920		2.58495		2.51337		2.46008		2.42408		2.40185
0.930		2.74755		2.66422		2.60002		2.55522		2.52680
0.940		2.93737		2.84035		2.76316		2.70762		2.67153
0.950		3.16447		3.05120		2.95822		2.88925		2.84336
0.960		3.44585		3.31260		3.19982		3.11349		3.05461
0.970		3.81345		3.65438		3.51557		3.40560		3.32849
0.980		4.33939		4.14383		3.96778		3.82268		3.71731
0.990		5.25547		4.99722		4.75689		4.54859		4.38927
0.995		6.18809		5.86667		5.56183		5.28831		5.06970
0.999		8.39809		7.92806		7.47283		7.04569		6.67778

TABLE 2.4 N= 3 (CONT.)

P ↓	λ: 40	10 50	45	10 45	15 15	15 70	20	15 65	25	15 60
0.001	0.00661		0.00663			0.00610	0.00655		0.00686	
0.005	0.01956		0.01964			0.01804	0.01936		0.02029	
0.010	0.03139		0.03152			0.02894	0.03104		0.03254	
0.020	0.05075		0.05096			0.04674	0.05009		0.05249	
0.030	0.06755		0.06783			0.06216	0.06658		0.06974	
0.040	0.08300		0.08334			0.07632	0.08170		0.08556	
0.050	0.09760		0.09801			0.08969	0.09596		0.10047	
0.060	0.11162		0.11208			0.10251	0.10962		0.11474	
0.070	0.12521		0.12573			0.11494	0.12285		0.12854	
0.080	0.13849		0.13906			0.12706	0.13574		0.14199	
0.090	0.15153		0.15216			0.13896	0.14838		0.15517	
0.100	0.16439		0.16507			0.15069	0.16082		0.16814	
0.110	0.17711		0.17784			0.16229	0.17311		0.18094	
0.120	0.18973		0.19052			0.17379	0.18529		0.19362	
0.130	0.20228		0.20312			0.18523	0.19738		0.20620	
0.140	0.21478		0.21567			0.19661	0.20941		0.21871	
0.150	0.22726		0.22819			0.20797	0.22139		0.23117	
0.160	0.23972		0.24070			0.21932	0.23336		0.24359	
0.170	0.25219		0.25322			0.23067	0.24532		0.25600	
0.180	0.26468		0.26576			0.24204	0.25728		0.26841	
0.190	0.27721		0.27833			0.25345	0.26926		0.28083	
0.200	0.28977		0.29095			0.26489	0.28128		0.29328	
0.250	0.35363		0.35504			0.32310	0.34217		0.35625	
0.300	0.42000		0.42164			0.38377	0.40527		0.42126	
0.350	0.48988		0.49174			0.44795	0.47160		0.48934	
0.400	0.56428		0.56636			0.51675	0.54222		0.56152	
0.450	0.64433		0.64660			0.59144	0.61832		0.63894	
0.500	0.73138		0.73382			0.67359	0.70136		0.72296	
0.550	0.82712		0.82970			0.76524	0.79319		0.81535	
0.600	0.93379		0.93646			0.86917	0.89631		0.91841	
0.650	1.05448		1.05716			0.98926	1.01424		1.03538	
0.700	1.19366		1.19624			1.13130	1.15217		1.17103	
0.750	1.35826		1.36054			1.30435	1.31821		1.33275	
0.800	1.55982		1.56149			1.52375	1.52619		1.53310	
0.810	1.60617		1.60767			1.57535	1.57477		1.57957	
0.820	1.65505		1.65635			1.63020	1.62629		1.62873	
0.830	1.70674		1.70780			1.68869	1.68111		1.68090	
0.840	1.76158		1.76237			1.75128	1.73964		1.73647	
0.850	1.81998		1.82047			1.81852	1.80240		1.79588	
0.860	1.88244		1.88257			1.89108	1.86998		1.85970	
0.870	1.94955		1.94927			1.96977	1.94315		1.92859	
0.880	2.02207		2.02131			2.05560	2.02282		2.00341	
0.890	2.10094		2.09963			2.14986	2.11018		2.08522	
0.900	2.18739		2.18541			2.25418	2.20674		2.17540	
0.910	2.28300		2.28024			2.37073	2.31450		2.27578	
0.920	2.38996		2.38624			2.50246	2.43619		2.38882	
0.930	2.51132		2.50642			2.65349	2.57564		2.51803	
0.940	2.65153		2.64516			2.82990	2.73846		2.66853	
0.950	2.81754		2.80925			3.04111	2.93341		2.84832	
0.960	3.02097		3.01008			3.30296	3.17518		3.07087	
0.970	3.28363		3.26899			3.64527	3.49146		3.36156	
0.980	3.65460		3.63391			4.13535	3.94471		3.77778	
0.990	4.29081		4.25774			4.98946	4.73566		4.50420	
0.995	4.92955		4.88158			5.85934	5.54206		5.24557	
0.999	6.42205		6.33007			7.92125	7.45500		7.00712	

TABLE 2.5 N= 3 (CONT.)

P ↓	λ: 30	15 55	35	15 50	40	15 45	20	20 60	25	20 55
0.001		0.00709		0.00723		0.00729		0.00701		0.00734
0.005		0.02095		0.02136		0.02156		0.02072		0.02168
0.010		0.03358		0.03424		0.03456		0.03322		0.03473
0.020		0.05416		0.05522		0.05573		0.05356		0.05597
0.030		0.07195		0.07335		0.07403		0.07113		0.07430
0.040		0.08825		0.08996		0.09079		0.08722		0.09108
0.050		0.10361		0.10561		0.10658		0.10238		0.10687
0.060		0.11831		0.12058		0.12169		0.11687		0.12196
0.070		0.13252		0.13506		0.13629		0.13089		0.13654
0.080		0.14637		0.14915		0.15051		0.14454		0.15073
0.090		0.15993		0.16296		0.16444		0.15790		0.16461
0.100		0.17327		0.17654		0.17814		0.17104		0.17825
0.110		0.18644		0.18994		0.19165		0.18401		0.19170
0.120		0.19947		0.20320		0.20502		0.19683		0.20500
0.130		0.21240		0.21635		0.21828		0.20956		0.21819
0.140		0.22524		0.22942		0.23146		0.22220		0.23128
0.150		0.23804		0.24243		0.24458		0.23479		0.24430
0.160		0.25079		0.25540		0.25765		0.24733		0.25727
0.170		0.26353		0.26835		0.27070		0.25986		0.27021
0.180		0.27626		0.28128		0.28374		0.27238		0.28314
0.190		0.28900		0.29423		0.29678		0.28490		0.29606
0.200		0.30176		0.30719		0.30984		0.29744		0.30900
0.250		0.36623		0.37264		0.37577		0.36079		0.37419
0.300		0.43265		0.43998		0.44358		0.42603		0.44111
0.350		0.50204		0.51025		0.51428		0.49420		0.51075
0.400		0.57542		0.58442		0.58885		0.56630		0.58413
0.450		0.65388		0.66361		0.66840		0.64347		0.66232
0.500		0.73876		0.74909		0.75419		0.72705		0.74662
0.550		0.83172		0.84249		0.84784		0.81878		0.83864
0.600		0.93497		0.94596		0.95143		0.92093		0.94054
0.650		1.05157		1.06244		1.06789		1.03670		1.05529
0.700		1.18599		1.19623		1.20141		1.17080		1.18724
0.750		1.34513		1.35393		1.35847		1.33056		1.34314
0.800		1.54066		1.54665		1.54989		1.52847		1.53440
0.810		1.58576		1.59094		1.59381		1.57439		1.57850
0.820		1.63338		1.63764		1.64007		1.62297		1.62505
0.830		1.68380		1.68701		1.68896		1.67455		1.67436
0.840		1.73739		1.73940		1.74078		1.72950		1.72675
0.850		1.79455		1.79519		1.79593		1.78827		1.78264
0.860		1.85579		1.85486		1.85486		1.85142		1.84255
0.870		1.92174		1.91900		1.91814		1.91963		1.90707
0.880		1.99317		1.98832		1.98647		1.99374		1.97698
0.890		2.07105		2.06375		2.06073		2.07483		2.05324
0.900		2.15664		2.14646		2.14205		2.16428		2.13711
0.910		2.25160		2.23799		2.23194		2.26391		2.23024
0.920		2.35821		2.34046		2.33241		2.37622		2.33488
0.930		2.47962		2.45682		2.44632		2.50470		2.45419
0.940		2.62053		2.59141		2.57782		2.65452		2.59285
0.950		2.78820		2.75096		2.73336		2.83368		2.75812
0.960		2.99489		2.94678		2.92377		3.05572		2.96224
0.970		3.26368		3.20015		3.16932		3.34610		3.22834
0.980		3.64676		3.55896		3.51558		3.76235		3.60872
0.990		4.31217		4.17693		4.10804		4.48949		4.27192
0.995		4.98927		4.80043		4.70117		5.23180		4.94884
0.999		6.59686		6.26757		6.08104		6.99499		6.55914

TABLE 2.6 N= 3 (CONT.)

P ↓	λ: 30 20 50	35 20 45	40 20 40	25 25 50	30 25 45
0.001	0.00755	0.00768	0.00772	0.00766	0.00785
0.005	0.02231	0.02267	0.02279	0.02261	0.02319
0.010	0.03574	0.03632	0.03651	0.03621	0.03713
0.020	0.05758	0.05851	0.05881	0.05833	0.05980
0.030	0.07642	0.07764	0.07804	0.07739	0.07932
0.040	0.09366	0.09514	0.09562	0.09483	0.09717
0.050	0.10987	0.11160	0.11216	0.11122	0.11394
0.060	0.12536	0.12732	0.12796	0.12687	0.12995
0.070	0.14032	0.14249	0.14321	0.14199	0.14540
0.080	0.15487	0.15726	0.15804	0.15668	0.16042
0.090	0.16910	0.17169	0.17254	0.17105	0.17510
0.100	0.18308	0.18587	0.18678	0.18516	0.18950
0.110	0.19686	0.19984	0.20082	0.19906	0.20369
0.120	0.21048	0.21365	0.21468	0.21280	0.21771
0.130	0.22398	0.22732	0.22842	0.22641	0.23159
0.140	0.23737	0.24090	0.24205	0.23991	0.24535
0.150	0.25070	0.25439	0.25560	0.25333	0.25903
0.160	0.26396	0.26783	0.26910	0.26670	0.27265
0.170	0.27719	0.28123	0.28255	0.28002	0.28621
0.180	0.29040	0.29460	0.29598	0.29331	0.29975
0.190	0.30360	0.30796	0.30939	0.30660	0.31327
0.200	0.31680	0.32132	0.32281	0.31988	0.32678
0.250	0.38328	0.38856	0.39030	0.38672	0.39468
0.300	0.45137	0.45735	0.45932	0.45507	0.46399
0.350	0.52209	0.52871	0.53089	0.52596	0.53573
0.400	0.59641	0.60360	0.60597	0.60036	0.61084
0.450	0.67539	0.68308	0.68562	0.67932	0.69039
0.500	0.76030	0.76839	0.77106	0.76409	0.77555
0.550	0.85269	0.86104	0.86381	0.85620	0.86784
0.600	0.95462	0.96305	0.96585	0.95768	0.96920
0.650	1.06891	1.07716	1.07992	1.07131	1.08231
0.700	1.19970	1.20738	1.20997	1.20116	1.21104
0.750	1.35334	1.35985	1.36207	1.35349	1.36137
0.800	1.54053	1.54485	1.54639	1.53885	1.54326
0.810	1.58348	1.58719	1.58853	1.58136	1.58482
0.820	1.62875	1.63175	1.63287	1.62614	1.62853
0.830	1.67660	1.67880	1.67967	1.67347	1.67467
0.840	1.72735	1.72865	1.72922	1.72366	1.72350
0.850	1.78139	1.78165	1.78189	1.77708	1.77541
0.860	1.83916	1.83824	1.83810	1.83420	1.83079
0.870	1.90125	1.89896	1.89839	1.89556	1.89019
0.880	1.96836	1.96449	1.96341	1.96188	1.95425
0.890	2.04137	2.03566	2.03398	2.03403	2.02378
0.900	2.12143	2.11355	2.11118	2.11315	2.09985
0.910	2.21006	2.19961	2.19640	2.20073	2.18383
0.920	2.30932	2.29575	2.29153	2.29881	2.27762
0.930	2.42209	2.40471	2.39924	2.41027	2.38385
0.940	2.55263	2.53046	2.52343	2.53931	2.50641
0.950	2.70755	2.67920	2.67013	2.69251	2.65132
0.960	2.89797	2.86130	2.84946	2.88091	2.82870
0.970	3.14485	3.09623	3.08040	3.12538	3.05756
0.980	3.49552	3.42784	3.40553	3.47307	3.38074
0.990	4.10229	3.99641	3.96076	4.07598	3.93572
0.995	4.71782	4.56741	4.51561	4.68913	4.49463
0.999	6.17605	5.90308	5.80341	6.14600	5.80918

TABLE 2.7 N= 3 (CONT.)

P ↓ λ:	25 35 40	30 30 40	30 35 35	All λ's = 100/3
0.001	0.00795	0.00802	0.00808	0.00810
0.005	0.02347	0.02369	0.02385	0.02391
0.010	0.03758	0.03792	0.03818	0.03828
0.020	0.06051	0.06105	0.06146	0.06161
0.030	0.08025	0.08097	0.08150	0.08170
0.040	0.09830	0.09916	0.09981	0.10005
0.050	0.11526	0.11625	0.11700	0.11728
0.060	0.13144	0.13256	0.13341	0.13372
0.070	0.14705	0.14829	0.14923	0.14958
0.080	0.16222	0.16357	0.16460	0.16498
0.090	0.17705	0.17851	0.17962	0.18003
0.100	0.19160	0.19316	0.19435	0.19479
0.110	0.20593	0.20759	0.20885	0.20932
0.120	0.22009	0.22183	0.22317	0.22367
0.130	0.23410	0.23593	0.23734	0.23787
0.140	0.24799	0.24991	0.25139	0.25194
0.150	0.26180	0.26380	0.26535	0.26592
0.160	0.27553	0.27761	0.27923	0.27983
0.170	0.28922	0.29138	0.29306	0.29368
0.180	0.30287	0.30510	0.30685	0.30749
0.190	0.31650	0.31881	0.32061	0.32128
0.200	0.33013	0.33250	0.33437	0.33506
0.250	0.39856	0.40125	0.40339	0.40418
0.300	0.46835	0.47129	0.47368	0.47455
0.350	0.54050	0.54365	0.54625	0.54719
0.400	0.61599	0.61928	0.62206	0.62306
0.450	0.69583	0.69921	0.70212	0.70316
0.500	0.78122	0.78461	0.78760	0.78866
0.550	0.87363	0.87693	0.87995	0.88100
0.600	0.97497	0.97807	0.98104	0.98205
0.650	1.08788	1.09063	1.09343	1.09437
0.700	1.21613	1.21832	1.22082	1.22162
0.750	1.36555	1.36691	1.36886	1.36945
0.800	1.54585	1.54592	1.54697	1.54721
0.810	1.58696	1.58671	1.58751	1.58765
0.820	1.63018	1.62957	1.63009	1.63014
0.830	1.67576	1.67474	1.67496	1.67489
0.840	1.72396	1.72250	1.72238	1.72218
0.850	1.77514	1.77320	1.77268	1.77235
0.860	1.82971	1.82723	1.82628	1.82578
0.870	1.88818	1.88508	1.88363	1.88296
0.880	1.95116	1.94738	1.94536	1.94448
0.890	2.01944	2.01490	2.01222	2.01111
0.900	2.09403	2.08862	2.08518	2.08379
0.910	2.17628	2.16985	2.16551	2.16382
0.920	2.26797	2.26038	2.25496	2.25290
0.930	2.37165	2.36267	2.35595	2.35344
0.940	2.49101	2.48036	2.47202	2.46896
0.950	2.63179	2.61908	2.60867	2.60491
0.960	2.80360	2.78824	2.77507	2.77039
0.970	3.02447	3.00551	2.98839	2.98243
0.980	3.33481	3.31052	3.28711	3.27913
0.990	3.86366	3.82966	3.79355	3.78162
0.995	4.39131	4.34718	4.29588	4.27938
0.999	5.61530	5.54750	5.45145	5.42207

TABLE 3.1 N= 4

P ↓	λ:	5 5 5 85	5 5 10 80	5 5 15 75	5 5 20 70	5 5 25 65
0.001		0.00930	0.01089	0.01186	0.01253	0.01301
0.005		0.02143	0.02507	0.02733	0.02890	0.03003
0.010		0.03104	0.03629	0.03957	0.04186	0.04351
0.020		0.04546	0.05310	0.05793	0.06133	0.06378
0.030		0.05727	0.06684	0.07295	0.07726	0.08039
0.040		0.06780	0.07905	0.08630	0.09144	0.09518
0.050		0.07754	0.09034	0.09864	0.10455	0.10886
0.060		0.08677	0.10100	0.11030	0.11694	0.12180
0.070		0.09563	0.11122	0.12147	0.12882	0.13421
0.080		0.10423	0.12111	0.13229	0.14033	0.14623
0.090		0.11264	0.13077	0.14284	0.15155	0.15796
0.100		0.12092	0.14025	0.15320	0.16257	0.16948
0.110		0.12910	0.14959	0.16341	0.17343	0.18083
0.120		0.13723	0.15885	0.17350	0.18417	0.19206
0.130		0.14531	0.16803	0.18353	0.19483	0.20321
0.140		0.15339	0.17719	0.19351	0.20544	0.21430
0.150		0.16147	0.18632	0.20346	0.21602	0.22536
0.160		0.16958	0.19546	0.21340	0.22660	0.23641
0.170		0.17774	0.20462	0.22336	0.23718	0.24747
0.180		0.18595	0.21381	0.23335	0.24778	0.25855
0.190		0.19423	0.22306	0.24337	0.25843	0.26968
0.200		0.20259	0.23236	0.25346	0.26912	0.28085
0.250		0.24606	0.28021	0.30510	0.32380	0.33789
0.300		0.29336	0.33127	0.35972	0.38138	0.39780
0.350		0.34603	0.38686	0.41855	0.44300	0.46167
0.400		0.40581	0.44844	0.48287	0.50985	0.53062
0.450		0.47481	0.51773	0.55414	0.58325	0.60588
0.500		0.55559	0.59689	0.63419	0.66481	0.68893
0.550		0.65129	0.68867	0.72536	0.75659	0.78165
0.600		0.76575	0.79672	0.83075	0.86134	0.88655
0.650		0.90398	0.92596	0.95476	0.98292	1.00712
0.700		1.07299	1.08342	1.10375	1.12701	1.14853
0.750		1.28352	1.27970	1.28764	1.30253	1.31885
0.800		1.55382	1.53235	1.52301	1.52454	1.53175
0.810		1.61762	1.59208	1.57858	1.57664	1.58136
0.820		1.68548	1.65564	1.63770	1.63198	1.63393
0.830		1.75785	1.72347	1.70080	1.69094	1.68980
0.840		1.83530	1.79609	1.76836	1.75398	1.74940
0.850		1.91849	1.87412	1.84097	1.82163	1.81321
0.860		2.00821	1.95831	1.91934	1.89457	1.88183
0.870		2.10544	2.04959	2.00435	1.97361	1.95602
0.880		2.21141	2.14910	2.09706	2.05973	2.03669
0.890		2.32766	2.25830	2.19885	2.15423	2.12500
0.900		2.45617	2.37904	2.31147	2.25873	2.22245
0.910		2.59956	2.51380	2.43723	2.37540	2.33104
0.920		2.76140	2.66592	2.57929	2.50717	2.45346
0.930		2.94668	2.84013	2.74205	2.65817	2.59351
0.940		3.16277	3.04333	2.93201	2.83447	2.75678
0.950		3.42112	3.28630	3.15927	3.04548	2.95197
0.960		3.74097	3.58715	3.44080	3.30704	3.19373
0.970		4.15853	3.97997	3.80855	3.64900	3.50965
0.980		4.75561	4.54171	4.33465	4.13864	3.96206
0.990		5.79499	5.51971	5.25091	4.99225	4.75142
0.995		6.85273	6.51506	6.18365	5.86184	5.55654
0.999		9.35848	8.87317	8.39382	7.92343	7.46777

TABLE 3.2 N= 4 (CONT.)

P ↓	λ: 5 30 5 60	5 35 5 55	5 40 5 50	5 45 5 45	5 10 10 75
0.001	0.01335	0.01358	0.01372	0.01376	0.01272
0.005	0.03083	0.03137	0.03168	0.03178	0.02925
0.010	0.04469	0.04548	0.04594	0.04609	0.04227
0.020	0.06554	0.06672	0.06740	0.06763	0.06171
0.030	0.08263	0.08414	0.08502	0.08531	0.07754
0.040	0.09786	0.09968	0.10073	0.10107	0.09155
0.050	0.11196	0.11406	0.11527	0.11567	0.10444
0.060	0.12530	0.12766	0.12904	0.12949	0.11658
0.070	0.13809	0.14072	0.14225	0.14276	0.12818
0.080	0.15049	0.15338	0.15506	0.15562	0.13937
0.090	0.16259	0.16574	0.16757	0.16817	0.15026
0.100	0.17447	0.17787	0.17985	0.18050	0.16092
0.110	0.18619	0.18984	0.19196	0.19266	0.17139
0.120	0.19778	0.20168	0.20395	0.20470	0.18172
0.130	0.20929	0.21343	0.21585	0.21664	0.19195
0.140	0.22074	0.22513	0.22769	0.22853	0.20210
0.150	0.23216	0.23679	0.23950	0.24039	0.21220
0.160	0.24356	0.24845	0.25130	0.25224	0.22228
0.170	0.25498	0.26011	0.26310	0.26409	0.23234
0.180	0.26642	0.27179	0.27493	0.27597	0.24240
0.190	0.27789	0.28352	0.28680	0.28788	0.25249
0.200	0.28942	0.29529	0.29872	0.29985	0.26261
0.250	0.34824	0.35534	0.35950	0.36086	0.31411
0.300	0.40990	0.41823	0.42311	0.42472	0.36809
0.350	0.47548	0.48501	0.49061	0.49245	0.42578
0.400	0.54606	0.55674	0.56302	0.56509	0.48849
0.450	0.62280	0.63453	0.64145	0.64373	0.55773
0.500	0.70709	0.71974	0.72721	0.72968	0.63539
0.550	0.80071	0.81407	0.82198	0.82460	0.72393
0.600	0.90599	0.91974	0.92793	0.93064	0.82666
0.650	1.02622	1.03990	1.04811	1.05084	0.94817
0.700	1.16619	1.17911	1.18695	1.18958	1.09515
0.750	1.33337	1.34446	1.35134	1.35367	1.27779
0.800	1.54038	1.54786	1.55279	1.55450	1.51287
0.810	1.58833	1.59477	1.59913	1.60066	1.56850
0.820	1.63903	1.64429	1.64800	1.64932	1.62772
0.830	1.69279	1.69671	1.69968	1.70077	1.69095
0.840	1.75000	1.75238	1.75451	1.75533	1.75868
0.850	1.81111	1.81175	1.81291	1.81341	1.83149
0.860	1.87667	1.87532	1.87536	1.87551	1.91009
0.870	1.94736	1.94371	1.94247	1.94220	1.99535
0.880	2.02402	2.01773	2.01499	2.01424	2.08834
0.890	2.10772	2.09835	2.09386	2.09255	2.19041
0.900	2.19983	2.18685	2.18031	2.17833	2.30333
0.910	2.30217	2.28491	2.27593	2.27316	2.42939
0.920	2.41720	2.39483	2.38289	2.37916	2.57175
0.930	2.54840	2.51982	2.50425	2.49934	2.73482
0.940	2.70088	2.66458	2.64448	2.63807	2.92509
0.950	2.88260	2.83645	2.81050	2.80216	3.15264
0.960	3.10694	3.04775	3.01393	3.00299	3.43445
0.970	3.39918	3.32169	3.27661	3.26191	3.80247
0.980	3.81641	3.71059	3.64760	3.62683	4.32885
0.990	4.54254	4.38268	4.28385	4.25066	5.24541
0.995	5.28243	5.06322	4.92263	4.87449	6.17835
0.999	7.04008	6.67153	6.41521	6.32298	8.38878

TABLE 3.3 N= 4 (CONT.)

P ↓	λ:	5 10 15 70	5 10 20 65	5 10 25 60	5 10 30 55	5 10 35 50
0.001		0.01384	0.01460	0.01514	0.01550	0.01574
0.005		0.03182	0.03358	0.03482	0.03568	0.03622
0.010		0.04598	0.04854	0.05034	0.05159	0.05238
0.020		0.06712	0.07087	0.07353	0.07537	0.07654
0.030		0.08432	0.08904	0.09240	0.09473	0.09621
0.040		0.09953	0.10512	0.10911	0.11187	0.11363
0.050		0.11353	0.11991	0.12448	0.12765	0.12967
0.060		0.12670	0.13383	0.13894	0.14249	0.14475
0.070		0.13927	0.14711	0.15274	0.15667	0.15916
0.080		0.15139	0.15992	0.16605	0.17033	0.17305
0.090		0.16317	0.17236	0.17898	0.18361	0.18655
0.100		0.17469	0.18452	0.19161	0.19658	0.19974
0.110		0.18599	0.19645	0.20402	0.20931	0.21269
0.120		0.19714	0.20822	0.21624	0.22186	0.22545
0.130		0.20817	0.21985	0.22832	0.23426	0.23806
0.140		0.21910	0.23137	0.24029	0.24655	0.25055
0.150		0.22997	0.24282	0.25218	0.25875	0.26296
0.160		0.24079	0.25422	0.26401	0.27090	0.27530
0.170		0.25158	0.26558	0.27580	0.28300	0.28760
0.180		0.26237	0.27693	0.28757	0.29508	0.29988
0.190		0.27316	0.28828	0.29935	0.30715	0.31216
0.200		0.28398	0.29964	0.31113	0.31924	0.32444
0.250		0.33878	0.35711	0.37063	0.38023	0.38640
0.300		0.39576	0.41658	0.43207	0.44311	0.45022
0.350		0.45607	0.47922	0.49658	0.50900	0.51702
0.400		0.52094	0.54616	0.56524	0.57896	0.58785
0.450		0.59169	0.61865	0.63925	0.65416	0.66385
0.500		0.67001	0.69820	0.72003	0.73595	0.74634
0.550		0.75802	0.78674	0.80937	0.82603	0.83696
0.600		0.85860	0.88684	0.90963	0.92664	0.93789
0.650		0.97578	1.00207	1.02410	1.04087	1.05209
0.700		1.11551	1.13773	1.15758	1.17321	1.18387
0.750		1.28705	1.30205	1.31754	1.33062	1.33986
0.800		1.50591	1.50898	1.51661	1.52478	1.53118
0.810		1.55754	1.55744	1.56289	1.56966	1.57524
0.820		1.61246	1.60888	1.61189	1.61707	1.62171
0.830		1.67106	1.66365	1.66393	1.66731	1.67087
0.840		1.73381	1.72218	1.71938	1.72073	1.72306
0.850		1.80126	1.78497	1.77871	1.77774	1.77867
0.860		1.87407	1.85263	1.84247	1.83885	1.83818
0.870		1.95306	1.92590	1.91134	1.90469	1.90216
0.880		2.03923	2.00573	1.98617	1.97603	1.97135
0.890		2.13385	2.09329	2.06802	2.05385	2.04666
0.900		2.23858	2.19009	2.15828	2.13940	2.12926
0.910		2.35557	2.29814	2.25877	2.23436	2.22071
0.920		2.48776	2.42017	2.37198	2.34098	2.32311
0.930		2.63928	2.56000	2.50139	2.46245	2.43943
0.940		2.81619	2.72325	2.65215	2.60345	2.57399
0.950		3.02791	2.91867	2.83225	2.77126	2.73354
0.960		3.29027	3.16097	3.05519	2.97813	2.92940
0.970		3.63311	3.47782	3.34635	3.24718	3.18283
0.980		4.12372	3.93170	3.76315	3.63063	3.54177
0.990		4.97842	4.72338	4.49035	4.29662	4.15998
0.995		5.84867	5.53024	5.23227	4.97421	4.78372
0.999		7.91108	7.44377	6.99457	6.58260	6.25136

TABLE 3.4 N= 4 (CONT.)

P ↓	λ:	5 10 40 45	5 15 15 65	5 15 20 60	5 15 25 55	5 15 30 50
0.001		0.01585	0.01503	0.01583	0.01638	0.01675
0.005		0.03648	0.03455	0.03640	0.03768	0.03852
0.010		0.05276	0.04992	0.05260	0.05445	0.05567
0.020		0.07710	0.07284	0.07675	0.07946	0.08126
0.030		0.09693	0.09146	0.09637	0.09978	0.10205
0.040		0.11448	0.10792	0.11371	0.11774	0.12042
0.050		0.13064	0.12305	0.12964	0.13424	0.13730
0.060		0.14585	0.13726	0.14460	0.14974	0.15316
0.070		0.16038	0.15081	0.15887	0.16451	0.16827
0.080		0.17438	0.16387	0.17260	0.17873	0.18282
0.090		0.18799	0.17654	0.18592	0.19252	0.19693
0.100		0.20128	0.18891	0.19892	0.20597	0.21069
0.110		0.21434	0.20104	0.21166	0.21916	0.22418
0.120		0.22720	0.21299	0.22420	0.23213	0.23744
0.130		0.23991	0.22479	0.23658	0.24493	0.25053
0.140		0.25250	0.23647	0.24883	0.25759	0.26348
0.150		0.26501	0.24807	0.26098	0.27015	0.27632
0.160		0.27745	0.25960	0.27306	0.28263	0.28907
0.170		0.28985	0.27109	0.28508	0.29504	0.30175
0.180		0.30223	0.28255	0.29706	0.30741	0.31439
0.190		0.31460	0.29400	0.30903	0.31976	0.32700
0.200		0.32698	0.30546	0.32099	0.33210	0.33960
0.250		0.38942	0.36325	0.38118	0.39410	0.40286
0.300		0.45371	0.42282	0.44294	0.45754	0.46749
0.350		0.52096	0.48531	0.50738	0.52354	0.53460
0.400		0.59223	0.55183	0.57559	0.59315	0.60523
0.450		0.66863	0.62363	0.64873	0.66748	0.68045
0.500		0.75147	0.70221	0.72817	0.74784	0.76156
0.550		0.84238	0.78945	0.81567	0.83587	0.85011
0.600		0.94349	0.88791	0.91353	0.93374	0.94816
0.650		1.05770	1.00119	1.02498	1.04443	1.05857
0.700		1.18925	1.13458	1.15479	1.17231	1.18546
0.750		1.34461	1.29641	1.31039	1.32420	1.33522
0.800		1.53463	1.50079	1.50442	1.51156	1.51857
0.810		1.57830	1.54876	1.54962	1.55491	1.56077
0.820		1.62433	1.59971	1.59751	1.60072	1.60529
0.830		1.67300	1.65402	1.64843	1.64930	1.65239
0.840		1.72462	1.71211	1.70275	1.70098	1.70240
0.850		1.77957	1.77448	1.76093	1.75619	1.75570
0.860		1.83832	1.84176	1.82354	1.81542	1.81275
0.870		1.90143	1.91470	1.89126	1.87931	1.87413
0.880		1.96960	1.99423	1.96495	1.94861	1.94054
0.890		2.04371	2.08155	2.04568	2.02430	2.01287
0.900		2.12490	2.17818	2.13485	2.10765	2.09227
0.910		2.21467	2.28613	2.23430	2.20031	2.18026
0.920		2.31504	2.40814	2.34654	2.30455	2.27892
0.930		2.42885	2.54804	2.47508	2.42354	2.39112
0.940		2.56027	2.71147	2.62510	2.56199	2.52115
0.950		2.71575	2.90718	2.80467	2.72718	2.67562
0.960		2.90611	3.14989	3.02733	2.93141	2.86569
0.970		3.15163	3.46730	3.31862	3.19787	3.11235
0.980		3.49787	3.92188	3.73614	3.57901	3.46305
0.990		4.09034	4.71445	4.46510	4.24368	4.07038
0.995		4.68350	5.52186	5.20869	4.92193	4.68670
0.999		6.06343	7.43605	6.97351	6.53426	6.14668

TABLE 3.5 N= 4 (CONT.)

P ↓	λ: 5 15 35 45	5 15 40 40	5 20 20 55	5 20 25 50	5 20 30 45
0.001	0.01695	0.01537	0.01665	0.01719	0.01753
0.005	0.03900	0.03550	0.03828	0.03953	0.04031
0.010	0.05636	0.05149	0.05531	0.05712	0.05825
0.020	0.08228	0.07556	0.08070	0.08336	0.08501
0.030	0.10334	0.09530	0.10132	0.10465	0.10673
0.040	0.12196	0.11289	0.11954	0.12347	0.12592
0.050	0.13906	0.12916	0.13626	0.14074	0.14353
0.060	0.15513	0.14454	0.15196	0.15695	0.16007
0.070	0.17044	0.15929	0.16692	0.17239	0.17581
0.080	0.18518	0.17356	0.18131	0.18724	0.19096
0.090	0.19947	0.18748	0.19527	0.20164	0.20564
0.100	0.21341	0.20112	0.20888	0.21567	0.21994
0.110	0.22707	0.21455	0.22221	0.22942	0.23395
0.120	0.24051	0.22782	0.23532	0.24293	0.24772
0.130	0.25377	0.24096	0.24825	0.25625	0.26129
0.140	0.26688	0.25402	0.26104	0.26942	0.27471
0.150	0.27988	0.26701	0.27371	0.28247	0.28799
0.160	0.29279	0.27996	0.28630	0.29542	0.30118
0.170	0.30563	0.29289	0.29882	0.30830	0.31429
0.180	0.31843	0.30582	0.31129	0.32112	0.32734
0.190	0.33120	0.31875	0.32374	0.33391	0.34035
0.200	0.34395	0.33171	0.33617	0.34668	0.35333
0.250	0.40795	0.39726	0.39853	0.41063	0.41833
0.300	0.47328	0.46489	0.46221	0.47577	0.48443
0.350	0.54106	0.53556	0.52832	0.54318	0.55271
0.400	0.61230	0.61016	0.59790	0.61387	0.62418
0.450	0.68808	0.68963	0.67204	0.68892	0.69989
0.500	0.76965	0.77508	0.75203	0.76956	0.78104
0.550	0.85855	0.86792	0.83950	0.85732	0.86910
0.600	0.95678	0.97000	0.93655	0.95419	0.96599
0.650	1.06711	1.08387	1.04614	1.06292	1.07436
0.700	1.19353	1.21327	1.17257	1.18752	1.19802
0.750	1.34220	1.36396	1.32256	1.33419	1.34286
0.800	1.52341	1.54558	1.50750	1.51337	1.51872
0.810	1.56500	1.58696	1.55028	1.55458	1.55899
0.820	1.60881	1.63045	1.59551	1.59803	1.60139
0.830	1.65511	1.67628	1.64346	1.64399	1.64617
0.840	1.70420	1.72475	1.69449	1.69279	1.69362
0.850	1.75645	1.77618	1.74901	1.74479	1.74410
0.860	1.81230	1.83101	1.80752	1.80045	1.79802
0.870	1.87228	1.88971	1.87065	1.86033	1.85590
0.880	1.93707	1.95293	1.93917	1.92513	1.91839
0.890	2.00751	2.02145	2.01403	1.99571	1.98630
0.900	2.08468	2.09628	2.09652	2.07323	2.06067
0.910	2.17001	2.17874	2.18828	2.15918	2.14290
0.920	2.26545	2.27065	2.29159	2.25560	2.23485
0.930	2.37371	2.37454	2.40964	2.36535	2.33915
0.940	2.49879	2.49411	2.54711	2.49267	2.45968
0.950	2.64686	2.63511	2.71135	2.64413	2.60243
0.960	2.82833	2.80717	2.91467	2.83083	2.77750
0.970	3.06268	3.02837	3.18034	3.07369	3.00387
0.980	3.39379	3.33930	3.56096	3.42004	3.32437
0.990	3.96202	3.86968	4.22582	4.02249	3.87650
0.995	4.53304	4.39978	4.90488	4.63655	4.43413
0.999	5.86922	5.63330	6.51917	6.09697	5.74901

TABLE 3.6 N= 4 (CONT.)

P ↓	λ: 5 20 35 40	5 25 25 45	5 25 30 40	5 25 35 35	5 30 30 35
0.001	0.01662	0.01771	0.01711	0.01483	0.01551
0.005	0.03830	0.04072	0.03941	0.03435	0.03585
0.010	0.05544	0.05884	0.05703	0.04994	0.05202
0.020	0.08112	0.08586	0.08339	0.07354	0.07641
0.030	0.10208	0.10780	0.10488	0.09305	0.09646
0.040	0.12067	0.12717	0.12391	0.11053	0.11437
0.050	0.13779	0.14495	0.14144	0.12680	0.13097
0.060	0.15393	0.16163	0.15792	0.14226	0.14670
0.070	0.16934	0.17752	0.17366	0.15716	0.16182
0.080	0.18422	0.19279	0.18883	0.17166	0.17648
0.090	0.19868	0.20759	0.20356	0.18586	0.19081
0.100	0.21281	0.22202	0.21795	0.19984	0.20488
0.110	0.22669	0.23614	0.23206	0.21366	0.21877
0.120	0.24036	0.25001	0.24596	0.22738	0.23252
0.130	0.25388	0.26369	0.25968	0.24102	0.24617
0.140	0.26727	0.27720	0.27326	0.25462	0.25976
0.150	0.28056	0.29058	0.28673	0.26821	0.27330
0.160	0.29378	0.30386	0.30012	0.28179	0.28683
0.170	0.30696	0.31706	0.31345	0.29540	0.30036
0.180	0.32010	0.33019	0.32674	0.30904	0.31390
0.190	0.33323	0.34328	0.34000	0.32272	0.32749
0.200	0.34636	0.35634	0.35326	0.33647	0.34111
0.250	0.41244	0.42169	0.41981	0.40641	0.41031
0.300	0.48015	0.48804	0.48777	0.47895	0.48201
0.350	0.55050	0.55652	0.55816	0.55472	0.55695
0.400	0.62443	0.62809	0.63195	0.63425	0.63579
0.450	0.70293	0.70382	0.71010	0.71816	0.71919
0.500	0.78711	0.78487	0.79375	0.80721	0.80794
0.550	0.87836	0.87270	0.88425	0.90248	0.90308
0.600	0.97847	0.96922	0.98338	1.00547	1.00606
0.650	1.08991	1.07700	1.09357	1.11831	1.11894
0.700	1.21625	1.19981	1.21831	1.24419	1.24481
0.750	1.36295	1.34343	1.36294	1.38803	1.38848
0.800	1.53916	1.51755	1.53637	1.55808	1.55801
0.810	1.57922	1.55738	1.57574	1.59639	1.59614
0.820	1.62128	1.59931	1.61707	1.63649	1.63604
0.830	1.66556	1.64358	1.66057	1.67858	1.67790
0.840	1.71235	1.69048	1.70649	1.72292	1.72196
0.850	1.76196	1.74034	1.75516	1.76979	1.76851
0.860	1.81477	1.79360	1.80695	1.81953	1.81788
0.870	1.87127	1.85075	1.86231	1.87259	1.87050
0.880	1.93204	1.91244	1.92183	1.92948	1.92689
0.890	1.99782	1.97946	1.98620	1.99089	1.98769
0.900	2.06956	2.05285	2.05635	2.05765	2.05374
0.910	2.14853	2.13397	2.13351	2.13092	2.12616
0.920	2.23641	2.22466	2.21929	2.21221	2.20644
0.930	2.33559	2.32753	2.31601	2.30367	2.29666
0.940	2.44957	2.44639	2.42704	2.40845	2.39988
0.950	2.58374	2.58717	2.55757	2.53141	2.52086
0.960	2.74717	2.75987	2.71636	2.68070	2.66751
0.970	2.95687	2.98326	2.91977	2.87162	2.85471
0.980	3.25104	3.29980	3.20463	3.13853	3.11579
0.990	3.75188	3.84605	3.68868	3.59131	3.55716
0.995	4.25211	4.39913	4.17172	4.04239	3.99519
0.999	5.41859	5.70831	5.30068	5.09306	5.01051

TABLE 3.7 N= 4 (CONT.)

P ↓	λ:	10 10 10 70	10 10 15 65	10 10 20 60	10 10 25 55	10 10 30 50
0.001		0.01483	0.01611	0.01696	0.01755	0.01794
0.005		0.03400	0.03689	0.03885	0.04020	0.04108
0.010		0.04901	0.05316	0.05597	0.05791	0.05919
0.020		0.07130	0.07726	0.08134	0.08416	0.08602
0.030		0.08932	0.09672	0.10180	0.10533	0.10767
0.040		0.10518	0.11383	0.11979	0.12394	0.12669
0.050		0.11972	0.12948	0.13624	0.14095	0.14408
0.060		0.13334	0.14414	0.15163	0.15687	0.16035
0.070		0.14630	0.15806	0.16625	0.17198	0.17580
0.080		0.15876	0.17142	0.18027	0.18648	0.19063
0.090		0.17083	0.18436	0.19384	0.20051	0.20496
0.100		0.18260	0.19695	0.20704	0.21415	0.21890
0.110		0.19412	0.20927	0.21995	0.22749	0.23253
0.120		0.20545	0.22137	0.23263	0.24057	0.24590
0.130		0.21663	0.23329	0.24511	0.25346	0.25907
0.140		0.22769	0.24507	0.25743	0.26618	0.27206
0.150		0.23866	0.25674	0.26963	0.27878	0.28492
0.160		0.24956	0.26832	0.28173	0.29126	0.29767
0.170		0.26041	0.27983	0.29376	0.30367	0.31034
0.180		0.27124	0.29130	0.30573	0.31601	0.32294
0.190		0.28204	0.30274	0.31766	0.32831	0.33549
0.200		0.29286	0.31417	0.32957	0.34058	0.34802
0.250		0.34737	0.37154	0.38925	0.40201	0.41067
0.300		0.40363	0.43032	0.45017	0.46457	0.47438
0.350		0.46286	0.49171	0.51349	0.52943	0.54035
0.400		0.52628	0.55684	0.58033	0.59769	0.60963
0.450		0.59527	0.62701	0.65191	0.67050	0.68337
0.500		0.67153	0.70373	0.72963	0.74922	0.76287
0.550		0.75726	0.78896	0.81528	0.83552	0.84975
0.600		0.85543	0.88530	0.91121	0.93159	0.94610
0.650		0.97022	0.99643	1.02073	1.04048	1.05481
0.700		1.10778	1.12777	1.14864	1.16662	1.18004
0.750		1.27763	1.28780	1.30252	1.31686	1.32824
0.800		1.49559	1.49088	1.49512	1.50279	1.51018
0.810		1.54715	1.53867	1.54008	1.54589	1.55212
0.820		1.60205	1.58948	1.58777	1.59146	1.59639
0.830		1.66068	1.64368	1.63849	1.63981	1.64326
0.840		1.72349	1.70169	1.69264	1.69129	1.69304
0.850		1.79105	1.76403	1.75069	1.74630	1.74613
0.860		1.86400	1.83131	1.81319	1.80537	1.80298
0.870		1.94317	1.90430	1.88084	1.86910	1.86417
0.880		2.02956	1.98393	1.95448	1.93828	1.93041
0.890		2.12445	2.07138	2.03521	2.01387	2.00258
0.900		2.22946	2.16819	2.12442	2.09715	2.08185
0.910		2.34676	2.27636	2.22395	2.18976	2.16973
0.920		2.47927	2.39863	2.33631	2.29399	2.26829
0.930		2.63113	2.53884	2.46502	2.41302	2.38042
0.940		2.80838	2.70260	2.61527	2.55154	2.51041
0.950		3.02045	2.89868	2.79512	2.71686	2.66487
0.960		3.28316	3.14179	3.01813	2.92128	2.85498
0.970		3.62633	3.45961	3.30981	3.18801	3.10173
0.980		4.11728	3.91462	3.72781	3.56951	3.45260
0.990		4.97235	4.70765	4.45734	4.23472	4.06023
0.995		5.84283	5.51534	5.20130	4.91338	4.67684
0.999		7.90553	7.42988	6.96657	6.52629	6.13736

TABLE 3.8 N= 4 (CONT.)

P ↓	λ: 10 10 35 45	10 10 40 40	10 15 15 60	10 15 20 55	10 15 25 50
0.001	0.01816	0.01823	0.01746	0.01835	0.01895
0.005	0.04159	0.04176	0.03995	0.04198	0.04334
0.010	0.05993	0.06017	0.05752	0.06043	0.06238
0.020	0.08709	0.08744	0.08351	0.08770	0.09051
0.030	0.10901	0.10945	0.10444	0.10964	0.11315
0.040	0.12827	0.12879	0.12281	0.12889	0.13299
0.050	0.14588	0.14647	0.13959	0.14646	0.15110
0.060	0.16236	0.16301	0.15527	0.16286	0.16800
0.070	0.17800	0.17872	0.17014	0.17841	0.18403
0.080	0.19301	0.19379	0.18440	0.19332	0.19937
0.090	0.20752	0.20836	0.19818	0.20770	0.21419
0.100	0.22164	0.22253	0.21158	0.22169	0.22858
0.110	0.23544	0.23639	0.22467	0.23533	0.24262
0.120	0.24898	0.24998	0.23750	0.24871	0.25638
0.130	0.26230	0.26336	0.25013	0.26186	0.26991
0.140	0.27546	0.27657	0.26259	0.27483	0.28324
0.150	0.28848	0.28964	0.27491	0.28765	0.29641
0.160	0.30138	0.30260	0.28713	0.30035	0.30946
0.170	0.31420	0.31546	0.29926	0.31295	0.32239
0.180	0.32695	0.32826	0.31132	0.32548	0.33525
0.190	0.33965	0.34102	0.32334	0.33794	0.34804
0.200	0.35233	0.35374	0.33533	0.35037	0.36078
0.250	0.41569	0.41734	0.39526	0.41235	0.42428
0.300	0.48010	0.48198	0.45621	0.47514	0.48846
0.350	0.54672	0.54882	0.51936	0.53992	0.55450
0.400	0.61662	0.61893	0.58582	0.60776	0.62347
0.450	0.69093	0.69343	0.65678	0.67980	0.69646
0.500	0.77093	0.77359	0.73364	0.75735	0.77473
0.550	0.85819	0.86099	0.81815	0.84203	0.85983
0.600	0.95477	0.95765	0.91265	0.93599	0.95377
0.650	1.06345	1.06633	1.02040	1.04217	1.05931
0.700	1.18826	1.19102	1.14621	1.16493	1.18046
0.750	1.33541	1.33786	1.29765	1.31102	1.32345
0.800	1.51524	1.51702	1.48756	1.49191	1.49874
0.810	1.55656	1.55815	1.53197	1.53388	1.53914
0.820	1.60013	1.60149	1.57910	1.57830	1.58179
0.830	1.64619	1.64730	1.62928	1.62545	1.62695
0.840	1.69505	1.69586	1.68289	1.67569	1.67494
0.850	1.74708	1.74756	1.74041	1.72943	1.72613
0.860	1.80271	1.80280	1.80240	1.78719	1.78099
0.870	1.86250	1.86214	1.86956	1.84959	1.84006
0.880	1.92709	1.92621	1.94274	1.91740	1.90406
0.890	1.99735	1.99585	2.02305	1.99161	1.97387
0.900	2.07435	2.07213	2.11189	2.07348	2.05062
0.910	2.15954	2.15644	2.21112	2.16469	2.13582
0.920	2.25484	2.25069	2.32326	2.26753	2.23152
0.930	2.36298	2.35752	2.45184	2.38521	2.34060
0.940	2.48795	2.48086	2.60208	2.52245	2.46731
0.950	2.63594	2.62672	2.78205	2.68662	2.61826
0.960	2.81734	2.80524	3.00535	2.89009	2.80457
0.970	3.05166	3.03539	3.29752	3.15622	3.04723
0.980	3.38276	3.35976	3.71625	3.53772	3.39373
0.990	3.95106	3.91428	4.44684	4.20419	3.99704
0.995	4.52216	4.46880	5.19151	4.88455	4.61216
0.999	5.85854	5.75635	6.95764	6.50063	6.07462

TABLE 3.9 N= 4 (CONT.)

P ↓	λ:	10 15 30 45	10 15 35 40	10 20 20 50	10 20 25 45	10 20 30 40
0.001		0.01931	0.01949	0.01925	0.01982	0.02014
0.005		0.04418	0.04458	0.04402	0.04532	0.04606
0.010		0.06359	0.06417	0.06334	0.06521	0.06626
0.020		0.09226	0.09310	0.09188	0.09456	0.09608
0.030		0.11533	0.11638	0.11481	0.11815	0.12003
0.040		0.13555	0.13678	0.13492	0.13881	0.14101
0.050		0.15400	0.15539	0.15325	0.15764	0.16013
0.060		0.17122	0.17277	0.17035	0.17521	0.17797
0.070		0.18754	0.18923	0.18656	0.19185	0.19485
0.080		0.20317	0.20500	0.20207	0.20777	0.21101
0.090		0.21825	0.22022	0.21704	0.22313	0.22659
0.100		0.23290	0.23500	0.23157	0.23804	0.24171
0.110		0.24720	0.24941	0.24575	0.25257	0.25645
0.120		0.26120	0.26354	0.25964	0.26680	0.27089
0.130		0.27497	0.27742	0.27328	0.28078	0.28506
0.140		0.28853	0.29110	0.28673	0.29455	0.29901
0.150		0.30193	0.30461	0.30001	0.30814	0.31279
0.160		0.31520	0.31799	0.31315	0.32159	0.32642
0.170		0.32836	0.33125	0.32619	0.33492	0.33993
0.180		0.34143	0.34443	0.33913	0.34816	0.35333
0.190		0.35443	0.35753	0.35200	0.36132	0.36666
0.200		0.36738	0.37058	0.36482	0.37442	0.37992
0.250		0.43187	0.43557	0.42864	0.43954	0.44581
0.300		0.49697	0.50113	0.49303	0.50508	0.51205
0.350		0.56387	0.56845	0.55916	0.57224	0.57984
0.400		0.63361	0.63858	0.62810	0.64206	0.65021
0.450		0.70727	0.71259	0.70093	0.71559	0.72420
0.500		0.78610	0.79171	0.77889	0.79403	0.80299
0.550		0.87158	0.87740	0.86350	0.87884	0.88799
0.600		0.96564	0.97156	0.95673	0.97188	0.98102
0.650		1.07094	1.07679	1.06129	1.07570	1.08454
0.700		1.19129	1.19682	1.18113	1.19401	1.20211
0.750		1.33260	1.33738	1.32238	1.33248	1.33917
0.800		1.50469	1.50803	1.49534	1.50065	1.50481
0.810		1.54419	1.54711	1.53520	1.53918	1.54263
0.820		1.58580	1.58825	1.57726	1.57976	1.58241
0.830		1.62978	1.63169	1.62180	1.62264	1.62438
0.840		1.67643	1.67772	1.66913	1.66809	1.66881
0.850		1.72610	1.72667	1.71962	1.71646	1.71601
0.860		1.77920	1.77895	1.77372	1.76816	1.76638
0.870		1.83625	1.83506	1.83200	1.82369	1.82039
0.880		1.89791	1.89562	1.89515	1.88369	1.87863
0.890		1.96499	1.96140	1.96405	1.94895	1.94183
0.900		2.03852	2.03342	2.03983	2.02050	2.01097
0.910		2.11991	2.11298	2.12399	2.09967	2.08728
0.920		2.21102	2.20189	2.21859	2.18833	2.17249
0.930		2.31449	2.30264	2.32649	2.28904	2.26899
0.940		2.43419	2.41892	2.45194	2.40562	2.38029
0.950		2.57614	2.55643	2.60154	2.54397	2.51184
0.960		2.75044	2.72473	2.78644	2.71406	2.67279
0.970		2.97611	2.94174	3.02767	2.93467	2.88032
0.980		3.29603	3.24773	3.37282	3.24825	3.17310
0.990		3.84791	3.77132	3.97524	3.79161	3.67495
0.995		4.40579	4.29563	4.59063	4.34374	4.17904
0.999		5.72183	5.51598	6.05505	5.65455	5.35952

TABLE 3.10 N= 4 (CONT.)

P ↓	λ: 10 20 35 35	10 25 25 40	10 25 30 35	10 30 30 30	15 15 15 55
0.001	0.01952	0.02035	0.02007	0.01866	0.01889
0.005	0.04470	0.04652	0.04592	0.04287	0.04316
0.010	0.06438	0.06691	0.06610	0.06190	0.06208
0.020	0.09353	0.09701	0.09595	0.09025	0.09000
0.030	0.11702	0.12118	0.11996	0.11324	0.11242
0.040	0.13764	0.14235	0.14102	0.13353	0.13205
0.050	0.15649	0.16163	0.16024	0.15215	0.14994
0.060	0.17409	0.17961	0.17818	0.16962	0.16663
0.070	0.19080	0.19663	0.19518	0.18625	0.18244
0.080	0.20681	0.21291	0.21146	0.20225	0.19757
0.090	0.22227	0.22862	0.22718	0.21776	0.21216
0.100	0.23730	0.24385	0.24244	0.23287	0.22632
0.110	0.25198	0.25869	0.25733	0.24767	0.24014
0.120	0.26636	0.27323	0.27191	0.26221	0.25367
0.130	0.28051	0.28749	0.28624	0.27655	0.26696
0.140	0.29446	0.30154	0.30036	0.29072	0.28005
0.150	0.30824	0.31540	0.31430	0.30476	0.29299
0.160	0.32190	0.32912	0.32811	0.31869	0.30579
0.170	0.33544	0.34270	0.34179	0.33254	0.31848
0.180	0.34890	0.35618	0.35538	0.34632	0.33108
0.190	0.36229	0.36958	0.36889	0.36006	0.34362
0.200	0.37563	0.38292	0.38235	0.37378	0.35611
0.250	0.44209	0.44911	0.44926	0.44238	0.41828
0.300	0.50918	0.51560	0.51662	0.51199	0.48105
0.350	0.57805	0.58357	0.58559	0.58363	0.54563
0.400	0.64971	0.65404	0.65719	0.65822	0.61307
0.450	0.72519	0.72805	0.73244	0.73664	0.68452
0.500	0.80564	0.80678	0.81248	0.81990	0.76126
0.550	0.89244	0.89160	0.89867	0.90921	0.84490
0.600	0.98736	0.98431	0.99275	1.00612	0.93756
0.650	1.09280	1.08733	1.09705	1.11278	1.04216
0.700	1.21220	1.20414	1.21492	1.23224	1.16301
0.750	1.35080	1.34011	1.35149	1.36922	1.30685
0.800	1.51731	1.50412	1.51518	1.53155	1.48514
0.810	1.55517	1.54153	1.55235	1.56814	1.52655
0.820	1.59493	1.58087	1.59135	1.60645	1.57039
0.830	1.63679	1.62235	1.63241	1.64668	1.61696
0.840	1.68103	1.66625	1.67576	1.68904	1.66661
0.850	1.72794	1.71287	1.72170	1.73381	1.71976
0.860	1.77788	1.76259	1.77058	1.78133	1.77692
0.870	1.83131	1.81588	1.82284	1.83199	1.83873
0.880	1.88878	1.87333	1.87900	1.88629	1.90595
0.890	1.95099	1.93565	1.93975	1.94487	1.97958
0.900	2.01884	2.00378	2.00595	2.00852	2.06090
0.910	2.09351	2.07896	2.07872	2.07830	2.15159
0.920	2.17661	2.16286	2.15962	2.15565	2.25395
0.930	2.27036	2.25784	2.25079	2.24257	2.37122
0.940	2.37806	2.36735	2.35537	2.34198	2.50814
0.950	2.50476	2.49673	2.47823	2.45841	2.67211
0.960	2.65898	2.65498	2.62748	2.59942	2.87556
0.970	2.85662	2.85899	2.81832	2.77916	3.14190
0.980	3.13334	3.14681	3.08475	3.02927	3.52396
0.990	3.60284	3.64050	3.53494	3.45036	4.19152
0.995	4.06955	4.13718	3.98051	3.86590	4.87278
0.999	5.14918	5.30449	5.00644	4.81973	6.49005

TABLE 3.11 N= 4 (CONT.)

P ↓	λ: 15 15 20 50	15 15 25 45	15 15 30 40	15 15 35 35	15 20 20 45
0.001	0.01981	0.02040	0.02073	0.02083	0.02072
0.005	0.04525	0.04658	0.04733	0.04758	0.04731
0.010	0.06506	0.06696	0.06804	0.06838	0.06799
0.020	0.09425	0.09699	0.09853	0.09903	0.09843
0.030	0.11767	0.12106	0.12298	0.12360	0.12281
0.040	0.13817	0.14212	0.14435	0.14508	0.14413
0.050	0.15683	0.16128	0.16380	0.16462	0.16351
0.060	0.17422	0.17913	0.18192	0.18282	0.18156
0.070	0.19067	0.19602	0.19905	0.20004	0.19863
0.080	0.20641	0.21216	0.21543	0.21649	0.21494
0.090	0.22157	0.22771	0.23120	0.23234	0.23064
0.100	0.23629	0.24279	0.24650	0.24770	0.24586
0.110	0.25062	0.25749	0.26139	0.26266	0.26068
0.120	0.26466	0.27186	0.27596	0.27730	0.27518
0.130	0.27843	0.28596	0.29025	0.29165	0.28939
0.140	0.29199	0.29984	0.30432	0.30578	0.30338
0.150	0.30538	0.31353	0.31819	0.31971	0.31717
0.160	0.31861	0.32707	0.33191	0.33348	0.33080
0.170	0.33173	0.34048	0.34549	0.34712	0.34430
0.180	0.34474	0.35378	0.35896	0.36064	0.35768
0.190	0.35768	0.36700	0.37234	0.37408	0.37098
0.200	0.37056	0.38015	0.38564	0.38744	0.38419
0.250	0.43452	0.44538	0.45163	0.45368	0.44971
0.300	0.49886	0.51085	0.51779	0.52006	0.51536
0.350	0.56476	0.57776	0.58531	0.58779	0.58234
0.400	0.63329	0.64715	0.65525	0.65791	0.65169
0.450	0.70552	0.72009	0.72865	0.73147	0.72447
0.500	0.78270	0.79777	0.80668	0.80962	0.80185
0.550	0.86632	0.88162	0.89073	0.89376	0.88525
0.600	0.95834	0.97349	0.98262	0.98567	0.97648
0.650	1.06143	1.07592	1.08479	1.08777	1.07803
0.700	1.17952	1.19257	1.20075	1.20353	1.19349
0.750	1.31868	1.32908	1.33592	1.33829	1.32841
0.800	1.48920	1.49493	1.49932	1.50092	1.49212
0.810	1.52852	1.53294	1.53664	1.53803	1.52962
0.820	1.57003	1.57300	1.57591	1.57705	1.56911
0.830	1.61400	1.61532	1.61735	1.61821	1.61084
0.840	1.66074	1.66020	1.66122	1.66176	1.65509
0.850	1.71062	1.70798	1.70785	1.70802	1.70219
0.860	1.76411	1.75907	1.75762	1.75738	1.75254
0.870	1.82176	1.81398	1.81100	1.81028	1.80665
0.880	1.88426	1.87332	1.86859	1.86730	1.86514
0.890	1.95250	1.93790	1.93111	1.92917	1.92879
0.900	2.02762	2.00874	1.99954	1.99682	1.99861
0.910	2.11112	2.08720	2.07511	2.07147	2.07596
0.920	2.20505	2.17510	2.15954	2.15477	2.16264
0.930	2.31231	2.27504	2.25522	2.24906	2.26123
0.940	2.43715	2.39083	2.36565	2.35775	2.37551
0.950	2.58619	2.52838	2.49628	2.48611	2.51138
0.960	2.77064	2.69767	2.65625	2.64299	2.67878
0.970	3.01158	2.91752	2.86274	2.84499	2.89646
0.980	3.35675	3.23045	3.15441	3.12937	3.20692
0.990	3.95983	3.77350	3.65509	3.61504	3.74727
0.995	4.57609	4.32593	4.15865	4.10043	4.29859
0.999	6.04207	5.63803	5.33907	5.22715	5.61146

TABLE 3.12 N= 4 (CONT.)

P ↓	λ:	15 20 25 40	15 20 30 35	20 20 20 40	20 20 25 35	20 20 30 30
0.001		0.02127	0.02153	0.02161	0.02209	0.02225
0.005		0.04855	0.04913	0.04930	0.05039	0.05074
0.010		0.06975	0.07059	0.07081	0.07236	0.07285
0.020		0.10095	0.10215	0.10243	0.10463	0.10534
0.030		0.12593	0.12741	0.12773	0.13043	0.13130
0.040		0.14775	0.14947	0.14981	0.15294	0.15395
0.050		0.16759	0.16952	0.16987	0.17339	0.17452
0.060		0.18605	0.18819	0.18853	0.19239	0.19364
0.070		0.20350	0.20582	0.20616	0.21034	0.21168
0.080		0.22016	0.22265	0.22298	0.22746	0.22890
0.090		0.23620	0.23886	0.23918	0.24393	0.24546
0.100		0.25174	0.25455	0.25486	0.25987	0.26149
0.110		0.26687	0.26983	0.27011	0.27538	0.27708
0.120		0.28166	0.28476	0.28502	0.29052	0.29230
0.130		0.29616	0.29940	0.29963	0.30536	0.30721
0.140		0.31042	0.31379	0.31399	0.31994	0.32186
0.150		0.32447	0.32798	0.32814	0.33430	0.33630
0.160		0.33836	0.34199	0.34212	0.34848	0.35054
0.170		0.35210	0.35585	0.35594	0.36250	0.36463
0.180		0.36573	0.36959	0.36964	0.37639	0.37858
0.190		0.37925	0.38323	0.38324	0.39017	0.39242
0.200		0.39270	0.39679	0.39675	0.40386	0.40617
0.250		0.45926	0.46387	0.46358	0.47149	0.47406
0.300		0.52582	0.53089	0.53030	0.53887	0.54167
0.350		0.59357	0.59904	0.59810	0.60723	0.61021
0.400		0.66356	0.66937	0.66804	0.67759	0.68073
0.450		0.73682	0.74290	0.74114	0.75097	0.75422
0.500		0.81449	0.82075	0.81851	0.82847	0.83178
0.550		0.89793	0.90426	0.90152	0.91140	0.91470
0.600		0.98889	0.99515	0.99187	1.00140	1.00461
0.650		1.08973	1.09571	1.09188	1.10071	1.10373
0.700		1.20384	1.20926	1.20489	1.21252	1.21517
0.750		1.33643	1.34082	1.33601	1.34168	1.34374
0.800		1.49620	1.49882	1.49378	1.49628	1.49734
0.810		1.53263	1.53475	1.52972	1.53137	1.53216
0.820		1.57093	1.57250	1.56750	1.56820	1.56870
0.830		1.61132	1.61226	1.60733	1.60698	1.60715
0.840		1.65405	1.65430	1.64946	1.64794	1.64773
0.850		1.69944	1.69889	1.69421	1.69136	1.69073
0.860		1.74786	1.74641	1.74192	1.73758	1.73648
0.870		1.79977	1.79727	1.79307	1.78703	1.78538
0.880		1.85572	1.85203	1.84819	1.84022	1.83795
0.890		1.91644	1.91136	1.90800	1.89781	1.89481
0.900		1.98286	1.97614	1.97341	1.96063	1.95679
0.910		2.05617	2.04753	2.04561	2.02979	2.02496
0.920		2.13805	2.12707	2.12623	2.10679	2.10078
0.930		2.23080	2.21696	2.21757	2.19373	2.18628
0.940		2.33783	2.32041	2.32299	2.29370	2.28443
0.950		2.46443	2.44238	2.44771	2.41145	2.39987
0.960		2.61949	2.59120	2.60056	2.55501	2.54029
0.970		2.81977	2.78247	2.79811	2.73937	2.72014
0.980		3.10304	3.05125	3.07788	2.99827	2.97179
0.990		3.59079	3.50945	3.56073	3.43964	3.39823
0.995		4.08358	3.96697	4.05006	3.88090	3.82136
0.999		5.24748	5.02978	5.21068	4.91034	4.79643

TABLE 3.13 N= 4 (CONT.)

P ↓	λ: 20 25 25 30	All λ's = 100/4
0.001	0.02247	0.02270
0.005	0.05124	0.05175
0.010	0.07356	0.07428
0.020	0.10634	0.10735
0.030	0.13253	0.13376
0.040	0.15536	0.15679
0.050	0.17609	0.17768
0.060	0.19536	0.19709
0.070	0.21354	0.21541
0.080	0.23088	0.23287
0.090	0.24756	0.24966
0.100	0.26369	0.26591
0.110	0.27938	0.28170
0.120	0.29470	0.29711
0.130	0.30970	0.31220
0.140	0.32444	0.32702
0.150	0.33895	0.34162
0.160	0.35328	0.35602
0.170	0.36744	0.37025
0.180	0.38146	0.38435
0.190	0.39537	0.39832
0.200	0.40918	0.41219
0.250	0.47735	0.48064
0.300	0.54517	0.54867
0.350	0.61387	0.61752
0.400	0.68447	0.68821
0.450	0.75798	0.76174
0.500	0.83548	0.83917
0.550	0.91824	0.92178
0.600	1.00789	1.01116
0.650	1.10657	1.10942
0.700	1.21738	1.21961
0.750	1.34501	1.34632
0.800	1.49722	1.49715
0.810	1.53168	1.53127
0.820	1.56783	1.56703
0.830	1.60584	1.60462
0.840	1.64595	1.64426
0.850	1.68842	1.68622
0.860	1.73359	1.73081
0.870	1.78184	1.77843
0.880	1.83368	1.82954
0.890	1.88971	1.88476
0.900	1.95074	1.94486
0.910	2.01782	2.01086
0.920	2.09236	2.08413
0.930	2.17634	2.16661
0.940	2.27265	2.26109
0.950	2.38579	2.37193
0.960	2.52323	2.50638
0.970	2.69897	2.67797
0.980	2.94435	2.91696
0.990	3.35892	3.31918
0.995	3.76890	3.71506
0.999	4.70957	4.61671

TABLE 4.1 N= 5

P ↓	λ:	5 5 5 80	5 5 10 75	5 5 15 70	5 5 20 65	5 5 25 60
0.001		0.01854	0.02101	0.02250	0.02350	0.02420
0.005		0.03677	0.04167	0.04465	0.04667	0.04809
0.010		0.04997	0.05661	0.06069	0.06347	0.06544
0.020		0.06870	0.07779	0.08345	0.08735	0.09011
0.030		0.08341	0.09439	0.10131	0.10611	0.10951
0.040		0.09615	0.10876	0.11678	0.12237	0.12634
0.050		0.10772	0.12178	0.13080	0.13712	0.14162
0.060		0.11850	0.13390	0.14385	0.15085	0.15585
0.070		0.12871	0.14536	0.15620	0.16385	0.16933
0.080		0.13851	0.15634	0.16803	0.17631	0.18225
0.090		0.14800	0.16694	0.17946	0.18835	0.19475
0.100		0.15724	0.17727	0.19058	0.20007	0.20691
0.110		0.16630	0.18737	0.20146	0.21154	0.21881
0.120		0.17523	0.19730	0.21215	0.22281	0.23052
0.130		0.18405	0.20709	0.22270	0.23392	0.24206
0.140		0.19279	0.21679	0.23313	0.24492	0.25348
0.150		0.20150	0.22641	0.24348	0.25583	0.26481
0.160		0.21017	0.23598	0.25377	0.26668	0.27608
0.170		0.21884	0.24553	0.26403	0.27748	0.28731
0.180		0.22752	0.25506	0.27426	0.28827	0.29851
0.190		0.23622	0.26460	0.28450	0.29905	0.30971
0.200		0.24496	0.27416	0.29475	0.30985	0.32091
0.250		0.28976	0.32275	0.34668	0.36446	0.37759
0.300		0.33741	0.37368	0.40077	0.42116	0.43632
0.350		0.38942	0.42829	0.45829	0.48118	0.49832
0.400		0.44740	0.48799	0.52051	0.54570	0.56473
0.450		0.51333	0.55443	0.58887	0.61605	0.63677
0.500		0.58963	0.62963	0.66510	0.69377	0.71589
0.550		0.67934	0.71619	0.75140	0.78081	0.80388
0.600		0.78630	0.81756	0.85070	0.87977	0.90311
0.650		0.91548	0.93842	0.96710	0.99422	1.01684
0.700		1.07369	1.08547	1.10652	1.12945	1.14985
0.750		1.27116	1.26877	1.27821	1.29366	1.30959
0.800		1.52506	1.50487	1.49762	1.50078	1.50866
0.810		1.58502	1.56072	1.54939	1.54931	1.55496
0.820		1.64880	1.62017	1.60446	1.60083	1.60398
0.830		1.71685	1.68362	1.66323	1.65569	1.65605
0.840		1.78967	1.75157	1.72615	1.71432	1.71155
0.850		1.86789	1.82460	1.79377	1.77721	1.77093
0.860		1.95227	1.90340	1.86676	1.84498	1.83475
0.870		2.04371	1.98886	1.94593	1.91838	1.90368
0.880		2.14339	2.08204	2.03228	1.99833	1.97858
0.890		2.25274	2.18430	2.12709	2.08603	2.06051
0.900		2.37363	2.29739	2.23200	2.18297	2.15085
0.910		2.50853	2.42362	2.34918	2.29118	2.25143
0.920		2.66079	2.56614	2.48155	2.41336	2.36474
0.930		2.83511	2.72936	2.63325	2.55335	2.49426
0.940		3.03843	2.91977	2.81033	2.71678	2.64514
0.950		3.28153	3.14746	3.02222	2.91238	2.82539
0.960		3.58251	3.42942	3.28476	3.15486	3.04847
0.970		3.97544	3.79758	3.62777	3.47191	3.33981
0.980		4.53734	4.32412	4.11856	3.92601	3.75683
0.990		5.51548	5.24086	4.97347	4.71794	4.48430
0.995		6.51095	6.17392	5.84386	5.52496	5.22642
0.999		8.86921	8.38451	7.90645	7.43872	6.98899

TABLE 4.2 N= 5 (CONT.)

P ↓	λ:	5 5 30	5 55	5 5 35	5 50	5 5 40	5 45	5 10 10	5 70	5 10 15	5 65
0.001		0.02467		0.02497		0.02512		0.02378		0.02542	
0.005		0.04907		0.04968		0.04998		0.04710		0.05038	
0.010		0.06679		0.06764		0.06805		0.06393		0.06839	
0.020		0.09201		0.09321		0.09380		0.08773		0.09387	
0.030		0.11186		0.11335		0.11407		0.10632		0.11378	
0.040		0.12909		0.13084		0.13168		0.12237		0.13097	
0.050		0.14474		0.14672		0.14769		0.13688		0.14650	
0.060		0.15933		0.16153		0.16261		0.15035		0.16091	
0.070		0.17315		0.17557		0.17675		0.16306		0.17451	
0.080		0.18640		0.18903		0.19032		0.17520		0.18751	
0.090		0.19921		0.20206		0.20344		0.18691		0.20002	
0.100		0.21169		0.21474		0.21623		0.19828		0.21217	
0.110		0.22391		0.22716		0.22874		0.20938		0.22402	
0.120		0.23592		0.23937		0.24105		0.22026		0.23564	
0.130		0.24777		0.25142		0.25319		0.23097		0.24706	
0.140		0.25950		0.26334		0.26521		0.24155		0.25834	
0.150		0.27113		0.27517		0.27714		0.25202		0.26949	
0.160		0.28270		0.28693		0.28900		0.26241		0.28055	
0.170		0.29423		0.29865		0.30082		0.27275		0.29154	
0.180		0.30573		0.31035		0.31261		0.28305		0.30248	
0.190		0.31723		0.32205		0.32440		0.29333		0.31339	
0.200		0.32873		0.33375		0.33620		0.30361		0.32429	
0.250		0.38691		0.39290		0.39583		0.35541		0.37903	
0.300		0.44713		0.45410		0.45752		0.40894		0.43525	
0.350		0.51060		0.51853		0.52242		0.46548		0.49417	
0.400		0.57841		0.58727		0.59163		0.52631		0.55699	
0.450		0.65175		0.66148		0.66628		0.59287		0.62502	
0.500		0.73199		0.74249		0.74767		0.66693		0.69981	
0.550		0.82082		0.83193		0.83742		0.75077		0.78335	
0.600		0.92047		0.93193		0.93763		0.84744		0.87830	
0.650		1.03400		1.04546		1.05118		0.96122		0.98838	
0.700		1.16587		1.17677		1.18226		1.09834		1.11905	
0.750		1.32300		1.33246		1.33732		1.26833		1.27884	
0.800		1.51705		1.52362		1.52715		1.48696		1.48210	
0.810		1.56193		1.56766		1.57080		1.53870		1.52998	
0.820		1.60934		1.61411		1.61681		1.59379		1.58089	
0.830		1.65958		1.66326		1.66546		1.65262		1.63521	
0.840		1.71301		1.71545		1.71706		1.71564		1.69336	
0.850		1.77003		1.77105		1.77200		1.78341		1.75584	
0.860		1.83117		1.83055		1.83073		1.85658		1.82328	
0.870		1.89703		1.89454		1.89383		1.93597		1.89644	
0.880		1.96840		1.96373		1.96200		2.02257		1.97624	
0.890		2.04625		2.03905		2.03610		2.11766		2.06388	
0.900		2.13184		2.12166		2.11729		2.22288		2.16088	
0.910		2.22684		2.21312		2.20706		2.34037		2.26926	
0.920		2.33351		2.31554		2.30743		2.47308		2.39173	
0.930		2.45504		2.43187		2.42124		2.62512		2.53214	
0.940		2.59610		2.56646		2.55266		2.80255		2.69611	
0.950		2.76398		2.72604		2.70814		3.01479		2.89239	
0.960		2.97095		2.92192		2.89850		3.27766		3.13570	
0.970		3.24010		3.17540		3.14402		3.62100		3.45373	
0.980		3.62369		3.53440		3.49028		4.11213		3.90895	
0.990		4.28990		4.15271		4.08275		4.96740		4.70222	
0.995		4.96767		4.77653		4.67593		5.83802		5.51007	
0.999		6.57635		6.24435		6.05589		7.90090		7.42483	

TABLE 4.3 N= 5 (CONT.)

P ↓	λ: 5 10 20 60	5 10 25 55	5 10 30 50	5 10 35 45	5 10 40 40
0.001	0.02652	0.02726	0.02775	0.02803	0.02812
0.005	0.05257	0.05408	0.05507	0.05563	0.05581
0.010	0.07140	0.07346	0.07482	0.07560	0.07585
0.020	0.09804	0.10092	0.10283	0.10392	0.10427
0.030	0.11888	0.12241	0.12475	0.12609	0.12652
0.040	0.13688	0.14098	0.14370	0.14526	0.14577
0.050	0.15314	0.15777	0.16084	0.16260	0.16317
0.060	0.16824	0.17335	0.17675	0.17870	0.17934
0.070	0.18248	0.18806	0.19177	0.19390	0.19460
0.080	0.19609	0.20211	0.20612	0.20843	0.20919
0.090	0.20921	0.21565	0.21996	0.22244	0.22325
0.100	0.22193	0.22880	0.23339	0.23603	0.23689
0.110	0.23434	0.24162	0.24648	0.24929	0.25021
0.120	0.24651	0.25418	0.25932	0.26229	0.26326
0.130	0.25847	0.26653	0.27194	0.27507	0.27609
0.140	0.27027	0.27872	0.28440	0.28768	0.28875
0.150	0.28194	0.29078	0.29671	0.30015	0.30127
0.160	0.29351	0.30273	0.30892	0.31251	0.31368
0.170	0.30501	0.31459	0.32105	0.32478	0.32601
0.180	0.31645	0.32640	0.33311	0.33700	0.33827
0.190	0.32785	0.33817	0.34514	0.34917	0.35049
0.200	0.33923	0.34992	0.35713	0.36131	0.36269
0.250	0.39633	0.40879	0.41724	0.42215	0.42376
0.300	0.45478	0.46895	0.47860	0.48422	0.48607
0.350	0.51578	0.53158	0.54239	0.54870	0.55078
0.400	0.58048	0.59780	0.60971	0.61669	0.61898
0.450	0.65010	0.66879	0.68170	0.68929	0.69179
0.500	0.72608	0.74589	0.75967	0.76780	0.77048
0.550	0.81021	0.83079	0.84523	0.85378	0.85662
0.600	0.90488	0.92569	0.94047	0.94929	0.95221
0.650	1.01339	1.03364	1.04828	1.05709	1.06003
0.700	1.14059	1.15906	1.17282	1.18123	1.18405
0.750	1.29405	1.30883	1.32051	1.32787	1.33037
0.800	1.48655	1.49451	1.50213	1.50735	1.50918
0.810	1.53153	1.53758	1.54403	1.54862	1.55025
0.820	1.57924	1.58314	1.58827	1.59214	1.59354
0.830	1.63001	1.63149	1.63510	1.63815	1.63930
0.840	1.68422	1.68297	1.68486	1.68698	1.68783
0.850	1.74233	1.73800	1.73793	1.73897	1.73949
0.860	1.80492	1.79709	1.79478	1.79458	1.79470
0.870	1.87265	1.86085	1.85597	1.85434	1.85401
0.880	1.94640	1.93007	1.92220	1.91892	1.91806
0.890	2.02725	2.00572	1.99439	1.98917	1.98768
0.900	2.11658	2.08905	2.07367	2.06616	2.06394
0.910	2.21625	2.18174	2.16157	2.15134	2.14823
0.920	2.32877	2.28605	2.26016	2.24663	2.24247
0.930	2.45765	2.40518	2.37233	2.35478	2.34930
0.940	2.60808	2.54381	2.50236	2.47975	2.47262
0.950	2.78812	2.70925	2.65688	2.62776	2.61848
0.960	3.01133	2.91382	2.84706	2.80917	2.79700
0.970	3.30324	3.18072	3.09390	3.04351	3.02714
0.980	3.72149	3.56245	3.44489	3.37465	3.35152
0.990	4.45132	4.22796	4.05271	3.94300	3.90603
0.995	5.19547	4.90683	4.66949	4.51416	4.46055
0.999	6.96099	6.52006	6.13030	5.85065	5.74810

TABLE 4.4 N= 5 (CONT.)

P ↓	λ:	5 15 15 60	5 15 20 55	5 15 25 50	5 15 30 45	5 15 35 40
0.001		0.02714	0.02827	0.02901	0.02947	0.02969
0.005		0.05379	0.05605	0.05754	0.05847	0.05891
0.010		0.07303	0.07611	0.07817	0.07944	0.08005
0.020		0.10024	0.10451	0.10736	0.10913	0.10999
0.030		0.12151	0.12670	0.13019	0.13236	0.13340
0.040		0.13985	0.14585	0.14990	0.15241	0.15362
0.050		0.15642	0.16315	0.16770	0.17053	0.17190
0.060		0.17179	0.17919	0.18420	0.18733	0.18884
0.070		0.18628	0.19432	0.19977	0.20318	0.20482
0.080		0.20011	0.20875	0.21462	0.21830	0.22008
0.090		0.21343	0.22265	0.22893	0.23286	0.23476
0.100		0.22634	0.23612	0.24279	0.24698	0.24900
0.110		0.23893	0.24925	0.25630	0.26073	0.26287
0.120		0.25125	0.26210	0.26952	0.27419	0.27645
0.130		0.26337	0.27473	0.28251	0.28741	0.28978
0.140		0.27531	0.28717	0.29530	0.30043	0.30292
0.150		0.28711	0.29946	0.30794	0.31330	0.31589
0.160		0.29881	0.31163	0.32046	0.32603	0.32873
0.170		0.31041	0.32371	0.33287	0.33866	0.34147
0.180		0.32196	0.33571	0.34521	0.35121	0.35413
0.190		0.33345	0.34766	0.35749	0.36371	0.36672
0.200		0.34492	0.35958	0.36973	0.37615	0.37927
0.250		0.40233	0.41910	0.43080	0.43824	0.44187
0.300		0.46089	0.47960	0.49275	0.50115	0.50525
0.350		0.52178	0.54225	0.55675	0.56605	0.57059
0.400		0.58615	0.60814	0.62385	0.63398	0.63895
0.450		0.65519	0.67840	0.69517	0.70603	0.71138
0.500		0.73031	0.75436	0.77194	0.78342	0.78908
0.550		0.81329	0.83763	0.85571	0.86762	0.87352
0.600		0.90646	0.93035	0.94849	0.96057	0.96659
0.650		1.01313	1.03550	1.05303	1.06490	1.07087
0.700		1.13812	1.15741	1.17335	1.18444	1.19009
0.750		1.28904	1.30286	1.31567	1.32507	1.32998
0.800		1.47876	1.48336	1.49047	1.49663	1.50008
0.810		1.52318	1.52528	1.53080	1.53604	1.53906
0.820		1.57033	1.56966	1.57338	1.57757	1.58011
0.830		1.62054	1.61679	1.61848	1.62147	1.62347
0.840		1.67421	1.66701	1.66641	1.66805	1.66941
0.850		1.73180	1.72076	1.71756	1.71764	1.71829
0.860		1.79387	1.77853	1.77238	1.77068	1.77050
0.870		1.86113	1.84095	1.83143	1.82769	1.82654
0.880		1.93444	1.90880	1.89541	1.88930	1.88704
0.890		2.01488	1.98306	1.96521	1.95633	1.95277
0.900		2.10387	2.06500	2.04196	2.02983	2.02473
0.910		2.20327	2.15631	2.12718	2.11119	2.10424
0.920		2.31558	2.25925	2.22291	2.20228	2.19310
0.930		2.44436	2.37706	2.33204	2.30574	2.29382
0.940		2.59481	2.51445	2.45882	2.42544	2.41007
0.950		2.77500	2.67878	2.60984	2.56739	2.54755
0.960		2.99853	2.88246	2.79626	2.74172	2.71583
0.970		3.29095	3.14882	3.03907	2.96744	2.93282
0.980		3.70994	3.53061	3.38577	3.28743	3.23880
0.990		4.44083	4.19742	3.98938	3.83945	3.76240
0.995		5.18569	4.87802	4.60474	4.39746	4.28672
0.999		6.95207	6.49441	6.06758	5.71376	5.50711

TABLE 4.5 N= 5 (CONT.)

P ↓	λ: 5 5 20 20 50	5 5 20 25 45	5 5 20 30 40	5 5 20 35 35	5 5 25 25 40
0.001	0.02939	0.03010	0.03050	0.03062	0.03075
0.005	0.05829	0.05972	0.06052	0.06078	0.06102
0.010	0.07917	0.08113	0.08223	0.08259	0.08291
0.020	0.10874	0.11145	0.11298	0.11347	0.11392
0.030	0.13184	0.13515	0.13702	0.13763	0.13816
0.040	0.15179	0.15562	0.15778	0.15848	0.15909
0.050	0.16980	0.17409	0.17652	0.17731	0.17799
0.060	0.18649	0.19122	0.19390	0.19477	0.19550
0.070	0.20223	0.20737	0.21028	0.21123	0.21201
0.080	0.21725	0.22277	0.22591	0.22693	0.22776
0.090	0.23170	0.23760	0.24095	0.24204	0.24291
0.100	0.24570	0.25196	0.25552	0.25668	0.25759
0.110	0.25934	0.26595	0.26971	0.27093	0.27189
0.120	0.27269	0.27963	0.28359	0.28488	0.28587
0.130	0.28580	0.29307	0.29722	0.29857	0.29959
0.140	0.29871	0.30630	0.31063	0.31204	0.31309
0.150	0.31146	0.31936	0.32387	0.32534	0.32642
0.160	0.32408	0.33228	0.33698	0.33850	0.33961
0.170	0.33659	0.34509	0.34996	0.35155	0.35268
0.180	0.34902	0.35782	0.36286	0.36450	0.36565
0.190	0.36138	0.37047	0.37568	0.37738	0.37855
0.200	0.37371	0.38308	0.38845	0.39020	0.39140
0.250	0.43514	0.44585	0.45202	0.45403	0.45529
0.300	0.49733	0.50926	0.51616	0.51842	0.51970
0.350	0.56145	0.57449	0.58205	0.58453	0.58579
0.400	0.62857	0.64255	0.65071	0.65339	0.65458
0.450	0.69975	0.71451	0.72318	0.72603	0.72709
0.500	0.77622	0.79155	0.80060	0.80359	0.80446
0.550	0.85951	0.87510	0.88438	0.88746	0.88808
0.600	0.95157	0.96702	0.97632	0.97942	0.97970
0.650	1.05512	1.06986	1.07888	1.08191	1.08175
0.700	1.17409	1.18730	1.19560	1.19841	1.19771
0.750	1.31461	1.32503	1.33191	1.33429	1.33290
0.800	1.48703	1.49257	1.49689	1.49847	1.49623
0.810	1.52679	1.53099	1.53459	1.53595	1.53352
0.820	1.56877	1.57147	1.57425	1.57536	1.57273
0.830	1.61323	1.61424	1.61611	1.61693	1.61409
0.840	1.66049	1.65959	1.66043	1.66092	1.65786
0.850	1.71092	1.70787	1.70753	1.70764	1.70436
0.860	1.76498	1.75949	1.75779	1.75748	1.75397
0.870	1.82322	1.81494	1.81170	1.81090	1.80716
0.880	1.88634	1.87487	1.86984	1.86847	1.86449
0.890	1.95522	1.94006	1.93296	1.93092	1.92671
0.900	2.03100	2.01155	2.00200	1.99918	1.99473
0.910	2.11517	2.09067	2.07824	2.07448	2.06981
0.920	2.20979	2.17929	2.16336	2.15848	2.15362
0.930	2.31773	2.27997	2.25979	2.25351	2.24851
0.940	2.44325	2.39652	2.37101	2.36299	2.35792
0.950	2.59294	2.53487	2.50250	2.49222	2.48722
0.960	2.77797	2.70498	2.66339	2.65004	2.64539
0.970	3.01937	2.92564	2.87087	2.85309	2.84933
0.980	3.36476	3.23932	3.16361	3.13865	3.13710
0.990	3.96754	3.78288	3.66546	3.62571	3.63078
0.995	4.58320	4.33522	4.16959	4.11195	4.12751
0.999	6.04802	5.64643	5.35022	5.23953	5.29502

TABLE 4.6 N= 5 (CONT.)

P ↓	λ:	5 25 30 35	5 30 30 30	10 10 10 65	10 10 15 60	10 10 20 55
0.001		0.03105	0.03123	0.02686	0.02866	0.02985
0.005		0.06164	0.06199	0.05308	0.05665	0.05900
0.010		0.08376	0.08425	0.07193	0.07675	0.07995
0.020		0.11509	0.11577	0.09847	0.10505	0.10945
0.030		0.13959	0.14041	0.11912	0.12705	0.13239
0.040		0.16074	0.16169	0.13688	0.14597	0.15210
0.050		0.17984	0.18090	0.15288	0.16299	0.16985
0.060		0.19754	0.19871	0.16768	0.17873	0.18625
0.070		0.21423	0.21550	0.18161	0.19354	0.20168
0.080		0.23014	0.23150	0.19489	0.20763	0.21637
0.090		0.24545	0.24690	0.20765	0.22117	0.23047
0.100		0.26029	0.26182	0.22001	0.23427	0.24412
0.110		0.27473	0.27634	0.23204	0.24702	0.25739
0.120		0.28885	0.29055	0.24381	0.25947	0.27035
0.130		0.30271	0.30448	0.25536	0.27169	0.28306
0.140		0.31635	0.31820	0.26674	0.28371	0.29557
0.150		0.32981	0.33173	0.27798	0.29558	0.30790
0.160		0.34313	0.34512	0.28910	0.30731	0.32010
0.170		0.35632	0.35838	0.30014	0.31894	0.33218
0.180		0.36942	0.37155	0.31111	0.33049	0.34417
0.190		0.38244	0.38463	0.32204	0.34197	0.35609
0.200		0.39540	0.39766	0.33293	0.35342	0.36797
0.250		0.45985	0.46240	0.38742	0.41047	0.42707
0.300		0.52476	0.52757	0.44302	0.46834	0.48683
0.350		0.59130	0.59432	0.50099	0.52827	0.54848
0.400		0.66046	0.66366	0.56254	0.59140	0.61315
0.450		0.73327	0.73660	0.62897	0.65895	0.68198
0.500		0.81085	0.81425	0.70188	0.73236	0.75631
0.550		0.89454	0.89794	0.78328	0.81341	0.83781
0.600		0.98609	0.98938	0.87588	0.90451	0.92864
0.650		1.08784	1.09089	0.98352	1.00900	1.03180
0.700		1.20316	1.20581	1.11183	1.13181	1.15172
0.750		1.33724	1.33920	1.26958	1.28068	1.29524
0.800		1.49866	1.49951	1.47145	1.46871	1.47400
0.810		1.53542	1.53597	1.51916	1.51286	1.51561
0.820		1.57405	1.57427	1.56996	1.55977	1.55970
0.830		1.61476	1.61461	1.62419	1.60977	1.60655
0.840		1.65779	1.65723	1.68231	1.66326	1.65653
0.850		1.70346	1.70245	1.74481	1.72071	1.71004
0.860		1.75213	1.75060	1.81231	1.78269	1.76761
0.870		1.80424	1.80212	1.88558	1.84990	1.82986
0.880		1.86034	1.85756	1.96555	1.92320	1.89757
0.890		1.92112	1.91758	2.05340	2.00370	1.97173
0.900		1.98749	1.98307	2.15065	2.09279	2.05361
0.910		2.06060	2.05515	2.25932	2.19234	2.14490
0.920		2.14206	2.13539	2.38212	2.30487	2.24789
0.930		2.23409	2.22595	2.52289	2.43392	2.36578
0.940		2.33994	2.32999	2.68724	2.58470	2.50333
0.950		2.46467	2.45243	2.88393	2.76528	2.66790
0.960		2.61672	2.60144	3.12764	2.98923	2.87188
0.970		2.81192	2.79237	3.44608	3.28214	3.13865
0.980		3.08577	3.05952	3.90172	3.70165	3.52096
0.990		3.55137	3.51190	4.69544	4.43312	4.18846
0.995		4.01484	3.96000	5.50356	5.17832	4.86952
0.999		5.08697	4.98861	7.41866	6.94513	6.48648

TABLE 4.7 N= 5 (CONT.)

P ↓	λ:	10 10 25 50 ⁵	10 10 30 45 ⁵	10 10 35 40 ⁵	10 15 15 55 ⁵	10 15 20 50 ⁵
0.001		0.03063	0.03111	0.03134	0.03054	0.03175
0.005		0.06056	0.06153	0.06199	0.06034	0.06273
0.010		0.08208	0.08340	0.08404	0.08174	0.08497
0.020		0.11239	0.11422	0.11510	0.11183	0.11625
0.030		0.13597	0.13819	0.13926	0.13519	0.14054
0.040		0.15623	0.15880	0.16004	0.15526	0.16140
0.050		0.17448	0.17736	0.17875	0.17330	0.18014
0.060		0.19134	0.19452	0.19605	0.18997	0.19745
0.070		0.20720	0.21065	0.21232	0.20563	0.21371
0.080		0.22230	0.22601	0.22780	0.22052	0.22917
0.090		0.23680	0.24076	0.24268	0.23482	0.24401
0.100		0.25082	0.25503	0.25707	0.24864	0.25834
0.110		0.26446	0.26891	0.27106	0.26207	0.27226
0.120		0.27779	0.28246	0.28472	0.27518	0.28585
0.130		0.29085	0.29575	0.29812	0.28803	0.29916
0.140		0.30370	0.30882	0.31130	0.30066	0.31223
0.150		0.31637	0.32171	0.32430	0.31311	0.32512
0.160		0.32890	0.33445	0.33714	0.32541	0.33784
0.170		0.34130	0.34707	0.34986	0.33759	0.35043
0.180		0.35362	0.35959	0.36249	0.34967	0.36291
0.190		0.36586	0.37203	0.37503	0.36167	0.37531
0.200		0.37804	0.38442	0.38752	0.37361	0.38764
0.250		0.43864	0.44600	0.44959	0.43295	0.44878
0.300		0.49982	0.50812	0.51216	0.49277	0.51023
0.350		0.56280	0.57199	0.57648	0.55430	0.57319
0.400		0.62869	0.63870	0.64362	0.61865	0.63876
0.450		0.69860	0.70937	0.71466	0.68696	0.70803
0.500		0.77381	0.78523	0.79086	0.76054	0.78224
0.550		0.85590	0.86780	0.87369	0.84102	0.86291
0.600		0.94689	0.95903	0.96508	0.93053	0.95199
0.650		1.04959	1.06161	1.06764	1.03205	1.05219
0.700		1.16805	1.17937	1.18512	1.14994	1.16746
0.750		1.30855	1.31825	1.32331	1.29104	1.30392
0.800		1.48165	1.48816	1.49178	1.46698	1.47195
0.810		1.52166	1.52725	1.53045	1.50799	1.51080
0.820		1.56393	1.56848	1.57119	1.55145	1.55186
0.830		1.60874	1.61208	1.61425	1.59768	1.59540
0.840		1.65639	1.65837	1.65990	1.64703	1.64173
0.850		1.70727	1.70768	1.70849	1.69991	1.69122
0.860		1.76184	1.76045	1.76042	1.75685	1.74435
0.870		1.82066	1.81719	1.81620	1.81848	1.80166
0.880		1.88443	1.87855	1.87643	1.88558	1.86387
0.890		1.95403	1.94535	1.94191	1.95915	1.93185
0.900		2.03062	2.01864	2.01363	2.04048	2.00674
0.910		2.11571	2.09980	2.09292	2.13126	2.09008
0.920		2.21134	2.19070	2.18157	2.23379	2.18391
0.930		2.32040	2.29401	2.28209	2.35132	2.29113
0.940		2.44716	2.41358	2.39816	2.48861	2.41602
0.950		2.59822	2.55543	2.53548	2.65306	2.56524
0.960		2.78475	2.72970	2.70361	2.85711	2.74999
0.970		3.02775	2.95540	2.92048	3.12419	2.99141
0.980		3.37477	3.27546	3.22637	3.50716	3.33730
0.990		3.97892	3.82768	3.74992	4.17583	3.94152
0.995		4.59474	4.38593	4.27425	4.85779	4.55864
0.999		6.05827	5.70272	5.49471	6.47593	6.02579

TABLE 4.8 N= 5 (CONT.)

P ↓	λ:	5 10 15 25 45	5 10 15 30 40	5 10 15 35 35	5 10 20 20 45	5 10 20 25 40
0.001		0.03251	0.03294	0.03307	0.03293	0.03363
0.005		0.06425	0.06510	0.06537	0.06506	0.06646
0.010		0.08704	0.08820	0.08857	0.08813	0.09004
0.020		0.11909	0.12069	0.12121	0.12057	0.12318
0.030		0.14398	0.14593	0.14655	0.14575	0.14890
0.040		0.16536	0.16759	0.16832	0.16736	0.17097
0.050		0.18457	0.18707	0.18788	0.18677	0.19081
0.060		0.20230	0.20505	0.20594	0.20469	0.20911
0.070		0.21896	0.22194	0.22290	0.22152	0.22629
0.080		0.23480	0.23799	0.23903	0.23751	0.24262
0.090		0.24999	0.25340	0.25450	0.25285	0.25827
0.100		0.26467	0.26827	0.26944	0.26766	0.27338
0.110		0.27893	0.28272	0.28396	0.28204	0.28806
0.120		0.29284	0.29682	0.29811	0.29607	0.30236
0.130		0.30646	0.31062	0.31198	0.30980	0.31637
0.140		0.31984	0.32418	0.32560	0.32329	0.33012
0.150		0.33302	0.33754	0.33901	0.33657	0.34365
0.160		0.34604	0.35072	0.35225	0.34968	0.35701
0.170		0.35891	0.36376	0.36534	0.36264	0.37022
0.180		0.37167	0.37669	0.37832	0.37549	0.38330
0.190		0.38434	0.38952	0.39121	0.38823	0.39628
0.200		0.39694	0.40227	0.40401	0.40090	0.40917
0.250		0.45936	0.46545	0.46744	0.46364	0.47296
0.300		0.52197	0.52876	0.53099	0.52646	0.53673
0.350		0.58601	0.59345	0.59589	0.59061	0.60171
0.400		0.65253	0.66055	0.66319	0.65713	0.66895
0.450		0.72260	0.73115	0.73396	0.72710	0.73948
0.500		0.79743	0.80639	0.80935	0.80167	0.81444
0.550		0.87845	0.88770	0.89077	0.88228	0.89520
0.600		0.96754	0.97688	0.97999	0.97074	0.98350
0.650		1.06721	1.07637	1.07944	1.06954	1.08169
0.700		1.18116	1.18971	1.19260	1.18229	1.19318
0.750		1.31505	1.32231	1.32482	1.31454	1.32315
0.800		1.47839	1.48322	1.48498	1.47564	1.48032
0.810		1.51593	1.52006	1.52160	1.51264	1.51623
0.820		1.55551	1.55885	1.56014	1.55163	1.55401
0.830		1.59737	1.59981	1.60082	1.59287	1.59388
0.840		1.64179	1.64321	1.64390	1.63663	1.63610
0.850		1.68913	1.68938	1.68969	1.68325	1.68097
0.860		1.73979	1.73869	1.73858	1.73314	1.72888
0.870		1.79428	1.79163	1.79102	1.78679	1.78027
0.880		1.85323	1.84877	1.84760	1.84484	1.83572
0.890		1.91743	1.91087	1.90902	1.90807	1.89593
0.900		1.98792	1.97887	1.97623	1.97749	1.96184
0.910		2.06604	2.05404	2.05045	2.05445	2.03466
0.920		2.15365	2.13808	2.13333	2.14079	2.11605
0.930		2.25334	2.23338	2.22721	2.23908	2.20832
0.940		2.36892	2.34346	2.33550	2.35311	2.31488
0.950		2.50633	2.47376	2.46347	2.48880	2.44103
0.960		2.67556	2.63345	2.61999	2.65611	2.59568
0.970		2.89547	2.83969	2.82164	2.87385	2.79559
0.980		3.20866	3.13119	3.10570	3.18461	3.07859
0.990		3.75235	3.63186	3.59109	3.72575	3.56630
0.995		4.30544	4.13557	4.07636	4.27788	4.05935
0.999		5.61876	5.31648	5.20301	5.59219	5.22419

TABLE 4.9 N= 5 (CONT.)

P ↓	λ:	5 10 20 30 35	5 10 25 25 35	5 10 25 30 30	5 15 15 15 50	5 15 15 20 45
0.001		0.03397	0.03424	0.03444	0.03248	0.03369
0.005		0.06713	0.06767	0.06806	0.06414	0.06652
0.010		0.09094	0.09168	0.09220	0.08684	0.09006
0.020		0.12441	0.12542	0.12614	0.11873	0.12312
0.030		0.15040	0.15161	0.15247	0.14346	0.14874
0.040		0.17269	0.17407	0.17507	0.16467	0.17070
0.050		0.19273	0.19426	0.19536	0.18371	0.19042
0.060		0.21121	0.21288	0.21409	0.20128	0.20859
0.070		0.22856	0.23035	0.23166	0.21777	0.22564
0.080		0.24505	0.24696	0.24836	0.23343	0.24184
0.090		0.26086	0.26288	0.26436	0.24845	0.25736
0.100		0.27612	0.27824	0.27980	0.26295	0.27233
0.110		0.29093	0.29315	0.29480	0.27702	0.28686
0.120		0.30538	0.30769	0.30941	0.29075	0.30102
0.130		0.31951	0.32192	0.32371	0.30418	0.31487
0.140		0.33339	0.33588	0.33774	0.31737	0.32847
0.150		0.34705	0.34962	0.35155	0.33036	0.34184
0.160		0.36053	0.36318	0.36518	0.34318	0.35504
0.170		0.37386	0.37658	0.37864	0.35586	0.36808
0.180		0.38705	0.38985	0.39197	0.36842	0.38100
0.190		0.40015	0.40302	0.40520	0.38088	0.39381
0.200		0.41316	0.41609	0.41833	0.39327	0.40653
0.250		0.47747	0.48070	0.48322	0.45461	0.46942
0.300		0.54171	0.54519	0.54795	0.51607	0.53223
0.350		0.60712	0.61078	0.61375	0.57888	0.59620
0.400		0.67472	0.67852	0.68166	0.64412	0.66238
0.450		0.74556	0.74941	0.75268	0.71288	0.73183
0.500		0.82075	0.82457	0.82793	0.78637	0.80571
0.550		0.90164	0.90533	0.90870	0.86610	0.88542
0.600		0.98992	0.99334	0.99665	0.95399	0.97276
0.650		1.08789	1.09088	1.09400	1.05270	1.07018
0.700		1.19884	1.20117	1.20395	1.16615	1.18123
0.750		1.32783	1.32916	1.33133	1.30039	1.31143
0.800		1.48323	1.48305	1.48420	1.46574	1.47004
0.810		1.51864	1.51808	1.51895	1.50399	1.50647
0.820		1.55587	1.55488	1.55544	1.54444	1.54488
0.830		1.59511	1.59365	1.59387	1.58735	1.58551
0.840		1.63662	1.63464	1.63448	1.63302	1.62863
0.850		1.68069	1.67814	1.67754	1.68185	1.67459
0.860		1.72767	1.72449	1.72339	1.73429	1.72379
0.870		1.77801	1.77412	1.77246	1.79090	1.77673
0.880		1.83223	1.82756	1.82525	1.85238	1.83403
0.890		1.89102	1.88546	1.88240	1.91964	1.89648
0.900		1.95526	1.94869	1.94475	1.99381	1.96510
0.910		2.02610	2.01836	2.01340	2.07642	2.04123
0.920		2.10509	2.09600	2.08981	2.16953	2.12671
0.930		2.19442	2.18374	2.17606	2.27608	2.22411
0.940		2.29731	2.28471	2.27518	2.40035	2.33724
0.950		2.41870	2.40375	2.39184	2.54903	2.47202
0.960		2.56693	2.54898	2.53387	2.73339	2.63844
0.970		2.75760	2.73562	2.71595	2.97466	2.85539
0.980		3.02576	2.99783	2.97088	3.32078	3.16556
0.990		3.48331	3.44481	3.40310	3.92595	3.70670
0.995		3.94050	3.89126	3.83187	4.54410	4.25949
0.999		5.00310	4.93005	4.81858	6.01288	5.57563

TABLE 4.10 N= 5 (CONT.)

P ↓	λ:	5 15 15 25 40	5 15 15 30 35	5 15 20 20 40	5 15 20 25 35	5 15 20 30 30
0.001		0.03441	0.03475	0.03485	0.03548	0.03568
0.005		0.06795	0.06863	0.06881	0.07006	0.07045
0.010		0.09200	0.09292	0.09315	0.09483	0.09537
0.020		0.12577	0.12702	0.12731	0.12960	0.13034
0.030		0.15193	0.15345	0.15376	0.15653	0.15742
0.040		0.17436	0.17610	0.17643	0.17959	0.18061
0.050		0.19449	0.19643	0.19677	0.20028	0.20141
0.060		0.21305	0.21518	0.21551	0.21934	0.22057
0.070		0.23046	0.23275	0.23308	0.23721	0.23854
0.080		0.24698	0.24944	0.24976	0.25416	0.25558
0.090		0.26281	0.26542	0.26573	0.27039	0.27190
0.100		0.27809	0.28084	0.28113	0.28604	0.28763
0.110		0.29290	0.29580	0.29607	0.30122	0.30288
0.120		0.30734	0.31037	0.31062	0.31599	0.31773
0.130		0.32146	0.32462	0.32485	0.33044	0.33224
0.140		0.33532	0.33860	0.33880	0.34460	0.34648
0.150		0.34895	0.35236	0.35253	0.35853	0.36047
0.160		0.36239	0.36592	0.36606	0.37226	0.37426
0.170		0.37567	0.37932	0.37942	0.38581	0.38788
0.180		0.38882	0.39258	0.39265	0.39922	0.40135
0.190		0.40186	0.40573	0.40576	0.41251	0.41470
0.200		0.41480	0.41878	0.41878	0.42570	0.42794
0.250		0.47873	0.48322	0.48298	0.49071	0.49322
0.300		0.54246	0.54742	0.54691	0.55532	0.55806
0.350		0.60724	0.61262	0.61178	0.62078	0.62373
0.400		0.67412	0.67987	0.67866	0.68814	0.69125
0.450		0.74414	0.75019	0.74855	0.75839	0.76164
0.500		0.81842	0.82469	0.82259	0.83263	0.83596
0.550		0.89830	0.90471	0.90209	0.91215	0.91550
0.600		0.98551	0.99192	0.98873	0.99856	1.00186
0.650		1.08238	1.08859	1.08481	1.09405	1.09719
0.700		1.19225	1.19798	1.19360	1.20176	1.20458
0.750		1.32028	1.32507	1.32014	1.32647	1.32873
0.800		1.47508	1.47816	1.47288	1.47611	1.47742
0.810		1.51045	1.51305	1.50775	1.51014	1.51119
0.820		1.54767	1.54973	1.54442	1.54588	1.54664
0.830		1.58696	1.58840	1.58312	1.58354	1.58396
0.840		1.62856	1.62932	1.62410	1.62333	1.62338
0.850		1.67280	1.67276	1.66765	1.66555	1.66518
0.860		1.72004	1.71908	1.71414	1.71054	1.70969
0.870		1.77073	1.76872	1.76402	1.75869	1.75730
0.880		1.82543	1.82222	1.81784	1.81054	1.80851
0.890		1.88487	1.88024	1.87630	1.86672	1.86396
0.900		1.94996	1.94366	1.94031	1.92807	1.92446
0.910		2.02191	2.01362	2.01107	1.99568	1.99105
0.920		2.10238	2.09168	2.09019	2.07104	2.06519
0.930		2.19367	2.18001	2.17996	2.15624	2.14889
0.940		2.29919	2.28180	2.28374	2.25432	2.24510
0.950		2.42423	2.40201	2.40676	2.37003	2.35838
0.960		2.57769	2.54892	2.55783	2.51133	2.49640
0.970		2.77632	2.73811	2.75356	2.69314	2.67348
0.980		3.05798	3.00454	3.03153	2.94908	2.92176
0.990		3.54438	3.45996	3.51293	3.38676	3.34367
0.995		4.03699	3.91579	4.00221	3.82571	3.76345
0.999		5.20245	4.97701	5.16503	4.85314	4.73378

TABLE 4.11 N= 5 (CONT.)

P ↓	λ:	5 15 25 25 30	5 20 20 20 35	5 20 20 25 30	5 20 25 25 25	10 10 10 10 60
0.001		0.03598	0.03594	0.03644	0.03673	0.03026
0.005		0.07103	0.07094	0.07192	0.07250	0.05961
0.010		0.09614	0.09601	0.09733	0.09812	0.08059
0.020		0.13138	0.13118	0.13297	0.13403	0.10997
0.030		0.15867	0.15840	0.16055	0.16182	0.13269
0.040		0.18203	0.18170	0.18416	0.18560	0.15215
0.050		0.20298	0.20260	0.20532	0.20691	0.16961
0.060		0.22228	0.22184	0.22480	0.22653	0.18571
0.070		0.24037	0.23988	0.24305	0.24491	0.20081
0.080		0.25753	0.25698	0.26036	0.26233	0.21515
0.090		0.27395	0.27335	0.27692	0.27900	0.22890
0.100		0.28978	0.28913	0.29288	0.29506	0.24218
0.110		0.30513	0.30442	0.30835	0.31062	0.25507
0.120		0.32007	0.31931	0.32340	0.32576	0.26764
0.130		0.33467	0.33386	0.33811	0.34055	0.27996
0.140		0.34899	0.34812	0.35252	0.35505	0.29205
0.150		0.36306	0.36214	0.36668	0.36929	0.30397
0.160		0.37693	0.37595	0.38063	0.38331	0.31574
0.170		0.39062	0.38958	0.39440	0.39715	0.32739
0.180		0.40416	0.40307	0.40801	0.41083	0.33894
0.190		0.41758	0.41643	0.42150	0.42439	0.35042
0.200		0.43089	0.42968	0.43487	0.43782	0.36183
0.250		0.49644	0.49496	0.50069	0.50392	0.41855
0.300		0.56152	0.55974	0.56593	0.56938	0.47580
0.350		0.62736	0.62528	0.63184	0.63545	0.53482
0.400		0.69500	0.69262	0.69946	0.70318	0.59678
0.450		0.76543	0.76275	0.76977	0.77354	0.66292
0.500		0.83973	0.83676	0.84383	0.84758	0.73466
0.550		0.91915	0.91590	0.92288	0.92652	0.81382
0.600		1.00528	1.00177	1.00848	1.01189	0.90281
0.650		1.10023	1.09652	1.10270	1.10574	1.00503
0.700		1.20702	1.20321	1.20850	1.21097	1.12552
0.750		1.33026	1.32653	1.33040	1.33201	1.27216
0.800		1.47756	1.47425	1.47585	1.47614	1.45832
0.810		1.51097	1.50781	1.50880	1.50875	1.50217
0.820		1.54602	1.54304	1.54336	1.54293	1.54882
0.830		1.58291	1.58014	1.57972	1.57887	1.59860
0.840		1.62185	1.61934	1.61808	1.61677	1.65190
0.850		1.66312	1.66091	1.65872	1.65690	1.70922
0.860		1.70702	1.70519	1.70194	1.69955	1.77111
0.870		1.75397	1.75257	1.74813	1.74509	1.83828
0.880		1.80443	1.80357	1.79777	1.79400	1.91161
0.890		1.85902	1.85880	1.85144	1.84684	1.99218
0.900		1.91853	1.91910	1.90993	1.90437	2.08142
0.910		1.98399	1.98553	1.97423	1.96756	2.18117
0.920		2.05679	2.05956	2.04571	2.03773	2.29397
0.930		2.13890	2.14321	2.12629	2.11675	2.42335
0.940		2.23317	2.23950	2.21877	2.20730	2.57450
0.950		2.34403	2.35307	2.32747	2.31357	2.75551
0.960		2.47889	2.49174	2.45964	2.44254	2.97994
0.970		2.65160	2.67019	2.62883	2.60726	3.27334
0.980		2.89321	2.92150	2.86543	2.83685	3.69339
0.990		3.30247	3.35184	3.26613	3.22378	4.42542
0.995		3.70828	3.78442	3.66360	3.60522	5.17097
0.999		4.64235	4.80155	4.57999	4.47605	6.93821

TABLE 4.12 N= 5 (CONT.)

P ↓	λ: 10 10 15 55	10 10 20 50	10 10 25 45	10 10 30 40	10 10 35 35
0.001	0.03223	0.03349	0.03429	0.03474	0.03489
0.005	0.06346	0.06594	0.06752	0.06840	0.06869
0.010	0.08574	0.08909	0.09123	0.09243	0.09282
0.020	0.11693	0.12148	0.12439	0.12603	0.12657
0.030	0.14101	0.14648	0.15000	0.15199	0.15263
0.040	0.16162	0.16787	0.17190	0.17417	0.17491
0.050	0.18008	0.18703	0.19151	0.19405	0.19487
0.060	0.19709	0.20466	0.20957	0.21235	0.21325
0.070	0.21302	0.22118	0.22648	0.22949	0.23046
0.080	0.22815	0.23686	0.24253	0.24574	0.24679
0.090	0.24263	0.25187	0.25789	0.26130	0.26241
0.100	0.25660	0.26634	0.27269	0.27630	0.27748
0.110	0.27016	0.28037	0.28705	0.29085	0.29208
0.120	0.28336	0.29404	0.30103	0.30501	0.30631
0.130	0.29629	0.30741	0.31470	0.31886	0.32021
0.140	0.30897	0.32052	0.32811	0.33245	0.33385
0.150	0.32145	0.33343	0.34131	0.34581	0.34727
0.160	0.33377	0.34615	0.35431	0.35898	0.36050
0.170	0.34595	0.35873	0.36717	0.37199	0.37357
0.180	0.35801	0.37118	0.37989	0.38488	0.38650
0.190	0.36998	0.38354	0.39251	0.39765	0.39933
0.200	0.38188	0.39581	0.40504	0.41034	0.41206
0.250	0.44082	0.45650	0.46698	0.47301	0.47498
0.300	0.49997	0.51724	0.52886	0.53558	0.53778
0.350	0.56060	0.57929	0.59197	0.59933	0.60174
0.400	0.62382	0.64374	0.65738	0.66533	0.66794
0.450	0.69079	0.71173	0.72619	0.73467	0.73746
0.500	0.76284	0.78448	0.79961	0.80853	0.81148
0.550	0.84160	0.86354	0.87909	0.88833	0.89139
0.600	0.92922	0.95086	0.96651	0.97589	0.97901
0.650	1.02870	1.04919	1.06441	1.07367	1.07677
0.700	1.14444	1.16248	1.17650	1.18522	1.18816
0.750	1.28338	1.29692	1.30847	1.31597	1.31855
0.800	1.45729	1.46295	1.46989	1.47501	1.47685
0.810	1.49794	1.50142	1.50705	1.51147	1.51310
0.820	1.54105	1.54211	1.54625	1.54989	1.55127
0.830	1.58696	1.58528	1.58774	1.59048	1.59158
0.840	1.63600	1.63126	1.63180	1.63352	1.63430
0.850	1.68861	1.68042	1.67878	1.67932	1.67973
0.860	1.74531	1.73322	1.72909	1.72827	1.72826
0.870	1.80673	1.79024	1.78324	1.78085	1.78034
0.880	1.87368	1.85217	1.84186	1.83765	1.83656
0.890	1.94714	1.91990	1.90575	1.89941	1.89764
0.900	2.02842	1.99459	1.97594	1.96708	1.96451
0.910	2.11921	2.07776	2.05379	2.04193	2.03839
0.920	2.22183	2.17147	2.14116	2.12566	2.12095
0.930	2.33951	2.27863	2.24063	2.22067	2.21452
0.940	2.47704	2.40353	2.35604	2.33048	2.32251
0.950	2.64182	2.55284	2.49332	2.46053	2.45019
0.960	2.84630	2.73779	2.66250	2.62000	2.60644
0.970	3.11390	2.97954	2.88245	2.82607	2.80785
0.980	3.49750	3.32594	3.19582	3.11746	3.09170
0.990	4.16691	3.93093	3.73994	3.61814	3.57691
0.995	4.84933	4.54863	4.29348	4.12195	4.06212
0.999	6.46801	6.01654	5.60762	5.30320	5.18872

TABLE 4.13 N= 5 (CONT.)

P ↓	λ:	10 10 15 15 50	10 10 15 20 45	10 10 15 25 40	10 10 15 30 35	10 10 20 20 40
0.001		0.03426	0.03553	0.03629	0.03664	0.03675
0.005		0.06740	0.06988	0.07136	0.07206	0.07225
0.010		0.09102	0.09435	0.09635	0.09729	0.09753
0.020		0.12401	0.12851	0.13122	0.13251	0.13279
0.030		0.14945	0.15484	0.15809	0.15964	0.15996
0.040		0.17118	0.17731	0.18103	0.18280	0.18313
0.050		0.19063	0.19742	0.20155	0.20351	0.20384
0.060		0.20851	0.21590	0.22040	0.22255	0.22288
0.070		0.22526	0.23320	0.23804	0.24035	0.24068
0.080		0.24113	0.24958	0.25475	0.25722	0.25754
0.090		0.25631	0.26525	0.27073	0.27335	0.27365
0.100		0.27094	0.28034	0.28611	0.28887	0.28916
0.110		0.28512	0.29496	0.30101	0.30390	0.30417
0.120		0.29892	0.30918	0.31550	0.31853	0.31877
0.130		0.31241	0.32308	0.32966	0.33281	0.33303
0.140		0.32564	0.33669	0.34353	0.34680	0.34700
0.150		0.33864	0.35008	0.35715	0.36055	0.36071
0.160		0.35146	0.36326	0.37057	0.37408	0.37422
0.170		0.36412	0.37628	0.38382	0.38745	0.38755
0.180		0.37665	0.38915	0.39692	0.40066	0.40072
0.190		0.38907	0.40191	0.40990	0.41374	0.41377
0.200		0.40140	0.41456	0.42277	0.42672	0.42671
0.250		0.46228	0.47696	0.48618	0.49063	0.49040
0.300		0.52305	0.53906	0.54919	0.55410	0.55360
0.350		0.58496	0.60213	0.61307	0.61840	0.61759
0.400		0.64912	0.66725	0.67890	0.68460	0.68343
0.450		0.71663	0.73549	0.74773	0.75375	0.75216
0.500		0.78871	0.80803	0.82070	0.82696	0.82491
0.550		0.86687	0.88625	0.89916	0.90557	0.90301
0.600		0.95303	0.97198	0.98482	0.99126	0.98813
0.650		1.04988	1.06766	1.08003	1.08631	1.08258
0.700		1.16133	1.17686	1.18814	1.19398	1.18961
0.750		1.29349	1.30510	1.31429	1.31923	1.31428
0.800		1.45674	1.46165	1.46710	1.47038	1.46501
0.810		1.49459	1.49767	1.50206	1.50487	1.49946
0.820		1.53463	1.53566	1.53887	1.54114	1.53572
0.830		1.57714	1.57588	1.57775	1.57940	1.57399
0.840		1.62243	1.61859	1.61894	1.61990	1.61454
0.850		1.67088	1.66414	1.66275	1.66292	1.65765
0.860		1.72296	1.71293	1.70957	1.70882	1.70370
0.870		1.77924	1.76547	1.75984	1.75804	1.75313
0.880		1.84042	1.82238	1.81412	1.81110	1.80651
0.890		1.90739	1.88445	1.87314	1.86868	1.86452
0.900		1.98132	1.95271	1.93781	1.93166	1.92807
0.910		2.06374	2.02849	2.00934	2.00118	1.99837
0.920		2.15672	2.11365	2.08940	2.07879	2.07704
0.930		2.26320	2.21078	2.18029	2.16668	2.16637
0.940		2.38749	2.32368	2.28543	2.26802	2.26971
0.950		2.53629	2.45829	2.41010	2.38778	2.39231
0.960		2.72089	2.62465	2.56324	2.53425	2.54300
0.970		2.96256	2.84166	2.76160	2.72299	2.73840
0.980		3.30929	3.15211	3.04310	2.98900	3.01616
0.990		3.91534	3.69392	3.52957	3.44403	3.49765
0.995		4.53411	4.24736	4.02246	3.89972	3.98730
0.999		6.00365	5.56451	5.18869	4.96093	5.15114

TABLE 4.14 N= 5 (CONT.)

P ↓ λ:	10 10 20 25 35	10 10 20 30 30	10 10 25 25 30	10 15 15 15 45	10 15 15 20 40
0.001	0.03741	0.03762	0.03792	0.03634	0.03759
0.005	0.07354	0.07396	0.07455	0.07141	0.07383
0.010	0.09927	0.09982	0.10062	0.09636	0.09960
0.020	0.13515	0.13590	0.13697	0.13114	0.13549
0.030	0.16277	0.16368	0.16495	0.15790	0.16309
0.040	0.18633	0.18736	0.18880	0.18072	0.18660
0.050	0.20740	0.20854	0.21013	0.20111	0.20760
0.060	0.22674	0.22799	0.22971	0.21984	0.22688
0.070	0.24483	0.24617	0.24801	0.23735	0.24490
0.080	0.26196	0.26338	0.26534	0.25393	0.26194
0.090	0.27832	0.27983	0.28189	0.26977	0.27822
0.100	0.29407	0.29566	0.29781	0.28501	0.29387
0.110	0.30932	0.31097	0.31322	0.29977	0.30902
0.120	0.32414	0.32587	0.32821	0.31412	0.32374
0.130	0.33861	0.34041	0.34283	0.32812	0.33811
0.140	0.35277	0.35464	0.35714	0.34184	0.35217
0.150	0.36669	0.36862	0.37120	0.35532	0.36597
0.160	0.38039	0.38238	0.38503	0.36858	0.37955
0.170	0.39390	0.39595	0.39868	0.38167	0.39295
0.180	0.40725	0.40937	0.41216	0.39461	0.40619
0.190	0.42047	0.42265	0.42550	0.40742	0.41929
0.200	0.43358	0.43581	0.43873	0.42013	0.43227
0.250	0.49806	0.50054	0.50374	0.48268	0.49609
0.300	0.56194	0.56466	0.56808	0.54478	0.55926
0.350	0.62651	0.62943	0.63304	0.60769	0.62307
0.400	0.69285	0.69594	0.69967	0.67251	0.68860
0.450	0.76196	0.76519	0.76898	0.74029	0.75688
0.500	0.83495	0.83827	0.84205	0.81219	0.82900
0.550	0.91310	0.91646	0.92014	0.88958	0.90630
0.600	0.99804	1.00136	1.00483	0.97424	0.99041
0.650	1.09196	1.09514	1.09824	1.06858	1.08360
0.700	1.19798	1.20086	1.20339	1.17611	1.18907
0.750	1.32087	1.32321	1.32485	1.30226	1.31180
0.800	1.46856	1.46997	1.47023	1.45621	1.46010
0.810	1.50218	1.50333	1.50323	1.49163	1.49399
0.820	1.53750	1.53836	1.53786	1.52901	1.52965
0.830	1.57473	1.57526	1.57432	1.56857	1.56731
0.840	1.61409	1.61425	1.61283	1.61060	1.60720
0.850	1.65587	1.65561	1.65365	1.65543	1.64962
0.860	1.70040	1.69966	1.69710	1.70347	1.69494
0.870	1.74810	1.74681	1.74357	1.75523	1.74361
0.880	1.79948	1.79756	1.79355	1.81132	1.79616
0.890	1.85518	1.85252	1.84764	1.87254	1.85330
0.900	1.91604	1.91251	1.90663	1.93991	1.91594
0.910	1.98315	1.97860	1.97156	2.01477	1.98525
0.920	2.05800	2.05220	2.04380	2.09898	2.06287
0.930	2.14266	2.13535	2.12532	2.19513	2.15106
0.940	2.24020	2.23098	2.21897	2.30704	2.25319
0.950	2.35535	2.34366	2.32917	2.44068	2.37449
0.960	2.49606	2.48103	2.46331	2.60611	2.52378
0.970	2.67727	2.65741	2.63523	2.82233	2.71769
0.980	2.93257	2.90492	2.87592	3.13231	2.99393
0.990	3.36963	3.32591	3.28401	3.67442	3.47410
0.995	3.80832	3.74512	3.68901	4.22878	3.96358
0.999	4.83584	4.71480	4.62199	5.54802	5.12886

TABLE 4.15 N= 5 (CONT.)

P ↓	λ:	10 15 15 25 35	10 15 15 30 30	10 15 20 20 35	10 15 20 25 30	10 20 20 20 30
0.001		0.03826	0.03848	0.03875	0.03928	0.03978
0.005		0.07515	0.07557	0.07608	0.07711	0.07806
0.010		0.10137	0.10193	0.10260	0.10398	0.10524
0.020		0.13787	0.13864	0.13950	0.14136	0.14302
0.030		0.16594	0.16685	0.16786	0.17007	0.17202
0.040		0.18984	0.19088	0.19200	0.19451	0.19670
0.050		0.21119	0.21234	0.21355	0.21632	0.21871
0.060		0.23078	0.23203	0.23331	0.23632	0.23888
0.070		0.24908	0.25042	0.25177	0.25499	0.25771
0.080		0.26639	0.26782	0.26923	0.27264	0.27551
0.090		0.28292	0.28443	0.28589	0.28949	0.29248
0.100		0.29881	0.30040	0.30190	0.30568	0.30879
0.110		0.31419	0.31585	0.31739	0.32133	0.32455
0.120		0.32912	0.33086	0.33243	0.33653	0.33986
0.130		0.34370	0.34550	0.34710	0.35135	0.35478
0.140		0.35796	0.35983	0.36146	0.36585	0.36937
0.150		0.37196	0.37389	0.37554	0.38007	0.38367
0.160		0.38573	0.38772	0.38939	0.39406	0.39774
0.170		0.39931	0.40136	0.40305	0.40784	0.41159
0.180		0.41272	0.41483	0.41653	0.42145	0.42527
0.190		0.42599	0.42816	0.42987	0.43490	0.43879
0.200		0.43914	0.44137	0.44309	0.44824	0.45219
0.250		0.50372	0.50620	0.50793	0.51360	0.51780
0.300		0.56756	0.57027	0.57194	0.57805	0.58241
0.350		0.63196	0.63487	0.63642	0.64290	0.64733
0.400		0.69797	0.70105	0.70244	0.70920	0.71363
0.450		0.76662	0.76984	0.77099	0.77795	0.78228
0.500		0.83900	0.84230	0.84315	0.85021	0.85433
0.550		0.91636	0.91971	0.92018	0.92719	0.93099
0.600		1.00032	1.00364	1.00365	1.01043	1.01376
0.650		1.09302	1.09621	1.09565	1.10198	1.10463
0.700		1.19754	1.20045	1.19920	1.20472	1.20646
0.750		1.31857	1.32098	1.31888	1.32308	1.32354
0.800		1.46393	1.46544	1.46233	1.46435	1.46302
0.810		1.49701	1.49826	1.49494	1.49637	1.49459
0.820		1.53177	1.53274	1.52918	1.52996	1.52769
0.830		1.56840	1.56905	1.56525	1.56530	1.56250
0.840		1.60713	1.60742	1.60338	1.60261	1.59924
0.850		1.64825	1.64812	1.64383	1.64214	1.63814
0.860		1.69207	1.69148	1.68693	1.68420	1.67950
0.870		1.73902	1.73788	1.73308	1.72916	1.72370
0.880		1.78960	1.78783	1.78277	1.77749	1.77119
0.890		1.84445	1.84195	1.83664	1.82979	1.82254
0.900		1.90438	1.90102	1.89548	1.88679	1.87848
0.910		1.97050	1.96612	1.96035	1.94951	1.93999
0.920		2.04427	2.03864	2.03271	2.01928	2.00837
0.930		2.12776	2.12061	2.11457	2.09798	2.08546
0.940		2.22400	2.21492	2.20890	2.18839	2.17394
0.950		2.33771	2.32613	2.32032	2.29476	2.27797
0.960		2.47679	2.46181	2.45658	2.42426	2.40452
0.970		2.65612	2.63620	2.63230	2.59028	2.56663
0.980		2.90919	2.88125	2.88040	2.82288	2.79356
0.990		3.34343	3.29889	3.30682	3.21792	3.17872
0.995		3.78036	3.71565	3.73711	3.61097	3.56193
0.999		4.80645	4.68201	4.75303	4.52069	4.45053

TABLE 4.16 N= 5 (CONT.)

		10	15	15	15	15
		20 20	15 15	15 15	15 15	15 20
P ↓	λ:	25 25	15 40	20 35	25 30	20 30
0.001		0.04010	0.03844	0.03963	0.04017	0.04068
0.005		0.07868	0.07544	0.07773	0.07878	0.07974
0.010		0.10606	0.10170	0.10474	0.10615	0.10742
0.020		0.14413	0.13821	0.14228	0.14417	0.14584
0.030		0.17334	0.16625	0.17107	0.17331	0.17528
0.040		0.19818	0.19010	0.19555	0.19809	0.20030
0.050		0.22034	0.21139	0.21738	0.22018	0.22258
0.060		0.24064	0.23091	0.23739	0.24041	0.24299
0.070		0.25959	0.24913	0.25605	0.25929	0.26202
0.080		0.27750	0.26635	0.27368	0.27712	0.27999
0.090		0.29458	0.28279	0.29050	0.29412	0.29712
0.100		0.31098	0.29860	0.30666	0.31045	0.31357
0.110		0.32684	0.31387	0.32227	0.32622	0.32946
0.120		0.34223	0.32871	0.33743	0.34153	0.34487
0.130		0.35723	0.34318	0.35220	0.35645	0.35988
0.140		0.37189	0.35733	0.36664	0.37104	0.37456
0.150		0.38627	0.37122	0.38080	0.38534	0.38894
0.160		0.40041	0.38488	0.39473	0.39939	0.40307
0.170		0.41433	0.39834	0.40844	0.41323	0.41698
0.180		0.42807	0.41164	0.42198	0.42689	0.43071
0.190		0.44166	0.42479	0.43536	0.44039	0.44428
0.200		0.45511	0.43782	0.44862	0.45376	0.45770
0.250		0.52099	0.50175	0.51356	0.51921	0.52340
0.300		0.58581	0.56489	0.57753	0.58362	0.58796
0.350		0.65090	0.62854	0.64183	0.64828	0.65270
0.400		0.71731	0.69377	0.70754	0.71427	0.71868
0.450		0.78602	0.76160	0.77565	0.78258	0.78690
0.500		0.85806	0.83312	0.84723	0.85425	0.85838
0.550		0.93464	0.90964	0.92351	0.93050	0.93431
0.600		1.01722	0.99277	1.00603	1.01283	1.01618
0.650		1.10778	1.08473	1.09687	1.10324	1.10595
0.700		1.20909	1.18867	1.19898	1.20458	1.20640
0.750		1.32538	1.30948	1.31686	1.32119	1.32176
0.800		1.46362	1.45534	1.45800	1.46023	1.45904
0.810		1.49486	1.48866	1.49007	1.49173	1.49009
0.820		1.52760	1.52373	1.52375	1.52477	1.52265
0.830		1.56202	1.56075	1.55922	1.55953	1.55688
0.840		1.59830	1.59998	1.59671	1.59621	1.59300
0.850		1.63671	1.64170	1.63648	1.63508	1.63124
0.860		1.67751	1.68627	1.67885	1.67643	1.67190
0.870		1.72108	1.73414	1.72423	1.72063	1.71534
0.880		1.76785	1.78584	1.77309	1.76815	1.76201
0.890		1.81838	1.84208	1.82605	1.81956	1.81246
0.900		1.87337	1.90375	1.88392	1.87560	1.86744
0.910		1.93376	1.97203	1.94774	1.93726	1.92787
0.920		2.00082	2.04853	2.01893	2.00586	1.99506
0.930		2.07631	2.13553	2.09951	2.08327	2.07082
0.940		2.16281	2.23637	2.19240	2.17221	2.15778
0.950		2.26433	2.35627	2.30221	2.27689	2.26006
0.960		2.38752	2.50406	2.43662	2.40440	2.38451
0.970		2.54485	2.69639	2.61015	2.56800	2.54402
0.980		2.76420	2.97101	2.85561	2.79746	2.76753
0.990		3.13404	3.44989	3.27866	3.18789	3.14751
0.995		3.49893	3.93936	3.70692	3.57726	3.52645
0.999		4.33341	5.10650	4.72173	4.48147	4.40838

TABLE 4.17 N= 5 (CONT.)

P ↓	λ:	15 15 20 25 25	15 20 20 20 25	All λ's = 100/5
0.001		0.04101	0.04152	0.04204
0.005		0.08038	0.08136	0.08235
0.010		0.10826	0.10956	0.11086
0.020		0.14696	0.14866	0.15038
0.030		0.17661	0.17860	0.18061
0.040		0.20179	0.20402	0.20627
0.050		0.22423	0.22665	0.22910
0.060		0.24477	0.24737	0.24998
0.070		0.26392	0.26667	0.26944
0.080		0.28200	0.28489	0.28780
0.090		0.29923	0.30225	0.30529
0.100		0.31577	0.31891	0.32206
0.110		0.33174	0.33499	0.33825
0.120		0.34724	0.35059	0.35395
0.130		0.36234	0.36578	0.36923
0.140		0.37708	0.38061	0.38415
0.150		0.39154	0.39515	0.39876
0.160		0.40574	0.40943	0.41311
0.170		0.41972	0.42348	0.42724
0.180		0.43351	0.43733	0.44116
0.190		0.44714	0.45102	0.45491
0.200		0.46062	0.46457	0.46851
0.250		0.52658	0.53075	0.53492
0.300		0.59134	0.59567	0.59998
0.350		0.65625	0.66064	0.66502
0.400		0.72234	0.72673	0.73110
0.450		0.79062	0.79491	0.79919
0.500		0.86210	0.86620	0.87029
0.550		0.93796	0.94176	0.94555
0.600		1.01965	1.02301	1.02637
0.650		1.10912	1.11185	1.11461
0.700		1.20908	1.21096	1.21289
0.750		1.32368	1.32437	1.32514
0.800		1.45975	1.45875	1.45786
0.810		1.49049	1.48906	1.48775
0.820		1.52269	1.52080	1.51904
0.830		1.55653	1.55414	1.55189
0.840		1.59221	1.58927	1.58648
0.850		1.62996	1.62642	1.62304
0.860		1.67007	1.66587	1.66183
0.870		1.71289	1.70795	1.70319
0.880		1.75885	1.75308	1.74752
0.890		1.80849	1.80180	1.79533
0.900		1.86251	1.85478	1.84727
0.910		1.92184	1.91291	1.90421
0.920		1.98771	1.97739	1.96732
0.930		2.06185	2.04991	2.03821
0.940		2.14682	2.13291	2.11925
0.950		2.24654	2.23020	2.21410
0.960		2.36758	2.34812	2.32887
0.970		2.52220	2.49851	2.47492
0.980		2.73789	2.70782	2.67764
0.990		3.10191	3.05999	3.01725
0.995		3.46158	3.40677	3.34992
0.999		4.28612	4.19835	4.10300

TABLE 5.1 N= 6

P ↓	λ:	5 5 5 5 5 75	5 5 5 5 10 70	5 5 5 5 15 65	5 5 5 5 20 60	5 5 5 5 25 55
0.001		0.03041	0.03374	0.03570	0.03700	0.03788
0.005		0.05464	0.06062	0.06420	0.06659	0.06822
0.010		0.07117	0.07895	0.08366	0.08683	0.08900
0.020		0.09379	0.10400	0.11029	0.11455	0.11749
0.030		0.11103	0.12309	0.13060	0.13572	0.13926
0.040		0.12572	0.13932	0.14788	0.15375	0.15782
0.050		0.13885	0.15382	0.16333	0.16988	0.17444
0.060		0.15096	0.16717	0.17756	0.18475	0.18975
0.070		0.16232	0.17969	0.19090	0.19869	0.20413
0.080		0.17313	0.19158	0.20358	0.21195	0.21780
0.090		0.18352	0.20299	0.21575	0.22467	0.23093
0.100		0.19358	0.21403	0.22752	0.23698	0.24364
0.110		0.20338	0.22476	0.23897	0.24896	0.25600
0.120		0.21297	0.23526	0.25016	0.26067	0.26809
0.130		0.22240	0.24556	0.26114	0.27217	0.27996
0.140		0.23171	0.25571	0.27195	0.28349	0.29165
0.150		0.24092	0.26574	0.28264	0.29467	0.30321
0.160		0.25006	0.27568	0.29322	0.30575	0.31465
0.170		0.25915	0.28555	0.30373	0.31674	0.32601
0.180		0.26821	0.29537	0.31417	0.32768	0.33730
0.190		0.27727	0.30516	0.32458	0.33857	0.34856
0.200		0.28632	0.31494	0.33497	0.34944	0.35979
0.250		0.33219	0.36412	0.38713	0.40396	0.41609
0.300		0.38011	0.41491	0.44072	0.45986	0.47374
0.350		0.43153	0.46864	0.49704	0.51839	0.53399
0.400		0.48797	0.52667	0.55735	0.58076	0.59800
0.450		0.55122	0.59052	0.62300	0.64823	0.66697
0.500		0.62353	0.66209	0.69564	0.72229	0.74231
0.550		0.70773	0.74377	0.77732	0.80477	0.82571
0.600		0.80754	0.83878	0.87078	0.89810	0.91939
0.650		0.92780	0.95153	0.97978	1.00561	1.02640
0.700		1.07518	1.08832	1.10985	1.13215	1.15117
0.750		1.25946	1.25867	1.26953	1.28529	1.30054
0.800		1.49683	1.47819	1.47315	1.47781	1.48607
0.810		1.55294	1.53015	1.52114	1.52283	1.52914
0.820		1.61263	1.58547	1.57219	1.57059	1.57470
0.830		1.67633	1.64452	1.62666	1.62142	1.62306
0.840		1.74450	1.70778	1.68496	1.67571	1.67456
0.850		1.81775	1.77578	1.74761	1.73391	1.72962
0.860		1.89677	1.84918	1.81523	1.79659	1.78875
0.870		1.98242	1.92879	1.88857	1.86444	1.85256
0.880		2.07578	2.01561	1.96857	1.93830	1.92183
0.890		2.17822	2.11092	2.05641	2.01928	1.99754
0.900		2.29148	2.21634	2.15361	2.10875	2.08094
0.910		2.41789	2.33403	2.26218	2.20857	2.17371
0.920		2.56056	2.46692	2.38486	2.32124	2.27811
0.930		2.72392	2.61914	2.52547	2.45029	2.39734
0.940		2.91446	2.79674	2.68964	2.60090	2.53608
0.950		3.14229	3.00914	2.88612	2.78113	2.70165
0.960		3.42440	3.27218	3.12963	3.00455	2.90637
0.970		3.79270	3.61568	3.44785	3.29668	3.17345
0.980		4.31938	4.10698	3.90329	3.71517	3.55539
0.990		5.23628	4.96245	4.69680	4.44530	4.22119
0.995		6.16950	5.83320	5.50480	5.18964	4.90029
0.999		8.38022	7.89628	7.41978	6.95542	6.51384

TABLE 5.2 N= 6 (CONT.)

P ↓	λ:	5 5 5 5 30 50	5 5 5 5 35 45	5 5 5 5 40 40	5 5 5 10 10 65	5 5 5 10 15 60
0.001		0.03846	0.03878	0.03889	0.03738	0.03949
0.005		0.06930	0.06991	0.07010	0.06710	0.07094
0.010		0.09042	0.09124	0.09150	0.08733	0.09237
0.020		0.11942	0.12053	0.12089	0.11493	0.12161
0.030		0.14160	0.14294	0.14338	0.13592	0.14385
0.040		0.16052	0.16206	0.16256	0.15372	0.16272
0.050		0.17746	0.17919	0.17976	0.16961	0.17957
0.060		0.19308	0.19499	0.19561	0.18420	0.19503
0.070		0.20775	0.20983	0.21051	0.19786	0.20951
0.080		0.22170	0.22395	0.22468	0.21081	0.22323
0.090		0.23511	0.23751	0.23830	0.22322	0.23638
0.100		0.24808	0.25064	0.25148	0.23520	0.24906
0.110		0.26071	0.26342	0.26431	0.24683	0.26138
0.120		0.27306	0.27592	0.27686	0.25817	0.27339
0.130		0.28519	0.28820	0.28919	0.26930	0.28515
0.140		0.29713	0.30030	0.30134	0.28023	0.29671
0.150		0.30894	0.31225	0.31334	0.29102	0.30811
0.160		0.32063	0.32410	0.32523	0.30168	0.31937
0.170		0.33224	0.33585	0.33703	0.31225	0.33052
0.180		0.34379	0.34754	0.34877	0.32275	0.34160
0.190		0.35529	0.35919	0.36047	0.33319	0.35260
0.200		0.36677	0.37082	0.37214	0.34360	0.36356
0.250		0.42431	0.42909	0.43066	0.39563	0.41821
0.300		0.48319	0.48870	0.49051	0.44873	0.47371
0.350		0.54466	0.55089	0.55294	0.50421	0.53132
0.400		0.60985	0.61678	0.61907	0.56330	0.59222
0.450		0.67992	0.68752	0.69003	0.62737	0.65765
0.500		0.75622	0.76442	0.76712	0.69805	0.72909
0.550		0.84037	0.84905	0.85192	0.77745	0.80836
0.600		0.93447	0.94345	0.94643	0.86836	0.89791
0.650		1.04139	1.05040	1.05340	0.97471	1.00113
0.700		1.16529	1.17390	1.17679	1.10224	1.12302
0.750		1.31256	1.32011	1.32267	1.25980	1.27138
0.800		1.49394	1.49931	1.50120	1.46209	1.45934
0.810		1.53582	1.54055	1.54223	1.50995	1.50354
0.820		1.58003	1.58403	1.58549	1.56091	1.55051
0.830		1.62685	1.63002	1.63121	1.61533	1.60059
0.840		1.67660	1.67882	1.67971	1.67364	1.65418
0.850		1.72967	1.73080	1.73135	1.73635	1.71175
0.860		1.78652	1.78639	1.78654	1.80407	1.77387
0.870		1.84771	1.84614	1.84583	1.87757	1.84123
0.880		1.91396	1.91071	1.90986	1.95777	1.91470
0.890		1.98616	1.98095	1.97947	2.04585	1.99538
0.900		2.06547	2.05794	2.05572	2.14333	2.08466
0.910		2.15340	2.14312	2.14001	2.25223	2.18442
0.920		2.25202	2.23842	2.23423	2.37525	2.29717
0.930		2.36424	2.34657	2.34106	2.51623	2.42644
0.940		2.49432	2.47155	2.46438	2.68080	2.57745
0.950		2.64889	2.61957	2.61024	2.87768	2.75826
0.960		2.83914	2.80100	2.78875	3.12160	2.98246
0.970		3.08608	3.03537	3.01890	3.44023	3.27560
0.980		3.43718	3.36653	3.34327	3.89608	3.69537
0.990		4.04520	3.93494	3.89779	4.69002	4.42712
0.995		4.66214	4.50615	4.45230	5.49830	5.17251
0.999		6.12324	5.84277	5.73985	7.41362	6.93957

TABLE 5.3 N= 6 (CONT.)

P ↓	λ: 5 5 5 10 20 55	5 5 5 10 25 50	5 5 5 10 30 45	5 5 5 10 35 40	5 5 5 15 15 55
0.001	0.04087	0.04178	0.04234	0.04261	0.04167
0.005	0.07347	0.07514	0.07617	0.07666	0.07489
0.010	0.09569	0.09791	0.09927	0.09993	0.09753
0.020	0.12606	0.12903	0.13087	0.13176	0.12845
0.030	0.14917	0.15274	0.15495	0.15601	0.15195
0.040	0.16879	0.17287	0.17541	0.17663	0.17190
0.050	0.18631	0.19085	0.19368	0.19505	0.18970
0.060	0.20240	0.20737	0.21048	0.21197	0.20604
0.070	0.21746	0.22284	0.22621	0.22783	0.22132
0.080	0.23174	0.23752	0.24113	0.24287	0.23581
0.090	0.24543	0.25157	0.25543	0.25729	0.24967
0.100	0.25863	0.26514	0.26922	0.27120	0.26305
0.110	0.27144	0.27831	0.28262	0.28470	0.27602
0.120	0.28394	0.29115	0.29569	0.29788	0.28867
0.130	0.29618	0.30373	0.30849	0.31079	0.30104
0.140	0.30821	0.31610	0.32106	0.32347	0.31320
0.150	0.32007	0.32828	0.33346	0.33597	0.32518
0.160	0.33178	0.34032	0.34571	0.34832	0.33700
0.170	0.34338	0.35224	0.35784	0.36055	0.34871
0.180	0.35489	0.36407	0.36987	0.37269	0.36031
0.190	0.36634	0.37583	0.38183	0.38475	0.37184
0.200	0.37773	0.38753	0.39374	0.39675	0.38331
0.250	0.43445	0.44577	0.45297	0.45647	0.44033
0.300	0.49191	0.50469	0.51285	0.51683	0.49790
0.350	0.55135	0.56553	0.57462	0.57907	0.55728
0.400	0.61392	0.62941	0.63939	0.64428	0.61959
0.450	0.68080	0.69747	0.70827	0.71357	0.68598
0.500	0.75334	0.77100	0.78251	0.78818	0.75778
0.550	0.83322	0.85158	0.86364	0.86960	0.83663
0.600	0.92263	0.94126	0.95362	0.95977	0.92470
0.650	1.02462	1.04285	1.05514	1.06129	1.02497
0.700	1.14361	1.16042	1.17204	1.17793	1.14184
0.750	1.28650	1.30026	1.31025	1.31544	1.28220
0.800	1.46494	1.47293	1.47968	1.48343	1.45771
0.810	1.50654	1.51288	1.51870	1.52201	1.49868
0.820	1.55061	1.55511	1.55986	1.56268	1.54212
0.830	1.59747	1.59987	1.60340	1.60567	1.58834
0.840	1.64746	1.64750	1.64964	1.65127	1.63770
0.850	1.70101	1.69836	1.69891	1.69980	1.69061
0.860	1.75863	1.75292	1.75164	1.75168	1.74759
0.870	1.82094	1.81174	1.80835	1.80742	1.80928
0.880	1.88873	1.87553	1.86969	1.86761	1.87647
0.890	1.96298	1.94516	1.93648	1.93306	1.95014
0.900	2.04498	2.02178	2.00975	2.00475	2.03159
0.910	2.13639	2.10692	2.09091	2.08402	2.12252
0.920	2.23952	2.20261	2.18183	2.17265	2.22522
0.930	2.35758	2.31175	2.28514	2.27316	2.34294
0.940	2.49530	2.43859	2.40474	2.38921	2.48044
0.950	2.66006	2.58977	2.54663	2.52653	2.64512
0.960	2.86427	2.77642	2.72094	2.69466	2.84943
0.970	3.13128	3.01959	2.94671	2.91153	3.11680
0.980	3.51387	3.36682	3.26686	3.21743	3.50008
0.990	4.18172	3.97127	3.81922	3.74100	4.16910
0.995	4.86300	4.58733	4.37760	4.26535	4.85129
0.999	6.48028	6.05123	5.69466	5.48584	6.46973

TABLE 5.4 N= 6 (CONT.)

P ↓	λ:	5 5 20	5 15 50	5 5 25	5 15 45	5 5 30	5 15 40	5 5 35	5 15 35	5 5 20	5 20 45
0.001		0.04306		0.04394		0.04443		0.04459		0.04442	
0.005		0.07743		0.07905		0.07995		0.08024		0.07991	
0.010		0.10088		0.10301		0.10421		0.10459		0.10414	
0.020		0.13290		0.13576		0.13737		0.13788		0.13724	
0.030		0.15727		0.16069		0.16262		0.16324		0.16244	
0.040		0.17796		0.18186		0.18406		0.18477		0.18383	
0.050		0.19641		0.20075		0.20320		0.20399		0.20291	
0.060		0.21335		0.21809		0.22078		0.22165		0.22043	
0.070		0.22921		0.23432		0.23722		0.23816		0.23682	
0.080		0.24423		0.24971		0.25282		0.25382		0.25235	
0.090		0.25861		0.26443		0.26774		0.26881		0.26722	
0.100		0.27248		0.27863		0.28213		0.28327		0.28155	
0.110		0.28593		0.29240		0.29609		0.29729		0.29544	
0.120		0.29903		0.30582		0.30969		0.31095		0.30898	
0.130		0.31186		0.31896		0.32300		0.32432		0.32222	
0.140		0.32446		0.33185		0.33607		0.33745		0.33522	
0.150		0.33686		0.34455		0.34894		0.35037		0.34802	
0.160		0.34911		0.35708		0.36164		0.36313		0.36065	
0.170		0.36122		0.36948		0.37420		0.37574		0.37314	
0.180		0.37323		0.38177		0.38665		0.38825		0.38551	
0.190		0.38515		0.39396		0.39901		0.40066		0.39779	
0.200		0.39701		0.40609		0.41130		0.41300		0.41000	
0.250		0.45587		0.46626		0.47223		0.47418		0.47050	
0.300		0.51515		0.52674		0.53345		0.53564		0.53124	
0.350		0.57606		0.58879		0.59617		0.59860		0.59343	
0.400		0.63971		0.65346		0.66148		0.66411		0.65814	
0.450		0.70720		0.72183		0.73041		0.73323		0.72643	
0.500		0.77976		0.79510		0.80414		0.80712		0.79946	
0.550		0.85892		0.87470		0.88407		0.88717		0.87866	
0.600		0.94665		0.96251		0.97201		0.97517		0.96584	
0.650		1.04567		1.06105		1.07041		1.07354		1.06350	
0.700		1.15993		1.17402		1.18279		1.18575		1.17524	
0.750		1.29558		1.30708		1.31456		1.31714		1.30661	
0.800		1.46300		1.46976		1.47478		1.47660		1.46699	
0.810		1.50177		1.50718		1.51150		1.51310		1.50385	
0.820		1.54275		1.54665		1.55017		1.55152		1.54273	
0.830		1.58622		1.58841		1.59102		1.59208		1.58385	
0.840		1.63249		1.63275		1.63432		1.63505		1.62751	
0.850		1.68195		1.68001		1.68038		1.68075		1.67403	
0.860		1.73504		1.73060		1.72961		1.72954		1.72383	
0.870		1.79234		1.78502		1.78245		1.78189		1.77740	
0.880		1.85454		1.84391		1.83952		1.83837		1.83538	
0.890		1.92253		1.90807		1.90154		1.89971		1.89854	
0.900		1.99747		1.97852		1.96948		1.96684		1.96791	
0.910		2.08085		2.05663		2.04458		2.04098		2.04484	
0.920		2.17475		2.14423		2.12856		2.12380		2.13116	
0.930		2.28207		2.24392		2.22382		2.21762		2.22944	
0.940		2.40708		2.35952		2.33386		2.32585		2.34349	
0.950		2.55644		2.49697		2.46414		2.45377		2.47922	
0.960		2.74138		2.66628		2.62380		2.61024		2.64660	
0.970		2.98304		2.88629		2.83005		2.81185		2.86446	
0.980		3.32922		3.19964		3.12157		3.09588		3.17542	
0.990		3.93384		3.74360		3.62230		3.58125		3.71689	
0.995		4.55124		4.29693		4.12609		4.06651		4.26932	
0.999		6.01877		5.61066		5.30718		5.19315		5.58409	

TABLE 5.5 N= 6 (CONT.)

P ↓	λ:	5 5 5 20 25 40	5 5 5 20 30 35	5 5 5 25 25 35	5 5 5 25 30 30	5 5 10 10 10 60
0.001		0.04523	0.04561	0.04593	0.04615	0.04132
0.005		0.08139	0.08209	0.08267	0.08308	0.07407
0.010		0.10609	0.10701	0.10777	0.10831	0.09630
0.020		0.13985	0.14109	0.14210	0.14282	0.12653
0.030		0.16556	0.16704	0.16824	0.16910	0.14944
0.040		0.18739	0.18908	0.19043	0.19141	0.16882
0.050		0.20686	0.20874	0.21024	0.21132	0.18607
0.060		0.22474	0.22680	0.22842	0.22961	0.20188
0.070		0.24147	0.24369	0.24543	0.24671	0.21664
0.080		0.25732	0.25969	0.26155	0.26291	0.23062
0.090		0.27249	0.27501	0.27698	0.27842	0.24397
0.100		0.28711	0.28977	0.29184	0.29336	0.25684
0.110		0.30129	0.30409	0.30625	0.30785	0.26930
0.120		0.31510	0.31804	0.32029	0.32196	0.28144
0.130		0.32861	0.33168	0.33402	0.33577	0.29332
0.140		0.34187	0.34506	0.34749	0.34931	0.30497
0.150		0.35492	0.35824	0.36075	0.36263	0.31643
0.160		0.36780	0.37123	0.37383	0.37577	0.32775
0.170		0.38053	0.38408	0.38675	0.38877	0.33894
0.180		0.39314	0.39681	0.39956	0.40163	0.35004
0.190		0.40566	0.40944	0.41226	0.41439	0.36105
0.200		0.41809	0.42199	0.42488	0.42707	0.37201
0.250		0.47968	0.48411	0.48731	0.48979	0.42643
0.300		0.54140	0.54633	0.54979	0.55253	0.48139
0.350		0.60448	0.60986	0.61353	0.61649	0.53816
0.400		0.66996	0.67574	0.67956	0.68271	0.59794
0.450		0.73888	0.74500	0.74890	0.75219	0.66195
0.500		0.81237	0.81875	0.82263	0.82603	0.73167
0.550		0.89179	0.89831	0.90208	0.90551	0.80895
0.600		0.97886	0.98539	0.98891	0.99228	0.89625
0.650		1.07594	1.08227	1.08536	1.08856	0.99705
0.700		1.18643	1.19225	1.19467	1.19752	1.11645
0.750		1.31552	1.32035	1.32176	1.32400	1.26246
0.800		1.47191	1.47495	1.47482	1.47604	1.44854
0.810		1.50767	1.51022	1.50969	1.51062	1.49245
0.820		1.54531	1.54730	1.54633	1.54696	1.53917
0.830		1.58505	1.58640	1.58496	1.58523	1.58906
0.840		1.62714	1.62776	1.62580	1.62568	1.64249
0.850		1.67189	1.67170	1.66915	1.66859	1.69995
0.860		1.71967	1.71855	1.71536	1.71429	1.76202
0.870		1.77095	1.76875	1.76485	1.76321	1.82938
0.880		1.82628	1.82284	1.81814	1.81584	1.90292
0.890		1.88639	1.88151	1.87590	1.87285	1.98371
0.900		1.95220	1.94563	1.93899	1.93505	2.07318
0.910		2.02493	2.01634	2.00852	2.00355	2.17318
0.920		2.10623	2.09521	2.08602	2.07981	2.28623
0.930		2.19841	2.18443	2.17363	2.16590	2.41586
0.940		2.30490	2.28720	2.27446	2.26485	2.56726
0.950		2.43099	2.40848	2.39337	2.38135	2.74852
0.960		2.58560	2.55660	2.53845	2.52321	2.97319
0.970		2.78548	2.74717	2.72494	2.70510	3.26683
0.980		3.06850	3.01521	2.98701	2.95981	3.68713
0.990		3.55631	3.47266	3.43383	3.39176	4.41944
0.995		4.04950	3.92980	3.88021	3.82033	5.16516
0.999		5.21469	4.99239	4.91898	4.80676	6.93264

TABLE 5.6 N= 6 (CONT.)

P ↓	λ:	5 5 10 10 15 55	5 5 10 10 20 50	5 5 10 10 25 45	5 5 10 10 30 40	5 5 10 10 35 35
0.001		0.04359	0.04504	0.04596	0.04647	0.04663
0.005		0.07816	0.08079	0.08246	0.08339	0.08369
0.010		0.10162	0.10507	0.10727	0.10850	0.10890
0.020		0.13354	0.13811	0.14103	0.14268	0.14321
0.030		0.15771	0.16314	0.16663	0.16860	0.16923
0.040		0.17816	0.18432	0.18829	0.19054	0.19126
0.050		0.19636	0.20317	0.20757	0.21006	0.21087
0.060		0.21302	0.22043	0.22523	0.22795	0.22883
0.070		0.22858	0.23655	0.24172	0.24465	0.24560
0.080		0.24330	0.25179	0.25731	0.26045	0.26146
0.090		0.25736	0.26635	0.27221	0.27554	0.27662
0.100		0.27089	0.28037	0.28655	0.29007	0.29121
0.110		0.28400	0.29394	0.30044	0.30413	0.30533
0.120		0.29676	0.30715	0.31395	0.31782	0.31908
0.130		0.30922	0.32005	0.32715	0.33120	0.33251
0.140		0.32145	0.33270	0.34009	0.34431	0.34568
0.150		0.33348	0.34514	0.35282	0.35720	0.35863
0.160		0.34534	0.35741	0.36536	0.36991	0.37139
0.170		0.35706	0.36953	0.37776	0.38246	0.38399
0.180		0.36867	0.38153	0.39002	0.39489	0.39647
0.190		0.38019	0.39343	0.40219	0.40721	0.40885
0.200		0.39163	0.40525	0.41427	0.41945	0.42114
0.250		0.44834	0.46375	0.47403	0.47995	0.48188
0.300		0.50532	0.52238	0.53385	0.54048	0.54265
0.350		0.56385	0.58242	0.59500	0.60230	0.60469
0.400		0.62505	0.64496	0.65857	0.66650	0.66910
0.450		0.69009	0.71114	0.72565	0.73415	0.73695
0.500		0.76031	0.78220	0.79745	0.80644	0.80941
0.550		0.83736	0.85967	0.87544	0.88479	0.88789
0.600		0.92342	0.94556	0.96149	0.97103	0.97421
0.650		1.02153	1.04259	1.05818	1.06763	1.07079
0.700		1.13613	1.15478	1.16919	1.17814	1.18115
0.750		1.27423	1.28832	1.30027	1.30800	1.31066
0.800		1.44767	1.45371	1.46098	1.46631	1.46823
0.810		1.48827	1.49208	1.49803	1.50266	1.50435
0.820		1.53137	1.53269	1.53712	1.54095	1.54240
0.830		1.57727	1.57580	1.57852	1.58144	1.58260
0.840		1.62634	1.62173	1.62249	1.62437	1.62521
0.850		1.67899	1.67085	1.66939	1.67008	1.67054
0.860		1.73576	1.72364	1.71964	1.71894	1.71898
0.870		1.79727	1.78065	1.77373	1.77145	1.77098
0.880		1.86433	1.84260	1.83231	1.82818	1.82712
0.890		1.93793	1.91037	1.89617	1.88987	1.88812
0.900		2.01936	1.98511	1.96635	1.95749	1.95493
0.910		2.11033	2.06836	2.04420	2.03229	2.02875
0.920		2.21315	2.16216	2.13157	2.11599	2.11126
0.930		2.33107	2.26945	2.23108	2.21097	2.20479
0.940		2.46884	2.39450	2.34653	2.32076	2.31273
0.950		2.63388	2.54399	2.48388	2.45081	2.44039
0.960		2.83863	2.72915	2.65316	2.61029	2.59662
0.970		3.10653	2.97116	2.87324	2.81638	2.79801
0.980		3.49044	3.31788	3.18679	3.10781	3.08185
0.990		4.16020	3.92327	3.73120	3.60857	3.56705
0.995		4.84284	4.54124	4.28498	4.11247	4.05226
0.999		6.46181	6.00953	5.59952	5.29390	5.17886

TABLE 5.7 N= 6 (CONT.)

P ↓	λ:	5 5 10 15 15 50	5 5 10 15 20 45	5 5 10 15 25 40	5 5 10 15 30 35	5 5 10 20 20 40
0.001		0.04592	0.04736	0.04821	0.04862	0.04873
0.005		0.08233	0.08493	0.08649	0.08722	0.08742
0.010		0.10704	0.11045	0.11249	0.11346	0.11369
0.020		0.14064	0.14514	0.14785	0.14914	0.14942
0.030		0.16608	0.17141	0.17464	0.17617	0.17647
0.040		0.18758	0.19362	0.19728	0.19902	0.19934
0.050		0.20670	0.21336	0.21741	0.21933	0.21966
0.060		0.22421	0.23143	0.23583	0.23793	0.23826
0.070		0.24053	0.24829	0.25302	0.25527	0.25560
0.080		0.25597	0.26422	0.26926	0.27166	0.27198
0.090		0.27070	0.27942	0.28476	0.28731	0.28761
0.100		0.28488	0.29404	0.29967	0.30235	0.30264
0.110		0.29860	0.30819	0.31409	0.31691	0.31718
0.120		0.31194	0.32194	0.32811	0.33105	0.33131
0.130		0.32497	0.33537	0.34179	0.34486	0.34510
0.140		0.33773	0.34853	0.35519	0.35839	0.35860
0.150		0.35028	0.36145	0.36836	0.37167	0.37186
0.160		0.36264	0.37418	0.38132	0.38475	0.38491
0.170		0.37485	0.38674	0.39412	0.39766	0.39779
0.180		0.38693	0.39917	0.40677	0.41043	0.41052
0.190		0.39891	0.41149	0.41931	0.42308	0.42314
0.200		0.41080	0.42371	0.43175	0.43562	0.43565
0.250		0.46953	0.48400	0.49307	0.49746	0.49727
0.300		0.52824	0.54410	0.55413	0.55898	0.55854
0.350		0.58819	0.60529	0.61617	0.62147	0.62072
0.400		0.65048	0.66864	0.68028	0.68597	0.68487
0.450		0.71621	0.73520	0.74750	0.75353	0.75201
0.500		0.78662	0.80616	0.81895	0.82526	0.82326
0.550		0.86319	0.88290	0.89598	0.90248	0.89995
0.600		0.94789	0.96725	0.98033	0.98688	0.98376
0.650		1.04340	1.06166	1.07432	1.08073	1.07698
0.700		1.15367	1.16971	1.18130	1.18729	1.18287
0.750		1.28484	1.29692	1.30643	1.31154	1.30649
0.800		1.44734	1.45261	1.45835	1.46178	1.45627
0.810		1.48508	1.48848	1.49315	1.49610	1.49055
0.820		1.52502	1.52633	1.52980	1.53221	1.52663
0.830		1.56744	1.56642	1.56852	1.57031	1.56474
0.840		1.61267	1.60901	1.60957	1.61065	1.60512
0.850		1.66107	1.65445	1.65325	1.65353	1.64808
0.860		1.71312	1.70314	1.69993	1.69928	1.69398
0.870		1.76938	1.75559	1.75007	1.74836	1.74327
0.880		1.83057	1.81243	1.80424	1.80129	1.79651
0.890		1.89758	1.87445	1.86315	1.85874	1.85439
0.900		1.97156	1.94266	1.92772	1.92160	1.91783
0.910		2.05406	2.01843	1.99917	1.99100	1.98802
0.920		2.14716	2.10359	2.07915	2.06850	2.06659
0.930		2.25379	2.20073	2.16998	2.15628	2.15583
0.940		2.37825	2.31369	2.27507	2.25753	2.25911
0.950		2.52727	2.44839	2.39972	2.37720	2.38166
0.960		2.71213	2.61488	2.55285	2.52359	2.53233
0.970		2.95411	2.83209	2.75124	2.71227	2.72776
0.980		3.30120	3.14281	3.03282	2.97823	3.00561
0.990		3.90769	3.68504	3.51948	3.43324	3.48734
0.995		4.52675	4.23881	4.01257	3.88894	3.97726
0.999		5.99665	5.55645	5.17919	4.95020	5.14160

TABLE 5.8 N= 6 (CONT.)

P ↓	λ:	5 5 10 20 25 35	5 5 10 20 30 30	5 5 10 25 25 30	5 5 15 15 15 45	5 5 15 15 20 40
0.001		0.04948	0.04972	0.05006	0.04827	0.04967
0.005		0.08877	0.08921	0.08983	0.08654	0.08906
0.010		0.11547	0.11603	0.11684	0.11250	0.11579
0.020		0.15177	0.15252	0.15358	0.14776	0.15210
0.030		0.17926	0.18015	0.18140	0.17444	0.17956
0.040		0.20249	0.20350	0.20492	0.19697	0.20275
0.050		0.22314	0.22426	0.22581	0.21699	0.22335
0.060		0.24204	0.24325	0.24493	0.23530	0.24218
0.070		0.25965	0.26096	0.26275	0.25236	0.25973
0.080		0.27630	0.27768	0.27959	0.26848	0.27630
0.090		0.29217	0.29364	0.29565	0.28385	0.29210
0.100		0.30744	0.30898	0.31109	0.29863	0.30728
0.110		0.32220	0.32382	0.32602	0.31291	0.32195
0.120		0.33654	0.33823	0.34052	0.32680	0.33620
0.130		0.35054	0.35230	0.35467	0.34034	0.35010
0.140		0.36424	0.36607	0.36852	0.35360	0.36370
0.150		0.37770	0.37959	0.38212	0.36662	0.37705
0.160		0.39094	0.39289	0.39550	0.37944	0.39019
0.170		0.40401	0.40602	0.40870	0.39209	0.40315
0.180		0.41692	0.41900	0.42174	0.40459	0.41595
0.190		0.42971	0.43184	0.43466	0.41697	0.42862
0.200		0.44240	0.44458	0.44746	0.42925	0.44119
0.250		0.50483	0.50728	0.51045	0.48973	0.50298
0.300		0.56681	0.56951	0.57292	0.54988	0.56426
0.350		0.62962	0.63253	0.63614	0.61095	0.62630
0.400		0.69430	0.69740	0.70115	0.67403	0.69017
0.450		0.76187	0.76511	0.76894	0.74016	0.75687
0.500		0.83340	0.83676	0.84060	0.81048	0.82751
0.550		0.91020	0.91360	0.91736	0.88639	0.90339
0.600		0.99386	0.99725	1.00080	0.96966	0.98618
0.650		1.08659	1.08984	1.09303	1.06269	1.07810
0.700		1.19148	1.19445	1.19706	1.16901	1.18239
0.750		1.31333	1.31575	1.31747	1.29406	1.30401
0.800		1.46004	1.46153	1.46185	1.44706	1.45128
0.810		1.49347	1.49470	1.49465	1.48232	1.48498
0.820		1.52861	1.52954	1.52909	1.51953	1.52045
0.830		1.56565	1.56625	1.56536	1.55894	1.55792
0.840		1.60484	1.60506	1.60368	1.60083	1.59763
0.850		1.64644	1.64624	1.64431	1.64553	1.63989
0.860		1.69080	1.69012	1.68757	1.69346	1.68504
0.870		1.73833	1.73709	1.73385	1.74512	1.73355
0.880		1.78955	1.78766	1.78364	1.80112	1.78595
0.890		1.84508	1.84245	1.83755	1.86227	1.84295
0.900		1.90578	1.90227	1.89635	1.92959	1.90545
0.910		1.97273	1.96817	1.96108	2.00443	1.97464
0.920		2.04743	2.04161	2.03313	2.08863	2.05215
0.930		2.13194	2.12458	2.11446	2.18481	2.14025
0.940		2.22933	2.22004	2.20790	2.29678	2.24231
0.950		2.34434	2.33255	2.31790	2.43053	2.36357
0.960		2.48491	2.46974	2.45182	2.59614	2.51285
0.970		2.66599	2.64594	2.62351	2.81260	2.70682
0.980		2.92119	2.89326	2.86394	3.12291	2.98320
0.990		3.35817	3.31403	3.27171	3.66552	3.46368
0.995		3.79687	3.73311	3.67648	4.22025	3.95349
0.999		4.82456	4.70265	4.60916	5.53997	5.11935

TABLE 5.9 N= 6 (CONT.)

P ↓	λ: 5 5 / 15 15 / 25 35	5 5 / 15 15 / 30 30	5 5 / 15 20 / 20 35	5 5 / 15 20 / 25 30	5 5 / 15 25 / 25 25
0.001	0.05043	0.05067	0.05098	0.05157	0.04798
0.005	0.09044	0.09088	0.09141	0.09248	0.08651
0.010	0.11759	0.11817	0.11884	0.12025	0.11289
0.020	0.15448	0.15524	0.15610	0.15795	0.14905
0.030	0.18238	0.18328	0.18427	0.18646	0.17664
0.040	0.20594	0.20696	0.20806	0.21053	0.20010
0.050	0.22686	0.22799	0.22917	0.23189	0.22106
0.060	0.24599	0.24722	0.24847	0.25141	0.24032
0.070	0.26382	0.26513	0.26645	0.26959	0.25834
0.080	0.28064	0.28204	0.28342	0.28675	0.27543
0.090	0.29669	0.29816	0.29959	0.30311	0.29178
0.100	0.31210	0.31365	0.31512	0.31881	0.30754
0.110	0.32699	0.32862	0.33013	0.33398	0.32282
0.120	0.34146	0.34315	0.34470	0.34871	0.33770
0.130	0.35556	0.35733	0.35891	0.36307	0.35225
0.140	0.36937	0.37120	0.37281	0.37711	0.36653
0.150	0.38291	0.38480	0.38644	0.39088	0.38057
0.160	0.39624	0.39819	0.39985	0.40442	0.39441
0.170	0.40937	0.41139	0.41307	0.41777	0.40809
0.180	0.42235	0.42443	0.42612	0.43095	0.42162
0.190	0.43520	0.43733	0.43904	0.44399	0.43504
0.200	0.44793	0.45012	0.45184	0.45691	0.44836
0.250	0.51052	0.51297	0.51470	0.52031	0.51405
0.300	0.57250	0.57519	0.57688	0.58296	0.57936
0.350	0.63516	0.63806	0.63965	0.64613	0.64545
0.400	0.69955	0.70263	0.70407	0.71085	0.71326
0.450	0.76667	0.76990	0.77111	0.77813	0.78368
0.500	0.83760	0.84094	0.84185	0.84899	0.85765
0.550	0.91360	0.91699	0.91753	0.92465	0.93627
0.600	0.99627	0.99964	0.99971	1.00663	1.02093
0.650	1.08774	1.09100	1.09048	1.09697	1.11351
0.700	1.19110	1.19409	1.19286	1.19855	1.21664
0.750	1.31103	1.31351	1.31141	1.31578	1.33438
0.800	1.45534	1.45692	1.45377	1.45595	1.47342
0.810	1.48822	1.48955	1.48617	1.48775	1.50471
0.820	1.52278	1.52382	1.52020	1.52112	1.53745
0.830	1.55921	1.55993	1.55606	1.55625	1.57180
0.840	1.59775	1.59810	1.59398	1.59334	1.60796
0.850	1.63867	1.63861	1.63423	1.63265	1.64617
0.860	1.68231	1.68177	1.67712	1.67448	1.68669
0.870	1.72908	1.72798	1.72307	1.71923	1.72988
0.880	1.77947	1.77774	1.77256	1.76734	1.77616
0.890	1.83414	1.83167	1.82623	1.81941	1.82605
0.900	1.89390	1.89056	1.88487	1.87619	1.88025
0.910	1.95985	1.95546	1.94955	1.93868	1.93966
0.920	2.03345	2.02781	2.02172	2.00822	2.00548
0.930	2.11678	2.10959	2.10338	2.08669	2.07943
0.940	2.21287	2.20372	2.19753	2.17685	2.16398
0.950	2.32643	2.31475	2.30877	2.28298	2.26299
0.960	2.46538	2.45026	2.44487	2.41221	2.38289
0.970	2.64459	2.62447	2.62044	2.57796	2.53571
0.980	2.89757	2.86935	2.86843	2.81027	2.74838
0.990	3.33180	3.28681	3.29482	3.20496	3.10660
0.995	3.76879	3.70349	3.72520	3.59780	3.46041
0.999	4.79513	4.66978	4.74151	4.50732	4.27383

TABLE 5.10 N= 6 (CONT.)

P ↓	λ:	5 5 20 20 20 30	5 5 20 20 25 25	5 10 10 10 10 55	5 10 10 10 15 50	5 10 10 10 20 45
0.001		0.05213	0.04931	0.04559	0.04800	0.04949
0.005		0.09347	0.08874	0.08152	0.08582	0.08851
0.010		0.12152	0.11566	0.10581	0.11138	0.11488
0.020		0.15960	0.15246	0.13871	0.14597	0.15057
0.030		0.18838	0.18046	0.16354	0.17205	0.17748
0.040		0.21267	0.20423	0.18448	0.19404	0.20016
0.050		0.23423	0.22541	0.20306	0.21353	0.22027
0.060		0.25392	0.24485	0.22003	0.23133	0.23863
0.070		0.27226	0.26302	0.23585	0.24790	0.25571
0.080		0.28956	0.28022	0.25078	0.26353	0.27183
0.090		0.30604	0.29666	0.26502	0.27843	0.28719
0.100		0.32187	0.31250	0.27871	0.29274	0.30193
0.110		0.33715	0.32784	0.29194	0.30657	0.31618
0.120		0.35198	0.34276	0.30481	0.32000	0.33001
0.130		0.36643	0.35735	0.31736	0.33309	0.34349
0.140		0.38057	0.37164	0.32965	0.34590	0.35668
0.150		0.39443	0.38569	0.34172	0.35848	0.36962
0.160		0.40805	0.39954	0.35361	0.37086	0.38235
0.170		0.42148	0.41321	0.36535	0.38307	0.39491
0.180		0.43473	0.42673	0.37696	0.39514	0.40731
0.190		0.44784	0.44012	0.38847	0.40709	0.41959
0.200		0.46082	0.45342	0.39989	0.41894	0.43177
0.250		0.52450	0.51891	0.45631	0.47732	0.49167
0.300		0.58733	0.58393	0.51274	0.53544	0.55115
0.350		0.65059	0.64967	0.57046	0.59458	0.61152
0.400		0.71533	0.71709	0.63062	0.65586	0.67386
0.450		0.78253	0.78707	0.69438	0.72038	0.73924
0.500		0.85320	0.86058	0.76308	0.78936	0.80884
0.550		0.92855	0.93870	0.83836	0.86432	0.88406
0.600		1.01006	1.02280	0.92241	0.94720	0.96671
0.650		1.09973	1.11472	1.01829	1.04071	1.05926
0.700		1.20038	1.21706	1.13051	1.14880	1.16529
0.750		1.31632	1.33377	1.26617	1.27767	1.29036
0.800		1.45466	1.47139	1.43735	1.43785	1.44382
0.810		1.48600	1.50233	1.47755	1.47513	1.47924
0.820		1.51888	1.53468	1.52027	1.51463	1.51664
0.830		1.55346	1.56862	1.56582	1.55663	1.55628
0.840		1.58996	1.60432	1.61457	1.60144	1.59843
0.850		1.62863	1.64202	1.66694	1.64945	1.64343
0.860		1.66976	1.68199	1.72348	1.70113	1.69170
0.870		1.71373	1.72456	1.78482	1.75705	1.74374
0.880		1.76098	1.77014	1.85175	1.81793	1.80017
0.890		1.81209	1.81925	1.92529	1.88468	1.86180
0.900		1.86778	1.87255	2.00674	1.95845	1.92966
0.910		1.92904	1.93092	2.09780	2.04080	2.00510
0.920		1.99717	1.99554	2.20078	2.13382	2.08997
0.930		2.07399	2.06806	2.31895	2.24044	2.18688
0.940		2.16220	2.15089	2.45706	2.36501	2.29966
0.950		2.26595	2.24775	2.62252	2.51424	2.43427
0.960		2.39220	2.36489	2.82778	2.69946	2.60078
0.970		2.55398	2.51392	3.09626	2.94194	2.81815
0.980		2.78056	2.72090	3.48082	3.28972	3.12927
0.990		3.16529	3.06862	4.15132	3.89714	3.67227
0.995		3.54825	3.41124	4.83440	4.51680	4.22672
0.999		4.43665	4.19778	6.45391	5.98743	5.54536

TABLE 5.11 N= 6 (CONT.)

P ↓	λ:	5 10 10 10 25 40	5 10 10 10 30 35	5 10 10 15 15 45	5 10 10 15 20 40	5 10 10 15 25 35
0.001		0.05038	0.05080	0.05044	0.05190	0.05269
0.005		0.09012	0.09088	0.09016	0.09277	0.09419
0.010		0.11698	0.11797	0.11697	0.12036	0.12220
0.020		0.15334	0.15465	0.15324	0.15767	0.16009
0.030		0.18076	0.18232	0.18055	0.18576	0.18862
0.040		0.20387	0.20564	0.20355	0.20941	0.21263
0.050		0.22436	0.22631	0.22392	0.23035	0.23390
0.060		0.24307	0.24518	0.24251	0.24945	0.25329
0.070		0.26048	0.26275	0.25980	0.26721	0.27133
0.080		0.27690	0.27932	0.27610	0.28396	0.28832
0.090		0.29255	0.29511	0.29162	0.29989	0.30449
0.100		0.30757	0.31027	0.30651	0.31518	0.32001
0.110		0.32208	0.32491	0.32089	0.32993	0.33498
0.120		0.33617	0.33912	0.33484	0.34424	0.34950
0.130		0.34990	0.35297	0.34844	0.35818	0.36364
0.140		0.36333	0.36652	0.36173	0.37181	0.37745
0.150		0.37651	0.37981	0.37477	0.38516	0.39100
0.160		0.38947	0.39289	0.38759	0.39829	0.40431
0.170		0.40225	0.40578	0.40022	0.41123	0.41742
0.180		0.41488	0.41851	0.41269	0.42400	0.43037
0.190		0.42737	0.43112	0.42504	0.43662	0.44316
0.200		0.43976	0.44361	0.43727	0.44913	0.45584
0.250		0.50066	0.50501	0.49736	0.51050	0.51798
0.300		0.56108	0.56589	0.55690	0.57115	0.57932
0.350		0.62229	0.62754	0.61717	0.63239	0.64117
0.400		0.68540	0.69104	0.67926	0.69529	0.70460
0.450		0.75145	0.75744	0.74424	0.76087	0.77062
0.500		0.82157	0.82785	0.81325	0.83025	0.84031
0.550		0.89713	0.90362	0.88767	0.90472	0.91494
0.600		0.97985	0.98643	0.96927	0.98594	0.99609
0.650		1.07207	1.07855	1.06047	1.07614	1.08591
0.700		1.17714	1.18325	1.16478	1.17855	1.18746
0.750		1.30024	1.30551	1.28765	1.29811	1.30541
0.800		1.45001	1.45365	1.43833	1.44315	1.44756
0.810		1.48437	1.48755	1.47311	1.47638	1.47999
0.820		1.52058	1.52322	1.50985	1.51138	1.51408
0.830		1.55886	1.56089	1.54878	1.54837	1.55004
0.840		1.59946	1.60079	1.59020	1.58760	1.58809
0.850		1.64270	1.64322	1.63443	1.62936	1.62852
0.860		1.68894	1.68853	1.68189	1.67401	1.67166
0.870		1.73864	1.73716	1.73309	1.72202	1.71791
0.880		1.79237	1.78964	1.78865	1.77391	1.76779
0.890		1.85085	1.84665	1.84937	1.83041	1.82192
0.900		1.91500	1.90907	1.91628	1.89240	1.88114
0.910		1.98604	1.97803	1.99075	1.96109	1.94654
0.920		2.06562	2.05509	2.07463	2.03811	2.01958
0.930		2.15607	2.14244	2.17053	2.12574	2.10234
0.940		2.26081	2.24326	2.28230	2.22735	2.19784
0.950		2.38515	2.36252	2.41595	2.34819	2.31082
0.960		2.53801	2.50851	2.58159	2.49712	2.44918
0.970		2.73622	2.69682	2.79827	2.69084	2.62781
0.980		3.01774	2.96244	3.10908	2.96716	2.88024
0.990		3.50460	3.41718	3.65264	3.44802	3.31402
0.995		3.99803	3.87280	4.20814	3.93841	3.75095
0.999		5.16546	4.93412	5.52892	5.10550	4.77770

TABLE 5.12 N= 6 (CONT.)

P ↓	λ: 5 10 30 30	5 10 20 35	5 10 25 30	5 10 25 25	5 10 15 40
	10 15	10 20	10 20	10 25	15 15
0.001	0.05261	0.05325	0.05387	0.04991	0.05289
0.005	0.09413	0.09518	0.09629	0.08985	0.09448
0.010	0.12221	0.12348	0.12492	0.11713	0.12253
0.020	0.16021	0.16174	0.16363	0.15441	0.16042
0.030	0.18886	0.19054	0.19276	0.18277	0.18892
0.040	0.21299	0.21477	0.21727	0.20681	0.21288
0.050	0.23437	0.23623	0.23897	0.22824	0.23409
0.060	0.25388	0.25579	0.25875	0.24788	0.25341
0.070	0.27202	0.27397	0.27713	0.26623	0.27138
0.080	0.28912	0.29110	0.29445	0.28358	0.28829
0.090	0.30540	0.30740	0.31093	0.30016	0.30438
0.100	0.32102	0.32303	0.32672	0.31612	0.31981
0.110	0.33609	0.33811	0.34196	0.33156	0.33469
0.120	0.35071	0.35273	0.35674	0.34657	0.34912
0.130	0.36495	0.36697	0.37112	0.36122	0.36317
0.140	0.37887	0.38088	0.38517	0.37558	0.37689
0.150	0.39251	0.39451	0.39894	0.38967	0.39033
0.160	0.40592	0.40791	0.41246	0.40355	0.40354
0.170	0.41912	0.42110	0.42578	0.41724	0.41654
0.180	0.43216	0.43411	0.43891	0.43077	0.42938
0.190	0.44504	0.44698	0.45190	0.44417	0.44206
0.200	0.45781	0.45972	0.46475	0.45745	0.45462
0.250	0.52038	0.52213	0.52770	0.52273	0.51615
0.300	0.58212	0.58367	0.58970	0.58731	0.57682
0.350	0.64432	0.64564	0.65206	0.65240	0.63795
0.400	0.70806	0.70910	0.71585	0.71897	0.70059
0.450	0.77433	0.77507	0.78205	0.78795	0.76578
0.500	0.84421	0.84459	0.85172	0.86032	0.83460
0.550	0.91893	0.91892	0.92606	0.93719	0.90832
0.600	1.00007	0.99961	1.00659	1.01998	0.98857
0.650	1.08971	1.08876	1.09534	1.11058	1.07754
0.700	1.19088	1.18935	1.19519	1.21165	1.17838
0.750	1.30814	1.30594	1.31051	1.32726	1.29594
0.800	1.44912	1.44613	1.44856	1.46412	1.43841
0.810	1.48123	1.47807	1.47990	1.49497	1.47104
0.820	1.51497	1.51163	1.51281	1.52728	1.50540
0.830	1.55053	1.54701	1.54745	1.56119	1.54172
0.840	1.58815	1.58443	1.58405	1.59692	1.58024
0.850	1.62808	1.62417	1.62286	1.63470	1.62124
0.860	1.67066	1.66655	1.66418	1.67479	1.66510
0.870	1.71628	1.71197	1.70838	1.71756	1.71225
0.880	1.76544	1.76092	1.75594	1.76342	1.76324
0.890	1.81875	1.81402	1.80744	1.81291	1.81877
0.900	1.87701	1.87209	1.86362	1.86671	1.87974
0.910	1.94129	1.93618	1.92549	1.92572	1.94733
0.920	2.01300	2.00774	1.99438	1.99118	2.02317
0.930	2.09415	2.08878	2.07216	2.06478	2.10953
0.940	2.18767	2.18228	2.16160	2.14903	2.20979
0.950	2.29811	2.29286	2.26695	2.24778	2.32920
0.960	2.43308	2.42828	2.39535	2.36748	2.47661
0.970	2.60688	2.60316	2.56017	2.52022	2.66875
0.980	2.85156	2.85047	2.79143	2.73299	2.94356
0.990	3.26938	3.27630	3.18484	3.09166	3.42337
0.995	3.68671	3.70661	3.57686	3.44589	3.91400
0.999	4.65351	4.72355	4.48550	4.25870	5.08319

TABLE 5.13 N= 6 (CONT.)

P ↓	λ:	5 10 15 15 20 35	5 10 15 15 25 30	5 10 15 20 20 30	5 10 15 20 25 25	5 10 20 20 20 25
0.001		0.05427	0.05490	0.05548	0.05257	0.05396
0.005		0.09693	0.09806	0.09909	0.09438	0.09669
0.010		0.12570	0.12716	0.12848	0.12281	0.12564
0.020		0.16454	0.16645	0.16815	0.16147	0.16492
0.030		0.19375	0.19599	0.19796	0.19075	0.19457
0.040		0.21830	0.22082	0.22300	0.21548	0.21958
0.050		0.24001	0.24277	0.24515	0.23746	0.24176
0.060		0.25980	0.26278	0.26532	0.25755	0.26201
0.070		0.27818	0.28136	0.28405	0.27627	0.28085
0.080		0.29548	0.29885	0.30167	0.29393	0.29862
0.090		0.31193	0.31547	0.31842	0.31078	0.31555
0.100		0.32770	0.33141	0.33447	0.32695	0.33179
0.110		0.34291	0.34677	0.34994	0.34258	0.34747
0.120		0.35764	0.36166	0.36492	0.35775	0.36267
0.130		0.37198	0.37614	0.37950	0.37253	0.37749
0.140		0.38598	0.39028	0.39373	0.38699	0.39196
0.150		0.39970	0.40413	0.40766	0.40116	0.40615
0.160		0.41316	0.41772	0.42133	0.41510	0.42010
0.170		0.42642	0.43110	0.43479	0.42884	0.43383
0.180		0.43949	0.44430	0.44805	0.44239	0.44738
0.190		0.45241	0.45733	0.46115	0.45580	0.46078
0.200		0.46520	0.47023	0.47411	0.46908	0.47404
0.250		0.52775	0.53331	0.53745	0.53414	0.53896
0.300		0.58929	0.59530	0.59962	0.59822	0.60283
0.350		0.65114	0.65753	0.66196	0.66259	0.66691
0.400		0.71434	0.72105	0.72551	0.72823	0.73223
0.450		0.77991	0.78686	0.79125	0.79610	0.79972
0.500		0.84888	0.85598	0.86021	0.86715	0.87035
0.550		0.92249	0.92961	0.93357	0.94250	0.94522
0.600		1.00226	1.00924	1.01277	1.02354	1.02569
0.650		1.09024	1.09685	1.09976	1.11209	1.11357
0.700		1.18935	1.19526	1.19730	1.21076	1.21140
0.750		1.30407	1.30877	1.30957	1.32345	1.32301
0.800		1.44186	1.44449	1.44349	1.45665	1.45473
0.810		1.47322	1.47529	1.47384	1.48664	1.48436
0.820		1.50618	1.50761	1.50567	1.51803	1.51535
0.830		1.54093	1.54164	1.53916	1.55098	1.54786
0.840		1.57768	1.57759	1.57451	1.58567	1.58208
0.850		1.61670	1.61569	1.61197	1.62232	1.61822
0.860		1.65831	1.65627	1.65183	1.66122	1.65654
0.870		1.70290	1.69967	1.69444	1.70268	1.69736
0.880		1.75096	1.74636	1.74025	1.74711	1.74108
0.890		1.80312	1.79692	1.78982	1.79502	1.78820
0.900		1.86015	1.85209	1.84387	1.84708	1.83934
0.910		1.92312	1.91284	1.90334	1.90415	1.89536
0.920		1.99345	1.98050	1.96952	1.96740	1.95737
0.930		2.07314	2.05692	2.04421	2.03845	2.02697
0.940		2.16513	2.14481	2.13004	2.11971	2.10645
0.950		2.27402	2.24839	2.23109	2.21486	2.19938
0.960		2.40751	2.37472	2.35421	2.33007	2.31171
0.970		2.58015	2.53703	2.51225	2.47687	2.45454
0.980		2.82479	2.76506	2.73407	2.68106	2.65265
0.990		3.24736	3.15385	3.11200	3.02464	2.98467
0.995		3.67582	3.54226	3.48969	3.36353	3.31070
0.999		4.69208	4.44568	4.37063	4.14124	4.05489

TABLE 5.14 N= 6 (CONT.)

P ↓	λ:	5 15 15 15 15 35	5 15 15 15 20 30	5 15 15 15 25 25	5 15 15 20 20 25	5 15 20 20 20 20
0.001		0.05530	0.05653	0.05385	0.05527	0.04302
0.005		0.09871	0.10090	0.09662	0.09895	0.07837
0.010		0.12794	0.13076	0.12564	0.12850	0.10309
0.020		0.16737	0.17102	0.16507	0.16852	0.13773
0.030		0.19698	0.20123	0.19486	0.19868	0.16481
0.040		0.22184	0.22659	0.22000	0.22408	0.18830
0.050		0.24382	0.24899	0.24230	0.24658	0.20967
0.060		0.26383	0.26938	0.26267	0.26711	0.22964
0.070		0.28240	0.28830	0.28162	0.28618	0.24860
0.080		0.29987	0.30609	0.29949	0.30415	0.26682
0.090		0.31647	0.32299	0.31651	0.32125	0.28448
0.100		0.33238	0.33917	0.33284	0.33764	0.30170
0.110		0.34771	0.35476	0.34860	0.35345	0.31856
0.120		0.36255	0.36985	0.36389	0.36878	0.33513
0.130		0.37699	0.38452	0.37877	0.38369	0.35147
0.140		0.39108	0.39884	0.39332	0.39825	0.36760
0.150		0.40487	0.41284	0.40757	0.41252	0.38357
0.160		0.41841	0.42659	0.42157	0.42652	0.39938
0.170		0.43173	0.44010	0.43535	0.44031	0.41506
0.180		0.44487	0.45342	0.44895	0.45390	0.43063
0.190		0.45783	0.46657	0.46238	0.46733	0.44609
0.200		0.47066	0.47957	0.47568	0.48061	0.46145
0.250		0.53335	0.54302	0.54071	0.54552	0.53706
0.300		0.59490	0.60519	0.60458	0.60918	0.61110
0.350		0.65661	0.66739	0.66855	0.67290	0.68414
0.400		0.71957	0.73068	0.73365	0.73770	0.75682
0.450		0.78475	0.79605	0.80082	0.80451	0.82993
0.500		0.85319	0.86449	0.87103	0.87431	0.90438
0.550		0.92610	0.93717	0.94537	0.94818	0.98123
0.600		1.00498	1.01551	1.02523	1.02748	1.06180
0.650		1.09182	1.10142	1.11242	1.11399	1.14777
0.700		1.18949	1.19759	1.20949	1.21022	1.24146
0.750		1.30237	1.30811	1.32031	1.31996	1.34633
0.800		1.43776	1.43976	1.45126	1.44941	1.46795
0.810		1.46856	1.46956	1.48075	1.47853	1.49505
0.820		1.50092	1.50082	1.51161	1.50899	1.52332
0.830		1.53502	1.53370	1.54400	1.54094	1.55289
0.840		1.57109	1.56840	1.57810	1.57457	1.58392
0.850		1.60938	1.60516	1.61414	1.61009	1.61660
0.860		1.65021	1.64427	1.65238	1.64775	1.65116
0.870		1.69396	1.68607	1.69315	1.68788	1.68787
0.880		1.74112	1.73099	1.73684	1.73085	1.72710
0.890		1.79229	1.77960	1.78397	1.77717	1.76926
0.900		1.84826	1.83259	1.83518	1.82745	1.81493
0.910		1.91006	1.89090	1.89132	1.88252	1.86483
0.920		1.97910	1.95578	1.95354	1.94351	1.91995
0.930		2.05736	2.02900	2.02347	2.01195	1.98168
0.940		2.14776	2.11316	2.10345	2.09013	2.05204
0.950		2.25484	2.21226	2.19712	2.18155	2.13416
0.960		2.38626	2.33305	2.31058	2.29208	2.23326
0.970		2.55648	2.48821	2.45520	2.43263	2.35908
0.980		2.79824	2.70622	2.65646	2.62764	2.53342
0.990		3.21731	3.07846	2.99536	2.95454	2.82544
0.995		3.64393	3.45155	3.32993	3.27561	3.11230
0.999		4.66000	4.32590	4.09893	4.00872	3.76772

TABLE 5.15 N= 6 (CONT.)

P ↓	λ:	10 10 10 10 10 50	10 10 10 10 15 45	10 10 10 10 20 40	10 10 10 10 25 35	10 10 10 10 30 30
0.001		0.05016	0.05269	0.05420	0.05502	0.05528
0.005		0.08940	0.09387	0.09656	0.09802	0.09848
0.010		0.11580	0.12154	0.12500	0.12690	0.12750
0.020		0.15137	0.15878	0.16329	0.16576	0.16656
0.030		0.17809	0.18671	0.19200	0.19491	0.19584
0.040		0.20053	0.21016	0.21608	0.21935	0.22040
0.050		0.22038	0.23087	0.23736	0.24094	0.24209
0.060		0.23846	0.24973	0.25672	0.26059	0.26184
0.070		0.25526	0.26723	0.27469	0.27882	0.28015
0.080		0.27108	0.28370	0.29159	0.29597	0.29738
0.090		0.28614	0.29936	0.30766	0.31227	0.31376
0.100		0.30057	0.31437	0.32304	0.32788	0.32944
0.110		0.31450	0.32883	0.33787	0.34292	0.34455
0.120		0.32801	0.34285	0.35224	0.35749	0.35918
0.130		0.34117	0.35649	0.36622	0.37166	0.37342
0.140		0.35402	0.36981	0.37986	0.38550	0.38732
0.150		0.36663	0.38286	0.39322	0.39904	0.40092
0.160		0.37902	0.39568	0.40635	0.41234	0.41428
0.170		0.39123	0.40830	0.41926	0.42543	0.42742
0.180		0.40328	0.42075	0.43200	0.43833	0.44039
0.190		0.41521	0.43306	0.44458	0.45108	0.45319
0.200		0.42703	0.44525	0.45704	0.46370	0.46586
0.250		0.48508	0.50497	0.51801	0.52542	0.52783
0.300		0.54265	0.56393	0.57807	0.58616	0.58880
0.350		0.60103	0.62344	0.63854	0.64724	0.65009
0.400		0.66134	0.68460	0.70051	0.70975	0.71278
0.450		0.72469	0.74847	0.76502	0.77471	0.77789
0.500		0.79231	0.81620	0.83316	0.84319	0.84650
0.550		0.86568	0.88916	0.90623	0.91645	0.91985
0.600		0.94675	0.96910	0.98589	0.99609	0.99950
0.650		1.03822	1.05844	1.07435	1.08424	1.08757
0.700		1.14406	1.16068	1.17482	1.18393	1.18704
0.750		1.27049	1.28127	1.29225	1.29983	1.30249
0.800		1.42817	1.42949	1.43494	1.43972	1.44153
0.810		1.46496	1.46376	1.46767	1.47166	1.47323
0.820		1.50398	1.49999	1.50217	1.50526	1.50655
0.830		1.54551	1.53841	1.53865	1.54072	1.54170
0.840		1.58987	1.57931	1.57736	1.57827	1.57888
0.850		1.63745	1.62304	1.61860	1.61818	1.61838
0.860		1.68873	1.67000	1.66273	1.66079	1.66052
0.870		1.74429	1.72071	1.71020	1.70650	1.70568
0.880		1.80485	1.77580	1.76156	1.75583	1.75436
0.890		1.87132	1.83607	1.81752	1.80940	1.80719
0.900		1.94489	1.90256	1.87899	1.86806	1.86495
0.910		2.02710	1.97664	1.94716	1.93288	1.92871
0.920		2.12006	2.06018	2.02367	2.00533	1.99987
0.930		2.22673	2.15582	2.11081	2.08750	2.08044
0.940		2.35146	2.26742	2.21197	2.18242	2.17333
0.950		2.50098	2.40102	2.33241	2.29481	2.28308
0.960		2.68662	2.56675	2.48102	2.43259	2.41727
0.970		2.92969	2.78374	2.67455	2.61067	2.59013
0.980		3.27824	3.09516	2.95091	2.86261	2.83360
0.990		3.88662	3.63976	3.43226	3.29607	3.24961
0.995		4.50689	4.19607	3.92332	3.73302	3.66553
0.999		5.97822	5.51788	5.09169	4.76025	4.63121

TABLE 5.16 N= 6 (CONT.)

P ↓	λ:	10 10 10 15 15 40	10 10 10 15 20 35	10 10 10 15 25 30	10 10 10 20 20 30	10 10 10 20 25 25
0.001		0.05523	0.05666	0.05731	0.05792	0.05831
0.005		0.09832	0.10084	0.10200	0.10305	0.10375
0.010		0.12722	0.13047	0.13196	0.13331	0.13420
0.020		0.16609	0.17028	0.17223	0.17395	0.17510
0.030		0.19519	0.20009	0.20236	0.20435	0.20569
0.040		0.21958	0.22506	0.22761	0.22981	0.23131
0.050		0.24112	0.24709	0.24988	0.25227	0.25390
0.060		0.26070	0.26712	0.27012	0.27267	0.27443
0.070		0.27886	0.28569	0.28888	0.29158	0.29345
0.080		0.29593	0.30314	0.30652	0.30934	0.31131
0.090		0.31215	0.31971	0.32326	0.32620	0.32827
0.100		0.32767	0.33556	0.33927	0.34233	0.34448
0.110		0.34262	0.35083	0.35469	0.35786	0.36009
0.120		0.35710	0.36561	0.36962	0.37288	0.37519
0.130		0.37118	0.37997	0.38412	0.38747	0.38987
0.140		0.38492	0.39399	0.39827	0.40171	0.40417
0.150		0.39837	0.40770	0.41211	0.41563	0.41816
0.160		0.41156	0.42115	0.42569	0.42928	0.43188
0.170		0.42455	0.43437	0.43903	0.44270	0.44537
0.180		0.43734	0.44741	0.45219	0.45592	0.45865
0.190		0.44999	0.46028	0.46517	0.46897	0.47176
0.200		0.46249	0.47300	0.47801	0.48186	0.48471
0.250		0.52362	0.53514	0.54065	0.54477	0.54787
0.300		0.58371	0.59609	0.60205	0.60634	0.60967
0.350		0.64408	0.65718	0.66353	0.66793	0.67144
0.400		0.70583	0.71950	0.72617	0.73061	0.73425
0.450		0.76997	0.78404	0.79096	0.79536	0.79909
0.500		0.83758	0.85186	0.85894	0.86320	0.86697
0.550		0.90995	0.92416	0.93129	0.93529	0.93903
0.600		0.98867	1.00247	1.00950	1.01310	1.01671
0.650		1.07592	1.08883	1.09552	1.09854	1.10190
0.700		1.17484	1.18612	1.19217	1.19434	1.19727
0.750		1.29025	1.29880	1.30370	1.30464	1.30686
0.800		1.43031	1.43428	1.43717	1.43632	1.43738
0.810		1.46243	1.46515	1.46748	1.46618	1.46693
0.820		1.49627	1.49760	1.49930	1.49751	1.49790
0.830		1.53205	1.53182	1.53281	1.53048	1.53049
0.840		1.57002	1.56802	1.56822	1.56529	1.56486
0.850		1.61048	1.60649	1.60578	1.60219	1.60127
0.860		1.65377	1.64752	1.64578	1.64147	1.63999
0.870		1.70035	1.69153	1.68859	1.68348	1.68136
0.880		1.75077	1.73898	1.73467	1.72866	1.72582
0.890		1.80572	1.79051	1.78459	1.77757	1.77389
0.900		1.86610	1.84690	1.83910	1.83093	1.82627
0.910		1.93311	1.90920	1.89915	1.88967	1.88385
0.920		2.00838	1.97884	1.96608	1.95508	1.94788
0.930		2.09420	2.05782	2.04174	2.02895	2.02006
0.940		2.19395	2.14908	2.12882	2.11390	2.10291
0.950		2.31290	2.25723	2.23154	2.21401	2.20030
0.960		2.45995	2.38997	2.35694	2.33609	2.31875
0.970		2.65190	2.56187	2.51825	2.49298	2.47041
0.980		2.92679	2.80585	2.74517	2.71348	2.68253
0.990		3.40728	3.22801	3.13270	3.08983	3.04176
0.995		3.89875	3.65664	3.52041	3.46658	3.39791
0.999		5.06941	4.67399	4.42330	4.34681	4.21732

TABLE 5.17 N= 6 (CONT.)

P ↓	λ: 10 10 15 15 15 35	10 10 15 15 20 30	10 10 15 15 25 25	10 10 15 20 20 25	10 10 20 20 20 20
0.001	0.05772	0.05900	0.05940	0.06003	0.05926
0.005	0.10266	0.10491	0.10561	0.10669	0.10552
0.010	0.13275	0.13563	0.13653	0.13792	0.13656
0.020	0.17315	0.17685	0.17802	0.17978	0.17828
0.030	0.20335	0.20765	0.20901	0.21104	0.20951
0.040	0.22863	0.23342	0.23493	0.23718	0.23567
0.050	0.25092	0.25613	0.25777	0.26020	0.25874
0.060	0.27116	0.27675	0.27851	0.28110	0.27970
0.070	0.28991	0.29584	0.29772	0.30044	0.29913
0.080	0.30753	0.31376	0.31574	0.31860	0.31738
0.090	0.32424	0.33077	0.33284	0.33581	0.33470
0.100	0.34023	0.34702	0.34918	0.35227	0.35126
0.110	0.35562	0.36266	0.36491	0.36809	0.36720
0.120	0.37050	0.37779	0.38011	0.38339	0.38262
0.130	0.38496	0.39248	0.39487	0.39824	0.39760
0.140	0.39906	0.40679	0.40926	0.41271	0.41220
0.150	0.41285	0.42079	0.42332	0.42685	0.42647
0.160	0.42637	0.43451	0.43711	0.44071	0.44047
0.170	0.43966	0.44799	0.45065	0.45433	0.45422
0.180	0.45275	0.46126	0.46399	0.46772	0.46776
0.190	0.46567	0.47435	0.47714	0.48094	0.48112
0.200	0.47844	0.48729	0.49014	0.49399	0.49432
0.250	0.54071	0.55031	0.55342	0.55751	0.55861
0.300	0.60167	0.61189	0.61520	0.61947	0.62136
0.350	0.66265	0.67336	0.67685	0.68122	0.68393
0.400	0.72474	0.73580	0.73943	0.74383	0.74735
0.450	0.78892	0.80019	0.80392	0.80827	0.81259
0.500	0.85624	0.86754	0.87130	0.87553	0.88061
0.550	0.92787	0.93899	0.94272	0.94671	0.95249
0.600	1.00532	1.01596	1.01958	1.02319	1.02956
0.650	1.09056	1.10033	1.10371	1.10679	1.11358
0.700	1.18643	1.19478	1.19775	1.20003	1.20698
0.750	1.29726	1.30334	1.30563	1.30679	1.31345
0.800	1.43031	1.43272	1.43389	1.43340	1.43904
0.810	1.46060	1.46203	1.46290	1.46198	1.46728
0.820	1.49243	1.49277	1.49331	1.49192	1.49683
0.830	1.52600	1.52512	1.52527	1.52339	1.52782
0.840	1.56150	1.55927	1.55899	1.55655	1.56044
0.850	1.59921	1.59545	1.59470	1.59163	1.59489
0.860	1.63944	1.63395	1.63265	1.62891	1.63142
0.870	1.68257	1.67512	1.67320	1.66869	1.67033
0.880	1.72908	1.71939	1.71676	1.71139	1.71200
0.890	1.77959	1.76730	1.76385	1.75751	1.75691
0.900	1.83486	1.81957	1.81514	1.80770	1.80565
0.910	1.89594	1.87710	1.87153	1.86282	1.85904
0.920	1.96424	1.94116	1.93421	1.92401	1.91813
0.930	2.04172	2.01350	2.00487	1.99290	1.98443
0.940	2.13132	2.09670	2.08596	2.07184	2.06013
0.950	2.23758	2.19478	2.18130	2.16451	2.14860
0.960	2.36816	2.31444	2.29726	2.27700	2.25548
0.970	2.53756	2.46832	2.44580	2.42077	2.39123
0.980	2.77859	2.68487	2.65368	2.62140	2.57922
0.990	3.19730	3.05538	3.00626	2.96033	2.89324
0.995	3.62422	3.42751	3.35654	3.29565	3.20005
0.999	4.64178	4.30140	4.16534	4.06621	3.89326

TABLE 5.18 N= 6 (CONT.)

P ↓	λ: 10 15 15 15 15 30	10 15 15 15 20 25	10 15 15 20 20 20	15 15 15 15 15 25	15 15 15 15 20 20
0.001	0.06011	0.06115	0.06076	0.06229	0.06294
0.005	0.10678	0.10860	0.10805	0.11052	0.11164
0.010	0.13798	0.14029	0.13970	0.14270	0.14412
0.020	0.17979	0.18274	0.18216	0.18572	0.18753
0.030	0.21098	0.21439	0.21388	0.21777	0.21984
0.040	0.23705	0.24083	0.24041	0.24451	0.24680
0.050	0.26001	0.26411	0.26377	0.26803	0.27049
0.060	0.28084	0.28521	0.28498	0.28934	0.29196
0.070	0.30011	0.30474	0.30461	0.30904	0.31180
0.080	0.31819	0.32305	0.32303	0.32751	0.33040
0.090	0.33534	0.34040	0.34049	0.34500	0.34800
0.100	0.35172	0.35698	0.35718	0.36170	0.36480
0.110	0.36747	0.37291	0.37323	0.37774	0.38094
0.120	0.38270	0.38831	0.38875	0.39323	0.39652
0.130	0.39748	0.40325	0.40380	0.40825	0.41163
0.140	0.41187	0.41780	0.41847	0.42288	0.42634
0.150	0.42594	0.43201	0.43280	0.43716	0.44070
0.160	0.43973	0.44593	0.44685	0.45115	0.45475
0.170	0.45326	0.45960	0.46064	0.46487	0.46854
0.180	0.46659	0.47305	0.47421	0.47837	0.48210
0.190	0.47973	0.48631	0.48759	0.49166	0.49545
0.200	0.49271	0.49940	0.50080	0.50479	0.50863
0.250	0.55585	0.56303	0.56504	0.56852	0.57259
0.300	0.61741	0.62497	0.62759	0.63045	0.63469
0.350	0.67877	0.68660	0.68981	0.69197	0.69629
0.400	0.74099	0.74899	0.75277	0.75413	0.75849
0.450	0.80503	0.81308	0.81740	0.81789	0.82221
0.500	0.87190	0.87986	0.88467	0.88422	0.88842
0.550	0.94272	0.95043	0.95566	0.95419	0.95818
0.600	1.01888	1.02613	1.03166	1.02913	1.03278
0.650	1.10222	1.10873	1.11440	1.11076	1.11392
0.700	1.19534	1.20071	1.20627	1.20149	1.20395
0.750	1.30219	1.30582	1.31089	1.30499	1.30642
0.800	1.42931	1.43024	1.43419	1.42728	1.42718
0.810	1.45807	1.45830	1.46191	1.45482	1.45433
0.820	1.48823	1.48769	1.49090	1.48365	1.48274
0.830	1.51996	1.51855	1.52131	1.51391	1.51253
0.840	1.55343	1.55107	1.55330	1.54579	1.54389
0.850	1.58889	1.58546	1.58709	1.57948	1.57700
0.860	1.62662	1.62198	1.62292	1.61525	1.61213
0.870	1.66694	1.66094	1.66107	1.65339	1.64955
0.880	1.71029	1.70274	1.70194	1.69429	1.68963
0.890	1.75719	1.74787	1.74596	1.73842	1.73283
0.900	1.80833	1.79696	1.79375	1.78640	1.77975
0.910	1.86462	1.85085	1.84609	1.83904	1.83115
0.920	1.92729	1.91065	1.90401	1.89743	1.88808
0.930	1.99805	1.97795	1.96901	1.96310	1.95199
0.940	2.07943	2.05505	2.04321	2.03828	2.02502
0.950	2.17536	2.14550	2.12993	2.12645	2.11044
0.960	2.29244	2.25528	2.23469	2.23338	2.21376
0.970	2.44309	2.39554	2.36776	2.36992	2.34518
0.980	2.65536	2.59126	2.55204	2.56036	2.52755
0.990	3.01945	2.92201	2.85985	2.88214	2.83310
0.995	3.38651	3.24960	3.16056	3.20107	3.13266
0.999	4.25381	4.00461	3.83967	3.93832	3.81252

TABLE 5.19 N= 6 (CONT.)

P ↓	All λ's = 100/6
0.001	0.06351
0.005	0.11262
0.010	0.14535
0.020	0.18907
0.030	0.22160
0.040	0.24873
0.050	0.27256
0.060	0.29415
0.070	0.31410
0.080	0.33279
0.090	0.35048
0.100	0.36736
0.110	0.38357
0.120	0.39921
0.130	0.41438
0.140	0.42914
0.150	0.44355
0.160	0.45765
0.170	0.47148
0.180	0.48508
0.190	0.49847
0.200	0.51168
0.250	0.57577
0.300	0.63793
0.350	0.69955
0.400	0.76169
0.450	0.82531
0.500	0.89136
0.550	0.96087
0.600	1.03513
0.650	1.11580
0.700	1.20519
0.750	1.30680
0.800	1.42635
0.810	1.45319
0.820	1.48127
0.830	1.51070
0.840	1.54167
0.850	1.57435
0.860	1.60900
0.870	1.64590
0.880	1.68539
0.890	1.72794
0.900	1.77411
0.910	1.82465
0.920	1.88059
0.930	1.94332
0.940	2.01493
0.950	2.09860
0.960	2.19964
0.970	2.32794
0.980	2.50554
0.990	2.80199
0.995	3.09127
0.999	3.74296

TABLE 6.1 N= 7

P ↓	λ:	5 5 5 5 5 5 70	5 5 5 5 5 10 65	5 5 5 5 5 15 60	5 5 5 5 5 20 55	5 5 5 5 5 25 50
0.001		0.04444	0.04856	0.05095	0.05249	0.05351
0.005		0.07442	0.08132	0.08539	0.08806	0.08983
0.010		0.09404	0.10274	0.10795	0.11139	0.11367
0.020		0.12017	0.13126	0.13802	0.14252	0.14551
0.030		0.13968	0.15255	0.16048	0.16580	0.16935
0.040		0.15607	0.17040	0.17934	0.18535	0.18939
0.050		0.17059	0.18620	0.19603	0.20268	0.20715
0.060		0.18385	0.20062	0.21128	0.21851	0.22339
0.070		0.19621	0.21405	0.22547	0.23326	0.23852
0.080		0.20789	0.22673	0.23888	0.24719	0.25283
0.090		0.21906	0.23884	0.25168	0.26050	0.26649
0.100		0.22981	0.25048	0.26400	0.27331	0.27964
0.110		0.24024	0.26176	0.27593	0.28571	0.29238
0.120		0.25040	0.27274	0.28754	0.29779	0.30479
0.130		0.26035	0.28348	0.29889	0.30960	0.31693
0.140		0.27013	0.29401	0.31003	0.32119	0.32884
0.150		0.27976	0.30438	0.32099	0.33260	0.34057
0.160		0.28930	0.31463	0.33181	0.34386	0.35215
0.170		0.29874	0.32477	0.34252	0.35500	0.36360
0.180		0.30813	0.33482	0.35314	0.36605	0.37496
0.190		0.31747	0.34482	0.36369	0.37703	0.38625
0.200		0.32679	0.35477	0.37419	0.38796	0.39748
0.250		0.37351	0.40442	0.42648	0.44232	0.45336
0.300		0.42160	0.45503	0.47957	0.49742	0.50996
0.350		0.47246	0.50793	0.53477	0.55455	0.56855
0.400		0.52751	0.56441	0.59328	0.61488	0.63028
0.450		0.58838	0.62590	0.65642	0.67964	0.69633
0.500		0.65710	0.69412	0.72569	0.75021	0.76803
0.550		0.73627	0.77126	0.80300	0.82833	0.84699
0.600		0.82935	0.86028	0.89086	0.91624	0.93527
0.650		0.94098	0.96523	0.99275	1.01700	1.03573
0.700		1.07758	1.09202	1.11373	1.13509	1.15242
0.750		1.24859	1.24956	1.26167	1.27742	1.29166
0.800		1.46930	1.45250	1.44973	1.45565	1.46399
0.810		1.52152	1.50055	1.49400	1.49725	1.50391
0.820		1.57710	1.55172	1.54106	1.54134	1.54611
0.830		1.63642	1.60637	1.59126	1.58823	1.59086
0.840		1.69993	1.66491	1.64499	1.63826	1.63847
0.850		1.76818	1.72786	1.70270	1.69187	1.68934
0.860		1.84182	1.79583	1.76498	1.74956	1.74391
0.870		1.92166	1.86957	1.83252	1.81195	1.80275
0.880		2.00870	1.95001	1.90618	1.87984	1.86656
0.890		2.10422	2.03833	1.98705	1.95420	1.93623
0.900		2.20984	2.13605	2.07655	2.03632	2.01291
0.910		2.32772	2.24518	2.17652	2.12787	2.09810
0.920		2.46080	2.36842	2.28949	2.23115	2.19386
0.930		2.61319	2.50962	2.41899	2.34937	2.30309
0.940		2.79094	2.67438	2.57023	2.48728	2.43003
0.950		3.00350	2.87146	2.75127	2.65224	2.58132
0.960		3.26671	3.11557	2.97570	2.85667	2.76811
0.970		3.61037	3.43439	3.26908	3.12393	3.01144
0.980		4.10183	3.89043	3.68910	3.50680	3.35888
0.990		4.95749	4.68461	4.42113	4.17498	3.96364
0.995		5.82843	5.49304	5.16671	4.85650	4.57993
0.999		7.89158	7.40855	6.93400	6.47407	6.04419

TABLE 6.2 N= 7 (CONT.)

		5 5 5 5	5 5 5 5	5 5 5 10	5 5 5 10	5 5 5 10
P ↓	λ:	30 45	35 40	10 60	15 55	20 50
0.001		0.05414	0.05444	0.05297	0.05550	0.05711
0.005		0.09092	0.09144	0.08866	0.09295	0.09571
0.010		0.11508	0.11576	0.11197	0.11744	0.12097
0.020		0.14737	0.14826	0.14295	0.15000	0.15459
0.030		0.17156	0.17262	0.16603	0.17427	0.17967
0.040		0.19190	0.19311	0.18536	0.19461	0.20069
0.050		0.20994	0.21128	0.20245	0.21258	0.21928
0.060		0.22643	0.22790	0.21801	0.22895	0.23622
0.070		0.24181	0.24339	0.23249	0.24418	0.25198
0.080		0.25635	0.25805	0.24614	0.25854	0.26683
0.090		0.27024	0.27205	0.25914	0.27222	0.28099
0.100		0.28361	0.28553	0.27164	0.28536	0.29459
0.110		0.29657	0.29859	0.28373	0.29806	0.30774
0.120		0.30919	0.31132	0.29548	0.31040	0.32052
0.130		0.32154	0.32377	0.30694	0.32245	0.33299
0.140		0.33366	0.33599	0.31818	0.33425	0.34521
0.150		0.34559	0.34802	0.32923	0.34585	0.35721
0.160		0.35737	0.35990	0.34012	0.35727	0.36903
0.170		0.36903	0.37166	0.35089	0.36856	0.38071
0.180		0.38059	0.38332	0.36155	0.37973	0.39227
0.190		0.39208	0.39491	0.37212	0.39081	0.40372
0.200		0.40351	0.40644	0.38264	0.40181	0.41510
0.250		0.46038	0.46380	0.43480	0.45631	0.47141
0.300		0.51796	0.52186	0.48746	0.51108	0.52788
0.350		0.57752	0.58190	0.54191	0.56742	0.58582
0.400		0.64019	0.64505	0.59936	0.62649	0.64635
0.450		0.70713	0.71244	0.66109	0.68946	0.71059
0.500		0.77962	0.78533	0.72862	0.75771	0.77982
0.550		0.85921	0.86525	0.80383	0.83291	0.85561
0.600		0.94787	0.95413	0.88929	0.91730	0.93996
0.650		1.04831	1.05460	0.98855	1.01395	1.03565
0.700		1.16436	1.17041	1.10684	1.12738	1.14670
0.750		1.30197	1.30732	1.25229	1.26467	1.27937
0.800		1.47101	1.47488	1.43845	1.43775	1.44419
0.810		1.50997	1.51341	1.48246	1.47834	1.48250
0.820		1.55108	1.55402	1.52930	1.52144	1.52304
0.830		1.59458	1.59696	1.57933	1.56737	1.56611
0.840		1.64078	1.64251	1.63293	1.61649	1.61201
0.850		1.69002	1.69100	1.69058	1.66921	1.66112
0.860		1.74274	1.74285	1.75285	1.72607	1.71391
0.870		1.79943	1.79855	1.82043	1.78770	1.77094
0.880		1.86077	1.85873	1.89420	1.85490	1.83293
0.890		1.92755	1.92415	1.97524	1.92866	1.90075
0.900		2.00083	1.99583	2.06496	2.01028	1.97557
0.910		2.08200	2.07508	2.16522	2.10145	2.05891
0.920		2.17293	2.16371	2.27852	2.20449	2.15284
0.930		2.27627	2.26420	2.40840	2.32263	2.26026
0.940		2.39589	2.38026	2.56006	2.46066	2.38548
0.950		2.53782	2.51757	2.74155	2.62596	2.53515
0.960		2.71218	2.68570	2.96646	2.83099	2.72054
0.970		2.93801	2.90258	3.26034	3.09918	2.96280
0.980		3.25825	3.20849	3.68087	3.48339	3.30983
0.990		3.81076	3.73208	4.41346	4.15350	3.91562
0.995		4.36928	4.25644	5.15936	4.83636	4.53386
0.999		5.68660	5.47698	6.92708	6.45562	6.00252

TABLE 6.3 N= 7 (CONT.)

P ↓	λ: 5 5 5 10 25 45	5 5 5 10 30 40	5 5 5 10 35 35	5 5 5 15 15 50	5 5 5 15 20 45
0.001	0.05812	0.05869	0.05887	0.05807	0.05966
0.005	0.09745	0.09843	0.09875	0.09731	0.10002
0.010	0.12322	0.12447	0.12488	0.12297	0.12645
0.020	0.15753	0.15918	0.15971	0.15712	0.16163
0.030	0.18314	0.18509	0.18572	0.18258	0.18787
0.040	0.20461	0.20682	0.20754	0.20391	0.20986
0.050	0.22361	0.22605	0.22684	0.22275	0.22930
0.060	0.24092	0.24358	0.24444	0.23992	0.24701
0.070	0.25703	0.25989	0.26082	0.25588	0.26347
0.080	0.27222	0.27528	0.27627	0.27093	0.27899
0.090	0.28671	0.28995	0.29100	0.28526	0.29377
0.100	0.30062	0.30404	0.30515	0.29902	0.30796
0.110	0.31407	0.31767	0.31884	0.31231	0.32167
0.120	0.32714	0.33091	0.33214	0.32522	0.33498
0.130	0.33990	0.34384	0.34512	0.33782	0.34797
0.140	0.35240	0.35650	0.35784	0.35015	0.36068
0.150	0.36468	0.36894	0.37033	0.36226	0.37316
0.160	0.37678	0.38120	0.38264	0.37419	0.38545
0.170	0.38872	0.39331	0.39480	0.38596	0.39758
0.180	0.40055	0.40528	0.40683	0.39760	0.40957
0.190	0.41227	0.41716	0.41876	0.40914	0.42145
0.200	0.42391	0.42896	0.43060	0.42060	0.43324
0.250	0.48148	0.48727	0.48917	0.47718	0.49140
0.300	0.53917	0.54570	0.54783	0.53378	0.54946
0.350	0.59828	0.60550	0.60787	0.59169	0.60868
0.400	0.65989	0.66778	0.67037	0.65200	0.67015
0.450	0.72512	0.73363	0.73643	0.71584	0.73493
0.500	0.79520	0.80424	0.80723	0.78445	0.80421
0.550	0.87159	0.88106	0.88419	0.85933	0.87939
0.600	0.95621	0.96591	0.96914	0.94248	0.96228
0.650	1.05162	1.06129	1.06452	1.03659	1.05537
0.700	1.16156	1.17074	1.17384	1.14565	1.16224
0.750	1.29176	1.29975	1.30249	1.27584	1.28844
0.800	1.45183	1.45739	1.45938	1.43765	1.44331
0.810	1.48877	1.49362	1.49539	1.47529	1.47904
0.820	1.52778	1.53182	1.53334	1.51515	1.51677
0.830	1.56909	1.57220	1.57343	1.55751	1.55674
0.840	1.61300	1.61505	1.61595	1.60269	1.59922
0.850	1.65984	1.66068	1.66120	1.65106	1.64456
0.860	1.71004	1.70948	1.70955	1.70310	1.69317
0.870	1.76410	1.76192	1.76149	1.75937	1.74556
0.880	1.82266	1.81859	1.81756	1.82059	1.80234
0.890	1.88650	1.88024	1.87851	1.88766	1.86432
0.900	1.95668	1.94782	1.94527	1.96172	1.93251
0.910	2.03454	2.02259	2.01905	2.04433	2.00827
0.920	2.12195	2.10626	2.10152	2.13756	2.09345
0.930	2.22149	2.20123	2.19501	2.24435	2.19064
0.940	2.33700	2.31102	2.30293	2.36901	2.30366
0.950	2.47444	2.44107	2.43057	2.51826	2.43847
0.960	2.64382	2.60056	2.58678	2.70339	2.60511
0.970	2.86403	2.80669	2.78816	2.94567	2.82252
0.980	3.17777	3.09816	3.07199	3.29312	3.13353
0.990	3.72247	3.59901	3.55720	3.90006	3.67618
0.995	4.27649	4.10300	4.04240	4.51939	4.23028
0.999	5.59143	5.28461	5.16901	5.98965	5.54838

TABLE 6.4 N= 7 (CONT.)

P ↓	λ:	5 5 5 15 25 40	5 5 5 5 15 30 35	5 5 5 5 20 20 40	5 5 5 5 20 25 35	5 5 5 5 20 30 30
0.001		0.06060	0.06105	0.06117	0.06199	0.06225
0.005		0.10164	0.10241	0.10261	0.10402	0.10446
0.010		0.12853	0.12952	0.12976	0.13156	0.13214
0.020		0.16435	0.16563	0.16591	0.16826	0.16901
0.030		0.19107	0.19259	0.19289	0.19565	0.19653
0.040		0.21347	0.21518	0.21550	0.21861	0.21960
0.050		0.23327	0.23516	0.23548	0.23890	0.24000
0.060		0.25132	0.25337	0.25369	0.25739	0.25858
0.070		0.26810	0.27030	0.27062	0.27459	0.27586
0.080		0.28391	0.28626	0.28658	0.29079	0.29215
0.090		0.29898	0.30146	0.30177	0.30622	0.30765
0.100		0.31344	0.31606	0.31636	0.32103	0.32254
0.110		0.32742	0.33016	0.33044	0.33534	0.33692
0.120		0.34099	0.34386	0.34413	0.34923	0.35088
0.130		0.35422	0.35722	0.35747	0.36278	0.36449
0.140		0.36718	0.37029	0.37052	0.37603	0.37781
0.150		0.37990	0.38313	0.38333	0.38904	0.39089
0.160		0.39242	0.39577	0.39595	0.40184	0.40375
0.170		0.40478	0.40824	0.40839	0.41447	0.41644
0.180		0.41699	0.42056	0.42069	0.42695	0.42898
0.190		0.42910	0.43278	0.43287	0.43931	0.44139
0.200		0.44110	0.44489	0.44495	0.45156	0.45371
0.250		0.50032	0.50462	0.50449	0.51192	0.51434
0.300		0.55935	0.56415	0.56376	0.57195	0.57462
0.350		0.61948	0.62474	0.62405	0.63291	0.63580
0.400		0.68177	0.68745	0.68641	0.69585	0.69894
0.450		0.74727	0.75332	0.75186	0.76178	0.76504
0.500		0.81711	0.82347	0.82153	0.83178	0.83517
0.550		0.89265	0.89923	0.89674	0.90715	0.91060
0.600		0.97561	0.98228	0.97918	0.98948	0.99293
0.650		1.06833	1.07489	1.07113	1.08098	1.08430
0.700		1.17418	1.18033	1.17586	1.18474	1.18779
0.750		1.29831	1.30358	1.29844	1.30554	1.30805
0.800		1.44936	1.45295	1.44730	1.45130	1.45286
0.810		1.48400	1.48711	1.48141	1.48455	1.48585
0.820		1.52051	1.52307	1.51733	1.51951	1.52052
0.830		1.55909	1.56102	1.55528	1.55638	1.55705
0.840		1.60000	1.60122	1.59551	1.59540	1.59569
0.850		1.64355	1.64395	1.63833	1.63684	1.63669
0.860		1.69012	1.68958	1.68409	1.68103	1.68040
0.870		1.74015	1.73852	1.73325	1.72841	1.72720
0.880		1.79422	1.79133	1.78636	1.77946	1.77761
0.890		1.85303	1.84867	1.84413	1.83485	1.83223
0.900		1.91752	1.91142	1.90746	1.89540	1.89189
0.910		1.98890	1.98072	1.97756	1.96220	1.95764
0.920		2.06882	2.05813	2.05605	2.03675	2.03091
0.930		2.15960	2.14582	2.14522	2.12113	2.11373
0.940		2.26465	2.24698	2.24844	2.21839	2.20903
0.950		2.38929	2.36658	2.37097	2.33326	2.32137
0.960		2.54243	2.51290	2.52163	2.47372	2.45840
0.970		2.74087	2.70153	2.71709	2.65468	2.63444
0.980		3.02254	2.96744	2.99506	2.90978	2.88157
0.990		3.50940	3.42244	3.47704	3.34670	3.30214
0.995		4.00269	3.87815	3.96722	3.78542	3.72110
0.999		5.16971	4.93947	5.13208	4.81328	4.69049

TABLE 6.5 N= 7 (CONT.)

P ↓	λ:	5 5 5 25 25 30	5 5 5 10 10 10 55	5 5 5 10 10 15 50	5 5 5 10 10 20 45	5 5 5 10 10 25 40
0.001		0.06263	0.05768	0.06034	0.06198	0.06295
0.005		0.10511	0.09643	0.10091	0.10370	0.10537
0.010		0.13296	0.12168	0.12736	0.13092	0.13305
0.020		0.17007	0.15516	0.16243	0.16703	0.16980
0.030		0.19777	0.18004	0.18850	0.19388	0.19713
0.040		0.22099	0.20084	0.21028	0.21632	0.21997
0.050		0.24152	0.21918	0.22948	0.23610	0.24012
0.060		0.26023	0.23585	0.24694	0.25409	0.25845
0.070		0.27762	0.25133	0.26314	0.27079	0.27545
0.080		0.29401	0.26591	0.27839	0.28650	0.29146
0.090		0.30962	0.27977	0.29288	0.30144	0.30667
0.100		0.32460	0.29307	0.30678	0.31576	0.32126
0.110		0.33907	0.30590	0.32020	0.32958	0.33534
0.120		0.35312	0.31836	0.33321	0.34298	0.34899
0.130		0.36681	0.33050	0.34588	0.35603	0.36229
0.140		0.38021	0.34238	0.35827	0.36880	0.37529
0.150		0.39336	0.35404	0.37043	0.38131	0.38804
0.160		0.40630	0.36551	0.38238	0.39362	0.40057
0.170		0.41906	0.37683	0.39417	0.40575	0.41293
0.180		0.43167	0.38802	0.40582	0.41773	0.42513
0.190		0.44416	0.39910	0.41735	0.42959	0.43721
0.200		0.45654	0.41010	0.42878	0.44135	0.44918
0.250		0.51747	0.46439	0.48509	0.49920	0.50804
0.300		0.57801	0.51869	0.54117	0.55670	0.56650
0.350		0.63941	0.57430	0.59832	0.61514	0.62583
0.400		0.70271	0.63237	0.65766	0.67563	0.68713
0.450		0.76891	0.69407	0.72028	0.73922	0.75145
0.500		0.83907	0.76076	0.78744	0.80711	0.81994
0.550		0.91443	0.83413	0.86065	0.88070	0.89394
0.600		0.99657	0.91640	0.94188	0.96182	0.97520
0.650		1.08760	1.01067	1.03388	1.05295	1.06606
0.700		1.19050	1.12156	1.14063	1.15770	1.16990
0.750		1.30986	1.25626	1.26839	1.28165	1.29191
0.800		1.45326	1.42699	1.42780	1.43422	1.44075
0.810		1.48587	1.46717	1.46498	1.46949	1.47495
0.820		1.52013	1.50990	1.50439	1.50676	1.51100
0.830		1.55621	1.55549	1.54632	1.54627	1.54914
0.840		1.59434	1.60431	1.59108	1.58831	1.58960
0.850		1.63479	1.65678	1.63907	1.63322	1.63271
0.860		1.67787	1.71344	1.69075	1.68141	1.67883
0.870		1.72397	1.77492	1.74670	1.73339	1.72843
0.880		1.77358	1.84204	1.80764	1.78978	1.78207
0.890		1.82731	1.91578	1.87446	1.85139	1.84046
0.900		1.88593	1.99745	1.94835	1.91924	1.90454
0.910		1.95048	2.08876	2.03084	1.99470	1.97552
0.920		2.02235	2.19201	2.12403	2.07962	2.05507
0.930		2.10349	2.31045	2.23086	2.17660	2.14549
0.940		2.19675	2.44885	2.35567	2.28948	2.25023
0.950		2.30655	2.61460	2.50517	2.42424	2.37458
0.960		2.44027	2.82016	2.69069	2.59093	2.52749
0.970		2.61173	3.08894	2.93351	2.80854	2.72578
0.980		2.85191	3.47380	3.28167	3.11998	3.00742
0.990		3.25938	4.14463	3.88953	3.66343	3.49451
0.995		3.66393	4.82792	4.50946	4.21820	3.98815
0.999		4.59632	6.44771	5.98044	5.53730	5.15598

TABLE 6.6 N= 7 (CONT.)

P ↓	λ: 5 5 10 10 30 35	5 5 5 10 15 15 45	5 5 5 10 15 20 40	5 5 5 10 15 25 35	5 5 5 10 15 30 30
0.001	0.06341	0.06301	0.06460	0.06546	0.06574
0.005	0.10615	0.10540	0.10810	0.10957	0.11003
0.010	0.13406	0.13304	0.13647	0.13835	0.13894
0.020	0.17111	0.16969	0.17411	0.17653	0.17730
0.030	0.19867	0.19692	0.20207	0.20490	0.20581
0.040	0.22171	0.21966	0.22543	0.22860	0.22962
0.050	0.24203	0.23970	0.24601	0.24949	0.25061
0.060	0.26052	0.25791	0.26472	0.26848	0.26969
0.070	0.27767	0.27480	0.28206	0.28609	0.28738
0.080	0.29382	0.29069	0.29838	0.30264	0.30402
0.090	0.30917	0.30579	0.31388	0.31838	0.31983
0.100	0.32389	0.32026	0.32873	0.33345	0.33497
0.110	0.33810	0.33421	0.34305	0.34798	0.34957
0.120	0.35187	0.34774	0.35693	0.36206	0.36372
0.130	0.36529	0.36091	0.37044	0.37577	0.37749
0.140	0.37840	0.37378	0.38363	0.38916	0.39094
0.150	0.39126	0.38639	0.39657	0.40228	0.40412
0.160	0.40391	0.39879	0.40928	0.41517	0.41707
0.170	0.41638	0.41100	0.42179	0.42786	0.42983
0.180	0.42869	0.42306	0.43415	0.44039	0.44241
0.190	0.44087	0.43499	0.44636	0.45278	0.45485
0.200	0.45294	0.44681	0.45846	0.46504	0.46718
0.250	0.51231	0.50489	0.51785	0.52522	0.52761
0.300	0.57124	0.56248	0.57661	0.58469	0.58733
0.350	0.63103	0.62087	0.63603	0.64476	0.64762
0.400	0.69275	0.68115	0.69719	0.70649	0.70954
0.450	0.75745	0.74438	0.76110	0.77089	0.77411
0.500	0.82626	0.81170	0.82888	0.83903	0.84238
0.550	0.90050	0.88450	0.90182	0.91218	0.91561
0.600	0.98188	0.96456	0.98158	0.99192	0.99538
0.650	1.07268	1.05431	1.07040	1.08040	1.08378
0.700	1.17617	1.15727	1.17151	1.18069	1.18383
0.750	1.29736	1.27895	1.28988	1.29746	1.30012
0.800	1.44457	1.42862	1.43383	1.43852	1.44030
0.810	1.47829	1.46323	1.46686	1.47073	1.47227
0.820	1.51381	1.49981	1.50168	1.50462	1.50587
0.830	1.55132	1.53859	1.53848	1.54038	1.54131
0.840	1.59107	1.57987	1.57753	1.57824	1.57880
0.850	1.63336	1.62399	1.61912	1.61848	1.61862
0.860	1.67854	1.67135	1.66362	1.66144	1.66109
0.870	1.72705	1.72246	1.71147	1.70751	1.70661
0.880	1.77942	1.77796	1.76323	1.75721	1.75566
0.890	1.83632	1.83865	1.81960	1.81118	1.80887
0.900	1.89863	1.90554	1.88148	1.87024	1.86704
0.910	1.96751	1.98002	1.95008	1.93549	1.93121
0.920	2.04449	2.06394	2.02702	2.00838	2.00281
0.930	2.13177	2.15993	2.11459	2.09101	2.08384
0.940	2.23253	2.27183	2.21618	2.18639	2.17721
0.950	2.35174	2.40565	2.33703	2.29926	2.28745
0.960	2.49770	2.57151	2.48601	2.43753	2.42216
0.970	2.68598	2.78849	2.67984	2.61609	2.59555
0.980	2.95159	3.09970	2.95637	2.86850	2.83958
0.990	3.40635	3.64378	3.43761	3.30233	3.25616
0.995	3.86200	4.19964	3.92835	3.73937	3.67238
0.999	4.92339	5.52088	5.09601	4.76639	4.63828

TABLE 6.7 N= 7 (CONT.)

P ↓	λ: 5 5 / 10 20 / 20 35	5 5 / 10 20 / 25 30	5 5 / 10 25 / 25 25	5 5 / 15 15 / 15 40	5 5 / 15 15 / 20 35
0.001	0.06607	0.06675	0.06715	0.06568	0.06717
0.005	0.11059	0.11173	0.11241	0.10986	0.11239
0.010	0.13964	0.14110	0.14196	0.13867	0.14187
0.020	0.17817	0.18005	0.18115	0.17684	0.18095
0.030	0.20680	0.20899	0.21028	0.20519	0.20996
0.040	0.23071	0.23317	0.23461	0.22885	0.23418
0.050	0.25178	0.25447	0.25604	0.24968	0.25550
0.060	0.27093	0.27383	0.27552	0.26861	0.27487
0.070	0.28868	0.29177	0.29357	0.28615	0.29281
0.080	0.30537	0.30865	0.31055	0.30264	0.30967
0.090	0.32123	0.32468	0.32667	0.31830	0.32569
0.100	0.33642	0.34003	0.34212	0.33329	0.34101
0.110	0.35106	0.35482	0.35700	0.34774	0.35578
0.120	0.36524	0.36916	0.37142	0.36174	0.37008
0.130	0.37905	0.38311	0.38544	0.37535	0.38399
0.140	0.39253	0.39673	0.39914	0.38865	0.39757
0.150	0.40573	0.41007	0.41256	0.40167	0.41086
0.160	0.41871	0.42317	0.42573	0.41447	0.42391
0.170	0.43148	0.43607	0.43870	0.42706	0.43676
0.180	0.44408	0.44880	0.45149	0.43948	0.44943
0.190	0.45654	0.46138	0.46414	0.45176	0.46195
0.200	0.46888	0.47383	0.47665	0.46392	0.47434
0.250	0.52935	0.53484	0.53795	0.52350	0.53498
0.300	0.58904	0.59502	0.59837	0.58232	0.59471
0.350	0.64925	0.65565	0.65920	0.64167	0.65483
0.400	0.71104	0.71780	0.72150	0.70261	0.71639
0.450	0.77541	0.78243	0.78624	0.76615	0.78038
0.500	0.84340	0.85060	0.85445	0.83339	0.84785
0.550	0.91627	0.92352	0.92733	0.90559	0.92000
0.600	0.99557	1.00269	1.00635	0.98438	0.99836
0.650	1.08338	1.09013	1.09351	1.07194	1.08499
0.700	1.18270	1.18872	1.19161	1.17144	1.18279
0.750	1.29808	1.30284	1.30494	1.28773	1.29625
0.800	1.43714	1.43974	1.44055	1.42903	1.43282
0.810	1.46885	1.47086	1.47133	1.46144	1.46395
0.820	1.50219	1.50354	1.50363	1.49559	1.49667
0.830	1.53736	1.53796	1.53763	1.53171	1.53118
0.840	1.57458	1.57434	1.57354	1.57002	1.56770
0.850	1.61411	1.61293	1.61160	1.61083	1.60650
0.860	1.65629	1.65402	1.65210	1.65451	1.64788
0.870	1.70150	1.69801	1.69542	1.70149	1.69226
0.880	1.75026	1.74535	1.74200	1.75232	1.74010
0.890	1.80317	1.79663	1.79240	1.80770	1.79204
0.900	1.86106	1.85260	1.84735	1.86853	1.84887
0.910	1.92497	1.91425	1.90782	1.93601	1.91164
0.920	1.99635	1.98292	1.97508	2.01176	1.98177
0.930	2.07723	2.06049	2.05095	2.09806	2.06128
0.940	2.17058	2.14971	2.13807	2.19829	2.15310
0.950	2.28101	2.25484	2.24054	2.31770	2.26183
0.960	2.41630	2.38301	2.36519	2.46518	2.39519
0.970	2.59108	2.54760	2.52482	2.65747	2.56773
0.980	2.83834	2.77860	2.74804	2.93255	2.81235
0.990	3.26422	3.17174	3.12581	3.41285	3.23504
0.995	3.69468	3.56360	3.49980	3.90390	3.66373
0.999	4.71205	4.47210	4.35787	5.07373	4.68056

TABLE 6.8 N= 7 (CONT.)

P ↓	λ: 5 5 / 15 15 / 25 30	5 5 / 15 20 / 20 30	5 5 / 15 20 / 25 25	5 5 / 20 20 / 20 25	5 10 / 10 10 / 10 50
0.001	0.06785	0.06848	0.06889	0.06953	0.06267
0.005	0.11355	0.11460	0.11530	0.11636	0.10459
0.010	0.14335	0.14468	0.14556	0.14690	0.13182
0.020	0.18285	0.18454	0.18566	0.18737	0.16781
0.030	0.21218	0.21412	0.21543	0.21740	0.19446
0.040	0.23666	0.23881	0.24027	0.24245	0.21668
0.050	0.25822	0.26055	0.26214	0.26450	0.23623
0.060	0.27779	0.28028	0.28200	0.28451	0.25397
0.070	0.29593	0.29857	0.30039	0.30305	0.27040
0.080	0.31297	0.31574	0.31766	0.32045	0.28584
0.090	0.32915	0.33204	0.33406	0.33697	0.30050
0.100	0.34464	0.34765	0.34976	0.35278	0.31453
0.110	0.35956	0.36267	0.36487	0.36800	0.32805
0.120	0.37401	0.37722	0.37950	0.38273	0.34115
0.130	0.38806	0.39137	0.39372	0.39704	0.35390
0.140	0.40178	0.40517	0.40760	0.41101	0.36635
0.150	0.41520	0.41868	0.42118	0.42468	0.37855
0.160	0.42839	0.43195	0.43451	0.43809	0.39053
0.170	0.44136	0.44499	0.44763	0.45127	0.40233
0.180	0.45415	0.45786	0.46056	0.46427	0.41398
0.190	0.46678	0.47056	0.47332	0.47711	0.42550
0.200	0.47929	0.48313	0.48595	0.48980	0.43692
0.250	0.54047	0.54458	0.54768	0.55180	0.49297
0.300	0.60067	0.60499	0.60832	0.61263	0.54856
0.350	0.66120	0.66565	0.66918	0.67360	0.60500
0.400	0.72311	0.72761	0.73128	0.73575	0.66340
0.450	0.78737	0.79183	0.79561	0.80004	0.72487
0.500	0.85501	0.85933	0.86315	0.86745	0.79064
0.550	0.92722	0.93128	0.93507	0.93911	0.86221
0.600	1.00547	1.00912	1.01278	1.01642	0.94155
0.650	1.09176	1.09479	1.09819	1.10124	1.03140
0.700	1.18888	1.19104	1.19399	1.19620	1.13576
0.750	1.30114	1.30204	1.30425	1.30526	1.26096
0.800	1.43564	1.43471	1.43572	1.43499	1.41775
0.810	1.46619	1.46480	1.46549	1.46432	1.45442
0.820	1.49828	1.49638	1.49672	1.49507	1.49334
0.830	1.53207	1.52962	1.52956	1.52738	1.53480
0.840	1.56777	1.56472	1.56422	1.56146	1.57911
0.850	1.60564	1.60193	1.60092	1.59754	1.62667
0.860	1.64597	1.64153	1.63996	1.63587	1.67796
0.870	1.68914	1.68389	1.68167	1.67681	1.73356
0.880	1.73559	1.72944	1.72649	1.72076	1.79419
0.890	1.78591	1.77875	1.77495	1.76825	1.86078
0.900	1.84084	1.83253	1.82775	1.81994	1.93449
0.910	1.90136	1.89174	1.88579	1.87671	2.01688
0.920	1.96878	1.95765	1.95031	1.93976	2.11006
0.930	2.04495	2.03206	2.02304	2.01075	2.21699
0.940	2.13261	2.11761	2.10649	2.09210	2.34201
0.950	2.23595	2.21838	2.20456	2.18759	2.49185
0.960	2.36203	2.34120	2.32378	2.30349	2.67784
0.970	2.52410	2.49893	2.47634	2.45154	2.92128
0.980	2.75189	2.72043	2.68957	2.65799	3.27022
0.990	3.14043	3.09803	3.05033	3.00619	3.87903
0.995	3.52874	3.47557	3.40760	3.34990	4.49956
0.999	4.43214	4.35651	4.22866	4.13664	5.97124

TABLE 6.9 N= 7 (CONT.)

P ↓	λ:	5 5 10 10 15 45	5 5 10 10 20 40	5 5 10 10 25 35	5 5 10 10 30 30	5 5 10 10 15 40
0.001		0.06544	0.06708	0.06796	0.06825	0.06818
0.005		0.10920	0.11197	0.11348	0.11395	0.11377
0.010		0.13763	0.14114	0.14306	0.14367	0.14338
0.020		0.17520	0.17969	0.18215	0.18294	0.18247
0.030		0.20301	0.20824	0.21111	0.21203	0.21139
0.040		0.22618	0.23202	0.23523	0.23626	0.23547
0.050		0.24655	0.25292	0.25644	0.25757	0.25662
0.060		0.26503	0.27189	0.27568	0.27690	0.27580
0.070		0.28214	0.28944	0.29349	0.29479	0.29354
0.080		0.29820	0.30593	0.31021	0.31159	0.31019
0.090		0.31345	0.32157	0.32608	0.32753	0.32598
0.100		0.32804	0.33653	0.34126	0.34278	0.34108
0.110		0.34209	0.35094	0.35587	0.35747	0.35562
0.120		0.35569	0.36488	0.37002	0.37168	0.36968
0.130		0.36892	0.37844	0.38377	0.38549	0.38334
0.140		0.38183	0.39167	0.39719	0.39897	0.39667
0.150		0.39447	0.40462	0.41032	0.41216	0.40971
0.160		0.40688	0.41734	0.42321	0.42511	0.42250
0.170		0.41910	0.42985	0.43589	0.43785	0.43509
0.180		0.43115	0.44218	0.44840	0.45041	0.44749
0.190		0.44306	0.45437	0.46075	0.46282	0.45974
0.200		0.45485	0.46643	0.47298	0.47510	0.47186
0.250		0.51262	0.52548	0.53279	0.53517	0.53110
0.300		0.56969	0.58369	0.59171	0.59432	0.58938
0.350		0.62736	0.64239	0.65104	0.65386	0.64801
0.400		0.68673	0.70263	0.71186	0.71488	0.70806
0.450		0.74885	0.76547	0.77518	0.77838	0.77055
0.500		0.81487	0.83199	0.84209	0.84543	0.83657
0.550		0.88616	0.90349	0.91384	0.91727	0.90737
0.600		0.96450	0.98163	0.99201	0.99548	0.98457
0.650		1.05230	1.06863	1.07875	1.08215	1.07035
0.700		1.15309	1.16772	1.17709	1.18029	1.16784
0.750		1.27237	1.28385	1.29172	1.29448	1.28189
0.800		1.41948	1.42537	1.43044	1.43235	1.42069
0.810		1.45356	1.45789	1.46216	1.46383	1.45257
0.820		1.48961	1.49219	1.49555	1.49693	1.48619
0.830		1.52787	1.52847	1.53080	1.53186	1.52175
0.840		1.56863	1.56699	1.56814	1.56884	1.55950
0.850		1.61224	1.60805	1.60785	1.60814	1.59975
0.860		1.65909	1.65201	1.65027	1.65007	1.64286
0.870		1.70972	1.69933	1.69580	1.69504	1.68926
0.880		1.76475	1.75055	1.74494	1.74354	1.73950
0.890		1.82499	1.80639	1.79836	1.79619	1.79430
0.900		1.89148	1.86775	1.85685	1.85378	1.85456
0.910		1.96560	1.93584	1.92153	1.91737	1.92146
0.920		2.04922	2.01228	1.99385	1.98838	1.99665
0.930		2.14498	2.09939	2.07590	2.06880	2.08242
0.940		2.25675	2.20055	2.17072	2.16156	2.18217
0.950		2.39056	2.32104	2.28302	2.27119	2.30116
0.960		2.55658	2.46974	2.42075	2.40527	2.44833
0.970		2.77392	2.66344	2.59880	2.57803	2.64048
0.980		3.08578	2.94005	2.85077	2.82143	2.91571
0.990		3.63093	3.42185	3.28433	3.23739	3.39676
0.995		4.18760	3.91328	3.72141	3.65330	3.88868
0.999		5.50986	5.08221	4.74896	4.61896	5.05996

TABLE 6.10 N= 7 (CONT.)

		5	5	5	5	5
		5 10	5 10	5 10	5 10	5 10
		10 15	10 15	10 20	10 20	15 15
P ↓	λ:	20 35	25 30	20 30	25 25	15 35
0.001		0.06972	0.07043	0.07108	0.07150	0.07087
0.005		0.11636	0.11755	0.11863	0.11934	0.11822
0.010		0.14665	0.14816	0.14951	0.15041	0.14895
0.020		0.18664	0.18858	0.19029	0.19143	0.18949
0.030		0.21623	0.21848	0.22044	0.22177	0.21945
0.040		0.24086	0.24336	0.24554	0.24701	0.24438
0.050		0.26249	0.26523	0.26758	0.26918	0.26626
0.060		0.28210	0.28504	0.28755	0.28927	0.28608
0.070		0.30024	0.30337	0.30602	0.30785	0.30440
0.080		0.31726	0.32057	0.32334	0.32527	0.32158
0.090		0.33339	0.33687	0.33976	0.34179	0.33786
0.100		0.34882	0.35245	0.35546	0.35757	0.35342
0.110		0.36366	0.36745	0.37055	0.37275	0.36838
0.120		0.37802	0.38195	0.38515	0.38742	0.38285
0.130		0.39197	0.39603	0.39933	0.40168	0.39689
0.140		0.40557	0.40977	0.41315	0.41557	0.41058
0.150		0.41887	0.42320	0.42666	0.42916	0.42397
0.160		0.43192	0.43637	0.43992	0.44248	0.43709
0.170		0.44475	0.44933	0.45294	0.45557	0.44999
0.180		0.45739	0.46209	0.46577	0.46846	0.46270
0.190		0.46987	0.47468	0.47844	0.48118	0.47523
0.200		0.48221	0.48714	0.49095	0.49376	0.48762
0.250		0.54249	0.54794	0.55203	0.55510	0.54807
0.300		0.60167	0.60758	0.61187	0.61518	0.60729
0.350		0.66107	0.66739	0.67181	0.67531	0.66661
0.400		0.72175	0.72842	0.73290	0.73656	0.72710
0.450		0.78471	0.79167	0.79612	0.79989	0.78972
0.500		0.85100	0.85814	0.86248	0.86630	0.85552
0.550		0.92181	0.92904	0.93314	0.93694	0.92568
0.600		0.99867	1.00582	1.00954	1.01323	1.00167
0.650		1.08361	1.09047	1.09360	1.09705	1.08548
0.700		1.17952	1.18576	1.18804	1.19107	1.17993
0.750		1.29086	1.29595	1.29701	1.29933	1.28938
0.800		1.42504	1.42813	1.42736	1.42852	1.42105
0.810		1.45565	1.45818	1.45695	1.45780	1.45107
0.820		1.48785	1.48975	1.48801	1.48851	1.48264
0.830		1.52182	1.52301	1.52072	1.52082	1.51593
0.840		1.55779	1.55816	1.55527	1.55492	1.55117
0.850		1.59601	1.59547	1.59190	1.59106	1.58862
0.860		1.63681	1.63522	1.63091	1.62950	1.62858
0.870		1.68058	1.67778	1.67265	1.67060	1.67146
0.880		1.72781	1.72361	1.71757	1.71477	1.71772
0.890		1.77912	1.77329	1.76621	1.76256	1.76798
0.900		1.83530	1.82755	1.81929	1.81465	1.82301
0.910		1.89740	1.88737	1.87777	1.87195	1.88386
0.920		1.96684	1.95406	1.94290	1.93568	1.95193
0.930		2.04565	2.02947	2.01650	2.00755	2.02921
0.940		2.13675	2.11632	2.10117	2.09008	2.11862
0.950		2.24476	2.21881	2.20100	2.18715	2.22472
0.960		2.37740	2.34399	2.32280	2.30526	2.35518
0.970		2.54925	2.50508	2.47940	2.45655	2.52452
0.980		2.79326	2.73180	2.69962	2.66825	2.76562
0.990		3.21562	3.11916	3.07570	3.02698	3.18462
0.995		3.64453	3.50681	3.45235	3.38278	3.61194
0.999		4.66248	4.40974	4.33266	4.20174	4.63027

TABLE 6.11 N= 7 (CONT.)

P ↓	λ:	5 5 10 15 15 20 30	5 5 10 15 15 25 25	5 5 10 15 20 20 25	5 5 10 20 20 20 20	5 5 15 15 15 15 30
0.001		0.07224	0.07267	0.07334	0.07402	0.07343
0.005		0.12052	0.12124	0.12235	0.12347	0.12244
0.010		0.15185	0.15276	0.15414	0.15554	0.15421
0.020		0.19317	0.19433	0.19607	0.19783	0.19608
0.030		0.22370	0.22504	0.22704	0.22906	0.22699
0.040		0.24909	0.25058	0.25279	0.25501	0.25267
0.050		0.27138	0.27300	0.27538	0.27778	0.27520
0.060		0.29156	0.29329	0.29583	0.29838	0.29559
0.070		0.31021	0.31205	0.31473	0.31742	0.31442
0.080		0.32769	0.32963	0.33244	0.33526	0.33206
0.090		0.34426	0.34629	0.34922	0.35215	0.34877
0.100		0.36008	0.36220	0.36523	0.36828	0.36472
0.110		0.37530	0.37750	0.38063	0.38377	0.38005
0.120		0.39000	0.39228	0.39551	0.39874	0.39485
0.130		0.40428	0.40663	0.40995	0.41327	0.40922
0.140		0.41818	0.42061	0.42401	0.42742	0.42322
0.150		0.43178	0.43427	0.43775	0.44124	0.43689
0.160		0.44510	0.44766	0.45122	0.45478	0.45028
0.170		0.45819	0.46082	0.46445	0.46808	0.46344
0.180		0.47108	0.47377	0.47747	0.48116	0.47638
0.190		0.48380	0.48655	0.49031	0.49407	0.48915
0.200		0.49636	0.49917	0.50299	0.50681	0.50176
0.250		0.55759	0.56066	0.56474	0.56881	0.56313
0.300		0.61746	0.62076	0.62502	0.62928	0.62303
0.350		0.67731	0.68080	0.68519	0.68956	0.68279
0.400		0.73820	0.74183	0.74627	0.75070	0.74348
0.450		0.80108	0.80483	0.80924	0.81364	0.80603
0.500		0.86696	0.87076	0.87507	0.87936	0.87145
0.550		0.93698	0.94077	0.94485	0.94892	0.94085
0.600		1.01255	1.01623	1.01996	1.02368	1.01561
0.650		1.09553	1.09900	1.10219	1.10539	1.09755
0.700		1.18859	1.19166	1.19407	1.19651	1.18927
0.750		1.29577	1.29816	1.29944	1.30078	1.29469
0.800		1.42376	1.42504	1.42463	1.42434	1.42036
0.810		1.45279	1.45376	1.45292	1.45221	1.44883
0.820		1.48325	1.48388	1.48257	1.48141	1.47869
0.830		1.51531	1.51556	1.51374	1.51207	1.51011
0.840		1.54917	1.54899	1.54660	1.54438	1.54328
0.850		1.58506	1.58440	1.58138	1.57855	1.57844
0.860		1.62328	1.62206	1.61835	1.61483	1.61585
0.870		1.66416	1.66231	1.65782	1.65355	1.65586
0.880		1.70813	1.70556	1.70020	1.69507	1.69889
0.890		1.75575	1.75234	1.74599	1.73990	1.74547
0.900		1.80772	1.80333	1.79584	1.78864	1.79628
0.910		1.86495	1.85939	1.85061	1.84213	1.85224
0.920		1.92870	1.92175	1.91144	1.90146	1.91456
0.930		2.00074	1.99207	1.97995	1.96819	1.98498
0.940		2.08363	2.07281	2.05850	2.04457	2.06601
0.950		2.18139	2.16779	2.15075	2.13409	2.16158
0.960		2.30073	2.28337	2.26279	2.24258	2.27830
0.970		2.45430	2.43149	2.40606	2.38091	2.42860
0.980		2.67055	2.63893	2.60612	2.57336	2.64052
0.990		3.04080	2.99098	2.94432	2.89684	3.00436
0.995		3.41289	3.34094	3.27913	3.21487	3.37144
0.999		4.28704	4.14939	4.04904	3.93834	4.23923

TABLE 6.12 N= 7 (CONT.)

		5	5	5	5	5
		5 15	5 15	10 10	10 10	10 10
		15 15	15 20	10 10	10 10	10 10
P ↓	λ:	20 25	20 20	10 45	15 40	20 35
0.001		0.07454	0.07522	0.06793	0.07076	0.07235
0.005		0.12429	0.12542	0.11308	0.11776	0.12042
0.010		0.15653	0.15794	0.14229	0.14816	0.15150
0.020		0.19901	0.20078	0.18076	0.18815	0.19239
0.030		0.23035	0.23238	0.20914	0.21763	0.22253
0.040		0.25639	0.25863	0.23273	0.24211	0.24755
0.050		0.27923	0.28164	0.25342	0.26357	0.26949
0.060		0.29988	0.30245	0.27215	0.28299	0.28933
0.070		0.31896	0.32166	0.28947	0.30093	0.30766
0.080		0.33683	0.33965	0.30570	0.31773	0.32482
0.090		0.35374	0.35668	0.32109	0.33365	0.34108
0.100		0.36989	0.37294	0.33579	0.34885	0.35660
0.110		0.38540	0.38854	0.34993	0.36346	0.37151
0.120		0.40037	0.40362	0.36361	0.37758	0.38592
0.130		0.41490	0.41823	0.37689	0.39129	0.39990
0.140		0.42905	0.43246	0.38984	0.40464	0.41352
0.150		0.44287	0.44636	0.40251	0.41770	0.42683
0.160		0.45640	0.45997	0.41493	0.43050	0.43988
0.170		0.46969	0.47332	0.42715	0.44307	0.45269
0.180		0.48277	0.48646	0.43919	0.45546	0.46531
0.190		0.49565	0.49941	0.45108	0.46768	0.47775
0.200		0.50838	0.51220	0.46284	0.47975	0.49005
0.250		0.57026	0.57433	0.52032	0.53867	0.54998
0.300		0.63057	0.63481	0.57690	0.59644	0.60864
0.350		0.69064	0.69500	0.63389	0.65439	0.66735
0.400		0.75152	0.75593	0.69239	0.71360	0.72720
0.450		0.81416	0.81855	0.75345	0.77508	0.78917
0.500		0.87953	0.88381	0.81822	0.83991	0.85431
0.550		0.94871	0.95279	0.88804	0.90935	0.92381
0.600		1.02304	1.02678	0.96467	0.98498	0.99918
0.650		1.10426	1.10750	1.05052	1.06897	1.08242
0.700		1.19485	1.19735	1.14908	1.16441	1.17641
0.750		1.29854	1.29999	1.26587	1.27615	1.28556
0.800		1.42150	1.42136	1.41024	1.41232	1.41725
0.810		1.44926	1.44871	1.44376	1.44364	1.44732
0.820		1.47834	1.47734	1.47924	1.47668	1.47897
0.830		1.50888	1.50740	1.51693	1.51166	1.51237
0.840		1.54109	1.53906	1.55713	1.54882	1.54774
0.850		1.57515	1.57253	1.60017	1.58846	1.58536
0.860		1.61134	1.60805	1.64648	1.63094	1.62554
0.870		1.64997	1.64593	1.69657	1.67672	1.66867
0.880		1.69143	1.68655	1.75110	1.72634	1.71525
0.890		1.73620	1.73037	1.81086	1.78052	1.76588
0.900		1.78493	1.77800	1.87692	1.84015	1.82137
0.910		1.83844	1.83023	1.95066	1.90645	1.88276
0.920		1.89785	1.88814	2.03398	1.98105	1.95148
0.930		1.96473	1.95324	2.12952	2.06627	2.02954
0.940		2.04139	2.02770	2.24120	2.16552	2.11991
0.950		2.13138	2.11493	2.37506	2.28411	2.22717
0.960		2.24066	2.22057	2.54132	2.43100	2.35909
0.970		2.38035	2.35518	2.75914	2.62309	2.53026
0.980		2.57543	2.54232	3.07177	2.89860	2.77374
0.990		2.90537	2.85663	3.61810	3.38059	3.19594
0.995		3.23242	3.16548	4.17559	3.87347	3.62519
0.999		3.98680	3.86806	5.49885	5.04624	4.64443

TABLE 6.13 N= 7 (CONT.)

P ↓	λ:	5 10 10 10 10 25 30	5 10 10 10 15 15 35	5 10 10 10 15 20 30	5 10 10 10 15 25 25	5 10 10 10 20 20 25
0.001		0.07307	0.07353	0.07494	0.07538	0.07607
0.005		0.12164	0.12232	0.12467	0.12541	0.12654
0.010		0.15304	0.15384	0.15679	0.15772	0.15913
0.020		0.19435	0.19527	0.19901	0.20018	0.20195
0.030		0.22481	0.22578	0.23008	0.23144	0.23346
0.040		0.25009	0.25110	0.25586	0.25736	0.25958
0.050		0.27225	0.27327	0.27843	0.28006	0.28246
0.060		0.29229	0.29332	0.29883	0.30058	0.30313
0.070		0.31080	0.31182	0.31766	0.31950	0.32219
0.080		0.32814	0.32915	0.33528	0.33722	0.34003
0.090		0.34456	0.34554	0.35195	0.35399	0.35691
0.100		0.36023	0.36119	0.36786	0.36998	0.37301
0.110		0.37529	0.37622	0.38313	0.38533	0.38846
0.120		0.38984	0.39073	0.39788	0.40016	0.40338
0.130		0.40396	0.40481	0.41218	0.41453	0.41784
0.140		0.41771	0.41852	0.42610	0.42852	0.43191
0.150		0.43115	0.43191	0.43970	0.44218	0.44565
0.160		0.44432	0.44503	0.45301	0.45556	0.45910
0.170		0.45725	0.45791	0.46608	0.46869	0.47230
0.180		0.46998	0.47059	0.47894	0.48161	0.48529
0.190		0.48254	0.48309	0.49161	0.49435	0.49808
0.200		0.49495	0.49543	0.50413	0.50692	0.51072
0.250		0.55539	0.55553	0.56499	0.56804	0.57209
0.300		0.61450	0.61424	0.62433	0.62761	0.63185
0.350		0.67362	0.67288	0.68351	0.68698	0.69134
0.400		0.73382	0.73255	0.74359	0.74721	0.75163
0.450		0.79608	0.79421	0.80553	0.80926	0.81367
0.500		0.86143	0.85890	0.87033	0.87412	0.87844
0.550		0.93104	0.92777	0.93911	0.94291	0.94702
0.600		1.00636	1.00230	1.01327	1.01699	1.02077
0.650		1.08936	1.08445	1.09467	1.09819	1.10146
0.700		1.18278	1.17701	1.18592	1.18907	1.19160
0.750		1.29086	1.28427	1.29102	1.29353	1.29495
0.800		1.42061	1.41343	1.41659	1.41801	1.41778
0.810		1.45014	1.44290	1.44508	1.44621	1.44555
0.820		1.48117	1.47389	1.47499	1.47578	1.47466
0.830		1.51387	1.50659	1.50647	1.50689	1.50526
0.840		1.54845	1.54122	1.53974	1.53973	1.53753
0.850		1.58515	1.57804	1.57501	1.57453	1.57169
0.860		1.62429	1.61736	1.61257	1.61154	1.60801
0.870		1.66622	1.65956	1.65278	1.65112	1.64680
0.880		1.71139	1.70512	1.69604	1.69366	1.68846
0.890		1.76038	1.75466	1.74292	1.73970	1.73349
0.900		1.81393	1.80894	1.79410	1.78989	1.78253
0.910		1.87301	1.86902	1.85051	1.84512	1.83644
0.920		1.93894	1.93630	1.91339	1.90658	1.89634
0.930		2.01355	2.01276	1.98449	1.97594	1.96384
0.940		2.09956	2.10133	2.06638	2.05564	2.04129
0.950		2.20117	2.20659	2.16306	2.14948	2.13231
0.960		2.32543	2.33622	2.28123	2.26379	2.24296
0.970		2.48554	2.50480	2.43350	2.41046	2.38459
0.980		2.71123	2.74528	2.64832	2.61618	2.58264
0.990		3.09749	3.16419	3.01698	2.96602	2.91806
0.995		3.48460	3.59206	3.38828	3.31446	3.25078
0.999		4.38724	4.61211	4.26235	4.12105	4.01764

TABLE 6.14 N= 7 (CONT.)

P ↓ λ:	5 10 10 15 15 15 30	5 10 10 15 15 20 25	5 10 10 15 20 20 20	5 10 15 15 15 15 25	5 10 15 15 15 20 20
0.001	0.07616	0.07731	0.07801	0.07855	0.07927
0.005	0.12662	0.12852	0.12967	0.13051	0.13168
0.010	0.15919	0.16155	0.16299	0.16400	0.16545
0.020	0.20195	0.20491	0.20671	0.20791	0.20972
0.030	0.23339	0.23680	0.23885	0.24016	0.24222
0.040	0.25945	0.26321	0.26546	0.26685	0.26912
0.050	0.28226	0.28632	0.28874	0.29019	0.29263
0.060	0.30287	0.30719	0.30976	0.31126	0.31385
0.070	0.32187	0.32642	0.32913	0.33067	0.33339
0.080	0.33964	0.34442	0.34725	0.34881	0.35166
0.090	0.35645	0.36143	0.36437	0.36596	0.36891
0.100	0.37248	0.37765	0.38070	0.38230	0.38535
0.110	0.38787	0.39321	0.39636	0.39797	0.40112
0.120	0.40272	0.40823	0.41146	0.41308	0.41632
0.130	0.41711	0.42278	0.42610	0.42772	0.43104
0.140	0.43111	0.43693	0.44033	0.44195	0.44535
0.150	0.44478	0.45075	0.45422	0.45584	0.45931
0.160	0.45817	0.46426	0.46781	0.46942	0.47296
0.170	0.47130	0.47752	0.48114	0.48274	0.48635
0.180	0.48421	0.49056	0.49424	0.49583	0.49950
0.190	0.49694	0.50341	0.50714	0.50872	0.51245
0.200	0.50949	0.51608	0.51987	0.52143	0.52522
0.250	0.57050	0.57759	0.58163	0.58308	0.58711
0.300	0.62988	0.63737	0.64159	0.64288	0.64709
0.350	0.68899	0.69679	0.70113	0.70221	0.70654
0.400	0.74888	0.75689	0.76129	0.76213	0.76651
0.450	0.81052	0.81863	0.82301	0.82358	0.82795
0.500	0.87488	0.88296	0.88726	0.88750	0.89178
0.550	0.94307	0.95097	0.95508	0.95495	0.95905
0.600	1.01646	1.02397	1.02776	1.02721	1.03102
0.650	1.09684	1.10368	1.10700	1.10597	1.10933
0.700	1.18677	1.19254	1.19516	1.19360	1.19629
0.750	1.29013	1.29423	1.29582	1.29367	1.29536
0.800	1.41336	1.41483	1.41486	1.41207	1.41226
0.810	1.44129	1.44205	1.44169	1.43877	1.43856
0.820	1.47059	1.47058	1.46977	1.46672	1.46609
0.830	1.50142	1.50055	1.49926	1.49607	1.49497
0.840	1.53398	1.53215	1.53032	1.52701	1.52538
0.850	1.56850	1.56558	1.56315	1.55972	1.55751
0.860	1.60524	1.60111	1.59801	1.59445	1.59159
0.870	1.64455	1.63904	1.63519	1.63152	1.62792
0.880	1.68683	1.67975	1.67506	1.67128	1.66686
0.890	1.73263	1.72374	1.71808	1.71422	1.70885
0.900	1.78263	1.77162	1.76485	1.76093	1.75446
0.910	1.83770	1.82422	1.81616	1.81221	1.80447
0.920	1.89909	1.88265	1.87306	1.86914	1.85988
0.930	1.96850	1.94846	1.93705	1.93322	1.92214
0.940	2.04843	2.02393	2.01027	2.00665	1.99332
0.950	2.14282	2.11259	2.09609	2.09286	2.07667
0.960	2.25823	2.22034	2.20010	2.19756	2.17757
0.970	2.40707	2.35823	2.33272	2.33146	2.30607
0.980	2.61736	2.55104	2.51728	2.51862	2.48466
0.990	2.97938	2.87782	2.82773	2.83583	2.78454
0.995	3.34564	3.20247	3.13333	3.15133	3.07926
0.999	4.21373	3.95350	3.83012	3.88404	3.75044

TABLE 6.15 N= 7 (CONT.)

P ↓	λ:	5 15 15 15 15 15 20	10 10 10 10 10 10 10 40	10 10 10 10 10 15 35	10 10 10 10 10 20 30	10 10 10 10 10 25 25
0.001		0.08054	0.07340	0.07625	0.07771	0.07817
0.005		0.13371	0.12183	0.12648	0.12889	0.12964
0.010		0.16793	0.15301	0.15879	0.16180	0.16274
0.020		0.21275	0.19388	0.20110	0.20488	0.20607
0.030		0.24562	0.22390	0.23215	0.23649	0.23786
0.040		0.27280	0.24877	0.25783	0.26263	0.26415
0.050		0.29654	0.27053	0.28029	0.28548	0.28712
0.060		0.31795	0.29018	0.30056	0.30610	0.30785
0.070		0.33766	0.30830	0.31924	0.32509	0.32694
0.080		0.35607	0.32526	0.33670	0.34284	0.34479
0.090		0.37346	0.34130	0.35320	0.35962	0.36165
0.100		0.39001	0.35659	0.36893	0.37560	0.37772
0.110		0.40588	0.37128	0.38402	0.39093	0.39313
0.120		0.42118	0.38545	0.39859	0.40572	0.40800
0.130		0.43598	0.39920	0.41270	0.42005	0.42239
0.140		0.45037	0.41258	0.42642	0.43398	0.43640
0.150		0.46440	0.42565	0.43982	0.44758	0.45006
0.160		0.47811	0.43845	0.45293	0.46088	0.46342
0.170		0.49155	0.45102	0.46580	0.47393	0.47653
0.180		0.50476	0.46338	0.47845	0.48676	0.48942
0.190		0.51775	0.47557	0.49091	0.49939	0.50211
0.200		0.53056	0.48761	0.50321	0.51186	0.51464
0.250		0.59257	0.54623	0.56299	0.57238	0.57541
0.300		0.65256	0.60351	0.62119	0.63122	0.63447
0.350		0.71193	0.66082	0.67920	0.68976	0.69320
0.400		0.77173	0.71922	0.73808	0.74906	0.75266
0.450		0.83288	0.77973	0.79882	0.81009	0.81380
0.500		0.89631	0.84342	0.86242	0.87383	0.87762
0.550		0.96305	0.91152	0.93004	0.94141	0.94521
0.600		1.03432	0.98561	1.00314	1.01419	1.01792
0.650		1.11173	1.06780	1.08362	1.09400	1.09756
0.700		1.19752	1.16118	1.17427	1.18343	1.18664
0.750		1.29505	1.27052	1.27930	1.28640	1.28902
0.800		1.40985	1.40395	1.40584	1.40947	1.41104
0.810		1.43565	1.43467	1.43473	1.43741	1.43869
0.820		1.46262	1.46710	1.46513	1.46674	1.46770
0.830		1.49091	1.50145	1.49721	1.49762	1.49821
0.840		1.52067	1.53797	1.53120	1.53026	1.53044
0.850		1.55210	1.57696	1.56735	1.56488	1.56458
0.860		1.58542	1.61878	1.60597	1.60176	1.60092
0.870		1.62091	1.66389	1.64746	1.64125	1.63979
0.880		1.65892	1.71284	1.69228	1.68376	1.68159
0.890		1.69988	1.76633	1.74104	1.72985	1.72683
0.900		1.74434	1.82529	1.79453	1.78020	1.77619
0.910		1.79303	1.89093	1.85378	1.83573	1.83054
0.920		1.84694	1.96491	1.92020	1.89768	1.89105
0.930		1.90744	2.04955	1.99579	1.96779	1.95940
0.940		1.97655	2.14829	2.08348	2.04862	2.03800
0.950		2.05735	2.26648	2.18785	2.14416	2.13063
0.960		2.15503	2.41313	2.31663	2.26110	2.24360
0.970		2.27922	2.60525	2.48444	2.41204	2.38877
0.980		2.45143	2.88120	2.72440	2.62540	2.59272
0.990		2.73974	3.36433	3.14342	2.99254	2.94037
0.995		3.02215	3.85828	3.57203	3.36321	3.28739
0.999		3.66264	5.03255	4.59399	4.23752	4.09237

TABLE 6.16 N= 7 (CONT.)

P ↓	λ: 10 10 10 15 15 30	10 10 10 15 20 25	10 10 10 20 20 20	10 10 15 15 15 25	10 10 15 15 20 20
0.001	0.07896	0.08014	0.08087	0.08142	0.08216
0.005	0.13088	0.13282	0.13400	0.13485	0.13605
0.010	0.16424	0.16664	0.16810	0.16912	0.17060
0.020	0.20786	0.21086	0.21268	0.21389	0.21572
0.030	0.23983	0.24326	0.24533	0.24665	0.24873
0.040	0.26625	0.27003	0.27230	0.27370	0.27598
0.050	0.28933	0.29341	0.29584	0.29730	0.29975
0.060	0.31014	0.31448	0.31706	0.31856	0.32116
0.070	0.32931	0.33387	0.33659	0.33812	0.34085
0.080	0.34721	0.35199	0.35482	0.35638	0.35923
0.090	0.36412	0.36910	0.37204	0.37362	0.37657
0.100	0.38022	0.38538	0.38843	0.39002	0.39308
0.110	0.39566	0.40100	0.40414	0.40574	0.40889
0.120	0.41055	0.41605	0.41927	0.42088	0.42412
0.130	0.42496	0.43062	0.43393	0.43554	0.43885
0.140	0.43898	0.44478	0.44817	0.44978	0.45317
0.150	0.45265	0.45859	0.46205	0.46365	0.46712
0.160	0.46601	0.47209	0.47562	0.47722	0.48075
0.170	0.47912	0.48532	0.48892	0.49051	0.49411
0.180	0.49201	0.49833	0.50199	0.50357	0.50723
0.190	0.50469	0.51113	0.51485	0.51641	0.52013
0.200	0.51720	0.52375	0.52753	0.52908	0.53285
0.250	0.57787	0.58491	0.58893	0.59037	0.59438
0.300	0.63675	0.64420	0.64839	0.64968	0.65387
0.350	0.69523	0.70298	0.70730	0.70840	0.71271
0.400	0.75436	0.76233	0.76672	0.76758	0.77195
0.450	0.81511	0.82320	0.82758	0.82818	0.83255
0.500	0.87844	0.88653	0.89083	0.89111	0.89541
0.550	0.94545	0.95339	0.95752	0.95744	0.96157
0.600	1.01749	1.02507	1.02892	1.02842	1.03228
0.650	1.09632	1.10328	1.10668	1.10571	1.10914
0.700	1.18445	1.19041	1.19314	1.19163	1.19442
0.750	1.28571	1.29007	1.29180	1.28970	1.29153
0.800	1.40643	1.40823	1.40845	1.40567	1.40604
0.810	1.43380	1.43491	1.43473	1.43182	1.43180
0.820	1.46251	1.46287	1.46225	1.45920	1.45876
0.830	1.49274	1.49224	1.49115	1.48795	1.48705
0.840	1.52466	1.52321	1.52159	1.51825	1.51683
0.850	1.55850	1.55599	1.55377	1.55029	1.54829
0.860	1.59454	1.59082	1.58793	1.58432	1.58168
0.870	1.63311	1.62801	1.62438	1.62063	1.61726
0.880	1.67462	1.66795	1.66346	1.65960	1.65539
0.890	1.71960	1.71111	1.70565	1.70168	1.69652
0.900	1.76871	1.75810	1.75152	1.74747	1.74121
0.910	1.82286	1.80974	1.80186	1.79775	1.79021
0.920	1.88325	1.86713	1.85770	1.85359	1.84451
0.930	1.95158	1.93181	1.92051	1.91646	1.90554
0.940	2.03036	2.00603	1.99243	1.98856	1.97534
0.950	2.12348	2.09328	2.07676	2.07324	2.05709
0.960	2.23751	2.19941	2.17903	2.17617	2.15611
0.970	2.38482	2.33539	2.30955	2.30795	2.28230
0.980	2.59342	2.52583	2.49139	2.49242	2.45784
0.990	2.95368	2.84935	2.79782	2.80581	2.75306
0.995	3.31926	3.17159	3.10005	3.11841	3.04375
0.999	4.18808	3.91948	3.79102	3.84740	3.70764

TABLE 6.17 N= 7 (CONT.)

P ↓	λ:	10 10 15 15 15 15 15 20	10 15 15 15 15 15 15 15	All λ's = 100/7
0.001		0.08346	0.08479	0.08550
0.005		0.13812	0.14021	0.14132
0.010		0.17312	0.17566	0.17701
0.020		0.21878	0.22186	0.22347
0.030		0.25215	0.25559	0.25738
0.040		0.27967	0.28338	0.28530
0.050		0.30366	0.30760	0.30962
0.060		0.32526	0.32938	0.33149
0.070		0.34512	0.34940	0.35158
0.080		0.36364	0.36806	0.37031
0.090		0.38111	0.38566	0.38795
0.100		0.39773	0.40238	0.40473
0.110		0.41364	0.41839	0.42078
0.120		0.42896	0.43381	0.43623
0.130		0.44378	0.44870	0.45116
0.140		0.45817	0.46317	0.46565
0.150		0.47219	0.47725	0.47975
0.160		0.48588	0.49100	0.49353
0.170		0.49929	0.50447	0.50701
0.180		0.51246	0.51768	0.52024
0.190		0.52541	0.53067	0.53324
0.200		0.53816	0.54346	0.54605
0.250		0.59982	0.60524	0.60784
0.300		0.65933	0.66477	0.66733
0.350		0.71809	0.72347	0.72595
0.400		0.77718	0.78239	0.78475
0.450		0.83750	0.84246	0.84465
0.500		0.89999	0.90458	0.90654
0.550		0.96565	0.96974	0.97142
0.600		1.03568	1.03913	1.04046
0.650		1.11167	1.11427	1.11516
0.700		1.19581	1.19731	1.19763
0.750		1.29140	1.29142	1.29102
0.800		1.40383	1.40182	1.40046
0.810		1.42908	1.42657	1.42499
0.820		1.45549	1.45244	1.45060
0.830		1.48318	1.47955	1.47745
0.840		1.51231	1.50805	1.50566
0.850		1.54307	1.53811	1.53541
0.860		1.57568	1.56996	1.56692
0.870		1.61042	1.60386	1.60045
0.880		1.64761	1.64012	1.63630
0.890		1.68769	1.67916	1.67489
0.900		1.73120	1.72149	1.71672
0.910		1.77885	1.76780	1.76246
0.920		1.83161	1.81901	1.81302
0.930		1.89082	1.87641	1.86967
0.940		1.95847	1.94188	1.93425
0.950		2.03758	2.01830	2.00959
0.960		2.13324	2.11051	2.10043
0.970		2.25490	2.22747	2.21558
0.980		2.42372	2.38918	2.37463
0.990		2.70662	2.65868	2.63933
0.995		2.98415	2.92125	2.89682
0.999		3.61512	3.51171	3.47455

TABLE 7.1 N= 8

P ↓	λ:	5 5 5 5 5 5 5 65	5 5 5 5 5 5 10 60	5 5 5 5 5 5 15 55	5 5 5 5 5 5 20 50	5 5 5 5 5 5 25 45
0.001		0.06024	0.06506	0.06782	0.06956	0.07066
0.005		0.09569	0.10334	0.10781	0.11068	0.11249
0.010		0.11815	0.12760	0.13320	0.13680	0.13909
0.020		0.14748	0.15925	0.16634	0.17095	0.17390
0.030		0.16904	0.18250	0.19071	0.19609	0.19954
0.040		0.18695	0.20180	0.21096	0.21698	0.22086
0.050		0.20268	0.21874	0.22874	0.23535	0.23961
0.060		0.21696	0.23410	0.24487	0.25201	0.25663
0.070		0.23020	0.24833	0.25980	0.26745	0.27240
0.080		0.24265	0.26169	0.27384	0.28196	0.28723
0.090		0.25449	0.27440	0.28718	0.29576	0.30134
0.100		0.26584	0.28657	0.29997	0.30899	0.31486
0.110		0.27681	0.29831	0.31230	0.32175	0.32792
0.120		0.28746	0.30970	0.32427	0.33413	0.34058
0.130		0.29785	0.32080	0.33593	0.34620	0.35293
0.140		0.30803	0.33167	0.34733	0.35801	0.36501
0.150		0.31803	0.34233	0.35853	0.36959	0.37686
0.160		0.32789	0.35283	0.36955	0.38100	0.38854
0.170		0.33763	0.36319	0.38042	0.39226	0.40006
0.180		0.34729	0.37344	0.39118	0.40339	0.41145
0.190		0.35687	0.38360	0.40183	0.41442	0.42274
0.200		0.36639	0.39369	0.41241	0.42537	0.43395
0.250		0.41379	0.44368	0.46474	0.47950	0.48935
0.300		0.46196	0.49405	0.51729	0.53380	0.54488
0.350		0.51225	0.54613	0.57139	0.58957	0.60186
0.400		0.56600	0.60115	0.62821	0.64795	0.66140
0.450		0.62471	0.66044	0.68897	0.71012	0.72465
0.500		0.69017	0.72556	0.75507	0.77739	0.79286
0.550		0.76475	0.79849	0.82825	0.85133	0.86754
0.600		0.85157	0.88188	0.91081	0.93403	0.95061
0.650		0.95497	0.97945	1.00590	1.02829	1.04469
0.700		1.08103	1.09659	1.11813	1.13819	1.15353
0.750		1.23877	1.24157	1.25465	1.27003	1.28290
0.800		1.44267	1.42803	1.42749	1.43437	1.44240
0.810		1.49097	1.47217	1.46809	1.47263	1.47927
0.820		1.54240	1.51919	1.51124	1.51314	1.51820
0.830		1.59732	1.56940	1.55724	1.55619	1.55946
0.840		1.65613	1.62322	1.60643	1.60208	1.60332
0.850		1.71936	1.68109	1.65927	1.65121	1.65012
0.860		1.78759	1.74361	1.71626	1.70403	1.70030
0.870		1.86160	1.81145	1.77804	1.76111	1.75434
0.880		1.94229	1.88548	1.84540	1.82316	1.81289
0.890		2.03086	1.96678	1.91935	1.89106	1.87675
0.900		2.12882	2.05677	2.00117	1.96598	1.94694
0.910		2.23817	2.15729	2.09256	2.04943	2.02483
0.920		2.36164	2.27085	2.19583	2.14349	2.11228
0.930		2.50304	2.40099	2.31422	2.25106	2.21188
0.940		2.66799	2.55288	2.45250	2.37645	2.32746
0.950		2.86525	2.73462	2.61807	2.52633	2.46499
0.960		3.10955	2.95976	2.82338	2.71193	2.63448
0.970		3.42856	3.25386	3.09185	2.95446	2.85483
0.980		3.88478	3.67463	3.47636	3.30179	3.16875
0.990		4.67918	4.40748	4.14680	3.90797	3.71374
0.995		5.48781	5.15356	4.82987	4.52649	4.26800
0.999		7.40344	6.92151	6.44942	5.99551	5.58334

TABLE 7.2 N= 8 (CONT.)

P ↓	λ: 5 5 5 5 / 30 40	5 5 5 5 / 35 35	5 5 5 5 / 10 55	5 5 5 5 / 15 50	5 5 5 5 / 20 45
0.001	0.07127	0.07147	0.07016	0.07302	0.07479
0.005	0.11350	0.11383	0.11139	0.11602	0.11890
0.010	0.14038	0.14079	0.13749	0.14326	0.14688
0.020	0.17556	0.17609	0.17150	0.17878	0.18338
0.030	0.20148	0.20211	0.19644	0.20484	0.21018
0.040	0.22305	0.22375	0.21712	0.22645	0.23242
0.050	0.24201	0.24279	0.23525	0.24540	0.25193
0.060	0.25924	0.26008	0.25167	0.26257	0.26959
0.070	0.27520	0.27611	0.26685	0.27844	0.28594
0.080	0.29022	0.29119	0.28111	0.29334	0.30128
0.090	0.30451	0.30553	0.29463	0.30747	0.31584
0.100	0.31820	0.31929	0.30758	0.32100	0.32977
0.110	0.33142	0.33256	0.32005	0.33403	0.34319
0.120	0.34425	0.34544	0.33213	0.34665	0.35619
0.130	0.35676	0.35801	0.34389	0.35893	0.36884
0.140	0.36900	0.37030	0.35539	0.37092	0.38120
0.150	0.38101	0.38236	0.36665	0.38268	0.39331
0.160	0.39284	0.39424	0.37773	0.39423	0.40521
0.170	0.40452	0.40597	0.38865	0.40562	0.41693
0.180	0.41606	0.41757	0.39944	0.41686	0.42851
0.190	0.42751	0.42906	0.41012	0.42798	0.43996
0.200	0.43887	0.44047	0.42071	0.43901	0.45131
0.250	0.49501	0.49686	0.47291	0.49327	0.50712
0.300	0.55128	0.55338	0.52508	0.54729	0.56260
0.350	0.60898	0.61132	0.57852	0.60239	0.61905
0.400	0.66923	0.67180	0.63440	0.65968	0.67759
0.450	0.73315	0.73594	0.69393	0.72030	0.73929
0.500	0.80196	0.80496	0.75847	0.78551	0.80535
0.550	0.87712	0.88029	0.82975	0.85684	0.87720
0.600	0.96049	0.96377	0.91007	0.93630	0.95668
0.650	1.05459	1.05789	1.00261	1.02668	1.04632
0.700	1.16299	1.16617	1.11208	1.13203	1.14974
0.750	1.29116	1.29399	1.24584	1.25869	1.27257
0.800	1.44821	1.45028	1.41623	1.41738	1.42429
0.810	1.48435	1.48620	1.45643	1.45448	1.45943
0.820	1.52246	1.52405	1.49921	1.49383	1.49658
0.830	1.56277	1.56407	1.54488	1.53571	1.53599
0.840	1.60555	1.60651	1.59380	1.58046	1.57794
0.850	1.65111	1.65169	1.64641	1.62845	1.62277
0.860	1.69986	1.69999	1.70323	1.68017	1.67091
0.870	1.75226	1.75187	1.76491	1.73617	1.72284
0.880	1.80889	1.80790	1.83224	1.79719	1.77922
0.890	1.87051	1.86881	1.90622	1.86413	1.84083
0.900	1.93807	1.93552	1.98814	1.93816	1.90870
0.910	2.01282	2.00928	2.07972	2.02082	1.98421
0.920	2.09649	2.09173	2.18325	2.11421	2.06919
0.930	2.19146	2.18520	2.30198	2.22127	2.16627
0.940	2.30125	2.29310	2.44067	2.34633	2.27928
0.950	2.43132	2.42073	2.60672	2.49612	2.41419
0.960	2.59083	2.57693	2.81257	2.68195	2.58108
0.970	2.79698	2.77831	3.08164	2.92511	2.79894
0.980	3.08851	3.06213	3.46680	3.27364	3.11070
0.990	3.58945	3.54734	4.13795	3.88192	3.65459
0.995	4.09352	4.03254	4.82145	4.50212	4.20969
0.999	5.27532	5.15915	6.44152	5.97345	5.52925

TABLE 7.3 N= 8 (CONT.)

P ↓	λ:	5 5 5 5 5 10 25 40	5 5 5 5 5 10 30 35	5 5 5 5 5 15 15 45	5 5 5 5 5 15 20 40	5 5 5 5 5 15 25 35
0.001		0.07584	0.07633	0.07590	0.07760	0.07852
0.005		0.12062	0.12143	0.12065	0.12342	0.12493
0.010		0.14905	0.15007	0.14903	0.15250	0.15440
0.020		0.18615	0.18746	0.18603	0.19044	0.19286
0.030		0.21341	0.21494	0.21319	0.21830	0.22111
0.040		0.23603	0.23775	0.23572	0.24141	0.24455
0.050		0.25588	0.25776	0.25547	0.26168	0.26511
0.060		0.27387	0.27590	0.27335	0.28003	0.28373
0.070		0.29051	0.29269	0.28988	0.29700	0.30095
0.080		0.30613	0.30844	0.30540	0.31293	0.31710
0.090		0.32096	0.32340	0.32011	0.32803	0.33243
0.100		0.33515	0.33772	0.33419	0.34248	0.34709
0.110		0.34882	0.35151	0.34775	0.35639	0.36121
0.120		0.36206	0.36487	0.36087	0.36986	0.37488
0.130		0.37495	0.37787	0.37364	0.38296	0.38817
0.140		0.38754	0.39058	0.38610	0.39575	0.40115
0.150		0.39988	0.40302	0.39831	0.40827	0.41386
0.160		0.41200	0.41526	0.41031	0.42057	0.42634
0.170		0.42394	0.42731	0.42212	0.43268	0.43862
0.180		0.43573	0.43920	0.43377	0.44463	0.45074
0.190		0.44740	0.45097	0.44530	0.45644	0.46273
0.200		0.45896	0.46264	0.45672	0.46814	0.47459
0.250		0.51579	0.51998	0.51279	0.52555	0.53279
0.300		0.57225	0.57692	0.56841	0.58237	0.59036
0.350		0.62963	0.63477	0.62486	0.63992	0.64858
0.400		0.68904	0.69463	0.68324	0.69925	0.70853
0.450		0.75153	0.75753	0.74461	0.76141	0.77122
0.500		0.81826	0.82462	0.81013	0.82749	0.83772
0.550		0.89061	0.89724	0.88121	0.89881	0.90930
0.600		0.97032	0.97712	0.95963	0.97703	0.98757
0.650		1.05976	1.06653	1.04784	1.06440	1.07466
0.700		1.16233	1.16878	1.14942	1.16417	1.17364
0.750		1.28324	1.28889	1.26988	1.28132	1.28921
0.800		1.43119	1.43520	1.41858	1.42422	1.42919
0.810		1.46524	1.46876	1.45302	1.45706	1.46121
0.820		1.50115	1.50413	1.48945	1.49169	1.49490
0.830		1.53915	1.54150	1.52811	1.52832	1.53047
0.840		1.57950	1.58112	1.56927	1.56720	1.56815
0.850		1.62250	1.62329	1.61329	1.60864	1.60821
0.860		1.66852	1.66836	1.66057	1.65299	1.65099
0.870		1.71803	1.71676	1.71162	1.70071	1.69690
0.880		1.77159	1.76902	1.76708	1.75235	1.74644
0.890		1.82993	1.82584	1.82775	1.80861	1.80026
0.900		1.89396	1.88807	1.89466	1.87040	1.85918
0.910		1.96490	1.95688	1.96918	1.93893	1.92429
0.920		2.04442	2.03380	2.05317	2.01581	1.99706
0.930		2.13485	2.12102	2.14926	2.10335	2.07956
0.940		2.23959	2.22175	2.26131	2.20493	2.17485
0.950		2.36398	2.34093	2.39533	2.32581	2.28762
0.960		2.51694	2.48686	2.56144	2.47486	2.42582
0.970		2.71532	2.67513	2.77873	2.66882	2.60434
0.980		2.99710	2.94075	3.09033	2.94557	2.85673
0.990		3.48442	3.39552	3.63493	3.42720	3.29063
0.995		3.97828	3.85120	4.19115	3.91830	3.72778
0.999		5.14651	4.91266	5.51285	5.08653	4.75509

TABLE 7.4 N= 8 (CONT.)

P ↓	λ:	5 5 5 15 30 30	5 5 5 20 20 35	5 5 5 20 25 30	5 5 5 25 25 25	5 5 5 5 10 10 10 50
0.001		0.07882	0.07918	0.07990	0.08032	0.07551
0.005		0.12541	0.12598	0.12716	0.12785	0.11978
0.010		0.15500	0.15570	0.15717	0.15804	0.14776
0.020		0.19363	0.19449	0.19637	0.19747	0.18413
0.030		0.22201	0.22299	0.22516	0.22643	0.21075
0.040		0.24555	0.24663	0.24905	0.25047	0.23278
0.050		0.26621	0.26736	0.27000	0.27155	0.25207
0.060		0.28491	0.28613	0.28898	0.29064	0.26950
0.070		0.30221	0.30349	0.30652	0.30829	0.28560
0.080		0.31845	0.31978	0.32299	0.32485	0.30069
0.090		0.33385	0.33523	0.33860	0.34056	0.31499
0.100		0.34858	0.35000	0.35354	0.35558	0.32865
0.110		0.36277	0.36423	0.36791	0.37004	0.34180
0.120		0.37650	0.37800	0.38183	0.38404	0.35452
0.130		0.38986	0.39139	0.39536	0.39765	0.36688
0.140		0.40289	0.40446	0.40857	0.41093	0.37895
0.150		0.41566	0.41725	0.42150	0.42394	0.39076
0.160		0.42820	0.42982	0.43419	0.43670	0.40235
0.170		0.44054	0.44218	0.44668	0.44926	0.41376
0.180		0.45272	0.45438	0.45901	0.46165	0.42502
0.190		0.46476	0.46644	0.47118	0.47390	0.43615
0.200		0.47668	0.47838	0.48324	0.48601	0.44716
0.250		0.53515	0.53689	0.54231	0.54537	0.50124
0.300		0.59296	0.59469	0.60061	0.60394	0.55485
0.350		0.65141	0.65309	0.65945	0.66299	0.60930
0.400		0.71157	0.71312	0.71988	0.72359	0.66571
0.450		0.77444	0.77581	0.78287	0.78670	0.72519
0.500		0.84110	0.84219	0.84946	0.85336	0.78900
0.550		0.91278	0.91352	0.92087	0.92476	0.85865
0.600		0.99108	0.99135	0.99861	1.00237	0.93614
0.650		1.07811	1.07777	1.08470	1.08818	1.02424
0.700		1.17687	1.17578	1.18201	1.18501	1.12705
0.750		1.29198	1.28994	1.29491	1.29711	1.25097
0.800		1.43108	1.42787	1.43066	1.43157	1.40691
0.810		1.46284	1.45937	1.46156	1.46212	1.44348
0.820		1.49624	1.49250	1.49402	1.49420	1.48232
0.830		1.53149	1.52746	1.52823	1.52797	1.52373
0.840		1.56879	1.56447	1.56439	1.56366	1.56802
0.850		1.60843	1.60381	1.60276	1.60149	1.61560
0.860		1.65072	1.64579	1.64365	1.64178	1.66693
0.870		1.69606	1.69082	1.68743	1.68488	1.72261
0.880		1.74494	1.73940	1.73457	1.73123	1.78336
0.890		1.79799	1.79214	1.78564	1.78141	1.85009
0.900		1.85600	1.84985	1.84141	1.83615	1.92399
0.910		1.92002	1.91360	1.90285	1.89638	2.00660
0.920		1.99147	1.98482	1.97132	1.96342	2.10003
0.930		2.07236	2.06555	2.04869	2.03906	2.20725
0.940		2.16559	2.15876	2.13770	2.12594	2.33257
0.950		2.27571	2.26907	2.24262	2.22816	2.48275
0.960		2.41029	2.40426	2.37058	2.35255	2.66909
0.970		2.58358	2.57895	2.53495	2.51188	2.91290
0.980		2.82750	2.82618	2.76573	2.73478	3.26222
0.990		3.24399	3.25214	3.15861	3.11212	3.87145
0.995		3.66016	3.68275	3.55032	3.48580	4.49224
0.999		4.62603	4.70056	4.45869	4.34339	5.96425

TABLE 7.5 N= 8 (CONT.)

P ↓ λ:	5 5 5 5 10 10 15 45	5 5 5 5 10 10 20 40	5 5 5 5 10 10 25 35	5 5 5 5 10 10 30 30	5 5 5 5 10 15 15 40
0.001	0.07847	0.08022	0.08117	0.08147	0.08140
0.005	0.12452	0.12736	0.12891	0.12940	0.12921
0.010	0.15364	0.15719	0.15912	0.15974	0.15944
0.020	0.19151	0.19599	0.19845	0.19923	0.19876
0.030	0.21923	0.22441	0.22725	0.22816	0.22753
0.040	0.24216	0.24792	0.25110	0.25211	0.25133
0.050	0.26223	0.26851	0.27197	0.27308	0.27215
0.060	0.28038	0.28711	0.29084	0.29204	0.29096
0.070	0.29712	0.30429	0.30826	0.30954	0.30833
0.080	0.31282	0.32039	0.32459	0.32594	0.32459
0.090	0.32769	0.33564	0.34006	0.34148	0.33999
0.100	0.34189	0.35020	0.35483	0.35632	0.35469
0.110	0.35555	0.36421	0.36904	0.37060	0.36882
0.120	0.36876	0.37776	0.38279	0.38441	0.38249
0.130	0.38160	0.39092	0.39614	0.39782	0.39575
0.140	0.39412	0.40376	0.40915	0.41090	0.40869
0.150	0.40637	0.41631	0.42189	0.42369	0.42133
0.160	0.41839	0.42863	0.43438	0.43624	0.43374
0.170	0.43021	0.44075	0.44667	0.44858	0.44593
0.180	0.44187	0.45269	0.45878	0.46075	0.45795
0.190	0.45339	0.46449	0.47074	0.47277	0.46981
0.200	0.46479	0.47616	0.48258	0.48466	0.48154
0.250	0.52063	0.53329	0.54048	0.54282	0.53890
0.300	0.57578	0.58963	0.59754	0.60012	0.59534
0.350	0.63156	0.64648	0.65507	0.65787	0.65217
0.400	0.68906	0.70494	0.71413	0.71714	0.71047
0.450	0.74934	0.76601	0.77575	0.77894	0.77124
0.500	0.81355	0.83082	0.84100	0.84435	0.83557
0.550	0.88308	0.90067	0.91114	0.91461	0.90473
0.600	0.95971	0.97720	0.98777	0.99129	0.98033
0.650	1.04588	1.06267	1.07303	1.07651	1.06455
0.700	1.14515	1.16031	1.16998	1.17327	1.16056
0.750	1.26309	1.27512	1.28332	1.28618	1.27322
0.800	1.40910	1.41547	1.42086	1.42287	1.41074
0.810	1.44300	1.44779	1.45236	1.45413	1.44239
0.820	1.47889	1.48188	1.48554	1.48703	1.47578
0.830	1.51700	1.51798	1.52058	1.52175	1.51113
0.840	1.55764	1.55632	1.55773	1.55853	1.54868
0.850	1.60113	1.59721	1.59726	1.59763	1.58873
0.860	1.64791	1.64103	1.63949	1.63938	1.63165
0.870	1.69848	1.68821	1.68485	1.68417	1.67789
0.880	1.75348	1.73931	1.73384	1.73250	1.72799
0.890	1.81372	1.79505	1.78710	1.78498	1.78265
0.900	1.88024	1.85633	1.84546	1.84242	1.84280
0.910	1.95443	1.92435	1.91001	1.90586	1.90962
0.920	2.03816	2.00076	1.98222	1.97674	1.98476
0.930	2.13407	2.08787	2.06417	2.05704	2.07051
0.940	2.24603	2.18905	2.15891	2.14968	2.17028
0.950	2.38010	2.30961	2.27116	2.25921	2.28935
0.960	2.54642	2.45843	2.40885	2.39320	2.43666
0.970	2.76413	2.65231	2.58690	2.56589	2.62905
0.980	3.07642	2.92921	2.83890	2.80923	2.90464
0.990	3.62211	3.41145	3.27259	3.22516	3.38627
0.995	4.17913	3.90324	3.70982	3.64105	3.87863
0.999	5.50184	5.07275	4.73766	4.60672	5.05052

TABLE 7.6 N= 8 (CONT.)

P ↓	λ:	5 5 5 5 10 15 20 35	5 5 5 5 10 15 25 30	5 5 5 5 10 20 20 30	5 5 5 5 10 20 25 25	5 5 5 5 15 15 15 35
0.001		0.08304	0.08379	0.08448	0.08493	0.08425
0.005		0.13186	0.13307	0.13418	0.13490	0.13375
0.010		0.16274	0.16426	0.16562	0.16653	0.16505
0.020		0.20292	0.20484	0.20654	0.20768	0.20574
0.030		0.23232	0.23454	0.23649	0.23780	0.23551
0.040		0.25664	0.25912	0.26126	0.26271	0.26012
0.050		0.27793	0.28062	0.28293	0.28451	0.28164
0.060		0.29716	0.30004	0.30251	0.30420	0.30107
0.070		0.31490	0.31797	0.32057	0.32237	0.31899
0.080		0.33152	0.33476	0.33749	0.33938	0.33577
0.090		0.34725	0.35065	0.35350	0.35548	0.35165
0.100		0.36227	0.36583	0.36878	0.37085	0.36680
0.110		0.37671	0.38041	0.38347	0.38562	0.38136
0.120		0.39066	0.39451	0.39766	0.39989	0.39543
0.130		0.40421	0.40819	0.41144	0.41374	0.40908
0.140		0.41741	0.42153	0.42486	0.42723	0.42237
0.150		0.43032	0.43457	0.43798	0.44042	0.43537
0.160		0.44298	0.44735	0.45084	0.45335	0.44810
0.170		0.45542	0.45991	0.46348	0.46606	0.46062
0.180		0.46767	0.47228	0.47592	0.47857	0.47294
0.190		0.47977	0.48450	0.48820	0.49091	0.48510
0.200		0.49173	0.49657	0.50034	0.50311	0.49711
0.250		0.55015	0.55552	0.55958	0.56262	0.55572
0.300		0.60752	0.61338	0.61766	0.62094	0.61318
0.350		0.66517	0.67146	0.67589	0.67938	0.67079
0.400		0.72416	0.73083	0.73534	0.73900	0.72961
0.450		0.78547	0.79245	0.79697	0.80075	0.79061
0.500		0.85015	0.85735	0.86178	0.86563	0.85483
0.550		0.91940	0.92672	0.93093	0.93479	0.92343
0.600		0.99473	1.00202	1.00586	1.00963	0.99790
0.650		1.07818	1.08521	1.08848	1.09203	1.08021
0.700		1.17266	1.17910	1.18152	1.18466	1.17320
0.750		1.28262	1.28794	1.28912	1.29156	1.28122
0.800		1.41550	1.41881	1.41813	1.41940	1.41150
0.810		1.44586	1.44860	1.44745	1.44840	1.44125
0.820		1.47781	1.47992	1.47825	1.47885	1.47254
0.830		1.51154	1.51293	1.51069	1.51089	1.50557
0.840		1.54727	1.54783	1.54498	1.54472	1.54055
0.850		1.58526	1.58489	1.58135	1.58059	1.57774
0.860		1.62583	1.62440	1.62009	1.61876	1.61745
0.870		1.66938	1.66672	1.66157	1.65959	1.66007
0.880		1.71640	1.71232	1.70623	1.70349	1.70609
0.890		1.76750	1.76176	1.75461	1.75100	1.75611
0.900		1.82348	1.81579	1.80744	1.80281	1.81092
0.910		1.88539	1.87537	1.86565	1.85983	1.87155
0.920		1.95467	1.94184	1.93053	1.92327	1.93941
0.930		2.03331	2.01704	2.00386	1.99486	2.01651
0.940		2.12428	2.10367	2.08827	2.07709	2.10575
0.950		2.23218	2.20595	2.18784	2.17385	2.21172
0.960		2.36474	2.33093	2.30938	2.29163	2.34209
0.970		2.53656	2.49183	2.46572	2.44257	2.51141
0.980		2.78063	2.71838	2.68568	2.65388	2.75260
0.990		3.20323	3.10558	3.06153	3.01214	3.17195
0.995		3.63242	3.49320	3.43810	3.36762	3.59967
0.999		4.65099	4.39617	4.31851	4.18614	4.61877

TABLE 7.7 N= 8 (CONT.)

P ↓	λ:	5 5 5 5 15 15 20 30	5 5 5 5 15 15 25 25	5 5 5 5 15 20 20 25	5 5 5 5 20 20 20 20	5 5 5 10 10 10 10 45
0.001		0.08571	0.08616	0.08687	0.08758	0.08110
0.005		0.13610	0.13683	0.13796	0.13910	0.12847
0.010		0.16797	0.16888	0.17027	0.17168	0.15832
0.020		0.20940	0.21055	0.21229	0.21403	0.19704
0.030		0.23971	0.24103	0.24301	0.24500	0.22530
0.040		0.26477	0.26623	0.26841	0.27060	0.24863
0.050		0.28667	0.28827	0.29061	0.29297	0.26902
0.060		0.30646	0.30816	0.31066	0.31317	0.28741
0.070		0.32470	0.32651	0.32914	0.33178	0.30437
0.080		0.34177	0.34367	0.34643	0.34920	0.32023
0.090		0.35793	0.35992	0.36279	0.36568	0.33525
0.100		0.37334	0.37542	0.37840	0.38139	0.34957
0.110		0.38815	0.39031	0.39339	0.39648	0.36333
0.120		0.40245	0.40469	0.40786	0.41105	0.37662
0.130		0.41632	0.41863	0.42190	0.42517	0.38952
0.140		0.42984	0.43222	0.43557	0.43893	0.40209
0.150		0.44304	0.44549	0.44892	0.45236	0.41438
0.160		0.45598	0.45850	0.46200	0.46552	0.42642
0.170		0.46869	0.47127	0.47485	0.47843	0.43826
0.180		0.48120	0.48384	0.48749	0.49114	0.44992
0.190		0.49353	0.49624	0.49996	0.50368	0.46143
0.200		0.50573	0.50849	0.51227	0.51605	0.47282
0.250		0.56514	0.56818	0.57223	0.57628	0.52843
0.300		0.62328	0.62655	0.63081	0.63506	0.58315
0.350		0.68146	0.68494	0.68934	0.69373	0.63830
0.400		0.74073	0.74438	0.74885	0.75331	0.69497
0.450		0.80205	0.80582	0.81029	0.81475	0.75421
0.500		0.86640	0.87024	0.87462	0.87899	0.81716
0.550		0.93492	0.93877	0.94295	0.94712	0.88518
0.600		1.00902	1.01278	1.01662	1.02046	0.96005
0.650		1.09056	1.09412	1.09743	1.10075	1.04417
0.700		1.18219	1.18537	1.18790	1.19047	1.14110
0.750		1.28795	1.29046	1.29186	1.29332	1.25639
0.800		1.41453	1.41592	1.41561	1.41542	1.39952
0.810		1.44327	1.44436	1.44361	1.44300	1.43283
0.820		1.47345	1.47419	1.47297	1.47189	1.46813
0.830		1.50523	1.50559	1.50384	1.50224	1.50565
0.840		1.53880	1.53873	1.53640	1.53424	1.54570
0.850		1.57441	1.57384	1.57088	1.56810	1.58863
0.860		1.61234	1.61121	1.60753	1.60406	1.63485
0.870		1.65293	1.65117	1.64669	1.64245	1.68488
0.880		1.69662	1.69412	1.68875	1.68364	1.73938
0.890		1.74395	1.74060	1.73422	1.72811	1.79916
0.900		1.79563	1.79127	1.78374	1.77650	1.86527
0.910		1.85257	1.84703	1.83816	1.82962	1.93911
0.920		1.91603	1.90906	1.89864	1.88857	2.02258
0.930		1.98777	1.97906	1.96679	1.95489	2.11832
0.940		2.07037	2.05947	2.04496	2.03084	2.23025
0.950		2.16783	2.15410	2.13680	2.11990	2.36444
0.960		2.28688	2.26931	2.24841	2.22788	2.53106
0.970		2.44016	2.41705	2.39120	2.36564	2.74932
0.980		2.65614	2.62407	2.59071	2.55740	3.06243
0.990		3.02618	2.97563	2.92822	2.87994	3.60932
0.995		3.39825	3.32529	3.26256	3.19725	4.16715
0.999		4.27269	4.13342	4.03184	3.91953	5.49084

TABLE 7.8 N= 8 (CONT.)

P ↓	λ:	5 5 5 10 10 10 15 40	5 5 5 10 10 10 20 35	5 5 5 10 10 10 25 30	5 5 5 10 10 15 15 35	5 5 5 10 10 15 20 30
0.001		0.08411	0.08579	0.08657	0.08704	0.08854
0.005		0.13326	0.13596	0.13721	0.13790	0.14030
0.010		0.16423	0.16759	0.16914	0.16994	0.17291
0.020		0.20440	0.20862	0.21057	0.21148	0.21519
0.030		0.23370	0.23856	0.24081	0.24177	0.24602
0.040		0.25789	0.26327	0.26576	0.26676	0.27145
0.050		0.27902	0.28484	0.28756	0.28857	0.29365
0.060		0.29807	0.30430	0.30721	0.30823	0.31365
0.070		0.31563	0.32224	0.32533	0.32634	0.33207
0.080		0.33205	0.33901	0.34226	0.34327	0.34929
0.090		0.34758	0.35487	0.35828	0.35927	0.36556
0.100		0.36239	0.36999	0.37355	0.37452	0.38106
0.110		0.37661	0.38451	0.38822	0.38915	0.39594
0.120		0.39035	0.39852	0.40237	0.40328	0.41030
0.130		0.40367	0.41212	0.41610	0.41697	0.42421
0.140		0.41664	0.42535	0.42946	0.43030	0.43775
0.150		0.42931	0.43828	0.44252	0.44331	0.45096
0.160		0.44173	0.45094	0.45530	0.45605	0.46390
0.170		0.45393	0.46338	0.46786	0.46856	0.47659
0.180		0.46594	0.47562	0.48022	0.48086	0.48908
0.190		0.47778	0.48770	0.49240	0.49299	0.50139
0.200		0.48949	0.49962	0.50444	0.50498	0.51355
0.250		0.54658	0.55775	0.56308	0.56330	0.57265
0.300		0.60256	0.61464	0.62045	0.62027	0.63030
0.350		0.65875	0.67165	0.67788	0.67724	0.68784
0.400		0.71623	0.72982	0.73643	0.73527	0.74632
0.450		0.77600	0.79015	0.79708	0.79532	0.80670
0.500		0.83914	0.85367	0.86084	0.85840	0.86996
0.550		0.90691	0.92158	0.92889	0.92570	0.93722
0.600		0.98089	0.99537	1.00268	0.99867	1.00988
0.650		1.06325	1.07706	1.08417	1.07926	1.08978
0.700		1.15711	1.16954	1.17611	1.17028	1.17953
0.750		1.26733	1.27722	1.28275	1.27602	1.28313
0.800		1.40211	1.40749	1.41109	1.40368	1.40719
0.810		1.43317	1.43729	1.44034	1.43285	1.43538
0.820		1.46597	1.46867	1.47110	1.46356	1.46498
0.830		1.50071	1.50180	1.50353	1.49597	1.49617
0.840		1.53764	1.53692	1.53784	1.53032	1.52913
0.850		1.57707	1.57429	1.57428	1.56686	1.56409
0.860		1.61936	1.61423	1.61316	1.60590	1.60135
0.870		1.66496	1.65712	1.65483	1.64783	1.64125
0.880		1.71442	1.70347	1.69975	1.69314	1.68421
0.890		1.76846	1.75389	1.74850	1.74242	1.73078
0.900		1.82798	1.80917	1.80181	1.79647	1.78166
0.910		1.89421	1.87037	1.86066	1.85632	1.83776
0.920		1.96877	1.93893	1.92636	1.92338	1.90034
0.930		2.05400	2.01684	2.00076	1.99966	1.97114
0.940		2.15331	2.10709	2.08656	2.08808	2.05273
0.950		2.27202	2.21427	2.18798	2.19323	2.14912
0.960		2.41911	2.34614	2.31206	2.32281	2.26701
0.970		2.61151	2.51735	2.47202	2.49141	2.41903
0.980		2.88747	2.76095	2.69759	2.73207	2.63362
0.990		3.37010	3.18348	3.08379	3.15144	3.00215
0.995		3.86345	3.61308	3.47092	3.57979	3.37352
0.999		5.03682	4.63295	4.37367	4.60064	4.24799

TABLE 7.9 N= 8 (CONT.)

P ↓ λ:	5 5 5 10 10 15 25 25	5 5 5 10 10 20 20 25	5 5 5 10 15 15 15 30	5 5 5 10 15 15 20 25	5 5 5 10 15 20 20 20
0.001	0.08900	0.08973	0.08981	0.09102	0.09176
0.005	0.14104	0.14220	0.14228	0.14420	0.14537
0.010	0.17384	0.17526	0.17532	0.17769	0.17913
0.020	0.21636	0.21811	0.21811	0.22106	0.22284
0.030	0.24736	0.24936	0.24930	0.25266	0.25469
0.040	0.27293	0.27513	0.27501	0.27871	0.28093
0.050	0.29525	0.29761	0.29743	0.30142	0.30380
0.060	0.31536	0.31787	0.31763	0.32188	0.32441
0.070	0.33389	0.33653	0.33622	0.34070	0.34337
0.080	0.35120	0.35396	0.35359	0.35829	0.36107
0.090	0.36756	0.37043	0.37000	0.37489	0.37779
0.100	0.38314	0.38613	0.38563	0.39071	0.39371
0.110	0.39810	0.40118	0.40062	0.40587	0.40897
0.120	0.41253	0.41570	0.41508	0.42050	0.42368
0.130	0.42652	0.42978	0.42908	0.43466	0.43793
0.140	0.44012	0.44346	0.44271	0.44844	0.45179
0.150	0.45341	0.45682	0.45600	0.46187	0.46530
0.160	0.46641	0.46990	0.46901	0.47502	0.47852
0.170	0.47917	0.48273	0.48177	0.48792	0.49149
0.180	0.49172	0.49535	0.49432	0.50059	0.50423
0.190	0.50409	0.50778	0.50669	0.51308	0.51678
0.200	0.51630	0.52006	0.51889	0.52540	0.52916
0.250	0.57567	0.57969	0.57817	0.58520	0.58922
0.300	0.63355	0.63778	0.63588	0.64334	0.64756
0.350	0.69129	0.69567	0.69338	0.70118	0.70553
0.400	0.74994	0.75440	0.75170	0.75974	0.76417
0.450	0.81045	0.81491	0.81180	0.81998	0.82442
0.500	0.87379	0.87818	0.87465	0.88284	0.88720
0.550	0.94107	0.94528	0.94133	0.94938	0.95357
0.600	1.01367	1.01756	1.01322	1.02091	1.02481
0.650	1.09338	1.09679	1.09210	1.09915	1.10259
0.700	1.18278	1.18544	1.18052	1.18652	1.18928
0.750	1.28575	1.28731	1.28234	1.28670	1.28842
0.800	1.40874	1.40862	1.40400	1.40571	1.40586
0.810	1.43663	1.43608	1.43161	1.43261	1.43236
0.820	1.46590	1.46487	1.46058	1.46081	1.46011
0.830	1.49670	1.49515	1.49110	1.49045	1.48926
0.840	1.52923	1.52710	1.52333	1.52171	1.51997
0.850	1.56372	1.56094	1.55752	1.55480	1.55245
0.860	1.60042	1.59693	1.59393	1.58997	1.58695
0.870	1.63968	1.63539	1.63290	1.62754	1.62375
0.880	1.68191	1.67672	1.67486	1.66789	1.66324
0.890	1.72763	1.72141	1.72032	1.71151	1.70587
0.900	1.77750	1.77010	1.76997	1.75901	1.75224
0.910	1.83240	1.82365	1.82470	1.81121	1.80312
0.920	1.89353	1.88318	1.88574	1.86924	1.85959
0.930	1.96256	1.95031	1.95480	1.93463	1.92311
0.940	2.04192	2.02737	2.03440	2.00966	1.99584
0.950	2.13541	2.11797	2.12844	2.09785	2.08113
0.960	2.24936	2.22818	2.24352	2.20510	2.18454
0.970	2.39567	2.36934	2.39204	2.34244	2.31650
0.980	2.60100	2.56686	2.60207	2.53464	2.50027
0.990	2.95042	2.90165	2.96399	2.86069	2.80967
0.995	3.29864	3.23395	3.33039	3.18489	3.11448
0.999	4.10500	4.00030	4.19915	3.93548	3.81005

TABLE 7.10 N= 8 (CONT.)

P ↓	λ:	5 5 5 15 15 15 15 25	5 5 5 15 15 15 20 20	5 5 10 10 10 10 10 40	5 5 10 10 10 10 15 35	5 5 10 10 10 10 20 30
0.001		0.09233	0.09308	0.08689	0.08989	0.09143
0.005		0.14622	0.14741	0.13737	0.14211	0.14456
0.010		0.18013	0.18159	0.16909	0.17490	0.17792
0.020		0.22403	0.22583	0.21009	0.21726	0.22103
0.030		0.25598	0.25802	0.23992	0.24807	0.25237
0.040		0.28230	0.28454	0.26448	0.27343	0.27816
0.050		0.30524	0.30764	0.28590	0.29551	0.30063
0.060		0.32589	0.32844	0.30518	0.31540	0.32085
0.070		0.34489	0.34757	0.32293	0.33368	0.33944
0.080		0.36262	0.36542	0.33950	0.35075	0.35679
0.090		0.37936	0.38227	0.35516	0.36687	0.37317
0.100		0.39530	0.39831	0.37007	0.38221	0.38876
0.110		0.41057	0.41368	0.38438	0.39692	0.40371
0.120		0.42530	0.42849	0.39818	0.41110	0.41811
0.130		0.43955	0.44283	0.41155	0.42483	0.43206
0.140		0.45341	0.45676	0.42456	0.43818	0.44562
0.150		0.46692	0.47035	0.43725	0.45121	0.45884
0.160		0.48014	0.48364	0.44968	0.46396	0.47178
0.170		0.49309	0.49667	0.46188	0.47646	0.48447
0.180		0.50583	0.50946	0.47388	0.48875	0.49694
0.190		0.51836	0.52206	0.48571	0.50086	0.50922
0.200		0.53073	0.53449	0.49739	0.51281	0.52134
0.250		0.59069	0.59470	0.55423	0.57085	0.58014
0.300		0.64888	0.65308	0.60978	0.62737	0.63733
0.350		0.70666	0.71100	0.66536	0.68374	0.69426
0.400		0.76507	0.76948	0.72207	0.74100	0.75199
0.450		0.82504	0.82946	0.78088	0.80013	0.81146
0.500		0.88750	0.89185	0.84288	0.86213	0.87366
0.550		0.95349	0.95768	0.90929	0.92816	0.93970
0.600		1.02430	1.02821	0.98168	0.99965	1.01094
0.650		1.10159	1.10507	1.06218	1.07854	1.08920
0.700		1.18772	1.19053	1.15389	1.16756	1.17706
0.750		1.28624	1.28807	1.26160	1.27098	1.27845
0.800		1.40301	1.40333	1.39349	1.39591	1.39992
0.810		1.42937	1.42929	1.42393	1.42448	1.42753
0.820		1.45698	1.45646	1.45608	1.45456	1.45653
0.830		1.48598	1.48499	1.49017	1.48633	1.48709
0.840		1.51656	1.51504	1.52643	1.52001	1.51940
0.850		1.54891	1.54679	1.56519	1.55585	1.55369
0.860		1.58327	1.58050	1.60679	1.59418	1.59025
0.870		1.61996	1.61644	1.65170	1.63537	1.62941
0.880		1.65933	1.65496	1.70047	1.67991	1.67160
0.890		1.70186	1.69653	1.75382	1.72840	1.71736
0.900		1.74816	1.74171	1.81267	1.78163	1.76739
0.910		1.79901	1.79126	1.87824	1.84064	1.82259
0.920		1.85549	1.84619	1.95219	1.90685	1.88422
0.930		1.91909	1.90793	2.03686	1.98225	1.95402
0.940		1.99203	1.97857	2.13570	2.06980	2.03455
0.950		2.07771	2.06132	2.25407	2.17407	2.12980
0.960		2.18183	2.16156	2.40101	2.30284	2.24647
0.970		2.31511	2.28930	2.59352	2.47075	2.39717
0.980		2.50155	2.46697	2.87002	2.71098	2.61037
0.990		2.81789	2.76560	3.35387	3.13059	2.97750
0.995		3.13286	3.05937	3.84830	3.55976	3.34832
0.999		3.86517	3.72907	5.02315	4.58256	4.22315

TABLE 7.11 N= 8 (CONT.)

P ↓	λ:	5 5 10 10 10 10 25 25	5 5 10 10 10 15 15 30	5 5 10 10 10 15 20 25	5 5 10 10 10 20 20 20	5 5 10 10 15 15 15 25
0.001		0.09191	0.09274	0.09398	0.09442	0.09532
0.005		0.14532	0.14657	0.14854	0.14933	0.15059
0.010		0.17886	0.18036	0.18277	0.18381	0.18525
0.020		0.22221	0.22398	0.22696	0.22834	0.22996
0.030		0.25372	0.25567	0.25906	0.26071	0.26241
0.040		0.27965	0.28173	0.28546	0.28732	0.28908
0.050		0.30224	0.30442	0.30844	0.31049	0.31227
0.060		0.32257	0.32483	0.32910	0.33132	0.33313
0.070		0.34126	0.34359	0.34809	0.35047	0.35228
0.080		0.35870	0.36110	0.36580	0.36833	0.37014
0.090		0.37517	0.37761	0.38251	0.38517	0.38697
0.100		0.39084	0.39332	0.39840	0.40119	0.40299
0.110		0.40587	0.40838	0.41363	0.41654	0.41832
0.120		0.42035	0.42288	0.42830	0.43132	0.43308
0.130		0.43436	0.43692	0.44249	0.44562	0.44736
0.140		0.44799	0.45057	0.45628	0.45952	0.46123
0.150		0.46128	0.46387	0.46972	0.47306	0.47475
0.160		0.47428	0.47687	0.48286	0.48629	0.48796
0.170		0.48703	0.48963	0.49574	0.49926	0.50090
0.180		0.49956	0.50215	0.50840	0.51200	0.51361
0.190		0.51190	0.51449	0.52085	0.52453	0.52611
0.200		0.52407	0.52665	0.53313	0.53689	0.53843
0.250		0.58314	0.58563	0.59262	0.59673	0.59808
0.300		0.64056	0.64289	0.65030	0.65468	0.65582
0.350		0.69769	0.69978	0.70753	0.71210	0.71301
0.400		0.75558	0.75737	0.76537	0.77005	0.77070
0.450		0.81519	0.81659	0.82474	0.82946	0.82982
0.500		0.87748	0.87840	0.88658	0.89124	0.89129
0.550		0.94355	0.94389	0.95196	0.95643	0.95615
0.600		1.01474	1.01439	1.02215	1.02630	1.02565
0.650		1.09285	1.09168	1.09885	1.10249	1.10144
0.700		1.18039	1.17824	1.18443	1.18730	1.18581
0.750		1.28120	1.27788	1.28250	1.28425	1.28225
0.800		1.40162	1.39694	1.39900	1.39906	1.39652
0.810		1.42894	1.42396	1.42534	1.42497	1.42231
0.820		1.45762	1.45233	1.45294	1.45210	1.44932
0.830		1.48781	1.48221	1.48196	1.48060	1.47770
0.840		1.51970	1.51378	1.51257	1.51064	1.50762
0.850		1.55352	1.54728	1.54498	1.54242	1.53928
0.860		1.58953	1.58296	1.57944	1.57618	1.57292
0.870		1.62806	1.62117	1.61626	1.61220	1.60882
0.880		1.66952	1.66232	1.65580	1.65087	1.64737
0.890		1.71442	1.70693	1.69856	1.69262	1.68903
0.900		1.76344	1.75568	1.74515	1.73806	1.73437
0.910		1.81744	1.80946	1.79638	1.78796	1.78419
0.920		1.87761	1.86948	1.85333	1.84335	1.83955
0.930		1.94560	1.93745	1.91756	1.90572	1.90193
0.940		2.02385	2.01587	1.99132	1.97718	1.97349
0.950		2.11613	2.10864	2.07808	2.06107	2.05762
0.960		2.22876	2.22233	2.18369	2.16290	2.15995
0.970		2.37357	2.36935	2.31912	2.29300	2.29107
0.980		2.57718	2.57774	2.50896	2.47447	2.47480
0.990		2.92450	2.93801	2.83180	2.78054	2.78736
0.995		3.27137	3.30386	3.15367	3.08242	3.09950
0.999		4.07625	4.17350	3.90128	3.77105	3.82829

TABLE 7.12 N= 8 (CONT.)

P ↓	λ:	5 5 10 10 15 15 20 20	5 5 10 15 15 15 15 20	5 5 15 15 15 15 15 15	5 10 10 10 10 10 10 35	5 10 10 10 10 10 15 30
0.001		0.09609	0.09745	0.08082	0.09281	0.09573
0.005		0.15180	0.15389	0.13010	0.14639	0.15094
0.010		0.18673	0.18924	0.16204	0.17991	0.18545
0.020		0.23178	0.23481	0.20456	0.22309	0.22988
0.030		0.26447	0.26785	0.23638	0.25440	0.26206
0.040		0.29133	0.29497	0.26315	0.28011	0.28847
0.050		0.31469	0.31855	0.28689	0.30247	0.31142
0.060		0.33568	0.33974	0.30859	0.32256	0.33204
0.070		0.35496	0.35918	0.32880	0.34102	0.35096
0.080		0.37294	0.37730	0.34788	0.35822	0.36858
0.090		0.38988	0.39437	0.36606	0.37445	0.38520
0.100		0.40599	0.41059	0.38351	0.38988	0.40099
0.110		0.42142	0.42612	0.40037	0.40465	0.41611
0.120		0.43627	0.44107	0.41671	0.41889	0.43066
0.130		0.45063	0.45551	0.43262	0.43266	0.44473
0.140		0.46458	0.46954	0.44815	0.44604	0.45839
0.150		0.47817	0.48320	0.46336	0.45908	0.47170
0.160		0.49145	0.49654	0.47827	0.47183	0.48470
0.170		0.50446	0.50961	0.49292	0.48433	0.49744
0.180		0.51723	0.52243	0.50735	0.49660	0.50995
0.190		0.52979	0.53504	0.52157	0.50869	0.52226
0.200		0.54217	0.54747	0.53560	0.52060	0.53439
0.250		0.60207	0.60751	0.60353	0.57838	0.59308
0.300		0.66000	0.66548	0.66887	0.63448	0.64990
0.350		0.71732	0.72276	0.73275	0.69026	0.70622
0.400		0.77509	0.78039	0.79609	0.74680	0.76310
0.450		0.83423	0.83929	0.85973	0.80505	0.82147
0.500		0.89565	0.90035	0.92452	0.86601	0.88228
0.550		0.96037	0.96458	0.99139	0.93080	0.94661
0.600		1.02961	1.03316	1.06145	1.00085	1.01576
0.650		1.10498	1.10767	1.13611	1.07804	1.09147
0.700		1.18873	1.19027	1.21733	1.16507	1.17617
0.750		1.28422	1.28423	1.30798	1.26612	1.27360
0.800		1.39702	1.39491	1.41273	1.38823	1.38999
0.810		1.42243	1.41979	1.43601	1.41617	1.41641
0.820		1.44901	1.44582	1.46027	1.44559	1.44415
0.830		1.47693	1.47312	1.48562	1.47668	1.47337
0.840		1.50632	1.50185	1.51219	1.50965	1.50425
0.850		1.53739	1.53220	1.54014	1.54476	1.53702
0.860		1.57037	1.56439	1.56966	1.58232	1.57194
0.870		1.60554	1.59868	1.60098	1.62272	1.60934
0.880		1.64325	1.63542	1.63440	1.66643	1.64963
0.890		1.68393	1.67503	1.67026	1.71407	1.69334
0.900		1.72815	1.71804	1.70904	1.76641	1.74113
0.910		1.77666	1.76516	1.75134	1.82451	1.79389
0.920		1.83045	1.81737	1.79797	1.88979	1.85282
0.930		1.89094	1.87599	1.85009	1.96424	1.91961
0.940		1.96016	1.94299	1.90935	2.05084	1.99676
0.950		2.04128	2.02140	1.97833	2.15419	2.08816
0.960		2.13960	2.11626	2.06131	2.28211	2.20038
0.970		2.26500	2.23700	2.16627	2.44933	2.34579
0.980		2.43958	2.40468	2.31099	2.68923	2.55249
0.990		2.73353	2.68601	2.55170	3.10938	2.91119
0.995		3.02328	2.96231	2.78624	3.53960	3.27670
0.999		3.68579	3.59134	3.31605	4.56454	4.14772

TABLE 7.13 N= 8 (CONT.)

P ↓ λ:	5 10 10 10 10 10 20 25	5 10 10 10 10 15 15 25	5 10 10 10 10 15 20 20	5 10 10 10 15 15 15 20	5 10 10 15 15 15 15 15
0.001	0.09700	0.09837	0.09892	0.10055	0.08535
0.005	0.15293	0.15503	0.15597	0.15838	0.13718
0.010	0.18790	0.19042	0.19162	0.19447	0.17064
0.020	0.23290	0.23593	0.23747	0.24082	0.21492
0.030	0.26549	0.26886	0.27066	0.27433	0.24786
0.040	0.29223	0.29586	0.29787	0.30179	0.27540
0.050	0.31546	0.31931	0.32150	0.32561	0.29970
0.060	0.33632	0.34036	0.34271	0.34698	0.32179
0.070	0.35546	0.35966	0.36216	0.36657	0.34226
0.080	0.37330	0.37764	0.38027	0.38480	0.36151
0.090	0.39010	0.39457	0.39732	0.40196	0.37977
0.100	0.40607	0.41065	0.41352	0.41825	0.39723
0.110	0.42136	0.42604	0.42902	0.43383	0.41403
0.120	0.43606	0.44084	0.44392	0.44880	0.43027
0.130	0.45029	0.45515	0.45833	0.46328	0.44602
0.140	0.46409	0.46903	0.47230	0.47731	0.46136
0.150	0.47754	0.48255	0.48591	0.49097	0.47633
0.160	0.49067	0.49575	0.49919	0.50430	0.49098
0.170	0.50354	0.50867	0.51220	0.51735	0.50534
0.180	0.51617	0.52136	0.52495	0.53014	0.51945
0.190	0.52859	0.53383	0.53750	0.54272	0.53333
0.200	0.54083	0.54611	0.54985	0.55510	0.54701
0.250	0.60003	0.60547	0.60951	0.61485	0.61302
0.300	0.65727	0.66277	0.66704	0.67239	0.67630
0.350	0.71393	0.71939	0.72384	0.72911	0.73815
0.400	0.77106	0.77639	0.78095	0.78607	0.79955
0.450	0.82959	0.83469	0.83929	0.84418	0.86137
0.500	0.89046	0.89521	0.89976	0.90432	0.92446
0.550	0.95470	0.95897	0.96337	0.96748	0.98975
0.600	1.02358	1.02719	1.03132	1.03484	1.05833
0.650	1.09875	1.10148	1.10518	1.10791	1.13158
0.700	1.18254	1.18409	1.18714	1.18883	1.21142
0.750	1.27849	1.27845	1.28050	1.28077	1.30069
0.800	1.39241	1.39016	1.39069	1.38897	1.40399
0.810	1.41817	1.41537	1.41550	1.41328	1.42697
0.820	1.44516	1.44177	1.44146	1.43871	1.45092
0.830	1.47354	1.46951	1.46872	1.46538	1.47594
0.840	1.50348	1.49875	1.49742	1.49344	1.50218
0.850	1.53518	1.52969	1.52777	1.52308	1.52978
0.860	1.56889	1.56256	1.55997	1.55451	1.55894
0.870	1.60491	1.59766	1.59432	1.58800	1.58988
0.880	1.64361	1.63535	1.63114	1.62386	1.62289
0.890	1.68547	1.67607	1.67089	1.66252	1.65832
0.900	1.73109	1.72042	1.71410	1.70451	1.69663
0.910	1.78127	1.76915	1.76151	1.75051	1.73842
0.920	1.83710	1.82332	1.81411	1.80147	1.78449
0.930	1.90009	1.88439	1.87327	1.85870	1.83597
0.940	1.97248	1.95449	1.94101	1.92411	1.89450
0.950	2.05771	2.03695	2.02046	2.00067	1.96261
0.960	2.16158	2.13735	2.11683	2.09332	2.04452
0.970	2.29496	2.26617	2.23989	2.21130	2.14808
0.980	2.48231	2.44700	2.41146	2.37526	2.29076
0.990	2.80182	2.75554	2.70095	2.65068	2.52779
0.995	3.12138	3.06480	2.98688	2.92167	2.75837
0.999	3.86633	3.79037	3.64152	3.54054	3.27792

TABLE 7.14 N= 8 (CONT.)

P ↓	λ:	10 10 10 10 10 10 10 30	10 10 10 10 10 10 15 25	10 10 10 10 10 10 20 20	10 10 10 10 10 15 15 20	10 10 10 10 15 15 15 15
0.001		0.09878	0.10148	0.10229	0.10372	0.10516
0.005		0.15536	0.15952	0.16078	0.16293	0.16511
0.010		0.19060	0.19563	0.19716	0.19974	0.20234
0.020		0.23582	0.24193	0.24378	0.24687	0.24997
0.030		0.26848	0.27533	0.27742	0.28084	0.28428
0.040		0.29523	0.30266	0.30493	0.30861	0.31231
0.050		0.31843	0.32635	0.32878	0.33267	0.33657
0.060		0.33924	0.34758	0.35015	0.35422	0.35830
0.070		0.35831	0.36703	0.36972	0.37394	0.37817
0.080		0.37606	0.38512	0.38793	0.39228	0.39665
0.090		0.39277	0.40214	0.40505	0.40952	0.41401
0.100		0.40864	0.41829	0.42130	0.42588	0.43047
0.110		0.42381	0.43373	0.43682	0.44151	0.44619
0.120		0.43840	0.44857	0.45175	0.45652	0.46129
0.130		0.45250	0.46290	0.46616	0.47101	0.47586
0.140		0.46618	0.47680	0.48013	0.48505	0.48998
0.150		0.47950	0.49032	0.49372	0.49871	0.50370
0.160		0.49250	0.50351	0.50698	0.51203	0.51708
0.170		0.50523	0.51642	0.51995	0.52506	0.53017
0.180		0.51772	0.52908	0.53267	0.53783	0.54299
0.190		0.53000	0.54152	0.54517	0.55037	0.55557
0.200		0.54209	0.55377	0.55747	0.56272	0.56796
0.250		0.60052	0.61284	0.61678	0.62218	0.62756
0.300		0.65692	0.66972	0.67385	0.67930	0.68474
0.350		0.71270	0.72581	0.73008	0.73550	0.74090
0.400		0.76889	0.78215	0.78651	0.79181	0.79710
0.450		0.82645	0.83966	0.84406	0.84915	0.85424
0.500		0.88630	0.89926	0.90362	0.90841	0.91319
0.550		0.94950	0.96194	0.96619	0.97053	0.97488
0.600		1.01733	1.02890	1.03294	1.03668	1.04046
0.650		1.09147	1.10172	1.10541	1.10835	1.11135
0.700		1.17432	1.18259	1.18572	1.18759	1.18955
0.750		1.26953	1.27485	1.27711	1.27752	1.27807
0.800		1.38319	1.38397	1.38486	1.38321	1.38177
0.810		1.40899	1.40858	1.40911	1.40695	1.40501
0.820		1.43608	1.43435	1.43448	1.43176	1.42928
0.830		1.46461	1.46143	1.46111	1.45779	1.45472
0.840		1.49477	1.48997	1.48915	1.48516	1.48145
0.850		1.52677	1.52017	1.51878	1.51407	1.50966
0.860		1.56089	1.55225	1.55022	1.54472	1.53953
0.870		1.59744	1.58651	1.58375	1.57736	1.57131
0.880		1.63683	1.62329	1.61970	1.61233	1.60531
0.890		1.67958	1.66304	1.65849	1.65001	1.64190
0.900		1.72634	1.70633	1.70066	1.69092	1.68158
0.910		1.77800	1.75392	1.74693	1.73575	1.72498
0.920		1.83575	1.80682	1.79825	1.78539	1.77298
0.930		1.90127	1.86649	1.85597	1.84115	1.82677
0.940		1.97705	1.93502	1.92208	1.90487	1.88812
0.950		2.06697	2.01569	1.99961	1.97946	1.95975
0.960		2.17757	2.11401	2.09368	2.06975	2.04619
0.970		2.32126	2.24033	2.21385	2.18475	2.15586
0.980		2.52619	2.41802	2.38156	2.34466	2.30758
0.990		2.88340	2.72227	2.66508	2.61364	2.56069
0.995		3.24885	3.02855	2.94602	2.87881	2.80769
0.999		4.12185	3.75129	3.59348	3.48669	3.36473

TABLE 7.15 N= 8 (CONT.)

P ↓	All λ's = 100/8
0.001	0.10714
0.005	0.16805
0.010	0.20581
0.020	0.25406
0.030	0.28876
0.040	0.31708
0.050	0.34158
0.060	0.36349
0.070	0.38353
0.080	0.40214
0.090	0.41963
0.100	0.43619
0.110	0.45200
0.120	0.46718
0.130	0.48182
0.140	0.49600
0.150	0.50977
0.160	0.52320
0.170	0.53632
0.180	0.54918
0.190	0.56179
0.200	0.57420
0.250	0.63383
0.300	0.69093
0.350	0.74691
0.400	0.80283
0.450	0.85958
0.500	0.91802
0.550	0.97906
0.600	1.04382
0.650	1.11367
0.700	1.19056
0.750	1.27736
0.800	1.37876
0.810	1.40144
0.820	1.42512
0.830	1.44991
0.840	1.47594
0.850	1.50338
0.860	1.53242
0.870	1.56330
0.880	1.59629
0.890	1.63177
0.900	1.67020
0.910	1.71218
0.920	1.75855
0.930	1.81045
0.940	1.86954
0.950	1.93841
0.960	2.02135
0.970	2.12631
0.980	2.27103
0.990	2.51128
0.995	2.74437
0.999	3.26556

TABLE 8.1 N= 9

P ↓	λ: 5 60	10 55	15 50	20 45	25 40
0.001	0.07750	0.08294	0.08599	0.08787	0.08899
0.005	0.11810	0.12639	0.13115	0.13412	0.13588
0.010	0.14321	0.15325	0.15911	0.16278	0.16497
0.020	0.17546	0.18774	0.19503	0.19964	0.20241
0.030	0.19886	0.21275	0.22110	0.22641	0.22962
0.040	0.21814	0.23333	0.24257	0.24848	0.25205
0.050	0.23496	0.25128	0.26131	0.26774	0.27164
0.060	0.25015	0.26747	0.27821	0.28512	0.28932
0.070	0.26415	0.28240	0.29378	0.30115	0.30563
0.080	0.27727	0.29636	0.30837	0.31615	0.32091
0.090	0.28970	0.30958	0.32217	0.33036	0.33538
0.100	0.30159	0.32221	0.33535	0.34394	0.34920
0.110	0.31302	0.33435	0.34803	0.35699	0.36249
0.120	0.32409	0.34609	0.36029	0.36962	0.37536
0.130	0.33486	0.35751	0.37221	0.38190	0.38786
0.140	0.34538	0.36864	0.38383	0.39387	0.40006
0.150	0.35569	0.37954	0.39521	0.40560	0.41200
0.160	0.36582	0.39025	0.40639	0.41711	0.42373
0.170	0.37581	0.40080	0.41739	0.42844	0.43528
0.180	0.38569	0.41120	0.42824	0.43962	0.44667
0.190	0.39546	0.42149	0.43898	0.45068	0.45794
0.200	0.40516	0.43169	0.44961	0.46163	0.46910
0.250	0.45307	0.48189	0.50186	0.51543	0.52392
0.300	0.50120	0.53195	0.55383	0.56888	0.57837
0.350	0.55089	0.58320	0.60684	0.62330	0.63375
0.400	0.60341	0.63682	0.66201	0.67981	0.69117
0.450	0.66010	0.69403	0.72051	0.73950	0.75173
0.500	0.72259	0.75625	0.78362	0.80361	0.81658
0.550	0.79296	0.82526	0.85291	0.87358	0.88715
0.600	0.87402	0.90342	0.93045	0.95130	0.96519
0.650	0.96967	0.99404	1.01908	1.03933	1.05312
0.700	1.08557	1.10199	1.12295	1.14136	1.15437
0.750	1.23019	1.23481	1.24849	1.26307	1.27418
0.800	1.41721	1.40501	1.40655	1.41399	1.42129
0.810	1.46156	1.44527	1.44358	1.44901	1.45521
0.820	1.50881	1.48815	1.48290	1.48607	1.49100
0.830	1.55927	1.53394	1.52477	1.52540	1.52889
0.840	1.61335	1.58302	1.56953	1.56728	1.56913
0.850	1.67151	1.63582	1.61757	1.61207	1.61204
0.860	1.73431	1.69285	1.66935	1.66017	1.65799
0.870	1.80244	1.75477	1.72546	1.71210	1.70744
0.880	1.87676	1.82235	1.78660	1.76848	1.76095
0.890	1.95835	1.89661	1.85369	1.83012	1.81924
0.900	2.04862	1.97883	1.92789	1.89805	1.88324
0.910	2.14941	2.07070	2.01076	1.97362	1.95418
0.920	2.26324	2.17453	2.10437	2.05870	2.03370
0.930	2.39362	2.29356	2.21168	2.15590	2.12414
0.940	2.54575	2.43255	2.33701	2.26905	2.22891
0.950	2.72771	2.59888	2.48710	2.40414	2.35335
0.960	2.95308	2.80502	2.67324	2.57125	2.50639
0.>70	3.24741	3.07436	2.91673	2.78936	2.70487
0.980	3.66838	3.45980	3.26563	3.10144	2.98679
0.990	4.40151	4.13128	3.87433	3.64578	3.47435
0.995	5.14780	4.81499	4.49480	4.20120	3.96842
0.999	6.91589	6.43531	5.96646	5.52120	5.13704

TABLE 8.2 N= 9 (CONT.)

P ↓	λ: 5 5 5 5 30 35	5 5 5 5 10 50	5 5 5 5 10 15 45	5 5 5 5 10 20 40	5 5 5 5 10 25 35
0.001	0.08951	0.08861	0.09174	0.09359	0.09459
0.005	0.13672	0.13499	0.13984	0.14275	0.14433
0.010	0.16601	0.16363	0.16958	0.17316	0.17512
0.020	0.20372	0.20037	0.20774	0.21221	0.21466
0.030	0.23114	0.22697	0.23538	0.24052	0.24334
0.040	0.25374	0.24884	0.25811	0.26381	0.26694
0.050	0.27349	0.26789	0.27792	0.28410	0.28751
0.060	0.29132	0.28505	0.29576	0.30239	0.30605
0.070	0.30777	0.30086	0.31218	0.31922	0.32312
0.080	0.32317	0.31563	0.32754	0.33497	0.33909
0.090	0.33777	0.32960	0.34206	0.34985	0.35418
0.100	0.35171	0.34292	0.35590	0.36405	0.36858
0.110	0.36512	0.35572	0.36920	0.37769	0.38242
0.120	0.37810	0.36809	0.38205	0.39086	0.39578
0.130	0.39071	0.38010	0.39452	0.40364	0.40874
0.140	0.40302	0.39180	0.40666	0.41610	0.42138
0.150	0.41508	0.40324	0.41854	0.42827	0.43373
0.160	0.42691	0.41446	0.43018	0.44021	0.44584
0.170	0.43856	0.42550	0.44163	0.45195	0.45775
0.180	0.45006	0.43638	0.45291	0.46351	0.46947
0.190	0.46143	0.44713	0.46405	0.47493	0.48106
0.200	0.47269	0.45777	0.47507	0.48622	0.49251
0.250	0.52802	0.50991	0.52900	0.54145	0.54852
0.300	0.58296	0.56154	0.58224	0.59590	0.60370
0.350	0.63882	0.61397	0.63609	0.65087	0.65937
0.400	0.69671	0.66832	0.69166	0.70746	0.71661
0.450	0.75771	0.72572	0.75000	0.76670	0.77643
0.500	0.82296	0.78744	0.81230	0.82970	0.83992
0.550	0.89385	0.85502	0.87994	0.89777	0.90836
0.600	0.97211	0.93050	0.95473	0.97260	0.98335
0.650	1.06006	1.01672	1.03916	1.05644	1.06705
0.700	1.16102	1.11785	1.13683	1.15257	1.16255
0.750	1.28005	1.24044	1.25336	1.26601	1.27456
0.800	1.42551	1.39559	1.39829	1.40521	1.41094
0.810	1.45893	1.43208	1.43202	1.43732	1.44223
0.820	1.49416	1.47087	1.46776	1.47123	1.47520
0.830	1.53141	1.51226	1.50575	1.50714	1.51006
0.840	1.57092	1.55657	1.54628	1.54532	1.54702
0.850	1.61298	1.60420	1.58969	1.58607	1.58637
0.860	1.65796	1.65562	1.63642	1.62975	1.62844
0.870	1.70627	1.71142	1.68696	1.67682	1.67365
0.880	1.75846	1.77233	1.74196	1.72782	1.72250
0.890	1.81521	1.83926	1.80223	1.78348	1.77563
0.900	1.87739	1.91338	1.86883	1.84470	1.83387
0.910	1.94614	1.99626	1.94312	1.91270	1.89832
0.920	2.02302	2.08998	2.02699	1.98910	1.97044
0.930	2.11021	2.19750	2.12309	2.07623	2.05232
0.940	2.21091	2.32316	2.23528	2.17748	2.14700
0.950	2.33007	2.47369	2.36962	2.29813	2.25921
0.960	2.47600	2.66039	2.53627	2.44710	2.39689
0.970	2.66427	2.90456	2.75436	2.64118	2.57496
0.980	2.92989	3.25425	3.06709	2.91837	2.82703
0.990	3.38470	3.86389	3.61332	3.40107	3.26085
0.995	3.84040	4.48493	4.17068	3.89323	3.69822
0.999	4.90193	5.95726	5.49382	5.06329	4.72638

TABLE 8.3 N= 9 (CONT.)

		5 5 5 5 10	5 5 5 5 15	5 5 5 5 15	5 5 5 5 15	5 5 5 5 20
P ↓	λ:	30 30	15 40	20 35	25 30	20 30
0.001		0.09491	0.09483	0.09656	0.09735	0.09807
0.005		0.14483	0.14463	0.14733	0.14857	0.14970
0.010		0.17574	0.17543	0.17876	0.18029	0.18166
0.020		0.21544	0.21497	0.21911	0.22103	0.22272
0.030		0.24425	0.24361	0.24836	0.25056	0.25249
0.040		0.26794	0.26717	0.27242	0.27486	0.27698
0.050		0.28861	0.28769	0.29339	0.29604	0.29831
0.060		0.30723	0.30618	0.31227	0.31511	0.31753
0.070		0.32437	0.32320	0.32965	0.33267	0.33522
0.080		0.34041	0.33910	0.34590	0.34908	0.35176
0.090		0.35558	0.35413	0.36125	0.36459	0.36739
0.100		0.37004	0.36846	0.37590	0.37939	0.38228
0.110		0.38394	0.38223	0.38995	0.39359	0.39659
0.120		0.39736	0.39551	0.40353	0.40730	0.41039
0.130		0.41039	0.40841	0.41669	0.42060	0.42378
0.140		0.42308	0.42096	0.42951	0.43355	0.43682
0.150		0.43549	0.43323	0.44204	0.44620	0.44956
0.160		0.44766	0.44526	0.45432	0.45860	0.46204
0.170		0.45962	0.45707	0.46638	0.47079	0.47430
0.180		0.47140	0.46871	0.47826	0.48278	0.48637
0.190		0.48304	0.48020	0.48998	0.49462	0.49828
0.200		0.49455	0.49156	0.50157	0.50632	0.51005
0.250		0.55081	0.54705	0.55813	0.56342	0.56744
0.300		0.60624	0.60163	0.61368	0.61947	0.62373
0.350		0.66214	0.65662	0.66954	0.67578	0.68022
0.400		0.71960	0.71310	0.72677	0.73341	0.73796
0.450		0.77963	0.77207	0.78635	0.79334	0.79792
0.500		0.84330	0.83462	0.84933	0.85659	0.86110
0.550		0.91187	0.90203	0.91693	0.92434	0.92866
0.600		0.98694	0.97593	0.99065	0.99808	1.00205
0.650		1.07062	1.05852	1.07254	1.07976	1.08317
0.700		1.16594	1.15296	1.16553	1.17218	1.17475
0.750		1.27754	1.26418	1.27407	1.27963	1.28094
0.800		1.41307	1.40043	1.40563	1.40917	1.40859
0.810		1.44411	1.43184	1.43574	1.43872	1.43765
0.820		1.47680	1.46501	1.46745	1.46978	1.46818
0.830		1.51133	1.50015	1.50094	1.50254	1.50036
0.840		1.54792	1.53751	1.53643	1.53721	1.53439
0.850		1.58684	1.57738	1.57420	1.57402	1.57051
0.860		1.62841	1.62014	1.61456	1.61330	1.60900
0.870		1.67304	1.66622	1.65790	1.65539	1.65023
0.880		1.72122	1.71620	1.70471	1.70076	1.69463
0.890		1.77356	1.77076	1.75562	1.74998	1.74276
0.900		1.83086	1.83082	1.81143	1.80379	1.79534
0.910		1.89418	1.89758	1.87317	1.86317	1.85331
0.920		1.96495	1.97269	1.94230	1.92943	1.91794
0.930		2.04514	2.05846	2.02081	2.00442	1.99103
0.940		2.13770	2.15829	2.11167	2.09087	2.07521
0.950		2.24715	2.27747	2.21948	2.19296	2.17453
0.960		2.38107	2.42495	2.35199	2.31777	2.29584
0.970		2.55372	2.61760	2.52382	2.47850	2.45195
0.980		2.79702	2.89358	2.76798	2.70490	2.67168
0.990		3.21292	3.37579	3.19084	3.09199	3.04732
0.995		3.62881	3.86859	3.62033	3.47957	3.42383
0.999		4.59447	5.04109	4.63951	4.38261	4.30437

TABLE 8.4 N= 9 (CONT.)

		5 5 5 5 5 20	5 5 5 5 10 10	5 5 5 5 10 10	5 5 5 5 10 10	5 5 5 5 10 10
P ↓	λ:	25 25	10 45	15 40	20 35	25 30
0.001		0.09854	0.09449	0.09766	0.09943	0.10024
0.005		0.15043	0.14385	0.14873	0.15149	0.15275
0.010		0.18257	0.17428	0.18023	0.18362	0.18518
0.020		0.22385	0.21324	0.22057	0.22478	0.22672
0.030		0.25379	0.24140	0.24973	0.25454	0.25677
0.040		0.27841	0.26452	0.27367	0.27897	0.28144
0.050		0.29987	0.28463	0.29448	0.30022	0.30290
0.060		0.31920	0.30271	0.31321	0.31934	0.32220
0.070		0.33699	0.31934	0.33042	0.33691	0.33995
0.080		0.35362	0.33487	0.34648	0.35332	0.35651
0.090		0.36933	0.34954	0.36165	0.36880	0.37215
0.100		0.38431	0.36351	0.37610	0.38355	0.38705
0.110		0.39870	0.37691	0.38996	0.39770	0.40134
0.120		0.41258	0.38985	0.40332	0.41134	0.41512
0.130		0.42604	0.40239	0.41628	0.42456	0.42847
0.140		0.43915	0.41459	0.42888	0.43743	0.44146
0.150		0.45196	0.42652	0.44119	0.44999	0.45414
0.160		0.46451	0.43820	0.45324	0.46228	0.46656
0.170		0.47683	0.44967	0.46507	0.47435	0.47874
0.180		0.48897	0.46096	0.47671	0.48623	0.49074
0.190		0.50094	0.47211	0.48819	0.49794	0.50256
0.200		0.51276	0.48312	0.49954	0.50950	0.51423
0.250		0.57044	0.53689	0.55482	0.56583	0.57108
0.300		0.62698	0.58975	0.60900	0.62095	0.62669
0.350		0.68369	0.64303	0.66340	0.67621	0.68239
0.400		0.74162	0.69781	0.71909	0.73265	0.73924
0.450		0.80172	0.75515	0.77708	0.79127	0.79820
0.500		0.86500	0.81619	0.83845	0.85309	0.86030
0.550		0.93258	0.88231	0.90444	0.91932	0.92672
0.600		1.00590	0.95528	0.97667	0.99145	0.99890
0.650		1.08682	1.03756	1.05731	1.07152	1.07880
0.700		1.17801	1.13274	1.14951	1.16240	1.16919
0.750		1.28350	1.24647	1.25815	1.26855	1.27432
0.800		1.40998	1.38834	1.39151	1.39738	1.40124
0.810		1.43872	1.42145	1.42231	1.42691	1.43021
0.820		1.46889	1.45657	1.45486	1.45802	1.46069
0.830		1.50067	1.49395	1.48936	1.49089	1.49286
0.840		1.53424	1.53387	1.52608	1.52576	1.52691
0.850		1.56984	1.57670	1.56531	1.56288	1.56309
0.860		1.60775	1.62286	1.60742	1.60258	1.60172
0.870		1.64831	1.67286	1.65285	1.64525	1.64314
0.880		1.69195	1.72737	1.70218	1.69139	1.68783
0.890		1.73919	1.78720	1.75611	1.74161	1.73634
0.900		1.79074	1.85341	1.81555	1.79671	1.78943
0.910		1.84749	1.92740	1.88173	1.85774	1.84806
0.920		1.91066	2.01106	1.95629	1.92615	1.91356
0.930		1.98197	2.10705	2.04157	2.00395	1.98776
0.940		2.06392	2.21927	2.14098	2.09410	2.07339
0.950		2.16039	2.35381	2.25986	2.20123	2.17465
0.960		2.27786	2.52083	2.40720	2.33311	2.29858
0.970		2.42848	2.73953	2.59994	2.50438	2.45843
0.980		2.63943	3.05313	2.87636	2.74815	2.68391
0.990		2.99724	3.60056	3.35965	3.17103	3.07007
0.995		3.35242	4.15873	3.85345	3.60098	3.45723
0.999		4.17052	5.48283	5.02741	4.62149	4.36010

TABLE 8.5 N= 9 (CONT.)

P ↓	λ:	5 5 5 10 15	5 5 5 10 15	5 5 5 10 15	5 5 5 15 15	5 5 5 15 15
		15 35	20 30	25 25	15 30	20 25
0.001		0.10073	0.10230	0.10279	0.10364	0.10490
0.005		0.15345	0.15589	0.15665	0.15789	0.15984
0.010		0.18598	0.18896	0.18989	0.19137	0.19375
0.020		0.22762	0.23132	0.23248	0.23422	0.23715
0.030		0.25773	0.26193	0.26326	0.26518	0.26850
0.040		0.28243	0.28706	0.28852	0.29057	0.29422
0.050		0.30390	0.30890	0.31048	0.31264	0.31657
0.060		0.32321	0.32855	0.33023	0.33247	0.33665
0.070		0.34096	0.34659	0.34838	0.35068	0.35509
0.080		0.35752	0.36343	0.36531	0.36767	0.37229
0.090		0.37314	0.37932	0.38128	0.38370	0.38851
0.100		0.38802	0.39445	0.39649	0.39895	0.40395
0.110		0.40228	0.40895	0.41107	0.41357	0.41873
0.120		0.41603	0.42293	0.42512	0.42765	0.43298
0.130		0.42936	0.43647	0.43873	0.44129	0.44677
0.140		0.44231	0.44963	0.45197	0.45454	0.46018
0.150		0.45496	0.46248	0.46488	0.46747	0.47325
0.160		0.46733	0.47505	0.47752	0.48012	0.48604
0.170		0.47948	0.48738	0.48991	0.49252	0.49857
0.180		0.49142	0.49951	0.50210	0.50471	0.51089
0.190		0.50319	0.51146	0.51411	0.51672	0.52302
0.200		0.51482	0.52326	0.52597	0.52857	0.53499
0.250		0.57136	0.58061	0.58359	0.58612	0.59308
0.300		0.62660	0.63655	0.63977	0.64215	0.64957
0.350		0.68186	0.69241	0.69585	0.69800	0.70579
0.400		0.73819	0.74925	0.75286	0.75471	0.76277
0.450		0.79655	0.80800	0.81176	0.81321	0.82145
0.500		0.85798	0.86965	0.87351	0.87447	0.88277
0.550		0.92362	0.93532	0.93922	0.93959	0.94778
0.600		0.99494	1.00639	1.01025	1.00991	1.01778
0.650		1.07390	1.08472	1.08842	1.08722	1.09449
0.700		1.16330	1.17291	1.17628	1.17405	1.18032
0.750		1.26746	1.27496	1.27771	1.27429	1.27892
0.800		1.39359	1.39748	1.39916	1.39433	1.39631
0.810		1.42246	1.42536	1.42674	1.42162	1.42288
0.820		1.45287	1.45466	1.45570	1.45027	1.45074
0.830		1.48500	1.48554	1.48620	1.48045	1.48005
0.840		1.51906	1.51819	1.51842	1.51236	1.51096
0.850		1.55532	1.55286	1.55260	1.54622	1.54371
0.860		1.59410	1.58981	1.58899	1.58230	1.57853
0.870		1.63577	1.62941	1.62794	1.62094	1.61575
0.880		1.68082	1.67207	1.66986	1.66256	1.65573
0.890		1.72987	1.71835	1.71526	1.70769	1.69897
0.900		1.78369	1.76893	1.76482	1.75700	1.74610
0.910		1.84334	1.82474	1.81941	1.81140	1.79791
0.920		1.91022	1.88703	1.88022	1.87211	1.85554
0.930		1.98634	1.95755	1.94893	1.94084	1.92051
0.940		2.07464	2.03887	2.02797	2.02010	1.99511
0.950		2.17970	2.13499	2.12113	2.11383	2.08285
0.960		2.30927	2.25263	2.23475	2.22860	2.18962
0.970		2.47795	2.40441	2.38072	2.37685	2.32644
0.980		2.71883	2.61882	2.58570	2.58667	2.51807
0.990		3.13870	2.98729	2.93476	2.94854	2.84345
0.995		3.56753	3.35875	3.28277	3.31513	3.16723
0.999		4.58919	4.23363	4.08895	4.18458	3.91743

TABLE 8.6 N= 9 (CONT.)

P ↓	λ: 5 5 5 5 5 15 20 20 20	5 5 5 5 10 10 10 10 40	5 5 5 5 10 10 10 15 35	5 5 5 5 10 10 10 20 30	5 5 5 5 10 10 10 25 25
0.001	0.10567	0.10055	0.10370	0.10530	0.10580
0.005	0.16103	0.15289	0.15770	0.16019	0.16096
0.010	0.19519	0.18509	0.19093	0.19396	0.19491
0.020	0.23892	0.22623	0.23337	0.23711	0.23828
0.030	0.27051	0.25589	0.26397	0.26822	0.26956
0.040	0.29641	0.28019	0.28902	0.29370	0.29517
0.050	0.31892	0.30129	0.31078	0.31581	0.31740
0.060	0.33914	0.32024	0.33030	0.33567	0.33736
0.070	0.35772	0.33764	0.34823	0.35389	0.35568
0.080	0.37503	0.35387	0.36493	0.37087	0.37275
0.090	0.39136	0.36916	0.38068	0.38687	0.38884
0.100	0.40690	0.38372	0.39566	0.40209	0.40414
0.110	0.42178	0.39767	0.41000	0.41667	0.41879
0.120	0.43612	0.41111	0.42382	0.43071	0.43290
0.130	0.45000	0.42412	0.43719	0.44429	0.44656
0.140	0.46348	0.43677	0.45018	0.45749	0.45982
0.150	0.47663	0.44911	0.46285	0.47036	0.47275
0.160	0.48949	0.46118	0.47524	0.48294	0.48540
0.170	0.50210	0.47303	0.48739	0.49527	0.49779
0.180	0.51448	0.48467	0.49933	0.50739	0.50997
0.190	0.52668	0.49615	0.51108	0.51932	0.52196
0.200	0.53871	0.50747	0.52268	0.53109	0.53379
0.250	0.59707	0.56256	0.57900	0.58819	0.59115
0.300	0.65377	0.61635	0.63384	0.64372	0.64691
0.350	0.71014	0.67020	0.68853	0.69901	0.70241
0.400	0.76722	0.72515	0.74413	0.75511	0.75870
0.450	0.82593	0.78221	0.80160	0.81297	0.81671
0.500	0.88720	0.84243	0.86193	0.87356	0.87741
0.550	0.95206	0.90706	0.92629	0.93799	0.94189
0.600	1.02179	0.97766	0.99611	1.00762	1.01149
0.650	1.09806	1.05637	1.07330	1.08426	1.08800
0.700	1.18321	1.14631	1.16063	1.17049	1.17393
0.750	1.28078	1.25231	1.26235	1.27023	1.27311
0.800	1.39660	1.38262	1.38562	1.39004	1.39189
0.810	1.42276	1.41276	1.41386	1.41732	1.41888
0.820	1.45017	1.44464	1.44362	1.44599	1.44723
0.830	1.47897	1.47846	1.47507	1.47622	1.47708
0.840	1.50933	1.51448	1.50844	1.50820	1.50864
0.850	1.54145	1.55300	1.54398	1.54216	1.54212
0.860	1.57559	1.59440	1.58201	1.57839	1.57779
0.870	1.61202	1.63912	1.62291	1.61722	1.61599
0.880	1.65113	1.68774	1.66718	1.65909	1.65711
0.890	1.69336	1.74097	1.71541	1.70453	1.70168
0.900	1.73933	1.79974	1.76840	1.75424	1.75037
0.910	1.78979	1.86527	1.82719	1.80914	1.80403
0.920	1.84582	1.93924	1.89321	1.87046	1.86387
0.930	1.90888	2.02399	1.96845	1.93997	1.93153
0.940	1.98113	2.12298	2.05589	2.02022	2.00945
0.950	2.06589	2.24158	2.16012	2.11521	2.10140
0.960	2.16874	2.38885	2.28892	2.23165	2.21371
0.970	2.30005	2.58180	2.45697	2.38216	2.35821
0.980	2.48307	2.85886	2.69753	2.59525	2.56152
0.990	2.79145	3.34345	3.11779	2.96243	2.90857
0.995	3.09550	3.83834	3.54752	3.33344	3.25532
0.999	3.78990	5.01376	4.57115	4.20880	4.06013

TABLE 8.7 N= 9 (CONT.)

P ↓	λ:	5 5 5 10 10 15 15 30	5 5 5 10 10 15 20 25	5 5 5 10 10 20 20 20	5 5 5 10 15 15 15 25	5 5 5 10 15 15 20 20
0.001		0.10667	0.10796	0.10875	0.10935	0.11015
0.005		0.16223	0.16421	0.16542	0.16629	0.16751
0.010		0.19640	0.19882	0.20029	0.20130	0.20278
0.020		0.24004	0.24301	0.24480	0.24598	0.24779
0.030		0.27149	0.27485	0.27687	0.27816	0.28019
0.040		0.29723	0.30091	0.30311	0.30448	0.30670
0.050		0.31956	0.32352	0.32588	0.32731	0.32968
0.060		0.33960	0.34380	0.34631	0.34778	0.35030
0.070		0.35799	0.36241	0.36504	0.36655	0.36919
0.080		0.37512	0.37975	0.38249	0.38403	0.38678
0.090		0.39126	0.39607	0.39893	0.40049	0.40335
0.100		0.40660	0.41160	0.41455	0.41612	0.41909
0.110		0.42128	0.42645	0.42950	0.43109	0.43414
0.120		0.43542	0.44075	0.44388	0.44548	0.44863
0.130		0.44910	0.45458	0.45780	0.45941	0.46263
0.140		0.46238	0.46801	0.47131	0.47292	0.47622
0.150		0.47533	0.48110	0.48447	0.48608	0.48946
0.160		0.48798	0.49389	0.49733	0.49894	0.50239
0.170		0.50038	0.50642	0.50993	0.51153	0.51505
0.180		0.51257	0.51872	0.52230	0.52390	0.52748
0.190		0.52455	0.53083	0.53447	0.53606	0.53970
0.200		0.53638	0.54277	0.54647	0.54805	0.55175
0.250		0.59368	0.60060	0.60456	0.60606	0.61002
0.300		0.64929	0.65666	0.66083	0.66220	0.66636
0.350		0.70458	0.71231	0.71664	0.71783	0.72215
0.400		0.76057	0.76858	0.77302	0.77399	0.77841
0.450		0.81820	0.82641	0.83088	0.83159	0.83605
0.500		0.87844	0.88671	0.89115	0.89155	0.89597
0.550		0.94234	0.95055	0.95485	0.95489	0.95919
0.600		1.01125	1.01919	1.02325	1.02285	1.02691
0.650		1.08692	1.09431	1.09796	1.09706	1.10074
0.700		1.17183	1.17829	1.18130	1.17982	1.18289
0.750		1.26979	1.27470	1.27673	1.27459	1.27671
0.800		1.38713	1.38948	1.38998	1.38709	1.38775
0.810		1.41381	1.41547	1.41556	1.41251	1.41278
0.820		1.44183	1.44271	1.44236	1.43915	1.43899
0.830		1.47135	1.47137	1.47052	1.46716	1.46652
0.840		1.50257	1.50162	1.50022	1.49669	1.49552
0.850		1.53570	1.53366	1.53165	1.52795	1.52619
0.860		1.57103	1.56774	1.56504	1.56119	1.55876
0.870		1.60888	1.60417	1.60069	1.59669	1.59350
0.880		1.64967	1.64333	1.63896	1.63481	1.63077
0.890		1.69392	1.68569	1.68031	1.67603	1.67100
0.900		1.74231	1.73187	1.72533	1.72093	1.71476
0.910		1.79573	1.78268	1.77477	1.77030	1.76278
0.920		1.85539	1.83921	1.82968	1.82518	1.81606
0.930		1.92301	1.90300	1.89151	1.88706	1.87600
0.940		2.00109	1.97630	1.96240	1.95810	1.94464
0.950		2.09353	2.06259	2.04563	2.04167	2.02514
0.960		2.20693	2.16771	2.14669	2.14342	2.12277
0.970		2.35370	2.30262	2.27586	2.27391	2.24739
0.980		2.56194	2.49190	2.45614	2.45695	2.42105
0.990		2.92229	2.81413	2.76053	2.76873	2.71377
0.995		3.28845	3.13567	3.06130	3.08048	3.00262
0.999		4.15895	3.88307	3.75011	3.80916	3.66382

TABLE 8.8 N= 9 (CONT.)

P ↓ λ:	5 5 5 5 15 15 15 15 20	5 5 5 10 10 10 10 10 35	5 5 5 10 10 10 10 15 30	5 5 5 10 10 10 10 20 25	5 5 5 10 10 10 15 15 25
0.001	0.11157	0.10672	0.10976	0.11108	0.11250
0.005	0.16962	0.16201	0.16662	0.16864	0.17075
0.010	0.20530	0.19594	0.20149	0.20395	0.20646
0.020	0.25080	0.23915	0.24590	0.24890	0.25191
0.030	0.28354	0.27023	0.27782	0.28121	0.28455
0.040	0.31030	0.29564	0.30390	0.30761	0.31120
0.050	0.33350	0.31766	0.32649	0.33047	0.33428
0.060	0.35430	0.33740	0.34674	0.35096	0.35494
0.070	0.37335	0.35549	0.36529	0.36973	0.37387
0.080	0.39109	0.37234	0.38255	0.38719	0.39147
0.090	0.40778	0.38820	0.39879	0.40362	0.40803
0.100	0.42363	0.40327	0.41422	0.41922	0.42375
0.110	0.43879	0.41769	0.42897	0.43414	0.43877
0.120	0.45337	0.43157	0.44317	0.44849	0.45322
0.130	0.46746	0.44499	0.45689	0.46236	0.46717
0.140	0.48113	0.45802	0.47020	0.47581	0.48071
0.150	0.49444	0.47071	0.48316	0.48891	0.49388
0.160	0.50744	0.48311	0.49582	0.50171	0.50674
0.170	0.52016	0.49526	0.50822	0.51423	0.51933
0.180	0.53265	0.50720	0.52039	0.52652	0.53168
0.190	0.54493	0.51894	0.53236	0.53861	0.54382
0.200	0.55702	0.53052	0.54415	0.55052	0.55577
0.250	0.61546	0.58662	0.60121	0.60809	0.61353
0.300	0.67187	0.64107	0.65643	0.66375	0.66927
0.350	0.72764	0.69522	0.71118	0.71886	0.72437
0.400	0.78378	0.75013	0.76649	0.77446	0.77986
0.450	0.84120	0.80674	0.82329	0.83146	0.83666
0.500	0.90079	0.86604	0.88254	0.89079	0.89567
0.550	0.96354	0.92915	0.94527	0.95349	0.95790
0.600	1.03062	0.99750	1.01280	1.02079	1.02456
0.650	1.10358	1.07295	1.08685	1.09435	1.09725
0.700	1.18458	1.15821	1.16984	1.17648	1.17820
0.750	1.27686	1.25746	1.26551	1.27069	1.27081
0.800	1.38574	1.37776	1.38007	1.38280	1.38066
0.810	1.41024	1.40534	1.40611	1.40817	1.40547
0.820	1.43587	1.43441	1.43347	1.43479	1.43148
0.830	1.46278	1.46515	1.46231	1.46278	1.45882
0.840	1.49110	1.49778	1.49280	1.49232	1.48764
0.850	1.52103	1.53255	1.52518	1.52362	1.51816
0.860	1.55279	1.56978	1.55971	1.55693	1.55061
0.870	1.58664	1.60985	1.59672	1.59253	1.58526
0.880	1.62291	1.65326	1.63662	1.63081	1.62250
0.890	1.66203	1.70060	1.67994	1.67224	1.66275
0.900	1.70454	1.75268	1.72733	1.71742	1.70662
0.910	1.75114	1.81054	1.77970	1.76716	1.75486
0.920	1.80278	1.87561	1.83824	1.82253	1.80851
0.930	1.86081	1.94990	1.90465	1.88505	1.86904
0.940	1.92716	2.03640	1.98144	1.95695	1.93858
0.950	2.00486	2.13973	2.07250	2.04168	2.02045
0.960	2.09893	2.26774	2.18442	2.14504	2.12022
0.970	2.21876	2.43520	2.32961	2.27789	2.24838
0.980	2.38533	2.67556	2.53622	2.46469	2.42851
0.990	2.66511	3.09652	2.89517	2.78367	2.73632
0.995	2.94022	3.52739	3.26114	3.10300	3.04528
0.999	3.56740	4.55317	4.13320	3.84795	3.77100

TABLE 8.9 N= 9 (CONT.)

P ↓	5 5 5 10 10 10 15 λ: 20 20	5 5 5 10 10 15 15 15 20	5 5 5 10 15 15 15 15 15	5 5 10 10 10 10 10 10 30	5 5 10 10 10 10 10 15 25
0.001	0.11332	0.11476	0.11622	0.11290	0.11570
0.005	0.17199	0.17414	0.17630	0.17107	0.17527
0.010	0.20796	0.21051	0.21308	0.20663	0.21167
0.020	0.25373	0.25677	0.25982	0.25180	0.25786
0.030	0.28660	0.28997	0.29335	0.28418	0.29096
0.040	0.31343	0.31706	0.32069	0.31059	0.31793
0.050	0.33667	0.34050	0.34434	0.33343	0.34125
0.060	0.35747	0.36148	0.36551	0.35387	0.36211
0.070	0.37652	0.38069	0.38486	0.37258	0.38118
0.080	0.39423	0.39854	0.40285	0.38997	0.39890
0.090	0.41090	0.41532	0.41976	0.40631	0.41556
0.100	0.42671	0.43125	0.43580	0.42182	0.43135
0.110	0.44182	0.44647	0.45111	0.43664	0.44643
0.120	0.45635	0.46109	0.46583	0.45089	0.46092
0.130	0.47039	0.47521	0.48003	0.46464	0.47491
0.140	0.48400	0.48890	0.49380	0.47798	0.48847
0.150	0.49725	0.50222	0.50718	0.49096	0.50165
0.160	0.51018	0.51521	0.52024	0.50362	0.51451
0.170	0.52283	0.52792	0.53302	0.51602	0.52709
0.180	0.53524	0.54039	0.54554	0.52818	0.53943
0.190	0.54744	0.55264	0.55784	0.54013	0.55154
0.200	0.55946	0.56470	0.56994	0.55190	0.56347
0.250	0.61746	0.62288	0.62828	0.60873	0.62098
0.300	0.67341	0.67890	0.68437	0.66357	0.67634
0.350	0.72866	0.73414	0.73960	0.71780	0.73093
0.400	0.78426	0.78963	0.79499	0.77246	0.78579
0.450	0.84110	0.84628	0.85144	0.82847	0.84182
0.500	0.90009	0.90495	0.90981	0.88676	0.89991
0.550	0.96221	0.96663	0.97105	0.94836	0.96106
0.600	1.02866	1.03246	1.03630	1.01455	1.02645
0.650	1.10099	1.10397	1.10701	1.08701	1.09764
0.700	1.18137	1.18324	1.18521	1.16810	1.17681
0.750	1.27308	1.27344	1.27395	1.26147	1.26725
0.800	1.38152	1.37976	1.37821	1.37318	1.37442
0.810	1.40596	1.40367	1.40161	1.39857	1.39861
0.820	1.43155	1.42869	1.42607	1.42525	1.42397
0.830	1.45842	1.45493	1.45172	1.45337	1.45061
0.840	1.48673	1.48257	1.47869	1.48311	1.47871
0.850	1.51666	1.51176	1.50716	1.51470	1.50846
0.860	1.54845	1.54273	1.53733	1.54838	1.54008
0.870	1.58236	1.57574	1.56946	1.58450	1.57386
0.880	1.61874	1.61111	1.60384	1.62346	1.61015
0.890	1.65801	1.64925	1.64088	1.66577	1.64940
0.900	1.70073	1.69070	1.68107	1.71210	1.69217
0.910	1.74762	1.73613	1.72507	1.76333	1.73921
0.920	1.79966	1.78649	1.77375	1.82065	1.79156
0.930	1.85822	1.84307	1.82836	1.88575	1.85063
0.940	1.92531	1.90779	1.89070	1.96113	1.91855
0.950	2.00403	1.98359	1.96354	2.05068	1.99858
0.960	2.09956	2.07539	2.05152	2.16098	2.09622
0.970	2.22162	2.19239	2.16327	2.30447	2.22183
0.980	2.39193	2.35515	2.31804	2.50938	2.39879
0.990	2.67962	2.62894	2.57659	2.86705	2.70235
0.995	2.96424	2.89870	2.82919	3.23315	3.00845
0.999	3.61818	3.51577	3.39930	4.10737	3.73169

TABLE 8.10 N= 9 (CONT.)

		5	5	5	5	5
		5 10	5 10	5 10	10 10	10 10
		10 10	10 10	10 10	10 10	10 10
		10 10	10 15	15 15	10 10	10 10
P ↓	λ:	20 20	15 20	15 15	10 25	15 20
0.001		0.11654	0.11801	0.11950	0.11897	0.12133
0.005		0.17653	0.17871	0.18090	0.17984	0.18333
0.010		0.21319	0.21577	0.21836	0.21692	0.22107
0.020		0.25970	0.26277	0.26585	0.26385	0.26880
0.030		0.29303	0.29642	0.29982	0.29739	0.30288
0.040		0.32018	0.32382	0.32747	0.32468	0.33059
0.050		0.34365	0.34750	0.35135	0.34823	0.35450
0.060		0.36464	0.36866	0.37269	0.36927	0.37584
0.070		0.38384	0.38801	0.39219	0.38849	0.39532
0.080		0.40167	0.40598	0.41029	0.40632	0.41340
0.090		0.41843	0.42285	0.42729	0.42307	0.43036
0.100		0.43431	0.43885	0.44339	0.43893	0.44643
0.110		0.44948	0.45412	0.45876	0.45407	0.46175
0.120		0.46406	0.46878	0.47351	0.46861	0.47645
0.130		0.47812	0.48293	0.48774	0.48262	0.49063
0.140		0.49175	0.49664	0.50152	0.49620	0.50435
0.150		0.50501	0.50996	0.51491	0.50939	0.51768
0.160		0.51794	0.52295	0.52797	0.52225	0.53067
0.170		0.53058	0.53566	0.54073	0.53483	0.54337
0.180		0.54298	0.54811	0.55323	0.54715	0.55580
0.190		0.55515	0.56033	0.56551	0.55924	0.56800
0.200		0.56714	0.57236	0.57758	0.57114	0.58000
0.250		0.62490	0.63029	0.63567	0.62842	0.63769
0.300		0.68046	0.68593	0.69138	0.68342	0.69296
0.350		0.73520	0.74066	0.74611	0.73752	0.74721
0.400		0.79016	0.79554	0.80089	0.79177	0.80149
0.450		0.84625	0.85143	0.85661	0.84706	0.85667
0.500		0.90433	0.90923	0.91412	0.90427	0.91361
0.550		0.96539	0.96987	0.97435	0.96437	0.97325
0.600		1.03059	1.03449	1.03841	1.02853	1.03669
0.650		1.10145	1.10456	1.10772	1.09825	1.10535
0.700		1.18008	1.18212	1.18426	1.17565	1.18124
0.750		1.26968	1.27025	1.27097	1.26394	1.26730
0.800		1.37548	1.37398	1.37269	1.36839	1.36843
0.810		1.39932	1.39729	1.39550	1.39196	1.39114
0.820		1.42427	1.42168	1.41934	1.41665	1.41488
0.830		1.45046	1.44726	1.44432	1.44259	1.43978
0.840		1.47806	1.47418	1.47059	1.46994	1.46597
0.850		1.50723	1.50261	1.49830	1.49889	1.49363
0.860		1.53821	1.53277	1.52767	1.52966	1.52296
0.870		1.57126	1.56491	1.55892	1.56253	1.55420
0.880		1.60670	1.59935	1.59237	1.59783	1.58767
0.890		1.64497	1.63647	1.62838	1.63601	1.62374
0.900		1.68660	1.67681	1.66745	1.67763	1.66292
0.910		1.73230	1.72102	1.71020	1.72341	1.70585
0.920		1.78302	1.77002	1.75750	1.77435	1.75342
0.930		1.84011	1.82508	1.81054	1.83188	1.80686
0.940		1.90554	1.88805	1.87106	1.89804	1.86797
0.950		1.98234	1.96181	1.94176	1.97606	1.93955
0.960		2.07561	2.05116	2.02714	2.07137	2.02626
0.970		2.19489	2.16509	2.13555	2.19417	2.13685
0.980		2.36157	2.32370	2.28567	2.36764	2.29089
0.990		2.64379	2.59092	2.53646	2.66657	2.55081
0.995		2.92388	2.85480	2.78154	2.96971	2.80813
0.999		3.57038	3.46094	3.33521	3.69109	3.40221

TABLE 8.11 N= 9 (CONT.)

P ↓	λ:	5 10 10 10 10 10 15 15 15	10 10 10 10 10 10 10 10 20	10 10 10 10 10 10 10 15 15	All λ's = 100/9
0.001		0.12284	0.12469	0.12624	0.12800
0.005		0.18555	0.18800	0.19026	0.19277
0.010		0.22369	0.22641	0.22907	0.23199
0.020		0.27190	0.27486	0.27799	0.28138
0.030		0.30631	0.30937	0.31281	0.31650
0.040		0.33426	0.33737	0.34106	0.34496
0.050		0.35836	0.36150	0.36538	0.36946
0.060		0.37988	0.38301	0.38705	0.39128
0.070		0.39951	0.40263	0.40682	0.41116
0.080		0.41772	0.42081	0.42513	0.42957
0.090		0.43479	0.43786	0.44229	0.44682
0.100		0.45096	0.45399	0.45852	0.46313
0.110		0.46638	0.46936	0.47398	0.47866
0.120		0.48117	0.48410	0.48881	0.49354
0.130		0.49543	0.49830	0.50309	0.50787
0.140		0.50922	0.51204	0.51689	0.52173
0.150		0.52262	0.52537	0.53030	0.53517
0.160		0.53567	0.53836	0.54335	0.54825
0.170		0.54842	0.55105	0.55609	0.56102
0.180		0.56091	0.56346	0.56855	0.57351
0.190		0.57315	0.57564	0.58078	0.58576
0.200		0.58520	0.58761	0.59279	0.59778
0.250		0.64305	0.64507	0.65041	0.65542
0.300		0.69839	0.70000	0.70541	0.71037
0.350		0.75265	0.75379	0.75921	0.76403
0.400		0.80684	0.80749	0.81284	0.81745
0.450		0.86185	0.86198	0.86717	0.87149
0.500		0.91853	0.91810	0.92305	0.92698
0.550		0.97778	0.97677	0.98135	0.98480
0.600		1.04069	1.03906	1.04314	1.04596
0.650		1.10864	1.10635	1.10975	1.11178
0.700		1.18353	1.18058	1.18303	1.18404
0.750		1.26823	1.26460	1.26574	1.26542
0.800		1.36741	1.36313	1.36238	1.36024
0.810		1.38962	1.38523	1.38399	1.38141
0.820		1.41282	1.40832	1.40656	1.40350
0.830		1.43713	1.43253	1.43019	1.42662
0.840		1.46268	1.45799	1.45502	1.45089
0.850		1.48963	1.48486	1.48119	1.47645
0.860		1.51817	1.51334	1.50889	1.50348
0.870		1.54854	1.54366	1.53835	1.53220
0.880		1.58102	1.57613	1.56984	1.56288
0.890		1.61598	1.61111	1.60371	1.59584
0.900		1.65388	1.64908	1.64041	1.63152
0.910		1.69534	1.69068	1.68053	1.67047
0.920		1.74118	1.73675	1.72487	1.71345
0.930		1.79257	1.78848	1.77452	1.76152
0.940		1.85117	1.84762	1.83110	1.81621
0.950		1.91960	1.91686	1.89711	1.87988
0.960		2.00218	2.00072	1.97670	1.95647
0.970		2.10698	2.10765	2.07760	2.05329
0.980		2.25202	2.25665	2.21706	2.18656
0.990		2.49417	2.50834	2.44951	2.40733
0.995		2.73073	2.75816	2.67628	2.62104
0.999		3.26535	3.33859	3.18818	3.09746

TABLE 9.1 N= 10

P ↓	λ: 5 55	10 50	15 45	20 40	25 35
	5 5 / 5 5 / 5 5 / 5 5	5 5 / 5 5 / 5 5 / 5 5	5 5 / 5 5 / 5 5 / 5 5	5 5 / 5 5 / 5 5 / 5 5	5 5 / 5 5 / 5 5 / 5 5
0.001	0.09595	0.10191	0.10519	0.10714	0.10818
0.005	0.14142	0.15020	0.15516	0.15812	0.15973
0.010	0.16897	0.17946	0.18546	0.18908	0.19105
0.020	0.20390	0.21652	0.22389	0.22836	0.23080
0.030	0.22897	0.24312	0.25148	0.25658	0.25938
0.040	0.24948	0.26485	0.27403	0.27966	0.28276
0.050	0.26728	0.28370	0.29360	0.29970	0.30306
0.060	0.28326	0.30062	0.31117	0.31769	0.32130
0.070	0.29795	0.31615	0.32730	0.33422	0.33805
0.080	0.31166	0.33063	0.34234	0.34964	0.35368
0.090	0.32461	0.34430	0.35654	0.36419	0.36843
0.100	0.33694	0.35732	0.37006	0.37804	0.38249
0.110	0.34878	0.36980	0.38302	0.39133	0.39596
0.120	0.36021	0.38184	0.39553	0.40415	0.40897
0.130	0.37131	0.39351	0.40765	0.41658	0.42158
0.140	0.38211	0.40487	0.41944	0.42868	0.43385
0.150	0.39268	0.41597	0.43097	0.44050	0.44583
0.160	0.40304	0.42685	0.44226	0.45207	0.45758
0.170	0.41324	0.43753	0.45334	0.46344	0.46912
0.180	0.42328	0.44806	0.46426	0.47464	0.48048
0.190	0.43321	0.45845	0.47504	0.48569	0.49169
0.200	0.44305	0.46872	0.48569	0.49661	0.50277
0.250	0.49131	0.51899	0.53776	0.54998	0.55691
0.300	0.53931	0.56866	0.58909	0.60254	0.61021
0.350	0.58837	0.61906	0.64098	0.65560	0.66399
0.400	0.63968	0.67130	0.69456	0.71027	0.71934
0.450	0.69447	0.72654	0.75089	0.76757	0.77728
0.500	0.75421	0.78604	0.81115	0.82865	0.83891
0.550	0.82072	0.85138	0.87676	0.89483	0.90553
0.600	0.89646	0.92466	0.94958	0.96782	0.97877
0.650	0.98492	1.00880	1.03212	1.04992	1.06081
0.700	1.09118	1.10811	1.12806	1.14444	1.15476
0.750	1.22324	1.22930	1.24314	1.25646	1.26540
0.800	1.39325	1.38370	1.38700	1.39451	1.40062
0.810	1.43363	1.42014	1.42058	1.42643	1.43171
0.820	1.47667	1.45893	1.45619	1.46017	1.46450
0.830	1.52265	1.50035	1.49407	1.49592	1.49917
0.840	1.57196	1.54472	1.53452	1.53396	1.53597
0.850	1.62500	1.59245	1.57789	1.57458	1.57516
0.860	1.68231	1.64401	1.62459	1.61816	1.61709
0.870	1.74452	1.69999	1.67514	1.66513	1.66217
0.880	1.81241	1.76112	1.73018	1.71607	1.71090
0.890	1.88698	1.82829	1.79054	1.77168	1.76392
0.900	1.96952	1.90269	1.85724	1.83288	1.82208
0.910	2.06171	1.98587	1.93168	1.90088	1.88645
0.920	2.16585	2.07991	2.01574	1.97731	1.95851
0.930	2.28519	2.18778	2.11205	2.06449	2.04035
0.940	2.42447	2.31379	2.22451	2.16583	2.13500
0.950	2.59109	2.46467	2.35915	2.28660	2.24720
0.960	2.79749	2.65172	2.52614	2.43574	2.38490
0.970	3.06712	2.89625	2.74462	2.63005	2.56300
0.980	3.45282	3.24630	3.05779	2.90755	2.81514
0.990	4.12461	3.85633	3.60454	3.39070	3.24911
0.995	4.80853	4.47761	4.16224	3.88322	3.68663
0.999	6.42912	5.95029	5.48581	5.05383	4.71510

TABLE 9.2 N= 10 (CONT.)

P ↓	λ:	5 5 5 5 30 30	5 5 5 5 10 45	5 5 5 5 15 40	5 5 5 5 20 35	5 5 5 5 25 30
0.001		0.10852	0.10806	0.11136	0.11321	0.11405
0.005		0.16024	0.15921	0.16418	0.16698	0.16827
0.010		0.19167	0.19018	0.19617	0.19958	0.20115
0.020		0.23158	0.22937	0.23668	0.24087	0.24281
0.030		0.26028	0.25745	0.26572	0.27048	0.27269
0.040		0.28375	0.28037	0.28943	0.29467	0.29710
0.050		0.30414	0.30023	0.30997	0.31563	0.31826
0.060		0.32246	0.31804	0.32839	0.33443	0.33724
0.070		0.33928	0.33438	0.34528	0.35167	0.35465
0.080		0.35498	0.34960	0.36102	0.36773	0.37087
0.090		0.36980	0.36394	0.37585	0.38287	0.38616
0.100		0.38392	0.37759	0.38996	0.39727	0.40071
0.110		0.39746	0.39066	0.40347	0.41107	0.41464
0.120		0.41052	0.40326	0.41649	0.42436	0.42806
0.130		0.42318	0.41546	0.42910	0.43723	0.44106
0.140		0.43551	0.42732	0.44135	0.44974	0.45369
0.150		0.44756	0.43890	0.45331	0.46194	0.46602
0.160		0.45936	0.45023	0.46501	0.47388	0.47807
0.170		0.47095	0.46136	0.47649	0.48560	0.48991
0.180		0.48236	0.47230	0.48778	0.49712	0.50154
0.190		0.49363	0.48309	0.49891	0.50847	0.51301
0.200		0.50476	0.49375	0.50989	0.51968	0.52433
0.250		0.55916	0.54572	0.56339	0.57423	0.57940
0.300		0.61271	0.59673	0.61578	0.62758	0.63325
0.350		0.66673	0.64812	0.66836	0.68106	0.68719
0.400		0.72231	0.70097	0.72222	0.73573	0.74228
0.450		0.78047	0.75633	0.77836	0.79256	0.79949
0.500		0.84230	0.81536	0.83786	0.85260	0.85985
0.550		0.90907	0.87944	0.90198	0.91705	0.92453
0.600		0.98241	0.95038	0.97234	0.98743	0.99500
0.650		1.06445	1.03067	1.05114	1.06576	1.07322
0.700		1.15826	1.12397	1.14157	1.15496	1.16199
0.750		1.26850	1.23603	1.24855	1.25951	1.26556
0.800		1.40288	1.37662	1.38045	1.38687	1.39101
0.810		1.43372	1.40954	1.41099	1.41612	1.41971
0.820		1.46622	1.44450	1.44330	1.44696	1.44992
0.830		1.50056	1.48174	1.47758	1.47958	1.48181
0.840		1.53697	1.52157	1.51409	1.51420	1.51560
0.850		1.57573	1.56433	1.55313	1.55109	1.55154
0.860		1.61715	1.61046	1.59507	1.59057	1.58992
0.870		1.66164	1.66048	1.64037	1.63303	1.63111
0.880		1.70968	1.71505	1.68959	1.67897	1.67557
0.890		1.76190	1.77498	1.74344	1.72901	1.72388
0.900		1.81910	1.84134	1.80284	1.78395	1.77676
0.910		1.88233	1.91552	1.86901	1.84485	1.83520
0.920		1.95300	1.99943	1.94361	1.91314	1.90051
0.930		2.03312	2.09571	2.02898	1.99086	1.97456
0.940		2.12562	2.20828	2.12853	2.08096	2.06004
0.950		2.23501	2.34319	2.24762	2.18807	2.16117
0.960		2.36890	2.51063	2.39525	2.31998	2.28499
0.970		2.54151	2.72978	2.58836	2.49135	2.44475
0.980		2.78479	3.04386	2.86528	2.73533	2.67018
0.990		3.20067	3.59182	3.34922	3.15860	3.05634
0.995		3.61656	4.15031	3.84347	3.58890	3.44354
0.999		4.58222	5.47483	5.01800	4.61005	4.34653

TABLE 9.3 N= 10 (CONT.)

P ↓	λ:	5 5 5 5 5 5 5 15 15 35	5 5 5 5 5 5 5 15 20 30	5 5 5 5 5 5 5 15 25 25	5 5 5 5 5 5 5 20 20 25	5 5 5 5 5 5 10 10 10 40
0.001		0.11456	0.11620	0.11670	0.11749	0.11435
0.005		0.16897	0.17144	0.17221	0.17340	0.16838
0.010		0.20194	0.20494	0.20588	0.20731	0.20104
0.020		0.24370	0.24738	0.24853	0.25027	0.24231
0.030		0.27364	0.27781	0.27912	0.28108	0.27182
0.040		0.29808	0.30266	0.30410	0.30625	0.29589
0.050		0.31926	0.32419	0.32575	0.32805	0.31670
0.060		0.33825	0.34350	0.34516	0.34759	0.33535
0.070		0.35566	0.36120	0.36296	0.36552	0.35243
0.080		0.37187	0.37769	0.37953	0.38221	0.36833
0.090		0.38715	0.39322	0.39515	0.39793	0.38329
0.100		0.40168	0.40799	0.41000	0.41288	0.39751
0.110		0.41559	0.42214	0.42422	0.42720	0.41112
0.120		0.42899	0.43576	0.43791	0.44098	0.42422
0.130		0.44196	0.44894	0.45117	0.45432	0.43689
0.140		0.45456	0.46175	0.46405	0.46728	0.44920
0.150		0.46685	0.47424	0.47660	0.47992	0.46120
0.160		0.47888	0.48646	0.48888	0.49228	0.47293
0.170		0.49067	0.49844	0.50093	0.50439	0.48443
0.180		0.50226	0.51022	0.51276	0.51630	0.49574
0.190		0.51368	0.52182	0.52442	0.52802	0.50687
0.200		0.52495	0.53326	0.53593	0.53959	0.51785
0.250		0.57975	0.58887	0.59181	0.59577	0.57121
0.300		0.63324	0.64309	0.64628	0.65047	0.62326
0.350		0.68676	0.69725	0.70066	0.70504	0.67534
0.400		0.74134	0.75239	0.75599	0.76049	0.72851
0.450		0.79797	0.80945	0.81322	0.81778	0.78375
0.500		0.85764	0.86942	0.87331	0.87784	0.84213
0.550		0.92154	0.93341	0.93737	0.94177	0.90487
0.600		0.99112	1.00282	1.00676	1.01090	0.97357
0.650		1.06835	1.07950	1.08330	1.08698	1.05038
0.700		1.15605	1.16605	1.16954	1.17251	1.13842
0.750		1.25855	1.26648	1.26938	1.27124	1.24261
0.800		1.38310	1.38742	1.38924	1.38939	1.37126
0.810		1.41167	1.41499	1.41651	1.41621	1.40111
0.820		1.44179	1.44398	1.44516	1.44436	1.43271
0.830		1.47363	1.47455	1.47535	1.47401	1.46626
0.840		1.50741	1.50690	1.50726	1.50532	1.50204
0.850		1.54340	1.54126	1.54113	1.53851	1.54034
0.860		1.58191	1.57792	1.57722	1.57385	1.58154
0.870		1.62334	1.61722	1.61586	1.61165	1.62611
0.880		1.66815	1.65959	1.65747	1.65231	1.67460
0.890		1.71698	1.70558	1.70257	1.69634	1.72774
0.900		1.77060	1.75588	1.75183	1.74435	1.78646
0.910		1.83006	1.81142	1.80612	1.79721	1.85200
0.920		1.89679	1.87344	1.86663	1.85605	1.92604
0.930		1.97278	1.94369	1.93504	1.92246	2.01092
0.940		2.06099	2.02477	2.01378	1.99878	2.11012
0.950		2.16602	2.12065	2.10664	2.08862	2.22901
0.960		2.29562	2.23807	2.21997	2.19803	2.37666
0.970		2.46442	2.38967	2.36563	2.33833	2.57010
0.980		2.70557	2.60394	2.57030	2.53490	2.84775
0.990		3.12597	2.97239	2.91904	2.86854	3.33306
0.995		3.55530	3.34396	3.26687	3.20010	3.82840
0.999		4.57775	4.21929	4.07288	3.96553	5.00438

TABLE 9.4 N= 10 (CONT.)

P ↓	λ:	5 5 5 5 5 5 10 10 15 35	5 5 5 5 5 5 10 10 20 30	5 5 5 5 5 5 10 10 25 25	5 5 5 5 5 5 10 15 15 30	5 5 5 5 5 5 10 15 20 25
0.001		0.11762	0.11929	0.11981	0.12070	0.12204
0.005		0.17326	0.17578	0.17656	0.17784	0.17985
0.010		0.20690	0.20994	0.21090	0.21239	0.21481
0.020		0.24942	0.25314	0.25431	0.25605	0.25900
0.030		0.27983	0.28404	0.28537	0.28728	0.29061
0.040		0.30462	0.30924	0.31069	0.31273	0.31636
0.050		0.32607	0.33103	0.33260	0.33474	0.33864
0.060		0.34527	0.35055	0.35222	0.35444	0.35858
0.070		0.36286	0.36843	0.37019	0.37248	0.37683
0.080		0.37922	0.38506	0.38691	0.38925	0.39381
0.090		0.39462	0.40072	0.40265	0.40504	0.40978
0.100		0.40926	0.41559	0.41760	0.42003	0.42495
0.110		0.42325	0.42981	0.43190	0.43437	0.43945
0.120		0.43673	0.44350	0.44566	0.44816	0.45340
0.130		0.44975	0.45674	0.45896	0.46149	0.46688
0.140		0.46240	0.46959	0.47188	0.47442	0.47996
0.150		0.47473	0.48211	0.48446	0.48703	0.49270
0.160		0.48677	0.49434	0.49676	0.49934	0.50515
0.170		0.49858	0.50633	0.50881	0.51140	0.51734
0.180		0.51017	0.51811	0.52064	0.52324	0.52931
0.190		0.52159	0.52970	0.53229	0.53489	0.54108
0.200		0.53285	0.54113	0.54378	0.54638	0.55268
0.250		0.58747	0.59654	0.59946	0.60201	0.60886
0.300		0.64061	0.65039	0.65355	0.65598	0.66329
0.350		0.69360	0.70402	0.70740	0.70963	0.71733
0.400		0.74750	0.75846	0.76203	0.76399	0.77201
0.450		0.80325	0.81465	0.81839	0.81999	0.82824
0.500		0.86185	0.87357	0.87744	0.87858	0.88695
0.550		0.92446	0.93632	0.94026	0.94084	0.94917
0.600		0.99250	1.00425	1.00819	1.00807	1.01618
0.650		1.06790	1.07918	1.08302	1.08203	1.08966
0.700		1.15344	1.16369	1.16725	1.16522	1.17195
0.750		1.25339	1.26171	1.26473	1.26143	1.26664
0.800		1.37492	1.37980	1.38181	1.37698	1.37965
0.810		1.40283	1.40674	1.40846	1.40329	1.40527
0.820		1.43226	1.43508	1.43646	1.43095	1.43215
0.830		1.46339	1.46497	1.46598	1.46012	1.46044
0.840		1.49644	1.49661	1.49720	1.49097	1.49031
0.850		1.53168	1.53025	1.53035	1.52375	1.52197
0.860		1.56941	1.56614	1.56568	1.55871	1.55567
0.870		1.61004	1.60465	1.60354	1.59620	1.59171
0.880		1.65404	1.64620	1.64434	1.63663	1.63048
0.890		1.70202	1.69133	1.68858	1.68052	1.67244
0.900		1.75479	1.74074	1.73694	1.72856	1.71822
0.910		1.81339	1.79534	1.79028	1.78163	1.76862
0.920		1.87924	1.85638	1.84980	1.84095	1.82474
0.930		1.95437	1.92561	1.91715	1.90824	1.88810
0.940		2.04174	2.00562	1.99477	1.98600	1.96096
0.950		2.14597	2.10038	2.08643	2.07815	2.04680
0.960		2.27486	2.21663	2.19845	2.19129	2.15146
0.970		2.44312	2.36700	2.34268	2.33786	2.28588
0.980		2.68406	2.58003	2.54574	2.54600	2.47465
0.990		3.10501	2.94732	2.89257	2.90653	2.79633
0.995		3.53530	3.31855	3.23924	3.27304	3.11759
0.999		4.55975	4.19446	4.04400	4.14441	3.86483

TABLE 9.5 N= 10 (CONT.)

P ↓	λ:	5 5 5 10 20 20 20	5 5 5 15 15 15 15 25	5 5 5 15 15 20 20	5 5 5 10 10 10 35	5 5 5 10 10 15 30
0.001		0.12286	0.12348	0.12431	0.12074	0.12388
0.005		0.18107	0.18194	0.18318	0.17760	0.18226
0.010		0.21628	0.21730	0.21878	0.21191	0.21747
0.020		0.26078	0.26196	0.26375	0.25517	0.26188
0.030		0.29261	0.29389	0.29590	0.28605	0.29356
0.040		0.31854	0.31990	0.32209	0.31117	0.31934
0.050		0.34097	0.34238	0.34473	0.33288	0.34161
0.060		0.36105	0.36250	0.36499	0.35230	0.36151
0.070		0.37943	0.38092	0.38352	0.37006	0.37971
0.080		0.39651	0.39804	0.40075	0.38657	0.39663
0.090		0.41259	0.41414	0.41696	0.40209	0.41252
0.100		0.42786	0.42942	0.43234	0.41682	0.42761
0.110		0.44245	0.44403	0.44704	0.43090	0.44202
0.120		0.45649	0.45808	0.46117	0.44444	0.45587
0.130		0.47005	0.47165	0.47483	0.45752	0.46924
0.140		0.48321	0.48482	0.48807	0.47021	0.48222
0.150		0.49603	0.49764	0.50097	0.48257	0.49484
0.160		0.50855	0.51016	0.51356	0.49463	0.50717
0.170		0.52081	0.52242	0.52588	0.50645	0.51924
0.180		0.53284	0.53444	0.53798	0.51805	0.53108
0.190		0.54468	0.54627	0.54987	0.52946	0.54272
0.200		0.55634	0.55793	0.56159	0.54071	0.55418
0.250		0.61279	0.61431	0.61823	0.59516	0.60962
0.300		0.66745	0.66885	0.67299	0.64797	0.66324
0.350		0.72166	0.72290	0.72721	0.70046	0.71639
0.400		0.77646	0.77749	0.78192	0.75370	0.77010
0.450		0.83275	0.83353	0.83802	0.80862	0.82530
0.500		0.89144	0.89191	0.89639	0.86620	0.88291
0.550		0.95356	0.95367	0.95805	0.92757	0.94399
0.600		1.02035	1.02002	1.02419	0.99412	1.00983
0.650		1.09344	1.09259	1.09640	1.06773	1.08213
0.700		1.17511	1.17367	1.17688	1.15112	1.16334
0.750		1.26884	1.26669	1.26898	1.24847	1.25716
0.800		1.38030	1.37737	1.37818	1.36688	1.36980
0.810		1.40551	1.40241	1.40284	1.39409	1.39545
0.820		1.43195	1.42866	1.42866	1.42280	1.42242
0.830		1.45973	1.45628	1.45579	1.45317	1.45086
0.840		1.48905	1.48542	1.48440	1.48545	1.48096
0.850		1.52009	1.51628	1.51466	1.51987	1.51294
0.860		1.55308	1.54910	1.54680	1.55677	1.54707
0.870		1.58833	1.58418	1.58112	1.59653	1.58368
0.880		1.62620	1.62189	1.61794	1.63963	1.62319
0.890		1.66713	1.66267	1.65772	1.68670	1.66610
0.900		1.71171	1.70712	1.70100	1.73852	1.71311
0.910		1.76070	1.75602	1.74853	1.79616	1.76509
0.920		1.81515	1.81043	1.80130	1.86105	1.82325
0.930		1.87651	1.87182	1.86070	1.93523	1.88930
0.940		1.94690	1.94235	1.92877	2.02167	1.96575
0.950		2.02959	2.02539	2.00865	2.12504	2.05651
0.960		2.13008	2.12657	2.10561	2.25320	2.16818
0.970		2.25862	2.25646	2.22947	2.42097	2.31320
0.980		2.43817	2.43886	2.40223	2.66187	2.51981
0.990		2.74162	2.74994	2.69378	3.08369	2.87911
0.995		3.04172	3.06135	2.98178	3.51521	3.24560
0.999		3.72955	3.79000	3.64175	4.54181	4.11870

TABLE 9.6 N= 10 (CONT.)

P ↓	λ:	5 5 5 5 5 10 10 10 20 25	5 5 5 5 5 10 10 15 15 25	5 5 5 5 5 10 10 15 20 20	5 5 5 5 5 10 15 15 15 20	5 5 5 5 5 15 15 15 15 15
0.001		0.12525	0.12671	0.12756	0.12904	0.13055
0.005		0.18430	0.18643	0.18768	0.18984	0.19202
0.010		0.21993	0.22244	0.22395	0.22649	0.22906
0.020		0.26486	0.26784	0.26965	0.27267	0.27571
0.030		0.29692	0.30023	0.30226	0.30559	0.30894
0.040		0.32300	0.32656	0.32877	0.33235	0.33595
0.050		0.34553	0.34929	0.35165	0.35544	0.35924
0.060		0.36567	0.36961	0.37210	0.37606	0.38004
0.070		0.38409	0.38818	0.39079	0.39491	0.39904
0.080		0.40119	0.40543	0.40815	0.41240	0.41667
0.090		0.41728	0.42163	0.42446	0.42883	0.43322
0.100		0.43253	0.43700	0.43992	0.44441	0.44891
0.110		0.44710	0.45168	0.45469	0.45928	0.46388
0.120		0.46111	0.46578	0.46887	0.47356	0.47825
0.130		0.47463	0.47940	0.48257	0.48734	0.49212
0.140		0.48775	0.49259	0.49584	0.50069	0.50555
0.150		0.50051	0.50543	0.50875	0.51368	0.51861
0.160		0.51297	0.51796	0.52135	0.52635	0.53134
0.170		0.52516	0.53022	0.53368	0.53873	0.54379
0.180		0.53712	0.54224	0.54576	0.55088	0.55599
0.190		0.54888	0.55406	0.55764	0.56281	0.56797
0.200		0.56047	0.56569	0.56933	0.57455	0.57976
0.250		0.61643	0.62186	0.62576	0.63117	0.63656
0.300		0.67051	0.67604	0.68016	0.68566	0.69115
0.350		0.72404	0.72959	0.73388	0.73940	0.74490
0.400		0.77807	0.78355	0.78795	0.79339	0.79882
0.450		0.83350	0.83880	0.84327	0.84854	0.85379
0.500		0.89125	0.89625	0.90072	0.90570	0.91067
0.550		0.95232	0.95689	0.96128	0.96584	0.97040
0.600		1.01798	1.02192	1.02613	1.03010	1.03409
0.650		1.08985	1.09294	1.09681	1.09997	1.10318
0.700		1.17025	1.17216	1.17548	1.17754	1.17969
0.750		1.26265	1.26294	1.26539	1.26593	1.26662
0.800		1.37287	1.37086	1.37190	1.37028	1.36889
0.810		1.39785	1.39527	1.39593	1.39378	1.39187
0.820		1.42407	1.42087	1.42111	1.41837	1.41589
0.830		1.45166	1.44779	1.44756	1.44418	1.44109
0.840		1.48081	1.47619	1.47544	1.47137	1.46760
0.850		1.51170	1.50627	1.50493	1.50010	1.49560
0.860		1.54458	1.53827	1.53627	1.53060	1.52528
0.870		1.57977	1.57248	1.56972	1.56312	1.55689
0.880		1.61762	1.60925	1.60562	1.59798	1.59075
0.890		1.65862	1.64903	1.64440	1.63560	1.62723
0.900		1.70336	1.69241	1.68660	1.67650	1.66683
0.910		1.75264	1.74015	1.73296	1.72135	1.71020
0.920		1.80756	1.79328	1.78445	1.77110	1.75822
0.930		1.86962	1.85327	1.84242	1.82703	1.81212
0.940		1.94105	1.92226	1.90889	1.89105	1.87368
0.950		2.02530	2.00355	1.98694	1.96608	1.94565
0.960		2.12817	2.10273	2.08176	2.05703	2.03265
0.970		2.26053	2.23025	2.20301	2.17305	2.14325
0.980		2.44685	2.40973	2.37238	2.33464	2.29657
0.990		2.76538	2.71690	2.65888	2.60685	2.55303
0.995		3.08454	3.02563	2.94269	2.87543	2.80391
0.999		3.82956	3.75162	3.59557	3.49081	3.37103

TABLE 9.7 N= 10 (CONT.)

P ↓	λ:	5 5 5 5 10 10 10 10 10 30	5 5 5 5 10 10 10 10 15 25	5 5 5 5 10 10 10 10 20 20	5 5 5 5 10 10 10 15 15 20	5 5 5 5 10 10 15 15 15 15
0.001		0.12711	0.13000	0.13086	0.13238	0.13391
0.005		0.18673	0.19097	0.19224	0.19443	0.19664
0.010		0.22260	0.22764	0.22916	0.23174	0.23433
0.020		0.26773	0.27376	0.27559	0.27863	0.28169
0.030		0.29987	0.30658	0.30863	0.31199	0.31536
0.040		0.32597	0.33323	0.33545	0.33905	0.34267
0.050		0.34848	0.35620	0.35858	0.36238	0.36619
0.060		0.36858	0.37671	0.37921	0.38318	0.38717
0.070		0.38695	0.39543	0.39805	0.40218	0.40631
0.080		0.40399	0.41280	0.41553	0.41979	0.42406
0.090		0.41999	0.42911	0.43194	0.43632	0.44070
0.100		0.43516	0.44456	0.44748	0.45197	0.45646
0.110		0.44964	0.45931	0.46231	0.46690	0.47149
0.120		0.46355	0.47346	0.47655	0.48123	0.48591
0.130		0.47697	0.48711	0.49028	0.49504	0.49981
0.140		0.48998	0.50034	0.50358	0.50842	0.51326
0.150		0.50263	0.51320	0.51651	0.52142	0.52634
0.160		0.51497	0.52573	0.52912	0.53409	0.53907
0.170		0.52704	0.53799	0.54144	0.54648	0.55152
0.180		0.53888	0.55001	0.55352	0.55861	0.56371
0.190		0.55051	0.56181	0.56538	0.57053	0.57567
0.200		0.56196	0.57342	0.57705	0.58225	0.58744
0.250		0.61721	0.62939	0.63327	0.63865	0.64402
0.300		0.67050	0.68322	0.68732	0.69280	0.69827
0.350		0.72317	0.73630	0.74056	0.74606	0.75155
0.400		0.77626	0.78965	0.79403	0.79947	0.80488
0.450		0.83069	0.84415	0.84861	0.85388	0.85914
0.500		0.88736	0.90070	0.90517	0.91018	0.91518
0.550		0.94731	0.96026	0.96467	0.96928	0.97390
0.600		1.01180	1.02402	1.02826	1.03232	1.03640
0.650		1.08248	1.09352	1.09745	1.10074	1.10408
0.700		1.16172	1.17090	1.17432	1.17656	1.17888
0.750		1.25315	1.25945	1.26205	1.26282	1.26373
0.800		1.36282	1.36458	1.36583	1.36450	1.36338
0.810		1.38780	1.38834	1.38924	1.38737	1.38575
0.820		1.41405	1.41326	1.41375	1.41131	1.40913
0.830		1.44174	1.43946	1.43950	1.43643	1.43364
0.840		1.47105	1.46710	1.46663	1.46287	1.45942
0.850		1.50219	1.49638	1.49534	1.49082	1.48664
0.860		1.53544	1.52753	1.52583	1.52048	1.51548
0.870		1.57112	1.56082	1.55838	1.55209	1.54619
0.880		1.60964	1.59660	1.59331	1.58598	1.57907
0.890		1.65151	1.63533	1.63104	1.62254	1.61448
0.900		1.69739	1.67756	1.67211	1.66228	1.65292
0.910		1.74818	1.72406	1.71723	1.70586	1.69500
0.920		1.80508	1.77583	1.76734	1.75420	1.74158
0.930		1.86978	1.83432	1.82380	1.80854	1.79384
0.940		1.94478	1.90162	1.88854	1.87075	1.85351
0.950		2.03400	1.98101	1.96462	1.94367	1.92327
0.960		2.14405	2.07799	2.05710	2.03209	2.00757
0.970		2.28740	2.20292	2.17550	2.14495	2.11471
0.980		2.49242	2.37921	2.34118	2.30227	2.26323
0.990		2.85066	2.68220	2.62219	2.56778	2.51169
0.995		3.21747	2.98821	2.90151	2.83044	2.75487
0.999		4.09292	3.71209	3.54717	3.43497	3.30526

TABLE 9.8 N= 10 (CONT.)

P ↓	λ:	5 5 5 10 10 10 10 10 10 25	5 5 5 10 10 10 10 10 15 20	5 5 5 10 10 10 10 15 15 15	5 5 10 10 10 10 10 10 10 20	5 5 10 10 10 10 10 10 15 15
0.001		0.13334	0.13576	0.13732	0.13920	0.14079
0.005		0.19556	0.19907	0.20130	0.20375	0.20602
0.010		0.23288	0.23702	0.23964	0.24235	0.24500
0.020		0.27971	0.28462	0.28770	0.29064	0.29375
0.030		0.31297	0.31841	0.32180	0.32484	0.32825
0.040		0.33992	0.34577	0.34940	0.35250	0.35614
0.050		0.36313	0.36932	0.37315	0.37627	0.38011
0.060		0.38381	0.39031	0.39430	0.39743	0.40143
0.070		0.40268	0.40944	0.41358	0.41670	0.42084
0.080		0.42017	0.42717	0.43143	0.43453	0.43880
0.090		0.43658	0.44379	0.44817	0.45124	0.45562
0.100		0.45211	0.45951	0.46400	0.46704	0.47152
0.110		0.46691	0.47450	0.47909	0.48208	0.48666
0.120		0.48112	0.48887	0.49355	0.49650	0.50117
0.130		0.49481	0.50272	0.50748	0.51038	0.51513
0.140		0.50806	0.51613	0.52096	0.52380	0.52862
0.150		0.52093	0.52914	0.53404	0.53683	0.54172
0.160		0.53348	0.54182	0.54678	0.54951	0.55446
0.170		0.54574	0.55420	0.55922	0.56189	0.56690
0.180		0.55775	0.56633	0.57140	0.57401	0.57907
0.190		0.56953	0.57822	0.58335	0.58589	0.59100
0.200		0.58113	0.58992	0.59509	0.59756	0.60272
0.250		0.63690	0.64612	0.65147	0.65357	0.65890
0.300		0.69041	0.69994	0.70538	0.70707	0.71250
0.350		0.74303	0.75275	0.75822	0.75945	0.76490
0.400		0.79579	0.80558	0.81099	0.81174	0.81714
0.450		0.84958	0.85930	0.86457	0.86479	0.87006
0.500		0.90527	0.91476	0.91979	0.91944	0.92449
0.550		0.96379	0.97287	0.97754	0.97659	0.98131
0.600		1.02631	1.03472	1.03888	1.03730	1.04153
0.650		1.09432	1.10172	1.10518	1.10292	1.10649
0.700		1.16990	1.17583	1.17832	1.17534	1.17799
0.750		1.25623	1.25997	1.26110	1.25740	1.25875
0.800		1.35855	1.35897	1.35814	1.35373	1.35318
0.810		1.38166	1.38122	1.37989	1.37535	1.37432
0.820		1.40588	1.40449	1.40262	1.39795	1.39639
0.830		1.43135	1.42890	1.42644	1.42165	1.41951
0.840		1.45821	1.45460	1.45147	1.44658	1.44380
0.850		1.48666	1.48174	1.47789	1.47290	1.46941
0.860		1.51691	1.51053	1.50588	1.50081	1.49653
0.870		1.54924	1.54121	1.53567	1.53053	1.52538
0.880		1.58400	1.57409	1.56754	1.56237	1.55623
0.890		1.62162	1.60955	1.60186	1.59670	1.58942
0.900		1.66264	1.64808	1.63909	1.63398	1.62539
0.910		1.70781	1.69033	1.67982	1.67483	1.66474
0.920		1.75813	1.73717	1.72489	1.72011	1.70823
0.930		1.81499	1.78983	1.77543	1.77098	1.75696
0.940		1.88047	1.85010	1.83310	1.82918	1.81253
0.950		1.95777	1.92075	1.90049	1.89738	1.87739
0.960		2.05234	2.00642	1.98188	1.98007	1.95565
0.970		2.17439	2.11581	2.08526	2.08565	2.05495
0.980		2.34713	2.26842	2.22848	2.23299	2.19233
0.990		2.64554	2.52644	2.46799	2.48250	2.42171
0.995		2.94878	2.78244	2.70237	2.73083	2.64588
0.999		3.67129	3.37501	3.23325	3.30983	3.15323

TABLE 9.9 N= 10 (CONT.)

P ↓	λ:	5 10 10 10 10 10 10 10 10 15	All λ's = 100/10
0.001		0.14430	0.14787
0.005		0.21078	0.21559
0.010		0.25039	0.25582
0.020		0.29981	0.30591
0.030		0.33472	0.34121
0.040		0.36290	0.36965
0.050		0.38707	0.39403
0.060		0.40855	0.41567
0.070		0.42809	0.43534
0.080		0.44616	0.45350
0.090		0.46306	0.47049
0.100		0.47903	0.48652
0.110		0.49422	0.50176
0.120		0.50876	0.51634
0.130		0.52275	0.53036
0.140		0.53627	0.54389
0.150		0.54937	0.55701
0.160		0.56212	0.56976
0.170		0.57456	0.58219
0.180		0.58671	0.59434
0.190		0.59863	0.60623
0.200		0.61032	0.61791
0.250		0.66632	0.67372
0.300		0.71961	0.72672
0.350		0.77161	0.77832
0.400		0.82333	0.82955
0.450		0.87562	0.88124
0.500		0.92929	0.93418
0.550		0.98520	0.98922
0.600		1.04435	1.04732
0.650		1.10800	1.10971
0.700		1.17791	1.17807
0.750		1.25668	1.25489
0.800		1.34853	1.34420
0.810		1.36905	1.36411
0.820		1.39047	1.38488
0.830		1.41289	1.40661
0.840		1.43643	1.42940
0.850		1.46124	1.45339
0.860		1.48748	1.47876
0.870		1.51537	1.50569
0.880		1.54517	1.53444
0.890		1.57721	1.56532
0.900		1.61191	1.59872
0.910		1.64982	1.63516
0.920		1.69168	1.67535
0.930		1.73853	1.72026
0.940		1.79189	1.77131
0.950		1.85408	1.83070
0.960		1.92900	1.90207
0.970		2.02387	1.99219
0.980		2.15481	2.11608
0.990		2.37269	2.32093
0.995		2.58484	2.51882
0.999		3.06280	2.95883

Selected Tables in Mathematical Statistics
Volume 10, 1986

CONFIDENCE LIMITS ON THE CORRELATION COEFFICIENT

ROBERT E. ODEH

Department of Mathematics

University of Victoria

Victoria, B.C., Canada

Dedicated to my family.

ABSTRACT

Tables are given for finding upper, lower, or two-sided confidence limits
and intervals on the population correlation coefficient for a bivariate normal
distribution. A review of existing and/or related tables is given. Examples
are given of the use of the tables. Interpolation in the tables is discussed.
An extensive section is given on the distribution theory and numerical methods
used to compute the tables.

Received by the Editors October,1983 and in revised form September, 1985.
MOS 1980 subject classification 62Q05 primary, 62E15, 62G15, 62J99 secondary.
Research for this project was sponsored by a grant from Natural Sciences and
Engineering Research Council, Canada.

CONTENTS

1. INTRODUCTION

1.1 NOTATION AND DESCRIPTION OF TABLES

Let (X,Y) have a joint bivariate normal distribution with means μ_x, μ_y, standard deviations σ_x, σ_y, respectively, and correlation ρ.

If $(X_1,Y_1),\ldots,(X_n,Y_n)$ denotes a random sample of size n from (X,Y), the sample correlation coefficient is defined by

$$R = \sum_1^n [(X_i-\overline{X})(Y_i-\overline{Y})]/[\sum_1^n(X_i-\overline{X})^2 \sum_1^n(Y_j-\overline{Y})^2]^{1/2}$$

where

$$\overline{X}= (\sum_1^n X_i)/n , \qquad \overline{Y}= (\sum_1^n Y_j)/n .$$

For given n, the distribution of R is independent of μ_x, μ_y, σ_x, σ_y, but depends upon ρ, $-1<\rho<1$.

For $-1\leq r\leq 1$ define $f_n(r;\rho)$ to be the probability density function for R. We denote the cumulative distribution function of R by

$$F_n(r;\rho) = Pr[R\leq r] = \int_{-1}^r f_n(x;\rho) \, dx .$$

We also denote $Pr[R>r]$ by

$$P_n(r;\rho) = 1 - F_n(r;\rho) .$$

For a given value of $\alpha_1<1/2$, n, and an observed value of R, say r_0 , a lower $100(1-\alpha_1)\%$ confidence limit on ρ is the unique value of ρ, say ρ_L , which satisfies $P_n(r_0;\rho_L) = \alpha_1$ (A one-sided lower $100(1-\alpha_1)\%$ confidence interval on ρ is given by $\rho_L\leq \rho \leq 1.0$). Alternatively, ρ_L satisfies $F_n(r_0;\rho_L) = 1-\alpha_1$.

For $\alpha_2>1/2$, an upper $100\alpha_2\%$ confidence limit on ρ is the unique value of ρ, say ρ_U , which satisfies $P_n(r_0;\rho_U) = \alpha_2$ (A one-sided upper $100\alpha_2\%$ confidence interval on ρ is given by $-1.0\leq \rho \leq \rho_U$).

For $\alpha_1 = 1-\alpha_2 = \alpha_0$, an equal-tailed two-sided $100(1-2\alpha_0)\%$ confidence interval on ρ is given by $\rho_L \leq \rho \leq \rho_U$.

In this paper we give values of $\rho_\alpha = \rho(\alpha, n, r_0)$ to 5 decimal places which satisfy $P_n(r_0; \rho_\alpha) = \alpha$. Tables are given for values of

$$r_0 = -0.98(0.02)0.98$$

$$n = 3(1)60(2)80(5)100(10)200(25)300(50)600(100)1000$$

and for $1-\alpha$ and

$$\alpha = 0.005, 0.01, 0.025, 0.05, 0.10, 0.25 \quad .$$

A review of existing and/or related tables is given in Section 1.2. Examples of use of the tables and interpolation are discussed in Section 2. An extensive discussion of the computational formulas used to compute the tables is given in Section 3.

1.2 REVIEW OF EXISTING TABLES

Tables of the cumulative distribution function and ordinates of the density function for the sample correlation coefficient were first given by F.N. David (1938,1954). There are, however, many errors in the body of the tables due to the difficulty of computing these tables by hand.

Tables of $F_n(r; \rho)$ are given for

$$n = 3(1)25, 50, 100, 200, 400$$

$$\rho = 0.0(0.1)0.9$$

and for values of r which depend upon n.

Charts for finding confidence intervals are also given for values of

$$n = 3(1)8, 10, 12, 15, 20, 25, 50, 100, 200, 400$$

$$\alpha = 0.005, 0.01, 0.025, 0.05 \quad \text{and} \quad 1-\alpha \quad .$$

David's tables are superseded by the heroic effort of Oksoy and Aroian (1982a), who give accurate 6 decimal place tables of $F_n(r;\rho)$, for values of

$$n = 3(1)10(2)20(2)30(5)40(10)100(100)500$$

$$\rho = 0.00(0.05)0.90, \ 0.92, \ 0.94, \ 0.95, \ 0.96, \ 0.98$$

and 41 selected values of r depending upon the value of ρ and n. Additional tables are given for $n=3$ and for all $n\rho \geq 0.65$.

In a related publication Öksoy and Aroian (1982b) give 6 decimal place tables of the percentage points of the distribution of the correlation coefficient. For the values of n and ρ listed above, values of r_γ are given such that $F_n(r_\gamma;\rho) = \gamma$, for 39 values of γ.

More accessible, but less extensive tables of percentage points are given by Odeh(1982). Five decimal place tables are given for

$$n = 4(1)30(2)40(5)50(10)100(20)200(100)1000$$

$$\rho = 0.0(0.10)0.90, \ 0.95$$

$$\gamma = 0.75, \ 0.90, \ 0.95, \ 0.975, \ 0.99, \ 0.995 \text{ and } 1-\gamma \ .$$

Some tables of the cumulative distribution function and percentage points are also given by Subrahmaniam and Subrahmaniam (1980). These tables are intended as a supplement to the tables of F.N. David. Five decimal place tables of $F_n(r;\rho)$ are given for $n = 26(1)49$ for values of r which depend upon n. Tables of r_γ are given for $n = 3(1)50$,

$$\gamma = 0.80, \ 0.90, \ 0.95, \ 0.975, \ 0.98, \ 0.99 \text{ and } 1-\gamma \ .$$

The only available tables of confidence limits are given in Table 10 of Odeh and Owen (1980). Four decimal place tables are given for values of

$$n = 3(1)30, \ 40, \ 60, \ 100, \ 120, \ 150, \ 200, \ 400$$

$$r_0 = -0.95(0.05)0.95$$

$$\alpha = 0.005, \ 0.01, \ 0.025, \ 0.05, \ 0.10, \ 0.25 \text{ and } 1-\alpha \ .$$

These tables were computed especially for use in screening problems.

2. USE OF THE TABLES

The tables may be used to find one-sided lower or upper confidence intervals, or two-sided confidence intervals for ρ. For given n, r_0, and α, the entry in the table is the value of ρ (given to 5 decimal places) which satisfies $P_n(r_0;\rho) = \alpha$.

For each value of n, there are two pages of tables. The tables make use of the relationship $P_n(r_0;-\rho)=1-P_n(-r_0;\rho)$.

To find a <u>lower</u> $100(1-\alpha)\%$ confidence limit, we enter the tables with the value of r_0 chosen from the left-hand column of the page labelled $r_0\downarrow$ at the top, and the value of α picked from the row at the top. The notation $\downarrow\downarrow$ means that we do not change the sign of the entry in the table. For a given value of n, we use the left-hand page of a table (TABLE *.1) if $r_0\leq 0.0$, and the right-hand page (TABLE *.2) if $r_0>0.0$.

To find an <u>upper</u> $100\alpha\%$ confidence limit, we enter the tables with the value of r_0 chosen from the right-hand column labelled $\uparrow r_0$ at the bottom of the page, and the value of α picked from the row at the bottom. The notation $\uparrow-$ means that we change the sign of the entry. For a given value of n, we use the right-hand page of a table (TABLE *.2) if $r_0\leq 0.0$, and the left-hand page (TABLE *.1) if $r_0>0.0$.

Example 1

As an example, suppose a sample of size 10 yields a value of $r_0=0.60$. Tables 8.1 and 8.2 give confidence limits for samples of size 10. To find a lower one-sided interval for ρ, we enter Table 8.2 (since $r_0>0.0$) under the column on the left labelled $r_0\downarrow$, until we find the row corresponding to 0.60, and read 0.07108 in the column labelled $\alpha=.05\downarrow\downarrow$ at the top of the page. Thus a lower $100(1-\alpha)\%=95\%$ confidence limit on ρ is 0.07108 . (A one-sided lower 95% confidence interval on ρ is $0.07108<\rho<1.0$).

To find an upper one-sided interval for ρ, we enter Table 8.1 under the column at the right labelled $\uparrow r_0$ (at the bottom of the page) until we find the row corresponding to $r_0=0.60$, and read -0.84520 in the column labelled $\alpha=.95\uparrow-$ at the bottom of the page. The notation $\uparrow-$ means we must change the sign of the entry from $-$ to $+$. Thus an upper $100\alpha\%=95\%$ confidence limit on ρ is 0.84520. (A one-sided upper 95% confidence interval on ρ is $-1<\rho<0.84520$).

Combining the above results, a two-sided 90% confidence interval on ρ is $0.07108<\rho<0.84520$. Similarly, we find a two-sided 95% confidence interval on ρ is $-0.04823<\rho<0.87589$.

Example 2

As a second example, a sample of size $n=25$ yields a value of $r_0=-.52$. Since r_0 is negative, we enter Table 23.1 under the column $r_0\downarrow$ on the left and read -0.80408 under the column labelled $\alpha=.005\downarrow+$ at the top of the page. We enter Table 23.2 under the column at the right labelled $\uparrow r_0$ at the bottom, and read 0.01967 in the column labelled $\alpha=.995\uparrow-$ at the bottom. We change the sign of this entry from $+$ to $-$.

A lower one-sided 99.5% confidence interval on ρ is $-0.80408<\rho<1.0$. An upper one-sided 99.5% confidence interval on ρ is $-1.0<\rho<-0.01967$. A two-sided 99% confidence interval on ρ is $-0.80408<\rho<-0.01967$.

Example 3 Interpolation in r_0

The tables give values of ρ_L (or $-\rho_U$) for $r_0=-0.98(0.02)0.98$. For a value of r_0 not in the table, linear interpolation through two surrounding tabled values generally overestimates the true value of ρ_L (underestimates the true value of ρ_U).

Figure 2.1 below gives a comparison of the true value of ρ_L with the value obtained by linear interpolation for samples of size 10, 25, 50, and 100. The last column gives ERROR = $10^5 \cdot \max|$true$-$linear$|$. We note the following properties of linear interpolation:

(a) The maximum error always occurs when $\alpha = 0.005$. The error decreases as α increases.

(b) The error increases as r_0 increases.

(c) The error decreases as n increases.

(d) For $25 \leq n \leq 50$, at least 3 decimals places are correct. For $n > 50$ at least 4 decimal places are correct.

We also note that interpolation through 4 points is normally accurate to all 5 decimal places but may be off by ± 0.00001 (limited by the 5 decimal place accuracy of the tables and round-off error).

FIGURE 2.1

Comparison of True and Interpolated Value in r_0 of ρ_L

$\alpha =$	0.005	0.01	0.025	0.05	0.10	0.25	ERROR
		$r_0 = -0.95$		$n = 10$			
TRUE	-0.99145	-0.98954	-0.98609	-0.98237	-0.97702	-0.96469	
LINEAR	-0.99144	-0.98953	-0.98608	-0.98236	-0.97701	-0.96469	1
		$r_0 = -0.65$					
TRUE	-0.93304	-0.91869	-0.89322	-0.86649	-0.82925	-0.74867	
LINEAR	-0.93302	-0.91868	-0.89321	-0.86648	-0.82923	-0.74866	2
		$r_0 = -0.25$					
TRUE	-0.82760	-0.79320	-0.73412	-0.67472	-0.59606	-0.44045	
LINEAR	-0.82758	-0.79318	-0.73410	-0.67470	-0.59604	-0.44044	2
		$r_0 = 0.25$					
TRUE	-0.60451	-0.53685	-0.42786	-0.32657	-0.20369	0.00770	
LINEAR	-0.60445	-0.53679	-0.42780	-0.32652	-0.20365	0.00772	6
		$r_0 = 0.65$					
TRUE	-0.21358	-0.11478	0.02888	0.14761	0.27554	0.46204	
LINEAR	-0.21335	-0.11457	0.02906	0.14776	0.27564	0.46209	23
		$r_0 = 0.95$					
TRUE	0.65844	0.71441	0.78031	0.82439	0.86399	0.91072	
LINEAR	0.66034	0.71575	0.78111	0.82488	0.86427	0.91081	190
		$r_0 = -0.95$		$n = 25$			
TRUE	-0.98228	-0.98023	-0.97686	-0.97355	-0.96921	-0.96045	
LINEAR	-0.98226	-0.98022	-0.97684	-0.97354	-0.96920	-0.96044	2
		$r_0 = -0.65$					
TRUE	-0.86347	-0.84882	-0.82506	-0.80230	-0.77316	-0.71686	
LINEAR	-0.86345	-0.84881	-0.82504	-0.80228	-0.77315	-0.71685	2

$\alpha =$	0.005	0.01	0.025	0.05	0.10	0.25	ERROR
		$r_0 = -0.25$		$n = 25$ (continued)			
TRUE	-0.66071	-0.62872	-0.57836	-0.53186	-0.47465	-0.37090	
LINEAR	-0.66068	-0.62869	-0.57834	-0.53183	-0.47463	-0.37089	3
		$r_0 = 0.25$					
TRUE	-0.28586	-0.23481	-0.15862	-0.09243	-0.01600	0.11015	
LINEAR	-0.28580	-0.23476	-0.15857	-0.09239	-0.01597	0.11016	6
		$r_0 = 0.65$					
TRUE	0.21141	0.26334	0.33583	0.39439	0.45742	0.55197	
LINEAR	0.21153	0.26344	0.33592	0.39446	0.45746	0.55199	12
		$r_0 = 0.95$					
TRUE	0.85106	0.86569	0.88451	0.89850	0.91247	0.93157	
LINEAR	0.85136	0.86593	0.88467	0.89861	0.91254	0.93161	30
		$r_0 = -0.95$		$n = 50$			
TRUE	-0.97561	-0.97376	-0.97081	-0.96804	-0.96453	-0.95785	
LINEAR	-0.97560	-0.97375	-0.97080	-0.96802	-0.96452	-0.95784	1
		$r_0 = -0.65$					
TRUE	-0.81552	-0.80275	-0.78272	-0.76422	-0.74134	-0.69916	
LINEAR	-0.81550	-0.80273	-0.78270	-0.76420	-0.74132	-0.69916	2
		$r_0 = -0.25$					
TRUE	-0.55598	-0.52993	-0.49009	-0.45435	-0.41150	-0.33618	
LINEAR	-0.55595	-0.52991	-0.49007	-0.45433	-0.41148	-0.33617	3
		$r_0 = 0.25$					
TRUE	-0.12077	-0.08422	-0.03054	0.01547	0.06812	0.15462	
LINEAR	-0.12073	-0.08418	-0.03050	0.01549	0.06814	0.15462	4
		$r_0 = 0.65$					
TRUE	0.37442	0.40581	0.44990	0.48592	0.52526	0.58590	
LINEAR	0.37449	0.40586	0.44995	0.48595	0.52528	0.58591	7
		$r_0 = 0.95$					
TRUE	0.89482	0.90199	0.91161	0.91911	0.92696	0.93840	
LINEAR	0.89494	0.90209	0.91168	0.91917	0.92700	0.93841	12
		$r_0 = -0.95$		$n = 100$			
TRUE	-0.96977	-0.96821	-0.96578	-0.96356	-0.96082	-0.95581	
LINEAR	-0.96975	-0.96819	-0.96577	-0.96354	-0.96081	-0.95581	2
		$r_0 = -0.65$					
TRUE	-0.77520	-0.76480	-0.74884	-0.73442	-0.71699	-0.68583	
LINEAR	-0.77518	-0.76479	-0.74882	-0.73441	-0.71698	-0.68582	2
		$r_0 = -0.25$					
TRUE	-0.47396	-0.45393	-0.42378	-0.39716	-0.36571	-0.31144	
LINEAR	-0.47394	-0.45392	-0.42376	-0.39715	-0.36570	-0.31143	2
		$r_0 = 0.25$					
TRUE	-0.00689	0.01863	0.05593	0.08778	0.12417	0.18402	
LINEAR	-0.00686	0.01866	0.05595	0.08779	0.12418	0.18403	3
		$r_0 = 0.65$					
TRUE	0.47048	0.49014	0.51802	0.54108	0.56662	0.60685	
LINEAR	0.47052	0.49017	0.51805	0.54110	0.56664	0.60686	4
		$r_0 = 0.95$					
TRUE	0.91626	0.92028	0.92582	0.93027	0.93507	0.94237	
LINEAR	0.91632	0.92032	0.92585	0.93030	0.93509	0.94237	6

Example 4   Interpolation in n

Since the tables given here cover all values of n from 3 to 60,

interpolation in n is necessary only if n>60 and only if no table exists.

For a value of n for which no table exists, linear interpolation through

two surrounding tabled values generally underestimates the true value of

ρ_L (overestimates the true value of ρ_U).

Figure 2.2 below gives a comparison of the true value of ρ_L with the

value obtained by linear interpolation for samples of size 63, 125, 235, and

635. The last column gives ERROR = $10^5 \cdot \max|\text{true-linear}|$. We denote by S the

set of values of n used for interpolation. We note that linear interpolation

gives at least 3 decimal places of accuracy and may be more accurate depending

upon the value of n and r_0 .

We also note that interpolation through 4 points is normally accurate

to all 5 decimal places but may be off by ±0.00001 (limited by the 5

decimal place accuracy of the tables and round-off error).

Figure 2.2

Comparison of True and Interpolated Value in n of ρ_L

α =	0.005	0.01	0.025	0.05	0.10	0.25	ERROR
		$r_0 = -0.94$		n = 63	S = {62,64}		
TRUE	-0.96820	-0.96609	-0.96276	-0.95965	-0.95578	-0.94851	
LINEAR	-0.96820	-0.96609	-0.96276	-0.95965	-0.95577	-0.94851	1
		$r_0 = -0.64$					
TRUE	-0.79497	-0.78265	-0.76349	-0.74595	-0.72446	-0.68535	
LINEAR	-0.79498	-0.78266	-0.76350	-0.74596	-0.72446	-0.68535	1
		$r_0 = -0.24$					
TRUE	-0.51850	-0.49427	-0.45746	-0.42466	-0.38561	-0.31753	
LINEAR	-0.51852	-0.49429	-0.45747	-0.42468	-0.38562	-0.31753	2
		$r_0 = 0.24$					
TRUE	-0.08843	-0.05601	-0.00846	0.03225	0.07883	0.15546	
LINEAR	-0.08846	-0.05604	-0.00849	0.03222	0.07882	0.15545	3
		$r_0 = 0.64$					
TRUE	0.39770	0.42483	0.46311	0.49454	0.52908	0.58282	
LINEAR	0.39767	0.42480	0.46308	0.49451	0.52907	0.58281	3
		$r_0 = 0.94$					
TRUE	0.88480	0.89171	0.90110	0.90852	0.91637	0.92802	
LINEAR	0.88479	0.89170	0.90109	0.90851	0.91636	0.92802	1

$\alpha =$	0.005	0.01	0.025	0.05	0.10	0.25	ERROR

$r_0 = -0.94$ $n = 125$ $S = \{120,130\}$

	0.005	0.01	0.025	0.05	0.10	0.25	ERROR
TRUE	-0.96165	-0.95990	-0.95720	-0.95474	-0.95173	-0.94629	
LINEAR	-0.96166	-0.95991	-0.95720	-0.95474	-0.95174	-0.94629	1

$r_0 = -0.64$

| TRUE | -0.75675 | -0.74687 | -0.73177 | -0.71822 | -0.70194 | -0.67307 | |
| LINEAR | -0.75680 | -0.74691 | -0.73181 | -0.71826 | -0.70197 | -0.67308 | 5 |

$r_0 = -0.24$

| TRUE | -0.44355 | -0.42512 | -0.39747 | -0.37317 | -0.34456 | -0.29544 | |
| LINEAR | -0.44366 | -0.42522 | -0.39756 | -0.37324 | -0.34462 | -0.29547 | 11 |

$r_0 = 0.24$

| TRUE | 0.01095 | 0.03367 | 0.06686 | 0.09521 | 0.12764 | 0.18106 | |
| LINEAR | 0.01080 | 0.03353 | 0.06674 | 0.09512 | 0.12756 | 0.18102 | 15 |

$r_0 = 0.64$

| TRUE | 0.47966 | 0.49698 | 0.52164 | 0.54213 | 0.56492 | 0.60106 | |
| LINEAR | 0.47953 | 0.49687 | 0.52155 | 0.54206 | 0.56486 | 0.60103 | 13 |

$r_0 = 0.94$

| TRUE | 0.90531 | 0.90934 | 0.91494 | 0.91948 | 0.92441 | 0.93200 | |
| LINEAR | 0.90528 | 0.90931 | 0.91492 | 0.91946 | 0.92440 | 0.93199 | 3 |

$r_0 = -0.94$ $n = 235$ $S = \{225,250\}$

| TRUE | -0.95668 | -0.95526 | -0.95310 | -0.95117 | -0.94884 | -0.94472 | |
| LINEAR | -0.95669 | -0.95527 | -0.95311 | -0.95117 | -0.94885 | -0.94472 | 4 |

$r_0 = -0.64$

| TRUE | -0.72865 | -0.72084 | -0.70906 | -0.69862 | -0.68621 | -0.66458 | |
| LINEAR | -0.72872 | -0.72091 | -0.70912 | -0.69867 | -0.68625 | -0.66460 | 7 |

$r_0 = -0.24$

| TRUE | -0.39122 | -0.37721 | -0.35636 | -0.33816 | -0.31689 | -0.28067 | |
| LINEAR | -0.39136 | -0.37733 | -0.35646 | -0.33825 | -0.31696 | -0.28071 | 14 |

$r_0 = 0.24$

| TRUE | 0.07514 | 0.09146 | 0.11530 | 0.13567 | 0.15898 | 0.19748 | |
| LINEAR | 0.07496 | 0.09130 | 0.11516 | 0.13556 | 0.15889 | 0.19744 | 18 |

$r_0 = 0.64$

| TRUE | 0.52828 | 0.54003 | 0.55688 | 0.57100 | 0.58685 | 0.61233 | |
| LINEAR | 0.52814 | 0.53991 | 0.55679 | 0.57092 | 0.58679 | 0.61230 | 14 |

$r_0 = 0.94$

| TRUE | 0.91653 | 0.91912 | 0.92278 | 0.92580 | 0.92914 | 0.93438 | |
| LINEAR | 0.91650 | 0.91910 | 0.92276 | 0.92579 | 0.92912 | 0.93438 | 3 |

$r_0 = -0.94$ $n = 635$ $S = \{600,700\}$

| TRUE | -0.95077 | -0.94981 | -0.94836 | -0.94709 | -0.94558 | -0.94296 | |
| LINEAR | -0.95079 | -0.94983 | -0.94837 | -0.94709 | -0.94559 | -0.94297 | 2 |

$r_0 = -0.64$

| TRUE | -0.69632 | -0.69116 | -0.68347 | -0.67673 | -0.66882 | -0.65527 | |
| LINEAR | -0.69642 | -0.69126 | -0.68355 | -0.67680 | -0.66887 | -0.65529 | 10 |

$r_0 = -0.24$

| TRUE | -0.33373 | -0.32487 | -0.31176 | -0.30040 | -0.28720 | -0.26491 | |
| LINEAR | -0.33391 | -0.32503 | -0.31190 | -0.30052 | -0.28730 | -0.26496 | 19 |

$r_0 = 0.24$

| TRUE | 0.14120 | 0.15093 | 0.16515 | 0.17731 | 0.19127 | 0.21441 | |
| LINEAR | 0.14099 | 0.15074 | 0.16499 | 0.17718 | 0.19117 | 0.21435 | 21 |

α =	0.005	0.01	0.025	0.05	0.10	0.25	ERROR
		r_0 = 0.64		n = 635 (continued)			
TRUE	0.57517	0.58179	0.59135	0.59945	0.60862	0.62359	
LINEAR	0.57502	0.58166	0.59125	0.59936	0.60855	0.62355	15
		r_0 = 0.94					
TRUE	0.92675	0.92814	0.93013	0.93180	0.93368	0.93671	
LINEAR	0.92672	0.92811	0.93011	0.93178	0.93367	0.93670	3

Example 5   An Approximation

Based on Fisher's z-transformation, Winterbottom (1980) develops the following asymptotic approximation for confidence limits on ρ .

If we let $x=x(\alpha)$ be the lower 100α percentage point for the standard normal distribution, $m=n-1$,

$$z = 1/2 \ln\{(1+r)/(1-r)\}$$

and

$$y = z + x/m^{1/2} - r/(2m) + \{x^3+3(1+r^2)x\}/(12m^{3/2})$$

$$- (4r^3x^2+5r^3+9r)/(24m^2)$$

$$+ \{x^5+(60r^4-30r^2+20)x^3 + (165r^4+30r^2+15)x\}/(480m^{5/2}) ,$$

then a lower $100(1-\alpha)\%$ confidence limit on ρ is given by

$$\rho_L = (e^{2y}-1)/(e^{2y}+1) \tag{2.1}$$

The high accuracy of this approximation is illustrated in Figure 2.3 given below. For n = 20 and n = 64,   the tables give the values of ρ_L computed from the approximation to 5 decimal places. If the last one or two digits are incorrect, the correct digits follow the entry and are underlined.

The maximum error over the table for n = 20 is 0.00011 for α = .005, r_0 = .96. However, when n = 64, the maximum error in the approximation is only 0.00001. We therefore suggest using the approximation in case interpolation in both n and r_0 is required.

Figure 2.3

Comparison of True and Approximate Values of ρ_L

r_0 ↓	α = 0.005	0.01	0.025	0.05	0.10	0.25
	n = 20					
-0.96	-0.987676	-0.986021	-0.983254	-0.980498	-0.9768079	-0.969154
-0.60	-0.86052	-0.843310	-0.815032	-0.787621	-0.752187	-0.682853
-0.20	-0.671842	-0.635965	-0.57895	-0.525865	-0.460143	-0.340310
0.20	-0.3966159	-0.342476	-0.26027	-0.18766	-0.10262	0.040054
0.60	0.0564959	0.1190613	0.20757	0.27986	0.358298	0.4767068
0.96	0.8595544	0.875398	0.8953025	0.909740	0.9238279	0.942487
	n = 64					
-0.96	-0.97879	-0.97738	-0.97517	-0.97310	-0.97052	-0.96567
-0.60	-0.76881	-0.75525	-0.73422	-0.71503	-0.69159	-0.649143
-0.20	-0.48528	-0.46019	-0.42218	-0.38844	-0.34839	-0.27892
0.20	-0.12693	-0.09501	-0.04806	-0.00771	0.03862	0.11523
0.60	0.34442	0.37263	0.41260	0.44559	0.48202	0.539065
0.96	0.92286	0.92754	0.93388	0.93887	0.94416	0.95198

3. DETAILS OF COMPUTATION

3.1 EXPRESSIONS FOR THE DENSITY FUNCTION

The probability density function of the sample correlation coefficient was derived by Fisher(1915), and can be written as

$$f_n(r;\rho) = [(n-2)/\pi](1-\rho^2)^{(n-1)/2}(1-r^2)^{(n-4)/2}I_{n-1}(x) \qquad (3.1)$$

where $-1<r<1$, $-1<\rho<1$, $x=\rho r$

and where

$$I_m(x) = \int_0^\infty [\cosh(w)-x]^{-m}\, dw \qquad \text{for} \quad |x| < 1 . \qquad (3.2)$$

If we let $\cosh(w) = (1-xz)/(1-z)$ then the integral in eqn (3.2) is
transformed to

$$I_m(x) = 2^{-1/2}(1-x)^{-m+1/2}\int_0^1 z^{-1/2} (1-z)^{m-1}[1-(1+x)z/2]^{-1/2} \, dz \; . \tag{3.3}$$

By expanding $[1-(1+x)z/2]^{-1/2}$ in a uniformly convergent series and integrating
term by term we obtain the expression

$$I_m(x) = [\sqrt{\pi/2} \; \Gamma(m)]/[\Gamma(m+1/2)](1-x)^{-m+1/2} \; F[1/2,1/2,m+1/2;(1+x)/2] \tag{3.4}$$

where $F[a,b,c;w] = 1 + [ab/c]w + [a(a+1)b(b+1)]/[2!c(c+1)]w^2 + \ldots$
is the hypergeometric series.

In particular   (3.5)

$$F[1/2,1/2,m+1/2;(1+x)/2] = 1 + [1+x]/[4(2m+1)] + [9(1+x)^2]/[16(2m+1)(2m+3)] + \ldots$$

Some useful special properties of $I_m(x)$ are given by the following:

$$I_1(x) = \arccos(-x)/\sqrt{1-x^2} \tag{3.6}$$

By repeated differentiation, with respect to x, under the integral sign in
eqn (3.2), it follows that

$$I_m(x) = [1/(m-1)!] \; \partial^{m-1}I_1(x)/\partial x^{m-1} \tag{3.7}$$

which gives for m=2

$$I_2(x) = \{1+xI_1(x)\}/(1-x^2) \; . \tag{3.8}$$

For $m \geq 1$ the following recurrence formula is satisfied (see pages 205-206
of Hotelling(1953) for details):

$$I_{m+2}(x) = \{(2m+1)xI_{m+1}(x)+mI_m(x)\}/\{(m+1)(1-x^2)\} \; . \tag{3.9}$$

For $0 \leq x \leq 1$, eqn (3.9) can be used to find a sequence of values for $I_3(x)$, $I_4(x), \ldots, I_{n-1}(x)$. However, for $-1 < x < 0$ the recurrence formula is numerically unstable, since the two terms on the right-hand side of eqn (3.9) are of opposite sign. In this case $(1+x) < 1$, so the series given by eqns (3.4) and (3.5) will converge extremely fast and can be used.

3.2 EXPRESSIONS FOR THE CUMULATIVE DISTRIBUTION FUNCTION

3.2.1 Recurrence Formulas

The cumulative distribution function $F_n(r; \rho)$ satisfies the following recurrence formula (see pages 207-208 of Hotelling (1953) for details):

$$(m+2)\rho^2 F_{m+5}(r; \rho) = -[m-2(m+1)\rho^2]F_{m+3}(r; \rho) + m(1-\rho^2)F_{m+1}(r; \rho)$$

$$+ [(1-\rho^2)^{(n+2)/2}(1-r^2)]/[\pi(m+1)]\{(m+2)\rho[(m+1)(1-r^2)(1-\rho^2) \qquad (3.10)$$

$$- (2m+1)(1-x^2)]I_{m+3}(x) + r[(m+1)r^2 + (3m^2+6m+2)\rho^2]I_{m+2}(x)\}$$

For n odd, the above formula can be used repeatedly for $m = 2, 4, \ldots, n-5$ to find F_7, $F_9, \ldots, F_n$ starting with values of F_3 and F_5, for which exact expressions are given by eqns (3.23) and (3.24) respectively. For n even, the formula can be used repeatedly for $m = 3, 5, \ldots, n-5$ to find F_8, $F_{10}, \ldots, F_n$ starting with values of F_4 and F_6, for which exact expressions are given by eqns (3.27) and (3.28) respectively.

However, the formula can only be applied if $\rho^2 \geq m/[2(m+1)]$, since if ρ^2 is less than this bound, the first two terms on the right hand side of equation (3.10) are of opposite sign and the formula is numerically unstable.

3.2.2 Hotelling's Series for the Cumulative Distribution Function

Hotelling (1953, pp 203-205) defines

$$Q_n(r;\rho) = \Pr[\rho < R < r] \tag{3.11}$$

and develops $Q_n(r;\rho)$ in the following uniformly convergent series for

$-1 < \rho < r < 1$. (In Hotelling's paper n is the sample size minus one.)

From the relationships

$$P_n(r;\rho) = 1 - F_n(r;\rho) = Q_n(1;\rho) - Q_n(r;\rho) ,$$

$$P_n(-1;\rho) = 1 , \tag{3.12}$$

$$F_n(r;\rho) = 1 - P_n(r;\rho) ,$$

$$F_n(-r;-\rho) = 1 - F_n(r;\rho) ,$$

we can compute $F_n(r;\rho)$ for any value of r and ρ with $-1 \le r \le 1$, $-1 < \rho < 1$.

By combining eqns (3.1) and (3.4) we may write

$$f_n(y;\rho) = \text{con}\cdot(1-\rho^2)^{m/2}(1-y^2)^{(m-3)/2}(1-\rho y)^{-m+1/2}F[1/2,1/2,m+1/2;(1+\rho y)/2]$$

where $\text{con} = [(m-1)\Gamma(m)]/[\sqrt{2\pi}\,\Gamma(m+1/2)]$ and m=n-1 . (3.13)

By using the relationship $(1+\rho y)=2-(1-\rho y)$, the term $(1+\rho y)^j$ in

$F[1/2,1/2,m+1/2;(1+\rho y)/2]$ can be expanded in terms of $1-\rho y$ by using the

binomial theorem to expand $[2-(1-\rho y)]^j$. If we define

$$N_k = \int_\rho^r (1-\rho^2)^{m/2}(1-y^2)^{(m-3)/2}(1-\rho y)^{-m+k+1/2} \, dy , \quad k = 0,1,\ldots \tag{3.14}$$

then $Q_n(r;\rho)$ can be expressed linearly in terms of $N_0, N_1,\ldots$.

The result is

$$Q_n(r;\rho) = \text{con} \cdot \{N_0 + [2N_0-N_1]/[4(2m+1)] + [9(4N_0-4N_1+N_2)]/[32(2m+1)(2m+3)] + \ldots\}$$

where $\text{con} = [(m-1)\Gamma(m)]/[\sqrt{2\pi}\Gamma(m+1/2)]$ and $m=n-1$.   (3.15)

Hotelling shows that the error committed by truncating the series at

any point is less than $2/(1-|\rho|)$ times the last term used. However, it is

important to note that the series converges very slowly for small values

of n , and in this case a large number of terms must be computed.

Hotelling also develops N_k in a series involving the Incomplete

Beta Function as

$$N_k = (1-\rho^2)^k/2 \sum_0^\infty \Gamma (3/2-k)/[\Gamma(3/2-k-s)s!]\rho^s B_b[(s+1)/2,(m-1)/2]$$ (3.16)

where $b = (r-\rho)^2/(1-\rho r)^2$, $0 \leq b \leq 1$

and where

$$B_b[p,q] = \int_0^b w^{p-1}(1-w)^{q-1} \, dw$$ (3.17)

is the Incomplete Beta Function. Hotelling also shows that for large s

the absolute value of the ratio of the term of order (s+1) to the term of

order s is bounded by $|\rho\sqrt{b}| < 1$, so that the series converges rapidly.

A FORTRAN subroutine for evaluating the Incomplete Beta Function Ratio,

$$I_b[p,q] = B_b[p,q]/B_1[p,q]$$ (3.18)

is given by Majumder and Bhattacharjee (1973). The input parameter BETA to this

routine is the value of $B_1[p,q] = \Gamma(p)\Gamma(q)/\Gamma(p+q)$. If BETA is set equal to 1.0,

then the routine returns $B_b[p,q]$. The routine is written in single precision

but when implemented in double precision on an IBM Computer it yields at least

12 decimal place accuracy if the parameter ACU is set to 10^{-14} .

3.2.3 Guenther's Series for the Cumulative Distribution Function

Guenther(1977) writes

$$P_n(r;\rho) = Pr[R>0] - Pr[0<R<r] \qquad (3.19)$$

and develops $Pr[0<R<r]$ in an infinite series involving the cumulative F-distribution. (This series was developed for use on a desk calculator, some models of which have programs to evaluate the cumulative F-distribution). His series is given here in terms of the Incomplete Beta Function Ratio defined by eqns (3.17) and (3.18).

The result is

$$Pr[0<R<r] = \sum_0^\infty K_1(j)I_{r^2}[(n-2)/2,(2j+1)/2] \qquad (3.20)$$

$$+ \sum_0^\infty K_2(j)I_{r^2}[(n-2)/2,(j+1)]$$

where $K_1(j)$, $K_2(j)$ are defined recursively by

$$K_1(0) = \frac{1}{2}(1-\rho^2)^{(n-1)/2}$$

$$K_1(j) = [(2j+n-3)/(2j)]\rho^2 K_1(j-1) \qquad j\geq 1 , \qquad (3.21)$$

$$K_2(0) = \Gamma(n/2)\rho(1-\rho^2)^{(n-1)/2}/\sqrt{\pi}\Gamma[(n-1)/2] ,$$

$$K_2(j) = [(2j+n-2)/(2j+1)]\rho^2 K_2(j-1) \qquad j\geq 1 .$$

Also

$$Pr[R>0]\} = (1/2)\{1 +sgn(\rho)I_{\rho^2}[(n-1)/2,1/2]\} . \qquad (3.22)$$

Eqns (3.20) and (3.22) are used together to give $P_n(r;\rho)$. Guenther also obtains error bounds for truncating the infinite series given by (3.20).

3.2.4 Exact Expressions for the Cumulative Distribution Function

Garwood(1933) gives exact expressions for $F_n(r;\rho)$ for n = 3(1)8
in terms of the density function for smaller values of n. Except
for n = 3, these expressions involve powers of ρ in the denominator.
For computational purposes it is desirable to rewrite the expressions
in a form where ρ only appears in the numerator.

For n = 4(1)8 the expressions of Garwood were rewritten by Öksoy
and Aroian (1982a,b) in terms of $x=\rho r$, ρ, U = arccos$(-x)/\sqrt{1-x^2}$ =
$I_1(x)$, and the derivatives of U with respect to x. The
derivatives of U can be expressed as polynomials in x and U itself,
with powers of $\sqrt{1-x^2}$ in the denominator. The first nine derivatives
of U are listed by Öksoy and Aroian (1982a,b). They also investigated
a method derived (but not used) by Garwood to derive new exact formulas
for n = 9, 10. For odd n , ρ does not appear in the denominator of these
expressions. For n = 6, 8, 10, the expressions do have powers of ρ in
the denominator.

The expressions given here for n = 7(1)10 were generated using the
recurrence formula given by eqn (3.10). The expressions for n = 5 ,6 were
obtained by rewriting the expressions given by Öksoy and Aroian (1982a,b).
The expressions for n = 3, 4 are given in Odeh and Owen (1980).

The expressions were obtained using the algebraic language REDUCE 2
(Hearn (1973)) on an IBM Computer. The terms involving U and those
not involving U were found separately. The expressions are also given
in Odeh (1983).

The form of the expressions for n odd and n even are different.
Hence, we list them separately.

We use the following notation in the expressions given below:

$$x = \rho r$$

$$A = 1 - \rho^2$$

$$B = 1 - x^2$$

$$C = 1 - r^2$$

$$U = \arccos(-x)/(B^{1/2})$$

$$D_1 = \arccos(-r)/\pi$$

$$D_2 = \arccos(\rho)/\pi$$

EXPRESSIONS FOR n ODD

For n = 3

$$F3 = 0$$

$$q_0 = -1$$

$$F3U = q_0 \rho C^{1/2} U$$

$$F_3(r;\rho) = D_1 + F3U/\pi \qquad\qquad (3.23)$$

For $\rho = 0$, $F_3(r;0) = D_1$

For n = 5

$$P_0 = x^2 + 2$$

$$P_2 = 3\rho^2$$

$$F5 = (P_0 - P_2)rC^{1/2}A$$

$$q_0 = \rho^2 - 3$$

$$q_2 = 2\rho^2 x^2$$

$$F5U = (q_0 + q_2)\rho C^{3/2}$$

$$F_5(r;\rho) = D_1 + F5/(2\pi B^2) + F5U/(2\pi B^2) \qquad\qquad (3.24)$$

For $\rho = 0$, $F_5(r;0) = D_1 + rC^{1/2}/\pi$

For $n = 7$

$$p_0 = 18x^6 - x^4(6r^2 - 139) - x^2(83r^2 + 118) - 16r^2 + 40$$

$$p_2 = \rho^2(92x^4 + 116x^2 + 107)$$

$$p_4 = 5\rho^4(10x^2 + 11)$$

$$F7 = (p_0 - p_2 + p_4)rc^{1/2}A$$

$$q_0 = -3\rho^4 + 10\rho^2 - 15$$

$$q_2 = 4x^2(-6\rho^4 + 15\rho^2 - 5)$$

$$q_4 = -8\rho^4x^4$$

$$F7U = (q_0 + q_2 + q_4)\rho c^{5/2}U$$

$$F_7(r;\rho) = D_1 + F7/(24\pi B^4) + F7U/(8\pi B^4) \tag{3.25}$$

For $\rho = 0$, $F_7(r;0) = D_1 + r(-2r^2 + 5)c^{1/2}/(3\pi)$

For $n = 9$

$$p_0 = 220x^{10} - x^8(140r^2 - 3884)$$
$$+ x^6(40r^4 - 4414r^2 + 12015)$$
$$+ x^4(1518r^4 - 7338r^2 + 13946)$$
$$+ x^2(1779r^4 - 5017r^2 + 4057) + 128r^4 - 416r^2 + 528$$

$$p_2 = \rho^2(1588x^8 + 6972x^6 + 16281x^4 - 7790x^2 - 2019)$$

$$p_4 = 3\rho^4(572x^6 + 2204x^4 + 2311x^2 + 688)$$

$$p_6 = 21\rho^6(28x^4 + 104x^2 + 33)$$

$$F9 = (p_0 - p_2 + p_4 - p_6)rc^{1/2}A$$

$$q_0 = 5\rho^6 - 21\rho^4 + 35\rho^2 - 35$$

$$q_2 = 2x^2(45\rho^6 - 168\rho^4 + 217\rho^2 - 70)$$

$$q_4 = 8x^4(15\rho^6 - 42\rho^4 + 28\rho^2 - 7)$$

$$q_6 = 16\rho^6x^6$$

$$F9U = (q_0 + q_2 + q_4 + q_6)\rho c^{7/2}U$$

$$F_9(r;\rho) = D_1 + F9/(240\pi B^6) + F9U/(16\pi B^6) \tag{3.26}$$

For $\rho = 0$, $F_9(r;0) = D_1 + r(8r^4 - 26r^2 + 33)c^{1/2}/(15\pi)$

EXPRESSIONS FOR n EVEN

For $n = 4$

$$p_0 = 0$$

$$p_1 = \rho$$

$$F4 = (p_0 - p_1)A^{1/2}C$$

$$q_0 = 1$$

$$F4U = q_0 rA^{3/2}U$$

$$F_4(r;\rho) = D_2 + F4/(\pi B) + F4U/(\pi B) \tag{3.27}$$

For $\rho = 0$, $F_4(r;0) = 1/2 + r/2$

For $n = 6$

$$p_0 = xr(2x^2 + 13)$$

$$p_1 = 2\rho(4x^4 + 6x^2 + 5)$$

$$p_3 = \rho^3(11x^2 + 4)$$

$$F6 = (p_0 - p_1 + p_3)A^{1/2}C$$

$$q_0 = -r^2 + 3$$

$$q_2 = 2x^2(-2r^2 + 1)$$

$$F6U = (q_0 + q_2)rA^{5/2}U$$

$$F_6(r;\rho) = D_2 + F6/(6\pi B^3) + F6U/(2\pi B^3) \tag{3.28}$$

For $\rho = 0$, $F_6(r;0) = 1/2 + r(-r^2 + 3)/4$

For $n = 8$

$$p_0 = xr[88x^6 - x^4(24r^2 - 1642) - x^2(582r^2 - 1387)$$
$$-339r^2 + 663]$$

$$p_1 = \rho(184x^8 + 1098x^6 + 2871x^4 + 1253x^2 + 264)$$

$$p_3 = \rho^3(518x^6 + 1377x^4 + 1677x^2 + 208)$$

$$p_5 = \rho^5(274x^4 + 607x^2 + 64)$$

$$F8 = (p_0 - p_1 + p_3 - p_5)A^{1/2}C$$

$$q_0 = 3r^4 - 10r^2 + 15$$

$$q_2 = 4x^2(9r^4 - 23r^2 + 10)$$

$$q_4 = 8x^4(3r^4 - 3r^2 + 1)$$

$$F8U = (q_0 + q_2 + q_4)rA^{7/2}U$$

$$F_8(r;\rho) = D_2 + F8/(120\pi B^5) + F8U/(8\pi B^5) \qquad (3.29)$$

For $\rho = 0$, $F_8(r;0) = 1/2 + r(3r^4 - 10r^2 + 15)/16$

For $n = 10$

$$p_0 = xr[1952x^{10} - x^8(1056r^2 - 74704)$$

$$+ x^6(240r^4 - 51984r^2 + 214044)$$

$$+ x^4(13224r^4 - 117516r^2 + 274654)$$

$$+ x^2(25962r^4 - 82962r^2 + 93884)$$

$$+ 5619r^4 - 16752r^2 + 16437]$$

$$p_1 = 4\rho(704x^{12} + 9488x^{10} + 54344x^8 + 87724x^6 + 60064x^4$$

$$+ 11785x^2 + 1116)$$

$$p_3 = \rho^3(12088x^{10} + 82474x^8 + 258739x^6 + 231464x^4$$

$$+ 85694x^2 + 5216)$$

$$p_5 = 2\rho^5(6394x^8 + 37955x^6 + 58968x^4 + 30218x^2 + 1600)$$

$$p_7 = 3\rho^7(1452x^6 + 8132x^4 + 5175x^2 + 256)$$

$$F10 = (p_0 - p_1 + p_3 - p_5 + p_7)A^{1/2}C$$

$$q_0 = -5r^6 + 21r^4 - 35r^2 + 35$$

$$q_2 = 6x^2(-20r^6 + 75r^4 - 98r^2 + 35)$$

$$q_4 = 24x^4(-10r^6 + 30r^4 - 25r^2 + 7)$$

$$q_6 = 16x^6(-4r^6 + 6r^4 - 4r^2 + 1)$$

$$F10U = (q_0 + q_2 + q_4 + q_6)rA^{9/2}U$$

$$F_{10}(r;\rho) = D_2 + F10/(1680\pi B^7) + F10U/(16\pi B^7) \qquad (3.30)$$

For $\rho = 0$, $F_{10}(r;0) = 1/2 + r(-5r^6 + 21r^4 - 35r^2 + 35)/32$

3.3   CONSTRUCTION OF THE TABLES AND CHECKS ON ACCURACY

   For a given n , $\alpha < 0.5$, and r_0 , the entry in the table is the value

of ρ (given to 5 decimal places) which satisfies $P_n(r_0;\rho) = \alpha$,

or equivalently $F_n(r_0;\rho) = \alpha' = 1-\alpha$. The values of ρ were found by

using a modified form of successive approximations which uses an estimate

of the derivative of $F_n(r_0;\rho)$ with respect to ρ . The method is equivalent

to using linear interpolation.

   If ρ_0 and ρ_1 are two intitial estimates of ρ , then we obtain a

sequence of estimates $\rho_2, \rho_3, \ldots$ from

$$\rho_{j+1} = \rho_j + a_j[\alpha' - F_n(r_0;\rho_j)]$$

where

$$a_j = (\rho_j - \rho_{j-1})/[F_n(r_0;\rho_j) - F_n(r_0;\rho_{j-1})] \qquad j = 1,2,\ldots$$

This process was repeated until two successive values of ρ agreed to

8 decimal places. Originally the series given in Section 3.2.2 was used to

evaluate $F_n(r_0;\rho)$ for $n>10$, using sufficient terms to obtain 12 decimal

place accuracy. The series was checked by comparing the results with the

recurrence formula given by eqn (3.10) if $\rho^2 \geq 0.5$, and the series given

in Section 3.2.3 otherwise. For $3 \leq n \leq 10$ the exact expressions given

in Section 3.2.4 were used and also checked by the series in Section 3.2.3.

There is excellent agreement between the above methods. The values of

$F_n(r_0;\rho)$ computed by these various formulas agreed to at least 10 decimal

places. The values of ρ obtained were rounded to 5 decimal places and

stored. Since $F_n(r_0;\rho)$ is a monotone decreasing function in ρ , the

values were then incremented (if $F_n(r_0;\rho)>\alpha'$) or decremented (if $F_n(r_0;\rho)<\alpha'$)

by 0.00001 to see if $|F_n(r_0;\rho)-\alpha'|$ could be reduced. No stored values were

changed and the tabled values are accurate to all 5 decimal places printed.

The initial estimate ρ_0 was obtained from the approximation given by eqn (2.1). Then

if $F_n(r_0;\rho_0)<\alpha'$ we set $\rho_1 = \rho_0 - 0.001$,

and if $F_n(r_0;\rho_0)>\alpha'$ we set $\rho_1 = \rho_0 + 0.001$.

All computations were done in double precision on an IBM Model 4341 Computer at the University of Victoria.

ACKNOWLEDGEMENT

I would like to express my appreciation to the IMS Committee on Mathematical Tables for their continuing assistance with the publication of this series of Tables. I am also indebted to my Co-Editor, Professor J. M. Davenport, and to our Managing-Editor, Dr. N. S. Pearson, for their continuing effort, fortitude, and encouragement.

I was particularly helped in the preperation of the manuscript by many useful discussions with Professor W. Q. Meeker of Iowa State University and Professor L. Aroian of Union College.

The research for this project was supported by Natural Sciences and Engineering Research Council Canada under Grant No. A5203. I am particularly grateful to this agency for their continuing support.

REFERENCES

DAVID, F.N. (1938,1954),   TABLES OF THE CORRELATION COEFFICIENT,
    Cambridge: Cambridge University Press.

FISHER, R.A. (1915),   Frequency Distribution of the Value of the
    Correlation Coefficient in Samples from an Indefinitely Large
    Population, Biometrika, 10, 507-521.

GARWOOD, F. (1933),   The Probability Integral of the Correlation
    Coefficient in Samples from a Normal Population.
    Biometrika, 25, 71-78.

GUENTHER, W.C. (1977),   Desk Calculations of Probabilities for the
    Distribution of the Sample Correlation Coefficient,
    The American Statistician, 31, 45-48.

HEARN, A.C. (1973).   REDUCE 2 USER'S MANUAL, Salt Lake City,
    Utah: University of Utah.

HOTELLING, H. (1953),   New Light on the Correlation Coefficient and
    Its Transforms, Journal of the Royal Statistical Society B, 15,
    193-225; (discussion 225-232).

MAJUMDER, K.L. & BHATTACHARJEE G.P. (1973),   Algorithm AS 63:
    The Incomplete Beta Integral, Applied Statistics,22(3), 409-411.

ODEH, R.E. (1982),   Critical Values of the Sample Product-Moment Correlation
    Coefficient in the Bivariate Normal Distribution,
    Communications in Statistics, B11(1), 1-26.

ODEH, R.E. (1983),   On Formulas for the Exact Distribution of the Correlation
    Coefficient Using Reduce, Proceedings of the Statistical Computing
    Section of the American Statistical Association, 306-308.

ODEH, R.E. & Owen, D.B. (1980),   TABLES FOR NORMAL TOLERANCE LIMITS, SAMPLING
    PLANS, AND SCREENING, New York: Marcel Dekker, Inc.

ÖKSOY, D. & AROIAN, L.A. (1982a),   DISTRIBUTION OF THE COEFFICIENT OF
    CORRELATION, Schenectady, New York: Institute of Administration
    and Management, Union College.

    ---- (1982b),   PERCENTAGE POINTS OF THE DISTRIBUTION OF THE COEFFICIENT
    OF CORRELATION, Schenectady, New York: Institute of Administration
    and Management, Union College.

SUBRAHMANIAM, Ka. & SUBRAHMANIAM, Ko. (1980).   SOME EXTENSIONS TO MISS
    F.N. DAVID'S TABLES OF SAMPLE CORRELATION COEFFICIENT;
    DISTRIBUTION FUNCTION AND PERCENTILES, TR No. 99, Winnipeg,
    Manitoba: Department of Statistics, University of Manitoba.

WINTERBOTTOM, A. (1980),   Estimation for the Bivariate Normal Correlation
    Coefficient Using Asymptotic Expansions,
    Communications in Statistics, B9(6), 599-609.

INDEX OF TABLES

TABLE 1.1   SAMPLE SIZE =   3

α =	.005 ↓+	.01 ↓+	.025 ↓+	.05 ↓+	.10 ↓+	.25 ↓+	
r₀ ↓							
-0.98	-0.99980	-0.99960	-0.99897	-0.99786	-0.99538	-0.98489	0.98
-0.96	-0.99960	-0.99920	-0.99796	-0.99577	-0.99087	-0.97053	0.96
-0.94	-0.99941	-0.99881	-0.99696	-0.99371	-0.98646	-0.95679	0.94
-0.92	-0.99922	-0.99842	-0.99598	-0.99168	-0.98214	-0.94358	0.92
-0.90	-0.99903	-0.99804	-0.99500	-0.98967	-0.97790	-0.93084	0.90
-0.88	-0.99884	-0.99766	-0.99404	-0.98769	-0.97372	-0.91850	0.88
-0.86	-0.99865	-0.99728	-0.99308	-0.98573	-0.96960	-0.90652	0.86
-0.84	-0.99846	-0.99691	-0.99213	-0.98379	-0.96554	-0.89487	0.84
-0.82	-0.99828	-0.99653	-0.99118	-0.98186	-0.96152	-0.88351	0.82
-0.80	-0.99809	-0.99616	-0.99025	-0.97994	-0.95754	-0.87241	0.80
-0.78	-0.99791	-0.99579	-0.98931	-0.97804	-0.95361	-0.86154	0.78
-0.76	-0.99772	-0.99542	-0.98838	-0.97615	-0.94971	-0.85090	0.76
-0.74	-0.99754	-0.99505	-0.98745	-0.97427	-0.94584	-0.84045	0.74
-0.72	-0.99736	-0.99469	-0.98652	-0.97239	-0.94199	-0.83017	0.72
-0.70	-0.99717	-0.99432	-0.98560	-0.97052	-0.93817	-0.82007	0.70
-0.68	-0.99699	-0.99395	-0.98468	-0.96866	-0.93437	-0.81011	0.68
-0.66	-0.99681	-0.99358	-0.98375	-0.96680	-0.93058	-0.80029	0.66
-0.64	-0.99662	-0.99322	-0.98283	-0.96494	-0.92682	-0.79059	0.64
-0.62	-0.99644	-0.99285	-0.98191	-0.96308	-0.92306	-0.78100	0.62
-0.60	-0.99625	-0.99248	-0.98098	-0.96122	-0.91931	-0.77152	0.60
-0.58	-0.99607	-0.99211	-0.98006	-0.95936	-0.91557	-0.76214	0.58
-0.56	-0.99588	-0.99174	-0.97913	-0.95750	-0.91184	-0.75283	0.56
-0.54	-0.99570	-0.99137	-0.97820	-0.95563	-0.90810	-0.74361	0.54
-0.52	-0.99551	-0.99099	-0.97726	-0.95377	-0.90437	-0.73445	0.52
-0.50	-0.99532	-0.99062	-0.97632	-0.95189	-0.90063	-0.72535	0.50
-0.48	-0.99514	-0.99024	-0.97538	-0.95001	-0.89689	-0.71631	0.48
-0.46	-0.99495	-0.98986	-0.97444	-0.94812	-0.89315	-0.70731	0.46
-0.44	-0.99475	-0.98948	-0.97348	-0.94623	-0.88939	-0.69836	0.44
-0.42	-0.99456	-0.98910	-0.97253	-0.94432	-0.88563	-0.68944	0.42
-0.40	-0.99437	-0.98871	-0.97156	-0.94240	-0.88185	-0.68055	0.40
-0.38	-0.99417	-0.98832	-0.97059	-0.94048	-0.87806	-0.67167	0.38
-0.36	-0.99398	-0.98793	-0.96961	-0.93854	-0.87425	-0.66282	0.36
-0.34	-0.99378	-0.98753	-0.96863	-0.93658	-0.87042	-0.65398	0.34
-0.32	-0.99358	-0.98713	-0.96763	-0.93461	-0.86657	-0.64514	0.32
-0.30	-0.99338	-0.98673	-0.96663	-0.93263	-0.86269	-0.63630	0.30
-0.28	-0.99317	-0.98632	-0.96561	-0.93063	-0.85879	-0.62746	0.28
-0.26	-0.99296	-0.98590	-0.96459	-0.92861	-0.85486	-0.61860	0.26
-0.24	-0.99275	-0.98549	-0.96356	-0.92656	-0.85090	-0.60973	0.24
-0.22	-0.99254	-0.98507	-0.96251	-0.92450	-0.84691	-0.60084	0.22
-0.20	-0.99233	-0.98464	-0.96145	-0.92242	-0.84288	-0.59192	0.20
-0.18	-0.99211	-0.98420	-0.96038	-0.92031	-0.83881	-0.58297	0.18
-0.16	-0.99189	-0.98377	-0.95929	-0.91818	-0.83471	-0.57398	0.16
-0.14	-0.99167	-0.98332	-0.95819	-0.91602	-0.83055	-0.56494	0.14
-0.12	-0.99144	-0.98287	-0.95708	-0.91383	-0.82635	-0.55586	0.12
-0.10	-0.99121	-0.98241	-0.95594	-0.91161	-0.82211	-0.54672	0.10
-0.08	-0.99098	-0.98195	-0.95479	-0.90936	-0.81780	-0.53752	0.08
-0.06	-0.99074	-0.98147	-0.95363	-0.90708	-0.81345	-0.52826	0.06
-0.04	-0.99050	-0.98099	-0.95244	-0.90476	-0.80903	-0.51892	0.04
-0.02	-0.99025	-0.98050	-0.95123	-0.90240	-0.80455	-0.50950	0.02

↑ r₀

α =	.995 ↑-	.99 ↑-	.975 ↑-	.95 ↑-	.90 ↑-	.75 ↑-

TABLE 1.2   SAMPLE SIZE = 3   (CONT.)

α =	.005 ↓+	.01 ↓+	.025 ↓+	.05 ↓+	.10 ↓+	.25 ↓+	
r_o ↓							
0.00	-0.99000	-0.98000	-0.95000	-0.90000	-0.80000	-0.50000	0.00
0.02	-0.98974	-0.97949	-0.94875	-0.89756	-0.79538	-0.49040	-0.02
0.04	-0.98948	-0.97897	-0.94747	-0.89507	-0.79068	-0.48071	-0.04
0.06	-0.98922	-0.97844	-0.94617	-0.89254	-0.78591	-0.47090	-0.06
0.08	-0.98894	-0.97790	-0.94484	-0.88996	-0.78104	-0.46098	-0.08
0.10	-0.98867	-0.97735	-0.94348	-0.88732	-0.77609	-0.45093	-0.10
0.12	-0.98838	-0.97678	-0.94209	-0.88462	-0.77104	-0.44075	-0.12
0.14	-0.98809	-0.97620	-0.94066	-0.88187	-0.76589	-0.43043	-0.14
0.16	-0.98779	-0.97560	-0.93921	-0.87905	-0.76063	-0.41996	-0.16
0.18	-0.98748	-0.97499	-0.93771	-0.87616	-0.75526	-0.40932	-0.18
0.20	-0.98717	-0.97437	-0.93618	-0.87320	-0.74976	-0.39851	-0.20
0.22	-0.98684	-0.97372	-0.93460	-0.87017	-0.74413	-0.38752	-0.22
0.24	-0.98651	-0.97306	-0.93298	-0.86705	-0.73836	-0.37633	-0.24
0.26	-0.98616	-0.97238	-0.93131	-0.86384	-0.73244	-0.36492	-0.26
0.28	-0.98581	-0.97167	-0.92960	-0.86054	-0.72636	-0.35329	-0.28
0.30	-0.98544	-0.97095	-0.92782	-0.85714	-0.72011	-0.34142	-0.30
0.32	-0.98506	-0.97020	-0.92599	-0.85363	-0.71368	-0.32929	-0.32
0.34	-0.98467	-0.96942	-0.92410	-0.85000	-0.70705	-0.31688	-0.34
0.36	-0.98427	-0.96861	-0.92213	-0.84625	-0.70022	-0.30417	-0.36
0.38	-0.98384	-0.96777	-0.92010	-0.84237	-0.69315	-0.29115	-0.38
0.40	-0.98340	-0.96690	-0.91798	-0.83833	-0.68584	-0.27778	-0.40
0.42	-0.98295	-0.96600	-0.91578	-0.83414	-0.67827	-0.26405	-0.42
0.44	-0.98247	-0.96505	-0.91349	-0.82978	-0.67041	-0.24992	-0.44
0.46	-0.98197	-0.96406	-0.91109	-0.82524	-0.66224	-0.23536	-0.46
0.48	-0.98144	-0.96302	-0.90858	-0.82049	-0.65373	-0.22034	-0.48
0.50	-0.98089	-0.96194	-0.90595	-0.81551	-0.64486	-0.20481	-0.50
0.52	-0.98031	-0.96079	-0.90319	-0.81029	-0.63557	-0.18875	-0.52
0.54	-0.97970	-0.95958	-0.90027	-0.80480	-0.62585	-0.17208	-0.54
0.56	-0.97906	-0.95831	-0.89719	-0.79902	-0.61563	-0.15477	-0.56
0.58	-0.97837	-0.95695	-0.89393	-0.79290	-0.60487	-0.13675	-0.58
0.60	-0.97764	-0.95551	-0.89047	-0.78640	-0.59350	-0.11795	-0.60
0.62	-0.97686	-0.95397	-0.88677	-0.77949	-0.58146	-0.09829	-0.62
0.64	-0.97602	-0.95232	-0.88281	-0.77211	-0.56866	-0.07768	-0.64
0.66	-0.97512	-0.95053	-0.87854	-0.76419	-0.55500	-0.05601	-0.66
0.68	-0.97414	-0.94861	-0.87394	-0.75566	-0.54036	-0.03314	-0.68
0.70	-0.97307	-0.94651	-0.86894	-0.74642	-0.52461	-0.00895	-0.70
0.72	-0.97190	-0.94421	-0.86347	-0.73636	-0.50756	0.01675	-0.72
0.74	-0.97062	-0.94168	-0.85745	-0.72532	-0.48900	0.04418	-0.74
0.76	-0.96918	-0.93886	-0.85078	-0.71313	-0.46866	0.07358	-0.76
0.78	-0.96757	-0.93569	-0.84331	-0.69954	-0.44619	0.10528	-0.78
0.80	-0.96573	-0.93210	-0.83485	-0.68424	-0.42115	0.13966	-0.80
0.82	-0.96362	-0.92796	-0.82515	-0.66679	-0.39291	0.17724	-0.82
0.84	-0.96115	-0.92312	-0.81385	-0.64661	-0.36068	0.21866	-0.84
0.86	-0.95818	-0.91734	-0.80042	-0.62281	-0.32326	0.26481	-0.86
0.88	-0.95455	-0.91025	-0.78404	-0.59408	-0.27892	0.31687	-0.88
0.90	-0.94991	-0.90125	-0.76339	-0.55829	-0.22495	0.37656	-0.90
0.92	-0.94369	-0.88921	-0.73607	-0.51169	-0.15679	0.44646	-0.92
0.94	-0.93469	-0.87187	-0.69723	-0.44693	-0.06594	0.53067	-0.94
0.96	-0.91982	-0.84341	-0.63488	-0.34650	0.06622	0.63647	-0.96
0.98	-0.88698	-0.78151	-0.50504	-0.15116	0.29400	0.77882	-0.98

↑ r_o

α =	.995 ↑-	.99 ↑-	.975 ↑-	.95 ↑-	.90 ↑-	.75 ↑-

TABLE 2.1   SAMPLE SIZE =   4

α =	.005 ↓+	.01 ↓+	.025 ↓+	.05 ↓+	.10 ↓+	.25 ↓+	
r₀ ↓							
-0.98	-0.99940	-0.99902	-0.99813	-0.99686	-0.99451	-0.98694	0.98
-0.96	-0.99879	-0.99804	-0.99625	-0.99371	-0.98904	-0.97406	0.96
-0.94	-0.99818	-0.99706	-0.99436	-0.99055	-0.98357	-0.96134	0.94
-0.92	-0.99756	-0.99606	-0.99246	-0.98738	-0.97810	-0.94878	0.92
-0.90	-0.99694	-0.99506	-0.99055	-0.98421	-0.97264	-0.93635	0.90
-0.88	-0.99631	-0.99405	-0.98863	-0.98102	-0.96717	-0.92405	0.88
-0.86	-0.99568	-0.99304	-0.98670	-0.97781	-0.96170	-0.91186	0.86
-0.84	-0.99505	-0.99202	-0.98475	-0.97460	-0.95622	-0.89978	0.84
-0.82	-0.99441	-0.99099	-0.98280	-0.97136	-0.95074	-0.88780	0.82
-0.80	-0.99376	-0.98995	-0.98083	-0.96811	-0.94524	-0.87590	0.80
-0.78	-0.99311	-0.98890	-0.97884	-0.96485	-0.93972	-0.86409	0.78
-0.76	-0.99245	-0.98784	-0.97684	-0.96156	-0.93420	-0.85234	0.76
-0.74	-0.99178	-0.98677	-0.97482	-0.95825	-0.92865	-0.84066	0.74
-0.72	-0.99111	-0.98569	-0.97279	-0.95492	-0.92308	-0.82904	0.72
-0.70	-0.99043	-0.98460	-0.97074	-0.95156	-0.91749	-0.81747	0.70
-0.68	-0.98974	-0.98350	-0.96867	-0.94818	-0.91187	-0.80595	0.68
-0.66	-0.98905	-0.98239	-0.96657	-0.94477	-0.90623	-0.79446	0.66
-0.64	-0.98835	-0.98127	-0.96446	-0.94134	-0.90056	-0.78301	0.64
-0.62	-0.98763	-0.98013	-0.96233	-0.93787	-0.89486	-0.77159	0.62
-0.60	-0.98691	-0.97898	-0.96017	-0.93438	-0.88912	-0.76018	0.60
-0.58	-0.98619	-0.97781	-0.95799	-0.93085	-0.88334	-0.74880	0.58
-0.56	-0.98545	-0.97663	-0.95579	-0.92729	-0.87753	-0.73743	0.56
-0.54	-0.98470	-0.97544	-0.95356	-0.92369	-0.87168	-0.72606	0.54
-0.52	-0.98394	-0.97423	-0.95131	-0.92006	-0.86578	-0.71470	0.52
-0.50	-0.98317	-0.97300	-0.94902	-0.91639	-0.85983	-0.70334	0.50
-0.48	-0.98239	-0.97175	-0.94671	-0.91268	-0.85384	-0.69196	0.48
-0.46	-0.98160	-0.97049	-0.94437	-0.90892	-0.84780	-0.68058	0.46
-0.44	-0.98079	-0.96921	-0.94199	-0.90512	-0.84170	-0.66918	0.44
-0.42	-0.97998	-0.96791	-0.93959	-0.90128	-0.83555	-0.65775	0.42
-0.40	-0.97915	-0.96659	-0.93715	-0.89739	-0.82933	-0.64630	0.40
-0.38	-0.97830	-0.96525	-0.93467	-0.89344	-0.82306	-0.63482	0.38
-0.36	-0.97745	-0.96389	-0.93216	-0.88945	-0.81672	-0.62330	0.36
-0.34	-0.97657	-0.96251	-0.92961	-0.88540	-0.81031	-0.61174	0.34
-0.32	-0.97569	-0.96110	-0.92701	-0.88129	-0.80382	-0.60014	0.32
-0.30	-0.97478	-0.95966	-0.92438	-0.87713	-0.79727	-0.58848	0.30
-0.28	-0.97386	-0.95821	-0.92170	-0.87290	-0.79063	-0.57677	0.28
-0.26	-0.97293	-0.95672	-0.91898	-0.86861	-0.78391	-0.56500	0.26
-0.24	-0.97197	-0.95521	-0.91621	-0.86425	-0.77711	-0.55316	0.24
-0.22	-0.97099	-0.95366	-0.91339	-0.85982	-0.77021	-0.54125	0.22
-0.20	-0.97000	-0.95209	-0.91051	-0.85531	-0.76322	-0.52927	0.20
-0.18	-0.96899	-0.95049	-0.90759	-0.85073	-0.75613	-0.51721	0.18
-0.16	-0.96795	-0.94885	-0.90460	-0.84607	-0.74894	-0.50505	0.16
-0.14	-0.96689	-0.94717	-0.90156	-0.84132	-0.74164	-0.49281	0.14
-0.12	-0.96581	-0.94547	-0.89845	-0.83649	-0.73423	-0.48047	0.12
-0.10	-0.96470	-0.94372	-0.89528	-0.83156	-0.72670	-0.46802	0.10
-0.08	-0.96356	-0.94193	-0.89205	-0.82654	-0.71904	-0.45546	0.08
-0.06	-0.96240	-0.94010	-0.88874	-0.82141	-0.71125	-0.44278	0.06
-0.04	-0.96121	-0.93822	-0.88535	-0.81618	-0.70333	-0.42998	0.04
-0.02	-0.95999	-0.93630	-0.88189	-0.81084	-0.69526	-0.41705	0.02

↑ r₀

α =	.995 ↑−	.99 ↑−	.975 ↑−	.95 ↑−	.90 ↑−	.75 ↑−

TABLE 2.2   SAMPLE SIZE = 4   (CONT.)

α =	.005 ↓+	.01 ↓+	.025 ↓+	.05 ↓+	.10 ↓+	.25 ↓+	
r_o ↓							
0.00	-0.95874	-0.93433	-0.87834	-0.80538	-0.68705	-0.40397	0.00
0.02	-0.95745	-0.93231	-0.87471	-0.79980	-0.67868	-0.39075	-0.02
0.04	-0.95613	-0.93024	-0.87098	-0.79409	-0.67014	-0.37738	-0.04
0.06	-0.95477	-0.92810	-0.86716	-0.78825	-0.66143	-0.36384	-0.06
0.08	-0.95337	-0.92591	-0.86323	-0.78226	-0.65253	-0.35013	-0.08
0.10	-0.95193	-0.92365	-0.85920	-0.77612	-0.64344	-0.33624	-0.10
0.12	-0.95044	-0.92133	-0.85506	-0.76982	-0.63416	-0.32216	-0.12
0.14	-0.94891	-0.91893	-0.85079	-0.76335	-0.62465	-0.30788	-0.14
0.16	-0.94733	-0.91646	-0.84640	-0.75671	-0.61493	-0.29339	-0.16
0.18	-0.94569	-0.91391	-0.84187	-0.74987	-0.60497	-0.27867	-0.18
0.20	-0.94400	-0.91127	-0.83720	-0.74284	-0.59475	-0.26372	-0.20
0.22	-0.94225	-0.90854	-0.83238	-0.73560	-0.58428	-0.24852	-0.22
0.24	-0.94044	-0.90571	-0.82740	-0.72814	-0.57352	-0.23306	-0.24
0.26	-0.93855	-0.90278	-0.82224	-0.72043	-0.56246	-0.21732	-0.26
0.28	-0.93660	-0.89974	-0.81690	-0.71248	-0.55110	-0.20129	-0.28
0.30	-0.93456	-0.89658	-0.81136	-0.70425	-0.53939	-0.18495	-0.30
0.32	-0.93244	-0.89329	-0.80561	-0.69573	-0.52733	-0.16829	-0.32
0.34	-0.93023	-0.88986	-0.79963	-0.68690	-0.51489	-0.15128	-0.34
0.36	-0.92792	-0.88628	-0.79341	-0.67775	-0.50205	-0.13391	-0.36
0.38	-0.92550	-0.88255	-0.78693	-0.66823	-0.48877	-0.11615	-0.38
0.40	-0.92297	-0.87863	-0.78016	-0.65833	-0.47503	-0.09798	-0.40
0.42	-0.92031	-0.87453	-0.77309	-0.64802	-0.46080	-0.07937	-0.42
0.44	-0.91752	-0.87023	-0.76568	-0.63726	-0.44603	-0.06030	-0.44
0.46	-0.91458	-0.86570	-0.75791	-0.62602	-0.43068	-0.04074	-0.46
0.48	-0.91147	-0.86092	-0.74974	-0.61425	-0.41471	-0.02065	-0.48
0.50	-0.90818	-0.85587	-0.74114	-0.60191	-0.39808	0.0	-0.50
0.52	-0.90469	-0.85053	-0.73206	-0.58893	-0.38071	0.02125	-0.52
0.54	-0.90097	-0.84485	-0.72245	-0.57528	-0.36255	0.04315	-0.54
0.56	-0.89701	-0.83880	-0.71226	-0.56086	-0.34353	0.06573	-0.56
0.58	-0.89278	-0.83234	-0.70142	-0.54561	-0.32356	0.08906	-0.58
0.60	-0.88823	-0.82542	-0.68986	-0.52942	-0.30255	0.11318	-0.60
0.62	-0.88332	-0.81798	-0.67749	-0.51221	-0.28040	0.13818	-0.62
0.64	-0.87802	-0.80995	-0.66420	-0.49383	-0.25698	0.16410	-0.64
0.66	-0.87225	-0.80124	-0.64987	-0.47415	-0.23214	0.19105	-0.66
0.68	-0.86594	-0.79174	-0.63435	-0.45299	-0.20573	0.21910	-0.68
0.70	-0.85901	-0.78134	-0.61746	-0.43014	-0.17754	0.24836	-0.70
0.72	-0.85135	-0.76987	-0.59897	-0.40536	-0.14735	0.27895	-0.72
0.74	-0.84281	-0.75714	-0.57860	-0.37833	-0.11488	0.31100	-0.74
0.76	-0.83322	-0.74290	-0.55602	-0.34866	-0.07978	0.34466	-0.76
0.78	-0.82233	-0.72681	-0.53077	-0.31589	-0.04166	0.38013	-0.78
0.80	-0.80983	-0.70843	-0.50226	-0.27939	0.0	0.41759	-0.80
0.82	-0.79529	-0.68718	-0.46972	-0.23836	0.04585	0.45731	-0.82
0.84	-0.77808	-0.66222	-0.43207	-0.19172	0.09670	0.49959	-0.84
0.86	-0.75728	-0.63231	-0.38780	-0.13801	0.15364	0.54477	-0.86
0.88	-0.73148	-0.59559	-0.33464	-0.07514	0.21813	0.59330	-0.88
0.90	-0.69832	-0.54904	-0.26911	0.0	0.29218	0.64573	-0.90
0.92	-0.65358	-0.48736	-0.18542	0.09222	0.37868	0.70274	-0.92
0.94	-0.58869	-0.40017	-0.07301	0.20964	0.48204	0.76522	-0.94
0.96	-0.48260	-0.26330	0.09014	0.36733	0.60935	0.83438	-0.96
0.98	-0.26142	0.0	0.36191	0.59821	0.77324	0.91184	-0.98

↑ r_o

α =	.995 ↑-	.99 ↑-	.975 ↑-	.95 ↑-	.90 ↑-	.75 ↑-

TABLE 3.1   SAMPLE SIZE =   5

α =	.005 ↓+	.01 ↓+	.025 ↓+	.05 ↓+	.10 ↓+	.25 ↓+	
r_0 ↓							
-0.98	-0.99889	-0.99839	-0.99732	-0.99594	-0.99363	-0.98701	0.98
-0.96	-0.99778	-0.99677	-0.99461	-0.99186	-0.98723	-0.97409	0.96
-0.94	-0.99665	-0.99514	-0.99188	-0.98774	-0.98081	-0.96125	0.94
-0.92	-0.99551	-0.99348	-0.98913	-0.98360	-0.97437	-0.94846	0.92
-0.90	-0.99435	-0.99181	-0.98634	-0.97942	-0.96790	-0.93573	0.90
-0.88	-0.99318	-0.99011	-0.98354	-0.97522	-0.96139	-0.92305	0.88
-0.86	-0.99200	-0.98840	-0.98070	-0.97098	-0.95486	-0.91041	0.86
-0.84	-0.99080	-0.98667	-0.97784	-0.96670	-0.94829	-0.89781	0.84
-0.82	-0.98959	-0.98492	-0.97495	-0.96239	-0.94169	-0.88525	0.82
-0.80	-0.98836	-0.98315	-0.97202	-0.95805	-0.93505	-0.87271	0.80
-0.78	-0.98712	-0.98136	-0.96907	-0.95366	-0.92837	-0.86020	0.78
-0.76	-0.98586	-0.97954	-0.96608	-0.94923	-0.92165	-0.84771	0.76
-0.74	-0.98458	-0.97770	-0.96306	-0.94476	-0.91488	-0.83524	0.74
-0.72	-0.98329	-0.97584	-0.96000	-0.94025	-0.90807	-0.82278	0.72
-0.70	-0.98198	-0.97395	-0.95691	-0.93569	-0.90121	-0.81032	0.70
-0.68	-0.98064	-0.97203	-0.95378	-0.93109	-0.89430	-0.79787	0.68
-0.66	-0.97929	-0.97009	-0.95061	-0.92643	-0.88734	-0.78542	0.66
-0.64	-0.97792	-0.96812	-0.94740	-0.92173	-0.88033	-0.77296	0.64
-0.62	-0.97653	-0.96613	-0.94415	-0.91697	-0.87325	-0.76049	0.62
-0.60	-0.97512	-0.96410	-0.94086	-0.91216	-0.86612	-0.74801	0.60
-0.58	-0.97368	-0.96204	-0.93752	-0.90729	-0.85892	-0.73552	0.58
-0.56	-0.97222	-0.95995	-0.93413	-0.90237	-0.85166	-0.72300	0.56
-0.54	-0.97074	-0.95783	-0.93070	-0.89739	-0.84434	-0.71046	0.54
-0.52	-0.96923	-0.95568	-0.92722	-0.89234	-0.83694	-0.69788	0.52
-0.50	-0.96770	-0.95349	-0.92369	-0.88723	-0.82947	-0.68528	0.50
-0.48	-0.96614	-0.95126	-0.92010	-0.88205	-0.82192	-0.67264	0.48
-0.46	-0.96455	-0.94900	-0.91646	-0.87680	-0.81430	-0.65996	0.46
-0.44	-0.96293	-0.94670	-0.91276	-0.87148	-0.80659	-0.64724	0.44
-0.42	-0.96129	-0.94435	-0.90901	-0.86609	-0.79880	-0.63447	0.42
-0.40	-0.95961	-0.94197	-0.90519	-0.86061	-0.79092	-0.62164	0.40
-0.38	-0.95790	-0.93954	-0.90131	-0.85506	-0.78295	-0.60876	0.38
-0.36	-0.95616	-0.93706	-0.89736	-0.84943	-0.77489	-0.59582	0.36
-0.34	-0.95439	-0.93454	-0.89335	-0.84371	-0.76673	-0.58281	0.34
-0.32	-0.95257	-0.93197	-0.88926	-0.83789	-0.75846	-0.56974	0.32
-0.30	-0.95073	-0.92935	-0.88510	-0.83199	-0.75009	-0.55659	0.30
-0.28	-0.94884	-0.92668	-0.88086	-0.82599	-0.74161	-0.54337	0.28
-0.26	-0.94691	-0.92395	-0.87654	-0.81989	-0.73301	-0.53006	0.26
-0.24	-0.94494	-0.92116	-0.87214	-0.81368	-0.72429	-0.51667	0.24
-0.22	-0.94292	-0.91831	-0.86766	-0.80737	-0.71545	-0.50318	0.22
-0.20	-0.94086	-0.91540	-0.86308	-0.80094	-0.70648	-0.48960	0.20
-0.18	-0.93875	-0.91243	-0.85841	-0.79440	-0.69737	-0.47592	0.18
-0.16	-0.93659	-0.90939	-0.85364	-0.78773	-0.68813	-0.46213	0.16
-0.14	-0.93438	-0.90627	-0.84877	-0.78093	-0.67874	-0.44822	0.14
-0.12	-0.93211	-0.90309	-0.84379	-0.77400	-0.66920	-0.43421	0.12
-0.10	-0.92979	-0.89982	-0.83870	-0.76694	-0.65950	-0.42006	0.10
-0.08	-0.92740	-0.89647	-0.83350	-0.75973	-0.64964	-0.40579	0.08
-0.06	-0.92495	-0.89304	-0.82817	-0.75236	-0.63960	-0.39139	0.06
-0.04	-0.92244	-0.88952	-0.82272	-0.74484	-0.62939	-0.37684	0.04
-0.02	-0.91986	-0.88591	-0.81713	-0.73716	-0.61899	-0.36215	0.02

↑ r_0

α =	.995 ↑-	.99 ↑-	.975 ↑-	.95 ↑-	.90 ↑-	.75 ↑-

TABLE 3.2   SAMPLE SIZE =   5   (CONT.)

α =	.005 ↓+	.01 ↓+	.025 ↓+	.05 ↓+	.10 ↓+	.25 ↓+	
r_0 ↓							
0.00	-0.91720	-0.88219	-0.81140	-0.72930	-0.60840	-0.34730	0.00
0.02	-0.91447	-0.87838	-0.80553	-0.72126	-0.59760	-0.33229	-0.02
0.04	-0.91165	-0.87445	-0.79950	-0.71303	-0.58660	-0.31711	-0.04
0.06	-0.90875	-0.87041	-0.79331	-0.70461	-0.57537	-0.30175	-0.06
0.08	-0.90576	-0.86625	-0.78695	-0.69597	-0.56391	-0.28621	-0.08
0.10	-0.90267	-0.86196	-0.78041	-0.68712	-0.55220	-0.27047	-0.10
0.12	-0.89948	-0.85753	-0.77368	-0.67803	-0.54025	-0.25454	-0.12
0.14	-0.89619	-0.85297	-0.76675	-0.66871	-0.52802	-0.23839	-0.14
0.16	-0.89278	-0.84825	-0.75961	-0.65913	-0.51552	-0.22202	-0.16
0.18	-0.88925	-0.84337	-0.75225	-0.64928	-0.50272	-0.20542	-0.18
0.20	-0.88559	-0.83832	-0.74465	-0.63915	-0.48961	-0.18858	-0.20
0.22	-0.88180	-0.83309	-0.73681	-0.62872	-0.47618	-0.17148	-0.22
0.24	-0.87786	-0.82767	-0.72869	-0.61797	-0.46241	-0.15412	-0.24
0.26	-0.87376	-0.82204	-0.72030	-0.60689	-0.44828	-0.13649	-0.26
0.28	-0.86950	-0.81619	-0.71161	-0.59545	-0.43377	-0.11856	-0.28
0.30	-0.86506	-0.81011	-0.70259	-0.58363	-0.41886	-0.10032	-0.30
0.32	-0.86043	-0.80378	-0.69324	-0.57142	-0.40352	-0.08177	-0.32
0.34	-0.85559	-0.79717	-0.68352	-0.55877	-0.38774	-0.06288	-0.34
0.36	-0.85053	-0.79028	-0.67340	-0.54567	-0.37147	-0.04364	-0.36
0.38	-0.84523	-0.78307	-0.66287	-0.53209	-0.35471	-0.02403	-0.38
0.40	-0.83967	-0.77552	-0.65189	-0.51798	-0.33741	-0.00402	-0.40
0.42	-0.83382	-0.76760	-0.64042	-0.50332	-0.31953	0.01639	-0.42
0.44	-0.82767	-0.75928	-0.62842	-0.48805	-0.30105	0.03724	-0.44
0.46	-0.82117	-0.75053	-0.61585	-0.47214	-0.28191	0.05854	-0.46
0.48	-0.81431	-0.74131	-0.60267	-0.45554	-0.26209	0.08033	-0.48
0.50	-0.80705	-0.73156	-0.58881	-0.43818	-0.24151	0.10262	-0.50
0.52	-0.79933	-0.72124	-0.57421	-0.42000	-0.22013	0.12545	-0.52
0.54	-0.79112	-0.71029	-0.55881	-0.40094	-0.19790	0.14886	-0.54
0.56	-0.78236	-0.69865	-0.54253	-0.38091	-0.17473	0.17287	-0.56
0.58	-0.77299	-0.68622	-0.52526	-0.35982	-0.15055	0.19752	-0.58
0.60	-0.76292	-0.67293	-0.50692	-0.33758	-0.12529	0.22285	-0.60
0.62	-0.75208	-0.65866	-0.48737	-0.31405	-0.09884	0.24891	-0.62
0.64	-0.74035	-0.64330	-0.46648	-0.28911	-0.07110	0.27575	-0.64
0.66	-0.72761	-0.62668	-0.44408	-0.26260	-0.04194	0.30342	-0.66
0.68	-0.71371	-0.60864	-0.41997	-0.23434	-0.01122	0.33198	-0.68
0.70	-0.69846	-0.58896	-0.39393	-0.20411	0.02120	0.36149	-0.70
0.72	-0.68164	-0.56737	-0.36567	-0.17168	0.05552	0.39202	-0.72
0.74	-0.66296	-0.54354	-0.33486	-0.13673	0.09196	0.42367	-0.74
0.76	-0.64205	-0.51707	-0.30107	-0.09891	0.13074	0.45650	-0.76
0.78	-0.61846	-0.48744	-0.26379	-0.05779	0.17218	0.49064	-0.78
0.80	-0.59156	-0.45396	-0.22237	-0.01284	0.21662	0.52618	-0.80
0.82	-0.56051	-0.41574	-0.17596	0.03662	0.26447	0.56326	-0.82
0.84	-0.52419	-0.37156	-0.12348	0.09141	0.31624	0.60202	-0.84
0.86	-0.48093	-0.31972	-0.06343	0.15263	0.37255	0.64263	-0.86
0.88	-0.42829	-0.25776	0.00621	0.22170	0.43417	0.68528	-0.88
0.90	-0.36242	-0.18193	0.08834	0.30052	0.50209	0.73019	-0.90
0.92	-0.27689	-0.08626	0.18726	0.39178	0.57757	0.77764	-0.92
0.94	-0.15987	0.03963	0.30976	0.49931	0.66228	0.82793	-0.94
0.96	0.01350	0.21572	0.46728	0.62898	0.75852	0.88143	-0.96
0.98	0.30846	0.48781	0.68137	0.79024	0.86949	0.93860	-0.98

↑ r_0

α =	.995 ↑-	.99 ↑-	.975 ↑-	.95 ↑-	.90 ↑-	.75 ↑-

TABLE 4.1   SAMPLE SIZE =   6

α =	.005 ↓+	.01 ↓+	.025 ↓+	.05 ↓+	.10 ↓+	.25 ↓+	
r_0 ↓							
-0.98	-0.99838	-0.99779	-0.99660	-0.99517	-0.99288	-0.98681	0.98
-0.96	-0.99675	-0.99556	-0.99316	-0.99029	-0.98572	-0.97365	0.96
-0.94	-0.99509	-0.99330	-0.98969	-0.98537	-0.97852	-0.96052	0.94
-0.92	-0.99341	-0.99101	-0.98618	-0.98041	-0.97128	-0.94741	0.92
-0.90	-0.99171	-0.98870	-0.98264	-0.97541	-0.96400	-0.93432	0.90
-0.88	-0.98998	-0.98635	-0.97905	-0.97036	-0.95667	-0.92124	0.88
-0.86	-0.98824	-0.98398	-0.97543	-0.96526	-0.94929	-0.90818	0.86
-0.84	-0.98646	-0.98157	-0.97176	-0.96012	-0.94186	-0.89512	0.84
-0.82	-0.98467	-0.97914	-0.96805	-0.95492	-0.93438	-0.88207	0.82
-0.80	-0.98285	-0.97667	-0.96430	-0.94967	-0.92685	-0.86903	0.80
-0.78	-0.98100	-0.97416	-0.96050	-0.94437	-0.91927	-0.85598	0.78
-0.76	-0.97912	-0.97162	-0.95666	-0.93902	-0.91162	-0.84292	0.76
-0.74	-0.97722	-0.96905	-0.95276	-0.93360	-0.90392	-0.82986	0.74
-0.72	-0.97529	-0.96644	-0.94882	-0.92813	-0.89616	-0.81678	0.72
-0.70	-0.97332	-0.96379	-0.94483	-0.92260	-0.88833	-0.80369	0.70
-0.68	-0.97133	-0.96110	-0.94078	-0.91700	-0.88044	-0.79058	0.68
-0.66	-0.96931	-0.95837	-0.93668	-0.91134	-0.87248	-0.77745	0.66
-0.64	-0.96725	-0.95560	-0.93252	-0.90562	-0.86444	-0.76429	0.64
-0.62	-0.96516	-0.95279	-0.92831	-0.89982	-0.85634	-0.75111	0.62
-0.60	-0.96303	-0.94993	-0.92404	-0.89395	-0.84816	-0.73789	0.60
-0.58	-0.96087	-0.94703	-0.91970	-0.88802	-0.83990	-0.72464	0.58
-0.56	-0.95867	-0.94407	-0.91530	-0.88200	-0.83156	-0.71135	0.56
-0.54	-0.95644	-0.94107	-0.91083	-0.87591	-0.82314	-0.69802	0.54
-0.52	-0.95416	-0.93802	-0.90630	-0.86974	-0.81463	-0.68464	0.52
-0.50	-0.95184	-0.93492	-0.90170	-0.86348	-0.80604	-0.67122	0.50
-0.48	-0.94948	-0.93176	-0.89702	-0.85714	-0.79735	-0.65774	0.48
-0.46	-0.94707	-0.92854	-0.89227	-0.85071	-0.78856	-0.64422	0.46
-0.44	-0.94462	-0.92527	-0.88744	-0.84419	-0.77968	-0.63063	0.44
-0.42	-0.94212	-0.92194	-0.88254	-0.83757	-0.77069	-0.61698	0.42
-0.40	-0.93958	-0.91854	-0.87755	-0.83085	-0.76160	-0.60326	0.40
-0.38	-0.93698	-0.91508	-0.87247	-0.82404	-0.75240	-0.58948	0.38
-0.36	-0.93432	-0.91155	-0.86730	-0.81712	-0.74309	-0.57562	0.36
-0.34	-0.93162	-0.90795	-0.86204	-0.81009	-0.73366	-0.56169	0.34
-0.32	-0.92885	-0.90428	-0.85669	-0.80295	-0.72411	-0.54768	0.32
-0.30	-0.92603	-0.90054	-0.85124	-0.79569	-0.71444	-0.53358	0.30
-0.28	-0.92314	-0.89671	-0.84568	-0.78831	-0.70464	-0.51940	0.28
-0.26	-0.92019	-0.89281	-0.84002	-0.78081	-0.69470	-0.50512	0.26
-0.24	-0.91717	-0.88882	-0.83425	-0.77317	-0.68462	-0.49075	0.24
-0.22	-0.91409	-0.88474	-0.82836	-0.76540	-0.67441	-0.47628	0.22
-0.20	-0.91093	-0.88057	-0.82235	-0.75750	-0.66404	-0.46170	0.20
-0.18	-0.90769	-0.87631	-0.81622	-0.74945	-0.65352	-0.44701	0.18
-0.16	-0.90438	-0.87194	-0.80996	-0.74124	-0.64284	-0.43221	0.16
-0.14	-0.90098	-0.86748	-0.80356	-0.73289	-0.63199	-0.41729	0.14
-0.12	-0.89749	-0.86290	-0.79702	-0.72436	-0.62097	-0.40224	0.12
-0.10	-0.89392	-0.85821	-0.79034	-0.71568	-0.60977	-0.38707	0.10
-0.08	-0.89025	-0.85340	-0.78350	-0.70681	-0.59839	-0.37176	0.08
-0.06	-0.88648	-0.84847	-0.77651	-0.69776	-0.58682	-0.35631	0.06
-0.04	-0.88261	-0.84341	-0.76935	-0.68852	-0.57504	-0.34072	0.04
-0.02	-0.87862	-0.83821	-0.76201	-0.67908	-0.56306	-0.32498	0.02

↑ r_0

α =	.995 ↑−	.99 ↑−	.975 ↑−	.95 ↑−	.90 ↑−	.75 ↑−

TABLE 4.2   SAMPLE SIZE = 6   (CONT.)

α =	.005 ↓+	.01 ↓+	.025 ↓+	.05 ↓+	.10 ↓+	.25 ↓+	
r_0 ↓							
0.00	−0.87453	−0.83287	−0.75449	−0.66944	−0.55086	−0.30907	0.00
0.02	−0.87031	−0.82738	−0.74678	−0.65958	−0.53844	−0.29301	−0.02
0.04	−0.86596	−0.82174	−0.73888	−0.64949	−0.52578	−0.27677	−0.04
0.06	−0.86149	−0.81593	−0.73076	−0.63917	−0.51288	−0.26036	−0.06
0.08	−0.85687	−0.80994	−0.72242	−0.62859	−0.49973	−0.24376	−0.08
0.10	−0.85210	−0.80377	−0.71385	−0.61776	−0.48631	−0.22697	−0.10
0.12	−0.84717	−0.79741	−0.70504	−0.60666	−0.47261	−0.20998	−0.12
0.14	−0.84208	−0.79085	−0.69598	−0.59527	−0.45862	−0.19278	−0.14
0.16	−0.83682	−0.78407	−0.68665	−0.58359	−0.44434	−0.17537	−0.16
0.18	−0.83136	−0.77706	−0.67703	−0.57159	−0.42973	−0.15773	−0.18
0.20	−0.82571	−0.76981	−0.66712	−0.55926	−0.41480	−0.13986	−0.20
0.22	−0.81985	−0.76230	−0.65689	−0.54659	−0.39952	−0.12174	−0.22
0.24	−0.81376	−0.75452	−0.64632	−0.53355	−0.38387	−0.10337	−0.24
0.26	−0.80744	−0.74645	−0.63540	−0.52012	−0.36785	−0.08473	−0.26
0.28	−0.80086	−0.73807	−0.62411	−0.50629	−0.35143	−0.06582	−0.28
0.30	−0.79401	−0.72937	−0.61242	−0.49203	−0.33459	−0.04662	−0.30
0.32	−0.78686	−0.72031	−0.60030	−0.47731	−0.31730	−0.02712	−0.32
0.34	−0.77940	−0.71087	−0.58774	−0.46211	−0.29955	−0.00731	−0.34
0.36	−0.77160	−0.70103	−0.57469	−0.44640	−0.28132	0.01283	−0.36
0.38	−0.76344	−0.69076	−0.56112	−0.43014	−0.26256	0.03332	−0.38
0.40	−0.75488	−0.68002	−0.54701	−0.41330	−0.24326	0.05416	−0.40
0.42	−0.74590	−0.66877	−0.53231	−0.39585	−0.22338	0.07537	−0.42
0.44	−0.73645	−0.65698	−0.51697	−0.37774	−0.20289	0.09698	−0.44
0.46	−0.72650	−0.64460	−0.50094	−0.35893	−0.18175	0.11900	−0.46
0.48	−0.71600	−0.63157	−0.48418	−0.33936	−0.15992	0.14145	−0.48
0.50	−0.70490	−0.61784	−0.46663	−0.31898	−0.13735	0.16435	−0.50
0.52	−0.69313	−0.60335	−0.44820	−0.29772	−0.11401	0.18772	−0.52
0.54	−0.68064	−0.58802	−0.42884	−0.27553	−0.08983	0.21160	−0.54
0.56	−0.66734	−0.57176	−0.40845	−0.25233	−0.06476	0.23599	−0.56
0.58	−0.65315	−0.55448	−0.38694	−0.22802	−0.03873	0.26095	−0.58
0.60	−0.63796	−0.53606	−0.36420	−0.20252	−0.01168	0.28648	−0.60
0.62	−0.62165	−0.51639	−0.34010	−0.17571	0.01647	0.31263	−0.62
0.64	−0.60408	−0.49531	−0.31451	−0.14749	0.04581	0.33942	−0.64
0.66	−0.58508	−0.47264	−0.28725	−0.11770	0.07643	0.36691	−0.66
0.68	−0.56446	−0.44818	−0.25814	−0.08619	0.10844	0.39512	−0.68
0.70	−0.54197	−0.42168	−0.22695	−0.05279	0.14197	0.42410	−0.70
0.72	−0.51732	−0.39285	−0.19343	−0.01729	0.17714	0.45391	−0.72
0.74	−0.49016	−0.36133	−0.15725	0.02056	0.21412	0.48459	−0.74
0.76	−0.46003	−0.32667	−0.11805	0.06103	0.25307	0.51619	−0.76
0.78	−0.42637	−0.28833	−0.07537	0.10446	0.29420	0.54879	−0.78
0.80	−0.38846	−0.24562	−0.02868	0.15125	0.33774	0.58246	−0.80
0.82	−0.34534	−0.19766	0.02273	0.20186	0.38396	0.61726	−0.82
0.84	−0.29576	−0.14330	0.07969	0.25687	0.43317	0.65330	−0.84
0.86	−0.23797	−0.08100	0.14331	0.31699	0.48574	0.69065	−0.86
0.88	−0.16952	−0.00866	0.21499	0.38309	0.54211	0.72943	−0.88
0.90	−0.08677	0.07669	0.29664	0.45628	0.60281	0.76975	−0.90
0.92	0.01586	0.17943	0.39082	0.53800	0.66847	0.81176	−0.92
0.94	0.14756	0.30630	0.50116	0.63012	0.73990	0.85560	−0.94
0.96	0.32481	0.46841	0.63303	0.73519	0.81806	0.90144	−0.96
0.98	0.58108	0.68581	0.79470	0.85676	0.90422	0.94950	−0.98

↑ r_0

α =	.995 ↑−	.99 ↑−	.975 ↑−	.95 ↑−	.90 ↑−	.75 ↑−

TABLE 5.1   SAMPLE SIZE = 7

α =	.005 ↓+	.01 ↓+	.025 ↓+	.05 ↓+	.10 ↓+	.25 ↓+	
r₀ ↓							
-0.98	-0.99789	-0.99724	-0.99597	-0.99451	-0.99225	-0.98656	0.98
-0.96	-0.99576	-0.99444	-0.99190	-0.98896	-0.98446	-0.97313	0.96
-0.94	-0.99359	-0.99161	-0.98778	-0.98336	-0.97661	-0.95971	0.94
-0.92	-0.99140	-0.98875	-0.98361	-0.97772	-0.96871	-0.94629	0.92
-0.90	-0.98918	-0.98584	-0.97940	-0.97201	-0.96076	-0.93286	0.90
-0.88	-0.98692	-0.98290	-0.97514	-0.96625	-0.95275	-0.91944	0.88
-0.86	-0.98463	-0.97992	-0.97083	-0.96044	-0.94468	-0.90601	0.86
-0.84	-0.98231	-0.97689	-0.96646	-0.95456	-0.93656	-0.89257	0.84
-0.82	-0.97996	-0.97383	-0.96205	-0.94863	-0.92837	-0.87912	0.82
-0.80	-0.97757	-0.97072	-0.95758	-0.94263	-0.92012	-0.86565	0.80
-0.78	-0.97515	-0.96757	-0.95305	-0.93657	-0.91180	-0.85217	0.78
-0.76	-0.97268	-0.96437	-0.94846	-0.93044	-0.90342	-0.83866	0.76
-0.74	-0.97018	-0.96113	-0.94382	-0.92424	-0.89497	-0.82513	0.74
-0.72	-0.96764	-0.95784	-0.93911	-0.91798	-0.88644	-0.81158	0.72
-0.70	-0.96506	-0.95450	-0.93435	-0.91164	-0.87785	-0.79800	0.70
-0.68	-0.96244	-0.95111	-0.92951	-0.90523	-0.86917	-0.78439	0.68
-0.66	-0.95978	-0.94766	-0.92461	-0.89874	-0.86042	-0.77074	0.66
-0.64	-0.95707	-0.94416	-0.91964	-0.89217	-0.85159	-0.75706	0.64
-0.62	-0.95431	-0.94061	-0.91461	-0.88553	-0.84267	-0.74334	0.62
-0.60	-0.95151	-0.93700	-0.90949	-0.87880	-0.83367	-0.72957	0.60
-0.58	-0.94866	-0.93332	-0.90431	-0.87198	-0.82458	-0.71576	0.58
-0.56	-0.94576	-0.92959	-0.89904	-0.86508	-0.81540	-0.70190	0.56
-0.54	-0.94281	-0.92579	-0.89370	-0.85809	-0.80612	-0.68799	0.54
-0.52	-0.93980	-0.92193	-0.88827	-0.85100	-0.79675	-0.67403	0.52
-0.50	-0.93674	-0.91800	-0.88276	-0.84382	-0.78728	-0.66001	0.50
-0.48	-0.93362	-0.91400	-0.87716	-0.83653	-0.77770	-0.64593	0.48
-0.46	-0.93044	-0.90993	-0.87147	-0.82915	-0.76802	-0.63179	0.46
-0.44	-0.92719	-0.90578	-0.86569	-0.82166	-0.75824	-0.61758	0.44
-0.42	-0.92389	-0.90156	-0.85981	-0.81406	-0.74833	-0.60330	0.42
-0.40	-0.92052	-0.89726	-0.85384	-0.80635	-0.73832	-0.58895	0.40
-0.38	-0.91708	-0.89287	-0.84775	-0.79852	-0.72818	-0.57452	0.38
-0.36	-0.91357	-0.88840	-0.84157	-0.79057	-0.71792	-0.56002	0.36
-0.34	-0.90998	-0.88384	-0.83527	-0.78250	-0.70753	-0.54543	0.34
-0.32	-0.90632	-0.87918	-0.82886	-0.77430	-0.69701	-0.53076	0.32
-0.30	-0.90258	-0.87443	-0.82233	-0.76597	-0.68635	-0.51599	0.30
-0.28	-0.89876	-0.86959	-0.81567	-0.75750	-0.67555	-0.50114	0.28
-0.26	-0.89485	-0.86463	-0.80889	-0.74889	-0.66461	-0.48619	0.26
-0.24	-0.89085	-0.85958	-0.80198	-0.74013	-0.65352	-0.47113	0.24
-0.22	-0.88676	-0.85441	-0.79494	-0.73123	-0.64228	-0.45598	0.22
-0.20	-0.88257	-0.84912	-0.78775	-0.72216	-0.63087	-0.44071	0.20
-0.18	-0.87828	-0.84372	-0.78041	-0.71293	-0.61930	-0.42533	0.18
-0.16	-0.87389	-0.83818	-0.77293	-0.70354	-0.60756	-0.40984	0.16
-0.14	-0.86938	-0.83252	-0.76528	-0.69397	-0.59565	-0.39422	0.14
-0.12	-0.86477	-0.82672	-0.75747	-0.68422	-0.58355	-0.37848	0.12
-0.10	-0.86003	-0.82078	-0.74948	-0.67428	-0.57126	-0.36261	0.10
-0.08	-0.85517	-0.81470	-0.74132	-0.66414	-0.55877	-0.34660	0.08
-0.06	-0.85017	-0.80845	-0.73297	-0.65380	-0.54608	-0.33046	0.06
-0.04	-0.84504	-0.80205	-0.72443	-0.64325	-0.53319	-0.31417	0.04
-0.02	-0.83977	-0.79547	-0.71569	-0.63249	-0.52007	-0.29773	0.02

↑ r₀

α =	.995 ↑-	.99 ↑-	.975 ↑-	.95 ↑-	.90 ↑-	.75 ↑-

TABLE 5.2   SAMPLE SIZE = 7   (CONT.)

α =	.005 ↓+	.01 ↓+	.025 ↓+	.05 ↓+	.10 ↓+	.25 ↓+	
r_0 ↓							
0.00	−0.83434	−0.78872	−0.70673	−0.62149	−0.50673	−0.28113	0.00
0.02	−0.82876	−0.78178	−0.69756	−0.61025	−0.49315	−0.26437	−0.02
0.04	−0.82301	−0.77465	−0.68815	−0.59877	−0.47933	−0.24744	−0.04
0.06	−0.81709	−0.76731	−0.67851	−0.58703	−0.46525	−0.23034	−0.06
0.08	−0.81098	−0.75976	−0.66861	−0.57502	−0.45092	−0.21306	−0.08
0.10	−0.80468	−0.75198	−0.65844	−0.56273	−0.43631	−0.19560	−0.10
0.12	−0.79817	−0.74397	−0.64800	−0.55014	−0.42141	−0.17794	−0.12
0.14	−0.79145	−0.73570	−0.63727	−0.53725	−0.40622	−0.16008	−0.14
0.16	−0.78450	−0.72717	−0.62623	−0.52404	−0.39072	−0.14202	−0.16
0.18	−0.77731	−0.71836	−0.61487	−0.51049	−0.37489	−0.12374	−0.18
0.20	−0.76987	−0.70926	−0.60317	−0.49658	−0.35874	−0.10523	−0.20
0.22	−0.76215	−0.69984	−0.59112	−0.48231	−0.34223	−0.08650	−0.22
0.24	−0.75414	−0.69010	−0.57869	−0.46765	−0.32536	−0.06752	−0.24
0.26	−0.74583	−0.68000	−0.56587	−0.45257	−0.30810	−0.04829	−0.26
0.28	−0.73719	−0.66953	−0.55262	−0.43707	−0.29044	−0.02881	−0.28
0.30	−0.72821	−0.65867	−0.53893	−0.42112	−0.27237	−0.00905	−0.30
0.32	−0.71885	−0.64738	−0.52478	−0.40469	−0.25386	0.01098	−0.32
0.34	−0.70909	−0.63564	−0.51012	−0.38776	−0.23489	0.03131	−0.34
0.36	−0.69891	−0.62342	−0.49493	−0.37029	−0.21544	0.05194	−0.36
0.38	−0.68826	−0.61069	−0.47918	−0.35226	−0.19548	0.07288	−0.38
0.40	−0.67712	−0.59741	−0.46283	−0.33364	−0.17498	0.09415	−0.40
0.42	−0.66545	−0.58353	−0.44584	−0.31439	−0.15393	0.11576	−0.42
0.44	−0.65320	−0.56901	−0.42816	−0.29446	−0.13229	0.13773	−0.44
0.46	−0.64032	−0.55380	−0.40975	−0.27383	−0.11002	0.16007	−0.46
0.48	−0.62677	−0.53785	−0.39055	−0.25243	−0.08710	0.18279	−0.48
0.50	−0.61247	−0.52108	−0.37050	−0.23022	−0.06348	0.20592	−0.50
0.52	−0.59737	−0.50344	−0.34954	−0.20714	−0.03913	0.22947	−0.52
0.54	−0.58137	−0.48484	−0.32760	−0.18314	−0.01400	0.25347	−0.54
0.56	−0.56441	−0.46519	−0.30459	−0.15814	0.01197	0.27792	−0.56
0.58	−0.54636	−0.44439	−0.28041	−0.13208	0.03881	0.30286	−0.58
0.60	−0.52713	−0.42233	−0.25498	−0.10486	0.06659	0.32830	−0.60
0.62	−0.50657	−0.39888	−0.22818	−0.07639	0.09537	0.35427	−0.62
0.64	−0.48452	−0.37387	−0.19987	−0.04658	0.12522	0.38080	−0.64
0.66	−0.46082	−0.34714	−0.16990	−0.01531	0.15622	0.40792	−0.66
0.68	−0.43523	−0.31849	−0.13811	0.01756	0.18844	0.43565	−0.68
0.70	−0.40752	−0.28767	−0.10430	0.05217	0.22199	0.46402	−0.70
0.72	−0.37736	−0.25440	−0.06824	0.08867	0.25696	0.49308	−0.72
0.74	−0.34441	−0.21834	−0.02968	0.12727	0.29346	0.52286	−0.74
0.76	−0.30820	−0.17910	0.01169	0.16818	0.33163	0.55340	−0.76
0.78	−0.26819	−0.13617	0.05625	0.21165	0.37161	0.58474	−0.78
0.80	−0.22368	−0.08896	0.10441	0.25797	0.41356	0.61693	−0.80
0.82	−0.17379	−0.03672	0.15670	0.30747	0.45767	0.65003	−0.82
0.84	−0.11739	0.02150	0.21374	0.36056	0.50413	0.68409	−0.84
0.86	−0.05296	0.08690	0.27632	0.41770	0.55321	0.71917	−0.86
0.88	0.02153	0.16107	0.34540	0.47947	0.60516	0.75534	−0.88
0.90	0.10889	0.24611	0.42219	0.54655	0.66031	0.79267	−0.90
0.92	0.21318	0.34494	0.50827	0.61979	0.71905	0.83125	−0.92
0.94	0.34052	0.46166	0.60569	0.70022	0.78180	0.87116	−0.94
0.96	0.50058	0.60237	0.71726	0.78918	0.84909	0.91251	−0.96
0.98	0.70991	0.77654	0.84683	0.88835	0.92157	0.95542	−0.98

↑ r_0

α =	.995 ↑−	.99 ↑−	.975 ↑−	.95 ↑−	.90 ↑−	.75 ↑−

TABLE 6.1   SAMPLE SIZE =   8

α =	.005 ↓+	.01 ↓+	.025 ↓+	.05 ↓+	.10 ↓+	.25 ↓+	
r_0 ↓							
−0.98	−0.99743	−0.99673	−0.99542	−0.99394	−0.99172	−0.98632	0.98
−0.96	−0.99483	−0.99343	−0.99078	−0.98782	−0.98338	−0.97264	0.96
−0.94	−0.99220	−0.99008	−0.98609	−0.98164	−0.97499	−0.95894	0.94
−0.92	−0.98952	−0.98668	−0.98135	−0.97540	−0.96654	−0.94524	0.92
−0.90	−0.98681	−0.98325	−0.97655	−0.96910	−0.95802	−0.93153	0.90
−0.88	−0.98406	−0.97976	−0.97170	−0.96273	−0.94944	−0.91779	0.88
−0.86	−0.98127	−0.97623	−0.96679	−0.95630	−0.94080	−0.90405	0.86
−0.84	−0.97844	−0.97265	−0.96182	−0.94980	−0.93209	−0.89028	0.84
−0.82	−0.97557	−0.96902	−0.95678	−0.94324	−0.92331	−0.87649	0.82
−0.80	−0.97265	−0.96533	−0.95169	−0.93660	−0.91447	−0.86267	0.80
−0.78	−0.96969	−0.96160	−0.94652	−0.92990	−0.90555	−0.84883	0.78
−0.76	−0.96668	−0.95781	−0.94130	−0.92312	−0.89655	−0.83496	0.76
−0.74	−0.96363	−0.95396	−0.93600	−0.91626	−0.88748	−0.82106	0.74
−0.72	−0.96053	−0.95006	−0.93063	−0.90932	−0.87833	−0.80712	0.72
−0.70	−0.95737	−0.94609	−0.92519	−0.90231	−0.86910	−0.79315	0.70
−0.68	−0.95417	−0.94207	−0.91968	−0.89521	−0.85979	−0.77913	0.68
−0.66	−0.95091	−0.93798	−0.91409	−0.88803	−0.85039	−0.76508	0.66
−0.64	−0.94760	−0.93383	−0.90842	−0.88076	−0.84090	−0.75098	0.64
−0.62	−0.94423	−0.92961	−0.90268	−0.87340	−0.83132	−0.73683	0.62
−0.60	−0.94080	−0.92532	−0.89684	−0.86595	−0.82165	−0.72264	0.60
−0.58	−0.93732	−0.92096	−0.89093	−0.85840	−0.81189	−0.70839	0.58
−0.56	−0.93377	−0.91653	−0.88492	−0.85076	−0.80203	−0.69409	0.56
−0.54	−0.93015	−0.91202	−0.87882	−0.84302	−0.79206	−0.67973	0.54
−0.52	−0.92648	−0.90744	−0.87264	−0.83517	−0.78199	−0.66531	0.52
−0.50	−0.92273	−0.90278	−0.86635	−0.82722	−0.77182	−0.65083	0.50
−0.48	−0.91891	−0.89803	−0.85996	−0.81916	−0.76154	−0.63629	0.48
−0.46	−0.91502	−0.89320	−0.85348	−0.81099	−0.75114	−0.62167	0.46
−0.44	−0.91105	−0.88828	−0.84688	−0.80270	−0.74063	−0.60699	0.44
−0.42	−0.90701	−0.88326	−0.84018	−0.79429	−0.72999	−0.59223	0.42
−0.40	−0.90289	−0.87816	−0.83337	−0.78576	−0.71924	−0.57740	0.40
−0.38	−0.89868	−0.87295	−0.82644	−0.77711	−0.70836	−0.56249	0.38
−0.36	−0.89438	−0.86765	−0.81939	−0.76832	−0.69735	−0.54749	0.36
−0.34	−0.89000	−0.86224	−0.81221	−0.75940	−0.68620	−0.53241	0.34
−0.32	−0.88552	−0.85672	−0.80491	−0.75034	−0.67491	−0.51724	0.32
−0.30	−0.88095	−0.85109	−0.79748	−0.74114	−0.66349	−0.50198	0.30
−0.28	−0.87627	−0.84534	−0.78991	−0.73179	−0.65191	−0.48662	0.28
−0.26	−0.87149	−0.83948	−0.78219	−0.72229	−0.64019	−0.47117	0.26
−0.24	−0.86660	−0.83348	−0.77433	−0.71263	−0.62831	−0.45561	0.24
−0.22	−0.86161	−0.82736	−0.76632	−0.70280	−0.61627	−0.43995	0.22
−0.20	−0.85649	−0.82110	−0.75815	−0.69281	−0.60406	−0.42418	0.20
−0.18	−0.85125	−0.81471	−0.74982	−0.68264	−0.59168	−0.40830	0.18
−0.16	−0.84588	−0.80816	−0.74132	−0.67230	−0.57913	−0.39229	0.16
−0.14	−0.84039	−0.80147	−0.73265	−0.66176	−0.56639	−0.37617	0.14
−0.12	−0.83475	−0.79461	−0.72379	−0.65104	−0.55346	−0.35992	0.12
−0.10	−0.82897	−0.78759	−0.71474	−0.64011	−0.54034	−0.34355	0.10
−0.08	−0.82304	−0.78041	−0.70549	−0.62898	−0.52702	−0.32704	0.08
−0.06	−0.81696	−0.77304	−0.69604	−0.61763	−0.51349	−0.31039	0.06
−0.04	−0.81071	−0.76548	−0.68638	−0.60605	−0.49975	−0.29360	0.04
−0.02	−0.80428	−0.75773	−0.67650	−0.59425	−0.48578	−0.27666	0.02

↑ r_0

α =	.995 ↑−	.99 ↑−	.975 ↑−	.95 ↑−	.90 ↑−	.75 ↑−

TABLE 6.2   SAMPLE SIZE =   8   (CONT.)

α =	.005 ↓+	.01 ↓+	.025 ↓+	.05 ↓+	.10 ↓+	.25 ↓+	
r_0 ↓							
0.00	−0.79768	−0.74978	−0.66638	−0.58221	−0.47159	−0.25957	0.00
0.02	−0.79089	−0.74161	−0.65603	−0.56991	−0.45715	−0.24233	−0.02
0.04	−0.78390	−0.73322	−0.64542	−0.55736	−0.44247	−0.22492	−0.04
0.06	−0.77671	−0.72459	−0.63455	−0.54453	−0.42754	−0.20734	−0.06
0.08	−0.76929	−0.71572	−0.62341	−0.53142	−0.41233	−0.18959	−0.08
0.10	−0.76165	−0.70659	−0.61198	−0.51802	−0.39686	−0.17166	−0.10
0.12	−0.75376	−0.69720	−0.60025	−0.50432	−0.38109	−0.15354	−0.12
0.14	−0.74562	−0.68751	−0.58821	−0.49029	−0.36504	−0.13523	−0.14
0.16	−0.73721	−0.67753	−0.57584	−0.47594	−0.34867	−0.11673	−0.16
0.18	−0.72852	−0.66724	−0.56313	−0.46123	−0.33198	−0.09802	−0.18
0.20	−0.71953	−0.65661	−0.55005	−0.44617	−0.31496	−0.07909	−0.20
0.22	−0.71022	−0.64563	−0.53660	−0.43072	−0.29760	−0.05995	−0.22
0.24	−0.70058	−0.63428	−0.52274	−0.41487	−0.27987	−0.04058	−0.24
0.26	−0.69057	−0.62254	−0.50847	−0.39861	−0.26177	−0.02098	−0.26
0.28	−0.68019	−0.61038	−0.49375	−0.38192	−0.24328	−0.00113	−0.28
0.30	−0.66940	−0.59779	−0.47856	−0.36476	−0.22437	0.01897	−0.30
0.32	−0.65819	−0.58472	−0.46288	−0.34712	−0.20505	0.03934	−0.32
0.34	−0.64651	−0.57116	−0.44668	−0.32898	−0.18527	0.05997	−0.34
0.36	−0.63434	−0.55707	−0.42993	−0.31030	−0.16503	0.08088	−0.36
0.38	−0.62165	−0.54241	−0.41259	−0.29107	−0.14430	0.10209	−0.38
0.40	−0.60839	−0.52715	−0.39463	−0.27124	−0.12307	0.12359	−0.40
0.42	−0.59453	−0.51125	−0.37602	−0.25078	−0.10129	0.14541	−0.42
0.44	−0.58001	−0.49465	−0.35669	−0.22967	−0.07896	0.16756	−0.44
0.46	−0.56479	−0.47731	−0.33662	−0.20785	−0.05603	0.19004	−0.46
0.48	−0.54880	−0.45916	−0.31575	−0.18530	−0.03249	0.21288	−0.48
0.50	−0.53199	−0.44015	−0.29402	−0.16195	−0.00829	0.23608	−0.50
0.52	−0.51428	−0.42021	−0.27138	−0.13777	0.01659	0.25966	−0.52
0.54	−0.49558	−0.39925	−0.24775	−0.11270	0.04220	0.28363	−0.54
0.56	−0.47581	−0.37719	−0.22306	−0.08668	0.06857	0.30802	−0.56
0.58	−0.45487	−0.35393	−0.19724	−0.05964	0.09576	0.33284	−0.58
0.60	−0.43263	−0.32937	−0.17017	−0.03152	0.12379	0.35810	−0.60
0.62	−0.40896	−0.30336	−0.14177	−0.00223	0.15274	0.38383	−0.62
0.64	−0.38371	−0.27578	−0.11192	0.02831	0.18265	0.41005	−0.64
0.66	−0.35669	−0.24645	−0.08048	0.06020	0.21358	0.43678	−0.66
0.68	−0.32770	−0.21519	−0.04732	0.09354	0.24561	0.46405	−0.68
0.70	−0.29649	−0.18178	−0.01226	0.12845	0.27879	0.49187	−0.70
0.72	−0.26277	−0.14596	0.02488	0.16507	0.31322	0.52027	−0.72
0.74	−0.22621	−0.10745	0.06432	0.20355	0.34899	0.54929	−0.74
0.76	−0.18638	−0.06588	0.10630	0.24405	0.38618	0.57896	−0.76
0.78	−0.14279	−0.02085	0.15113	0.28677	0.42490	0.60930	−0.78
0.80	−0.09483	0.02816	0.19913	0.33192	0.46528	0.64035	−0.80
0.82	−0.04174	0.08174	0.25070	0.37976	0.50745	0.67215	−0.82
0.84	0.01742	0.14064	0.30631	0.43057	0.55156	0.70474	−0.84
0.86	0.08387	0.20578	0.36651	0.48468	0.59777	0.73816	−0.86
0.88	0.15918	0.27833	0.43199	0.54249	0.64627	0.77246	−0.88
0.90	0.24545	0.35978	0.50358	0.60445	0.69729	0.80770	−0.90
0.92	0.34553	0.45208	0.58228	0.67110	0.75105	0.84392	−0.92
0.94	0.46341	0.55784	0.66937	0.74310	0.80786	0.88119	−0.94
0.96	0.60492	0.68062	0.76650	0.82124	0.86802	0.91958	−0.96
0.98	0.77896	0.82550	0.87577	0.90647	0.93192	0.95915	−0.98

↑ r_0

α =	.995 ↑−	.99 ↑−	.975 ↑−	.95 ↑−	.90 ↑−	.75 ↑−

TABLE 7.1   SAMPLE SIZE =   9

α =	.005 ↓+	.01 ↓+	.025 ↓+	.05 ↓+	.10 ↓+	.25 ↓+	
r_0 ↓							
-0.98	-0.99701	-0.99628	-0.99492	-0.99344	-0.99126	-0.98610	0.98
-0.96	-0.99398	-0.99250	-0.98979	-0.98682	-0.98246	-0.97218	0.96
-0.94	-0.99090	-0.98868	-0.98460	-0.98013	-0.97359	-0.95825	0.94
-0.92	-0.98779	-0.98481	-0.97935	-0.97337	-0.96466	-0.94429	0.92
-0.90	-0.98462	-0.98089	-0.97403	-0.96655	-0.95567	-0.93032	0.90
-0.88	-0.98142	-0.97691	-0.96866	-0.95966	-0.94660	-0.91632	0.88
-0.86	-0.97816	-0.97288	-0.96322	-0.95270	-0.93747	-0.90230	0.86
-0.84	-0.97486	-0.96880	-0.95771	-0.94567	-0.92826	-0.88825	0.84
-0.82	-0.97151	-0.96465	-0.95213	-0.93856	-0.91898	-0.87417	0.82
-0.80	-0.96810	-0.96045	-0.94648	-0.93138	-0.90963	-0.86006	0.80
-0.78	-0.96465	-0.95619	-0.94076	-0.92411	-0.90020	-0.84592	0.78
-0.76	-0.96114	-0.95186	-0.93497	-0.91677	-0.89068	-0.83174	0.76
-0.74	-0.95758	-0.94747	-0.92910	-0.90934	-0.88109	-0.81752	0.74
-0.72	-0.95396	-0.94301	-0.92316	-0.90183	-0.87141	-0.80327	0.72
-0.70	-0.95028	-0.93849	-0.91713	-0.89424	-0.86165	-0.78897	0.70
-0.68	-0.94654	-0.93390	-0.91103	-0.88655	-0.85180	-0.77462	0.68
-0.66	-0.94274	-0.92924	-0.90484	-0.87878	-0.84186	-0.76023	0.66
-0.64	-0.93887	-0.92450	-0.89856	-0.87091	-0.83182	-0.74579	0.64
-0.62	-0.93494	-0.91968	-0.89219	-0.86294	-0.82170	-0.73130	0.62
-0.60	-0.93094	-0.91479	-0.88574	-0.85488	-0.81147	-0.71675	0.60
-0.58	-0.92687	-0.90982	-0.87918	-0.84671	-0.80114	-0.70215	0.58
-0.56	-0.92273	-0.90477	-0.87254	-0.83844	-0.79071	-0.68749	0.56
-0.54	-0.91852	-0.89963	-0.86579	-0.83006	-0.78018	-0.67277	0.54
-0.52	-0.91423	-0.89440	-0.85894	-0.82158	-0.76954	-0.65799	0.52
-0.50	-0.90986	-0.88908	-0.85198	-0.81298	-0.75878	-0.64314	0.50
-0.48	-0.90541	-0.88367	-0.84492	-0.80426	-0.74791	-0.62822	0.48
-0.46	-0.90087	-0.87817	-0.83775	-0.79543	-0.73693	-0.61323	0.46
-0.44	-0.89625	-0.87256	-0.83046	-0.78647	-0.72582	-0.59817	0.44
-0.42	-0.89154	-0.86685	-0.82305	-0.77739	-0.71459	-0.58304	0.42
-0.40	-0.88673	-0.86103	-0.81552	-0.76817	-0.70323	-0.56782	0.40
-0.38	-0.88183	-0.85511	-0.80786	-0.75883	-0.69174	-0.55253	0.38
-0.36	-0.87682	-0.84907	-0.80008	-0.74934	-0.68012	-0.53714	0.36
-0.34	-0.87172	-0.84291	-0.79216	-0.73971	-0.66836	-0.52168	0.34
-0.32	-0.86650	-0.83664	-0.78410	-0.72994	-0.65645	-0.50612	0.32
-0.30	-0.86118	-0.83023	-0.77590	-0.72002	-0.64440	-0.49047	0.30
-0.28	-0.85574	-0.82370	-0.76755	-0.70994	-0.63220	-0.47472	0.28
-0.26	-0.85018	-0.81703	-0.75905	-0.69970	-0.61985	-0.45888	0.26
-0.24	-0.84450	-0.81022	-0.75039	-0.68929	-0.60733	-0.44293	0.24
-0.22	-0.83869	-0.80327	-0.74157	-0.67872	-0.59465	-0.42688	0.22
-0.20	-0.83274	-0.79617	-0.73257	-0.66797	-0.58180	-0.41072	0.20
-0.18	-0.82666	-0.78891	-0.72341	-0.65703	-0.56878	-0.39444	0.18
-0.16	-0.82043	-0.78149	-0.71406	-0.64591	-0.55558	-0.37805	0.16
-0.14	-0.81405	-0.77390	-0.70452	-0.63460	-0.54219	-0.36154	0.14
-0.12	-0.80751	-0.76614	-0.69479	-0.62308	-0.52862	-0.34491	0.12
-0.10	-0.80081	-0.75819	-0.68486	-0.61136	-0.51484	-0.32815	0.10
-0.08	-0.79394	-0.75006	-0.67472	-0.59942	-0.50087	-0.31126	0.08
-0.06	-0.78689	-0.74173	-0.66436	-0.58726	-0.48668	-0.29423	0.06
-0.04	-0.77965	-0.73319	-0.65378	-0.57486	-0.47228	-0.27707	0.04
-0.02	-0.77222	-0.72444	-0.64296	-0.56223	-0.45765	-0.25976	0.02

↑ r_0

α =	.995 ↑-	.99 ↑-	.975 ↑-	.95 ↑-	.90 ↑-	.75 ↑-

TABLE 7.2 SAMPLE SIZE = 9 (CONT.)

α =	.005 ↓+	.01 ↓+	.025 ↓+	.05 ↓+	.10 ↓+	.25 ↓+	
r_o ↓							
0.00	−0.76459	−0.71546	−0.63190	−0.54936	−0.44280	−0.24230	0.00
0.02	−0.75675	−0.70625	−0.62058	−0.53622	−0.42770	−0.22469	−0.02
0.04	−0.74868	−0.69679	−0.60900	−0.52282	−0.41236	−0.20693	−0.04
0.06	−0.74038	−0.68708	−0.59714	−0.50914	−0.39677	−0.18900	−0.06
0.08	−0.73183	−0.67711	−0.58500	−0.49518	−0.38091	−0.17090	−0.08
0.10	−0.72303	−0.66685	−0.57255	−0.48092	−0.36478	−0.15263	−0.10
0.12	−0.71396	−0.65630	−0.55979	−0.46634	−0.34836	−0.13419	−0.12
0.14	−0.70460	−0.64544	−0.54671	−0.45144	−0.33165	−0.11556	−0.14
0.16	−0.69494	−0.63426	−0.53328	−0.43621	−0.31464	−0.09674	−0.16
0.18	−0.68497	−0.62274	−0.51950	−0.42063	−0.29732	−0.07772	−0.18
0.20	−0.67467	−0.61086	−0.50534	−0.40468	−0.27967	−0.05850	−0.20
0.22	−0.66401	−0.59860	−0.49079	−0.38834	−0.26168	−0.03908	−0.22
0.24	−0.65298	−0.58595	−0.47583	−0.37161	−0.24334	−0.01944	−0.24
0.26	−0.64156	−0.57287	−0.46043	−0.35447	−0.22463	0.00043	−0.26
0.28	−0.62972	−0.55936	−0.44459	−0.33689	−0.20554	0.02052	−0.28
0.30	−0.61744	−0.54538	−0.42826	−0.31885	−0.18606	0.04085	−0.30
0.32	−0.60469	−0.53090	−0.41143	−0.30034	−0.16616	0.06142	−0.32
0.34	−0.59143	−0.51589	−0.39407	−0.28132	−0.14584	0.08225	−0.34
0.36	−0.57765	−0.50033	−0.37615	−0.26178	−0.12507	0.10334	−0.36
0.38	−0.56329	−0.48417	−0.35764	−0.24170	−0.10383	0.12470	−0.38
0.40	−0.54832	−0.46738	−0.33851	−0.22103	−0.08210	0.14634	−0.40
0.42	−0.53270	−0.44992	−0.31871	−0.19975	−0.05987	0.16827	−0.42
0.44	−0.51638	−0.43173	−0.29821	−0.17783	−0.03710	0.19050	−0.44
0.46	−0.49930	−0.41278	−0.27697	−0.15523	−0.01378	0.21304	−0.46
0.48	−0.48142	−0.39300	−0.25494	−0.13192	0.01013	0.23590	−0.48
0.50	−0.46265	−0.37233	−0.23206	−0.10786	0.03464	0.25910	−0.50
0.52	−0.44294	−0.35071	−0.20828	−0.08299	0.05980	0.28264	−0.52
0.54	−0.42220	−0.32806	−0.18354	−0.05728	0.08564	0.30655	−0.54
0.56	−0.40033	−0.30430	−0.15777	−0.03067	0.11218	0.33082	−0.56
0.58	−0.37725	−0.27933	−0.13090	−0.00310	0.13946	0.35549	−0.58
0.60	−0.35283	−0.25305	−0.10285	0.02548	0.16753	0.38055	−0.60
0.62	−0.32694	−0.22535	−0.07352	0.05515	0.19643	0.40604	−0.62
0.64	−0.29945	−0.19609	−0.04281	0.08597	0.22620	0.43196	−0.64
0.66	−0.27016	−0.16513	−0.01061	0.11804	0.25689	0.45833	−0.66
0.68	−0.23890	−0.13229	0.02320	0.15142	0.28857	0.48518	−0.68
0.70	−0.20544	−0.09738	0.05877	0.18624	0.32128	0.51252	−0.70
0.72	−0.16950	−0.06018	0.09625	0.22259	0.35509	0.54037	−0.72
0.74	−0.13080	−0.02044	0.13582	0.26059	0.39007	0.56876	−0.74
0.76	−0.08895	0.02215	0.17768	0.30038	0.42630	0.59770	−0.76
0.78	−0.04354	0.06793	0.22208	0.34212	0.46387	0.62724	−0.78
0.80	0.00597	0.11732	0.26927	0.38596	0.50285	0.65738	−0.80
0.82	0.06020	0.17080	0.31956	0.43211	0.54336	0.68817	−0.82
0.84	0.11992	0.22896	0.37332	0.48077	0.58550	0.71962	−0.84
0.86	0.18611	0.29252	0.43095	0.53219	0.62941	0.75178	−0.86
0.88	0.25997	0.36233	0.49295	0.58666	0.67521	0.78469	−0.88
0.90	0.34304	0.43949	0.55991	0.64450	0.72306	0.81837	−0.90
0.92	0.43737	0.52535	0.63253	0.70609	0.77314	0.85287	−0.92
0.94	0.54566	0.62164	0.71167	0.77187	0.82564	0.88823	−0.94
0.96	0.67161	0.73064	0.79837	0.84235	0.88077	0.92451	−0.96
0.98	0.82048	0.85538	0.89393	0.91815	0.93880	0.96174	−0.98

↑ r_o

α =	.995 ↑−	.99 ↑−	.975 ↑−	.95 ↑−	.90 ↑−	.75 ↑−

TABLE 8.1   SAMPLE SIZE =   10

α =	.005 ↓+	.01 ↓+	.025 ↓+	.05 ↓+	.10 ↓+	.25 ↓+	
r_0 ↓							
-0.98	-0.99662	-0.99586	-0.99448	-0.99300	-0.99086	-0.98589	0.98
-0.96	-0.99318	-0.99166	-0.98891	-0.98593	-0.98165	-0.97177	0.96
-0.94	-0.98971	-0.98741	-0.98326	-0.97880	-0.97237	-0.95762	0.94
-0.92	-0.98618	-0.98311	-0.97756	-0.97159	-0.96303	-0.94344	0.92
-0.90	-0.98260	-0.97874	-0.97178	-0.96431	-0.95361	-0.92924	0.90
-0.88	-0.97897	-0.97432	-0.96594	-0.95696	-0.94412	-0.91501	0.88
-0.86	-0.97529	-0.96984	-0.96002	-0.94953	-0.93456	-0.90075	0.86
-0.84	-0.97155	-0.96529	-0.95404	-0.94203	-0.92493	-0.88645	0.84
-0.82	-0.96776	-0.96068	-0.94798	-0.93444	-0.91521	-0.87213	0.82
-0.80	-0.96391	-0.95601	-0.94184	-0.92678	-0.90542	-0.85776	0.80
-0.78	-0.96000	-0.95127	-0.93563	-0.91903	-0.89555	-0.84336	0.78
-0.76	-0.95603	-0.94646	-0.92934	-0.91119	-0.88559	-0.82892	0.76
-0.74	-0.95200	-0.94158	-0.92297	-0.90327	-0.87555	-0.81443	0.74
-0.72	-0.94790	-0.93663	-0.91651	-0.89526	-0.86542	-0.79990	0.72
-0.70	-0.94374	-0.93160	-0.90997	-0.88716	-0.85520	-0.78533	0.70
-0.68	-0.93951	-0.92650	-0.90334	-0.87897	-0.84489	-0.77070	0.68
-0.66	-0.93521	-0.92131	-0.89662	-0.87068	-0.83449	-0.75603	0.66
-0.64	-0.93084	-0.91605	-0.88980	-0.86229	-0.82398	-0.74130	0.64
-0.62	-0.92640	-0.91070	-0.88289	-0.85380	-0.81339	-0.72652	0.62
-0.60	-0.92188	-0.90527	-0.87589	-0.84520	-0.80269	-0.71168	0.60
-0.58	-0.91728	-0.89975	-0.86878	-0.83650	-0.79188	-0.69679	0.58
-0.56	-0.91260	-0.89414	-0.86157	-0.82769	-0.78097	-0.68183	0.56
-0.54	-0.90784	-0.88843	-0.85425	-0.81877	-0.76996	-0.66681	0.54
-0.52	-0.90299	-0.88263	-0.84683	-0.80973	-0.75883	-0.65173	0.52
-0.50	-0.89806	-0.87673	-0.83929	-0.80058	-0.74758	-0.63657	0.50
-0.48	-0.89303	-0.87073	-0.83164	-0.79130	-0.73622	-0.62135	0.48
-0.46	-0.88791	-0.86462	-0.82386	-0.78190	-0.72474	-0.60605	0.46
-0.44	-0.88269	-0.85840	-0.81597	-0.77237	-0.71313	-0.59068	0.44
-0.42	-0.87738	-0.85207	-0.80795	-0.76271	-0.70140	-0.57524	0.42
-0.40	-0.87195	-0.84563	-0.79980	-0.75291	-0.68954	-0.55971	0.40
-0.38	-0.86642	-0.83906	-0.79152	-0.74298	-0.67754	-0.54410	0.38
-0.36	-0.86078	-0.83238	-0.78310	-0.73290	-0.66541	-0.52841	0.36
-0.34	-0.85503	-0.82556	-0.77453	-0.72268	-0.65314	-0.51263	0.34
-0.32	-0.84916	-0.81862	-0.76582	-0.71230	-0.64072	-0.49675	0.32
-0.30	-0.84316	-0.81153	-0.75696	-0.70177	-0.62815	-0.48079	0.30
-0.28	-0.83703	-0.80431	-0.74795	-0.69108	-0.61543	-0.46473	0.28
-0.26	-0.83078	-0.79694	-0.73877	-0.68022	-0.60256	-0.44857	0.26
-0.24	-0.82438	-0.78942	-0.72943	-0.66919	-0.58952	-0.43231	0.24
-0.22	-0.81785	-0.78174	-0.71991	-0.65798	-0.57632	-0.41594	0.22
-0.20	-0.81116	-0.77390	-0.71022	-0.64660	-0.56294	-0.39947	0.20
-0.18	-0.80433	-0.76589	-0.70035	-0.63503	-0.54940	-0.38289	0.18
-0.16	-0.79733	-0.75771	-0.69028	-0.62326	-0.53567	-0.36619	0.16
-0.14	-0.79017	-0.74935	-0.68002	-0.61130	-0.52175	-0.34937	0.14
-0.12	-0.78284	-0.74079	-0.66956	-0.59913	-0.50765	-0.33243	0.12
-0.10	-0.77532	-0.73205	-0.65888	-0.58675	-0.49334	-0.31537	0.10
-0.08	-0.76762	-0.72310	-0.64799	-0.57415	-0.47884	-0.29818	0.08
-0.06	-0.75973	-0.71394	-0.63687	-0.56132	-0.46412	-0.28086	0.06
-0.04	-0.75163	-0.70455	-0.62552	-0.54826	-0.44919	-0.26340	0.04
-0.02	-0.74332	-0.69494	-0.61392	-0.53496	-0.43404	-0.24581	0.02

↑ r_0

α =	.995 ↑-	.99 ↑-	.975 ↑-	.95 ↑-	.90 ↑-	.75 ↑-

TABLE 8.2   SAMPLE SIZE =   10   (CONT.)

α =	.005 ↓+	.01 ↓+	.025 ↓+	.05 ↓+	.10 ↓+	.25 ↓+	
r_o ↓							
0.00	−0.73479	−0.68510	−0.60207	−0.52140	−0.41866	−0.22807	0.00
0.02	−0.72602	−0.67500	−0.58996	−0.50759	−0.40305	−0.21018	−0.02
0.04	−0.71702	−0.66464	−0.57757	−0.49351	−0.38718	−0.19214	−0.04
0.06	−0.70776	−0.65401	−0.56490	−0.47914	−0.37107	−0.17394	−0.06
0.08	−0.69824	−0.64310	−0.55194	−0.46449	−0.35470	−0.15558	−0.08
0.10	−0.68843	−0.63189	−0.53866	−0.44954	−0.33806	−0.13706	−0.10
0.12	−0.67834	−0.62038	−0.52507	−0.43428	−0.32114	−0.11836	−0.12
0.14	−0.66794	−0.60853	−0.51114	−0.41869	−0.30393	−0.09949	−0.14
0.16	−0.65722	−0.59635	−0.49686	−0.40277	−0.28643	−0.08044	−0.16
0.18	−0.64616	−0.58381	−0.48222	−0.38650	−0.26862	−0.06120	−0.18
0.20	−0.63474	−0.57090	−0.46720	−0.36986	−0.25049	−0.04176	−0.20
0.22	−0.62295	−0.55759	−0.45178	−0.35284	−0.23203	−0.02213	−0.22
0.24	−0.61076	−0.54387	−0.43594	−0.33543	−0.21323	−0.00230	−0.24
0.26	−0.59815	−0.52971	−0.41967	−0.31761	−0.19407	0.01775	−0.26
0.28	−0.58510	−0.51510	−0.40293	−0.29936	−0.17455	0.03801	−0.28
0.30	−0.57157	−0.49999	−0.38572	−0.28066	−0.15464	0.05850	−0.30
0.32	−0.55755	−0.48438	−0.36801	−0.26149	−0.13434	0.07921	−0.32
0.34	−0.54301	−0.46822	−0.34976	−0.24183	−0.11363	0.10017	−0.34
0.36	−0.52790	−0.45149	−0.33096	−0.22166	−0.09248	0.12137	−0.36
0.38	−0.51219	−0.43416	−0.31157	−0.20096	−0.07089	0.14282	−0.38
0.40	−0.49584	−0.41618	−0.29156	−0.17969	−0.04884	0.16454	−0.40
0.42	−0.47882	−0.39751	−0.27089	−0.15783	−0.02630	0.18652	−0.42
0.44	−0.46106	−0.37811	−0.24954	−0.13536	−0.00326	0.20879	−0.44
0.46	−0.44253	−0.35793	−0.22746	−0.11223	0.02030	0.23134	−0.46
0.48	−0.42316	−0.33692	−0.20460	−0.08842	0.04442	0.25419	−0.48
0.50	−0.40289	−0.31502	−0.18092	−0.06389	0.06911	0.27735	−0.50
0.52	−0.38164	−0.29217	−0.15636	−0.03859	0.09440	0.30083	−0.52
0.54	−0.35935	−0.26830	−0.13088	−0.01250	0.12032	0.32464	−0.54
0.56	−0.33593	−0.24333	−0.10441	0.01445	0.14690	0.34879	−0.56
0.58	−0.31127	−0.21716	−0.07689	0.04229	0.17416	0.37330	−0.58
0.60	−0.28528	−0.18972	−0.04823	0.07108	0.20215	0.39817	−0.60
0.62	−0.25782	−0.16089	−0.01837	0.10088	0.23090	0.42342	−0.62
0.64	−0.22876	−0.13055	0.01279	0.13175	0.26045	0.44907	−0.64
0.66	−0.19795	−0.09858	0.04534	0.16377	0.29084	0.47512	−0.66
0.68	−0.16520	−0.06481	0.07939	0.19700	0.32212	0.50160	−0.68
0.70	−0.13032	−0.02908	0.11507	0.23152	0.35433	0.52852	−0.70
0.72	−0.09306	0.00880	0.15250	0.26744	0.38753	0.55590	−0.72
0.74	−0.05316	0.04906	0.19184	0.30484	0.42178	0.58376	−0.74
0.76	−0.01030	0.09194	0.23325	0.34385	0.45713	0.61211	−0.76
0.78	0.03590	0.13775	0.27692	0.38457	0.49366	0.64098	−0.78
0.80	0.08587	0.18681	0.32308	0.42715	0.53143	0.67039	−0.80
0.82	0.14013	0.23954	0.37196	0.47173	0.57054	0.70036	−0.82
0.84	0.19933	0.29638	0.42384	0.51849	0.61105	0.73092	−0.84
0.86	0.26422	0.35792	0.47905	0.56761	0.65308	0.76209	−0.86
0.88	0.33575	0.42480	0.53796	0.61931	0.69672	0.79390	−0.88
0.90	0.41509	0.49783	0.60101	0.67383	0.74210	0.82638	−0.90
0.92	0.50373	0.57800	0.66870	0.73145	0.78933	0.85956	−0.92
0.94	0.60357	0.66654	0.74165	0.79248	0.83857	0.89347	−0.94
0.96	0.71711	0.76497	0.82057	0.85729	0.88997	0.92816	−0.96
0.98	0.84771	0.87527	0.90633	0.92630	0.94372	0.96365	−0.98

↑ r_o

α =	.995 ↑−	.99 ↑−	.975 ↑−	.95 ↑−	.90 ↑−	.75 ↑−

TABLE 9.1   SAMPLE SIZE =   11

α =	.005 ↓+	.01 ↓+	.025 ↓+	.05 ↓+	.10 ↓+	.25 ↓+	
r_0 ↓							
-0.98	-0.99625	-0.99548	-0.99409	-0.99261	-0.99050	-0.98571	0.98
-0.96	-0.99245	-0.99089	-0.98811	-0.98514	-0.98093	-0.97139	0.96
-0.94	-0.98860	-0.98625	-0.98206	-0.97761	-0.97129	-0.95705	0.94
-0.92	-0.98469	-0.98155	-0.97594	-0.97000	-0.96158	-0.94267	0.92
-0.90	-0.98073	-0.97678	-0.96975	-0.96231	-0.95180	-0.92827	0.90
-0.88	-0.97671	-0.97195	-0.96349	-0.95455	-0.94194	-0.91383	0.88
-0.86	-0.97263	-0.96706	-0.95715	-0.94671	-0.93200	-0.89936	0.86
-0.84	-0.96849	-0.96209	-0.95074	-0.93879	-0.92199	-0.88485	0.84
-0.82	-0.96429	-0.95706	-0.94425	-0.93078	-0.91189	-0.87030	0.82
-0.80	-0.96003	-0.95196	-0.93768	-0.92269	-0.90172	-0.85572	0.80
-0.78	-0.95570	-0.94678	-0.93102	-0.91451	-0.89146	-0.84109	0.78
-0.76	-0.95131	-0.94153	-0.92428	-0.90624	-0.88111	-0.82642	0.76
-0.74	-0.94685	-0.93621	-0.91746	-0.89789	-0.87068	-0.81170	0.74
-0.72	-0.94232	-0.93081	-0.91054	-0.88944	-0.86015	-0.79694	0.72
-0.70	-0.93771	-0.92532	-0.90354	-0.88089	-0.84954	-0.78212	0.70
-0.68	-0.93304	-0.91975	-0.89644	-0.87225	-0.83883	-0.76726	0.68
-0.66	-0.92828	-0.91410	-0.88925	-0.86350	-0.82802	-0.75234	0.66
-0.64	-0.92345	-0.90836	-0.88196	-0.85466	-0.81712	-0.73737	0.64
-0.62	-0.91854	-0.90253	-0.87457	-0.84571	-0.80611	-0.72235	0.62
-0.60	-0.91354	-0.89661	-0.86708	-0.83665	-0.79500	-0.70726	0.60
-0.58	-0.90846	-0.89060	-0.85948	-0.82748	-0.78379	-0.69211	0.58
-0.56	-0.90329	-0.88448	-0.85177	-0.81820	-0.77246	-0.67690	0.56
-0.54	-0.89803	-0.87827	-0.84395	-0.80880	-0.76103	-0.66163	0.54
-0.52	-0.89268	-0.87195	-0.83602	-0.79928	-0.74948	-0.64629	0.52
-0.50	-0.88723	-0.86553	-0.82797	-0.78965	-0.73782	-0.63088	0.50
-0.48	-0.88168	-0.85900	-0.81979	-0.77988	-0.72603	-0.61540	0.48
-0.46	-0.87603	-0.85235	-0.81150	-0.76999	-0.71413	-0.59984	0.46
-0.44	-0.87027	-0.84559	-0.80307	-0.75997	-0.70210	-0.58422	0.44
-0.42	-0.86440	-0.83871	-0.79451	-0.74981	-0.68994	-0.56851	0.42
-0.40	-0.85842	-0.83170	-0.78582	-0.73951	-0.67764	-0.55272	0.40
-0.38	-0.85233	-0.82457	-0.77699	-0.72907	-0.66522	-0.53685	0.38
-0.36	-0.84611	-0.81731	-0.76802	-0.71848	-0.65265	-0.52090	0.36
-0.34	-0.83977	-0.80991	-0.75889	-0.70774	-0.63995	-0.50485	0.34
-0.32	-0.83331	-0.80237	-0.74962	-0.69685	-0.62709	-0.48872	0.32
-0.30	-0.82671	-0.79468	-0.74019	-0.68580	-0.61409	-0.47250	0.30
-0.28	-0.81997	-0.78685	-0.73059	-0.67458	-0.60093	-0.45618	0.28
-0.26	-0.81308	-0.77886	-0.72083	-0.66319	-0.58762	-0.43976	0.26
-0.24	-0.80605	-0.77071	-0.71090	-0.65163	-0.57415	-0.42324	0.24
-0.22	-0.79887	-0.76239	-0.70079	-0.63990	-0.56050	-0.40662	0.22
-0.20	-0.79153	-0.75390	-0.69049	-0.62797	-0.54669	-0.38989	0.20
-0.18	-0.78402	-0.74524	-0.68001	-0.61586	-0.53270	-0.37305	0.18
-0.16	-0.77635	-0.73639	-0.66933	-0.60355	-0.51854	-0.35610	0.16
-0.14	-0.76849	-0.72735	-0.65845	-0.59104	-0.50418	-0.33904	0.14
-0.12	-0.76045	-0.71811	-0.64736	-0.57833	-0.48964	-0.32185	0.12
-0.10	-0.75222	-0.70867	-0.63605	-0.56539	-0.47490	-0.30455	0.10
-0.08	-0.74379	-0.69901	-0.62452	-0.55224	-0.45996	-0.28712	0.08
-0.06	-0.73515	-0.68913	-0.61275	-0.53886	-0.44481	-0.26956	0.06
-0.04	-0.72629	-0.67902	-0.60075	-0.52524	-0.42945	-0.25187	0.04
-0.02	-0.71721	-0.66867	-0.58849	-0.51138	-0.41387	-0.23404	0.02

↑ r_0

α =	.995 ↑-	.99 ↑-	.975 ↑-	.95 ↑-	.90 ↑-	.75 ↑-

TABLE 9.2   SAMPLE SIZE = 11 (CONT.)

α =	.005 ↓+	.01 ↓+	.025 ↓+	.05 ↓+	.10 ↓+	.25 ↓+	
r_0 ↓							
0.00	-0.70789	-0.65807	-0.57598	-0.49726	-0.39806	-0.21607	0.00
0.02	-0.69832	-0.64721	-0.56320	-0.48289	-0.38202	-0.19796	-0.02
0.04	-0.68850	-0.63608	-0.55015	-0.46825	-0.36574	-0.17970	-0.04
0.06	-0.67841	-0.62467	-0.53680	-0.45332	-0.34921	-0.16130	-0.06
0.08	-0.66804	-0.61296	-0.52315	-0.43811	-0.33243	-0.14273	-0.08
0.10	-0.65738	-0.60095	-0.50919	-0.42260	-0.31538	-0.12401	-0.10
0.12	-0.64640	-0.58861	-0.49491	-0.40678	-0.29806	-0.10512	-0.12
0.14	-0.63511	-0.57594	-0.48029	-0.39064	-0.28046	-0.08606	-0.14
0.16	-0.62348	-0.56292	-0.46531	-0.37416	-0.26256	-0.06683	-0.16
0.18	-0.61149	-0.54953	-0.44997	-0.35734	-0.24437	-0.04741	-0.18
0.20	-0.59912	-0.53576	-0.43425	-0.34015	-0.22587	-0.02782	-0.20
0.22	-0.58637	-0.52158	-0.41812	-0.32260	-0.20705	-0.00804	-0.22
0.24	-0.57320	-0.50697	-0.40158	-0.30465	-0.18789	0.01194	-0.24
0.26	-0.55959	-0.49192	-0.38461	-0.28630	-0.16840	0.03212	-0.26
0.28	-0.54552	-0.47641	-0.36717	-0.26753	-0.14855	0.05251	-0.28
0.30	-0.53097	-0.46039	-0.34926	-0.24832	-0.12832	0.07311	-0.30
0.32	-0.51590	-0.44386	-0.33085	-0.22865	-0.10772	0.09392	-0.32
0.34	-0.50028	-0.42678	-0.31192	-0.20851	-0.08672	0.11496	-0.34
0.36	-0.48409	-0.40912	-0.29243	-0.18787	-0.06532	0.13623	-0.36
0.38	-0.46728	-0.39084	-0.27237	-0.16671	-0.04348	0.15774	-0.38
0.40	-0.44982	-0.37192	-0.25169	-0.14500	-0.02120	0.17950	-0.40
0.42	-0.43167	-0.35231	-0.23038	-0.12273	0.00154	0.20151	-0.42
0.44	-0.41277	-0.33197	-0.20839	-0.09986	0.02475	0.22378	-0.44
0.46	-0.39308	-0.31085	-0.18569	-0.07637	0.04847	0.24632	-0.46
0.48	-0.37255	-0.28891	-0.16224	-0.05222	0.07270	0.26914	-0.48
0.50	-0.35110	-0.26609	-0.13799	-0.02738	0.09748	0.29225	-0.50
0.52	-0.32869	-0.24233	-0.11290	-0.00182	0.12282	0.31565	-0.52
0.54	-0.30522	-0.21757	-0.08692	0.02450	0.14874	0.33936	-0.54
0.56	-0.28063	-0.19173	-0.06000	0.05162	0.17528	0.36338	-0.56
0.58	-0.25481	-0.16473	-0.03207	0.07958	0.20247	0.38773	-0.58
0.60	-0.22768	-0.13649	-0.00306	0.10843	0.23032	0.41242	-0.60
0.62	-0.19911	-0.10692	0.02708	0.13823	0.25888	0.43746	-0.62
0.64	-0.16898	-0.07589	0.05844	0.16902	0.28818	0.46286	-0.64
0.66	-0.13714	-0.04331	0.09111	0.20087	0.31824	0.48863	-0.66
0.68	-0.10344	-0.00902	0.12518	0.23384	0.34912	0.51479	-0.68
0.70	-0.06769	0.02711	0.16075	0.26800	0.38085	0.54135	-0.70
0.72	-0.02969	0.06526	0.19794	0.30343	0.41348	0.56832	-0.72
0.74	0.01081	0.10561	0.23686	0.34021	0.44705	0.59573	-0.74
0.76	0.05409	0.14839	0.27768	0.37842	0.48162	0.62358	-0.76
0.78	0.10045	0.19384	0.32053	0.41818	0.51724	0.65190	-0.78
0.80	0.15028	0.24224	0.36561	0.45960	0.55397	0.68070	-0.80
0.82	0.20400	0.29393	0.41310	0.50279	0.59188	0.71000	-0.82
0.84	0.26215	0.34927	0.46324	0.54789	0.63103	0.73982	-0.84
0.86	0.32534	0.40872	0.51627	0.59505	0.67150	0.77019	-0.86
0.88	0.39432	0.47280	0.57250	0.64444	0.71339	0.80112	-0.88
0.90	0.46999	0.54213	0.63226	0.69625	0.75676	0.83264	-0.90
0.92	0.55347	0.61745	0.69593	0.75068	0.80174	0.86477	-0.92
0.94	0.64616	0.69966	0.76396	0.80799	0.84842	0.89754	-0.94
0.96	0.74984	0.78985	0.83689	0.86843	0.89694	0.93098	-0.96
0.98	0.86678	0.88940	0.91532	0.93231	0.94741	0.96513	-0.98

↑ r_0

α =	.995 ↑-	.99 ↑-	.975 ↑-	.95 ↑-	.90 ↑-	.75 ↑-

TABLE 10.1   SAMPLE SIZE =   12

α =	.005 ↓+	.01 ↓+	.025 ↓+	.05 ↓+	.10 ↓+	.25 ↓+	
r_0 ↓							
-0.98	-0.99591	-0.99512	-0.99373	-0.99225	-0.99018	-0.98554	0.98
-0.96	-0.99177	-0.99018	-0.98738	-0.98443	-0.98029	-0.97105	0.96
-0.94	-0.98757	-0.98518	-0.98096	-0.97654	-0.97033	-0.95653	0.94
-0.92	-0.98331	-0.98011	-0.97447	-0.96857	-0.96029	-0.94198	0.92
-0.90	-0.97899	-0.97498	-0.96791	-0.96052	-0.95017	-0.92739	0.90
-0.88	-0.97461	-0.96977	-0.96127	-0.95239	-0.93998	-0.91277	0.88
-0.86	-0.97017	-0.96450	-0.95455	-0.94417	-0.92972	-0.89811	0.86
-0.84	-0.96566	-0.95916	-0.94775	-0.93587	-0.91937	-0.88341	0.84
-0.82	-0.96108	-0.95374	-0.94087	-0.92749	-0.90893	-0.86867	0.82
-0.80	-0.95644	-0.94824	-0.93390	-0.91902	-0.89842	-0.85388	0.80
-0.78	-0.95173	-0.94267	-0.92685	-0.91046	-0.88782	-0.83906	0.78
-0.76	-0.94694	-0.93702	-0.91971	-0.90181	-0.87713	-0.82419	0.76
-0.74	-0.94209	-0.93129	-0.91248	-0.89306	-0.86635	-0.80927	0.74
-0.72	-0.93715	-0.92548	-0.90515	-0.88422	-0.85548	-0.79430	0.72
-0.70	-0.93214	-0.91958	-0.89773	-0.87527	-0.84451	-0.77928	0.70
-0.68	-0.92705	-0.91359	-0.89022	-0.86623	-0.83345	-0.76420	0.68
-0.66	-0.92188	-0.90751	-0.88260	-0.85709	-0.82229	-0.74908	0.66
-0.64	-0.91662	-0.90134	-0.87489	-0.84784	-0.81103	-0.73389	0.64
-0.62	-0.91128	-0.89507	-0.86707	-0.83848	-0.79967	-0.71865	0.62
-0.60	-0.90585	-0.88871	-0.85914	-0.82901	-0.78820	-0.70335	0.60
-0.58	-0.90032	-0.88224	-0.85110	-0.81943	-0.77663	-0.68799	0.58
-0.56	-0.89471	-0.87568	-0.84295	-0.80973	-0.76494	-0.67256	0.56
-0.54	-0.88899	-0.86900	-0.83468	-0.79991	-0.75314	-0.65707	0.54
-0.52	-0.88317	-0.86222	-0.82629	-0.78998	-0.74123	-0.64151	0.52
-0.50	-0.87726	-0.85533	-0.81778	-0.77991	-0.72920	-0.62588	0.50
-0.48	-0.87123	-0.84832	-0.80915	-0.76972	-0.71705	-0.61018	0.48
-0.46	-0.86510	-0.84119	-0.80039	-0.75940	-0.70477	-0.59440	0.46
-0.44	-0.85885	-0.83394	-0.79149	-0.74894	-0.69237	-0.57855	0.44
-0.42	-0.85249	-0.82656	-0.78246	-0.73835	-0.67984	-0.56262	0.42
-0.40	-0.84600	-0.81905	-0.77329	-0.72761	-0.66718	-0.54661	0.40
-0.38	-0.83940	-0.81141	-0.76398	-0.71673	-0.65438	-0.53052	0.38
-0.36	-0.83266	-0.80363	-0.75451	-0.70569	-0.64144	-0.51435	0.36
-0.34	-0.82579	-0.79571	-0.74490	-0.69451	-0.62836	-0.49808	0.34
-0.32	-0.81879	-0.78764	-0.73512	-0.68317	-0.61514	-0.48173	0.32
-0.30	-0.81164	-0.77942	-0.72519	-0.67166	-0.60176	-0.46529	0.30
-0.28	-0.80435	-0.77104	-0.71509	-0.65999	-0.58823	-0.44875	0.28
-0.26	-0.79690	-0.76250	-0.70482	-0.64815	-0.57454	-0.43211	0.26
-0.24	-0.78930	-0.75379	-0.69437	-0.63613	-0.56069	-0.41538	0.24
-0.22	-0.78154	-0.74491	-0.68374	-0.62393	-0.54668	-0.39854	0.22
-0.20	-0.77361	-0.73585	-0.67292	-0.61155	-0.53249	-0.38160	0.20
-0.18	-0.76551	-0.72660	-0.66191	-0.59897	-0.51814	-0.36455	0.18
-0.16	-0.75722	-0.71716	-0.65069	-0.58620	-0.50360	-0.34739	0.16
-0.14	-0.74875	-0.70753	-0.63928	-0.57322	-0.48887	-0.33012	0.14
-0.12	-0.74008	-0.69769	-0.62764	-0.56004	-0.47396	-0.31273	0.12
-0.10	-0.73121	-0.68763	-0.61579	-0.54664	-0.45886	-0.29522	0.10
-0.08	-0.72213	-0.67736	-0.60371	-0.53301	-0.44355	-0.27759	0.08
-0.06	-0.71283	-0.66685	-0.59139	-0.51916	-0.42804	-0.25984	0.06
-0.04	-0.70330	-0.65610	-0.57883	-0.50507	-0.41232	-0.24195	0.04
-0.02	-0.69354	-0.64511	-0.56602	-0.49074	-0.39638	-0.22394	0.02

↑ r_0

α =	.995 ↑-	.99 ↑-	.975 ↑-	.95 ↑-	.90 ↑-	.75 ↑-

TABLE 10.2   SAMPLE SIZE = 12   (CONT.)

α =	.005 ↓+	.01 ↓+	.025 ↓+	.05 ↓+	.10 ↓+	.25 ↓+	
r_o ↓							
0.00	−0.68353	−0.63386	−0.55294	−0.47616	−0.38022	−0.20579	0.00
0.02	−0.67326	−0.62235	−0.53960	−0.46131	−0.36382	−0.18750	−0.02
0.04	−0.66272	−0.61055	−0.52597	−0.44620	−0.34720	−0.16906	−0.04
0.06	−0.65191	−0.59846	−0.51205	−0.43081	−0.33033	−0.15048	−0.06
0.08	−0.64080	−0.58608	−0.49783	−0.41513	−0.31320	−0.13175	−0.08
0.10	−0.62939	−0.57337	−0.48330	−0.39916	−0.29582	−0.11287	−0.10
0.12	−0.61765	−0.56034	−0.46843	−0.38288	−0.27818	−0.09382	−0.12
0.14	−0.60559	−0.54697	−0.45323	−0.36628	−0.26026	−0.07462	−0.14
0.16	−0.59317	−0.53323	−0.43768	−0.34935	−0.24205	−0.05525	−0.16
0.18	−0.58038	−0.51912	−0.42176	−0.33208	−0.22356	−0.03570	−0.18
0.20	−0.56721	−0.50462	−0.40546	−0.31445	−0.20476	−0.01598	−0.20
0.22	−0.55363	−0.48971	−0.38876	−0.29646	−0.18565	0.00392	−0.22
0.24	−0.53963	−0.47437	−0.37165	−0.27809	−0.16622	0.02401	−0.24
0.26	−0.52518	−0.45858	−0.35410	−0.25932	−0.14646	0.04429	−0.26
0.28	−0.51026	−0.44232	−0.33610	−0.24014	−0.12635	0.06477	−0.28
0.30	−0.49484	−0.42556	−0.31763	−0.22053	−0.10589	0.08544	−0.30
0.32	−0.47889	−0.40827	−0.29867	−0.20047	−0.08506	0.10633	−0.32
0.34	−0.46239	−0.39044	−0.27919	−0.17996	−0.06386	0.12743	−0.34
0.36	−0.44530	−0.37202	−0.25916	−0.15895	−0.04225	0.14875	−0.36
0.38	−0.42759	−0.35300	−0.23858	−0.13745	−0.02024	0.17029	−0.38
0.40	−0.40923	−0.33332	−0.21739	−0.11542	0.00219	0.19207	−0.40
0.42	−0.39016	−0.31297	−0.19558	−0.09284	0.02507	0.21408	−0.42
0.44	−0.37034	−0.29189	−0.17312	−0.06969	0.04839	0.23634	−0.44
0.46	−0.34973	−0.27005	−0.14996	−0.04594	0.07220	0.25886	−0.46
0.48	−0.32828	−0.24739	−0.12608	−0.02157	0.09649	0.28163	−0.48
0.50	−0.30593	−0.22387	−0.10143	0.00346	0.12130	0.30468	−0.50
0.52	−0.28260	−0.19943	−0.07597	0.02918	0.14663	0.32800	−0.52
0.54	−0.25824	−0.17401	−0.04966	0.05562	0.17252	0.35161	−0.54
0.56	−0.23277	−0.14755	−0.02244	0.08282	0.19899	0.37551	−0.56
0.58	−0.20610	−0.11997	0.00574	0.11082	0.22606	0.39971	−0.58
0.60	−0.17814	−0.09119	0.03493	0.13965	0.25375	0.42423	−0.60
0.62	−0.14878	−0.06113	0.06520	0.16936	0.28210	0.44907	−0.62
0.64	−0.11792	−0.02968	0.09662	0.20001	0.31113	0.47425	−0.64
0.66	−0.08541	0.00325	0.12927	0.23163	0.34088	0.49977	−0.66
0.68	−0.05111	0.03778	0.16322	0.26430	0.37138	0.52564	−0.68
0.70	−0.01487	0.07406	0.19856	0.29806	0.40266	0.55189	−0.70
0.72	0.02351	0.11222	0.23540	0.33299	0.43476	0.57851	−0.72
0.74	0.06424	0.15244	0.27384	0.36915	0.46772	0.60553	−0.74
0.76	0.10755	0.19489	0.31401	0.40662	0.50159	0.63296	−0.76
0.78	0.15373	0.23979	0.35603	0.44550	0.53641	0.66081	−0.78
0.80	0.20308	0.28738	0.40005	0.48586	0.57224	0.68909	−0.80
0.82	0.25599	0.33794	0.44625	0.52781	0.60912	0.71783	−0.82
0.84	0.31288	0.39176	0.49480	0.57147	0.64711	0.74704	−0.84
0.86	0.37426	0.44923	0.54592	0.61695	0.68629	0.77674	−0.86
0.88	0.44072	0.51075	0.59983	0.66440	0.72670	0.80695	−0.88
0.90	0.51300	0.57683	0.65681	0.71395	0.76844	0.83768	−0.90
0.92	0.59196	0.64802	0.71715	0.76578	0.81158	0.86896	−0.92
0.94	0.67865	0.72503	0.78121	0.82008	0.85620	0.90080	−0.94
0.96	0.77438	0.80866	0.84938	0.87706	0.90241	0.93324	−0.96
0.98	0.88081	0.89991	0.92213	0.93694	0.95030	0.96630	−0.98

↑ r_o

α =	.995 ↑−	.99 ↑−	.975 ↑−	.95 ↑−	.90 ↑−	.75 ↑−

TABLE 11.1   SAMPLE SIZE =   13

α =	.005 ↓+	.01 ↓+	.025 ↓+	.05 ↓+	.10 ↓+	.25 ↓+	
r_0 ↓							
−0.98	−0.99560	−0.99480	−0.99340	−0.99193	−0.98989	−0.98539	0.98
−0.96	−0.99113	−0.98953	−0.98672	−0.98379	−0.97971	−0.97074	0.96
−0.94	−0.98661	−0.98419	−0.97996	−0.97557	−0.96945	−0.95606	0.94
−0.92	−0.98202	−0.97879	−0.97314	−0.96727	−0.95912	−0.94135	0.92
−0.90	−0.97737	−0.97331	−0.96623	−0.95889	−0.94871	−0.92659	0.90
−0.88	−0.97266	−0.96777	−0.95924	−0.95042	−0.93823	−0.91180	0.88
−0.86	−0.96788	−0.96214	−0.95217	−0.94187	−0.92766	−0.89698	0.86
−0.84	−0.96302	−0.95645	−0.94502	−0.93324	−0.91701	−0.88210	0.84
−0.82	−0.95810	−0.95067	−0.93778	−0.92451	−0.90628	−0.86719	0.82
−0.80	−0.95311	−0.94482	−0.93046	−0.91570	−0.89546	−0.85223	0.80
−0.78	−0.94804	−0.93889	−0.92304	−0.90679	−0.88455	−0.83723	0.78
−0.76	−0.94289	−0.93287	−0.91554	−0.89779	−0.87355	−0.82218	0.76
−0.74	−0.93767	−0.92676	−0.90794	−0.88870	−0.86246	−0.80708	0.74
−0.72	−0.93236	−0.92057	−0.90024	−0.87950	−0.85128	−0.79193	0.72
−0.70	−0.92698	−0.91429	−0.89245	−0.87021	−0.84001	−0.77673	0.70
−0.68	−0.92151	−0.90792	−0.88456	−0.86081	−0.82863	−0.76147	0.68
−0.66	−0.91595	−0.90145	−0.87656	−0.85130	−0.81716	−0.74616	0.66
−0.64	−0.91030	−0.89489	−0.86846	−0.84169	−0.80558	−0.73079	0.64
−0.62	−0.90456	−0.88822	−0.86025	−0.83197	−0.79391	−0.71536	0.62
−0.60	−0.89873	−0.88145	−0.85193	−0.82213	−0.78212	−0.69987	0.60
−0.58	−0.89280	−0.87458	−0.84349	−0.81218	−0.77023	−0.68431	0.58
−0.56	−0.88677	−0.86760	−0.83494	−0.80211	−0.75822	−0.66870	0.56
−0.54	−0.88064	−0.86051	−0.82627	−0.79192	−0.74610	−0.65301	0.54
−0.52	−0.87440	−0.85331	−0.81748	−0.78161	−0.73387	−0.63726	0.52
−0.50	−0.86805	−0.84599	−0.80856	−0.77117	−0.72152	−0.62144	0.50
−0.48	−0.86159	−0.83854	−0.79952	−0.76060	−0.70904	−0.60555	0.48
−0.46	−0.85502	−0.83098	−0.79034	−0.74989	−0.69645	−0.58958	0.46
−0.44	−0.84832	−0.82328	−0.78102	−0.73905	−0.68372	−0.57354	0.44
−0.42	−0.84150	−0.81546	−0.77157	−0.72807	−0.67087	−0.55742	0.42
−0.40	−0.83456	−0.80749	−0.76197	−0.71695	−0.65788	−0.54122	0.40
−0.38	−0.82749	−0.79939	−0.75222	−0.70567	−0.64475	−0.52493	0.38
−0.36	−0.82028	−0.79115	−0.74233	−0.69425	−0.63149	−0.50857	0.36
−0.34	−0.81293	−0.78276	−0.73227	−0.68267	−0.61809	−0.49211	0.34
−0.32	−0.80544	−0.77421	−0.72206	−0.67094	−0.60453	−0.47557	0.32
−0.30	−0.79780	−0.76551	−0.71168	−0.65903	−0.59083	−0.45894	0.30
−0.28	−0.79001	−0.75664	−0.70113	−0.64697	−0.57698	−0.44221	0.28
−0.26	−0.78206	−0.74761	−0.69041	−0.63473	−0.56297	−0.42539	0.26
−0.24	−0.77394	−0.73841	−0.67951	−0.62231	−0.54879	−0.40847	0.24
−0.22	−0.76566	−0.72902	−0.66842	−0.60971	−0.53446	−0.39145	0.22
−0.20	−0.75720	−0.71945	−0.65714	−0.59692	−0.51995	−0.37433	0.20
−0.18	−0.74855	−0.70969	−0.64567	−0.58394	−0.50527	−0.35710	0.18
−0.16	−0.73972	−0.69973	−0.63399	−0.57077	−0.49041	−0.33976	0.16
−0.14	−0.73070	−0.68957	−0.62210	−0.55739	−0.47537	−0.32232	0.14
−0.12	−0.72147	−0.67919	−0.60999	−0.54380	−0.46015	−0.30476	0.12
−0.10	−0.71203	−0.66860	−0.59767	−0.52999	−0.44473	−0.28708	0.10
−0.08	−0.70237	−0.65778	−0.58511	−0.51596	−0.42911	−0.26928	0.08
−0.06	−0.69248	−0.64672	−0.57231	−0.50171	−0.41329	−0.25136	0.06
−0.04	−0.68236	−0.63542	−0.55927	−0.48722	−0.39726	−0.23332	0.04
−0.02	−0.67200	−0.62386	−0.54597	−0.47248	−0.38102	−0.21515	0.02

↑ r_0

α =	.995 ↑−	.99 ↑−	.975 ↑−	.95 ↑−	.90 ↑−	.75 ↑−

TABLE 11.2   SAMPLE SIZE =   13   (CONT.)

α =	.005 ↓+	.01 ↓+	.025 ↓+	.05 ↓+	.10 ↓+	.25 ↓+	
r_o ↓							
0.00	−0.66138	−0.61205	−0.53241	−0.45750	−0.36456	−0.19684	0.00
0.02	−0.65049	−0.59996	−0.51858	−0.44226	−0.34788	−0.17840	−0.02
0.04	−0.63932	−0.58758	−0.50447	−0.42675	−0.33096	−0.15982	−0.04
0.06	−0.62787	−0.57491	−0.49006	−0.41096	−0.31380	−0.14110	−0.06
0.08	−0.61612	−0.56193	−0.47535	−0.39490	−0.29640	−0.12224	−0.08
0.10	−0.60405	−0.54863	−0.46033	−0.37853	−0.27874	−0.10322	−0.10
0.12	−0.59166	−0.53500	−0.44498	−0.36187	−0.26083	−0.08405	−0.12
0.14	−0.57892	−0.52102	−0.42929	−0.34489	−0.24264	−0.06473	−0.14
0.16	−0.56582	−0.50668	−0.41325	−0.32758	−0.22419	−0.04524	−0.16
0.18	−0.55234	−0.49196	−0.39685	−0.30994	−0.20544	−0.02559	−0.18
0.20	−0.53848	−0.47685	−0.38007	−0.29196	−0.18641	−0.00577	−0.20
0.22	−0.52420	−0.46132	−0.36289	−0.27361	−0.16707	0.01423	−0.22
0.24	−0.50948	−0.44536	−0.34530	−0.25489	−0.14742	0.03440	−0.24
0.26	−0.49432	−0.42894	−0.32728	−0.23578	−0.12744	0.05476	−0.26
0.28	−0.47867	−0.41206	−0.30882	−0.21627	−0.10714	0.07530	−0.28
0.30	−0.46252	−0.39467	−0.28990	−0.19634	−0.08649	0.09604	−0.30
0.32	−0.44584	−0.37677	−0.27049	−0.17598	−0.06549	0.11697	−0.32
0.34	−0.42860	−0.35831	−0.25057	−0.15517	−0.04412	0.13811	−0.34
0.36	−0.41076	−0.33928	−0.23012	−0.13389	−0.02238	0.15946	−0.36
0.38	−0.39231	−0.31964	−0.20912	−0.11212	−0.00024	0.18102	−0.38
0.40	−0.37319	−0.29937	−0.18753	−0.08985	0.02230	0.20280	−0.40
0.42	−0.35338	−0.27842	−0.16534	−0.06705	0.04526	0.22481	−0.42
0.44	−0.33282	−0.25676	−0.14252	−0.04369	0.06866	0.24705	−0.44
0.46	−0.31148	−0.23434	−0.11902	−0.01977	0.09251	0.26953	−0.46
0.48	−0.28929	−0.21113	−0.09482	0.00476	0.11683	0.29226	−0.48
0.50	−0.26622	−0.18708	−0.06988	0.02992	0.14163	0.31524	−0.50
0.52	−0.24219	−0.16213	−0.04417	0.05573	0.16693	0.33848	−0.52
0.54	−0.21714	−0.13623	−0.01763	0.08222	0.19276	0.36199	−0.54
0.56	−0.19101	−0.10932	0.00977	0.10944	0.21913	0.38577	−0.56
0.58	−0.16371	−0.08133	0.03808	0.13740	0.24606	0.40984	−0.58
0.60	−0.13515	−0.05219	0.06736	0.16616	0.27359	0.43421	−0.60
0.62	−0.10524	−0.02182	0.09766	0.19575	0.30173	0.45887	−0.62
0.64	−0.07388	0.00987	0.12905	0.22620	0.33050	0.48384	−0.64
0.66	−0.04094	0.04297	0.16158	0.25758	0.35994	0.50914	−0.66
0.68	−0.00630	0.07759	0.19534	0.28993	0.39008	0.53477	−0.68
0.70	0.03020	0.11385	0.23040	0.32329	0.42094	0.56073	−0.70
0.72	0.06871	0.15187	0.26684	0.35773	0.45257	0.58705	−0.72
0.74	0.10943	0.19181	0.30477	0.39331	0.48498	0.61373	−0.74
0.76	0.15256	0.23382	0.34428	0.43010	0.51823	0.64079	−0.76
0.78	0.19835	0.27809	0.38550	0.46816	0.55235	0.66824	−0.78
0.80	0.24707	0.32482	0.42854	0.50758	0.58738	0.69608	−0.80
0.82	0.29903	0.37424	0.47355	0.54844	0.62337	0.72435	−0.82
0.84	0.35459	0.42661	0.52067	0.59084	0.66037	0.75304	−0.84
0.86	0.41419	0.48224	0.57009	0.63487	0.69843	0.78217	−0.86
0.88	0.47830	0.54147	0.62200	0.68065	0.73762	0.81177	−0.88
0.90	0.54752	0.60470	0.67661	0.72831	0.77798	0.84184	−0.90
0.92	0.62254	0.67238	0.73417	0.77797	0.81958	0.87241	−0.92
0.94	0.70417	0.74505	0.79495	0.82980	0.86251	0.90349	−0.94
0.96	0.79343	0.82336	0.85927	0.88395	0.90683	0.93510	−0.96
0.98	0.89153	0.90804	0.92748	0.94062	0.95263	0.96726	−0.98

↑ r_o

α =	.995 ↑−	.99 ↑−	.975 ↑−	.95 ↑−	.90 ↑−	.75 ↑−

TABLE 12.1   SAMPLE SIZE =   14

α =	.005 ↓+	.01 ↓+	.025 ↓+	.05 ↓+	.10 ↓+	.25 ↓+	
r_0 ↓							
-0.98	-0.99530	-0.99450	-0.99309	-0.99164	-0.98963	-0.98525	0.98
-0.96	-0.99054	-0.98892	-0.98611	-0.98320	-0.97918	-0.97046	0.96
-0.94	-0.98572	-0.98328	-0.97905	-0.97468	-0.96866	-0.95563	0.94
-0.92	-0.98083	-0.97756	-0.97190	-0.96608	-0.95807	-0.94077	0.92
-0.90	-0.97587	-0.97177	-0.96468	-0.95740	-0.94739	-0.92587	0.90
-0.88	-0.97084	-0.96591	-0.95738	-0.94863	-0.93663	-0.91093	0.88
-0.86	-0.96574	-0.95996	-0.94999	-0.93978	-0.92580	-0.89595	0.86
-0.84	-0.96057	-0.95394	-0.94252	-0.93084	-0.91487	-0.88092	0.84
-0.82	-0.95532	-0.94784	-0.93495	-0.92180	-0.90387	-0.86585	0.82
-0.80	-0.95000	-0.94165	-0.92730	-0.91268	-0.89277	-0.85074	0.80
-0.78	-0.94460	-0.93538	-0.91956	-0.90346	-0.88159	-0.83557	0.78
-0.76	-0.93912	-0.92903	-0.91172	-0.89414	-0.87032	-0.82036	0.76
-0.74	-0.93356	-0.92258	-0.90378	-0.88473	-0.85895	-0.80510	0.74
-0.72	-0.92791	-0.91604	-0.89575	-0.87521	-0.84749	-0.78979	0.72
-0.70	-0.92218	-0.90941	-0.88762	-0.86560	-0.83594	-0.77442	0.70
-0.68	-0.91635	-0.90268	-0.87938	-0.85588	-0.82428	-0.75900	0.68
-0.66	-0.91044	-0.89586	-0.87103	-0.84605	-0.81253	-0.74352	0.66
-0.64	-0.90443	-0.88893	-0.86258	-0.83611	-0.80067	-0.72798	0.64
-0.62	-0.89833	-0.88190	-0.85402	-0.82606	-0.78871	-0.71239	0.62
-0.60	-0.89212	-0.87477	-0.84534	-0.81589	-0.77664	-0.69673	0.60
-0.58	-0.88582	-0.86752	-0.83655	-0.80561	-0.76446	-0.68101	0.58
-0.56	-0.87941	-0.86017	-0.82764	-0.79521	-0.75217	-0.66523	0.56
-0.54	-0.87289	-0.85269	-0.81861	-0.78468	-0.73977	-0.64937	0.54
-0.52	-0.86626	-0.84510	-0.80945	-0.77403	-0.72725	-0.63345	0.52
-0.50	-0.85952	-0.83739	-0.80016	-0.76325	-0.71461	-0.61746	0.50
-0.48	-0.85266	-0.82955	-0.79074	-0.75234	-0.70185	-0.60140	0.48
-0.46	-0.84568	-0.82159	-0.78119	-0.74130	-0.68896	-0.58527	0.46
-0.44	-0.83858	-0.81349	-0.77149	-0.73011	-0.67595	-0.56905	0.44
-0.42	-0.83135	-0.80526	-0.76166	-0.71879	-0.66281	-0.55277	0.42
-0.40	-0.82398	-0.79689	-0.75168	-0.70732	-0.64953	-0.53640	0.40
-0.38	-0.81648	-0.78837	-0.74155	-0.69570	-0.63612	-0.51995	0.38
-0.36	-0.80884	-0.77971	-0.73126	-0.68393	-0.62257	-0.50341	0.36
-0.34	-0.80106	-0.77089	-0.72082	-0.67200	-0.60888	-0.48680	0.34
-0.32	-0.79313	-0.76192	-0.71021	-0.65991	-0.59504	-0.47009	0.32
-0.30	-0.78504	-0.75278	-0.69943	-0.64766	-0.58106	-0.45329	0.30
-0.28	-0.77679	-0.74348	-0.68849	-0.63524	-0.56692	-0.43640	0.28
-0.26	-0.76838	-0.73400	-0.67736	-0.62265	-0.55262	-0.41942	0.26
-0.24	-0.75980	-0.72435	-0.66606	-0.60988	-0.53817	-0.40234	0.24
-0.22	-0.75104	-0.71451	-0.65456	-0.59693	-0.52355	-0.38517	0.22
-0.20	-0.74210	-0.70448	-0.64288	-0.58379	-0.50876	-0.36789	0.20
-0.18	-0.73297	-0.69426	-0.63099	-0.57046	-0.49380	-0.35050	0.18
-0.16	-0.72365	-0.68384	-0.61890	-0.55693	-0.47867	-0.33301	0.16
-0.14	-0.71413	-0.67321	-0.60660	-0.54319	-0.46335	-0.31542	0.14
-0.12	-0.70440	-0.66236	-0.59408	-0.52925	-0.44785	-0.29771	0.12
-0.10	-0.69445	-0.65128	-0.58133	-0.51509	-0.43216	-0.27989	0.10
-0.08	-0.68427	-0.63998	-0.56836	-0.50072	-0.41628	-0.26195	0.08
-0.06	-0.67386	-0.62843	-0.55514	-0.48611	-0.40019	-0.24389	0.06
-0.04	-0.66321	-0.61664	-0.54168	-0.47127	-0.38390	-0.22571	0.04
-0.02	-0.65231	-0.60459	-0.52796	-0.45619	-0.36740	-0.20740	0.02

↑ r_0

α =	.995 ↑-	.99 ↑-	.975 ↑-	.95 ↑-	.90 ↑-	.75 ↑-

TABLE 12.1   SAMPLE SIZE = 14   (CONT.)

α =	.005 ↓+	.01 ↓+	.025 ↓+	.05 ↓+	.10 ↓+	.25 ↓+	
r_o ↓							
0.00	−0.64114	−0.59227	−0.51398	−0.44086	−0.35069	−0.18897	0.00
0.02	−0.62971	−0.57967	−0.49972	−0.42528	−0.33375	−0.17040	−0.02
0.04	−0.61799	−0.56679	−0.48519	−0.40943	−0.31659	−0.15170	−0.04
0.06	−0.60597	−0.55361	−0.47036	−0.39330	−0.29919	−0.13286	−0.06
0.08	−0.59365	−0.54012	−0.45523	−0.37690	−0.28155	−0.11388	−0.08
0.10	−0.58101	−0.52630	−0.43979	−0.36021	−0.26366	−0.09476	−0.10
0.12	−0.56803	−0.51215	−0.42402	−0.34322	−0.24552	−0.07549	−0.12
0.14	−0.55471	−0.49765	−0.40791	−0.32592	−0.22712	−0.05606	−0.14
0.16	−0.54102	−0.48278	−0.39146	−0.30830	−0.20845	−0.03648	−0.16
0.18	−0.52695	−0.46754	−0.37465	−0.29035	−0.18950	−0.01675	−0.18
0.20	−0.51248	−0.45189	−0.35746	−0.27206	−0.17027	0.00315	−0.20
0.22	−0.49759	−0.43584	−0.33988	−0.25342	−0.15074	0.02322	−0.22
0.24	−0.48226	−0.41935	−0.32190	−0.23441	−0.13091	0.04346	−0.24
0.26	−0.46648	−0.40241	−0.30349	−0.21502	−0.11077	0.06388	−0.26
0.28	−0.45021	−0.38500	−0.28465	−0.19524	−0.09031	0.08448	−0.28
0.30	−0.43344	−0.36709	−0.26535	−0.17506	−0.06951	0.10526	−0.30
0.32	−0.41614	−0.34867	−0.24557	−0.15445	−0.04838	0.12623	−0.32
0.34	−0.39827	−0.32970	−0.22530	−0.13341	−0.02689	0.14739	−0.34
0.36	−0.37982	−0.31016	−0.20451	−0.11191	−0.00504	0.16876	−0.36
0.38	−0.36074	−0.29002	−0.18318	−0.08994	0.01719	0.19032	−0.38
0.40	−0.34101	−0.26926	−0.16128	−0.06748	0.03981	0.21210	−0.40
0.42	−0.32058	−0.24783	−0.13879	−0.04451	0.06283	0.23410	−0.42
0.44	−0.29942	−0.22570	−0.11568	−0.02101	0.08627	0.25631	−0.44
0.46	−0.27748	−0.20283	−0.09193	0.00304	0.11013	0.27876	−0.46
0.48	−0.25471	−0.17919	−0.06750	0.02766	0.13445	0.30143	−0.48
0.50	−0.23106	−0.15473	−0.04235	0.05289	0.15922	0.32435	−0.50
0.52	−0.20648	−0.12939	−0.01646	0.07874	0.18447	0.34751	−0.52
0.54	−0.18091	−0.10314	0.01022	0.10525	0.21022	0.37093	−0.54
0.56	−0.15427	−0.07591	0.03772	0.13244	0.23649	0.39460	−0.56
0.58	−0.12650	−0.04763	0.06610	0.16034	0.26328	0.41855	−0.58
0.60	−0.09752	−0.01826	0.09539	0.18899	0.29064	0.44277	−0.60
0.62	−0.06723	0.01230	0.12566	0.21843	0.31857	0.46727	−0.62
0.64	−0.03554	0.04411	0.15695	0.24868	0.34709	0.49206	−0.64
0.66	−0.00234	0.07726	0.18932	0.27980	0.37625	0.51716	−0.66
0.68	0.03248	0.11186	0.22285	0.31183	0.40605	0.54256	−0.68
0.70	0.06906	0.14799	0.25759	0.34480	0.43653	0.56828	−0.70
0.72	0.10755	0.18579	0.29363	0.37878	0.46771	0.59433	−0.72
0.74	0.14812	0.22537	0.33104	0.41382	0.49964	0.62071	−0.74
0.76	0.19094	0.26689	0.36993	0.44997	0.53233	0.64745	−0.76
0.78	0.23623	0.31049	0.41038	0.48729	0.56582	0.67454	−0.78
0.80	0.28422	0.35635	0.45250	0.52586	0.60016	0.70201	−0.80
0.82	0.33520	0.40468	0.49643	0.56575	0.63537	0.72986	−0.82
0.84	0.38946	0.45570	0.54228	0.60704	0.67151	0.75810	−0.84
0.86	0.44736	0.50965	0.59020	0.64982	0.70862	0.78676	−0.86
0.88	0.50932	0.56683	0.64036	0.69417	0.74674	0.81583	−0.88
0.90	0.57580	0.62757	0.69294	0.74020	0.78593	0.84534	−0.90
0.92	0.64739	0.69223	0.74813	0.78804	0.82624	0.87530	−0.92
0.94	0.72473	0.76126	0.80616	0.83779	0.86774	0.90573	−0.94
0.96	0.80861	0.83516	0.86728	0.88960	0.91048	0.93665	−0.96
0.98	0.89999	0.91450	0.93179	0.94361	0.95454	0.96806	−0.98

↑ r_o

α =	.995 ↑−	.99 ↑−	.975 ↑−	.95 ↑−	.90 ↑−	.75 ↑−

TABLE 13.1   SAMPLE SIZE =   15

α =	.005 ↓+	.01 ↓+	.025 ↓+	.05 ↓+	.10 ↓+	.25 ↓+	
r_0 ↓							
-0.98	-0.99503	-0.99421	-0.99281	-0.99137	-0.98939	-0.98512	0.98
-0.96	-0.98999	-0.98836	-0.98554	-0.98266	-0.97870	-0.97019	0.96
-0.94	-0.98488	-0.98242	-0.97820	-0.97387	-0.96794	-0.95524	0.94
-0.92	-0.97970	-0.97642	-0.97077	-0.96499	-0.95710	-0.94024	0.92
-0.90	-0.97446	-0.97034	-0.96326	-0.95604	-0.94618	-0.92520	0.90
-0.88	-0.96914	-0.96418	-0.95566	-0.94699	-0.93518	-0.91012	0.88
-0.86	-0.96374	-0.95793	-0.94798	-0.93786	-0.92409	-0.89500	0.86
-0.84	-0.95827	-0.95161	-0.94021	-0.92863	-0.91293	-0.87984	0.84
-0.82	-0.95273	-0.94520	-0.93235	-0.91932	-0.90167	-0.86463	0.82
-0.80	-0.94710	-0.93871	-0.92439	-0.90991	-0.89033	-0.84937	0.80
-0.78	-0.94139	-0.93213	-0.91634	-0.90040	-0.87889	-0.83406	0.78
-0.76	-0.93560	-0.92546	-0.90820	-0.89080	-0.86737	-0.81871	0.76
-0.74	-0.92972	-0.91870	-0.89996	-0.88110	-0.85575	-0.80330	0.74
-0.72	-0.92375	-0.91184	-0.89161	-0.87129	-0.84404	-0.78784	0.72
-0.70	-0.91769	-0.90489	-0.88317	-0.86138	-0.83223	-0.77232	0.70
-0.68	-0.91155	-0.89783	-0.87461	-0.85137	-0.82032	-0.75676	0.68
-0.66	-0.90530	-0.89068	-0.86595	-0.84124	-0.80832	-0.74113	0.66
-0.64	-0.89896	-0.88342	-0.85718	-0.83101	-0.79620	-0.72544	0.64
-0.62	-0.89251	-0.87605	-0.84830	-0.82066	-0.78399	-0.70970	0.62
-0.60	-0.88597	-0.86858	-0.83930	-0.81020	-0.77166	-0.69389	0.60
-0.58	-0.87932	-0.86099	-0.83018	-0.79961	-0.75923	-0.67802	0.58
-0.56	-0.87256	-0.85329	-0.82094	-0.78891	-0.74668	-0.66209	0.56
-0.54	-0.86569	-0.84546	-0.81157	-0.77808	-0.73402	-0.64608	0.54
-0.52	-0.85870	-0.83752	-0.80208	-0.76713	-0.72124	-0.63001	0.52
-0.50	-0.85159	-0.82945	-0.79246	-0.75604	-0.70835	-0.61387	0.50
-0.48	-0.84437	-0.82125	-0.78270	-0.74482	-0.69533	-0.59766	0.48
-0.46	-0.83702	-0.81293	-0.77281	-0.73347	-0.68219	-0.58138	0.46
-0.44	-0.82954	-0.80446	-0.76277	-0.72198	-0.66892	-0.56502	0.44
-0.42	-0.82192	-0.79586	-0.75259	-0.71034	-0.65552	-0.54858	0.42
-0.40	-0.81417	-0.78711	-0.74226	-0.69856	-0.64199	-0.53206	0.40
-0.38	-0.80628	-0.77821	-0.73178	-0.68664	-0.62832	-0.51546	0.38
-0.36	-0.79825	-0.76917	-0.72115	-0.67455	-0.61452	-0.49878	0.36
-0.34	-0.79006	-0.75997	-0.71035	-0.66231	-0.60057	-0.48202	0.34
-0.32	-0.78173	-0.75060	-0.69939	-0.64991	-0.58648	-0.46517	0.32
-0.30	-0.77323	-0.74107	-0.68826	-0.63735	-0.57224	-0.44823	0.30
-0.28	-0.76457	-0.73137	-0.67695	-0.62462	-0.55785	-0.43120	0.28
-0.26	-0.75574	-0.72150	-0.66547	-0.61171	-0.54330	-0.41407	0.26
-0.24	-0.74673	-0.71144	-0.65380	-0.59863	-0.52860	-0.39685	0.24
-0.22	-0.73754	-0.70119	-0.64195	-0.58536	-0.51373	-0.37954	0.22
-0.20	-0.72817	-0.69076	-0.62990	-0.57191	-0.49869	-0.36212	0.20
-0.18	-0.71860	-0.68012	-0.61764	-0.55826	-0.48349	-0.34461	0.18
-0.16	-0.70884	-0.66928	-0.60519	-0.54442	-0.46811	-0.32698	0.16
-0.14	-0.69886	-0.65822	-0.59252	-0.53038	-0.45256	-0.30926	0.14
-0.12	-0.68868	-0.64695	-0.57963	-0.51612	-0.43682	-0.29142	0.12
-0.10	-0.67827	-0.63545	-0.56651	-0.50166	-0.42089	-0.27347	0.10
-0.08	-0.66763	-0.62371	-0.55317	-0.48697	-0.40477	-0.25541	0.08
-0.06	-0.65675	-0.61173	-0.53958	-0.47206	-0.38846	-0.23723	0.06
-0.04	-0.64562	-0.59950	-0.52575	-0.45692	-0.37194	-0.21893	0.04
-0.02	-0.63424	-0.58701	-0.51166	-0.44153	-0.35522	-0.20051	0.02

↑ r_0

α =	.995 ↑-	.99 ↑-	.975 ↑-	.95 ↑-	.90 ↑-	.75 ↑-

TABLE 13.2   SAMPLE SIZE = 15   (CONT.)

α =	.005 ↓+	.01 ↓+	.025 ↓+	.05 ↓+	.10 ↓+	.25 ↓+	
r_0 ↓							
0.00	−0.62259	−0.57425	−0.49731	−0.42590	−0.33828	−0.18197	0.00
0.02	−0.61066	−0.56120	−0.48269	−0.41002	−0.32113	−0.16329	−0.02
0.04	−0.59845	−0.54787	−0.46778	−0.39388	−0.30375	−0.14449	−0.04
0.06	−0.58594	−0.53424	−0.45259	−0.37746	−0.28615	−0.12555	−0.06
0.08	−0.57311	−0.52030	−0.43709	−0.36077	−0.26830	−0.10648	−0.08
0.10	−0.55996	−0.50603	−0.42128	−0.34380	−0.25022	−0.08726	−0.10
0.12	−0.54647	−0.49142	−0.40516	−0.32653	−0.23189	−0.06790	−0.12
0.14	−0.53263	−0.47646	−0.38869	−0.30895	−0.21330	−0.04840	−0.14
0.16	−0.51842	−0.46114	−0.37189	−0.29107	−0.19445	−0.02874	−0.16
0.18	−0.50383	−0.44544	−0.35472	−0.27286	−0.17533	−0.00893	−0.18
0.20	−0.48883	−0.42934	−0.33719	−0.25431	−0.15594	0.01103	−0.20
0.22	−0.47341	−0.41283	−0.31927	−0.23542	−0.13625	0.03116	−0.22
0.24	−0.45756	−0.39589	−0.30095	−0.21617	−0.11628	0.05146	−0.24
0.26	−0.44124	−0.37850	−0.28222	−0.19655	−0.09600	0.07192	−0.26
0.28	−0.42445	−0.36065	−0.26305	−0.17655	−0.07541	0.09256	−0.28
0.30	−0.40714	−0.34230	−0.24344	−0.15616	−0.05450	0.11337	−0.30
0.32	−0.38931	−0.32344	−0.22336	−0.13535	−0.03326	0.13437	−0.32
0.34	−0.37092	−0.30404	−0.20280	−0.11412	−0.01168	0.15555	−0.34
0.36	−0.35194	−0.28408	−0.18173	−0.09245	0.01026	0.17692	−0.36
0.38	−0.33234	−0.26353	−0.16013	−0.07033	0.03255	0.19849	−0.38
0.40	−0.31209	−0.24236	−0.13798	−0.04772	0.05522	0.22026	−0.40
0.42	−0.29116	−0.22054	−0.11526	−0.02463	0.07828	0.24224	−0.42
0.44	−0.26950	−0.19803	−0.09194	−0.00103	0.10173	0.26442	−0.44
0.46	−0.24707	−0.17481	−0.06799	0.02311	0.12560	0.28683	−0.46
0.48	−0.22383	−0.15083	−0.04339	0.04780	0.14989	0.30945	−0.48
0.50	−0.19973	−0.12605	−0.01810	0.07306	0.17463	0.33230	−0.50
0.52	−0.17471	−0.10042	0.00791	0.09892	0.19982	0.35539	−0.52
0.54	−0.14873	−0.07391	0.03467	0.12541	0.22548	0.37872	−0.54
0.56	−0.12172	−0.04644	0.06223	0.15255	0.25163	0.40229	−0.56
0.58	−0.09360	−0.01798	0.09062	0.18037	0.27829	0.42612	−0.58
0.60	−0.06431	0.01154	0.11989	0.20889	0.30547	0.45021	−0.60
0.62	−0.03376	0.04219	0.15008	0.23817	0.33320	0.47456	−0.62
0.64	−0.00187	0.07404	0.18124	0.26821	0.36150	0.49919	−0.64
0.66	0.03147	0.10717	0.21342	0.29907	0.39038	0.52410	−0.66
0.68	0.06635	0.14167	0.24669	0.33078	0.41987	0.54931	−0.68
0.70	0.10291	0.17762	0.28111	0.36339	0.44999	0.57481	−0.70
0.72	0.14128	0.21514	0.31674	0.39693	0.48078	0.60062	−0.72
0.74	0.18159	0.25433	0.35365	0.43146	0.51226	0.62674	−0.74
0.76	0.22403	0.29532	0.39194	0.46703	0.54445	0.65319	−0.76
0.78	0.26876	0.33825	0.43168	0.50368	0.57739	0.67998	−0.78
0.80	0.31601	0.38328	0.47296	0.54149	0.61110	0.70711	−0.80
0.82	0.36600	0.43058	0.51590	0.58052	0.64564	0.73460	−0.82
0.84	0.41901	0.48034	0.56061	0.62083	0.68102	0.76245	−0.84
0.86	0.47533	0.53277	0.60721	0.66249	0.71729	0.79068	−0.86
0.88	0.53532	0.58813	0.65583	0.70560	0.75449	0.81931	−0.88
0.90	0.59938	0.64667	0.70664	0.75024	0.79267	0.84833	−0.90
0.92	0.66797	0.70873	0.75979	0.79649	0.83187	0.87778	−0.92
0.94	0.74163	0.77465	0.81548	0.84448	0.87215	0.90765	−0.94
0.96	0.82099	0.84484	0.87392	0.89431	0.91356	0.93797	−0.96
0.98	0.90682	0.91977	0.93534	0.94610	0.95615	0.96875	−0.98

↑ r_0

α =	.995 ↑−	.99 ↑−	.975 ↑−	.95 ↑−	.90 ↑−	.75 ↑−

TABLE 14.1   SAMPLE SIZE =   16

α =	.005 ↓+	.01 ↓+	.025 ↓+	.05 ↓+	.10 ↓+	.25 ↓+	
r_0 ↓							
−0.98	−0.99477	−0.99395	−0.99255	−0.99112	−0.98917	−0.98500	0.98
−0.96	−0.98947	−0.98783	−0.98502	−0.98216	−0.97826	−0.96995	0.96
−0.94	−0.98409	−0.98163	−0.97741	−0.97312	−0.96728	−0.95487	0.94
−0.92	−0.97865	−0.97535	−0.96971	−0.96399	−0.95621	−0.93975	0.92
−0.90	−0.97313	−0.96900	−0.96193	−0.95478	−0.94507	−0.92459	0.90
−0.88	−0.96754	−0.96256	−0.95407	−0.94547	−0.93384	−0.90938	0.88
−0.86	−0.96187	−0.95604	−0.94611	−0.93608	−0.92253	−0.89413	0.86
−0.84	−0.95612	−0.94944	−0.93807	−0.92660	−0.91114	−0.87884	0.84
−0.82	−0.95029	−0.94275	−0.92993	−0.91702	−0.89965	−0.86350	0.82
−0.80	−0.94438	−0.93597	−0.92170	−0.90735	−0.88808	−0.84811	0.80
−0.78	−0.93838	−0.92910	−0.91337	−0.89759	−0.87642	−0.83267	0.78
−0.76	−0.93230	−0.92214	−0.90494	−0.88772	−0.86467	−0.81719	0.76
−0.74	−0.92613	−0.91508	−0.89642	−0.87775	−0.85282	−0.80165	0.74
−0.72	−0.91986	−0.90793	−0.88779	−0.86768	−0.84088	−0.78605	0.72
−0.70	−0.91350	−0.90067	−0.87905	−0.85750	−0.82884	−0.77041	0.70
−0.68	−0.90705	−0.89332	−0.87021	−0.84722	−0.81670	−0.75470	0.68
−0.66	−0.90050	−0.88586	−0.86126	−0.83683	−0.80446	−0.73894	0.66
−0.64	−0.89384	−0.87829	−0.85219	−0.82632	−0.79212	−0.72312	0.64
−0.62	−0.88709	−0.87061	−0.84301	−0.81570	−0.77967	−0.70724	0.62
−0.60	−0.88022	−0.86282	−0.83372	−0.80496	−0.76711	−0.69130	0.60
−0.58	−0.87325	−0.85492	−0.82430	−0.79411	−0.75445	−0.67530	0.58
−0.56	−0.86616	−0.84690	−0.81476	−0.78313	−0.74167	−0.65923	0.56
−0.54	−0.85896	−0.83875	−0.80509	−0.77203	−0.72878	−0.64309	0.54
−0.52	−0.85164	−0.83048	−0.79529	−0.76079	−0.71577	−0.62688	0.52
−0.50	−0.84420	−0.82209	−0.78536	−0.74943	−0.70264	−0.61061	0.50
−0.48	−0.83663	−0.81356	−0.77530	−0.73794	−0.68939	−0.59426	0.48
−0.46	−0.82894	−0.80489	−0.76509	−0.72631	−0.67602	−0.57784	0.46
−0.44	−0.82111	−0.79609	−0.75475	−0.71453	−0.66252	−0.56135	0.44
−0.42	−0.81315	−0.78715	−0.74425	−0.70262	−0.64889	−0.54478	0.42
−0.40	−0.80504	−0.77806	−0.73361	−0.69056	−0.63512	−0.52813	0.40
−0.38	−0.79679	−0.76882	−0.72281	−0.67835	−0.62123	−0.51140	0.38
−0.36	−0.78840	−0.75942	−0.71186	−0.66599	−0.60719	−0.49459	0.36
−0.34	−0.77984	−0.74987	−0.70074	−0.65347	−0.59302	−0.47770	0.34
−0.32	−0.77113	−0.74015	−0.68946	−0.64078	−0.57870	−0.46072	0.32
−0.30	−0.76226	−0.73026	−0.67801	−0.62794	−0.56423	−0.44365	0.30
−0.28	−0.75322	−0.72020	−0.66638	−0.61492	−0.54962	−0.42649	0.28
−0.26	−0.74401	−0.70996	−0.65457	−0.60174	−0.53484	−0.40924	0.26
−0.24	−0.73461	−0.69953	−0.64258	−0.58837	−0.51992	−0.39190	0.24
−0.22	−0.72504	−0.68892	−0.63040	−0.57483	−0.50483	−0.37446	0.22
−0.20	−0.71527	−0.67811	−0.61802	−0.56109	−0.48958	−0.35693	0.20
−0.18	−0.70530	−0.66710	−0.60544	−0.54717	−0.47415	−0.33929	0.18
−0.16	−0.69513	−0.65588	−0.59265	−0.53305	−0.45856	−0.32155	0.16
−0.14	−0.68475	−0.64444	−0.57965	−0.51873	−0.44279	−0.30371	0.14
−0.12	−0.67415	−0.63278	−0.56644	−0.50420	−0.42684	−0.28576	0.12
−0.10	−0.66332	−0.62090	−0.55299	−0.48946	−0.41071	−0.26770	0.10
−0.08	−0.65226	−0.60877	−0.53932	−0.47450	−0.39438	−0.24954	0.08
−0.06	−0.64096	−0.59640	−0.52540	−0.45932	−0.37787	−0.23125	0.06
−0.04	−0.62940	−0.58378	−0.51124	−0.44390	−0.36115	−0.21285	0.04
−0.02	−0.61759	−0.57089	−0.49682	−0.42825	−0.34423	−0.19433	0.02

↑ r_0

α =	.995 ↑−	.99 ↑−	.975 ↑−	.95 ↑−	.90 ↑−	.75 ↑−

TABLE 14.2   SAMPLE SIZE = 16   (CONT.)

α =	.005 ↓+	.01 ↓+	.025 ↓+	.05 ↓+	.10 ↓+	.25 ↓+	
r_0 ↓							
0.00	−0.60551	−0.55774	−0.48215	−0.41236	−0.32710	−0.17569	0.00
0.02	−0.59314	−0.54430	−0.46720	−0.39622	−0.30976	−0.15692	−0.02
0.04	−0.58048	−0.53057	−0.45197	−0.37982	−0.29220	−0.13803	−0.04
0.06	−0.56752	−0.51654	−0.43645	−0.36315	−0.27441	−0.11901	−0.06
0.08	−0.55425	−0.50219	−0.42063	−0.34621	−0.25640	−0.09985	−0.08
0.10	−0.54065	−0.48753	−0.40451	−0.32899	−0.23814	−0.08056	−0.10
0.12	−0.52670	−0.47252	−0.38807	−0.31148	−0.21965	−0.06113	−0.12
0.14	−0.51241	−0.45716	−0.37130	−0.29367	−0.20090	−0.04155	−0.14
0.16	−0.49774	−0.44144	−0.35418	−0.27555	−0.18190	−0.02183	−0.16
0.18	−0.48268	−0.42534	−0.33672	−0.25712	−0.16264	−0.00196	−0.18
0.20	−0.46723	−0.40885	−0.31888	−0.23835	−0.14310	0.01806	−0.20
0.22	−0.45135	−0.39194	−0.30067	−0.21925	−0.12329	0.03824	−0.22
0.24	−0.43503	−0.37461	−0.28207	−0.19980	−0.10319	0.05857	−0.24
0.26	−0.41826	−0.35684	−0.26306	−0.17999	−0.08280	0.07907	−0.26
0.28	−0.40100	−0.33860	−0.24362	−0.15981	−0.06211	0.09974	−0.28
0.30	−0.38324	−0.31987	−0.22375	−0.13924	−0.04110	0.12058	−0.30
0.32	−0.36495	−0.30064	−0.20342	−0.11827	−0.01978	0.14159	−0.32
0.34	−0.34610	−0.28088	−0.18261	−0.09689	0.00187	0.16279	−0.34
0.36	−0.32668	−0.26056	−0.16131	−0.07508	0.02387	0.18416	−0.36
0.38	−0.30664	−0.23967	−0.13950	−0.05283	0.04621	0.20573	−0.38
0.40	−0.28596	−0.21817	−0.11715	−0.03012	0.06891	0.22748	−0.40
0.42	−0.26460	−0.19603	−0.09425	−0.00694	0.09199	0.24944	−0.42
0.44	−0.24253	−0.17322	−0.07076	0.01674	0.11545	0.27159	−0.44
0.46	−0.21970	−0.14971	−0.04666	0.04093	0.13930	0.29396	−0.46
0.48	−0.19608	−0.12546	−0.02194	0.06566	0.16357	0.31653	−0.48
0.50	−0.17162	−0.10043	0.00346	0.09093	0.18825	0.33932	−0.50
0.52	−0.14626	−0.07459	0.02954	0.11678	0.21337	0.36234	−0.52
0.54	−0.11996	−0.04788	0.05635	0.14323	0.23894	0.38558	−0.54
0.56	−0.09266	−0.02026	0.08392	0.17030	0.26498	0.40906	−0.56
0.58	−0.06429	0.00833	0.11229	0.19802	0.29150	0.43278	−0.58
0.60	−0.03479	0.03793	0.14150	0.22642	0.31852	0.45675	−0.60
0.62	−0.00407	0.06862	0.17158	0.25552	0.34606	0.48097	−0.62
0.64	0.02794	0.10045	0.20259	0.28536	0.37413	0.50545	−0.64
0.66	0.06132	0.13350	0.23457	0.31596	0.40276	0.53020	−0.66
0.68	0.09620	0.16785	0.26758	0.34737	0.43196	0.55522	−0.68
0.70	0.13266	0.20358	0.30167	0.37963	0.46177	0.58052	−0.70
0.72	0.17083	0.24079	0.33690	0.41276	0.49219	0.60611	−0.72
0.74	0.21084	0.27957	0.37334	0.44682	0.52326	0.63201	−0.74
0.76	0.25284	0.32004	0.41106	0.48185	0.55500	0.65820	−0.76
0.78	0.29701	0.36233	0.45013	0.51790	0.58744	0.68471	−0.78
0.80	0.34350	0.40656	0.49065	0.55502	0.62060	0.71155	−0.80
0.82	0.39255	0.45289	0.53269	0.59327	0.65453	0.73872	−0.82
0.84	0.44437	0.50149	0.57637	0.63271	0.68924	0.76623	−0.84
0.86	0.49924	0.55255	0.62178	0.67340	0.72478	0.79410	−0.86
0.88	0.55744	0.60626	0.66905	0.71541	0.76117	0.82232	−0.88
0.90	0.61933	0.66288	0.71831	0.75882	0.79847	0.85093	−0.90
0.92	0.68528	0.72265	0.76970	0.80371	0.83671	0.87992	−0.92
0.94	0.75576	0.78589	0.82337	0.85017	0.87593	0.90931	−0.94
0.96	0.83128	0.85292	0.87952	0.89830	0.91619	0.93911	−0.96
0.98	0.91245	0.92415	0.93832	0.94821	0.95753	0.96934	−0.98

↑ r_0

α =	.995 ↑−	.99 ↑−	.975 ↑−	.95 ↑−	.90 ↑−	.75 ↑−

TABLE 15.1   SAMPLE SIZE = 17

$a =$	.005 ↓+	.01 ↓+	.025 ↓+	.05 ↓+	.10 ↓+	.25 ↓+	
r_0 ↓							
-0.98	-0.99452	-0.99370	-0.99231	-0.99089	-0.98897	-0.98488	0.98
-0.96	-0.98898	-0.98733	-0.98453	-0.98170	-0.97785	-0.96973	0.96
-0.94	-0.98335	-0.98088	-0.97667	-0.97242	-0.96666	-0.95453	0.94
-0.92	-0.97766	-0.97435	-0.96873	-0.96306	-0.95539	-0.93930	0.92
-0.90	-0.97189	-0.96774	-0.96070	-0.95361	-0.94404	-0.92402	0.90
-0.88	-0.96604	-0.96105	-0.95258	-0.94407	-0.93261	-0.90870	0.88
-0.86	-0.96011	-0.95427	-0.94438	-0.93444	-0.92109	-0.89333	0.86
-0.84	-0.95410	-0.94740	-0.93608	-0.92472	-0.90949	-0.87792	0.84
-0.82	-0.94801	-0.94045	-0.92768	-0.91490	-0.89780	-0.86246	0.82
-0.80	-0.94183	-0.93340	-0.91919	-0.90499	-0.88602	-0.84695	0.80
-0.78	-0.93556	-0.92627	-0.91061	-0.89498	-0.87415	-0.83140	0.78
-0.76	-0.92920	-0.91903	-0.90192	-0.88487	-0.86218	-0.81579	0.76
-0.74	-0.92275	-0.91170	-0.89313	-0.87466	-0.85013	-0.80013	0.74
-0.72	-0.91621	-0.90427	-0.88423	-0.86434	-0.83797	-0.78441	0.72
-0.70	-0.90957	-0.89674	-0.87523	-0.85392	-0.82572	-0.76865	0.70
-0.68	-0.90283	-0.88910	-0.86612	-0.84339	-0.81337	-0.75282	0.68
-0.66	-0.89599	-0.88136	-0.85690	-0.83275	-0.80092	-0.73694	0.66
-0.64	-0.88905	-0.87351	-0.84757	-0.82199	-0.78836	-0.72099	0.64
-0.62	-0.88200	-0.86554	-0.83811	-0.81112	-0.77570	-0.70499	0.62
-0.60	-0.87484	-0.85746	-0.82854	-0.80014	-0.76293	-0.68893	0.60
-0.58	-0.86756	-0.84926	-0.81885	-0.78903	-0.75006	-0.67280	0.58
-0.56	-0.86018	-0.84094	-0.80903	-0.77780	-0.73707	-0.65661	0.56
-0.54	-0.85267	-0.83250	-0.79908	-0.76644	-0.72396	-0.64035	0.54
-0.52	-0.84504	-0.82393	-0.78901	-0.75496	-0.71074	-0.62402	0.52
-0.50	-0.83729	-0.81523	-0.77880	-0.74334	-0.69740	-0.60762	0.50
-0.48	-0.82941	-0.80639	-0.76845	-0.73160	-0.68394	-0.59116	0.48
-0.46	-0.82139	-0.79742	-0.75796	-0.71971	-0.67036	-0.57462	0.46
-0.44	-0.81324	-0.78831	-0.74733	-0.70768	-0.65665	-0.55800	0.44
-0.42	-0.80495	-0.77905	-0.73655	-0.69552	-0.64281	-0.54131	0.42
-0.40	-0.79652	-0.76965	-0.72562	-0.68320	-0.62884	-0.52454	0.40
-0.38	-0.78794	-0.76009	-0.71453	-0.67073	-0.61473	-0.50770	0.38
-0.36	-0.77921	-0.75037	-0.70329	-0.65812	-0.60049	-0.49077	0.36
-0.34	-0.77032	-0.74049	-0.69188	-0.64534	-0.58611	-0.47376	0.34
-0.32	-0.76126	-0.73045	-0.68031	-0.63240	-0.57159	-0.45666	0.32
-0.30	-0.75205	-0.72023	-0.66856	-0.61930	-0.55692	-0.43948	0.30
-0.28	-0.74266	-0.70984	-0.65664	-0.60604	-0.54210	-0.42221	0.28
-0.26	-0.73309	-0.69927	-0.64454	-0.59259	-0.52713	-0.40485	0.26
-0.24	-0.72334	-0.68851	-0.63225	-0.57898	-0.51200	-0.38740	0.24
-0.22	-0.71340	-0.67756	-0.61977	-0.56518	-0.49671	-0.36985	0.22
-0.20	-0.70327	-0.66641	-0.60710	-0.55119	-0.48126	-0.35221	0.20
-0.18	-0.69294	-0.65505	-0.59422	-0.53702	-0.46565	-0.33447	0.18
-0.16	-0.68240	-0.64349	-0.58114	-0.52265	-0.44986	-0.31663	0.16
-0.14	-0.67165	-0.63171	-0.56784	-0.50808	-0.43390	-0.29869	0.14
-0.12	-0.66067	-0.61970	-0.55433	-0.49330	-0.41776	-0.28064	0.12
-0.10	-0.64947	-0.60747	-0.54059	-0.47832	-0.40144	-0.26248	0.10
-0.08	-0.63802	-0.59499	-0.52662	-0.46312	-0.38494	-0.24422	0.08
-0.06	-0.62634	-0.58227	-0.51241	-0.44769	-0.36824	-0.22584	0.06
-0.04	-0.61440	-0.56930	-0.49796	-0.43204	-0.35135	-0.20735	0.04
-0.02	-0.60219	-0.55606	-0.48325	-0.41615	-0.33426	-0.18875	0.02

↑ r_0

$a =$	.995 ↑-	.99 ↑-	.975 ↑-	.95 ↑-	.90 ↑-	.75 ↑-

TABLE 15.2   SAMPLE SIZE =   17   (CONT.)

α =	.005 ↓+	.01 ↓+	.025 ↓+	.05 ↓+	.10 ↓+	.25 ↓+	
r_0 ↓							
0.00	−0.58971	−0.54255	−0.46828	−0.40003	−0.31696	−0.17002	0.00
0.02	−0.57695	−0.52876	−0.45304	−0.38365	−0.29945	−0.15117	−0.02
0.04	−0.56390	−0.51467	−0.43752	−0.36702	−0.28173	−0.13220	−0.04
0.06	−0.55054	−0.50028	−0.42172	−0.35013	−0.26379	−0.11311	−0.06
0.08	−0.53686	−0.48558	−0.40562	−0.33298	−0.24562	−0.09388	−0.08
0.10	−0.52286	−0.47056	−0.38922	−0.31554	−0.22722	−0.07452	−0.10
0.12	−0.50851	−0.45520	−0.37250	−0.29782	−0.20858	−0.05502	−0.12
0.14	−0.49381	−0.43949	−0.35546	−0.27981	−0.18970	−0.03539	−0.14
0.16	−0.47873	−0.42342	−0.33807	−0.26149	−0.17057	−0.01561	−0.16
0.18	−0.46327	−0.40697	−0.32034	−0.24286	−0.15118	0.00430	−0.18
0.20	−0.44741	−0.39013	−0.30225	−0.22391	−0.13153	0.02437	−0.20
0.22	−0.43112	−0.37288	−0.28379	−0.20463	−0.11161	0.04459	−0.22
0.24	−0.41440	−0.35521	−0.26494	−0.18501	−0.09141	0.06496	−0.24
0.26	−0.39722	−0.33710	−0.24569	−0.16504	−0.07092	0.08549	−0.26
0.28	−0.37956	−0.31853	−0.22603	−0.14470	−0.05014	0.10618	−0.28
0.30	−0.36140	−0.29948	−0.20593	−0.12398	−0.02906	0.12704	−0.30
0.32	−0.34272	−0.27993	−0.18539	−0.10288	−0.00767	0.14807	−0.32
0.34	−0.32349	−0.25986	−0.16438	−0.08138	0.01404	0.16927	−0.34
0.36	−0.30368	−0.23924	−0.14289	−0.05946	0.03608	0.19064	−0.36
0.38	−0.28327	−0.21806	−0.12091	−0.03711	0.05845	0.21220	−0.38
0.40	−0.26222	−0.19628	−0.09840	−0.01432	0.08118	0.23394	−0.40
0.42	−0.24051	−0.17388	−0.07534	0.00894	0.10426	0.25587	−0.42
0.44	−0.21809	−0.15082	−0.05173	0.03267	0.12771	0.27799	−0.44
0.46	−0.19494	−0.12708	−0.02752	0.05689	0.15154	0.30032	−0.46
0.48	−0.17101	−0.10262	−0.00270	0.08163	0.17577	0.32284	−0.48
0.50	−0.14625	−0.07740	0.02276	0.10690	0.20040	0.34558	−0.50
0.52	−0.12063	−0.05139	0.04889	0.13272	0.22545	0.36852	−0.52
0.54	−0.09408	−0.02454	0.07571	0.15912	0.25093	0.39169	−0.54
0.56	−0.06656	0.00318	0.10327	0.18611	0.27686	0.41508	−0.56
0.58	−0.03801	0.03184	0.13160	0.21372	0.30324	0.43870	−0.58
0.60	−0.00836	0.06148	0.16072	0.24198	0.33011	0.46255	−0.60
0.62	0.02245	0.09216	0.19068	0.27091	0.35746	0.48665	−0.62
0.64	0.05451	0.12393	0.22153	0.30055	0.38532	0.51099	−0.64
0.66	0.08789	0.15687	0.25330	0.33091	0.41372	0.53559	−0.66
0.68	0.12269	0.19104	0.28605	0.36203	0.44265	0.56045	−0.68
0.70	0.15900	0.22653	0.31981	0.39396	0.47216	0.58557	−0.70
0.72	0.19694	0.26341	0.35466	0.42671	0.50225	0.61097	−0.72
0.74	0.23662	0.30178	0.39065	0.46034	0.53295	0.63665	−0.74
0.76	0.27817	0.34174	0.42784	0.49488	0.56428	0.66262	−0.76
0.78	0.32176	0.38340	0.46630	0.53037	0.59627	0.68889	−0.78
0.80	0.36753	0.42689	0.50610	0.56686	0.62894	0.71546	−0.80
0.82	0.41567	0.47232	0.54733	0.60441	0.66232	0.74234	−0.82
0.84	0.46638	0.51986	0.59008	0.64307	0.69643	0.76955	−0.84
0.86	0.51991	0.56966	0.63443	0.68288	0.73132	0.79709	−0.86
0.88	0.57649	0.62191	0.68049	0.72392	0.76700	0.82497	−0.88
0.90	0.63643	0.67680	0.72838	0.76625	0.80352	0.85320	−0.90
0.92	0.70006	0.73457	0.77822	0.80995	0.84091	0.88179	−0.92
0.94	0.76776	0.79547	0.83014	0.85508	0.87921	0.91076	−0.94
0.96	0.83996	0.85978	0.88430	0.90174	0.91847	0.94011	−0.96
0.98	0.91718	0.92784	0.94086	0.95001	0.95871	0.96985	−0.98

↑ r_0

α =	.995 ↑−	.99 ↑−	.975 ↑−	.95 ↑−	.90 ↑−	.75 ↑−

TABLE 16.1   SAMPLE SIZE =   18

α =	.005 ↓+	.01 ↓+	.025 ↓+	.05 ↓+	.10 ↓+	.25 ↓+	
r_0 ↓							
-0.98	-0.99429	-0.99347	-0.99208	-0.99067	-0.98878	-0.98478	0.98
-0.96	-0.98851	-0.98687	-0.98408	-0.98127	-0.97747	-0.96952	0.96
-0.94	-0.98266	-0.98018	-0.97599	-0.97177	-0.96609	-0.95422	0.94
-0.92	-0.97672	-0.97341	-0.96781	-0.96219	-0.95463	-0.93888	0.92
-0.90	-0.97071	-0.96656	-0.95955	-0.95252	-0.94309	-0.92349	0.90
-0.88	-0.96462	-0.95963	-0.95120	-0.94276	-0.93146	-0.90806	0.88
-0.86	-0.95845	-0.95261	-0.94275	-0.93291	-0.91975	-0.89259	0.86
-0.84	-0.95219	-0.94549	-0.93422	-0.92297	-0.90796	-0.87707	0.84
-0.82	-0.94585	-0.93829	-0.92558	-0.91293	-0.89607	-0.86150	0.82
-0.80	-0.93942	-0.93100	-0.91686	-0.90279	-0.88410	-0.84588	0.80
-0.78	-0.93290	-0.92361	-0.90803	-0.89256	-0.87204	-0.83021	0.78
-0.76	-0.92628	-0.91612	-0.89910	-0.88222	-0.85988	-0.81449	0.76
-0.74	-0.91958	-0.90853	-0.89006	-0.87179	-0.84763	-0.79872	0.74
-0.72	-0.91277	-0.90084	-0.88092	-0.86124	-0.83528	-0.78290	0.72
-0.70	-0.90587	-0.89305	-0.87167	-0.85059	-0.82284	-0.76702	0.70
-0.68	-0.89886	-0.88515	-0.86231	-0.83984	-0.81029	-0.75108	0.68
-0.66	-0.89176	-0.87715	-0.85284	-0.82896	-0.79765	-0.73508	0.66
-0.64	-0.88454	-0.86903	-0.84326	-0.81798	-0.78490	-0.71903	0.64
-0.62	-0.87722	-0.86079	-0.83355	-0.80688	-0.77204	-0.70292	0.62
-0.60	-0.86978	-0.85244	-0.82373	-0.79566	-0.75908	-0.68674	0.60
-0.58	-0.86223	-0.84397	-0.81378	-0.78433	-0.74600	-0.67050	0.58
-0.56	-0.85455	-0.83538	-0.80371	-0.77286	-0.73282	-0.65420	0.56
-0.54	-0.84676	-0.82666	-0.79350	-0.76128	-0.71952	-0.63782	0.54
-0.52	-0.83885	-0.81781	-0.78317	-0.74956	-0.70611	-0.62139	0.52
-0.50	-0.83080	-0.80882	-0.77270	-0.73771	-0.69258	-0.60488	0.50
-0.48	-0.82263	-0.79971	-0.76209	-0.72573	-0.67892	-0.58830	0.48
-0.46	-0.81432	-0.79045	-0.75134	-0.71361	-0.66515	-0.57165	0.46
-0.44	-0.80587	-0.78105	-0.74044	-0.70135	-0.65125	-0.55493	0.44
-0.42	-0.79728	-0.77150	-0.72940	-0.68895	-0.63722	-0.53813	0.42
-0.40	-0.78854	-0.76180	-0.71821	-0.67640	-0.62306	-0.52126	0.40
-0.38	-0.77965	-0.75195	-0.70686	-0.66370	-0.60876	-0.50430	0.38
-0.36	-0.77061	-0.74194	-0.69535	-0.65085	-0.59433	-0.48727	0.36
-0.34	-0.76140	-0.73176	-0.68367	-0.63784	-0.57976	-0.47015	0.34
-0.32	-0.75204	-0.72142	-0.67183	-0.62468	-0.56505	-0.45295	0.32
-0.30	-0.74250	-0.71090	-0.65982	-0.61134	-0.55020	-0.43567	0.30
-0.28	-0.73279	-0.70020	-0.64763	-0.59784	-0.53520	-0.41830	0.28
-0.26	-0.72290	-0.68933	-0.63526	-0.58417	-0.52004	-0.40084	0.26
-0.24	-0.71282	-0.67826	-0.62270	-0.57032	-0.50473	-0.38329	0.24
-0.22	-0.70255	-0.66700	-0.60995	-0.55630	-0.48927	-0.36564	0.22
-0.20	-0.69209	-0.65554	-0.59701	-0.54208	-0.47364	-0.34790	0.20
-0.18	-0.68142	-0.64388	-0.58387	-0.52768	-0.45785	-0.33007	0.18
-0.16	-0.67054	-0.63200	-0.57051	-0.51309	-0.44189	-0.31214	0.16
-0.14	-0.65945	-0.61990	-0.55695	-0.49829	-0.42576	-0.29410	0.14
-0.12	-0.64813	-0.60758	-0.54317	-0.48330	-0.40946	-0.27597	0.12
-0.10	-0.63658	-0.59503	-0.52916	-0.46809	-0.39297	-0.25772	0.10
-0.08	-0.62479	-0.58224	-0.51493	-0.45267	-0.37630	-0.23938	0.08
-0.06	-0.61275	-0.56920	-0.50045	-0.43703	-0.35944	-0.22092	0.06
-0.04	-0.60046	-0.55591	-0.48573	-0.42116	-0.34240	-0.20235	0.04
-0.02	-0.58790	-0.54235	-0.47076	-0.40507	-0.32515	-0.18367	0.02

↑ r_0

α =	.995 ↑-	.99 ↑-	.975 ↑-	.95 ↑-	.90 ↑-	.75 ↑-

TABLE 16.2   SAMPLE SIZE = 18   (CONT.)

α =	.005 ↓+	.01 ↓+	.025 ↓+	.05 ↓+	.10 ↓+	.25 ↓+	
r_0 ↓							
0.00	-0.57507	-0.52852	-0.45553	-0.38873	-0.30770	-0.16487	0.00
0.02	-0.56195	-0.51440	-0.44003	-0.37215	-0.29005	-0.14595	-0.02
0.04	-0.54853	-0.50000	-0.42426	-0.35532	-0.27218	-0.12691	-0.04
0.06	-0.53481	-0.48529	-0.40820	-0.33824	-0.25410	-0.10775	-0.06
0.08	-0.52078	-0.47028	-0.39185	-0.32088	-0.23580	-0.08846	-0.08
0.10	-0.50641	-0.45493	-0.37520	-0.30326	-0.21727	-0.06904	-0.10
0.12	-0.49170	-0.43926	-0.35824	-0.28535	-0.19851	-0.04949	-0.12
0.14	-0.47663	-0.42323	-0.34096	-0.26716	-0.17951	-0.02981	-0.14
0.16	-0.46119	-0.40685	-0.32334	-0.24867	-0.16027	-0.00998	-0.16
0.18	-0.44537	-0.39009	-0.30538	-0.22987	-0.14077	0.00998	-0.18
0.20	-0.42914	-0.37295	-0.28706	-0.21076	-0.12102	0.03008	-0.20
0.22	-0.41250	-0.35540	-0.26838	-0.19133	-0.10100	0.05034	-0.22
0.24	-0.39542	-0.33743	-0.24932	-0.17156	-0.08072	0.07074	-0.24
0.26	-0.37789	-0.31902	-0.22986	-0.15145	-0.06015	0.09129	-0.26
0.28	-0.35988	-0.30017	-0.21000	-0.13098	-0.03931	0.11200	-0.28
0.30	-0.34137	-0.28084	-0.18972	-0.11014	-0.01816	0.13287	-0.30
0.32	-0.32235	-0.26102	-0.16900	-0.08893	0.00328	0.15391	-0.32
0.34	-0.30278	-0.24068	-0.14782	-0.06732	0.02503	0.17511	-0.34
0.36	-0.28264	-0.21981	-0.12618	-0.04531	0.04711	0.19648	-0.36
0.38	-0.26191	-0.19838	-0.10404	-0.02289	0.06950	0.21802	-0.38
0.40	-0.24055	-0.17637	-0.08140	-0.00003	0.09224	0.23975	-0.40
0.42	-0.21854	-0.15375	-0.05823	0.02327	0.11532	0.26165	-0.42
0.44	-0.19584	-0.13049	-0.03451	0.04704	0.13876	0.28375	-0.44
0.46	-0.17241	-0.10656	-0.01022	0.07128	0.16256	0.30603	-0.46
0.48	-0.14823	-0.08193	0.01466	0.09602	0.18675	0.32851	-0.48
0.50	-0.12324	-0.05657	0.04016	0.12127	0.21132	0.35119	-0.50
0.52	-0.09740	-0.03044	0.06631	0.14705	0.23630	0.37407	-0.52
0.54	-0.07067	-0.00349	0.09313	0.17338	0.26168	0.39716	-0.54
0.56	-0.04299	0.02430	0.12066	0.20029	0.28750	0.42047	-0.56
0.58	-0.01431	0.05300	0.14892	0.22779	0.31376	0.44400	-0.58
0.60	0.01543	0.08263	0.17795	0.25592	0.34047	0.46775	-0.60
0.62	0.04629	0.11327	0.20778	0.28468	0.36765	0.49173	-0.62
0.64	0.07835	0.14496	0.23846	0.31411	0.39532	0.51595	-0.64
0.66	0.11168	0.17776	0.27002	0.34424	0.42349	0.54041	-0.66
0.68	0.14637	0.21174	0.30250	0.37509	0.45218	0.56512	-0.68
0.70	0.18250	0.24697	0.33596	0.40671	0.48141	0.59008	-0.70
0.72	0.22018	0.28353	0.37044	0.43910	0.51120	0.61530	-0.72
0.74	0.25951	0.32149	0.40600	0.47233	0.54156	0.64079	-0.74
0.76	0.30062	0.36096	0.44270	0.50641	0.57252	0.66655	-0.76
0.78	0.34363	0.40202	0.48058	0.54140	0.60409	0.69260	-0.78
0.80	0.38870	0.44480	0.51974	0.57732	0.63632	0.71893	-0.80
0.82	0.43598	0.48940	0.56022	0.61424	0.66920	0.74556	-0.82
0.84	0.48567	0.53596	0.60212	0.65219	0.70278	0.77250	-0.84
0.86	0.53795	0.58462	0.64551	0.69122	0.73708	0.79974	-0.86
0.88	0.59306	0.63554	0.69049	0.73139	0.77213	0.82731	-0.88
0.90	0.65125	0.68890	0.73716	0.77277	0.80796	0.85521	-0.90
0.92	0.71281	0.74489	0.78563	0.81540	0.84460	0.88345	-0.92
0.94	0.77807	0.80374	0.83601	0.85937	0.88209	0.91204	-0.94
0.96	0.84739	0.86568	0.88843	0.90473	0.92046	0.94099	-0.96
0.98	0.92120	0.93100	0.94305	0.95158	0.95975	0.97030	-0.98

↑ r_0

α =	.995 ↑-	.99 ↑-	.975 ↑-	.95 ↑-	.90 ↑-	.75 ↑-

TABLE 17.1   SAMPLE SIZE =   19

α =	.005 ↓+	.01 ↓+	.025 ↓+	.05 ↓+	.10 ↓+	.25 ↓+	
r_0 ↓							
-0.98	-0.99408	-0.99325	-0.99187	-0.99047	-0.98860	-0.98468	0.98
-0.96	-0.98807	-0.98643	-0.98365	-0.98086	-0.97712	-0.96933	0.96
-0.94	-0.98200	-0.97952	-0.97534	-0.97116	-0.96556	-0.95393	0.94
-0.92	-0.97584	-0.97253	-0.96695	-0.96138	-0.95392	-0.93849	0.92
-0.90	-0.96960	-0.96545	-0.95847	-0.95151	-0.94220	-0.92300	0.90
-0.88	-0.96328	-0.95829	-0.94990	-0.94154	-0.93040	-0.90747	0.88
-0.86	-0.95688	-0.95104	-0.94123	-0.93149	-0.91851	-0.89190	0.86
-0.84	-0.95039	-0.94370	-0.93248	-0.92134	-0.90654	-0.87627	0.84
-0.82	-0.94381	-0.93626	-0.92362	-0.91109	-0.89447	-0.86060	0.82
-0.80	-0.93715	-0.92873	-0.91467	-0.90075	-0.88232	-0.84488	0.80
-0.78	-0.93039	-0.92111	-0.90561	-0.89030	-0.87008	-0.82911	0.78
-0.76	-0.92353	-0.91338	-0.89645	-0.87976	-0.85774	-0.81329	0.76
-0.74	-0.91658	-0.90555	-0.88719	-0.86911	-0.84531	-0.79742	0.74
-0.72	-0.90953	-0.89762	-0.87782	-0.85836	-0.83278	-0.78149	0.72
-0.70	-0.90238	-0.88959	-0.86834	-0.84749	-0.82016	-0.76551	0.70
-0.68	-0.89512	-0.88144	-0.85875	-0.83652	-0.80743	-0.74947	0.68
-0.66	-0.88776	-0.87319	-0.84905	-0.82544	-0.79461	-0.73337	0.66
-0.64	-0.88029	-0.86482	-0.83923	-0.81424	-0.78168	-0.71721	0.64
-0.62	-0.87271	-0.85633	-0.82929	-0.80293	-0.76864	-0.70099	0.62
-0.60	-0.86501	-0.84773	-0.81923	-0.79150	-0.75550	-0.68471	0.60
-0.58	-0.85720	-0.83900	-0.80905	-0.77995	-0.74225	-0.66837	0.58
-0.56	-0.84926	-0.83015	-0.79873	-0.76827	-0.72889	-0.65197	0.56
-0.54	-0.84121	-0.82118	-0.78829	-0.75647	-0.71541	-0.63549	0.54
-0.52	-0.83302	-0.81207	-0.77772	-0.74454	-0.70182	-0.61895	0.52
-0.50	-0.82471	-0.80282	-0.76701	-0.73248	-0.68811	-0.60235	0.50
-0.48	-0.81626	-0.79344	-0.75616	-0.72028	-0.67428	-0.58567	0.48
-0.46	-0.80767	-0.78392	-0.74517	-0.70795	-0.66033	-0.56892	0.46
-0.44	-0.79894	-0.77425	-0.73403	-0.69547	-0.64625	-0.55209	0.44
-0.42	-0.79007	-0.76443	-0.72274	-0.68286	-0.63205	-0.53520	0.42
-0.40	-0.78105	-0.75447	-0.71131	-0.67010	-0.61771	-0.51822	0.40
-0.38	-0.77187	-0.74434	-0.69971	-0.65718	-0.60324	-0.50117	0.38
-0.36	-0.76254	-0.73405	-0.68796	-0.64412	-0.58864	-0.48404	0.36
-0.34	-0.75305	-0.72360	-0.67604	-0.63090	-0.57390	-0.46683	0.34
-0.32	-0.74339	-0.71298	-0.66395	-0.61752	-0.55902	-0.44954	0.32
-0.30	-0.73355	-0.70219	-0.65169	-0.60397	-0.54400	-0.43216	0.30
-0.28	-0.72354	-0.69121	-0.63926	-0.59026	-0.52883	-0.41470	0.28
-0.26	-0.71335	-0.68005	-0.62664	-0.57638	-0.51351	-0.39715	0.26
-0.24	-0.70297	-0.66870	-0.61384	-0.56232	-0.49804	-0.37951	0.24
-0.22	-0.69240	-0.65716	-0.60084	-0.54809	-0.48241	-0.36178	0.22
-0.20	-0.68163	-0.64542	-0.58765	-0.53367	-0.46662	-0.34395	0.20
-0.18	-0.67065	-0.63346	-0.57426	-0.51906	-0.45068	-0.32603	0.18
-0.16	-0.65946	-0.62130	-0.56067	-0.50426	-0.43456	-0.30801	0.16
-0.14	-0.64806	-0.60892	-0.54686	-0.48926	-0.41828	-0.28990	0.14
-0.12	-0.63642	-0.59631	-0.53284	-0.47406	-0.40182	-0.27168	0.12
-0.10	-0.62456	-0.58347	-0.51859	-0.45866	-0.38518	-0.25336	0.10
-0.08	-0.61245	-0.57039	-0.50411	-0.44304	-0.36837	-0.23494	0.08
-0.06	-0.60009	-0.55706	-0.48939	-0.42721	-0.35137	-0.21641	0.06
-0.04	-0.58748	-0.54347	-0.47444	-0.41115	-0.33418	-0.19777	0.04
-0.02	-0.57459	-0.52963	-0.45923	-0.39486	-0.31679	-0.17902	0.02

↑ r_0

α =   .995 ↑−   .99 ↑−   .975 ↑−   .95 ↑−   .90 ↑−   .75 ↑−

TABLE 17.2   SAMPLE SIZE =   19   (CONT.)

$\alpha =$	.005 ↓+	.01 ↓+	.025 ↓+	.05 ↓+	.10 ↓+	.25 ↓+	
r_0 ↓							
0.00	−0.56144	−0.51550	−0.44376	−0.37834	−0.29921	−0.16015	0.00
0.02	−0.54799	−0.50110	−0.42803	−0.36158	−0.28143	−0.14117	−0.02
0.04	−0.53425	−0.48641	−0.41203	−0.34457	−0.26344	−0.12207	−0.04
0.06	−0.52021	−0.47142	−0.39575	−0.32731	−0.24523	−0.10285	−0.06
0.08	−0.50584	−0.45611	−0.37917	−0.30978	−0.22681	−0.08351	−0.08
0.10	−0.49115	−0.44049	−0.36230	−0.29199	−0.20817	−0.06404	−0.10
0.12	−0.47611	−0.42453	−0.34512	−0.27392	−0.18930	−0.04444	−0.12
0.14	−0.46072	−0.40822	−0.32762	−0.25557	−0.17019	−0.02472	−0.14
0.16	−0.44495	−0.39156	−0.30980	−0.23692	−0.15085	−0.00485	−0.16
0.18	−0.42881	−0.37453	−0.29164	−0.21798	−0.13126	0.01514	−0.18
0.20	−0.41226	−0.35711	−0.27312	−0.19873	−0.11143	0.03528	−0.20
0.22	−0.39530	−0.33929	−0.25425	−0.17916	−0.09133	0.05556	−0.22
0.24	−0.37790	−0.32106	−0.23500	−0.15926	−0.07097	0.07599	−0.24
0.26	−0.36005	−0.30240	−0.21537	−0.13903	−0.05034	0.09656	−0.26
0.28	−0.34173	−0.28330	−0.19533	−0.11845	−0.02943	0.11729	−0.28
0.30	−0.32292	−0.26372	−0.17489	−0.09751	−0.00824	0.13817	−0.30
0.32	−0.30360	−0.24366	−0.15401	−0.07620	0.01325	0.15921	−0.32
0.34	−0.28374	−0.22310	−0.13270	−0.05452	0.03503	0.18041	−0.34
0.36	−0.26332	−0.20201	−0.11092	−0.03244	0.05713	0.20177	−0.36
0.38	−0.24231	−0.18038	−0.08866	−0.00995	0.07954	0.22331	−0.38
0.40	−0.22069	−0.15817	−0.06592	0.01296	0.10228	0.24501	−0.40
0.42	−0.19842	−0.13537	−0.04265	0.03629	0.12535	0.26689	−0.42
0.44	−0.17548	−0.11194	−0.01886	0.06008	0.14877	0.28895	−0.44
0.46	−0.15183	−0.08786	0.00549	0.08433	0.17255	0.31120	−0.46
0.48	−0.12743	−0.06310	0.03042	0.10906	0.19668	0.33363	−0.48
0.50	−0.10225	−0.03763	0.05594	0.13428	0.22120	0.35626	−0.50
0.52	−0.07625	−0.01140	0.08209	0.16001	0.24610	0.37908	−0.52
0.54	−0.04937	0.01561	0.10889	0.18628	0.27140	0.40211	−0.54
0.56	−0.02158	0.04344	0.13637	0.21310	0.29711	0.42533	−0.56
0.58	0.00719	0.07214	0.16456	0.24049	0.32324	0.44878	−0.58
0.60	0.03698	0.10175	0.19349	0.26847	0.34981	0.47243	−0.60
0.62	0.06785	0.13232	0.22318	0.29707	0.37683	0.49631	−0.62
0.64	0.09988	0.16391	0.25369	0.32631	0.40432	0.52041	−0.64
0.66	0.13313	0.19656	0.28504	0.35622	0.43228	0.54474	−0.66
0.68	0.16768	0.23033	0.31727	0.38682	0.46074	0.56931	−0.68
0.70	0.20361	0.26530	0.35043	0.41813	0.48971	0.59413	−0.70
0.72	0.24101	0.30153	0.38457	0.45020	0.51922	0.61919	−0.72
0.74	0.27999	0.33910	0.41972	0.48305	0.54927	0.64450	−0.74
0.76	0.32065	0.37810	0.45595	0.51672	0.57988	0.67008	−0.76
0.78	0.36311	0.41860	0.49331	0.55124	0.61109	0.69592	−0.78
0.80	0.40750	0.46071	0.53186	0.58664	0.64290	0.72204	−0.80
0.82	0.45398	0.50454	0.57166	0.62298	0.67535	0.74844	−0.82
0.84	0.50271	0.55019	0.61279	0.66029	0.70844	0.77513	−0.84
0.86	0.55385	0.59781	0.65532	0.69861	0.74222	0.80212	−0.86
0.88	0.60762	0.64754	0.69933	0.73801	0.77670	0.82940	−0.88
0.90	0.66423	0.69951	0.74490	0.77852	0.81191	0.85701	−0.90
0.92	0.72394	0.75392	0.79215	0.82021	0.84788	0.88493	−0.92
0.94	0.78703	0.81095	0.84116	0.86314	0.88464	0.91318	−0.94
0.96	0.85382	0.87081	0.89205	0.90736	0.92222	0.94177	−0.96
0.98	0.92467	0.93374	0.94496	0.95296	0.96066	0.97071	−0.98

↑ r_0

$\alpha =$	.995 ↑−	.99 ↑−	.975 ↑−	.95 ↑−	.90 ↑−	.75 ↑−

TABLE 18.1   SAMPLE SIZE =   20

α =	.005 ↓+	.01 ↓+	.025 ↓+	.05 ↓+	.10 ↓+	.25 ↓+	
r_0 ↓							
−0.98	−0.99387	−0.99305	−0.99167	−0.99028	−0.98843	−0.98459	0.98
−0.96	−0.98766	−0.98601	−0.98324	−0.98048	−0.97679	−0.96914	0.96
−0.94	−0.98137	−0.97890	−0.97474	−0.97060	−0.96507	−0.95365	0.94
−0.92	−0.97500	−0.97169	−0.96614	−0.96062	−0.95326	−0.93812	0.92
−0.90	−0.96855	−0.96440	−0.95746	−0.95056	−0.94138	−0.92254	0.90
−0.88	−0.96202	−0.95703	−0.94868	−0.94040	−0.92941	−0.90692	0.88
−0.86	−0.95540	−0.94956	−0.93981	−0.93015	−0.91735	−0.89125	0.86
−0.84	−0.94869	−0.94200	−0.93084	−0.91981	−0.90521	−0.87553	0.84
−0.82	−0.94189	−0.93435	−0.92177	−0.90937	−0.89298	−0.85977	0.82
−0.80	−0.93500	−0.92660	−0.91261	−0.89883	−0.88066	−0.84395	0.80
−0.78	−0.92801	−0.91875	−0.90334	−0.88819	−0.86825	−0.82809	0.78
−0.76	−0.92093	−0.91080	−0.89397	−0.87745	−0.85574	−0.81217	0.76
−0.74	−0.91375	−0.90275	−0.88450	−0.86660	−0.84315	−0.79620	0.74
−0.72	−0.90647	−0.89459	−0.87491	−0.85565	−0.83045	−0.78018	0.72
−0.70	−0.89908	−0.88632	−0.86522	−0.84460	−0.81766	−0.76410	0.70
−0.68	−0.89159	−0.87795	−0.85542	−0.83343	−0.80477	−0.74796	0.68
−0.66	−0.88399	−0.86946	−0.84549	−0.82215	−0.79178	−0.73177	0.66
−0.64	−0.87628	−0.86086	−0.83546	−0.81075	−0.77868	−0.71552	0.64
−0.62	−0.86846	−0.85214	−0.82530	−0.79924	−0.76548	−0.69921	0.62
−0.60	−0.86052	−0.84330	−0.81502	−0.78761	−0.75217	−0.68283	0.60
−0.58	−0.85246	−0.83433	−0.80461	−0.77586	−0.73875	−0.66640	0.58
−0.56	−0.84427	−0.82524	−0.79408	−0.76399	−0.72523	−0.64989	0.56
−0.54	−0.83596	−0.81602	−0.78342	−0.75199	−0.71158	−0.63333	0.54
−0.52	−0.82753	−0.80667	−0.77262	−0.73986	−0.69783	−0.61670	0.52
−0.50	−0.81896	−0.79718	−0.76169	−0.72760	−0.68396	−0.59999	0.50
−0.48	−0.81025	−0.78756	−0.75061	−0.71520	−0.66996	−0.58322	0.48
−0.46	−0.80141	−0.77779	−0.73940	−0.70267	−0.65585	−0.56638	0.46
−0.44	−0.79242	−0.76787	−0.72804	−0.69000	−0.64161	−0.54947	0.44
−0.42	−0.78328	−0.75780	−0.71652	−0.67718	−0.62724	−0.53248	0.42
−0.40	−0.77400	−0.74758	−0.70486	−0.66422	−0.61275	−0.51542	0.40
−0.38	−0.76455	−0.73720	−0.69304	−0.65111	−0.59812	−0.49828	0.38
−0.36	−0.75495	−0.72666	−0.68106	−0.63785	−0.58336	−0.48106	0.36
−0.34	−0.74519	−0.71596	−0.66891	−0.62444	−0.56847	−0.46376	0.34
−0.32	−0.73525	−0.70508	−0.65660	−0.61086	−0.55343	−0.44638	0.32
−0.30	−0.72515	−0.69403	−0.64411	−0.59712	−0.53825	−0.42892	0.30
−0.28	−0.71486	−0.68279	−0.63145	−0.58321	−0.52293	−0.41137	0.28
−0.26	−0.70439	−0.67137	−0.61861	−0.56914	−0.50746	−0.39374	0.26
−0.24	−0.69373	−0.65976	−0.60558	−0.55489	−0.49184	−0.37602	0.24
−0.22	−0.68288	−0.64796	−0.59236	−0.54046	−0.47606	−0.35821	0.22
−0.20	−0.67182	−0.63595	−0.57895	−0.52585	−0.46013	−0.34030	0.20
−0.18	−0.66056	−0.62374	−0.56533	−0.51106	−0.44404	−0.32231	0.18
−0.16	−0.64908	−0.61131	−0.55151	−0.49607	−0.42778	−0.30422	0.16
−0.14	−0.63739	−0.59866	−0.53748	−0.48089	−0.41136	−0.28603	0.14
−0.12	−0.62546	−0.58579	−0.52323	−0.46551	−0.39476	−0.26774	0.12
−0.10	−0.61330	−0.57268	−0.50876	−0.44992	−0.37799	−0.24935	0.10
−0.08	−0.60090	−0.55934	−0.49407	−0.43413	−0.36104	−0.23086	0.08
−0.06	−0.58825	−0.54574	−0.47913	−0.41812	−0.34391	−0.21226	0.06
−0.04	−0.57534	−0.53189	−0.46396	−0.40188	−0.32659	−0.19356	0.04
−0.02	−0.56217	−0.51778	−0.44854	−0.38543	−0.30909	−0.17475	0.02

↑ r_0

| α = | .995 ↑− | .99 ↑− | .975 ↑− | .95 ↑− | .90 ↑− | .75 ↑− |

TABLE 18.2   SAMPLE SIZE = 20   (CONT.)

α =	.005 ↓+	.01 ↓+	.025 ↓+	.05 ↓+	.10 ↓+	.25 ↓+	
r_o ↓							
0.00	-0.54871	-0.50340	-0.43286	-0.36874	-0.29138	-0.15582	0.00
0.02	-0.53497	-0.48873	-0.41692	-0.35181	-0.27348	-0.13679	-0.02
0.04	-0.52093	-0.47378	-0.40071	-0.33464	-0.25538	-0.11763	-0.04
0.06	-0.50659	-0.45852	-0.38422	-0.31722	-0.23707	-0.09836	-0.06
0.08	-0.49193	-0.44296	-0.36744	-0.29954	-0.21854	-0.07897	-0.08
0.10	-0.47694	-0.42708	-0.35037	-0.28159	-0.19980	-0.05946	-0.10
0.12	-0.46161	-0.41086	-0.33300	-0.26338	-0.18083	-0.03982	-0.12
0.14	-0.44592	-0.39431	-0.31531	-0.24489	-0.16163	-0.02005	-0.14
0.16	-0.42986	-0.37740	-0.29730	-0.22611	-0.14221	-0.00015	-0.16
0.18	-0.41342	-0.36012	-0.27896	-0.20704	-0.12254	0.01988	-0.18
0.20	-0.39659	-0.34246	-0.26027	-0.18766	-0.10262	0.04004	-0.20
0.22	-0.37934	-0.32441	-0.24123	-0.16798	-0.08246	0.06035	-0.22
0.24	-0.36166	-0.30594	-0.22182	-0.14797	-0.06204	0.08079	-0.24
0.26	-0.34353	-0.28706	-0.20203	-0.12763	-0.04135	0.10138	-0.26
0.28	-0.32494	-0.26773	-0.18185	-0.10696	-0.02039	0.12212	-0.28
0.30	-0.30586	-0.24794	-0.16126	-0.08593	0.00084	0.14301	-0.30
0.32	-0.28628	-0.22768	-0.14025	-0.06455	0.02236	0.16405	-0.32
0.34	-0.26616	-0.20692	-0.11882	-0.04279	0.04418	0.18524	-0.34
0.36	-0.24550	-0.18564	-0.09693	-0.02065	0.06629	0.20660	-0.36
0.38	-0.22425	-0.16383	-0.07457	0.00188	0.08871	0.22812	-0.38
0.40	-0.20240	-0.14146	-0.05174	0.02483	0.11145	0.24981	-0.40
0.42	-0.17992	-0.11851	-0.02840	0.04819	0.13451	0.27166	-0.42
0.44	-0.15678	-0.09494	-0.00455	0.07198	0.15790	0.29370	-0.44
0.46	-0.13294	-0.07074	0.01985	0.09623	0.18164	0.31590	-0.46
0.48	-0.10837	-0.04587	0.04480	0.12094	0.20573	0.33829	-0.48
0.50	-0.08304	-0.02031	0.07033	0.14613	0.23019	0.36087	-0.50
0.52	-0.05690	0.00597	0.09647	0.17181	0.25501	0.38363	-0.52
0.54	-0.02991	0.03302	0.12324	0.19801	0.28022	0.40659	-0.54
0.56	-0.00204	0.06087	0.15066	0.22473	0.30583	0.42975	-0.56
0.58	0.02678	0.08956	0.17877	0.25201	0.33185	0.45311	-0.58
0.60	0.05659	0.11913	0.20759	0.27986	0.35828	0.47668	-0.60
0.62	0.08745	0.14962	0.23714	0.30830	0.38515	0.50045	-0.62
0.64	0.11942	0.18108	0.26748	0.33736	0.41246	0.52445	-0.64
0.66	0.15256	0.21357	0.29862	0.36705	0.44023	0.54867	-0.66
0.68	0.18695	0.24714	0.33061	0.39741	0.46848	0.57311	-0.68
0.70	0.22267	0.28185	0.36349	0.42845	0.49721	0.59779	-0.70
0.72	0.25979	0.31776	0.39729	0.46021	0.52645	0.62270	-0.72
0.74	0.29841	0.35495	0.43207	0.49271	0.55622	0.64786	-0.74
0.76	0.33864	0.39349	0.46786	0.52599	0.58652	0.67326	-0.76
0.78	0.38057	0.43346	0.50473	0.56007	0.61738	0.69892	-0.78
0.80	0.42433	0.47495	0.54272	0.59500	0.64882	0.72485	-0.80
0.82	0.47005	0.51805	0.58190	0.63081	0.68086	0.75104	-0.82
0.84	0.51787	0.56288	0.62232	0.66754	0.71352	0.77750	-0.84
0.86	0.56797	0.60955	0.66406	0.70522	0.74682	0.80425	-0.86
0.88	0.62051	0.65818	0.70718	0.74391	0.78078	0.83129	-0.88
0.90	0.67569	0.70891	0.75178	0.78365	0.81543	0.85862	-0.90
0.92	0.73373	0.76189	0.79792	0.82449	0.85080	0.88625	-0.92
0.94	0.79489	0.81730	0.84571	0.86649	0.88691	0.91420	-0.94
0.96	0.85944	0.87531	0.89525	0.90970	0.92379	0.94247	-0.96
0.98	0.92769	0.93613	0.94664	0.95418	0.96148	0.97107	-0.98

↑ r_o

α =	.995 ↑−	.99 ↑−	.975 ↑−	.95 ↑−	.90 ↑−	.75 ↑−

TABLE 19.1   SAMPLE SIZE =   21

α =	.005 ↓+	.01 ↓+	.025 ↓+	.05 ↓+	.10 ↓+	.25 ↓+	
r_0 ↓							
-0.98	-0.99367	-0.99285	-0.99148	-0.99011	-0.98828	-0.98451	0.98
-0.96	-0.98727	-0.98562	-0.98286	-0.98013	-0.97648	-0.96897	0.96
-0.94	-0.98078	-0.97830	-0.97417	-0.97006	-0.96460	-0.95340	0.94
-0.92	-0.97421	-0.97090	-0.96538	-0.95991	-0.95264	-0.93777	0.92
-0.90	-0.96755	-0.96341	-0.95650	-0.94966	-0.94060	-0.92211	0.90
-0.88	-0.96081	-0.95583	-0.94753	-0.93932	-0.92847	-0.90640	0.88
-0.86	-0.95399	-0.94816	-0.93846	-0.92889	-0.91626	-0.89064	0.86
-0.84	-0.94707	-0.94040	-0.92930	-0.91837	-0.90396	-0.87484	0.84
-0.82	-0.94006	-0.93253	-0.92003	-0.90775	-0.89158	-0.85899	0.82
-0.80	-0.93296	-0.92458	-0.91067	-0.89703	-0.87910	-0.84308	0.80
-0.78	-0.92576	-0.91652	-0.90121	-0.88620	-0.86654	-0.82713	0.78
-0.76	-0.91846	-0.90836	-0.89164	-0.87528	-0.85388	-0.81112	0.76
-0.74	-0.91106	-0.90009	-0.88196	-0.86425	-0.84112	-0.79507	0.74
-0.72	-0.90356	-0.89172	-0.87218	-0.85312	-0.82827	-0.77895	0.72
-0.70	-0.89596	-0.88324	-0.86228	-0.84188	-0.81532	-0.76279	0.70
-0.68	-0.88824	-0.87465	-0.85227	-0.83053	-0.80228	-0.74656	0.68
-0.66	-0.88042	-0.86594	-0.84215	-0.81906	-0.78913	-0.73028	0.66
-0.64	-0.87248	-0.85712	-0.83191	-0.80748	-0.77588	-0.71394	0.64
-0.62	-0.86443	-0.84818	-0.82155	-0.79579	-0.76252	-0.69754	0.62
-0.60	-0.85626	-0.83912	-0.81106	-0.78397	-0.74906	-0.68108	0.60
-0.58	-0.84797	-0.82993	-0.80045	-0.77204	-0.73549	-0.66455	0.58
-0.56	-0.83955	-0.82061	-0.78971	-0.75998	-0.72181	-0.64796	0.56
-0.54	-0.83101	-0.81117	-0.77884	-0.74779	-0.70801	-0.63131	0.54
-0.52	-0.82233	-0.80159	-0.76783	-0.73548	-0.69411	-0.61459	0.52
-0.50	-0.81352	-0.79187	-0.75669	-0.72303	-0.68008	-0.59781	0.50
-0.48	-0.80458	-0.78201	-0.74541	-0.71045	-0.66594	-0.58095	0.48
-0.46	-0.79549	-0.77201	-0.73399	-0.69774	-0.65167	-0.56402	0.46
-0.44	-0.78626	-0.76186	-0.72242	-0.68488	-0.63728	-0.54703	0.44
-0.42	-0.77688	-0.75156	-0.71069	-0.67188	-0.62277	-0.52995	0.42
-0.40	-0.76734	-0.74111	-0.69882	-0.65874	-0.60813	-0.51281	0.40
-0.38	-0.75765	-0.73049	-0.68679	-0.64545	-0.59336	-0.49559	0.38
-0.36	-0.74780	-0.71972	-0.67460	-0.63200	-0.57845	-0.47829	0.36
-0.34	-0.73778	-0.70877	-0.66224	-0.61840	-0.56341	-0.46091	0.34
-0.32	-0.72759	-0.69766	-0.64972	-0.60465	-0.54823	-0.44345	0.32
-0.30	-0.71723	-0.68636	-0.63702	-0.59073	-0.53291	-0.42591	0.30
-0.28	-0.70669	-0.67489	-0.62415	-0.57664	-0.51745	-0.40829	0.28
-0.26	-0.69596	-0.66323	-0.61110	-0.56239	-0.50184	-0.39058	0.26
-0.24	-0.68504	-0.65138	-0.59786	-0.54797	-0.48608	-0.37278	0.24
-0.22	-0.67392	-0.63933	-0.58444	-0.53336	-0.47017	-0.35490	0.22
-0.20	-0.66260	-0.62708	-0.57081	-0.51858	-0.45410	-0.33693	0.20
-0.18	-0.65107	-0.61462	-0.55699	-0.50361	-0.43787	-0.31886	0.18
-0.16	-0.63933	-0.60195	-0.54297	-0.48845	-0.42149	-0.30070	0.16
-0.14	-0.62737	-0.58906	-0.52873	-0.47310	-0.40493	-0.28244	0.14
-0.12	-0.61518	-0.57594	-0.51428	-0.45756	-0.38821	-0.26409	0.12
-0.10	-0.60275	-0.56259	-0.49961	-0.44180	-0.37132	-0.24564	0.10
-0.08	-0.59008	-0.54900	-0.48471	-0.42584	-0.35425	-0.22709	0.08
-0.06	-0.57716	-0.53517	-0.46957	-0.40967	-0.33700	-0.20843	0.06
-0.04	-0.56397	-0.52107	-0.45420	-0.39328	-0.31957	-0.18967	0.04
-0.02	-0.55053	-0.50672	-0.43858	-0.37667	-0.30195	-0.17080	0.02

↑ r_0

| α = | .995 ↑- | .99 ↑- | .975 ↑- | .95 ↑- | .90 ↑- | .75 ↑- |

TABLE 19.2   SAMPLE SIZE =   21   (CONT.)

α =	.005 ↓+	.01 ↓+	.025 ↓+	.05 ↓+	.10 ↓+	.25 ↓+	
r_0 ↓							
0.00	−0.53680	−0.49209	−0.42271	−0.35983	−0.28414	−0.15183	0.00
0.02	−0.52279	−0.47719	−0.40658	−0.34275	−0.26613	−0.13274	−0.02
0.04	−0.50848	−0.46200	−0.39018	−0.32543	−0.24793	−0.11354	−0.04
0.06	−0.49386	−0.44651	−0.37351	−0.30787	−0.22952	−0.09422	−0.06
0.08	−0.47893	−0.43071	−0.35655	−0.29005	−0.21090	−0.07479	−0.08
0.10	−0.46367	−0.41459	−0.33930	−0.27197	−0.19206	−0.05523	−0.10
0.12	−0.44807	−0.39815	−0.32176	−0.25363	−0.17301	−0.03556	−0.12
0.14	−0.43212	−0.38136	−0.30390	−0.23501	−0.15373	−0.01576	−0.14
0.16	−0.41580	−0.36423	−0.28572	−0.21611	−0.13423	0.00417	−0.16
0.18	−0.39910	−0.34673	−0.26722	−0.19693	−0.11449	0.02423	−0.18
0.20	−0.38200	−0.32886	−0.24837	−0.17744	−0.09451	0.04442	−0.20
0.22	−0.36450	−0.31059	−0.22918	−0.15765	−0.07428	0.06474	−0.22
0.24	−0.34656	−0.29192	−0.20963	−0.13755	−0.05381	0.08521	−0.24
0.26	−0.32819	−0.27283	−0.18970	−0.11712	−0.03307	0.10581	−0.26
0.28	−0.30935	−0.25331	−0.16939	−0.09636	−0.01207	0.12656	−0.28
0.30	−0.29004	−0.23334	−0.14868	−0.07526	0.00920	0.14745	−0.30
0.32	−0.27022	−0.21289	−0.12757	−0.05381	0.03074	0.16849	−0.32
0.34	−0.24988	−0.19196	−0.10602	−0.03200	0.05258	0.18968	−0.34
0.36	−0.22900	−0.17053	−0.08404	−0.00981	0.07470	0.21103	−0.36
0.38	−0.20755	−0.14856	−0.06161	0.01276	0.09712	0.23253	−0.38
0.40	−0.18550	−0.12605	−0.03870	0.03573	0.11985	0.25420	−0.40
0.42	−0.16284	−0.10297	−0.01531	0.05910	0.14290	0.27603	−0.42
0.44	−0.13952	−0.07929	0.00859	0.08290	0.16627	0.29804	−0.44
0.46	−0.11553	−0.05499	0.03302	0.10714	0.18997	0.32021	−0.46
0.48	−0.09082	−0.03004	0.05798	0.13182	0.21401	0.34256	−0.48
0.50	−0.06536	−0.00442	0.08352	0.15697	0.23841	0.36508	−0.50
0.52	−0.03912	0.02191	0.10963	0.18260	0.26316	0.38779	−0.52
0.54	−0.01206	0.04898	0.13636	0.20872	0.28829	0.41069	−0.54
0.56	0.01588	0.07683	0.16372	0.23536	0.31380	0.43378	−0.56
0.58	0.04472	0.10548	0.19174	0.26253	0.33970	0.45706	−0.58
0.60	0.07453	0.13499	0.22045	0.29025	0.36600	0.48055	−0.60
0.62	0.10535	0.16539	0.24987	0.31853	0.39272	0.50424	−0.62
0.64	0.13724	0.19673	0.28003	0.34741	0.41987	0.52813	−0.64
0.66	0.17026	0.22905	0.31098	0.37690	0.44746	0.55224	−0.66
0.68	0.20449	0.26241	0.34273	0.40703	0.47551	0.57657	−0.68
0.70	0.23998	0.29686	0.37533	0.43781	0.50402	0.60112	−0.70
0.72	0.27682	0.33247	0.40882	0.46928	0.53302	0.62589	−0.72
0.74	0.31509	0.36929	0.44324	0.50146	0.56252	0.65090	−0.74
0.76	0.35489	0.40739	0.47863	0.53437	0.59254	0.67615	−0.76
0.78	0.39631	0.44686	0.51504	0.56806	0.62308	0.70165	−0.78
0.80	0.43947	0.48777	0.55251	0.60255	0.65418	0.72739	−0.80
0.82	0.48448	0.53021	0.59111	0.63787	0.68585	0.75339	−0.82
0.84	0.53147	0.57427	0.63089	0.67407	0.71810	0.77965	−0.84
0.86	0.58059	0.62006	0.67190	0.71117	0.75097	0.80618	−0.86
0.88	0.63201	0.66769	0.71423	0.74922	0.78446	0.83299	−0.88
0.90	0.68588	0.71728	0.75793	0.78826	0.81861	0.86008	−0.90
0.92	0.74243	0.76898	0.80308	0.82833	0.85343	0.88745	−0.92
0.94	0.80185	0.82293	0.84977	0.86949	0.88895	0.91513	−0.94
0.96	0.86440	0.87929	0.89809	0.91178	0.92520	0.94310	−0.96
0.98	0.93034	0.93825	0.94813	0.95527	0.96221	0.97139	−0.98

↑ r_0

α =	.995 ↑−	.99 ↑−	.975 ↑−	.95 ↑−	.90 ↑−	.75 ↑−

TABLE 20.1   SAMPLE SIZE = 22

α =	.005 ↓+	.01 ↓+	.025 ↓+	.05 ↓+	.10 ↓+	.25 ↓+	
r_0 ↓							
−0.98	−0.99349	−0.99267	−0.99130	−0.98994	−0.98813	−0.98443	0.98
−0.96	−0.98689	−0.98525	−0.98251	−0.97979	−0.97619	−0.96881	0.96
−0.94	−0.98021	−0.97774	−0.97363	−0.96956	−0.96416	−0.95315	0.94
−0.92	−0.97345	−0.97015	−0.96466	−0.95923	−0.95206	−0.93745	0.92
−0.90	−0.96660	−0.96247	−0.95559	−0.94882	−0.93987	−0.92170	0.90
−0.88	−0.95967	−0.95470	−0.94644	−0.93831	−0.92760	−0.90591	0.88
−0.86	−0.95265	−0.94683	−0.93719	−0.92771	−0.91524	−0.89007	0.86
−0.84	−0.94553	−0.93887	−0.92784	−0.91701	−0.90280	−0.87419	0.84
−0.82	−0.93832	−0.93082	−0.91839	−0.90622	−0.89026	−0.85825	0.82
−0.80	−0.93102	−0.92266	−0.90884	−0.89533	−0.87764	−0.84227	0.80
−0.78	−0.92362	−0.91440	−0.89919	−0.88434	−0.86493	−0.82623	0.78
−0.76	−0.91611	−0.90604	−0.88943	−0.87324	−0.85212	−0.81014	0.76
−0.74	−0.90851	−0.89758	−0.87957	−0.86204	−0.83922	−0.79400	0.74
−0.72	−0.90080	−0.88900	−0.86960	−0.85073	−0.82623	−0.77780	0.72
−0.70	−0.89299	−0.88032	−0.85951	−0.83932	−0.81313	−0.76155	0.70
−0.68	−0.88507	−0.87152	−0.84931	−0.82780	−0.79994	−0.74525	0.68
−0.66	−0.87703	−0.86261	−0.83900	−0.81616	−0.78665	−0.72888	0.66
−0.64	−0.86888	−0.85358	−0.82856	−0.80441	−0.77325	−0.71246	0.64
−0.62	−0.86061	−0.84444	−0.81801	−0.79254	−0.75975	−0.69598	0.62
−0.60	−0.85223	−0.83516	−0.80733	−0.78055	−0.74615	−0.67943	0.60
−0.58	−0.84371	−0.82577	−0.79653	−0.76844	−0.73243	−0.66283	0.58
−0.56	−0.83508	−0.81624	−0.78559	−0.75621	−0.71861	−0.64616	0.56
−0.54	−0.82631	−0.80658	−0.77453	−0.74385	−0.70467	−0.62943	0.54
−0.52	−0.81742	−0.79679	−0.76333	−0.73137	−0.69062	−0.61263	0.52
−0.50	−0.80838	−0.78685	−0.75199	−0.71875	−0.67645	−0.59576	0.50
−0.48	−0.79921	−0.77678	−0.74052	−0.70600	−0.66217	−0.57883	0.48
−0.46	−0.78989	−0.76656	−0.72890	−0.69311	−0.64777	−0.56182	0.46
−0.44	−0.78043	−0.75620	−0.71713	−0.68008	−0.63324	−0.54475	0.44
−0.42	−0.77082	−0.74568	−0.70522	−0.66691	−0.61859	−0.52760	0.42
−0.40	−0.76105	−0.73500	−0.69315	−0.65360	−0.60381	−0.51038	0.40
−0.38	−0.75113	−0.72417	−0.68092	−0.64014	−0.58890	−0.49308	0.38
−0.36	−0.74104	−0.71317	−0.66853	−0.62652	−0.57386	−0.47571	0.36
−0.34	−0.73078	−0.70200	−0.65598	−0.61276	−0.55868	−0.45826	0.34
−0.32	−0.72036	−0.69066	−0.64326	−0.59883	−0.54337	−0.44072	0.32
−0.30	−0.70976	−0.67915	−0.63037	−0.58475	−0.52792	−0.42311	0.30
−0.28	−0.69897	−0.66745	−0.61731	−0.57050	−0.51233	−0.40542	0.28
−0.26	−0.68800	−0.65556	−0.60406	−0.55608	−0.49659	−0.38764	0.26
−0.24	−0.67684	−0.64349	−0.59063	−0.54149	−0.48071	−0.36978	0.24
−0.22	−0.66548	−0.63121	−0.57701	−0.52673	−0.46467	−0.35182	0.22
−0.20	−0.65391	−0.61874	−0.56320	−0.51178	−0.44848	−0.33378	0.20
−0.18	−0.64214	−0.60606	−0.54918	−0.49666	−0.43213	−0.31565	0.18
−0.16	−0.63015	−0.59316	−0.53497	−0.48134	−0.41563	−0.29743	0.16
−0.14	−0.61794	−0.58004	−0.52054	−0.46584	−0.39896	−0.27912	0.14
−0.12	−0.60550	−0.56670	−0.50590	−0.45013	−0.38212	−0.26070	0.12
−0.10	−0.59282	−0.55313	−0.49104	−0.43423	−0.36511	−0.24219	0.10
−0.08	−0.57990	−0.53931	−0.47596	−0.41812	−0.34793	−0.22359	0.08
−0.06	−0.56673	−0.52525	−0.46064	−0.40180	−0.33058	−0.20488	0.06
−0.04	−0.55329	−0.51093	−0.44509	−0.38527	−0.31304	−0.18606	0.04
−0.02	−0.53960	−0.49636	−0.42929	−0.36851	−0.29532	−0.16715	0.02

↑ r_0

α =	.995 ↑−	.99 ↑−	.975 ↑−	.95 ↑−	.90 ↑−	.75 ↑−

TABLE 20.2   SAMPLE SIZE = 22   (CONT.)

α =	.005 ↓+	.01 ↓+	.025 ↓+	.05 ↓+	.10 ↓+	.25 ↓+	
r_o ↓							
0.00	−0.52562	−0.48151	−0.41325	−0.35153	−0.27741	−0.14812	0.00
0.02	−0.51136	−0.46639	−0.39694	−0.33432	−0.25931	−0.12899	−0.02
0.04	−0.49680	−0.45098	−0.38037	−0.31687	−0.24101	−0.10974	−0.04
0.06	−0.48194	−0.43527	−0.36353	−0.29918	−0.22251	−0.09039	−0.06
0.08	−0.46676	−0.41926	−0.34641	−0.28123	−0.20380	−0.07092	−0.08
0.10	−0.45125	−0.40293	−0.32900	−0.26303	−0.18489	−0.05133	−0.10
0.12	−0.43541	−0.38628	−0.31129	−0.24457	−0.16576	−0.03162	−0.12
0.14	−0.41921	−0.36929	−0.29328	−0.22584	−0.14641	−0.01179	−0.14
0.16	−0.40265	−0.35195	−0.27495	−0.20684	−0.12684	0.00817	−0.16
0.18	−0.38571	−0.33425	−0.25631	−0.18755	−0.10704	0.02825	−0.18
0.20	−0.36838	−0.31619	−0.23732	−0.16797	−0.08700	0.04846	−0.20
0.22	−0.35065	−0.29773	−0.21800	−0.14808	−0.06672	0.06881	−0.22
0.24	−0.33249	−0.27888	−0.19832	−0.12790	−0.04620	0.08928	−0.24
0.26	−0.31389	−0.25961	−0.17827	−0.10739	−0.02542	0.10989	−0.26
0.28	−0.29484	−0.23991	−0.15785	−0.08656	−0.00439	0.13065	−0.28
0.30	−0.27531	−0.21977	−0.13703	−0.06540	0.01691	0.15154	−0.30
0.32	−0.25529	−0.19917	−0.11582	−0.04389	0.03848	0.17258	−0.32
0.34	−0.23475	−0.17809	−0.09419	−0.02203	0.06033	0.19377	−0.34
0.36	−0.21368	−0.15652	−0.07213	0.00019	0.08246	0.21511	−0.36
0.38	−0.19205	−0.13442	−0.04962	0.02279	0.10488	0.23660	−0.38
0.40	−0.16983	−0.11179	−0.02666	0.04578	0.12760	0.25825	−0.40
0.42	−0.14701	−0.08860	−0.00322	0.06917	0.15062	0.28006	−0.42
0.44	−0.12355	−0.06483	0.02071	0.09296	0.17397	0.30203	−0.44
0.46	−0.09942	−0.04045	0.04516	0.11718	0.19763	0.32417	−0.46
0.48	−0.07460	−0.01544	0.07013	0.14183	0.22162	0.34647	−0.48
0.50	−0.04905	0.01023	0.09565	0.16694	0.24596	0.36896	−0.50
0.52	−0.02273	0.03659	0.12174	0.19251	0.27065	0.39161	−0.52
0.54	0.00440	0.06366	0.14842	0.21856	0.29569	0.41445	−0.54
0.56	0.03236	0.09149	0.17571	0.24511	0.32110	0.43748	−0.56
0.58	0.06121	0.12011	0.20364	0.27217	0.34689	0.46069	−0.58
0.60	0.09100	0.14954	0.23223	0.29976	0.37308	0.48410	−0.60
0.62	0.12176	0.17985	0.26152	0.32790	0.39966	0.50770	−0.62
0.64	0.15356	0.21105	0.29152	0.35661	0.42666	0.53150	−0.64
0.66	0.18646	0.24321	0.32227	0.38590	0.45408	0.55551	−0.66
0.68	0.22051	0.27636	0.35379	0.41581	0.48194	0.57973	−0.68
0.70	0.25578	0.31056	0.38614	0.44635	0.51025	0.60416	−0.70
0.72	0.29234	0.34586	0.41933	0.47755	0.53902	0.62881	−0.72
0.74	0.33027	0.38233	0.45341	0.50942	0.56827	0.65369	−0.74
0.76	0.36966	0.42003	0.48842	0.54201	0.59802	0.67879	−0.76
0.78	0.41059	0.45902	0.52440	0.57532	0.62828	0.70413	−0.78
0.80	0.45317	0.49938	0.56139	0.60941	0.65906	0.72971	−0.80
0.82	0.49751	0.54119	0.59945	0.64428	0.69038	0.75554	−0.82
0.84	0.54373	0.58455	0.63863	0.67999	0.72227	0.78161	−0.84
0.86	0.59196	0.62953	0.67899	0.71655	0.75473	0.80794	−0.86
0.88	0.64233	0.67624	0.72058	0.75401	0.78780	0.83454	−0.88
0.90	0.69502	0.72481	0.76347	0.79241	0.82148	0.86140	−0.90
0.92	0.75019	0.77533	0.80772	0.83179	0.85581	0.88854	−0.92
0.94	0.80805	0.82797	0.85342	0.87219	0.89080	0.91597	−0.94
0.96	0.86880	0.88284	0.90064	0.91366	0.92648	0.94368	−0.96
0.98	0.93270	0.94013	0.94947	0.95624	0.96287	0.97169	−0.98

↑ r_o

α =	.995 ↑−	.99 ↑−	.975 ↑−	.95 ↑−	.90 ↑−	.75 ↑−

TABLE 21.1   SAMPLE SIZE =   23

α =	.005 ↓+	.01 ↓+	.025 ↓+	.05 ↓+	.10 ↓+	.25 ↓+	
r_0 ↓							
-0.98	-0.99331	-0.99249	-0.99113	-0.98978	-0.98800	-0.98435	0.98
-0.96	-0.98653	-0.98489	-0.98216	-0.97947	-0.97591	-0.96866	0.96
-0.94	-0.97967	-0.97721	-0.97311	-0.96908	-0.96375	-0.95292	0.94
-0.92	-0.97273	-0.96944	-0.96397	-0.95860	-0.95151	-0.93715	0.92
-0.90	-0.96570	-0.96158	-0.95474	-0.94802	-0.93918	-0.92132	0.90
-0.88	-0.95858	-0.95362	-0.94541	-0.93735	-0.92677	-0.90545	0.88
-0.86	-0.95137	-0.94557	-0.93598	-0.92659	-0.91428	-0.88954	0.86
-0.84	-0.94406	-0.93743	-0.92646	-0.91574	-0.90169	-0.87357	0.84
-0.82	-0.93667	-0.92918	-0.91684	-0.90478	-0.88902	-0.85756	0.82
-0.80	-0.92917	-0.92084	-0.90711	-0.89373	-0.87627	-0.84150	0.80
-0.78	-0.92158	-0.91240	-0.89728	-0.88257	-0.86341	-0.82538	0.78
-0.76	-0.91388	-0.90385	-0.88735	-0.87132	-0.85047	-0.80922	0.76
-0.74	-0.90608	-0.89519	-0.87731	-0.85995	-0.83743	-0.79300	0.74
-0.72	-0.89818	-0.88643	-0.86716	-0.84849	-0.82430	-0.77672	0.72
-0.70	-0.89017	-0.87755	-0.85689	-0.83691	-0.81107	-0.76040	0.70
-0.68	-0.88204	-0.86856	-0.84651	-0.82522	-0.79774	-0.74401	0.68
-0.66	-0.87381	-0.85946	-0.83602	-0.81342	-0.78431	-0.72757	0.66
-0.64	-0.86546	-0.85023	-0.82540	-0.80151	-0.77078	-0.71107	0.64
-0.62	-0.85698	-0.84089	-0.81467	-0.78948	-0.75715	-0.69451	0.62
-0.60	-0.84839	-0.83142	-0.80381	-0.77733	-0.74341	-0.67789	0.60
-0.58	-0.83968	-0.82182	-0.79282	-0.76506	-0.72956	-0.66121	0.58
-0.56	-0.83083	-0.81209	-0.78171	-0.75266	-0.71560	-0.64447	0.56
-0.54	-0.82186	-0.80223	-0.77046	-0.74014	-0.70153	-0.62766	0.54
-0.52	-0.81275	-0.79224	-0.75908	-0.72749	-0.68735	-0.61079	0.52
-0.50	-0.80350	-0.78211	-0.74756	-0.71472	-0.67305	-0.59385	0.50
-0.48	-0.79412	-0.77183	-0.73590	-0.70180	-0.65864	-0.57684	0.48
-0.46	-0.78459	-0.76141	-0.72410	-0.68876	-0.64410	-0.55976	0.46
-0.44	-0.77491	-0.75084	-0.71215	-0.67557	-0.62944	-0.54261	0.44
-0.42	-0.76508	-0.74011	-0.70005	-0.66224	-0.61466	-0.52539	0.42
-0.40	-0.75509	-0.72923	-0.68780	-0.64877	-0.59976	-0.50810	0.40
-0.38	-0.74495	-0.71819	-0.67539	-0.63515	-0.58472	-0.49073	0.38
-0.36	-0.73464	-0.70698	-0.66282	-0.62138	-0.56956	-0.47329	0.36
-0.34	-0.72416	-0.69561	-0.65009	-0.60745	-0.55426	-0.45577	0.34
-0.32	-0.71351	-0.68406	-0.63719	-0.59337	-0.53882	-0.43817	0.32
-0.30	-0.70268	-0.67234	-0.62412	-0.57913	-0.52325	-0.42050	0.30
-0.28	-0.69167	-0.66043	-0.61087	-0.56473	-0.50754	-0.40274	0.28
-0.26	-0.68048	-0.64834	-0.59744	-0.55016	-0.49168	-0.38489	0.26
-0.24	-0.66909	-0.63605	-0.58383	-0.53542	-0.47568	-0.36697	0.24
-0.22	-0.65750	-0.62357	-0.57003	-0.52051	-0.45953	-0.34895	0.22
-0.20	-0.64570	-0.61088	-0.55604	-0.50541	-0.44322	-0.33085	0.20
-0.18	-0.63370	-0.59799	-0.54185	-0.49014	-0.42676	-0.31267	0.18
-0.16	-0.62148	-0.58488	-0.52746	-0.47468	-0.41015	-0.29439	0.16
-0.14	-0.60904	-0.57156	-0.51286	-0.45903	-0.39337	-0.27601	0.14
-0.12	-0.59637	-0.55800	-0.49805	-0.44319	-0.37643	-0.25755	0.12
-0.10	-0.58346	-0.54422	-0.48302	-0.42714	-0.35932	-0.23898	0.10
-0.08	-0.57030	-0.53020	-0.46776	-0.41090	-0.34204	-0.22033	0.08
-0.06	-0.55690	-0.51593	-0.45228	-0.39444	-0.32458	-0.20157	0.06
-0.04	-0.54324	-0.50141	-0.43656	-0.37778	-0.30695	-0.18271	0.04
-0.02	-0.52931	-0.48663	-0.42060	-0.36089	-0.28914	-0.16374	0.02

↑ r_0

| α = | .995 ↑- | .99 ↑- | .975 ↑- | .95 ↑- | .90 ↑- | .75 ↑- |

TABLE 21.2   SAMPLE SIZE = 23   (CONT.)

α =	.005 ↓+	.01 ↓+	.025 ↓+	.05 ↓+	.10 ↓+	.25 ↓+	
r_0 ↓							
0.00	−0.51510	−0.47158	−0.40439	−0.34378	−0.27114	−0.14468	0.00
0.02	−0.50061	−0.45625	−0.38792	−0.32645	−0.25295	−0.12550	−0.02
0.04	−0.48582	−0.44064	−0.37120	−0.30888	−0.23456	−0.10622	−0.04
0.06	−0.47073	−0.42474	−0.35420	−0.29106	−0.21598	−0.08682	−0.06
0.08	−0.45532	−0.40853	−0.33693	−0.27301	−0.19720	−0.06732	−0.08
0.10	−0.43959	−0.39201	−0.31937	−0.25470	−0.17822	−0.04770	−0.10
0.12	−0.42352	−0.37517	−0.30152	−0.23613	−0.15902	−0.02796	−0.12
0.14	−0.40711	−0.35799	−0.28337	−0.21730	−0.13960	−0.00810	−0.14
0.16	−0.39033	−0.34047	−0.26491	−0.19820	−0.11997	0.01188	−0.16
0.18	−0.37317	−0.32259	−0.24613	−0.17882	−0.10011	0.03198	−0.18
0.20	−0.35563	−0.30435	−0.22703	−0.15915	−0.08002	0.05221	−0.20
0.22	−0.33769	−0.28572	−0.20758	−0.13919	−0.05970	0.07257	−0.22
0.24	−0.31932	−0.26670	−0.18779	−0.11893	−0.03913	0.09306	−0.24
0.26	−0.30053	−0.24727	−0.16763	−0.09835	−0.01832	0.11368	−0.26
0.28	−0.28128	−0.22743	−0.14711	−0.07746	0.00274	0.13444	−0.28
0.30	−0.26156	−0.20714	−0.12620	−0.05624	0.02406	0.15533	−0.30
0.32	−0.24136	−0.18640	−0.10490	−0.03469	0.04565	0.17637	−0.32
0.34	−0.22065	−0.16519	−0.08320	−0.01279	0.06750	0.19755	−0.34
0.36	−0.19941	−0.14349	−0.06107	0.00947	0.08964	0.21888	−0.36
0.38	−0.17762	−0.12128	−0.03851	0.03209	0.11205	0.24036	−0.38
0.40	−0.15526	−0.09855	−0.01550	0.05509	0.13476	0.26198	−0.40
0.42	−0.13230	−0.07527	0.00798	0.07848	0.15777	0.28377	−0.42
0.44	−0.10872	−0.05142	0.03194	0.10226	0.18108	0.30571	−0.44
0.46	−0.08448	−0.02698	0.05639	0.12646	0.20471	0.32782	−0.46
0.48	−0.05956	−0.00192	0.08136	0.15108	0.22865	0.35009	−0.48
0.50	−0.03393	0.02378	0.10686	0.17614	0.25293	0.37253	−0.50
0.52	−0.00755	0.05015	0.13291	0.20165	0.27755	0.39514	−0.52
0.54	0.01961	0.07722	0.15954	0.22763	0.30251	0.41792	−0.54
0.56	0.04759	0.10503	0.18676	0.25409	0.32783	0.44089	−0.56
0.58	0.07643	0.13359	0.21460	0.28104	0.35352	0.46403	−0.58
0.60	0.10618	0.16295	0.24308	0.30851	0.37959	0.48736	−0.60
0.62	0.13688	0.19315	0.27223	0.33651	0.40604	0.51089	−0.62
0.64	0.16858	0.22422	0.30207	0.36506	0.43289	0.53460	−0.64
0.66	0.20134	0.25620	0.33263	0.39417	0.46015	0.55852	−0.66
0.68	0.23521	0.28915	0.36394	0.42387	0.48784	0.58263	−0.68
0.70	0.27025	0.32311	0.39604	0.45418	0.51595	0.60696	−0.70
0.72	0.30654	0.35812	0.42895	0.48512	0.54452	0.63149	−0.72
0.74	0.34414	0.39425	0.46271	0.51672	0.57354	0.65624	−0.74
0.76	0.38313	0.43156	0.49736	0.54899	0.60304	0.68121	−0.76
0.78	0.42360	0.47010	0.53294	0.58196	0.63303	0.70641	−0.78
0.80	0.46565	0.50995	0.56949	0.61567	0.66352	0.73184	−0.80
0.82	0.50936	0.55118	0.60705	0.65013	0.69453	0.75750	−0.82
0.84	0.55485	0.59387	0.64568	0.68538	0.72607	0.78340	−0.84
0.86	0.60224	0.63811	0.68543	0.72145	0.75817	0.80955	−0.86
0.88	0.65166	0.68398	0.72634	0.75838	0.79084	0.83596	−0.88
0.90	0.70326	0.73160	0.76848	0.79619	0.82410	0.86262	−0.90
0.92	0.75718	0.78106	0.81192	0.83493	0.85798	0.88954	−0.92
0.94	0.81362	0.83250	0.85671	0.87464	0.89248	0.91673	−0.94
0.96	0.87275	0.88603	0.90294	0.91536	0.92763	0.94420	−0.96
0.98	0.93480	0.94182	0.95067	0.95713	0.96347	0.97196	−0.98

↑ r_0

α =	.995 ↑−	.99 ↑−	.975 ↑−	.95 ↑−	.90 ↑−	.75 ↑−

TABLE 22.1   SAMPLE SIZE =   24

α =	.005 ↓+	.01 ↓+	.025 ↓+	.05 ↓+	.10 ↓+	.25 ↓+	
r_0 ↓							
-0.98	-0.99314	-0.99232	-0.99096	-0.98963	-0.98787	-0.98428	0.98
-0.96	-0.98619	-0.98455	-0.98184	-0.97917	-0.97565	-0.96851	0.96
-0.94	-0.97916	-0.97670	-0.97263	-0.96863	-0.96336	-0.95271	0.94
-0.92	-0.97204	-0.96876	-0.96332	-0.95799	-0.95099	-0.93686	0.92
-0.90	-0.96483	-0.96072	-0.95392	-0.94727	-0.93853	-0.92096	0.90
-0.88	-0.95754	-0.95259	-0.94443	-0.93645	-0.92599	-0.90502	0.88
-0.86	-0.95015	-0.94437	-0.93484	-0.92553	-0.91337	-0.88903	0.86
-0.84	-0.94267	-0.93605	-0.92515	-0.91452	-0.90065	-0.87299	0.84
-0.82	-0.93509	-0.92763	-0.91536	-0.90342	-0.88785	-0.85690	0.82
-0.80	-0.92741	-0.91911	-0.90547	-0.89221	-0.87497	-0.84077	0.80
-0.78	-0.91964	-0.91049	-0.89547	-0.88090	-0.86199	-0.82458	0.78
-0.76	-0.91176	-0.90176	-0.88537	-0.86949	-0.84891	-0.80834	0.76
-0.74	-0.90377	-0.89292	-0.87516	-0.85798	-0.83575	-0.79205	0.74
-0.72	-0.89568	-0.88398	-0.86484	-0.84636	-0.82249	-0.77571	0.72
-0.70	-0.88748	-0.87492	-0.85441	-0.83463	-0.80913	-0.75931	0.70
-0.68	-0.87917	-0.86575	-0.84386	-0.82279	-0.79567	-0.74285	0.68
-0.66	-0.87074	-0.85646	-0.83319	-0.81084	-0.78211	-0.72634	0.66
-0.64	-0.86220	-0.84705	-0.82241	-0.79877	-0.76846	-0.70977	0.64
-0.62	-0.85353	-0.83752	-0.81150	-0.78659	-0.75469	-0.69313	0.62
-0.60	-0.84474	-0.82786	-0.80047	-0.77428	-0.74083	-0.67644	0.60
-0.58	-0.83583	-0.81808	-0.78932	-0.76186	-0.72685	-0.65969	0.58
-0.56	-0.82679	-0.80816	-0.77803	-0.74931	-0.71277	-0.64288	0.56
-0.54	-0.81762	-0.79811	-0.76661	-0.73664	-0.69857	-0.62600	0.54
-0.52	-0.80831	-0.78793	-0.75506	-0.72384	-0.68427	-0.60905	0.52
-0.50	-0.79886	-0.77760	-0.74337	-0.71091	-0.66985	-0.59205	0.50
-0.48	-0.78928	-0.76714	-0.73154	-0.69785	-0.65531	-0.57497	0.48
-0.46	-0.77954	-0.75652	-0.71957	-0.68465	-0.64065	-0.55782	0.46
-0.44	-0.76966	-0.74576	-0.70744	-0.67131	-0.62587	-0.54061	0.44
-0.42	-0.75963	-0.73484	-0.69517	-0.65784	-0.61097	-0.52332	0.42
-0.40	-0.74943	-0.72377	-0.68275	-0.64422	-0.59595	-0.50596	0.40
-0.38	-0.73908	-0.71253	-0.67017	-0.63045	-0.58080	-0.48853	0.38
-0.36	-0.72856	-0.70113	-0.65743	-0.61653	-0.56551	-0.47103	0.36
-0.34	-0.71788	-0.68956	-0.64453	-0.60246	-0.55010	-0.45344	0.34
-0.32	-0.70702	-0.67782	-0.63146	-0.58824	-0.53455	-0.43578	0.32
-0.30	-0.69598	-0.66590	-0.61821	-0.57385	-0.51887	-0.41804	0.30
-0.28	-0.68476	-0.65379	-0.60480	-0.55931	-0.50304	-0.40022	0.28
-0.26	-0.67335	-0.64150	-0.59120	-0.54459	-0.48707	-0.38232	0.26
-0.24	-0.66174	-0.62902	-0.57742	-0.52971	-0.47096	-0.36434	0.24
-0.22	-0.64994	-0.61634	-0.56346	-0.51466	-0.45470	-0.34627	0.22
-0.20	-0.63793	-0.60346	-0.54930	-0.49943	-0.43829	-0.32811	0.20
-0.18	-0.62572	-0.59037	-0.53495	-0.48402	-0.42173	-0.30987	0.18
-0.16	-0.61328	-0.57707	-0.52039	-0.46842	-0.40501	-0.29154	0.16
-0.14	-0.60062	-0.56355	-0.50563	-0.45264	-0.38814	-0.27311	0.14
-0.12	-0.58774	-0.54980	-0.49066	-0.43667	-0.37110	-0.25460	0.12
-0.10	-0.57461	-0.53583	-0.47547	-0.42049	-0.35389	-0.23599	0.10
-0.08	-0.56124	-0.52161	-0.46005	-0.40412	-0.33652	-0.21728	0.08
-0.06	-0.54763	-0.50715	-0.44441	-0.38754	-0.31897	-0.19847	0.06
-0.04	-0.53375	-0.49244	-0.42854	-0.37075	-0.30125	-0.17957	0.04
-0.02	-0.51960	-0.47747	-0.41243	-0.35375	-0.28335	-0.16057	0.02

↑ r_0

α =	.995 ↑-	.99 ↑-	.975 ↑-	.95 ↑-	.90 ↑-	.75 ↑-

TABLE 22.2   SAMPLE SIZE =   24   (CONT.)

α =	.005 ↓+	.01 ↓+	.025 ↓+	.05 ↓+	.10 ↓+	.25 ↓+	
r_o ↓							
0.00	−0.50518	−0.46223	−0.39607	−0.33652	−0.26527	−0.14146	0.00
0.02	−0.49048	−0.44672	−0.37946	−0.31907	−0.24700	−0.12225	−0.02
0.04	−0.47548	−0.43092	−0.36259	−0.30139	−0.22854	−0.10293	−0.04
0.06	−0.46018	−0.41484	−0.34545	−0.28347	−0.20989	−0.08350	−0.06
0.08	−0.44456	−0.39845	−0.32804	−0.26531	−0.19104	−0.06396	−0.08
0.10	−0.42862	−0.38175	−0.31035	−0.24691	−0.17198	−0.04431	−0.10
0.12	−0.41235	−0.36473	−0.29237	−0.22824	−0.15272	−0.02455	−0.12
0.14	−0.39573	−0.34739	−0.27410	−0.20932	−0.13325	−0.00467	−0.14
0.16	−0.37875	−0.32970	−0.25551	−0.19013	−0.11356	0.01534	−0.16
0.18	−0.36140	−0.31166	−0.23662	−0.17067	−0.09365	0.03546	−0.18
0.20	−0.34366	−0.29325	−0.21740	−0.15092	−0.07352	0.05570	−0.20
0.22	−0.32553	−0.27447	−0.19785	−0.13089	−0.05315	0.07607	−0.22
0.24	−0.30698	−0.25530	−0.17795	−0.11056	−0.03255	0.09657	−0.24
0.26	−0.28800	−0.23573	−0.15770	−0.08993	−0.01171	0.11720	−0.26
0.28	−0.26858	−0.21575	−0.13709	−0.06898	0.00938	0.13796	−0.28
0.30	−0.24869	−0.19533	−0.11610	−0.04771	0.03072	0.15886	−0.30
0.32	−0.22833	−0.17447	−0.09473	−0.02612	0.05232	0.17989	−0.32
0.34	−0.20746	−0.15314	−0.07296	−0.00419	0.07418	0.20106	−0.34
0.36	−0.18608	−0.13134	−0.05077	0.01809	0.09631	0.22238	−0.36
0.38	−0.16415	−0.10903	−0.02816	0.04073	0.11872	0.24384	−0.38
0.40	−0.14166	−0.08621	−0.00512	0.06374	0.14141	0.26545	−0.40
0.42	−0.11859	−0.06286	0.01839	0.08712	0.16440	0.28721	−0.42
0.44	−0.09490	−0.03895	0.04236	0.11090	0.18768	0.30913	−0.44
0.46	−0.07057	−0.01445	0.06681	0.13507	0.21127	0.33120	−0.46
0.48	−0.04557	0.01064	0.09177	0.15966	0.23517	0.35344	−0.48
0.50	−0.01988	0.03636	0.11725	0.18467	0.25939	0.37583	−0.50
0.52	0.00654	0.06274	0.14326	0.21012	0.28394	0.39840	−0.52
0.54	0.03373	0.08979	0.16984	0.23603	0.30883	0.42113	−0.54
0.56	0.06171	0.11756	0.19699	0.26240	0.33406	0.44404	−0.56
0.58	0.09053	0.14606	0.22473	0.28925	0.35965	0.46712	−0.58
0.60	0.12023	0.17534	0.25310	0.31660	0.38561	0.49038	−0.60
0.62	0.15086	0.20543	0.28212	0.34446	0.41194	0.51383	−0.62
0.64	0.18245	0.23637	0.31180	0.37285	0.43865	0.53747	−0.64
0.66	0.21506	0.26819	0.34218	0.40179	0.46576	0.56129	−0.66
0.68	0.24875	0.30093	0.37329	0.43130	0.49328	0.58532	−0.68
0.70	0.28358	0.33465	0.40515	0.46139	0.52121	0.60954	−0.70
0.72	0.31959	0.36939	0.43779	0.49209	0.54958	0.63397	−0.72
0.74	0.35687	0.40520	0.47125	0.52342	0.57839	0.65860	−0.74
0.76	0.39549	0.44213	0.50557	0.55540	0.60766	0.68345	−0.76
0.78	0.43552	0.48026	0.54077	0.58805	0.63739	0.70851	−0.78
0.80	0.47705	0.51962	0.57690	0.62140	0.66761	0.73380	−0.80
0.82	0.52017	0.56031	0.61401	0.65548	0.69833	0.75931	−0.82
0.84	0.56498	0.60238	0.65212	0.69032	0.72956	0.78505	−0.84
0.86	0.61159	0.64593	0.69130	0.72593	0.76132	0.81104	−0.86
0.88	0.66013	0.69102	0.73159	0.76236	0.79363	0.83726	−0.88
0.90	0.71072	0.73777	0.77305	0.79964	0.82650	0.86373	−0.90
0.92	0.76350	0.78625	0.81573	0.83779	0.85996	0.89045	−0.92
0.94	0.81864	0.83660	0.85970	0.87687	0.89401	0.91744	−0.94
0.96	0.87630	0.88892	0.90502	0.91690	0.92869	0.94468	−0.96
0.98	0.93669	0.94334	0.95176	0.95793	0.96401	0.97220	−0.98

↑ r_o

α =	.995 ↑−	.99 ↑−	.975 ↑−	.95 ↑−	.90 ↑−	.75 ↑−

TABLE 23.1   SAMPLE SIZE =   25

α =	.005 ↓+	.01 ↓+	.025 ↓+	.05 ↓+	.10 ↓+	.25 ↓+	
r_0 ↓							
-0.98	-0.99297	-0.99216	-0.99081	-0.98949	-0.98774	-0.98421	0.98
-0.96	-0.98586	-0.98423	-0.98153	-0.97889	-0.97541	-0.96838	0.96
-0.94	-0.97867	-0.97622	-0.97216	-0.96820	-0.96299	-0.95250	0.94
-0.92	-0.97138	-0.96811	-0.96270	-0.95742	-0.95050	-0.93658	0.92
-0.90	-0.96401	-0.95991	-0.95315	-0.94655	-0.93792	-0.92062	0.90
-0.88	-0.95654	-0.95162	-0.94350	-0.93559	-0.92525	-0.90461	0.88
-0.86	-0.94898	-0.94323	-0.93375	-0.92453	-0.91250	-0.88855	0.86
-0.84	-0.94133	-0.93474	-0.92391	-0.91338	-0.89967	-0.87244	0.84
-0.82	-0.93358	-0.92615	-0.91396	-0.90212	-0.88675	-0.85629	0.82
-0.80	-0.92573	-0.91746	-0.90391	-0.89077	-0.87374	-0.84008	0.80
-0.78	-0.91778	-0.90867	-0.89375	-0.87932	-0.86063	-0.82383	0.78
-0.76	-0.90972	-0.89777	-0.88349	-0.86777	-0.84744	-0.80752	0.76
-0.74	-0.90156	-0.89076	-0.87312	-0.85611	-0.83415	-0.79116	0.74
-0.72	-0.89329	-0.88164	-0.86264	-0.84434	-0.82077	-0.77474	0.72
-0.70	-0.88491	-0.87241	-0.85205	-0.83247	-0.80729	-0.75828	0.70
-0.68	-0.87642	-0.86307	-0.84134	-0.82048	-0.79371	-0.74175	0.68
-0.66	-0.86781	-0.85360	-0.83052	-0.80839	-0.78004	-0.72517	0.66
-0.64	-0.85909	-0.84402	-0.81957	-0.79618	-0.76626	-0.70853	0.64
-0.62	-0.85024	-0.83431	-0.80850	-0.78385	-0.75238	-0.69183	0.62
-0.60	-0.84127	-0.82448	-0.79731	-0.77140	-0.73839	-0.67508	0.60
-0.58	-0.83217	-0.81451	-0.78599	-0.75884	-0.72430	-0.65826	0.58
-0.56	-0.82294	-0.80442	-0.77455	-0.74615	-0.71010	-0.64138	0.56
-0.54	-0.81358	-0.79419	-0.76297	-0.73333	-0.69578	-0.62443	0.54
-0.52	-0.80408	-0.78383	-0.75125	-0.72039	-0.68136	-0.60742	0.52
-0.50	-0.79445	-0.77332	-0.73940	-0.70732	-0.66682	-0.59035	0.50
-0.48	-0.78467	-0.76268	-0.72741	-0.69411	-0.65217	-0.57321	0.48
-0.46	-0.77474	-0.75188	-0.71527	-0.68077	-0.63740	-0.55600	0.46
-0.44	-0.76467	-0.74093	-0.70299	-0.66730	-0.62251	-0.53872	0.44
-0.42	-0.75444	-0.72983	-0.69056	-0.65368	-0.60750	-0.52137	0.42
-0.40	-0.74405	-0.71858	-0.67797	-0.63992	-0.59236	-0.50395	0.40
-0.38	-0.73351	-0.70716	-0.66523	-0.62601	-0.57710	-0.48646	0.38
-0.36	-0.72279	-0.69557	-0.65233	-0.61196	-0.56171	-0.46890	0.36
-0.34	-0.71191	-0.68382	-0.63927	-0.59775	-0.54618	-0.45125	0.34
-0.32	-0.70085	-0.67189	-0.62604	-0.58339	-0.53053	-0.43354	0.32
-0.30	-0.68961	-0.65979	-0.61264	-0.56887	-0.51474	-0.41574	0.30
-0.28	-0.67819	-0.64750	-0.59906	-0.55419	-0.49881	-0.39786	0.28
-0.26	-0.66658	-0.63503	-0.58531	-0.53934	-0.48274	-0.37991	0.26
-0.24	-0.65478	-0.62236	-0.57137	-0.52433	-0.46652	-0.36187	0.24
-0.22	-0.64278	-0.60950	-0.55725	-0.50915	-0.45017	-0.34375	0.22
-0.20	-0.63057	-0.59644	-0.54294	-0.49379	-0.43366	-0.32554	0.20
-0.18	-0.61815	-0.58316	-0.52843	-0.47825	-0.41700	-0.30724	0.18
-0.16	-0.60551	-0.56968	-0.51372	-0.46253	-0.40019	-0.28886	0.16
-0.14	-0.59265	-0.55598	-0.49881	-0.44663	-0.38322	-0.27039	0.14
-0.12	-0.57956	-0.54205	-0.48369	-0.43053	-0.36609	-0.25183	0.12
-0.10	-0.56624	-0.52789	-0.46835	-0.41424	-0.34879	-0.23317	0.10
-0.08	-0.55267	-0.51350	-0.45279	-0.39775	-0.33134	-0.21442	0.08
-0.06	-0.53885	-0.49886	-0.43701	-0.38105	-0.31371	-0.19558	0.06
-0.04	-0.52477	-0.48397	-0.42099	-0.36415	-0.29590	-0.17663	0.04
-0.02	-0.51043	-0.46882	-0.40474	-0.34704	-0.27793	-0.15759	0.02

↑ r_0

α =	.995 ↑−	.99 ↑−	.975 ↑−	.95 ↑−	.90 ↑−	.75 ↑−

TABLE 23.2   SAMPLE SIZE = 25   (CONT.)

α =	.005 ↓+	.01 ↓+	.025 ↓+	.05 ↓+	.10 ↓+	.25 ↓+	
r_0 ↓							
0.00	-0.49581	-0.45341	-0.38824	-0.32970	-0.25977	-0.13845	0.00
0.02	-0.48091	-0.43773	-0.37150	-0.31215	-0.24143	-0.11920	-0.02
0.04	-0.46571	-0.42176	-0.35450	-0.29437	-0.22290	-0.09985	-0.04
0.06	-0.45021	-0.40551	-0.33723	-0.27635	-0.20418	-0.08039	-0.06
0.08	-0.43441	-0.38896	-0.31969	-0.25810	-0.18526	-0.06083	-0.08
0.10	-0.41828	-0.37209	-0.30188	-0.23960	-0.16615	-0.04115	-0.10
0.12	-0.40181	-0.35492	-0.28378	-0.22085	-0.14683	-0.02136	-0.12
0.14	-0.38501	-0.33741	-0.26539	-0.20184	-0.12730	-0.00146	-0.14
0.16	-0.36784	-0.31957	-0.24670	-0.18257	-0.10757	0.01856	-0.16
0.18	-0.35031	-0.30138	-0.22769	-0.16303	-0.08761	0.03870	-0.18
0.20	-0.33240	-0.28283	-0.20837	-0.14322	-0.06744	0.05896	-0.20
0.22	-0.31409	-0.26391	-0.18873	-0.12312	-0.04704	0.07934	-0.22
0.24	-0.29538	-0.24461	-0.16874	-0.10273	-0.02641	0.09985	-0.24
0.26	-0.27623	-0.22491	-0.14841	-0.08205	-0.00554	0.12048	-0.26
0.28	-0.25665	-0.20480	-0.12772	-0.06105	0.01557	0.14124	-0.28
0.30	-0.23662	-0.18427	-0.10666	-0.03975	0.03693	0.16214	-0.30
0.32	-0.21611	-0.16329	-0.08522	-0.01812	0.05854	0.18317	-0.32
0.34	-0.19510	-0.14187	-0.06339	0.00384	0.08040	0.20434	-0.34
0.36	-0.17359	-0.11997	-0.04116	0.02614	0.10253	0.22564	-0.36
0.38	-0.15154	-0.09758	-0.01851	0.04879	0.12493	0.24709	-0.38
0.40	-0.12894	-0.07469	0.00457	0.07180	0.14761	0.26868	-0.40
0.42	-0.10576	-0.05127	0.02809	0.09518	0.17057	0.29042	-0.42
0.44	-0.08198	-0.02730	0.05207	0.11894	0.19383	0.31231	-0.44
0.46	-0.05758	-0.00277	0.07653	0.14308	0.21737	0.33435	-0.46
0.48	-0.03252	0.02235	0.10147	0.16764	0.24123	0.35655	-0.48
0.50	-0.00678	0.04808	0.12692	0.19260	0.26539	0.37891	-0.50
0.52	0.01967	0.07445	0.15289	0.21799	0.28988	0.40143	-0.52
0.54	0.04686	0.10148	0.17941	0.24383	0.31469	0.42411	-0.54
0.56	0.07484	0.12920	0.20648	0.27011	0.33984	0.44696	-0.56
0.58	0.10363	0.15765	0.23414	0.29686	0.36534	0.46999	-0.58
0.60	0.13328	0.18684	0.26240	0.32410	0.39119	0.49319	-0.60
0.62	0.16382	0.21682	0.29128	0.35183	0.41740	0.51656	-0.62
0.64	0.19530	0.24762	0.32082	0.38007	0.44398	0.54012	-0.64
0.66	0.22777	0.27927	0.35102	0.40885	0.47095	0.56387	-0.66
0.68	0.26128	0.31183	0.38193	0.43817	0.49832	0.58780	-0.68
0.70	0.29588	0.34531	0.41356	0.46806	0.52608	0.61193	-0.70
0.72	0.33164	0.37978	0.44595	0.49853	0.55426	0.63625	-0.72
0.74	0.36861	0.41529	0.47913	0.52961	0.58287	0.66078	-0.74
0.76	0.40686	0.45187	0.51313	0.56131	0.61192	0.68551	-0.76
0.78	0.44647	0.48959	0.54798	0.59367	0.64142	0.71045	-0.78
0.80	0.48751	0.52851	0.58372	0.62669	0.67139	0.73561	-0.80
0.82	0.53008	0.56868	0.62039	0.66041	0.70184	0.76098	-0.82
0.84	0.57425	0.61018	0.65804	0.69486	0.73278	0.78658	-0.84
0.86	0.62014	0.65308	0.69669	0.73005	0.76422	0.81240	-0.86
0.88	0.66786	0.69746	0.73641	0.76602	0.79620	0.83846	-0.88
0.90	0.71752	0.74339	0.77723	0.80280	0.82871	0.86476	-0.90
0.92	0.76925	0.79098	0.81922	0.84042	0.86178	0.89130	-0.92
0.94	0.82320	0.84033	0.86243	0.87891	0.89542	0.91809	-0.94
0.96	0.87952	0.89153	0.90692	0.91831	0.92966	0.94513	-0.96
0.98	0.93839	0.94471	0.95275	0.95866	0.96451	0.97243	-0.98

↑ r_0

α =	.995 ↑−	.99 ↑−	.975 ↑−	.95 ↑−	.90 ↑−	.75 ↑−

TABLE 24.1   SAMPLE SIZE =   26

α =	.005 ↓+	.01 ↓+	.025 ↓+	.05 ↓+	.10 ↓+	.25 ↓+	
r_0 ↓							
-0.98	-0.99282	-0.99201	-0.99066	-0.98935	-0.98763	-0.98415	0.98
-0.96	-0.98555	-0.98392	-0.98124	-0.97861	-0.97517	-0.96825	0.96
-0.94	-0.97819	-0.97575	-0.97172	-0.96779	-0.96264	-0.95231	0.94
-0.92	-0.97075	-0.96749	-0.96211	-0.95687	-0.95003	-0.93632	0.92
-0.90	-0.96322	-0.95913	-0.95241	-0.94587	-0.93733	-0.92029	0.90
-0.88	-0.95559	-0.95068	-0.94261	-0.93477	-0.92455	-0.90422	0.88
-0.86	-0.94787	-0.94213	-0.93272	-0.92357	-0.91169	-0.88809	0.86
-0.84	-0.94005	-0.93349	-0.92272	-0.91228	-0.89874	-0.87192	0.84
-0.82	-0.93214	-0.92474	-0.91262	-0.90090	-0.88570	-0.85570	0.82
-0.80	-0.92412	-0.91589	-0.90242	-0.88941	-0.87257	-0.83943	0.80
-0.78	-0.91600	-0.90693	-0.89212	-0.87782	-0.85935	-0.82311	0.78
-0.76	-0.90778	-0.89787	-0.88171	-0.86613	-0.84604	-0.80674	0.76
-0.74	-0.89945	-0.88870	-0.87118	-0.85433	-0.83264	-0.79031	0.74
-0.72	-0.89101	-0.87942	-0.86055	-0.84243	-0.81914	-0.77383	0.72
-0.70	-0.88246	-0.87002	-0.84981	-0.83042	-0.80555	-0.75730	0.70
-0.68	-0.87380	-0.86051	-0.83895	-0.81830	-0.79186	-0.74071	0.68
-0.66	-0.86502	-0.85088	-0.82797	-0.80606	-0.77807	-0.72407	0.66
-0.64	-0.85612	-0.84113	-0.81687	-0.79372	-0.76417	-0.70736	0.64
-0.62	-0.84709	-0.83125	-0.80565	-0.78125	-0.75018	-0.69060	0.62
-0.60	-0.83795	-0.82125	-0.79431	-0.76867	-0.73608	-0.67378	0.60
-0.58	-0.82867	-0.81112	-0.78283	-0.75597	-0.72188	-0.65690	0.58
-0.56	-0.81927	-0.80086	-0.77123	-0.74314	-0.70757	-0.63996	0.56
-0.54	-0.80973	-0.79046	-0.75950	-0.73019	-0.69314	-0.62295	0.54
-0.52	-0.80005	-0.77993	-0.74763	-0.71712	-0.67861	-0.60588	0.52
-0.50	-0.79023	-0.76925	-0.73563	-0.70391	-0.66396	-0.58875	0.50
-0.48	-0.78027	-0.75843	-0.72348	-0.69057	-0.64920	-0.57155	0.48
-0.46	-0.77017	-0.74746	-0.71119	-0.67710	-0.63433	-0.55428	0.46
-0.44	-0.75991	-0.73635	-0.69876	-0.66349	-0.61933	-0.53694	0.44
-0.42	-0.74950	-0.72507	-0.68618	-0.64974	-0.60421	-0.51954	0.42
-0.40	-0.73893	-0.71365	-0.67344	-0.63585	-0.58897	-0.50206	0.40
-0.38	-0.72819	-0.70205	-0.66055	-0.62181	-0.57360	-0.48451	0.38
-0.36	-0.71730	-0.69030	-0.64750	-0.60763	-0.55811	-0.46689	0.36
-0.34	-0.70623	-0.67837	-0.63428	-0.59329	-0.54248	-0.44919	0.34
-0.32	-0.69498	-0.66627	-0.62090	-0.57880	-0.52673	-0.43142	0.32
-0.30	-0.68356	-0.65399	-0.60735	-0.56416	-0.51084	-0.41357	0.30
-0.28	-0.67195	-0.64153	-0.59363	-0.54935	-0.49481	-0.39564	0.28
-0.26	-0.66015	-0.62889	-0.57973	-0.53438	-0.47865	-0.37763	0.26
-0.24	-0.64816	-0.61605	-0.56565	-0.51925	-0.46234	-0.35954	0.24
-0.22	-0.63597	-0.60301	-0.55138	-0.50394	-0.44589	-0.34137	0.22
-0.20	-0.62357	-0.58978	-0.53692	-0.48846	-0.42929	-0.32312	0.20
-0.18	-0.61096	-0.57633	-0.52227	-0.47281	-0.41254	-0.30478	0.18
-0.16	-0.59813	-0.56268	-0.50742	-0.45697	-0.39564	-0.28635	0.16
-0.14	-0.58509	-0.54880	-0.49237	-0.44095	-0.37858	-0.26783	0.14
-0.12	-0.57181	-0.53471	-0.47710	-0.42474	-0.36137	-0.24923	0.12
-0.10	-0.55829	-0.52038	-0.46163	-0.40834	-0.34400	-0.23053	0.10
-0.08	-0.54454	-0.50582	-0.44593	-0.39174	-0.32646	-0.21174	0.08
-0.06	-0.53053	-0.49102	-0.43002	-0.37494	-0.30875	-0.19286	0.06
-0.04	-0.51627	-0.47596	-0.41387	-0.35793	-0.29087	-0.17388	0.04
-0.02	-0.50174	-0.46065	-0.39749	-0.34071	-0.27282	-0.15480	0.02

↑ r_0

| α = | .995 ↑− | .99 ↑− | .975 ↑− | .95 ↑− | .90 ↑− | .75 ↑− | |

TABLE 24.2   SAMPLE SIZE =   26   (CONT.)

α =	.005 ↓+	.01 ↓+	.025 ↓+	.05 ↓+	.10 ↓+	.25 ↓+	
r_0 ↓							
0.00	−0.48693	−0.44508	−0.38086	−0.32328	−0.25459	−0.13562	0.00
0.02	−0.47185	−0.42923	−0.36399	−0.30563	−0.23618	−0.11634	−0.02
0.04	−0.45647	−0.41311	−0.34687	−0.28776	−0.21759	−0.09696	−0.04
0.06	−0.44079	−0.39670	−0.32948	−0.26965	−0.19881	−0.07748	−0.06
0.08	−0.42480	−0.37999	−0.31183	−0.25131	−0.17984	−0.05788	−0.08
0.10	−0.40850	−0.36298	−0.29390	−0.23273	−0.16067	−0.03818	−0.10
0.12	−0.39186	−0.34566	−0.27570	−0.21390	−0.14130	−0.01837	−0.12
0.14	−0.37488	−0.32801	−0.25720	−0.19481	−0.12172	0.00155	−0.14
0.16	−0.35755	−0.31003	−0.23841	−0.17547	−0.10194	0.02159	−0.16
0.18	−0.33985	−0.29170	−0.21931	−0.15587	−0.08195	0.04174	−0.18
0.20	−0.32178	−0.27302	−0.19989	−0.13599	−0.06174	0.06201	−0.20
0.22	−0.30331	−0.25398	−0.18016	−0.11583	−0.04131	0.08240	−0.22
0.24	−0.28444	−0.23455	−0.16009	−0.09539	−0.02065	0.10291	−0.24
0.26	−0.26516	−0.21473	−0.13968	−0.07466	0.00024	0.12355	−0.26
0.28	−0.24543	−0.19451	−0.11892	−0.05362	0.02137	0.14432	−0.28
0.30	−0.22526	−0.17387	−0.09780	−0.03228	0.04274	0.16521	−0.30
0.32	−0.20462	−0.15280	−0.07631	−0.01063	0.06436	0.18624	−0.32
0.34	−0.18349	−0.13129	−0.05443	0.01136	0.08622	0.20739	−0.34
0.36	−0.16186	−0.10931	−0.03215	0.03367	0.10835	0.22869	−0.36
0.38	−0.13971	−0.08685	−0.00947	0.05633	0.13074	0.25012	−0.38
0.40	−0.11701	−0.06389	0.01363	0.07934	0.15340	0.27170	−0.40
0.42	−0.09374	−0.04042	0.03717	0.10271	0.17634	0.29341	−0.42
0.44	−0.06989	−0.01641	0.06115	0.12645	0.19956	0.31528	−0.44
0.46	−0.04542	0.00815	0.08560	0.15057	0.22307	0.33729	−0.46
0.48	−0.02032	0.03329	0.11052	0.17508	0.24688	0.35945	−0.48
0.50	0.00546	0.05902	0.13594	0.20000	0.27099	0.38177	−0.50
0.52	0.03193	0.08538	0.16187	0.22533	0.29542	0.40425	−0.52
0.54	0.05913	0.11238	0.18832	0.25109	0.32016	0.42689	−0.54
0.56	0.08709	0.14006	0.21533	0.27730	0.34523	0.44969	−0.56
0.58	0.11584	0.16844	0.24289	0.30395	0.37063	0.47266	−0.58
0.60	0.14542	0.19754	0.27104	0.33108	0.39638	0.49580	−0.60
0.62	0.17587	0.22741	0.29980	0.35868	0.42248	0.51911	−0.62
0.64	0.20724	0.25808	0.32919	0.38678	0.44894	0.54259	−0.64
0.66	0.23957	0.28957	0.35923	0.41540	0.47578	0.56626	−0.66
0.68	0.27290	0.32193	0.38995	0.44455	0.50299	0.59011	−0.68
0.70	0.30729	0.35520	0.42137	0.47424	0.53060	0.61415	−0.70
0.72	0.34279	0.38941	0.45352	0.50450	0.55861	0.63838	−0.72
0.74	0.37946	0.42462	0.48643	0.53534	0.58703	0.66281	−0.74
0.76	0.41736	0.46087	0.52012	0.56679	0.61588	0.68743	−0.76
0.78	0.45657	0.49821	0.55464	0.59886	0.64516	0.71225	−0.78
0.80	0.49715	0.53670	0.59002	0.63158	0.67489	0.73729	−0.80
0.82	0.53920	0.57639	0.62629	0.66497	0.70508	0.76253	−0.82
0.84	0.58278	0.61736	0.66348	0.69905	0.73575	0.78799	−0.84
0.86	0.62799	0.65965	0.70165	0.73385	0.76691	0.81367	−0.86
0.88	0.67494	0.70336	0.74083	0.76939	0.79857	0.83957	−0.88
0.90	0.72374	0.74855	0.78108	0.80571	0.83074	0.86571	−0.90
0.92	0.77450	0.79532	0.82243	0.84283	0.86346	0.89208	−0.92
0.94	0.82736	0.84374	0.86494	0.88079	0.89672	0.91869	−0.94
0.96	0.88245	0.89392	0.90866	0.91961	0.93056	0.94554	−0.96
0.98	0.93994	0.94597	0.95366	0.95934	0.96498	0.97264	−0.98

↑ r_0

α =	.995 ↑−	.99 ↑−	.975 ↑−	.95 ↑−	.90 ↑−	.75 ↑−

TABLE 25.1   SAMPLE SIZE =   27

α =	.005 ↓+	.01 ↓+	.025 ↓+	.05 ↓+	.10 ↓+	.25 ↓+	
r_0 ↓							
-0.98	-0.99267	-0.99186	-0.99052	-0.98922	-0.98751	-0.98408	0.98
-0.96	-0.98525	-0.98363	-0.98096	-0.97836	-0.97495	-0.96812	0.96
-0.94	-0.97774	-0.97531	-0.97130	-0.96740	-0.96231	-0.95212	0.94
-0.92	-0.97015	-0.96689	-0.96155	-0.95636	-0.94958	-0.93608	0.92
-0.90	-0.96246	-0.95839	-0.95171	-0.94522	-0.93678	-0.91998	0.90
-0.88	-0.95468	-0.94979	-0.94177	-0.93399	-0.92388	-0.90384	0.88
-0.86	-0.94680	-0.94109	-0.93173	-0.92266	-0.91091	-0.88766	0.86
-0.84	-0.93883	-0.93229	-0.92159	-0.91124	-0.89785	-0.87143	0.84
-0.82	-0.93076	-0.92339	-0.91135	-0.89973	-0.88470	-0.85514	0.82
-0.80	-0.92258	-0.91438	-0.90101	-0.88811	-0.87146	-0.83881	0.80
-0.78	-0.91431	-0.90527	-0.89056	-0.87639	-0.85813	-0.82243	0.78
-0.76	-0.90592	-0.89605	-0.88000	-0.86457	-0.84472	-0.80600	0.76
-0.74	-0.89743	-0.88673	-0.86934	-0.85264	-0.83120	-0.78951	0.74
-0.72	-0.88883	-0.87729	-0.85856	-0.84061	-0.81760	-0.77297	0.72
-0.70	-0.88012	-0.86774	-0.84767	-0.82847	-0.80390	-0.75638	0.70
-0.68	-0.87129	-0.85807	-0.83666	-0.81622	-0.79010	-0.73973	0.68
-0.66	-0.86234	-0.84828	-0.82554	-0.80385	-0.77620	-0.72302	0.66
-0.64	-0.85328	-0.83837	-0.81430	-0.79138	-0.76220	-0.70626	0.64
-0.62	-0.84409	-0.82833	-0.80293	-0.77878	-0.74810	-0.68944	0.62
-0.60	-0.83477	-0.81817	-0.79144	-0.76607	-0.73389	-0.67256	0.60
-0.58	-0.82533	-0.80788	-0.77983	-0.75324	-0.71958	-0.65562	0.58
-0.56	-0.81575	-0.79746	-0.76808	-0.74029	-0.70517	-0.63862	0.56
-0.54	-0.80605	-0.78690	-0.75620	-0.72721	-0.69064	-0.62155	0.54
-0.52	-0.79620	-0.77620	-0.74419	-0.71401	-0.67600	-0.60442	0.52
-0.50	-0.78621	-0.76537	-0.73204	-0.70067	-0.66126	-0.58723	0.50
-0.48	-0.77608	-0.75439	-0.71975	-0.68721	-0.64639	-0.56997	0.48
-0.46	-0.76580	-0.74326	-0.70732	-0.67361	-0.63141	-0.55265	0.46
-0.44	-0.75537	-0.73198	-0.69474	-0.65988	-0.61632	-0.53526	0.44
-0.42	-0.74478	-0.72054	-0.68201	-0.64600	-0.60110	-0.51780	0.42
-0.40	-0.73404	-0.70895	-0.66913	-0.63199	-0.58576	-0.50027	0.40
-0.38	-0.72313	-0.69719	-0.65610	-0.61783	-0.57029	-0.48266	0.38
-0.36	-0.71205	-0.68527	-0.64290	-0.60352	-0.55470	-0.46499	0.36
-0.34	-0.70081	-0.67318	-0.62955	-0.58907	-0.53899	-0.44724	0.34
-0.32	-0.68939	-0.66092	-0.61603	-0.57446	-0.52314	-0.42942	0.32
-0.30	-0.67778	-0.64848	-0.60234	-0.55969	-0.50715	-0.41152	0.30
-0.28	-0.66600	-0.63585	-0.58848	-0.54477	-0.49104	-0.39354	0.28
-0.26	-0.65402	-0.62304	-0.57444	-0.52969	-0.47478	-0.37549	0.26
-0.24	-0.64185	-0.61004	-0.56022	-0.51444	-0.45838	-0.35735	0.24
-0.22	-0.62948	-0.59685	-0.54581	-0.49902	-0.44184	-0.33913	0.22
-0.20	-0.61691	-0.58345	-0.53122	-0.48342	-0.42516	-0.32083	0.20
-0.18	-0.60412	-0.56984	-0.51643	-0.46766	-0.40833	-0.30245	0.18
-0.16	-0.59112	-0.55603	-0.50145	-0.45171	-0.39135	-0.28398	0.16
-0.14	-0.57789	-0.54200	-0.48626	-0.43558	-0.37421	-0.26542	0.14
-0.12	-0.56444	-0.52774	-0.47087	-0.41927	-0.35692	-0.24678	0.12
-0.10	-0.55075	-0.51326	-0.45527	-0.40276	-0.33947	-0.22804	0.10
-0.08	-0.53681	-0.49854	-0.43945	-0.38606	-0.32185	-0.20921	0.08
-0.06	-0.52263	-0.48358	-0.42340	-0.36916	-0.30408	-0.19029	0.06
-0.04	-0.50819	-0.46837	-0.40713	-0.35206	-0.28613	-0.17128	0.04
-0.02	-0.49349	-0.45291	-0.39063	-0.33475	-0.26801	-0.15217	0.02

↑ r_0

α =	.995 ↑-	.99 ↑-	.975 ↑-	.95 ↑-	.90 ↑-	.75 ↑-

TABLE 25.2   SAMPLE SIZE = 27   (CONT.)

α =	.005 ↓+	.01 ↓+	.025 ↓+	.05 ↓+	.10 ↓+	.25 ↓+	
r_0 ↓							
0.00	−0.47851	−0.43718	−0.37389	−0.31722	−0.24972	−0.13296	0.00
0.02	−0.46325	−0.42119	−0.35690	−0.29948	−0.23125	−0.11365	−0.02
0.04	−0.44771	−0.40492	−0.33966	−0.28152	−0.21259	−0.09424	−0.04
0.06	−0.43186	−0.38837	−0.32217	−0.26333	−0.19375	−0.07473	−0.06
0.08	−0.41571	−0.37152	−0.30441	−0.24491	−0.17473	−0.05512	−0.08
0.10	−0.39924	−0.35437	−0.28638	−0.22625	−0.15551	−0.03540	−0.10
0.12	−0.38244	−0.33691	−0.26807	−0.20735	−0.13609	−0.01557	−0.12
0.14	−0.36530	−0.31913	−0.24947	−0.18819	−0.11647	0.00438	−0.14
0.16	−0.34782	−0.30102	−0.23059	−0.16879	−0.09665	0.02443	−0.16
0.18	−0.32997	−0.28257	−0.21140	−0.14912	−0.07662	0.04459	−0.18
0.20	−0.31174	−0.26377	−0.19190	−0.12919	−0.05638	0.06487	−0.20
0.22	−0.29314	−0.24460	−0.17209	−0.10898	−0.03592	0.08527	−0.22
0.24	−0.27413	−0.22507	−0.15195	−0.08849	−0.01524	0.10579	−0.24
0.26	−0.25470	−0.20514	−0.13147	−0.06771	0.00567	0.12643	−0.26
0.28	−0.23485	−0.18482	−0.11065	−0.04664	0.02682	0.14720	−0.28
0.30	−0.21455	−0.16409	−0.08947	−0.02527	0.04820	0.16809	−0.30
0.32	−0.19379	−0.14293	−0.06793	−0.00359	0.06982	0.18911	−0.32
0.34	−0.17256	−0.12133	−0.04601	0.01841	0.09168	0.21026	−0.34
0.36	−0.15082	−0.09928	−0.02370	0.04074	0.11381	0.23154	−0.36
0.38	−0.12857	−0.07676	−0.00099	0.06340	0.13619	0.25296	−0.38
0.40	−0.10579	−0.05375	0.02213	0.08640	0.15883	0.27452	−0.40
0.42	−0.08245	−0.03023	0.04568	0.10976	0.18174	0.29621	−0.42
0.44	−0.05853	−0.00619	0.06966	0.13348	0.20493	0.31805	−0.44
0.46	−0.03401	0.01839	0.09410	0.15757	0.22841	0.34004	−0.46
0.48	−0.00887	0.04354	0.11900	0.18205	0.25217	0.36217	−0.48
0.50	0.01693	0.06927	0.14438	0.20692	0.27623	0.38445	−0.50
0.52	0.04341	0.09561	0.17026	0.23219	0.30059	0.40689	−0.52
0.54	0.07061	0.12258	0.19666	0.25789	0.32527	0.42948	−0.54
0.56	0.09854	0.15020	0.22359	0.28401	0.35026	0.45224	−0.56
0.58	0.12724	0.17851	0.25107	0.31057	0.37558	0.47515	−0.58
0.60	0.15676	0.20753	0.27911	0.33759	0.40123	0.49823	−0.60
0.62	0.18713	0.23729	0.30775	0.36507	0.42722	0.52148	−0.62
0.64	0.21838	0.26782	0.33700	0.39304	0.45357	0.54490	−0.64
0.66	0.25056	0.29916	0.36687	0.42151	0.48028	0.56850	−0.66
0.68	0.28371	0.33133	0.39741	0.45049	0.50735	0.59227	−0.68
0.70	0.31789	0.36438	0.42862	0.47999	0.53481	0.61622	−0.70
0.72	0.35314	0.39835	0.46055	0.51005	0.56265	0.64036	−0.72
0.74	0.38953	0.43328	0.49320	0.54067	0.59090	0.66469	−0.74
0.76	0.42710	0.46921	0.52661	0.57188	0.61955	0.68921	−0.76
0.78	0.46593	0.50619	0.56082	0.60368	0.64863	0.71393	−0.78
0.80	0.50607	0.54428	0.59585	0.63611	0.67814	0.73885	−0.80
0.82	0.54762	0.58352	0.63174	0.66919	0.70810	0.76397	−0.82
0.84	0.59064	0.62398	0.66853	0.70293	0.73851	0.78930	−0.84
0.86	0.63522	0.66572	0.70624	0.73736	0.76939	0.81485	−0.86
0.88	0.68146	0.70880	0.74492	0.77251	0.80076	0.84061	−0.88
0.90	0.72946	0.75330	0.78462	0.80840	0.83263	0.86659	−0.90
0.92	0.77932	0.79930	0.82538	0.84506	0.86501	0.89280	−0.92
0.94	0.83117	0.84687	0.86724	0.88252	0.89793	0.91924	−0.94
0.96	0.88514	0.89611	0.91026	0.92081	0.93138	0.94592	−0.96
0.98	0.94136	0.94712	0.95450	0.95996	0.96540	0.97284	−0.98

↑ r_0

α =	.995 ↑−	.99 ↑−	.975 ↑−	.95 ↑−	.90 ↑−	.75 ↑−

TABLE 26.1   SAMPLE SIZE =   28

α =	.005 ↓+	.01 ↓+	.025 ↓+	.05 ↓+	.10 ↓+	.25 ↓+	
r_0 ↓							
-0.98	-0.99252	-0.99172	-0.99039	-0.98910	-0.98741	-0.98403	0.98
-0.96	-0.98496	-0.98334	-0.98069	-0.97811	-0.97474	-0.96801	0.96
-0.94	-0.97731	-0.97488	-0.97090	-0.96703	-0.96199	-0.95195	0.94
-0.92	-0.96957	-0.96633	-0.96101	-0.95586	-0.94916	-0.93584	0.92
-0.90	-0.96173	-0.95767	-0.95103	-0.94460	-0.93625	-0.91969	0.90
-0.88	-0.95380	-0.94893	-0.94096	-0.93325	-0.92325	-0.90349	0.88
-0.86	-0.94578	-0.94008	-0.93078	-0.92180	-0.91017	-0.88725	0.86
-0.84	-0.93765	-0.93114	-0.92051	-0.91025	-0.89700	-0.87095	0.84
-0.82	-0.92943	-0.92209	-0.91013	-0.89861	-0.88375	-0.85461	0.82
-0.80	-0.92110	-0.91294	-0.89965	-0.88687	-0.87041	-0.83822	0.80
-0.78	-0.91268	-0.90368	-0.88907	-0.87503	-0.85698	-0.82178	0.78
-0.76	-0.90414	-0.89432	-0.87837	-0.86308	-0.84345	-0.80529	0.76
-0.74	-0.89549	-0.88484	-0.86757	-0.85103	-0.82984	-0.78875	0.74
-0.72	-0.88674	-0.87525	-0.85666	-0.83887	-0.81613	-0.77215	0.72
-0.70	-0.87787	-0.86555	-0.84563	-0.82661	-0.80232	-0.75550	0.70
-0.68	-0.86888	-0.85573	-0.83449	-0.81423	-0.78842	-0.73879	0.68
-0.66	-0.85978	-0.84579	-0.82322	-0.80175	-0.77442	-0.72203	0.66
-0.64	-0.85056	-0.83573	-0.81184	-0.78915	-0.76032	-0.70521	0.64
-0.62	-0.84121	-0.82554	-0.80034	-0.77643	-0.74612	-0.68833	0.62
-0.60	-0.83173	-0.81523	-0.78871	-0.76360	-0.73181	-0.67139	0.60
-0.58	-0.82213	-0.80479	-0.77696	-0.75065	-0.71740	-0.65440	0.58
-0.56	-0.81239	-0.79421	-0.76508	-0.73757	-0.70289	-0.63734	0.56
-0.54	-0.80252	-0.78350	-0.75306	-0.72438	-0.68826	-0.62022	0.54
-0.52	-0.79251	-0.77265	-0.74091	-0.71105	-0.67353	-0.60304	0.52
-0.50	-0.78236	-0.76166	-0.72863	-0.69760	-0.65868	-0.58579	0.50
-0.48	-0.77207	-0.75052	-0.71620	-0.68401	-0.64372	-0.56848	0.48
-0.46	-0.76162	-0.73924	-0.70363	-0.67030	-0.62865	-0.55111	0.46
-0.44	-0.75103	-0.72780	-0.69091	-0.65644	-0.61346	-0.53366	0.44
-0.42	-0.74027	-0.71621	-0.67805	-0.64245	-0.59815	-0.51615	0.42
-0.40	-0.72936	-0.70447	-0.66503	-0.62832	-0.58271	-0.49857	0.40
-0.38	-0.71829	-0.69256	-0.65186	-0.61405	-0.56716	-0.48092	0.38
-0.36	-0.70705	-0.68048	-0.63854	-0.59963	-0.55148	-0.46319	0.36
-0.34	-0.69563	-0.66824	-0.62505	-0.58506	-0.53567	-0.44540	0.34
-0.32	-0.68404	-0.65582	-0.61139	-0.57033	-0.51973	-0.42752	0.32
-0.30	-0.67228	-0.64322	-0.59757	-0.55546	-0.50366	-0.40958	0.30
-0.28	=0.66032	-0.63045	-0.58358	-0.54042	-0.48746	-0.39155	0.28
-0.26	-0.64818	-0.61748	-0.56941	-0.52523	-0.47112	-0.37345	0.26
-0.24	-0.63584	-0.60433	-0.55506	-0.50987	-0.45464	-0.35527	0.24
-0.22	-0.62330	-0.59098	-0.54052	-0.49434	-0.43802	-0.33701	0.22
-0.20	-0.61056	-0.57743	-0.52580	-0.47865	-0.42125	-0.31867	0.20
-0.18	-0.59760	-0.56367	-0.51089	-0.46278	-0.40434	-0.30025	0.18
-0.16	-0.58443	-0.54970	-0.49578	-0.44673	-0.38728	-0.28174	0.16
-0.14	-0.57104	-0.53552	-0.48047	-0.43050	-0.37007	-0.26314	0.14
-0.12	-0.55742	-0.52112	-0.46496	-0.41408	-0.35271	-0.24446	0.12
-0.10	-0.54356	-0.50648	-0.44923	-0.39748	-0.33518	-0.22569	0.10
-0.08	-0.52946	-0.49162	-0.43329	-0.38068	-0.31750	-0.20683	0.08
-0.06	-0.51512	-0.47651	-0.41713	-0.36369	-0.29966	-0.18787	0.06
-0.04	-0.50051	-0.46116	-0.40075	-0.34650	-0.28164	-0.16883	0.04
-0.02	-0.48565	-0.44556	-0.38413	-0.32910	-0.26346	-0.14969	0.02

↑ r_0

α =	.995 ↑−	.99 ↑−	.975 ↑−	.95 ↑−	.90 ↑−	.75 ↑−

TABLE 26.2   SAMPLE SIZE =   28   (CONT.)

α =	.005 ↓+	.01 ↓+	.025 ↓+	.05 ↓+	.10 ↓+	.25 ↓+	
r_o ↓							
0.00	−0.47051	−0.42969	−0.36728	−0.31149	−0.24511	−0.13045	0.00
0.02	−0.45509	−0.41356	−0.35018	−0.29367	−0.22658	−0.11112	−0.02
0.04	−0.43939	−0.39716	−0.33284	−0.27562	−0.20787	−0.09168	−0.04
0.06	−0.42339	−0.38047	−0.31524	−0.25736	−0.18898	−0.07215	−0.06
0.08	−0.40708	−0.36349	−0.29738	−0.23886	−0.16991	−0.05251	−0.08
0.10	−0.39046	−0.34621	−0.27926	−0.22013	−0.15064	−0.03277	−0.10
0.12	−0.37351	−0.32862	−0.26086	−0.20116	−0.13118	−0.01292	−0.12
0.14	−0.35622	−0.31072	−0.24217	−0.18195	−0.11152	0.00704	−0.14
0.16	−0.33859	−0.29249	−0.22320	−0.16248	−0.09167	0.02710	−0.16
0.18	−0.32060	−0.27393	−0.20393	−0.14276	−0.07160	0.04728	−0.18
0.20	−0.30224	−0.25501	−0.18436	−0.12277	−0.05133	0.06757	−0.20
0.22	−0.28350	−0.23574	−0.16448	−0.10252	−0.03084	0.08798	−0.22
0.24	−0.26436	−0.21610	−0.14427	−0.08199	−0.01014	0.10850	−0.24
0.26	−0.24482	−0.19609	−0.12373	−0.06117	0.01079	0.12914	−0.26
0.28	−0.22485	−0.17567	−0.10285	−0.04007	0.03194	0.14991	−0.28
0.30	−0.20444	−0.15486	−0.08162	−0.01867	0.05333	0.17080	−0.30
0.32	−0.18357	−0.13362	−0.06003	0.00303	0.07495	0.19181	−0.32
0.34	−0.16224	−0.11195	−0.03808	0.02505	0.09682	0.21295	−0.34
0.36	−0.14041	−0.08984	−0.01574	0.04738	0.11893	0.23423	−0.36
0.38	−0.11808	−0.06726	0.00699	0.07004	0.14130	0.25563	−0.38
0.40	−0.09522	−0.04420	0.03013	0.09304	0.16393	0.27717	−0.40
0.42	−0.07182	−0.02065	0.05368	0.11639	0.18682	0.29884	−0.42
0.44	−0.04784	0.00342	0.07766	0.14009	0.20998	0.32066	−0.44
0.46	−0.02328	0.02802	0.10208	0.16415	0.23341	0.34262	−0.46
0.48	0.00190	0.05317	0.12695	0.18859	0.25713	0.36472	−0.48
0.50	0.02771	0.07889	0.15230	0.21341	0.28114	0.38697	−0.50
0.52	0.05420	0.10521	0.17814	0.23863	0.30545	0.40937	−0.52
0.54	0.08137	0.13214	0.20447	0.26425	0.33005	0.43192	−0.54
0.56	0.10928	0.15971	0.23133	0.29029	0.35497	0.45462	−0.56
0.58	0.13793	0.18795	0.25872	0.31676	0.38021	0.47749	−0.58
0.60	0.16738	0.21688	0.28666	0.34368	0.40577	0.50051	−0.60
0.62	0.19765	0.24653	0.31518	0.37105	0.43166	0.52370	−0.62
0.64	0.22878	0.27693	0.34429	0.39889	0.45790	0.54706	−0.64
0.66	0.26082	0.30811	0.37401	0.42721	0.48448	0.57059	−0.66
0.68	0.29380	0.34011	0.40438	0.45603	0.51143	0.59428	−0.68
0.70	0.32778	0.37295	0.43539	0.48537	0.53874	0.61816	−0.70
0.72	0.36279	0.40668	0.46710	0.51523	0.56643	0.64222	−0.72
0.74	0.39890	0.44134	0.49951	0.54564	0.59451	0.66646	−0.74
0.76	0.43615	0.47697	0.53266	0.57662	0.62298	0.69088	−0.76
0.78	0.47462	0.51361	0.56657	0.60817	0.65187	0.71550	−0.78
0.80	0.51435	0.55132	0.60128	0.64033	0.68117	0.74031	−0.80
0.82	0.55543	0.59014	0.63681	0.67312	0.71090	0.76532	−0.82
0.84	0.59792	0.63012	0.67320	0.70654	0.74108	0.79053	−0.84
0.86	0.64191	0.67133	0.71049	0.74063	0.77171	0.81595	−0.86
0.88	0.68748	0.71383	0.74871	0.77541	0.80281	0.84158	−0.88
0.90	0.73473	0.75769	0.78790	0.81090	0.83439	0.86742	−0.90
0.92	0.78376	0.80297	0.82811	0.84713	0.86646	0.89348	−0.92
0.94	0.83467	0.84975	0.86937	0.88413	0.89904	0.91976	−0.94
0.96	0.88760	0.89813	0.91174	0.92192	0.93215	0.94627	−0.96
0.98	0.94266	0.94818	0.95527	0.96053	0.96580	0.97302	−0.98

↑ r_o

α =	.995 ↑−	.99 ↑−	.975 ↑−	.95 ↑−	.90 ↑−	.75 ↑−

TABLE 27.1 SAMPLE SIZE = 29

α =	.005 ↓+	.01 ↓+	.025 ↓+	.05 ↓+	.10 ↓+	.25 ↓+	
r_0 ↓							
−0.98	−0.99239	−0.99158	−0.99026	−0.98898	−0.98731	−0.98397	0.98
−0.96	−0.98468	−0.98307	−0.98043	−0.97787	−0.97454	−0.96789	0.96
−0.94	−0.97689	−0.97447	−0.97051	−0.96667	−0.96169	−0.95178	0.94
−0.92	−0.96901	−0.96578	−0.96050	−0.95539	−0.94875	−0.93562	0.92
−0.90	−0.96103	−0.95699	−0.95039	−0.94401	−0.93574	−0.91941	0.90
−0.88	−0.95296	−0.94811	−0.94018	−0.93254	−0.92264	−0.90315	0.88
−0.86	−0.94479	−0.93912	−0.92988	−0.92097	−0.90946	−0.88685	0.86
−0.84	−0.93652	−0.93004	−0.91947	−0.90931	−0.89620	−0.87051	0.84
−0.82	−0.92816	−0.92085	−0.90897	−0.89754	−0.88284	−0.85411	0.82
−0.80	−0.91969	−0.91156	−0.89836	−0.88568	−0.86940	−0.83766	0.80
−0.78	−0.91111	−0.90216	−0.88764	−0.87372	−0.85587	−0.82117	0.78
−0.76	−0.90243	−0.89265	−0.87681	−0.86166	−0.84225	−0.80462	0.76
−0.74	−0.89363	−0.88303	−0.86588	−0.84949	−0.82853	−0.78802	0.74
−0.72	−0.88473	−0.87330	−0.85484	−0.83722	−0.81473	−0.77137	0.72
−0.70	−0.87571	−0.86346	−0.84368	−0.82483	−0.80082	−0.75466	0.70
−0.68	−0.86658	−0.85349	−0.83240	−0.81234	−0.78682	−0.73790	0.68
−0.66	−0.85732	−0.84341	−0.82101	−0.79974	−0.77273	−0.72108	0.66
−0.64	−0.84795	−0.83320	−0.80950	−0.78702	−0.75853	−0.70421	0.64
−0.62	−0.83845	−0.82287	−0.79787	−0.77419	−0.74423	−0.68728	0.62
−0.60	−0.82882	−0.81241	−0.78611	−0.76124	−0.72983	−0.67029	0.60
−0.58	−0.81906	−0.80183	−0.77422	−0.74817	−0.71533	−0.65324	0.58
−0.56	−0.80917	−0.79110	−0.76221	−0.73499	−0.70072	−0.63613	0.56
−0.54	−0.79915	−0.78025	−0.75006	−0.72167	−0.68600	−0.61896	0.54
−0.52	−0.78898	−0.76925	−0.73778	−0.70823	−0.67117	−0.60173	0.52
−0.50	−0.77868	−0.75811	−0.72536	−0.69467	−0.65624	−0.58443	0.50
−0.48	−0.76822	−0.74683	−0.71281	−0.68097	−0.64118	−0.56707	0.48
−0.46	−0.75762	−0.73540	−0.70011	−0.66714	−0.62602	−0.54964	0.46
−0.44	−0.74687	−0.72382	−0.68726	−0.65317	−0.61074	−0.53215	0.44
−0.42	−0.73596	−0.71208	−0.67427	−0.63907	−0.59534	−0.51459	0.42
−0.40	−0.72489	−0.70018	−0.66113	−0.62483	−0.57982	−0.49696	0.40
−0.38	−0.71366	−0.68813	−0.64783	−0.61045	−0.56418	−0.47926	0.38
−0.36	−0.70226	−0.67591	−0.63437	−0.59592	−0.54841	−0.46149	0.36
−0.34	−0.69069	−0.66351	−0.62076	−0.58124	−0.53252	−0.44364	0.34
−0.32	−0.67894	−0.65095	−0.60698	−0.56641	−0.51649	−0.42573	0.32
−0.30	−0.66701	−0.63821	−0.59303	−0.55143	−0.50034	−0.40774	0.30
−0.28	−0.65490	−0.62528	−0.57891	−0.53629	−0.48406	−0.38967	0.28
−0.26	−0.64259	−0.61217	−0.56462	−0.52099	−0.46764	−0.37153	0.26
−0.24	−0.63010	−0.59888	−0.55014	−0.50553	−0.45108	−0.35331	0.24
−0.22	−0.61740	−0.58538	−0.53549	−0.48990	−0.43438	−0.33501	0.22
−0.20	−0.60450	−0.57169	−0.52065	−0.47411	−0.41754	−0.31663	0.20
−0.18	−0.59138	−0.55779	−0.50562	−0.45814	−0.40056	−0.29816	0.18
−0.16	−0.57806	−0.54368	−0.49039	−0.44199	−0.38343	−0.27962	0.16
−0.14	−0.56450	−0.52935	−0.47496	−0.42567	−0.36615	−0.26098	0.14
−0.12	−0.55073	−0.51481	−0.45933	−0.40916	−0.34871	−0.24227	0.12
−0.10	−0.53671	−0.50004	−0.44350	−0.39247	−0.33112	−0.22346	0.10
−0.08	−0.52246	−0.48503	−0.42745	−0.37558	−0.31338	−0.20457	0.08
−0.06	−0.50796	−0.46979	−0.41118	−0.35851	−0.29547	−0.18559	0.06
−0.04	−0.49320	−0.45431	−0.39468	−0.34123	−0.27740	−0.16651	0.04
−0.02	−0.47818	−0.43857	−0.37796	−0.32375	−0.25916	−0.14734	0.02

↑ r_0

| α = | .995 ↑− | .99 ↑− | .975 ↑− | .95 ↑− | .90 ↑− | .75 ↑− |

TABLE 27.2 SAMPLE SIZE = 29 (CONT.)

α =	.005 ↓+	.01 ↓+	.025 ↓+	.05 ↓+	.10 ↓+	.25 ↓+	
r_o ↓							
0.00	−0.46289	−0.42257	−0.36101	−0.30606	−0.24075	−0.12808	0.00
0.02	−0.44733	−0.40631	−0.34381	−0.28816	−0.22217	−0.10872	−0.02
0.04	−0.43147	−0.38978	−0.32637	−0.27004	−0.20341	−0.08926	−0.04
0.06	−0.41533	−0.37296	−0.30868	−0.25170	−0.18447	−0.06970	−0.06
0.08	−0.39887	−0.35586	−0.29073	−0.23314	−0.16535	−0.05005	−0.08
0.10	−0.38211	−0.33846	−0.27251	−0.21434	−0.14604	−0.03029	−0.10
0.12	−0.36502	−0.32076	−0.25403	−0.19531	−0.12654	−0.01042	−0.12
0.14	−0.34760	−0.30275	−0.23526	−0.17603	−0.10685	0.00955	−0.14
0.16	−0.32984	−0.28441	−0.21621	−0.15651	−0.08695	0.02963	−0.16
0.18	−0.31172	−0.26573	−0.19687	−0.13674	−0.06686	0.04982	−0.18
0.20	−0.29323	−0.24672	−0.17723	−0.11671	−0.04656	0.07011	−0.20
0.22	−0.27437	−0.22735	−0.15727	−0.09641	−0.02605	0.09053	−0.22
0.24	−0.25511	−0.20762	−0.13701	−0.07584	−0.00533	0.11106	−0.24
0.26	−0.23545	−0.18751	−0.11641	−0.05499	0.01561	0.13170	−0.26
0.28	−0.21538	−0.16702	−0.09548	−0.03386	0.03678	0.15246	−0.28
0.30	−0.19486	−0.14613	−0.07421	−0.01244	0.05817	0.17335	−0.30
0.32	−0.17390	−0.12482	−0.05258	0.00928	0.07980	0.19436	−0.32
0.34	−0.15248	−0.10309	−0.03060	0.03131	0.10166	0.21549	−0.34
0.36	−0.13057	−0.08092	−0.00823	0.05365	0.12377	0.23675	−0.36
0.38	−0.10817	−0.05829	0.01452	0.07631	0.14612	0.25814	−0.38
0.40	−0.08524	−0.03519	0.03766	0.09930	0.16873	0.27966	−0.40
0.42	−0.06178	−0.01161	0.06121	0.12263	0.19159	0.30132	−0.42
0.44	−0.03776	0.01248	0.08518	0.14630	0.21472	0.32311	−0.44
0.46	−0.01317	0.03709	0.10959	0.17034	0.23812	0.34504	−0.46
0.48	0.01204	0.06224	0.13444	0.19474	0.26180	0.36712	−0.48
0.50	0.03786	0.08794	0.15975	0.21951	0.28576	0.38933	−0.50
0.52	0.06434	0.11423	0.18554	0.24467	0.31001	0.41169	−0.52
0.54	0.09150	0.14113	0.21181	0.27023	0.33455	0.43420	−0.54
0.56	0.11937	0.16864	0.23860	0.29619	0.35940	0.45687	−0.56
0.58	0.14797	0.19681	0.26590	0.32258	0.38455	0.47968	−0.58
0.60	0.17734	0.22565	0.29375	0.34939	0.41003	0.50266	−0.60
0.62	0.20752	0.25519	0.32215	0.37665	0.43582	0.52579	−0.62
0.64	0.23854	0.28546	0.35113	0.40437	0.46195	0.54908	−0.64
0.66	0.27043	0.31649	0.38070	0.43255	0.48842	0.57255	−0.66
0.68	0.30325	0.34832	0.41090	0.46122	0.51524	0.59618	−0.68
0.70	0.33702	0.38096	0.44173	0.49039	0.54242	0.61998	−0.70
0.72	0.37180	0.41447	0.47322	0.52007	0.56997	0.64395	−0.72
0.74	0.40764	0.44887	0.50541	0.55029	0.59789	0.66811	−0.74
0.76	0.44459	0.48421	0.53830	0.58105	0.62619	0.69245	−0.76
0.78	0.48271	0.52053	0.57193	0.61237	0.65489	0.71697	−0.78
0.80	0.52205	0.55787	0.60634	0.64427	0.68400	0.74168	−0.80
0.82	0.56269	0.59629	0.64153	0.67678	0.71353	0.76658	−0.82
0.84	0.60468	0.63583	0.67756	0.70991	0.74348	0.79168	−0.84
0.86	0.64812	0.67655	0.71445	0.74367	0.77387	0.81698	−0.86
0.88	0.69306	0.71850	0.75223	0.77811	0.80471	0.84248	−0.88
0.90	0.73961	0.76175	0.79095	0.81323	0.83602	0.86819	−0.90
0.92	0.78786	0.80637	0.83064	0.84905	0.86781	0.89411	−0.92
0.94	0.83791	0.85242	0.87135	0.88562	0.90008	0.92025	−0.94
0.96	0.88987	0.89999	0.91311	0.92295	0.93286	0.94660	−0.96
0.98	0.94385	0.94915	0.95598	0.96106	0.96616	0.97319	−0.98

↑ r_o

α =	.995 ↑−	.99 ↑−	.975 ↑−	.95 ↑−	.90 ↑−	.75 ↑−

TABLE 28.1   SAMPLE SIZE =   30

α =	.005 ↓+	.01 ↓+	.025 ↓+	.05 ↓+	.10 ↓+	.25 ↓+	
r_0 ↓							
-0.98	-0.99225	-0.99145	-0.99014	-0.98887	-0.98721	-0.98392	0.98
-0.96	-0.98442	-0.98281	-0.98019	-0.97764	-0.97434	-0.96779	0.96
-0.94	-0.97649	-0.97408	-0.97014	-0.96633	-0.96140	-0.95162	0.94
-0.92	-0.96847	-0.96525	-0.96000	-0.95493	-0.94837	-0.93540	0.92
-0.90	-0.96036	-0.95633	-0.94977	-0.94344	-0.93526	-0.91914	0.90
-0.88	-0.95215	-0.94732	-0.93944	-0.93186	-0.92207	-0.90283	0.88
-0.86	-0.94384	-0.93820	-0.92901	-0.92017	-0.90879	-0.88648	0.86
-0.84	-0.93544	-0.92898	-0.91848	-0.90840	-0.89543	-0.87008	0.84
-0.82	-0.92693	-0.91966	-0.90785	-0.89652	-0.88198	-0.85363	0.82
-0.80	-0.91832	-0.91023	-0.89711	-0.88455	-0.86844	-0.83713	0.80
-0.78	-0.90960	-0.90070	-0.88627	-0.87248	-0.85482	-0.82058	0.78
-0.76	-0.90078	-0.89105	-0.87532	-0.86030	-0.84110	-0.80398	0.76
-0.74	-0.89185	-0.88130	-0.86426	-0.84802	-0.82729	-0.78732	0.74
-0.72	-0.88280	-0.87143	-0.85309	-0.83563	-0.81339	-0.77062	0.72
-0.70	-0.87364	-0.86145	-0.84181	-0.82314	-0.79939	-0.75386	0.70
-0.68	-0.86436	-0.85135	-0.83041	-0.81054	-0.78530	-0.73705	0.68
-0.66	-0.85496	-0.84113	-0.81889	-0.79782	-0.77111	-0.72018	0.66
-0.64	-0.84544	-0.83078	-0.80726	-0.78499	-0.75682	-0.70325	0.64
-0.62	-0.83579	-0.82031	-0.79550	-0.77205	-0.74244	-0.68627	0.62
-0.60	-0.82602	-0.80971	-0.78361	-0.75899	-0.72794	-0.66923	0.60
-0.58	-0.81612	-0.79899	-0.77160	-0.74581	-0.71335	-0.65213	0.58
-0.56	-0.80608	-0.78813	-0.75946	-0.73251	-0.69865	-0.63498	0.56
-0.54	-0.79591	-0.77713	-0.74719	-0.71909	-0.68384	-0.61776	0.54
-0.52	-0.78559	-0.76599	-0.73479	-0.70554	-0.66893	-0.60047	0.52
-0.50	-0.77514	-0.75471	-0.72225	-0.69187	-0.65390	-0.58313	0.50
-0.48	-0.76454	-0.74329	-0.70957	-0.67806	-0.63877	-0.56572	0.48
-0.46	-0.75379	-0.73172	-0.69674	-0.66413	-0.62352	-0.54825	0.46
-0.44	-0.74289	-0.72000	-0.68378	-0.65006	-0.60815	-0.53071	0.44
-0.42	-0.73183	-0.70812	-0.67066	-0.63585	-0.59266	-0.51310	0.42
-0.40	-0.72061	-0.69609	-0.65740	-0.62150	-0.57706	-0.49542	0.40
-0.38	-0.70922	-0.68389	-0.64398	-0.60701	-0.56134	-0.47768	0.38
-0.36	-0.69767	-0.67153	-0.63040	-0.59238	-0.54549	-0.45987	0.36
-0.34	-0.68595	-0.65900	-0.61666	-0.57760	-0.52952	-0.44198	0.34
-0.32	-0.67405	-0.64630	-0.60276	-0.56267	-0.51342	-0.42402	0.32
-0.30	-0.66197	-0.63341	-0.58870	-0.54759	-0.49719	-0.40599	0.30
-0.28	-0.64971	-0.62035	-0.57446	-0.53235	-0.48083	-0.38788	0.28
-0.26	-0.63725	-0.60711	-0.56005	-0.51696	-0.46433	-0.36970	0.26
-0.24	-0.62460	-0.59367	-0.54546	-0.50140	-0.44770	-0.35144	0.24
-0.22	-0.61176	-0.58004	-0.53069	-0.48568	-0.43093	-0.33310	0.22
-0.20	-0.59870	-0.56621	-0.51574	-0.46979	-0.41402	-0.31468	0.20
-0.18	-0.58544	-0.55217	-0.50059	-0.45373	-0.39697	-0.29618	0.18
-0.16	-0.57196	-0.53793	-0.48526	-0.43749	-0.37977	-0.27760	0.16
-0.14	-0.55826	-0.52347	-0.46972	-0.42108	-0.36242	-0.25894	0.14
-0.12	-0.54434	-0.50879	-0.45398	-0.40449	-0.34492	-0.24019	0.12
-0.10	-0.53017	-0.49389	-0.43804	-0.38771	-0.32727	-0.22135	0.10
-0.08	-0.51577	-0.47876	-0.42188	-0.37074	-0.30946	-0.20243	0.08
-0.06	-0.50113	-0.46339	-0.40551	-0.35358	-0.29150	-0.18342	0.06
-0.04	-0.48622	-0.44777	-0.38892	-0.33622	-0.27337	-0.16431	0.04
-0.02	-0.47106	-0.43191	-0.37210	-0.31866	-0.25507	-0.14512	0.02

↑ r_0

α =	.995 ↑-	.99 ↑-	.975 ↑-	.95 ↑-	.90 ↑-	.75 ↑-

TABLE 28.2   SAMPLE SIZE = 30 (CONT.)

α =	.005 ↓+	.01 ↓+	.025 ↓+	.05 ↓+	.10 ↓+	.25 ↓+	
r_0 ↓							
0.00	-0.45563	-0.41579	-0.35505	-0.30090	-0.23661	-0.12583	0.00
0.02	-0.43992	-0.39941	-0.33776	-0.28292	-0.21798	-0.10645	-0.02
0.04	-0.42393	-0.38276	-0.32022	-0.26474	-0.19917	-0.08697	-0.04
0.06	-0.40765	-0.36583	-0.30244	-0.24633	-0.18019	-0.06739	-0.06
0.08	-0.39106	-0.34861	-0.28441	-0.22770	-0.16103	-0.04772	-0.08
0.10	-0.37417	-0.33110	-0.26611	-0.20885	-0.14168	-0.02794	-0.10
0.12	-0.35695	-0.31329	-0.24754	-0.18976	-0.12214	-0.00806	-0.12
0.14	-0.33940	-0.29517	-0.22870	-0.17043	-0.10242	0.01193	-0.14
0.16	-0.32151	-0.27673	-0.20958	-0.15086	-0.08249	0.03201	-0.16
0.18	-0.30327	-0.25796	-0.19017	-0.13104	-0.06237	0.05221	-0.18
0.20	-0.28467	-0.23885	-0.17046	-0.11096	-0.04205	0.07252	-0.20
0.22	-0.26570	-0.21939	-0.15045	-0.09063	-0.02152	0.09294	-0.22
0.24	-0.24633	-0.19958	-0.13013	-0.07002	-0.00078	0.11347	-0.24
0.26	-0.22657	-0.17939	-0.10948	-0.04915	0.02017	0.13411	-0.26
0.28	-0.20639	-0.15882	-0.08851	-0.02799	0.04135	0.15488	-0.28
0.30	-0.18579	-0.13786	-0.06720	-0.00655	0.06275	0.17576	-0.30
0.32	-0.16474	-0.11649	-0.04554	0.01518	0.08437	0.19676	-0.32
0.34	-0.14323	-0.09470	-0.02352	0.03722	0.10623	0.21789	-0.34
0.36	-0.12125	-0.07248	-0.00114	0.05956	0.12833	0.23914	-0.36
0.38	-0.09878	-0.04981	0.02163	0.08222	0.15067	0.26051	-0.38
0.40	-0.07580	-0.02668	0.04478	0.10520	0.17326	0.28202	-0.40
0.42	-0.05229	-0.00307	0.06833	0.12852	0.19610	0.30365	-0.42
0.44	-0.02824	0.02103	0.09229	0.15217	0.21920	0.32543	-0.44
0.46	-0.00361	0.04565	0.11668	0.17617	0.24256	0.34733	-0.46
0.48	0.02161	0.07079	0.14150	0.20053	0.26620	0.36937	-0.48
0.50	0.04744	0.09648	0.16677	0.22526	0.29011	0.39156	-0.50
0.52	0.07391	0.12274	0.19251	0.25037	0.31430	0.41389	-0.52
0.54	0.10104	0.14959	0.21873	0.27586	0.33879	0.43636	-0.54
0.56	0.12887	0.17705	0.24544	0.30175	0.36356	0.45898	-0.56
0.58	0.15742	0.20515	0.27266	0.32805	0.38864	0.48175	-0.58
0.60	0.18672	0.23390	0.30041	0.35477	0.41403	0.50467	-0.60
0.62	0.21680	0.26333	0.32870	0.38192	0.43974	0.52775	-0.62
0.64	0.24770	0.29348	0.35755	0.40952	0.46577	0.55099	-0.64
0.66	0.27946	0.32436	0.38698	0.43757	0.49213	0.57439	-0.66
0.68	0.31210	0.35602	0.41701	0.46610	0.51883	0.59795	-0.68
0.70	0.34569	0.38847	0.44767	0.49511	0.54588	0.62168	-0.70
0.72	0.38025	0.42176	0.47897	0.52462	0.57329	0.64559	-0.72
0.74	0.41583	0.45592	0.51093	0.55464	0.60106	0.66966	-0.74
0.76	0.45249	0.49098	0.54358	0.58520	0.62920	0.69391	-0.76
0.78	0.49027	0.52700	0.57695	0.61630	0.65773	0.71834	-0.78
0.80	0.52924	0.56400	0.61106	0.64796	0.68666	0.74296	-0.80
0.82	0.56946	0.60203	0.64595	0.68021	0.71598	0.76776	-0.82
0.84	0.61098	0.64115	0.68163	0.71305	0.74572	0.79275	-0.84
0.86	0.65389	0.68141	0.71814	0.74652	0.77589	0.81794	-0.86
0.88	0.69825	0.72285	0.75552	0.78063	0.80650	0.84332	-0.88
0.90	0.74415	0.76553	0.79379	0.81539	0.83755	0.86891	-0.90
0.92	0.79167	0.80953	0.83300	0.85085	0.86907	0.89470	-0.92
0.94	0.84091	0.85490	0.87318	0.88701	0.90105	0.92070	-0.94
0.96	0.89197	0.90172	0.91438	0.92391	0.93353	0.94691	-0.96
0.98	0.94496	0.95005	0.95664	0.96156	0.96651	0.97335	-0.98

↑ r_0

α =	.995 ↑-	.99 ↑-	.975 ↑-	.95 ↑-	.90 ↑-	.75 ↑-

TABLE 29.1　SAMPLE SIZE = 31

α =	.005 ↓+	.01 ↓+	.025 ↓+	.05 ↓+	.10 ↓+	.25 ↓+	
r_0 ↓							
−0.98	−0.99212	−0.99133	−0.99002	−0.98876	−0.98712	−0.98386	0.98
−0.96	−0.98416	−0.98256	−0.97995	−0.97743	−0.97416	−0.96768	0.96
−0.94	−0.97610	−0.97370	−0.96979	−0.96601	−0.96112	−0.95146	0.94
−0.92	−0.96795	−0.96475	−0.95953	−0.95450	−0.94800	−0.93519	0.92
−0.90	−0.95971	−0.95570	−0.94918	−0.94290	−0.93480	−0.91888	0.90
−0.88	−0.95137	−0.94656	−0.93873	−0.93120	−0.92151	−0.90252	0.88
−0.86	−0.94293	−0.93731	−0.92818	−0.91941	−0.90814	−0.88612	0.86
−0.84	−0.93439	−0.92796	−0.91753	−0.90753	−0.89469	−0.86967	0.84
−0.82	−0.92575	−0.91851	−0.90678	−0.89555	−0.88115	−0.85317	0.82
−0.80	−0.91701	−0.90895	−0.89592	−0.88347	−0.86752	−0.83662	0.80
−0.78	−0.90816	−0.89929	−0.88496	−0.87128	−0.85381	−0.82002	0.78
−0.76	−0.89920	−0.88952	−0.87389	−0.85900	−0.84000	−0.80336	0.76
−0.74	−0.89013	−0.87963	−0.86271	−0.84661	−0.82610	−0.78666	0.74
−0.72	−0.88094	−0.86963	−0.85142	−0.83412	−0.81211	−0.76991	0.72
−0.70	−0.87164	−0.85952	−0.84002	−0.82152	−0.79803	−0.75310	0.70
−0.68	−0.86223	−0.84929	−0.82850	−0.80881	−0.78385	−0.73624	0.68
−0.66	−0.85269	−0.83893	−0.81686	−0.79598	−0.76957	−0.71932	0.66
−0.64	−0.84303	−0.82846	−0.80511	−0.78305	−0.75519	−0.70235	0.64
−0.62	−0.83325	−0.81785	−0.79323	−0.77000	−0.74072	−0.68532	0.62
−0.60	−0.82333	−0.80712	−0.78122	−0.75684	−0.72614	−0.66823	0.60
−0.58	−0.81329	−0.79626	−0.76910	−0.74355	−0.71146	−0.65108	0.58
−0.56	−0.80311	−0.78527	−0.75684	−0.73015	−0.69668	−0.63387	0.56
−0.54	−0.79280	−0.77414	−0.74445	−0.71662	−0.68178	−0.61661	0.54
−0.52	−0.78234	−0.76287	−0.73192	−0.70297	−0.66679	−0.59928	0.52
−0.50	−0.77174	−0.75146	−0.71926	−0.68919	−0.65168	−0.58189	0.50
−0.48	−0.76100	−0.73990	−0.70647	−0.67529	−0.63646	−0.56444	0.48
−0.46	−0.75011	−0.72820	−0.69353	−0.66125	−0.62113	−0.54692	0.46
−0.44	−0.73906	−0.71634	−0.68044	−0.64708	−0.60568	−0.52933	0.44
−0.42	−0.72786	−0.70433	−0.66721	−0.63277	−0.59012	−0.51168	0.42
−0.40	−0.71650	−0.69216	−0.65383	−0.61832	−0.57443	−0.49396	0.40
−0.38	−0.70497	−0.67983	−0.64029	−0.60374	−0.55863	−0.47618	0.38
−0.36	−0.69327	−0.66734	−0.62660	−0.58901	−0.54271	−0.45832	0.36
−0.34	−0.68141	−0.65468	−0.61275	−0.57413	−0.52666	−0.44040	0.34
−0.32	−0.66937	−0.64184	−0.59874	−0.55910	−0.51048	−0.42240	0.32
−0.30	−0.65714	−0.62883	−0.58456	−0.54393	−0.49418	−0.40433	0.30
−0.28	−0.64473	−0.61563	−0.57021	−0.52860	−0.47775	−0.38618	0.28
−0.26	−0.63214	−0.60226	−0.55569	−0.51311	−0.46118	−0.36796	0.26
−0.24	−0.61934	−0.58869	−0.54099	−0.49746	−0.44448	−0.34966	0.24
−0.22	−0.60635	−0.57493	−0.52611	−0.48165	−0.42764	−0.33129	0.22
−0.20	−0.59316	−0.56097	−0.51105	−0.46567	−0.41066	−0.31284	0.20
−0.18	−0.57975	−0.54681	−0.49580	−0.44952	−0.39355	−0.29430	0.18
−0.16	−0.56613	−0.53244	−0.48036	−0.43320	−0.37628	−0.27569	0.16
−0.14	−0.55229	−0.51785	−0.46472	−0.41671	−0.35887	−0.25699	0.14
−0.12	−0.53822	−0.50305	−0.44888	−0.40003	−0.34131	−0.23821	0.12
−0.10	−0.52392	−0.48802	−0.43284	−0.38317	−0.32360	−0.21935	0.10
−0.08	−0.50938	−0.47277	−0.41658	−0.36612	−0.30574	−0.20039	0.08
−0.06	−0.49460	−0.45728	−0.40011	−0.34889	−0.28772	−0.18135	0.06
−0.04	−0.47956	−0.44154	−0.38342	−0.33145	−0.26954	−0.16223	0.04
−0.02	−0.46426	−0.42556	−0.36651	−0.31382	−0.25119	−0.14301	0.02

↑ r_0

α =	.995 ↑−	.99 ↑−	.975 ↑−	.95 ↑−	.90 ↑−	.75 ↑−

TABLE 29.2 SAMPLE SIZE = 31 (CONT.)

α =	.005 ↓+	.01 ↓+	.025 ↓+	.05 ↓+	.10 ↓+	.25 ↓+	
r_0 ↓							
0.00	−0.44870	−0.40933	−0.34937	−0.29599	−0.23268	−0.12370	0.00
0.02	−0.43286	−0.39283	−0.33199	−0.27795	−0.21400	−0.10429	−0.02
0.04	−0.41674	−0.37607	−0.31438	−0.25970	−0.19515	−0.08479	−0.04
0.06	−0.40033	−0.35903	−0.29651	−0.24123	−0.17613	−0.06520	−0.06
0.08	−0.38361	−0.34170	−0.27839	−0.22254	−0.15692	−0.04550	−0.08
0.10	−0.36659	−0.32409	−0.26002	−0.20363	−0.13754	−0.02571	−0.10
0.12	−0.34926	−0.30618	−0.24138	−0.18449	−0.11797	−0.00582	−0.12
0.14	−0.33159	−0.28796	−0.22247	−0.16511	−0.09821	0.01418	−0.14
0.16	−0.31359	−0.26942	−0.20328	−0.14549	−0.07826	0.03428	−0.16
0.18	−0.29524	−0.25056	−0.18381	−0.12563	−0.05812	0.05448	−0.18
0.20	−0.27653	−0.23137	−0.16404	−0.10552	−0.03777	0.07480	−0.20
0.22	−0.25744	−0.21183	−0.14397	−0.08514	−0.01722	0.09522	−0.22
0.24	−0.23798	−0.19193	−0.12360	−0.06451	0.00353	0.11575	−0.24
0.26	−0.21812	−0.17167	−0.10291	−0.04361	0.02450	0.13640	−0.26
0.28	−0.19785	−0.15104	−0.08189	−0.02243	0.04568	0.15716	−0.28
0.30	−0.17716	−0.13001	−0.06055	−0.00097	0.06708	0.17804	−0.30
0.32	−0.15604	−0.10859	−0.03886	0.02078	0.08870	0.19904	−0.32
0.34	−0.13446	−0.08675	−0.01682	0.04282	0.11056	0.22015	−0.34
0.36	−0.11241	−0.06448	0.00558	0.06516	0.13265	0.24139	−0.36
0.38	−0.08989	−0.04178	0.02836	0.08782	0.15497	0.26275	−0.38
0.40	−0.06685	−0.01862	0.05151	0.11079	0.17754	0.28424	−0.40
0.42	−0.04330	0.00501	0.07506	0.13409	0.20036	0.30586	−0.42
0.44	−0.01922	0.02913	0.09901	0.15772	0.22343	0.32761	−0.44
0.46	0.00543	0.05374	0.12337	0.18169	0.24676	0.34949	−0.46
0.48	0.03066	0.07888	0.14817	0.20601	0.27035	0.37151	−0.48
0.50	0.05649	0.10455	0.17340	0.23069	0.29422	0.39366	−0.50
0.52	0.08295	0.13078	0.19909	0.25574	0.31836	0.41596	−0.52
0.54	0.11006	0.15758	0.22525	0.28117	0.34278	0.43839	−0.54
0.56	0.13784	0.18499	0.25189	0.30699	0.36749	0.46097	−0.56
0.58	0.16633	0.21301	0.27903	0.33321	0.39250	0.48370	−0.58
0.60	0.19555	0.24167	0.30668	0.35984	0.41781	0.50657	−0.60
0.62	0.22554	0.27100	0.33486	0.38689	0.44343	0.52960	−0.62
0.64	0.25633	0.30103	0.36359	0.41437	0.46936	0.55279	−0.64
0.66	0.28794	0.33177	0.39289	0.44230	0.49562	0.57613	−0.66
0.68	0.32043	0.36326	0.42277	0.47069	0.52221	0.59963	−0.68
0.70	0.35383	0.39553	0.45325	0.49955	0.54914	0.62329	−0.70
0.72	0.38817	0.42861	0.48436	0.52889	0.57641	0.64712	−0.72
0.74	0.42351	0.46253	0.51612	0.55874	0.60404	0.67112	−0.74
0.76	0.45989	0.49733	0.54854	0.58909	0.63203	0.69530	−0.76
0.78	0.49736	0.53305	0.58166	0.61998	0.66040	0.71964	−0.78
0.80	0.53597	0.56973	0.61550	0.65142	0.68915	0.74417	−0.80
0.82	0.57579	0.60741	0.65008	0.68342	0.71829	0.76887	−0.82
0.84	0.61687	0.64613	0.68544	0.71600	0.74783	0.79377	−0.84
0.86	0.65928	0.68594	0.72160	0.74918	0.77779	0.81885	−0.86
0.88	0.70308	0.72690	0.75859	0.78298	0.80817	0.84412	−0.88
0.90	0.74837	0.76906	0.79645	0.81742	0.83899	0.86959	−0.90
0.92	0.79521	0.81247	0.83520	0.85253	0.87025	0.89526	−0.92
0.94	0.84369	0.85720	0.87490	0.88831	0.90197	0.92113	−0.94
0.96	0.89391	0.90332	0.91557	0.92480	0.93416	0.94721	−0.96
0.98	0.94598	0.95089	0.95726	0.96202	0.96683	0.97350	−0.98

↑ r_0

α =	.995 ↑−	.99 ↑−	.975 ↑−	.95 ↑−	.90 ↑−	.75 ↑−

TABLE 30.1   SAMPLE SIZE =   32

α =	.005 ↓+	.01 ↓+	.025 ↓+	.05 ↓+	.10 ↓+	.25 ↓+	
r₀ ↓							
-0.98	-0.99200	-0.99120	-0.98991	-0.98865	-0.98703	-0.98381	0.98
-0.96	-0.98391	-0.98232	-0.97972	-0.97722	-0.97398	-0.96759	0.96
-0.94	-0.97573	-0.97334	-0.96945	-0.96570	-0.96085	-0.95131	0.94
-0.92	-0.96745	-0.96426	-0.95907	-0.95408	-0.94764	-0.93500	0.92
-0.90	-0.95908	-0.95509	-0.94861	-0.94238	-0.93435	-0.91864	0.90
-0.88	-0.95062	-0.94582	-0.93804	-0.93058	-0.92098	-0.90223	0.88
-0.86	-0.94205	-0.93645	-0.92738	-0.91869	-0.90752	-0.88577	0.86
-0.84	-0.93339	-0.92698	-0.91662	-0.90670	-0.89398	-0.86927	0.84
-0.82	-0.92462	-0.91741	-0.90575	-0.89461	-0.88036	-0.85272	0.82
-0.80	-0.91574	-0.90773	-0.89478	-0.88243	-0.86664	-0.83613	0.80
-0.78	-0.90676	-0.89794	-0.88370	-0.87014	-0.85284	-0.81948	0.78
-0.76	-0.89767	-0.88804	-0.87252	-0.85775	-0.83895	-0.80278	0.76
-0.74	-0.88847	-0.87803	-0.86123	-0.84526	-0.82496	-0.78603	0.74
-0.72	-0.87916	-0.86790	-0.84982	-0.83266	-0.81089	-0.76922	0.72
-0.70	-0.86972	-0.85766	-0.83830	-0.81996	-0.79672	-0.75237	0.70
-0.68	-0.86018	-0.84730	-0.82667	-0.80715	-0.78245	-0.73546	0.68
-0.66	-0.85051	-0.83682	-0.81492	-0.79423	-0.76809	-0.71850	0.66
-0.64	-0.84071	-0.82622	-0.80304	-0.78119	-0.75363	-0.70148	0.64
-0.62	-0.83079	-0.81549	-0.79105	-0.76804	-0.73907	-0.68440	0.62
-0.60	-0.82075	-0.80463	-0.77893	-0.75477	-0.72442	-0.66727	0.60
-0.58	-0.81057	-0.79365	-0.76669	-0.74139	-0.70965	-0.65007	0.58
-0.56	-0.80025	-0.78252	-0.75432	-0.72789	-0.69479	-0.63282	0.56
-0.54	-0.78980	-0.77127	-0.74181	-0.71426	-0.67982	-0.61551	0.54
-0.52	-0.77921	-0.75987	-0.72918	-0.70051	-0.66474	-0.59814	0.52
-0.50	-0.76848	-0.74833	-0.71641	-0.68664	-0.64955	-0.58071	0.50
-0.48	-0.75760	-0.73665	-0.70349	-0.67263	-0.63425	-0.56321	0.48
-0.46	-0.74657	-0.72482	-0.69044	-0.65850	-0.61884	-0.54565	0.46
-0.44	-0.73539	-0.71283	-0.67725	-0.64423	-0.60332	-0.52802	0.44
-0.42	-0.72405	-0.70069	-0.66390	-0.62982	-0.58768	-0.51033	0.42
-0.40	-0.71255	-0.68840	-0.65041	-0.61528	-0.57192	-0.49257	0.40
-0.38	-0.70088	-0.67594	-0.63676	-0.60060	-0.55605	-0.47475	0.38
-0.36	-0.68905	-0.66332	-0.62296	-0.58578	-0.54005	-0.45685	0.36
-0.34	-0.67705	-0.65053	-0.60901	-0.57081	-0.52393	-0.43889	0.34
-0.32	-0.66487	-0.63757	-0.59488	-0.55569	-0.50768	-0.42085	0.32
-0.30	-0.65251	-0.62443	-0.58060	-0.54043	-0.49131	-0.40274	0.30
-0.28	-0.63996	-0.61111	-0.56614	-0.52501	-0.47481	-0.38456	0.28
-0.26	-0.62723	-0.59761	-0.55152	-0.50943	-0.45818	-0.36630	0.26
-0.24	-0.61430	-0.58392	-0.53672	-0.49370	-0.44141	-0.34797	0.24
-0.22	-0.60117	-0.57004	-0.52173	-0.47780	-0.42451	-0.32956	0.22
-0.20	-0.58784	-0.55595	-0.50657	-0.46174	-0.40747	-0.31108	0.20
-0.18	-0.57430	-0.54167	-0.49122	-0.44551	-0.39028	-0.29251	0.18
-0.16	-0.56055	-0.52718	-0.47568	-0.42911	-0.37296	-0.27387	0.16
-0.14	-0.54657	-0.51248	-0.45994	-0.41253	-0.35549	-0.25514	0.14
-0.12	-0.53237	-0.49755	-0.44401	-0.39578	-0.33788	-0.23633	0.12
-0.10	-0.51794	-0.48241	-0.42787	-0.37884	-0.32011	-0.21744	0.10
-0.08	-0.50327	-0.46704	-0.41152	-0.36172	-0.30219	-0.19846	0.08
-0.06	-0.48835	-0.45143	-0.39496	-0.34441	-0.28412	-0.17939	0.06
-0.04	-0.47319	-0.43559	-0.37818	-0.32691	-0.26589	-0.16024	0.04
-0.02	-0.45776	-0.41950	-0.36118	-0.30921	-0.24750	-0.14100	0.02

↑ r₀

α =	.995 ↑-	.99 ↑-	.975 ↑-	.95 ↑-	.90 ↑-	.75 ↑-

TABLE 30.2   SAMPLE SIZE =   32   (CONT.)

α =	.005 ↓+	.01 ↓+	.025 ↓+	.05 ↓+	.10 ↓+	.25 ↓+	
r_0 ↓							
0.00	-0.44207	-0.40315	-0.34396	-0.29132	-0.22894	-0.12167	0.00
0.02	-0.42611	-0.38655	-0.32650	-0.27321	-0.21022	-0.10224	-0.02
0.04	-0.40986	-0.36968	-0.30880	-0.25490	-0.19133	-0.08273	-0.04
0.06	-0.39333	-0.35254	-0.29086	-0.23638	-0.17226	-0.06311	-0.06
0.08	-0.37650	-0.33512	-0.27267	-0.21763	-0.15302	-0.04340	-0.08
0.10	-0.35937	-0.31740	-0.25422	-0.19867	-0.13360	-0.02360	-0.10
0.12	-0.34191	-0.29940	-0.23551	-0.17948	-0.11400	-0.00369	-0.12
0.14	-0.32414	-0.28109	-0.21654	-0.16005	-0.09422	0.01632	-0.14
0.16	-0.30603	-0.26246	-0.19728	-0.14039	-0.07424	0.03643	-0.16
0.18	-0.28757	-0.24352	-0.17775	-0.12049	-0.05407	0.05664	-0.18
0.20	-0.26876	-0.22424	-0.15793	-0.10034	-0.03371	0.07696	-0.20
0.22	-0.24958	-0.20463	-0.13781	-0.07993	-0.01314	0.09739	-0.22
0.24	-0.23003	-0.18466	-0.11739	-0.05927	0.00762	0.11792	-0.24
0.26	-0.21008	-0.16433	-0.09666	-0.03834	0.02860	0.13857	-0.26
0.28	-0.18973	-0.14363	-0.07561	-0.01715	0.04979	0.15933	-0.28
0.30	-0.16896	-0.12255	-0.05423	0.00433	0.07119	0.18020	-0.30
0.32	-0.14776	-0.10107	-0.03252	0.02608	0.09281	0.20119	-0.32
0.34	-0.12612	-0.07919	-0.01046	0.04813	0.11466	0.22230	-0.34
0.36	-0.10402	-0.05689	0.01196	0.07048	0.13674	0.24353	-0.36
0.38	-0.08144	-0.03415	0.03474	0.09313	0.15905	0.26488	-0.38
0.40	-0.05836	-0.01097	0.05789	0.11609	0.18160	0.28635	-0.40
0.42	-0.03477	0.01268	0.08144	0.13936	0.20439	0.30795	-0.42
0.44	-0.01066	0.03680	0.10537	0.16297	0.22744	0.32968	-0.44
0.46	0.01401	0.06141	0.12972	0.18691	0.25073	0.35154	-0.46
0.48	0.03924	0.08654	0.15448	0.21120	0.27429	0.37353	-0.48
0.50	0.06507	0.11219	0.17968	0.23583	0.29811	0.39565	-0.50
0.52	0.09151	0.13838	0.20532	0.26083	0.32220	0.41791	-0.52
0.54	0.11858	0.16514	0.23142	0.28620	0.34656	0.44031	-0.54
0.56	0.14632	0.19249	0.25799	0.31195	0.37121	0.46286	-0.56
0.58	0.17475	0.22043	0.28505	0.33808	0.39615	0.48554	-0.58
0.60	0.20390	0.24901	0.31261	0.36462	0.42138	0.50837	-0.60
0.62	0.23379	0.27824	0.34069	0.39158	0.44691	0.53135	-0.62
0.64	0.26446	0.30814	0.36930	0.41895	0.47276	0.55448	-0.64
0.66	0.29595	0.33875	0.39846	0.44676	0.49891	0.57777	-0.66
0.68	0.32828	0.37008	0.42819	0.47501	0.52540	0.60121	-0.68
0.70	0.36149	0.40217	0.45851	0.50373	0.55221	0.62481	-0.70
0.72	0.39563	0.43505	0.48944	0.53292	0.57936	0.64858	-0.72
0.74	0.43073	0.46875	0.52100	0.56259	0.60685	0.67250	-0.74
0.76	0.46684	0.50330	0.55321	0.59276	0.63470	0.69660	-0.76
0.78	0.50401	0.53874	0.58609	0.62345	0.66291	0.72087	-0.78
0.80	0.54228	0.57511	0.61966	0.65468	0.69150	0.74531	-0.80
0.82	0.58172	0.61245	0.65396	0.68644	0.72046	0.76992	-0.82
0.84	0.62238	0.65079	0.68901	0.71877	0.74982	0.79472	-0.84
0.86	0.66432	0.69019	0.72484	0.75169	0.77958	0.81970	-0.86
0.88	0.70760	0.73070	0.76147	0.78520	0.80974	0.84487	-0.88
0.90	0.75231	0.77235	0.79893	0.81933	0.84033	0.87023	-0.90
0.92	0.79851	0.81522	0.83727	0.85410	0.87135	0.89578	-0.92
0.94	0.84629	0.85935	0.87650	0.88953	0.90282	0.92153	-0.94
0.96	0.89573	0.90482	0.91668	0.92564	0.93474	0.94748	-0.96
0.98	0.94693	0.95167	0.95783	0.96246	0.96713	0.97364	-0.98

↑ r_0

α =	.995 ↑−	.99 ↑−	.975 ↑−	.95 ↑−	.90 ↑−	.75 ↑−

TABLE 31.1   SAMPLE SIZE =   33

α =	.005 ↓+	.01 ↓+	.025 ↓+	.05 ↓+	.10 ↓+	.25 ↓+	
r_0 ↓							
-0.98	-0.99188	-0.99109	-0.98980	-0.98855	-0.98694	-0.98377	0.98
-0.96	-0.98367	-0.98208	-0.97950	-0.97702	-0.97381	-0.96749	0.96
-0.94	-0.97537	-0.97299	-0.96912	-0.96539	-0.96060	-0.95117	0.94
-0.92	-0.96697	-0.96380	-0.95864	-0.95368	-0.94730	-0.93481	0.92
-0.90	-0.95848	-0.95451	-0.94806	-0.94188	-0.93393	-0.91840	0.90
-0.88	-0.94989	-0.94512	-0.93738	-0.92998	-0.92047	-0.90195	0.88
-0.86	-0.94120	-0.93563	-0.92661	-0.91798	-0.90693	-0.88544	0.86
-0.84	-0.93241	-0.92604	-0.91574	-0.90590	-0.89331	-0.86890	0.84
-0.82	-0.92352	-0.91634	-0.90476	-0.89371	-0.87960	-0.85230	0.82
-0.80	-0.91452	-0.90654	-0.89368	-0.88143	-0.86580	-0.83566	0.80
-0.78	-0.90542	-0.89663	-0.88249	-0.86904	-0.85191	-0.81896	0.78
-0.76	-0.89620	-0.88662	-0.87120	-0.85655	-0.83794	-0.80222	0.76
-0.74	-0.88687	-0.87648	-0.85979	-0.84396	-0.82387	-0.78542	0.74
-0.72	-0.87743	-0.86624	-0.84828	-0.83127	-0.80972	-0.76857	0.72
-0.70	-0.86787	-0.85588	-0.83665	-0.81847	-0.79546	-0.75167	0.70
-0.68	-0.85820	-0.84540	-0.82491	-0.80556	-0.78112	-0.73472	0.68
-0.66	-0.84840	-0.83480	-0.81304	-0.79254	-0.76668	-0.71771	0.66
-0.64	-0.83848	-0.82407	-0.80106	-0.77940	-0.75214	-0.70064	0.64
-0.62	-0.82843	-0.81322	-0.78896	-0.76616	-0.73750	-0.68352	0.62
-0.60	-0.81826	-0.80224	-0.77673	-0.75280	-0.72276	-0.66635	0.60
-0.58	-0.80795	-0.79113	-0.76438	-0.73932	-0.70792	-0.64911	0.58
-0.56	-0.79750	-0.77989	-0.75190	-0.72572	-0.69298	-0.63182	0.56
-0.54	-0.78692	-0.76851	-0.73929	-0.71200	-0.67793	-0.61446	0.54
-0.52	-0.77620	-0.75699	-0.72654	-0.69815	-0.66278	-0.59705	0.52
-0.50	-0.76534	-0.74533	-0.71366	-0.68418	-0.64751	-0.57957	0.50
-0.48	-0.75433	-0.73352	-0.70064	-0.67009	-0.63214	-0.56204	0.48
-0.46	-0.74317	-0.72157	-0.68748	-0.65586	-0.61666	-0.54443	0.46
-0.44	-0.73186	-0.70946	-0.67418	-0.64150	-0.60106	-0.52677	0.44
-0.42	-0.72038	-0.69720	-0.66073	-0.62700	-0.58535	-0.50904	0.42
-0.40	-0.70875	-0.68478	-0.64713	-0.61237	-0.56952	-0.49124	0.40
-0.38	-0.69696	-0.67221	-0.63338	-0.59760	-0.55358	-0.47338	0.38
-0.36	-0.68500	-0.65946	-0.61948	-0.58269	-0.53751	-0.45544	0.36
-0.34	-0.67286	-0.64655	-0.60542	-0.56764	-0.52132	-0.43744	0.34
-0.32	-0.66055	-0.63347	-0.59119	-0.55243	-0.50501	-0.41937	0.32
-0.30	-0.64806	-0.62021	-0.57681	-0.53708	-0.48857	-0.40123	0.30
-0.28	-0.63538	-0.60678	-0.56225	-0.52158	-0.47200	-0.38301	0.28
-0.26	-0.62252	-0.59316	-0.54752	-0.50592	-0.45530	-0.36472	0.26
-0.24	-0.60946	-0.57935	-0.53262	-0.49010	-0.43847	-0.34636	0.24
-0.22	-0.59620	-0.56535	-0.51754	-0.47412	-0.42151	-0.32792	0.22
-0.20	-0.58274	-0.55115	-0.50228	-0.45798	-0.40441	-0.30940	0.20
-0.18	-0.56907	-0.53675	-0.48684	-0.44167	-0.38717	-0.29080	0.18
-0.16	-0.55519	-0.52214	-0.47120	-0.42520	-0.36979	-0.27213	0.16
-0.14	-0.54109	-0.50733	-0.45537	-0.40855	-0.35227	-0.25338	0.14
-0.12	-0.52676	-0.49229	-0.43935	-0.39172	-0.33460	-0.23454	0.12
-0.10	-0.51220	-0.47704	-0.42312	-0.37471	-0.31678	-0.21562	0.10
-0.08	-0.49741	-0.46156	-0.40668	-0.35752	-0.29881	-0.19662	0.08
-0.06	-0.48237	-0.44584	-0.39004	-0.34015	-0.28069	-0.17753	0.06
-0.04	-0.46708	-0.42989	-0.37318	-0.32258	-0.26241	-0.15835	0.04
-0.02	-0.45154	-0.41370	-0.35609	-0.30482	-0.24397	-0.13909	0.02
							↑ r_0

α =	.995 ↑−	.99 ↑−	.975 ↑−	.95 ↑−	.90 ↑−	.75 ↑−

TABLE 31.2   SAMPLE SIZE =   33   (CONT.)

α =	.005 ↓+	.01 ↓+	.025 ↓+	.05 ↓+	.10 ↓+	.25 ↓+	
r_0 ↓							
0.00	−0.43573	−0.39725	−0.33879	−0.28686	−0.22537	−0.11974	0.00
0.02	−0.41965	−0.38054	−0.32125	−0.26869	−0.20661	−0.10029	−0.02
0.04	−0.40329	−0.36358	−0.30348	−0.25032	−0.18768	−0.08076	−0.04
0.06	−0.38664	−0.34634	−0.28546	−0.23174	−0.16858	−0.06113	−0.06
0.08	−0.36970	−0.32882	−0.26720	−0.21295	−0.14931	−0.04140	−0.08
0.10	−0.35246	−0.31102	−0.24869	−0.19394	−0.12986	−0.02158	−0.10
0.12	−0.33490	−0.29293	−0.22992	−0.17470	−0.11023	−0.00167	−0.12
0.14	−0.31702	−0.27453	−0.21088	−0.15523	−0.09041	0.01835	−0.14
0.16	−0.29881	−0.25582	−0.19157	−0.13553	−0.07041	0.03847	−0.16
0.18	−0.28026	−0.23680	−0.17198	−0.11559	−0.05022	0.05869	−0.18
0.20	−0.26135	−0.21745	−0.15211	−0.09541	−0.02984	0.07902	−0.20
0.22	−0.24208	−0.19776	−0.13195	−0.07497	−0.00926	0.09945	−0.22
0.24	−0.22244	−0.17773	−0.11149	−0.05429	0.01152	0.11998	−0.24
0.26	−0.20241	−0.15734	−0.09072	−0.03334	0.03250	0.14063	−0.26
0.28	−0.18199	−0.13658	−0.06963	−0.01212	0.05369	0.16139	−0.28
0.30	−0.16115	−0.11545	−0.04823	0.00936	0.07510	0.18226	−0.30
0.32	−0.13989	−0.09393	−0.02649	0.03113	0.09672	0.20324	−0.32
0.34	−0.11818	−0.07200	−0.00441	0.05318	0.11856	0.22434	−0.34
0.36	−0.09603	−0.04967	0.01802	0.07552	0.14063	0.24556	−0.36
0.38	−0.07340	−0.02690	0.04080	0.09817	0.16292	0.26689	−0.38
0.40	−0.05028	−0.00370	0.06396	0.12111	0.18545	0.28835	−0.40
0.42	−0.02667	0.01996	0.08749	0.14437	0.20822	0.30993	−0.42
0.44	−0.00253	0.04408	0.11141	0.16795	0.23124	0.33164	−0.44
0.46	0.02215	0.06869	0.13574	0.19186	0.25450	0.35348	−0.46
0.48	0.04738	0.09380	0.16047	0.21611	0.27802	0.37544	−0.48
0.50	0.07320	0.11943	0.18563	0.24070	0.30179	0.39754	−0.50
0.52	0.09962	0.14559	0.21122	0.26565	0.32583	0.41977	−0.52
0.54	0.12667	0.17230	0.23726	0.29096	0.35014	0.44214	−0.54
0.56	0.15436	0.19959	0.26377	0.31664	0.37473	0.46464	−0.56
0.58	0.18273	0.22747	0.29075	0.34270	0.39960	0.48729	−0.58
0.60	0.21180	0.25596	0.31822	0.36915	0.42476	0.51008	−0.60
0.62	0.24160	0.28509	0.34619	0.39601	0.45021	0.53301	−0.62
0.64	0.27216	0.31487	0.37469	0.42328	0.47597	0.55609	−0.64
0.66	0.30351	0.34534	0.40373	0.45097	0.50203	0.57932	−0.66
0.68	0.33568	0.37653	0.43332	0.47910	0.52841	0.60271	−0.68
0.70	0.36872	0.40845	0.46348	0.50768	0.55511	0.62625	−0.70
0.72	0.40266	0.44113	0.49423	0.53672	0.58214	0.64995	−0.72
0.74	0.43753	0.47461	0.52560	0.56623	0.60951	0.67381	−0.74
0.76	0.47338	0.50893	0.55760	0.59623	0.63722	0.69783	−0.76
0.78	0.51026	0.54410	0.59026	0.62673	0.66528	0.72202	−0.78
0.80	0.54821	0.58017	0.62359	0.65774	0.69371	0.74638	−0.80
0.82	0.58729	0.61719	0.65762	0.68929	0.72251	0.77091	−0.82
0.84	0.62755	0.65517	0.69237	0.72138	0.75169	0.79562	−0.84
0.86	0.66904	0.69418	0.72788	0.75404	0.78126	0.82051	−0.86
0.88	0.71184	0.73426	0.76417	0.78728	0.81123	0.84558	−0.88
0.90	0.75600	0.77544	0.80127	0.82112	0.84160	0.87083	−0.90
0.92	0.80160	0.81780	0.83920	0.85558	0.87240	0.89627	−0.92
0.94	0.84871	0.86137	0.87801	0.89067	0.90363	0.92191	−0.94
0.96	0.89742	0.90622	0.91772	0.92643	0.93529	0.94774	−0.96
0.98	0.94782	0.95240	0.95837	0.96286	0.96741	0.97377	−0.98

↑ r_0

α =	.995 ↑−	.99 ↑−	.975 ↑−	.95 ↑−	.90 ↑−	.75 ↑−

TABLE 32.1   SAMPLE SIZE =   34

α =	.005 ↓+	.01 ↓+	.025 ↓+	.05 ↓+	.10 ↓+	.25 ↓+	
r_0 ↓							
−0.98	−0.99177	−0.99098	−0.98969	−0.98846	−0.98686	−0.98372	0.98
−0.96	−0.98344	−0.98186	−0.97929	−0.97683	−0.97364	−0.96740	0.96
−0.94	−0.97502	−0.97265	−0.96880	−0.96511	−0.96035	−0.95104	0.94
−0.92	−0.96651	−0.96334	−0.95821	−0.95330	−0.94698	−0.93463	0.92
−0.90	−0.95790	−0.95394	−0.94753	−0.94139	−0.93352	−0.91817	0.90
−0.88	−0.94919	−0.94444	−0.93675	−0.92940	−0.91998	−0.90167	0.88
−0.86	−0.94038	−0.93484	−0.92587	−0.91731	−0.90636	−0.88513	0.86
−0.84	−0.93147	−0.92513	−0.91489	−0.90513	−0.89266	−0.86854	0.84
−0.82	−0.92246	−0.91532	−0.90381	−0.89285	−0.87887	−0.85189	0.82
−0.80	−0.91334	−0.90540	−0.89262	−0.88047	−0.86499	−0.83520	0.80
−0.78	−0.90412	−0.89538	−0.88133	−0.86799	−0.85102	−0.81847	0.78
−0.76	−0.89478	−0.88524	−0.86993	−0.85540	−0.83697	−0.80168	0.76
−0.74	−0.88533	−0.87500	−0.85842	−0.84272	−0.82283	−0.78484	0.74
−0.72	−0.87577	−0.86464	−0.84680	−0.82993	−0.80859	−0.76795	0.72
−0.70	−0.86609	−0.85416	−0.83506	−0.81703	−0.79426	−0.75100	0.70
−0.68	−0.85629	−0.84356	−0.82321	−0.80403	−0.77984	−0.73400	0.68
−0.66	−0.84637	−0.83284	−0.81124	−0.79092	−0.76532	−0.71695	0.66
−0.64	−0.83632	−0.82200	−0.79916	−0.77769	−0.75070	−0.69985	0.64
−0.62	−0.82615	−0.81103	−0.78695	−0.76435	−0.73599	−0.68268	0.62
−0.60	−0.81585	−0.79993	−0.77462	−0.75090	−0.72118	−0.66546	0.60
−0.58	−0.80542	−0.78871	−0.76216	−0.73733	−0.70626	−0.64819	0.58
−0.56	−0.79485	−0.77735	−0.74957	−0.72363	−0.69125	−0.63085	0.56
−0.54	−0.78415	−0.76585	−0.73686	−0.70982	−0.67612	−0.61346	0.54
−0.52	−0.77330	−0.75421	−0.72401	−0.69589	−0.66090	−0.59601	0.52
−0.50	−0.76231	−0.74243	−0.71103	−0.68183	−0.64556	−0.57849	0.50
−0.48	−0.75118	−0.73051	−0.69790	−0.66764	−0.63012	−0.56091	0.48
−0.46	−0.73989	−0.71844	−0.68464	−0.65333	−0.61456	−0.54327	0.46
−0.44	−0.72845	−0.70622	−0.67124	−0.63888	−0.59890	−0.52557	0.44
−0.42	−0.71686	−0.69384	−0.65769	−0.62430	−0.58312	−0.50780	0.42
−0.40	−0.70510	−0.68131	−0.64399	−0.60958	−0.56722	−0.48997	0.40
−0.38	−0.69318	−0.66862	−0.63014	−0.59473	−0.55121	−0.47207	0.38
−0.36	−0.68109	−0.65576	−0.61614	−0.57973	−0.53508	−0.45410	0.36
−0.34	−0.66883	−0.64273	−0.60197	−0.56459	−0.51882	−0.43606	0.34
−0.32	−0.65639	−0.62954	−0.58765	−0.54931	−0.50245	−0.41796	0.32
−0.30	−0.64378	−0.61616	−0.57317	−0.53387	−0.48594	−0.39978	0.30
−0.28	−0.63098	−0.60261	−0.55852	−0.51829	−0.46931	−0.38153	0.28
−0.26	−0.61799	−0.58888	−0.54369	−0.50255	−0.45256	−0.36321	0.26
−0.24	−0.60481	−0.57496	−0.52870	−0.48666	−0.43567	−0.34481	0.24
−0.22	−0.59143	−0.56084	−0.51353	−0.47060	−0.41865	−0.32634	0.22
−0.20	−0.57784	−0.54654	−0.49818	−0.45439	−0.40149	−0.30780	0.20
−0.18	−0.56405	−0.53203	−0.48264	−0.43800	−0.38419	−0.28917	0.18
−0.16	−0.55005	−0.51731	−0.46691	−0.42145	−0.36676	−0.27047	0.16
−0.14	−0.53583	−0.50239	−0.45100	−0.40473	−0.34918	−0.25169	0.14
−0.12	−0.52138	−0.48725	−0.43488	−0.38784	−0.33146	−0.23283	0.12
−0.10	−0.50670	−0.47189	−0.41857	−0.37076	−0.31360	−0.21388	0.10
−0.08	−0.49179	−0.45630	−0.40205	−0.35350	−0.29558	−0.19486	0.08
−0.06	−0.47663	−0.44049	−0.38533	−0.33606	−0.27741	−0.17574	0.06
−0.04	−0.46123	−0.42444	−0.36839	−0.31843	−0.25909	−0.15655	0.04
−0.02	−0.44557	−0.40814	−0.35123	−0.30061	−0.24061	−0.13726	0.02

↑ r_0

α =	.995 ↑−	.99 ↑−	.975 ↑−	.95 ↑−	.90 ↑−	.75 ↑−

TABLE 32.2   SAMPLE SIZE = 34 (CONT.)

α =	.005 ↓+	.01 ↓+	.025 ↓+	.05 ↓+	.10 ↓+	.25 ↓+	
r₀ ↓							
0.00	-0.42965	-0.39160	-0.33384	-0.28259	-0.22197	-0.11789	0.00
0.02	-0.41346	-0.37480	-0.31623	-0.26438	-0.20317	-0.09843	-0.02
0.04	-0.39699	-0.35774	-0.29839	-0.24595	-0.18420	-0.07888	-0.04
0.06	-0.38024	-0.34041	-0.28031	-0.22732	-0.16507	-0.05924	-0.06
0.08	-0.36319	-0.32280	-0.26198	-0.20848	-0.14576	-0.03950	-0.08
0.10	-0.34584	-0.30491	-0.24340	-0.18942	-0.12628	-0.01966	-0.10
0.12	-0.32818	-0.28674	-0.22457	-0.17014	-0.10662	0.00026	-0.12
0.14	-0.31021	-0.26826	-0.20548	-0.15063	-0.08679	0.02029	-0.14
0.16	-0.29190	-0.24948	-0.18612	-0.13089	-0.06676	0.04042	-0.16
0.18	-0.27326	-0.23038	-0.16648	-0.11092	-0.04656	0.06064	-0.18
0.20	-0.25427	-0.21096	-0.14656	-0.09071	-0.02616	0.08097	-0.20
0.22	-0.23492	-0.19121	-0.12635	-0.07025	-0.00557	0.10141	-0.22
0.24	-0.21520	-0.17111	-0.10585	-0.04954	0.01522	0.12195	-0.24
0.26	-0.19509	-0.15067	-0.08505	-0.02857	0.03621	0.14259	-0.26
0.28	-0.17460	-0.12986	-0.06394	-0.00734	0.05741	0.16335	-0.28
0.30	-0.15369	-0.10868	-0.04250	0.01416	0.07882	0.18421	-0.30
0.32	-0.13237	-0.08711	-0.02074	0.03593	0.10043	0.20519	-0.32
0.34	-0.11061	-0.06515	0.00135	0.05799	0.12227	0.22628	-0.34
0.36	-0.08841	-0.04279	0.02379	0.08033	0.14432	0.24749	-0.36
0.38	-0.06574	-0.02000	0.04658	0.10296	0.16661	0.26881	-0.38
0.40	-0.04259	0.00322	0.06973	0.12589	0.18912	0.29025	-0.40
0.42	-0.01895	0.02689	0.09325	0.14914	0.21186	0.31182	-0.42
0.44	0.00520	0.05101	0.11716	0.17269	0.23485	0.33351	-0.44
0.46	0.02989	0.07562	0.14146	0.19657	0.25808	0.35532	-0.46
0.48	0.05513	0.10071	0.16616	0.22078	0.28156	0.37726	-0.48
0.50	0.08093	0.12631	0.19128	0.24533	0.30529	0.39933	-0.50
0.52	0.10733	0.15243	0.21682	0.27022	0.32929	0.42153	-0.52
0.54	0.13434	0.17910	0.24281	0.29547	0.35354	0.44387	-0.54
0.56	0.16198	0.20633	0.26925	0.32109	0.37807	0.46634	-0.56
0.58	0.19029	0.23413	0.29615	0.34708	0.40288	0.48894	-0.58
0.60	0.21928	0.26254	0.32353	0.37345	0.42797	0.51169	-0.60
0.62	0.24899	0.29157	0.35141	0.40021	0.45334	0.53458	-0.62
0.64	0.27944	0.32124	0.37980	0.42738	0.47901	0.55762	-0.64
0.66	0.31066	0.35159	0.40871	0.45496	0.50498	0.58080	-0.66
0.68	0.34269	0.38262	0.43816	0.48297	0.53126	0.60413	-0.68
0.70	0.37556	0.41437	0.46818	0.51142	0.55786	0.62761	-0.70
0.72	0.40930	0.44687	0.49877	0.54031	0.58477	0.65125	-0.72
0.74	0.44395	0.48015	0.52995	0.56967	0.61202	0.67504	-0.74
0.76	0.47956	0.51423	0.56176	0.59950	0.63960	0.69900	-0.76
0.78	0.51616	0.54915	0.59419	0.62982	0.66753	0.72312	-0.78
0.80	0.55380	0.58495	0.62729	0.66064	0.69581	0.74740	-0.80
0.82	0.59254	0.62165	0.66106	0.69198	0.72445	0.77185	-0.82
0.84	0.63241	0.65930	0.69554	0.72385	0.75346	0.79647	-0.84
0.86	0.67349	0.69794	0.73075	0.75627	0.78285	0.82127	-0.86
0.88	0.71581	0.73760	0.76672	0.78925	0.81263	0.84625	-0.88
0.90	0.75946	0.77835	0.80346	0.82281	0.84280	0.87140	-0.90
0.92	0.80449	0.82021	0.84102	0.85697	0.87338	0.89674	-0.92
0.94	0.85098	0.86325	0.87942	0.89175	0.90438	0.92227	-0.94
0.96	0.89901	0.90753	0.91869	0.92717	0.93581	0.94798	-0.96
0.98	0.94865	0.95309	0.95888	0.96324	0.96768	0.97389	-0.98

↑ r₀

α =	.995 ↑-	.99 ↑-	.975 ↑-	.95 ↑-	.90 ↑-	.75 ↑-

TABLE 33.1   SAMPLE SIZE =   35

α =	.005 ↓+	.01 ↓+	.025 ↓+	.05 ↓+	.10 ↓+	.25 ↓+	
r_0 ↓							
−0.98	−0.99165	−0.99087	−0.98959	−0.98836	−0.98678	−0.98368	0.98
−0.96	−0.98322	−0.98164	−0.97909	−0.97664	−0.97349	−0.96731	0.96
−0.94	−0.97468	−0.97232	−0.96849	−0.96483	−0.96011	−0.95091	0.94
−0.92	−0.96606	−0.96291	−0.95781	−0.95292	−0.94666	−0.93445	0.92
−0.90	−0.95733	−0.95339	−0.94702	−0.94093	−0.93313	−0.91796	0.90
−0.88	−0.94851	−0.94378	−0.93614	−0.92884	−0.91951	−0.90141	0.88
−0.86	−0.93959	−0.93407	−0.92516	−0.91666	−0.90581	−0.88482	0.86
−0.84	−0.93057	−0.92425	−0.91407	−0.90439	−0.89203	−0.86819	0.84
−0.82	−0.92144	−0.91433	−0.90289	−0.89201	−0.87816	−0.85150	0.82
−0.80	−0.91220	−0.90430	−0.89160	−0.87954	−0.86421	−0.83477	0.80
−0.78	−0.90286	−0.89417	−0.88021	−0.86697	−0.85017	−0.81799	0.78
−0.76	−0.89341	−0.88392	−0.86870	−0.85430	−0.83604	−0.80116	0.76
−0.74	−0.88384	−0.87356	−0.85709	−0.84152	−0.82182	−0.78428	0.74
−0.72	−0.87416	−0.86309	−0.84537	−0.82864	−0.80751	−0.76734	0.72
−0.70	−0.86436	−0.85250	−0.83353	−0.81565	−0.79311	−0.75036	0.70
−0.68	−0.85445	−0.84179	−0.82158	−0.80256	−0.77861	−0.73332	0.68
−0.66	−0.84441	−0.83096	−0.80951	−0.78936	−0.76402	−0.71623	0.66
−0.64	−0.83425	−0.82000	−0.79732	−0.77604	−0.74933	−0.69908	0.64
−0.62	−0.82396	−0.80892	−0.78501	−0.76261	−0.73454	−0.68188	0.62
−0.60	−0.81354	−0.79771	−0.77258	−0.74907	−0.71966	−0.66462	0.60
−0.58	−0.80298	−0.78637	−0.76002	−0.73541	−0.70467	−0.64730	0.58
−0.56	−0.79230	−0.77490	−0.74734	−0.72163	−0.68958	−0.62993	0.56
−0.54	−0.78147	−0.76329	−0.73452	−0.70773	−0.67439	−0.61250	0.54
−0.52	−0.77051	−0.75154	−0.72157	−0.69371	−0.65909	−0.59500	0.52
−0.50	−0.75940	−0.73965	−0.70849	−0.67957	−0.64369	−0.57745	0.50
−0.48	−0.74814	−0.72761	−0.69527	−0.66530	−0.62818	−0.55984	0.48
−0.46	−0.73673	−0.71543	−0.68191	−0.65090	−0.61256	−0.54216	0.46
−0.44	−0.72518	−0.70310	−0.66841	−0.63637	−0.59682	−0.52442	0.44
−0.42	−0.71346	−0.69061	−0.65476	−0.62170	−0.58098	−0.50662	0.42
−0.40	−0.70158	−0.67796	−0.64097	−0.60690	−0.56502	−0.48875	0.40
−0.38	−0.68954	−0.66516	−0.62702	−0.59197	−0.54894	−0.47081	0.38
−0.36	−0.67733	−0.65219	−0.61292	−0.57689	−0.53274	−0.45281	0.36
−0.34	−0.66495	−0.63906	−0.59867	−0.56167	−0.51643	−0.43474	0.34
−0.32	−0.65239	−0.62575	−0.58425	−0.54631	−0.49999	−0.41660	0.32
−0.30	−0.63966	−0.61227	−0.56967	−0.53080	−0.48343	−0.39839	0.30
−0.28	−0.62674	−0.59861	−0.55493	−0.51513	−0.46674	−0.38011	0.28
−0.26	−0.61363	−0.58477	−0.54002	−0.49932	−0.44992	−0.36176	0.26
−0.24	−0.60033	−0.57074	−0.52493	−0.48335	−0.43298	−0.34334	0.24
−0.22	−0.58683	−0.55652	−0.50967	−0.46723	−0.41590	−0.32484	0.22
−0.20	−0.57313	−0.54210	−0.49423	−0.45094	−0.39869	−0.30626	0.20
−0.18	−0.55922	−0.52749	−0.47861	−0.43449	−0.38134	−0.28761	0.18
−0.16	−0.54510	−0.51267	−0.46280	−0.41787	−0.36386	−0.26888	0.16
−0.14	−0.53077	−0.49765	−0.44680	−0.40108	−0.34623	−0.25008	0.14
−0.12	−0.51621	−0.48241	−0.43061	−0.38412	−0.32846	−0.23119	0.12
−0.10	−0.50141	−0.46695	−0.41421	−0.36698	−0.31055	−0.21222	0.10
−0.08	−0.48639	−0.45126	−0.39762	−0.34966	−0.29249	−0.19317	0.08
−0.06	−0.47112	−0.43535	−0.38081	−0.33216	−0.27427	−0.17404	0.06
−0.04	−0.45561	−0.41920	−0.36380	−0.31447	−0.25591	−0.15482	0.04
−0.02	−0.43984	−0.40281	−0.34656	−0.29659	−0.23739	−0.13552	0.02

↑ r_0

α =	.995 ↑−	.99 ↑−	.975 ↑−	.95 ↑−	.90 ↑−	.75 ↑−

TABLE 33.2   SAMPLE SIZE = 35 (CONT.)

α =	.005 ↓+	.01 ↓+	.025 ↓+	.05 ↓+	.10 ↓+	.25 ↓+	
r_0 ↓							
0.00	−0.42381	−0.38618	−0.32911	−0.27852	−0.21871	−0.11613	0.00
0.02	−0.40752	−0.36929	−0.31143	−0.26025	−0.19988	−0.09665	−0.02
0.04	−0.39095	−0.35214	−0.29352	−0.24177	−0.18088	−0.07709	−0.04
0.06	−0.37409	−0.33472	−0.27537	−0.22309	−0.16171	−0.05743	−0.06
0.08	−0.35695	−0.31704	−0.25699	−0.20421	−0.14237	−0.03768	−0.08
0.10	−0.33950	−0.29907	−0.23835	−0.18510	−0.12287	−0.01783	−0.10
0.12	−0.32175	−0.28081	−0.21946	−0.16578	−0.10318	0.00211	−0.12
0.14	−0.30369	−0.26226	−0.20031	−0.14624	−0.08332	0.02214	−0.14
0.16	−0.28529	−0.24341	−0.18090	−0.12647	−0.06328	0.04227	−0.16
0.18	−0.26657	−0.22424	−0.16122	−0.10646	−0.04306	0.06251	−0.18
0.20	−0.24749	−0.20476	−0.14126	−0.08622	−0.02264	0.08284	−0.20
0.22	−0.22806	−0.18494	−0.12101	−0.06574	−0.00204	0.10328	−0.22
0.24	−0.20827	−0.16479	−0.10047	−0.04500	0.01876	0.12382	−0.24
0.26	−0.18810	−0.14429	−0.07964	−0.02402	0.03976	0.14446	−0.26
0.28	−0.16754	−0.12344	−0.05850	−0.00277	0.06096	0.16521	−0.28
0.30	−0.14657	−0.10221	−0.03704	0.01873	0.08236	0.18607	−0.30
0.32	−0.12519	−0.08061	−0.01527	0.04051	0.10397	0.20704	−0.32
0.34	−0.10339	−0.05862	0.00684	0.06256	0.12580	0.22813	−0.34
0.36	−0.08114	−0.03622	0.02928	0.08490	0.14785	0.24932	−0.36
0.38	−0.05844	−0.01342	0.05208	0.10753	0.17011	0.27064	−0.38
0.40	−0.03526	0.00982	0.07522	0.13045	0.19261	0.29207	−0.40
0.42	−0.01159	0.03349	0.09874	0.15367	0.21533	0.31361	−0.42
0.44	0.01257	0.05761	0.12263	0.17720	0.23829	0.33528	−0.44
0.46	0.03726	0.08221	0.14690	0.20105	0.26149	0.35708	−0.46
0.48	0.06250	0.10728	0.17157	0.22522	0.28493	0.37899	−0.48
0.50	0.08829	0.13286	0.19665	0.24973	0.30862	0.40104	−0.50
0.52	0.11466	0.15894	0.22215	0.27457	0.33257	0.42321	−0.52
0.54	0.14163	0.18556	0.24808	0.29977	0.35678	0.44551	−0.54
0.56	0.16923	0.21273	0.27445	0.32532	0.38125	0.46795	−0.56
0.58	0.19748	0.24047	0.30128	0.35123	0.40599	0.49052	−0.58
0.60	0.22639	0.26879	0.32858	0.37753	0.43101	0.51323	−0.60
0.62	0.25601	0.29773	0.35636	0.40420	0.45631	0.53607	−0.62
0.64	0.28635	0.32729	0.38464	0.43128	0.48190	0.55906	−0.64
0.66	0.31745	0.35750	0.41344	0.45875	0.50779	0.58220	−0.66
0.68	0.34933	0.38840	0.44276	0.48664	0.53397	0.60548	−0.68
0.70	0.38203	0.41999	0.47263	0.51496	0.56047	0.62890	−0.70
0.72	0.41558	0.45231	0.50306	0.54372	0.58727	0.65248	−0.72
0.74	0.45003	0.48539	0.53407	0.57293	0.61440	0.67622	−0.74
0.76	0.48539	0.51925	0.56569	0.60260	0.64186	0.70011	−0.76
0.78	0.52173	0.55393	0.59792	0.63275	0.66965	0.72416	−0.78
0.80	0.55908	0.58945	0.63079	0.66338	0.69779	0.74837	−0.80
0.82	0.59749	0.62586	0.66432	0.69452	0.72628	0.77274	−0.82
0.84	0.63700	0.66319	0.69854	0.72618	0.75513	0.79728	−0.84
0.86	0.67767	0.70147	0.73346	0.75837	0.78435	0.82199	−0.86
0.88	0.71956	0.74076	0.76912	0.79110	0.81395	0.84688	−0.88
0.90	0.76272	0.78108	0.80553	0.82440	0.84393	0.87194	−0.90
0.92	0.80721	0.82249	0.84274	0.85829	0.87432	0.89718	−0.92
0.94	0.85311	0.86503	0.88075	0.89277	0.90510	0.92260	−0.94
0.96	0.90049	0.90876	0.91961	0.92787	0.93630	0.94821	−0.96
0.98	0.94943	0.95373	0.95935	0.96361	0.96793	0.97401	−0.98

↑ r_0

α =	.995 ↑−	.99 ↑−	.975 ↑−	.95 ↑−	.90 ↑−	.75 ↑−

TABLE 34.1   SAMPLE SIZE =   36

α =	.005 ↓+	.01 ↓+	.025 ↓+	.05 ↓+	.10 ↓+	.25 ↓+	
r_0 ↓							
-0.98	-0.99155	-0.99076	-0.98949	-0.98827	-0.98671	-0.98364	0.98
-0.96	-0.98300	-0.98143	-0.97889	-0.97646	-0.97334	-0.96723	0.96
-0.94	-0.97436	-0.97201	-0.96820	-0.96456	-0.95989	-0.95078	0.94
-0.92	-0.96562	-0.96248	-0.95741	-0.95257	-0.94636	-0.93428	0.92
-0.90	-0.95679	-0.95286	-0.94653	-0.94048	-0.93275	-0.91775	0.90
-0.88	-0.94786	-0.94315	-0.93555	-0.92831	-0.91906	-0.90116	0.88
-0.86	-0.93882	-0.93333	-0.92447	-0.91604	-0.90529	-0.88453	0.86
-0.84	-0.92969	-0.92340	-0.91329	-0.90367	-0.89143	-0.86785	0.84
-0.82	-0.92045	-0.91337	-0.90200	-0.89121	-0.87749	-0.85113	0.82
-0.80	-0.91110	-0.90324	-0.89062	-0.87865	-0.86346	-0.83436	0.80
-0.78	-0.90165	-0.89300	-0.87912	-0.86599	-0.84935	-0.81753	0.78
-0.76	-0.89208	-0.88264	-0.86752	-0.85323	-0.83514	-0.80066	0.76
-0.74	-0.88240	-0.87217	-0.85581	-0.84037	-0.82085	-0.78374	0.74
-0.72	-0.87261	-0.86159	-0.84399	-0.82740	-0.80647	-0.76677	0.72
-0.70	-0.86270	-0.85089	-0.83206	-0.81433	-0.79199	-0.74974	0.70
-0.68	-0.85267	-0.84008	-0.82001	-0.80114	-0.77743	-0.73266	0.68
-0.66	-0.84251	-0.82914	-0.80784	-0.78785	-0.76276	-0.71553	0.66
-0.64	-0.83224	-0.81807	-0.79556	-0.77445	-0.74800	-0.69834	0.64
-0.62	-0.82183	-0.80688	-0.78315	-0.76094	-0.73315	-0.68110	0.62
-0.60	-0.81130	-0.79557	-0.77062	-0.74731	-0.71819	-0.66381	0.60
-0.58	-0.80063	-0.78412	-0.75797	-0.73357	-0.70314	-0.64645	0.58
-0.56	-0.78983	-0.77254	-0.74518	-0.71971	-0.68798	-0.62904	0.56
-0.54	-0.77889	-0.76082	-0.73227	-0.70573	-0.67272	-0.61157	0.54
-0.52	-0.76781	-0.74896	-0.71923	-0.69162	-0.65736	-0.59404	0.52
-0.50	-0.75658	-0.73696	-0.70605	-0.67740	-0.64189	-0.57645	0.50
-0.48	-0.74521	-0.72482	-0.69274	-0.66304	-0.62631	-0.55881	0.48
-0.46	-0.73369	-0.71253	-0.67928	-0.64856	-0.61063	-0.54109	0.46
-0.44	-0.72201	-0.70009	-0.66569	-0.63395	-0.59483	-0.52332	0.44
-0.42	-0.71018	-0.68749	-0.65195	-0.61921	-0.57892	-0.50548	0.42
-0.40	-0.69819	-0.67474	-0.63806	-0.60433	-0.56290	-0.48758	0.40
-0.38	-0.68603	-0.66183	-0.62402	-0.58931	-0.54676	-0.46961	0.38
-0.36	-0.67371	-0.64876	-0.60983	-0.57416	-0.53051	-0.45158	0.36
-0.34	-0.66121	-0.63552	-0.59549	-0.55887	-0.51413	-0.43347	0.34
-0.32	-0.64854	-0.62210	-0.58098	-0.54343	-0.49763	-0.41530	0.32
-0.30	-0.63569	-0.60852	-0.56632	-0.52784	-0.48101	-0.39707	0.30
-0.28	-0.62266	-0.59476	-0.55149	-0.51211	-0.46427	-0.37876	0.28
-0.26	-0.60943	-0.58081	-0.53649	-0.49622	-0.44740	-0.36037	0.26
-0.24	-0.59602	-0.56668	-0.52132	-0.48018	-0.43040	-0.34192	0.24
-0.22	-0.58241	-0.55236	-0.50597	-0.46399	-0.41327	-0.32340	0.22
-0.20	-0.56860	-0.53784	-0.49045	-0.44763	-0.39601	-0.30479	0.20
-0.18	-0.55458	-0.52313	-0.47474	-0.43111	-0.37861	-0.28612	0.18
-0.16	-0.54035	-0.50821	-0.45885	-0.41443	-0.36108	-0.26736	0.16
-0.14	-0.52590	-0.49309	-0.44277	-0.39757	-0.34340	-0.24853	0.14
-0.12	-0.51123	-0.47775	-0.42650	-0.38055	-0.32559	-0.22962	0.12
-0.10	-0.49633	-0.46220	-0.41003	-0.36335	-0.30763	-0.21063	0.10
-0.08	-0.48120	-0.44642	-0.39336	-0.34597	-0.28953	-0.19156	0.08
-0.06	-0.46583	-0.43042	-0.37648	-0.32841	-0.27127	-0.17241	0.06
-0.04	-0.45021	-0.41418	-0.35940	-0.31067	-0.25287	-0.15318	0.04
-0.02	-0.43434	-0.39770	-0.34209	-0.29274	-0.23431	-0.13385	0.02

↑ r_0

α =	.995 ↑-	.99 ↑-	.975 ↑-	.95 ↑-	.90 ↑-	.75 ↑-

TABLE 34.2   SAMPLE SIZE = 36   (CONT.)

α =	.005 ↓+	.01 ↓+	.025 ↓+	.05 ↓+	.10 ↓+	.25 ↓+	
r_0 ↓							
0.00	-0.41821	-0.38098	-0.32457	-0.27461	-0.21560	-0.11445	0.00
0.02	-0.40182	-0.36400	-0.30683	-0.25629	-0.19673	-0.09495	-0.02
0.04	-0.38515	-0.34677	-0.28885	-0.23777	-0.17769	-0.07537	-0.04
0.06	-0.36820	-0.32927	-0.27065	-0.21905	-0.15850	-0.05570	-0.06
0.08	-0.35096	-0.31151	-0.25220	-0.20012	-0.13913	-0.03594	-0.08
0.10	-0.33343	-0.29346	-0.23351	-0.18097	-0.11960	-0.01608	-0.10
0.12	-0.31559	-0.27514	-0.21457	-0.16161	-0.09989	0.00387	-0.12
0.14	-0.29743	-0.25651	-0.19537	-0.14203	-0.08001	0.02391	-0.14
0.16	-0.27896	-0.23759	-0.17591	-0.12223	-0.05995	0.04405	-0.16
0.18	-0.26015	-0.21836	-0.15618	-0.10220	-0.03971	0.06429	-0.18
0.20	-0.24100	-0.19882	-0.13618	-0.08193	-0.01928	0.08462	-0.20
0.22	-0.22150	-0.17895	-0.11590	-0.06142	0.00133	0.10506	-0.22
0.24	-0.20164	-0.15874	-0.09533	-0.04067	0.02214	0.12560	-0.24
0.26	-0.18140	-0.13820	-0.07447	-0.01967	0.04314	0.14625	-0.26
0.28	-0.16078	-0.11730	-0.05330	0.00159	0.06434	0.16699	-0.28
0.30	-0.13976	-0.09603	-0.03183	0.02310	0.08575	0.18785	-0.30
0.32	-0.11833	-0.07440	-0.01003	0.04488	0.10736	0.20882	-0.32
0.34	-0.09648	-0.05237	0.01208	0.06694	0.12917	0.22989	-0.34
0.36	-0.07419	-0.02996	0.03453	0.08927	0.15121	0.25108	-0.36
0.38	-0.05146	-0.00713	0.05732	0.11188	0.17346	0.27238	-0.38
0.40	-0.02825	0.01611	0.08047	0.13479	0.19594	0.29379	-0.40
0.42	-0.00457	0.03979	0.10397	0.15799	0.21864	0.31532	-0.42
0.44	0.01961	0.06391	0.12784	0.18150	0.24157	0.33698	-0.44
0.46	0.04430	0.08849	0.15209	0.20532	0.26474	0.35875	-0.46
0.48	0.06952	0.11355	0.17673	0.22946	0.28814	0.38064	-0.48
0.50	0.09530	0.13909	0.20177	0.25392	0.31180	0.40266	-0.50
0.52	0.12164	0.16514	0.22723	0.27872	0.33570	0.42480	-0.52
0.54	0.14858	0.19171	0.25310	0.30386	0.35986	0.44708	-0.54
0.56	0.17613	0.21883	0.27941	0.32934	0.38427	0.46948	-0.56
0.58	0.20431	0.24649	0.30617	0.35519	0.40896	0.49202	-0.58
0.60	0.23315	0.27474	0.33338	0.38141	0.43391	0.51469	-0.60
0.62	0.26268	0.30358	0.36107	0.40800	0.45914	0.53750	-0.62
0.64	0.29291	0.33303	0.38925	0.43498	0.48465	0.56044	-0.64
0.66	0.32389	0.36312	0.41793	0.46235	0.51045	0.58353	-0.66
0.68	0.35563	0.39388	0.44713	0.49013	0.53655	0.60676	-0.68
0.70	0.38817	0.42532	0.47686	0.51833	0.56294	0.63013	-0.70
0.72	0.42154	0.45747	0.50714	0.54696	0.58965	0.65366	-0.72
0.74	0.45578	0.49035	0.53798	0.57602	0.61667	0.67733	-0.74
0.76	0.49092	0.52401	0.56941	0.60554	0.64401	0.70116	-0.76
0.78	0.52701	0.55845	0.60145	0.63552	0.67167	0.72514	-0.78
0.80	0.56407	0.59372	0.63410	0.66598	0.69968	0.74928	-0.80
0.82	0.60217	0.62985	0.66740	0.69693	0.72802	0.77359	-0.82
0.84	0.64133	0.66687	0.70137	0.72838	0.75672	0.79805	-0.84
0.86	0.68162	0.70482	0.73603	0.76036	0.78578	0.82268	-0.86
0.88	0.72309	0.74373	0.77139	0.79286	0.81521	0.84748	-0.88
0.90	0.76579	0.78366	0.80749	0.82591	0.84501	0.87245	-0.90
0.92	0.80978	0.82463	0.84436	0.85953	0.87520	0.89760	-0.92
0.94	0.85512	0.86670	0.88201	0.89373	0.90578	0.92293	-0.94
0.96	0.90189	0.90992	0.92048	0.92853	0.93677	0.94843	-0.96
0.98	0.95016	0.95433	0.95980	0.96395	0.96817	0.97412	-0.98

↑ r_0

α =	.995 ↑-	.99 ↑-	.975 ↑-	.95 ↑-	.90 ↑-	.75 ↑-

TABLE 35.1   SAMPLE SIZE = 37

α =	.005 ↓+	.01 ↓+	.025 ↓+	.05 ↓+	.10 ↓+	.25 ↓+	
r_0 ↓							
-0.98	-0.99144	-0.99066	-0.98940	-0.98819	-0.98663	-0.98360	0.98
-0.96	-0.98279	-0.98123	-0.97870	-0.97629	-0.97319	-0.96715	0.96
-0.94	-0.97404	-0.97170	-0.96791	-0.96430	-0.95967	-0.95066	0.94
-0.92	-0.96520	-0.96208	-0.95703	-0.95222	-0.94607	-0.93412	0.92
-0.90	-0.95626	-0.95235	-0.94605	-0.94005	-0.93238	-0.91754	0.90
-0.88	-0.94722	-0.94253	-0.93498	-0.92779	-0.91862	-0.90092	0.88
-0.86	-0.93808	-0.93261	-0.92380	-0.91543	-0.90478	-0.88425	0.86
-0.84	-0.92884	-0.92258	-0.91253	-0.90298	-0.89085	-0.86753	0.84
-0.82	-0.91949	-0.91245	-0.90115	-0.89044	-0.87684	-0.85077	0.82
-0.80	-0.91004	-0.90221	-0.88967	-0.87779	-0.86274	-0.83395	0.80
-0.78	-0.90047	-0.89186	-0.87808	-0.86505	-0.84855	-0.81709	0.78
-0.76	-0.89080	-0.88141	-0.86638	-0.85220	-0.83428	-0.80018	0.76
-0.74	-0.88101	-0.87083	-0.85458	-0.83925	-0.81992	-0.78322	0.74
-0.72	-0.87111	-0.86015	-0.84266	-0.82620	-0.80547	-0.76621	0.72
-0.70	-0.86109	-0.84935	-0.83063	-0.81305	-0.79092	-0.74915	0.70
-0.68	-0.85095	-0.83842	-0.81849	-0.79978	-0.77629	-0.73203	0.68
-0.66	-0.84068	-0.82738	-0.80623	-0.78641	-0.76156	-0.71486	0.66
-0.64	-0.83029	-0.81621	-0.79385	-0.77293	-0.74673	-0.69764	0.64
-0.62	-0.81978	-0.80492	-0.78135	-0.75933	-0.73181	-0.68036	0.62
-0.60	-0.80913	-0.79350	-0.76873	-0.74562	-0.71679	-0.66303	0.60
-0.58	-0.79836	-0.78194	-0.75598	-0.73180	-0.70167	-0.64564	0.58
-0.56	-0.78744	-0.77026	-0.74311	-0.71785	-0.68644	-0.62819	0.56
-0.54	-0.77639	-0.75843	-0.73010	-0.70379	-0.67112	-0.61068	0.54
-0.52	-0.76520	-0.74647	-0.71697	-0.68961	-0.65569	-0.59312	0.52
-0.50	-0.75386	-0.73437	-0.70370	-0.67530	-0.64016	-0.57550	0.50
-0.48	-0.74238	-0.72212	-0.69029	-0.66087	-0.62452	-0.55781	0.48
-0.46	-0.73075	-0.70973	-0.67675	-0.64631	-0.60877	-0.54007	0.46
-0.44	-0.71896	-0.69719	-0.66306	-0.63162	-0.59292	-0.52226	0.44
-0.42	-0.70702	-0.68449	-0.64924	-0.61680	-0.57695	-0.50439	0.42
-0.40	-0.69491	-0.67164	-0.63526	-0.60185	-0.56087	-0.48646	0.40
-0.38	-0.68264	-0.65862	-0.62113	-0.58676	-0.54467	-0.46846	0.38
-0.36	-0.67021	-0.64545	-0.60686	-0.57153	-0.52835	-0.45039	0.36
-0.34	-0.65760	-0.63210	-0.59243	-0.55617	-0.51192	-0.43226	0.34
-0.32	-0.64482	-0.61859	-0.57784	-0.54066	-0.49537	-0.41406	0.32
-0.30	-0.63186	-0.60491	-0.56308	-0.52500	-0.47870	-0.39579	0.30
-0.28	-0.61872	-0.59104	-0.54817	-0.50920	-0.46190	-0.37745	0.28
-0.26	-0.60539	-0.57700	-0.53309	-0.49324	-0.44498	-0.35904	0.26
-0.24	-0.59187	-0.56277	-0.51784	-0.47714	-0.42793	-0.34056	0.24
-0.22	-0.57815	-0.54835	-0.50241	-0.46087	-0.41075	-0.32201	0.22
-0.20	-0.56423	-0.53374	-0.48681	-0.44445	-0.39343	-0.30339	0.20
-0.18	-0.55010	-0.51893	-0.47103	-0.42787	-0.37599	-0.28468	0.18
-0.16	-0.53577	-0.50392	-0.45506	-0.41112	-0.35841	-0.26591	0.16
-0.14	-0.52121	-0.48870	-0.43890	-0.39421	-0.34069	-0.24705	0.14
-0.12	-0.50644	-0.47328	-0.42256	-0.37712	-0.32283	-0.22812	0.12
-0.10	-0.49144	-0.45763	-0.40601	-0.35987	-0.30483	-0.20911	0.10
-0.08	-0.47620	-0.44176	-0.38927	-0.34243	-0.28668	-0.19002	0.08
-0.06	-0.46073	-0.42567	-0.37232	-0.32482	-0.26839	-0.17085	0.06
-0.04	-0.44502	-0.40935	-0.35517	-0.30702	-0.24995	-0.15160	0.04
-0.02	-0.42905	-0.39278	-0.33780	-0.28904	-0.23136	-0.13226	0.02

↑ r_0

α =	.995 ↑-	.99 ↑-	.975 ↑-	.95 ↑-	.90 ↑-	.75 ↑-

TABLE 35.2   SAMPLE SIZE = 37   (CONT.)

α =	.005 ↓+	.01 ↓+	.025 ↓+	.05 ↓+	.10 ↓+	.25 ↓+	
r_0 ↓							
0.00	−0.41282	−0.37598	−0.32022	−0.27086	−0.21261	−0.11284	0.00
0.02	−0.39633	−0.35892	−0.30241	−0.25250	−0.19371	−0.09333	−0.02
0.04	−0.37957	−0.34161	−0.28438	−0.23393	−0.17464	−0.07373	−0.04
0.06	−0.36253	−0.32404	−0.26611	−0.21517	−0.15542	−0.05405	−0.06
0.08	−0.34521	−0.30620	−0.24761	−0.19619	−0.13603	−0.03427	−0.08
0.10	−0.32759	−0.28809	−0.22887	−0.17701	−0.11647	−0.01441	−0.10
0.12	−0.30966	−0.26969	−0.20988	−0.15762	−0.09674	0.00555	−0.12
0.14	−0.29143	−0.25100	−0.19063	−0.13801	−0.07684	0.02560	−0.14
0.16	−0.27288	−0.23202	−0.17113	−0.11817	−0.05677	0.04575	−0.16
0.18	−0.25399	−0.21273	−0.15136	−0.09811	−0.03651	0.06599	−0.18
0.20	−0.23477	−0.19313	−0.13132	−0.07782	−0.01607	0.08633	−0.20
0.22	−0.21520	−0.17320	−0.11101	−0.05729	0.00456	0.10677	−0.22
0.24	−0.19528	−0.15295	−0.09041	−0.03653	0.02537	0.12731	−0.24
0.26	−0.17498	−0.13236	−0.06952	−0.01551	0.04638	0.14795	−0.26
0.28	−0.15430	−0.11142	−0.04833	0.00576	0.06758	0.16870	−0.28
0.30	−0.13323	−0.09012	−0.02684	0.02728	0.08898	0.18955	−0.30
0.32	−0.11176	−0.06845	−0.00503	0.04906	0.11059	0.21051	−0.32
0.34	−0.08987	−0.04640	0.01710	0.07112	0.13240	0.23157	−0.34
0.36	−0.06755	−0.02396	0.03955	0.09344	0.15442	0.25275	−0.36
0.38	−0.04478	−0.00112	0.06234	0.11605	0.17666	0.27404	−0.38
0.40	−0.02156	0.02213	0.08548	0.13894	0.19911	0.29544	−0.40
0.42	0.00214	0.04581	0.10897	0.16212	0.22179	0.31696	−0.42
0.44	0.02633	0.06992	0.13282	0.18561	0.24470	0.33859	−0.44
0.46	0.05102	0.09449	0.15705	0.20939	0.26784	0.36034	−0.46
0.48	0.07623	0.11953	0.18166	0.23350	0.29121	0.38222	−0.48
0.50	0.10199	0.14504	0.20666	0.25792	0.31482	0.40421	−0.50
0.52	0.12831	0.17106	0.23206	0.28267	0.33868	0.42633	−0.52
0.54	0.15521	0.19758	0.25789	0.30775	0.36279	0.44857	−0.54
0.56	0.18270	0.22463	0.28413	0.33318	0.38716	0.47095	−0.56
0.58	0.21082	0.25223	0.31082	0.35896	0.41178	0.49345	−0.58
0.60	0.23959	0.28040	0.33796	0.38510	0.43667	0.51608	−0.60
0.62	0.26903	0.30914	0.36556	0.41161	0.46183	0.53885	−0.62
0.64	0.29916	0.33850	0.39363	0.43850	0.48727	0.56175	−0.64
0.66	0.33001	0.36847	0.42220	0.46578	0.51299	0.58480	−0.66
0.68	0.36162	0.39909	0.45128	0.49345	0.53900	0.60798	−0.68
0.70	0.39400	0.43038	0.48087	0.52153	0.56530	0.63131	−0.70
0.72	0.42720	0.46237	0.51101	0.55004	0.59191	0.65478	−0.72
0.74	0.46124	0.49507	0.54169	0.57897	0.61882	0.67839	−0.74
0.76	0.49616	0.52852	0.57295	0.60834	0.64605	0.70216	−0.76
0.78	0.53200	0.56274	0.60480	0.63816	0.67359	0.72608	−0.78
0.80	0.56880	0.59776	0.63725	0.66845	0.70147	0.75016	−0.80
0.82	0.60659	0.63362	0.67033	0.69922	0.72968	0.77439	−0.82
0.84	0.64543	0.67035	0.70406	0.73048	0.75823	0.79878	−0.84
0.86	0.68536	0.70798	0.73845	0.76224	0.78713	0.82334	−0.86
0.88	0.72643	0.74655	0.77354	0.79453	0.81640	0.84805	−0.88
0.90	0.76868	0.78609	0.80935	0.82734	0.84603	0.87294	−0.90
0.92	0.81219	0.82666	0.84589	0.86071	0.87603	0.89800	−0.92
0.94	0.85701	0.86828	0.88320	0.89464	0.90643	0.92323	−0.94
0.96	0.90321	0.91101	0.92130	0.92915	0.93721	0.94864	−0.96
0.98	0.95084	0.95490	0.96022	0.96427	0.96840	0.97423	−0.98

↑ r_0

α =	.995 ↑−	.99 ↑−	.975 ↑−	.95 ↑−	.90 ↑−	.75 ↑−

TABLE 36.1   SAMPLE SIZE =   38

α =	.005 ↓+	.01 ↓+	.025 ↓+	.05 ↓+	.10 ↓+	.25 ↓+	
r_0 ↓							
-0.98	-0.99134	-0.99056	-0.98931	-0.98810	-0.98656	-0.98356	0.98
-0.96	-0.98258	-0.98103	-0.97852	-0.97612	-0.97305	-0.96707	0.96
-0.94	-0.97374	-0.97140	-0.96764	-0.96405	-0.95945	-0.95054	0.94
-0.92	-0.96479	-0.96168	-0.95666	-0.95189	-0.94578	-0.93397	0.92
-0.90	-0.95575	-0.95186	-0.94559	-0.93963	-0.93203	-0.91735	0.90
-0.88	-0.94661	-0.94194	-0.93443	-0.92729	-0.91820	-0.90069	0.88
-0.86	-0.93736	-0.93191	-0.92316	-0.91485	-0.90429	-0.88398	0.86
-0.84	-0.92802	-0.92179	-0.91179	-0.90232	-0.89029	-0.86722	0.84
-0.82	-0.91856	-0.91156	-0.90032	-0.88969	-0.87621	-0.85042	0.82
-0.80	-0.90900	-0.90122	-0.88875	-0.87696	-0.86204	-0.83357	0.80
-0.78	-0.89934	-0.89077	-0.87707	-0.86414	-0.84779	-0.81667	0.78
-0.76	-0.88956	-0.88021	-0.86528	-0.85121	-0.83345	-0.79972	0.76
-0.74	-0.87966	-0.86954	-0.85338	-0.83818	-0.81902	-0.78272	0.74
-0.72	-0.86965	-0.85875	-0.84138	-0.82505	-0.80450	-0.76567	0.72
-0.70	-0.85953	-0.84785	-0.82926	-0.81181	-0.78989	-0.74857	0.70
-0.68	-0.84928	-0.83682	-0.81702	-0.79847	-0.77519	-0.73142	0.68
-0.66	-0.83891	-0.82568	-0.80467	-0.78501	-0.76039	-0.71422	0.66
-0.64	-0.82841	-0.81441	-0.79220	-0.77145	-0.74550	-0.69696	0.64
-0.62	-0.81779	-0.80302	-0.77962	-0.75778	-0.73052	-0.67964	0.62
-0.60	-0.80704	-0.79149	-0.76690	-0.74399	-0.71543	-0.66227	0.60
-0.58	-0.79615	-0.77984	-0.75407	-0.73009	-0.70025	-0.64485	0.58
-0.56	-0.78513	-0.76805	-0.74110	-0.71607	-0.68496	-0.62737	0.56
-0.54	-0.77398	-0.75613	-0.72801	-0.70193	-0.66958	-0.60983	0.54
-0.52	-0.76268	-0.74407	-0.71479	-0.68767	-0.65409	-0.59223	0.52
-0.50	-0.75123	-0.73187	-0.70143	-0.67329	-0.63850	-0.57458	0.50
-0.48	-0.73964	-0.71952	-0.68794	-0.65878	-0.62280	-0.55686	0.48
-0.46	-0.72790	-0.70703	-0.67431	-0.64415	-0.60699	-0.53908	0.46
-0.44	-0.71601	-0.69438	-0.66054	-0.62938	-0.59107	-0.52124	0.44
-0.42	-0.70396	-0.68159	-0.64662	-0.61449	-0.57505	-0.50334	0.42
-0.40	-0.69175	-0.66864	-0.63256	-0.59947	-0.55891	-0.48538	0.40
-0.38	-0.67937	-0.65553	-0.61835	-0.58430	-0.54265	-0.46735	0.38
-0.36	-0.66683	-0.64225	-0.60399	-0.56901	-0.52629	-0.44925	0.36
-0.34	-0.65412	-0.62881	-0.58948	-0.55357	-0.50980	-0.43109	0.34
-0.32	-0.64123	-0.61520	-0.57480	-0.53799	-0.49319	-0.41286	0.32
-0.30	-0.62817	-0.60142	-0.55997	-0.52227	-0.47647	-0.39457	0.30
-0.28	-0.61492	-0.58746	-0.54498	-0.50640	-0.45962	-0.37620	0.28
-0.26	-0.60148	-0.57333	-0.52982	-0.49038	-0.44265	-0.35777	0.26
-0.24	-0.58786	-0.55900	-0.51449	-0.47420	-0.42555	-0.33926	0.24
-0.22	-0.57404	-0.54449	-0.49898	-0.45788	-0.40832	-0.32068	0.22
-0.20	-0.56001	-0.52979	-0.48331	-0.44139	-0.39096	-0.30203	0.20
-0.18	-0.54579	-0.51489	-0.46745	-0.42475	-0.37347	-0.28331	0.18
-0.16	-0.53135	-0.49979	-0.45141	-0.40794	-0.35584	-0.26451	0.16
-0.14	-0.51670	-0.48448	-0.43518	-0.39097	-0.33808	-0.24563	0.14
-0.12	-0.50182	-0.46897	-0.41876	-0.37383	-0.32018	-0.22668	0.12
-0.10	-0.48672	-0.45323	-0.40215	-0.35652	-0.30214	-0.20765	0.10
-0.08	-0.47139	-0.43728	-0.38534	-0.33903	-0.28396	-0.18854	0.08
-0.06	-0.45582	-0.42111	-0.36833	-0.32137	-0.26563	-0.16935	0.06
-0.04	-0.44001	-0.40470	-0.35111	-0.30352	-0.24715	-0.15008	0.04
-0.02	-0.42395	-0.38806	-0.33368	-0.28549	-0.22852	-0.13073	0.02

↑ r_0

α =	.995 ↑-	.99 ↑-	.975 ↑-	.95 ↑-	.90 ↑-	.75 ↑-

TABLE 36.2   SAMPLE SIZE = 38   (CONT.)

α =	.005 ↓+	.01 ↓+	.025 ↓+	.05 ↓+	.10 ↓+	.25 ↓+	
r_o ↓							
0.00	−0.40764	−0.37117	−0.31603	−0.26727	−0.20975	−0.11129	0.00
0.02	−0.39106	−0.35404	−0.29817	−0.24886	−0.19081	−0.09177	−0.02
0.04	−0.37421	−0.33666	−0.28008	−0.23025	−0.17172	−0.07216	−0.04
0.06	−0.35708	−0.31901	−0.26176	−0.21144	−0.15247	−0.05246	−0.06
0.08	−0.33967	−0.30110	−0.24321	−0.19243	−0.13305	−0.03268	−0.08
0.10	−0.32197	−0.28292	−0.22441	−0.17322	−0.11347	−0.01280	−0.10
0.12	−0.30397	−0.26446	−0.20538	−0.15379	−0.09372	0.00716	−0.12
0.14	−0.28566	−0.24571	−0.18609	−0.13415	−0.07380	0.02722	−0.14
0.16	−0.26703	−0.22666	−0.16654	−0.11428	−0.05371	0.04737	−0.16
0.18	−0.24808	−0.20732	−0.14674	−0.09420	−0.03344	0.06762	−0.18
0.20	−0.22879	−0.18766	−0.12666	−0.07389	−0.01299	0.08796	−0.20
0.22	−0.20916	−0.16769	−0.10632	−0.05334	0.00764	0.10840	−0.22
0.24	−0.18918	−0.14739	−0.08569	−0.03255	0.02847	0.12894	−0.24
0.26	−0.16882	−0.12676	−0.06477	−0.01153	0.04948	0.14958	−0.26
0.28	−0.14809	−0.10578	−0.04357	0.00975	0.07068	0.17033	−0.28
0.30	−0.12698	−0.08445	−0.02206	0.03128	0.09208	0.19117	−0.30
0.32	−0.10546	−0.06275	−0.00024	0.05306	0.11368	0.21213	−0.32
0.34	−0.08353	−0.04068	0.02189	0.07511	0.13548	0.23319	−0.34
0.36	−0.06118	−0.01822	0.04435	0.09743	0.15749	0.25435	−0.36
0.38	−0.03839	0.00463	0.06714	0.12003	0.17972	0.27563	−0.38
0.40	−0.01515	0.02789	0.09027	0.14291	0.20215	0.29702	−0.40
0.42	0.00856	0.05156	0.11375	0.16607	0.22481	0.31852	−0.42
0.44	0.03275	0.07567	0.13758	0.18953	0.24769	0.34014	−0.44
0.46	0.05744	0.10023	0.16178	0.21329	0.27080	0.36187	−0.46
0.48	0.08265	0.12524	0.18636	0.23736	0.29414	0.38372	−0.48
0.50	0.10839	0.15073	0.21132	0.26174	0.31771	0.40569	−0.50
0.52	0.13467	0.17670	0.23669	0.28644	0.34153	0.42778	−0.52
0.54	0.16153	0.20318	0.26245	0.31147	0.36559	0.45000	−0.54
0.56	0.18898	0.23018	0.28864	0.33684	0.38991	0.47234	−0.56
0.58	0.21704	0.25771	0.31526	0.36256	0.41448	0.49481	−0.58
0.60	0.24573	0.28580	0.34232	0.38863	0.43931	0.51741	−0.60
0.62	0.27508	0.31445	0.36983	0.41506	0.46440	0.54014	−0.62
0.64	0.30511	0.34370	0.39781	0.44186	0.48977	0.56301	−0.64
0.66	0.33585	0.37356	0.42628	0.46904	0.51541	0.58601	−0.66
0.68	0.36732	0.40405	0.45523	0.49661	0.54134	0.60915	−0.68
0.70	0.39955	0.43520	0.48470	0.52459	0.56755	0.63242	−0.70
0.72	0.43258	0.46702	0.51469	0.55297	0.59406	0.65584	−0.72
0.74	0.46643	0.49955	0.54523	0.58177	0.62087	0.67941	−0.74
0.76	0.50114	0.53281	0.57632	0.61100	0.64799	0.70312	−0.76
0.78	0.53675	0.56681	0.60798	0.64067	0.67542	0.72698	−0.78
0.80	0.57329	0.60160	0.64024	0.67080	0.70317	0.75099	−0.80
0.82	0.61079	0.63721	0.67311	0.70140	0.73125	0.77516	−0.82
0.84	0.64932	0.67366	0.70661	0.73247	0.75967	0.79948	−0.84
0.86	0.68890	0.71098	0.74076	0.76404	0.78842	0.82396	−0.86
0.88	0.72958	0.74922	0.77558	0.79611	0.81753	0.84860	−0.88
0.90	0.77143	0.78840	0.81110	0.82870	0.84700	0.87341	−0.90
0.92	0.81448	0.82857	0.84734	0.86183	0.87683	0.89838	−0.92
0.94	0.85880	0.86978	0.88433	0.89550	0.90704	0.92352	−0.94
0.96	0.90445	0.91205	0.92208	0.92975	0.93763	0.94884	−0.96
0.98	0.95149	0.95544	0.96063	0.96457	0.96861	0.97433	−0.98

↑ r_o

α =	.995 ↑−	.99 ↑−	.975 ↑−	.95 ↑−	.90 ↑−	.75 ↑−

TABLE 37.1   SAMPLE SIZE =   39

$\alpha =$	.005 ↓+	.01 ↓+	.025 ↓+	.05 ↓+	.10 ↓+	.25 ↓+	
r_0 ↓							
-0.98	-0.99124	-0.99047	-0.98922	-0.98802	-0.98649	-0.98352	0.98
-0.96	-0.98239	-0.98084	-0.97834	-0.97596	-0.97291	-0.96699	0.96
-0.94	-0.97344	-0.97112	-0.96737	-0.96381	-0.95925	-0.95043	0.94
-0.92	-0.96439	-0.96130	-0.95631	-0.95156	-0.94551	-0.93382	0.92
-0.90	-0.95525	-0.95138	-0.94515	-0.93923	-0.93169	-0.91716	0.90
-0.88	-0.94601	-0.94136	-0.93389	-0.92681	-0.91779	-0.90046	0.88
-0.86	-0.93667	-0.93124	-0.92254	-0.91429	-0.90381	-0.88371	0.86
-0.84	-0.92722	-0.92102	-0.91108	-0.90168	-0.88975	-0.86692	0.84
-0.82	-0.91766	-0.91069	-0.89952	-0.88897	-0.87560	-0.85008	0.82
-0.80	-0.90800	-0.90025	-0.88786	-0.87616	-0.86137	-0.83319	0.80
-0.78	-0.89823	-0.88971	-0.87609	-0.86325	-0.84705	-0.81626	0.78
-0.76	-0.88835	-0.87905	-0.86422	-0.85025	-0.83265	-0.79928	0.76
-0.74	-0.87836	-0.86828	-0.85223	-0.83714	-0.81815	-0.78224	0.74
-0.72	-0.86824	-0.85740	-0.84014	-0.82393	-0.80357	-0.76516	0.72
-0.70	-0.85801	-0.84640	-0.82793	-0.81062	-0.78890	-0.74802	0.70
-0.68	-0.84766	-0.83528	-0.81561	-0.79720	-0.77413	-0.73083	0.68
-0.66	-0.83719	-0.82404	-0.80317	-0.78367	-0.75927	-0.71359	0.66
-0.64	-0.82659	-0.81267	-0.79061	-0.77003	-0.74432	-0.69630	0.64
-0.62	-0.81587	-0.80118	-0.77794	-0.75628	-0.72927	-0.67895	0.62
-0.60	-0.80501	-0.78956	-0.76514	-0.74241	-0.71412	-0.66155	0.60
-0.58	-0.79402	-0.77781	-0.75222	-0.72844	-0.69888	-0.64409	0.58
-0.56	-0.78290	-0.76592	-0.73917	-0.71434	-0.68354	-0.62658	0.56
-0.54	-0.77164	-0.75391	-0.72599	-0.70013	-0.66809	-0.60901	0.54
-0.52	-0.76024	-0.74175	-0.71268	-0.68579	-0.65254	-0.59138	0.52
-0.50	-0.74869	-0.72945	-0.69924	-0.67134	-0.63689	-0.57369	0.50
-0.48	-0.73700	-0.71701	-0.68566	-0.65676	-0.62113	-0.55594	0.48
-0.46	-0.72515	-0.70442	-0.67195	-0.64206	-0.60527	-0.53813	0.46
-0.44	-0.71315	-0.69168	-0.65810	-0.62722	-0.58930	-0.52026	0.44
-0.42	-0.70100	-0.67879	-0.64410	-0.61226	-0.57322	-0.50233	0.42
-0.40	-0.68869	-0.66574	-0.62996	-0.59717	-0.55702	-0.48434	0.40
-0.38	-0.67621	-0.65253	-0.61567	-0.58194	-0.54072	-0.46628	0.38
-0.36	-0.66357	-0.63917	-0.60122	-0.56657	-0.52429	-0.44815	0.36
-0.34	-0.65075	-0.62563	-0.58663	-0.55107	-0.50776	-0.42997	0.34
-0.32	-0.63776	-0.61193	-0.57188	-0.53542	-0.49110	-0.41171	0.32
-0.30	-0.62460	-0.59806	-0.55697	-0.51963	-0.47432	-0.39339	0.30
-0.28	-0.61125	-0.58401	-0.54190	-0.50370	-0.45743	-0.37500	0.28
-0.26	-0.59771	-0.56978	-0.52666	-0.48762	-0.44040	-0.35654	0.26
-0.24	-0.58399	-0.55537	-0.51126	-0.47138	-0.42326	-0.33801	0.24
-0.22	-0.57007	-0.54077	-0.49568	-0.45499	-0.40598	-0.31941	0.22
-0.20	-0.55595	-0.52598	-0.47993	-0.43845	-0.38858	-0.30073	0.20
-0.18	-0.54162	-0.51099	-0.46400	-0.42175	-0.37105	-0.28199	0.18
-0.16	-0.52709	-0.49580	-0.44789	-0.40489	-0.35338	-0.26317	0.16
-0.14	-0.51234	-0.48041	-0.43159	-0.38786	-0.33558	-0.24427	0.14
-0.12	-0.49737	-0.46481	-0.41511	-0.37066	-0.31764	-0.22530	0.12
-0.10	-0.48218	-0.44900	-0.39843	-0.35330	-0.29956	-0.20625	0.10
-0.08	-0.46675	-0.43296	-0.38155	-0.33576	-0.28134	-0.18712	0.08
-0.06	-0.45109	-0.41671	-0.36448	-0.31805	-0.26297	-0.16791	0.06
-0.04	-0.43519	-0.40022	-0.34720	-0.30015	-0.24446	-0.14863	0.04
-0.02	-0.41904	-0.38350	-0.32971	-0.28207	-0.22580	-0.12926	0.02

↑ r_0

$\alpha =$	.995 ↑-	.99 ↑-	.975 ↑-	.95 ↑-	.90 ↑-	.75 ↑-

TABLE 37.2   SAMPLE SIZE = 39   (CONT.)

α =	.005 ↓+	.01 ↓+	.025 ↓+	.05 ↓+	.10 ↓+	.25 ↓+	
r_0 ↓							
0.00	−0.40264	−0.36655	−0.31201	−0.26381	−0.20699	−0.10980	0.00
0.02	−0.38598	−0.34934	−0.29409	−0.24536	−0.18803	−0.09027	−0.02
0.04	−0.36904	−0.33189	−0.27594	−0.22671	−0.16891	−0.07065	−0.04
0.06	−0.35184	−0.31417	−0.25757	−0.20786	−0.14963	−0.05094	−0.06
0.08	−0.33435	−0.29620	−0.23897	−0.18882	−0.13019	−0.03115	−0.08
0.10	−0.31657	−0.27795	−0.22013	−0.16957	−0.11059	−0.01126	−0.10
0.12	−0.29849	−0.25943	−0.20105	−0.15011	−0.09083	0.00871	−0.12
0.14	−0.28011	−0.24062	−0.18172	−0.13044	−0.07089	0.02877	−0.14
0.16	−0.26142	−0.22152	−0.16214	−0.11055	−0.05078	0.04893	−0.16
0.18	−0.24240	−0.20212	−0.14230	−0.09044	−0.03050	0.06918	−0.18
0.20	−0.22305	−0.18242	−0.12219	−0.07011	−0.01004	0.08953	−0.20
0.22	−0.20336	−0.16240	−0.10181	−0.04954	0.01061	0.10997	−0.22
0.24	−0.18331	−0.14206	−0.08116	−0.02874	0.03143	0.13051	−0.24
0.26	−0.16291	−0.12138	−0.06022	−0.00770	0.05245	0.15115	−0.26
0.28	−0.14213	−0.10037	−0.03900	0.01358	0.07365	0.17189	−0.28
0.30	−0.12097	−0.07901	−0.01747	0.03511	0.09505	0.19273	−0.30
0.32	−0.09942	−0.05728	0.00436	0.05690	0.11664	0.21368	−0.32
0.34	−0.07746	−0.03519	0.02649	0.07895	0.13844	0.23473	−0.34
0.36	−0.05507	−0.01272	0.04895	0.10126	0.16044	0.25589	−0.36
0.38	−0.03226	0.01014	0.07174	0.12384	0.18264	0.27715	−0.38
0.40	−0.00900	0.03340	0.09486	0.14671	0.20506	0.29853	−0.40
0.42	0.01472	0.05708	0.11832	0.16985	0.22770	0.32001	−0.42
0.44	0.03891	0.08118	0.14214	0.19329	0.25055	0.34161	−0.44
0.46	0.06360	0.10572	0.16631	0.21702	0.27363	0.36333	−0.46
0.48	0.08879	0.13071	0.19086	0.24105	0.29694	0.38516	−0.48
0.50	0.11451	0.15617	0.21579	0.26539	0.32048	0.40711	−0.50
0.52	0.14076	0.18211	0.24110	0.29005	0.34426	0.42918	−0.52
0.54	0.16758	0.20854	0.26682	0.31503	0.36827	0.45137	−0.54
0.56	0.19498	0.23548	0.29295	0.34035	0.39254	0.47368	−0.56
0.58	0.22298	0.26294	0.31950	0.36600	0.41705	0.49612	−0.58
0.60	0.25159	0.29095	0.34648	0.39200	0.44182	0.51868	−0.60
0.62	0.28086	0.31952	0.37391	0.41835	0.46686	0.54138	−0.62
0.64	0.31079	0.34867	0.40180	0.44507	0.49215	0.56421	−0.64
0.66	0.34141	0.37842	0.43016	0.47216	0.51772	0.58716	−0.66
0.68	0.37275	0.40879	0.45900	0.49963	0.54357	0.61026	−0.68
0.70	0.40484	0.43979	0.48835	0.52750	0.56970	0.63349	−0.70
0.72	0.43771	0.47146	0.51821	0.55576	0.59612	0.65686	−0.72
0.74	0.47138	0.50382	0.54859	0.58444	0.62283	0.68037	−0.74
0.76	0.50588	0.53689	0.57952	0.61353	0.64984	0.70403	−0.76
0.78	0.54126	0.57069	0.61101	0.64306	0.67716	0.72783	−0.78
0.80	0.57755	0.60526	0.64308	0.67304	0.70480	0.75178	−0.80
0.82	0.61479	0.64061	0.67575	0.70347	0.73275	0.77589	−0.82
0.84	0.65301	0.67679	0.70903	0.73436	0.76103	0.80014	−0.84
0.86	0.69225	0.71383	0.74295	0.76574	0.78965	0.82455	−0.86
0.88	0.73258	0.75175	0.77752	0.79761	0.81861	0.84912	−0.88
0.90	0.77402	0.79059	0.81277	0.82999	0.84792	0.87385	−0.90
0.92	0.81664	0.83039	0.84872	0.86289	0.87759	0.89874	−0.92
0.94	0.86049	0.87119	0.88539	0.89632	0.90762	0.92380	−0.94
0.96	0.90563	0.91303	0.92281	0.93031	0.93803	0.94903	−0.96
0.98	0.95211	0.95595	0.96101	0.96486	0.96882	0.97443	−0.98

↑ r_0

α =	.995 ↑−	.99 ↑−	.975 ↑−	.95 ↑−	.90 ↑−	.75 ↑−

TABLE 38.1   SAMPLE SIZE =   40

α =	.005 ↓+	.01 ↓+	.025 ↓+	.05 ↓+	.10 ↓+	.25 ↓+	
r_o ↓							
-0.98	-0.99114	-0.99037	-0.98913	-0.98794	-0.98643	-0.98348	0.98
-0.96	-0.98220	-0.98065	-0.97817	-0.97580	-0.97278	-0.96692	0.96
-0.94	-0.97315	-0.97084	-0.96711	-0.96357	-0.95905	-0.95032	0.94
-0.92	-0.96401	-0.96092	-0.95596	-0.95125	-0.94525	-0.93367	0.92
-0.90	-0.95477	-0.95091	-0.94472	-0.93884	-0.93136	-0.91698	0.90
-0.88	-0.94543	-0.94080	-0.93338	-0.92634	-0.91740	-0.90024	0.88
-0.86	-0.93599	-0.93059	-0.92193	-0.91374	-0.90336	-0.88346	0.86
-0.84	-0.92644	-0.92027	-0.91039	-0.90105	-0.88923	-0.86663	0.84
-0.82	-0.91679	-0.90985	-0.89875	-0.88827	-0.87502	-0.84976	0.82
-0.80	-0.90703	-0.89932	-0.88700	-0.87539	-0.86072	-0.83283	0.80
-0.78	-0.89716	-0.88868	-0.87515	-0.86240	-0.84634	-0.81586	0.78
-0.76	-0.88718	-0.87793	-0.86319	-0.84932	-0.83187	-0.79885	0.76
-0.74	-0.87709	-0.86707	-0.85112	-0.83614	-0.81732	-0.78178	0.74
-0.72	-0.86688	-0.85609	-0.83894	-0.82285	-0.80267	-0.76466	0.72
-0.70	-0.85655	-0.84499	-0.82664	-0.80946	-0.78794	-0.74749	0.70
-0.68	-0.84610	-0.83378	-0.81424	-0.79597	-0.77311	-0.73027	0.68
-0.66	-0.83553	-0.82244	-0.80172	-0.78236	-0.75819	-0.71299	0.66
-0.64	-0.82483	-0.81098	-0.78908	-0.76865	-0.74318	-0.69567	0.64
-0.62	-0.81400	-0.79940	-0.77632	-0.75483	-0.72807	-0.67829	0.62
-0.60	-0.80305	-0.78768	-0.76343	-0.74089	-0.71286	-0.66085	0.60
-0.58	-0.79196	-0.77584	-0.75043	-0.72684	-0.69756	-0.64336	0.58
-0.56	-0.78074	-0.76386	-0.73730	-0.71267	-0.68216	-0.62581	0.56
-0.54	-0.76938	-0.75175	-0.72404	-0.69839	-0.66665	-0.60821	0.54
-0.52	-0.75787	-0.73950	-0.71065	-0.68399	-0.65105	-0.59055	0.52
-0.50	-0.74623	-0.72711	-0.69712	-0.66946	-0.63534	-0.57283	0.50
-0.48	-0.73443	-0.71457	-0.68347	-0.65481	-0.61953	-0.55505	0.48
-0.46	-0.72249	-0.70189	-0.66967	-0.64004	-0.60361	-0.53722	0.46
-0.44	-0.71039	-0.68906	-0.65574	-0.62514	-0.58759	-0.51932	0.44
-0.42	-0.69814	-0.67608	-0.64166	-0.61011	-0.57145	-0.50136	0.42
-0.40	-0.68573	-0.66294	-0.62744	-0.59495	-0.55520	-0.48333	0.40
-0.38	-0.67315	-0.64964	-0.61307	-0.57965	-0.53885	-0.46525	0.38
-0.36	-0.66041	-0.63619	-0.59856	-0.56422	-0.52237	-0.44710	0.36
-0.34	-0.64750	-0.62256	-0.58388	-0.54865	-0.50579	-0.42888	0.34
-0.32	-0.63441	-0.60877	-0.56906	-0.53294	-0.48908	-0.41060	0.32
-0.30	-0.62115	-0.59481	-0.55407	-0.51709	-0.47226	-0.39225	0.30
-0.28	-0.60770	-0.58067	-0.53893	-0.50110	-0.45531	-0.37384	0.28
-0.26	-0.59407	-0.56636	-0.52362	-0.48495	-0.43824	-0.35536	0.26
-0.24	-0.58025	-0.55186	-0.50814	-0.46866	-0.42105	-0.33680	0.24
-0.22	-0.56623	-0.53717	-0.49250	-0.45222	-0.40373	-0.31818	0.22
-0.20	-0.55202	-0.52230	-0.47667	-0.43562	-0.38629	-0.29948	0.20
-0.18	-0.53760	-0.50723	-0.46068	-0.41886	-0.36871	-0.28071	0.18
-0.16	-0.52297	-0.49196	-0.44450	-0.40194	-0.35101	-0.26187	0.16
-0.14	-0.50813	-0.47648	-0.42814	-0.38486	-0.33316	-0.24296	0.14
-0.12	-0.49307	-0.46080	-0.41159	-0.36761	-0.31519	-0.22397	0.12
-0.10	-0.47779	-0.44491	-0.39484	-0.35020	-0.29707	-0.20490	0.10
-0.08	-0.46228	-0.42880	-0.37791	-0.33261	-0.27881	-0.18575	0.08
-0.06	-0.44653	-0.41247	-0.36077	-0.31485	-0.26042	-0.16653	0.06
-0.04	-0.43054	-0.39591	-0.34343	-0.29691	-0.24187	-0.14723	0.04
-0.02	-0.41431	-0.37912	-0.32589	-0.27879	-0.22318	-0.12784	0.02

↑ r_o

α =	.995 ↑−	.99 ↑−	.975 ↑−	.95 ↑−	.90 ↑−	.75 ↑−

TABLE 38.2   SAMPLE SIZE =   40   (CONT.)

α =	.005 ↓+	.01 ↓+	.025 ↓+	.05 ↓+	.10 ↓+	.25 ↓+	
r_0 ↓							
0.00	−0.39782	−0.36209	−0.30813	−0.26048	−0.20434	−0.10838	0.00
0.02	−0.38108	−0.34481	−0.29016	−0.24199	−0.18535	−0.08883	−0.02
0.04	−0.36407	−0.32729	−0.27196	−0.22330	−0.16621	−0.06920	−0.04
0.06	−0.34678	−0.30951	−0.25355	−0.20442	−0.14691	−0.04948	−0.06
0.08	−0.32922	−0.29147	−0.23490	−0.18534	−0.12745	−0.02968	−0.08
0.10	−0.31137	−0.27317	−0.21601	−0.16606	−0.10783	−0.00978	−0.10
0.12	−0.29322	−0.25458	−0.19689	−0.14657	−0.08804	0.01020	−0.12
0.14	−0.27477	−0.23572	−0.17752	−0.12688	−0.06809	0.03027	−0.14
0.16	−0.25601	−0.21657	−0.15790	−0.10696	−0.04796	0.05043	−0.16
0.18	−0.23693	−0.19712	−0.13803	−0.08683	−0.02767	0.07068	−0.18
0.20	−0.21752	−0.17737	−0.11789	−0.06648	−0.00720	0.09103	−0.20
0.22	−0.19777	−0.15731	−0.09749	−0.04590	0.01345	0.11147	−0.22
0.24	−0.17768	−0.13693	−0.07681	−0.02509	0.03428	0.13201	−0.24
0.26	−0.15722	−0.11622	−0.05585	−0.00404	0.05530	0.15265	−0.26
0.28	−0.13640	−0.09517	−0.03461	0.01725	0.07650	0.17339	−0.28
0.30	−0.11520	−0.07378	−0.01307	0.03879	0.09789	0.19423	−0.30
0.32	−0.09361	−0.05204	0.00876	0.06058	0.11948	0.21517	−0.32
0.34	−0.07162	−0.02993	0.03091	0.08262	0.14127	0.23621	−0.34
0.36	−0.04921	−0.00744	0.05337	0.10493	0.16326	0.25736	−0.36
0.38	−0.02638	0.01543	0.07615	0.12750	0.18545	0.27861	−0.38
0.40	−0.00311	0.03869	0.09926	0.15035	0.20785	0.29997	−0.40
0.42	0.02062	0.06237	0.12271	0.17347	0.23047	0.32145	−0.42
0.44	0.04482	0.08646	0.14650	0.19689	0.25330	0.34303	−0.44
0.46	0.06949	0.11098	0.17066	0.22059	0.27635	0.36473	−0.46
0.48	0.09467	0.13595	0.19517	0.24459	0.29963	0.38654	−0.48
0.50	0.12037	0.16138	0.22006	0.26889	0.32313	0.40846	−0.50
0.52	0.14659	0.18728	0.24533	0.29350	0.34687	0.43051	−0.52
0.54	0.17337	0.21366	0.27100	0.31844	0.37084	0.45267	−0.54
0.56	0.20072	0.24055	0.29707	0.34370	0.39506	0.47496	−0.56
0.58	0.22865	0.26795	0.32356	0.36929	0.41952	0.49737	−0.58
0.60	0.25720	0.29588	0.35047	0.39522	0.44423	0.51990	−0.60
0.62	0.28638	0.32437	0.37781	0.42150	0.46920	0.54256	−0.62
0.64	0.31621	0.35342	0.40561	0.44814	0.49444	0.56535	−0.64
0.66	0.34673	0.38306	0.43387	0.47514	0.51993	0.58827	−0.66
0.68	0.37794	0.41330	0.46261	0.50252	0.54570	0.61132	−0.68
0.70	0.40989	0.44418	0.49183	0.53028	0.57175	0.63451	−0.70
0.72	0.44259	0.47570	0.52156	0.55843	0.59808	0.65784	−0.72
0.74	0.47609	0.50789	0.55180	0.58699	0.62470	0.68130	−0.74
0.76	0.51040	0.54078	0.58258	0.61595	0.65161	0.70490	−0.76
0.78	0.54556	0.57438	0.61391	0.64535	0.67883	0.72865	−0.78
0.80	0.58161	0.60873	0.64580	0.67517	0.70635	0.75254	−0.80
0.82	0.61858	0.64386	0.67827	0.70544	0.73418	0.77658	−0.82
0.84	0.65651	0.67978	0.71134	0.73617	0.76234	0.80077	−0.84
0.86	0.69545	0.71654	0.74504	0.76737	0.79082	0.82512	−0.86
0.88	0.73542	0.75416	0.77937	0.79905	0.81964	0.84962	−0.88
0.90	0.77649	0.79267	0.81436	0.83122	0.84880	0.87427	−0.90
0.92	0.81870	0.83212	0.85003	0.86390	0.87831	0.89909	−0.92
0.94	0.86210	0.87253	0.88641	0.89711	0.90817	0.92406	−0.94
0.96	0.90674	0.91396	0.92351	0.93085	0.93841	0.94921	−0.96
0.98	0.95269	0.95643	0.96137	0.96514	0.96901	0.97452	−0.98

↑ r_0

α =	.995 ↑−	.99 ↑−	.975 ↑−	.95 ↑−	.90 ↑−	.75 ↑−

TABLE 39.1   SAMPLE SIZE =   41

α =	.005 ↓+	.01 ↓+	.025 ↓+	.05 ↓+	.10 ↓+	.25 ↓+	
r_o ↓							
-0.98	-0.99105	-0.99028	-0.98905	-0.98787	-0.98636	-0.98345	0.98
-0.96	-0.98201	-0.98047	-0.97800	-0.97565	-0.97265	-0.96685	0.96
-0.94	-0.97287	-0.97057	-0.96686	-0.96334	-0.95886	-0.95021	0.94
-0.92	-0.96364	-0.96056	-0.95563	-0.95095	-0.94499	-0.93353	0.92
-0.90	-0.95430	-0.95046	-0.94430	-0.93846	-0.93105	-0.91680	0.90
-0.88	-0.94487	-0.94026	-0.93288	-0.92589	-0.91702	-0.90003	0.88
-0.86	-0.93533	-0.92996	-0.92135	-0.91322	-0.90291	-0.88321	0.86
-0.84	-0.92569	-0.91955	-0.90973	-0.90045	-0.88872	-0.86635	0.84
-0.82	-0.91594	-0.90903	-0.89800	-0.88759	-0.87445	-0.84944	0.82
-0.80	-0.90609	-0.89841	-0.88617	-0.87464	-0.86009	-0.83249	0.80
-0.78	-0.89613	-0.88768	-0.87423	-0.86158	-0.84565	-0.81548	0.78
-0.76	-0.88605	-0.87684	-0.86219	-0.84843	-0.83112	-0.79843	0.76
-0.74	-0.87586	-0.86589	-0.85004	-0.83517	-0.81651	-0.78133	0.74
-0.72	-0.86555	-0.85482	-0.83777	-0.82181	-0.80180	-0.76418	0.72
-0.70	-0.85513	-0.84363	-0.82540	-0.80835	-0.78701	-0.74697	0.70
-0.68	-0.84458	-0.83233	-0.81291	-0.79478	-0.77212	-0.72972	0.68
-0.66	-0.83391	-0.82090	-0.80031	-0.78111	-0.75714	-0.71241	0.66
-0.64	-0.82312	-0.80935	-0.78759	-0.76732	-0.74207	-0.69505	0.64
-0.62	-0.81220	-0.79767	-0.77475	-0.75343	-0.72691	-0.67764	0.62
-0.60	-0.80115	-0.78587	-0.76178	-0.73942	-0.71164	-0.66018	0.60
-0.58	-0.78996	-0.77393	-0.74870	-0.72530	-0.69628	-0.64266	0.58
-0.56	-0.77864	-0.76187	-0.73549	-0.71106	-0.68083	-0.62508	0.56
-0.54	-0.76718	-0.74966	-0.72215	-0.69671	-0.66527	-0.60745	0.54
-0.52	-0.75558	-0.73732	-0.70868	-0.68224	-0.64961	-0.58976	0.52
-0.50	-0.74384	-0.72484	-0.69508	-0.66765	-0.63385	-0.57201	0.50
-0.48	-0.73195	-0.71222	-0.68134	-0.65293	-0.61798	-0.55420	0.48
-0.46	-0.71991	-0.69945	-0.66747	-0.63809	-0.60201	-0.53633	0.46
-0.44	-0.70772	-0.68653	-0.65346	-0.62313	-0.58593	-0.51841	0.44
-0.42	-0.69537	-0.67346	-0.63931	-0.60803	-0.56975	-0.50042	0.42
-0.40	-0.68286	-0.66023	-0.62501	-0.59280	-0.55345	-0.48237	0.40
-0.38	-0.67019	-0.64685	-0.61057	-0.57745	-0.53704	-0.46426	0.38
-0.36	-0.65735	-0.63330	-0.59598	-0.56195	-0.52052	-0.44608	0.36
-0.34	-0.64435	-0.61959	-0.58123	-0.54632	-0.50389	-0.42784	0.34
-0.32	-0.63116	-0.60572	-0.56633	-0.53055	-0.48713	-0.40953	0.32
-0.30	-0.61781	-0.59167	-0.55128	-0.51464	-0.47026	-0.39116	0.30
-0.28	-0.60427	-0.57745	-0.53606	-0.49859	-0.45327	-0.37272	0.28
-0.26	-0.59054	-0.56305	-0.52068	-0.48239	-0.43616	-0.35422	0.26
-0.24	-0.57663	-0.54847	-0.50514	-0.46604	-0.41893	-0.33564	0.24
-0.22	-0.56252	-0.53370	-0.48942	-0.44954	-0.40157	-0.31699	0.22
-0.20	-0.54822	-0.51874	-0.47353	-0.43288	-0.38408	-0.29828	0.20
-0.18	-0.53371	-0.50359	-0.45747	-0.41607	-0.36646	-0.27949	0.18
-0.16	-0.51899	-0.48824	-0.44123	-0.39910	-0.34872	-0.26063	0.16
-0.14	-0.50406	-0.47269	-0.42480	-0.38197	-0.33084	-0.24169	0.14
-0.12	-0.48892	-0.45693	-0.40819	-0.36467	-0.31283	-0.22268	0.12
-0.10	-0.47355	-0.44097	-0.39139	-0.34721	-0.29467	-0.20360	0.10
-0.08	-0.45795	-0.42478	-0.37439	-0.32958	-0.27639	-0.18444	0.08
-0.06	-0.44212	-0.40838	-0.35720	-0.31177	-0.25796	-0.16520	0.06
-0.04	-0.42605	-0.39175	-0.33980	-0.29379	-0.23938	-0.14588	0.04
-0.02	-0.40974	-0.37488	-0.32221	-0.27562	-0.22066	-0.12648	0.02

↑ r_o

α =   .995 ↑-   .99 ↑-   .975 ↑-   .95 ↑-   .90 ↑-   .75 ↑-

TABLE 39.2   SAMPLE SIZE =   41   (CONT.)

α =	.005 ↓+	.01 ↓+	.025 ↓+	.05 ↓+	.10 ↓+	.25 ↓+	
r_o ↓							
0.00	−0.39317	−0.35779	−0.30440	−0.25728	−0.20180	−0.10701	0.00
0.02	−0.37635	−0.34045	−0.28637	−0.23875	−0.18278	−0.08745	−0.02
0.04	−0.35926	−0.32286	−0.26813	−0.22002	−0.16361	−0.06780	−0.04
0.06	−0.34191	−0.30502	−0.24967	−0.20111	−0.14429	−0.04807	−0.06
0.08	−0.32427	−0.28692	−0.23097	−0.18200	−0.12481	−0.02826	−0.08
0.10	−0.30635	−0.26856	−0.21205	−0.16269	−0.10517	−0.00836	−0.10
0.12	−0.28814	−0.24992	−0.19289	−0.14317	−0.08536	0.01162	−0.12
0.14	−0.26962	−0.23100	−0.17348	−0.12345	−0.06539	0.03170	−0.14
0.16	−0.25080	−0.21180	−0.15383	−0.10351	−0.04526	0.05187	−0.16
0.18	−0.23166	−0.19231	−0.13392	−0.08336	−0.02495	0.07212	−0.18
0.20	−0.21220	−0.17251	−0.11376	−0.06299	−0.00447	0.09247	−0.20
0.22	−0.19239	−0.15241	−0.09333	−0.04239	0.01618	0.11292	−0.22
0.24	−0.17225	−0.13199	−0.07263	−0.02157	0.03702	0.13346	−0.24
0.26	−0.15175	−0.11125	−0.05165	−0.00051	0.05804	0.15409	−0.26
0.28	−0.13089	−0.09018	−0.03039	0.02079	0.07924	0.17483	−0.28
0.30	−0.10965	−0.06876	−0.00884	0.04233	0.10063	0.19566	−0.30
0.32	−0.08803	−0.04699	0.01300	0.06411	0.12221	0.21659	−0.32
0.34	−0.06601	−0.02487	0.03515	0.08615	0.14399	0.23763	−0.34
0.36	−0.04358	−0.00237	0.05761	0.10845	0.16597	0.25877	−0.36
0.38	−0.02073	0.02051	0.08038	0.13101	0.18815	0.28001	−0.38
0.40	0.00256	0.04377	0.10348	0.15385	0.21053	0.30136	−0.40
0.42	0.02629	0.06744	0.12692	0.17695	0.23312	0.32282	−0.42
0.44	0.05048	0.09153	0.15070	0.20034	0.25593	0.34439	−0.44
0.46	0.07515	0.11603	0.17482	0.22401	0.27896	0.36607	−0.46
0.48	0.10032	0.14098	0.19931	0.24798	0.30220	0.38786	−0.48
0.50	0.12599	0.16637	0.22416	0.27224	0.32567	0.40977	−0.50
0.52	0.15218	0.19223	0.24939	0.29682	0.34937	0.43179	−0.52
0.54	0.17892	0.21857	0.27501	0.32170	0.37330	0.45393	−0.54
0.56	0.20622	0.24540	0.30102	0.34691	0.39747	0.47618	−0.56
0.58	0.23409	0.27274	0.32744	0.37244	0.42188	0.49856	−0.58
0.60	0.26257	0.30060	0.35428	0.39830	0.44654	0.52107	−0.60
0.62	0.29166	0.32900	0.38155	0.42451	0.47145	0.54369	−0.62
0.64	0.32140	0.35796	0.40926	0.45107	0.49662	0.56645	−0.64
0.66	0.35181	0.38749	0.43742	0.47799	0.52205	0.58933	−0.66
0.68	0.38290	0.41762	0.46605	0.50527	0.54775	0.61234	−0.68
0.70	0.41471	0.44837	0.49516	0.53294	0.57371	0.63549	−0.70
0.72	0.44726	0.47974	0.52476	0.56098	0.59996	0.65877	−0.72
0.74	0.48058	0.51178	0.55487	0.58942	0.62649	0.68218	−0.74
0.76	0.51471	0.54449	0.58550	0.61827	0.65331	0.70574	−0.76
0.78	0.54966	0.57791	0.61667	0.64753	0.68042	0.72943	−0.78
0.80	0.58548	0.61205	0.64838	0.67721	0.70783	0.75327	−0.80
0.82	0.62220	0.64695	0.68067	0.70733	0.73555	0.77725	−0.82
0.84	0.65985	0.68263	0.71355	0.73789	0.76359	0.80138	−0.84
0.86	0.69848	0.71912	0.74703	0.76892	0.79194	0.82566	−0.86
0.88	0.73813	0.75645	0.78113	0.80041	0.82062	0.85009	−0.88
0.90	0.77884	0.79465	0.81587	0.83239	0.84964	0.87468	−0.90
0.92	0.82065	0.83376	0.85128	0.86487	0.87900	0.89942	−0.92
0.94	0.86362	0.87381	0.88738	0.89785	0.90870	0.92432	−0.94
0.96	0.90780	0.91484	0.92418	0.93136	0.93877	0.94938	−0.96
0.98	0.95324	0.95689	0.96171	0.96540	0.96920	0.97461	−0.98

↑ r_o

α =	.995 ↑−	.99 ↑−	.975 ↑−	.95 ↑−	.90 ↑−	.75 ↑−

TABLE 40.1   SAMPLE SIZE =   42

α =	.005 ↓+	.01 ↓+	.025 ↓+	.05 ↓+	.10 ↓+	.25 ↓+	
r_o ↓							
-0.98	-0.99096	-0.99020	-0.98897	-0.98780	-0.98630	-0.98341	0.98
-0.96	-0.98183	-0.98030	-0.97784	-0.97550	-0.97253	-0.96678	0.96
-0.94	-0.97260	-0.97030	-0.96662	-0.96312	-0.95868	-0.95011	0.94
-0.92	-0.96327	-0.96021	-0.95531	-0.95066	-0.94475	-0.93339	0.92
-0.90	-0.95385	-0.95002	-0.94390	-0.93810	-0.93074	-0.91663	0.90
-0.88	-0.94432	-0.93973	-0.93239	-0.92545	-0.91665	-0.89983	0.88
-0.86	-0.93469	-0.92934	-0.92079	-0.91271	-0.90248	-0.88298	0.86
-0.84	-0.92496	-0.91885	-0.90908	-0.89987	-0.88824	-0.86608	0.84
-0.82	-0.91512	-0.90824	-0.89727	-0.88694	-0.87390	-0.84914	0.82
-0.80	-0.90518	-0.89754	-0.88536	-0.87391	-0.85949	-0.83215	0.80
-0.78	-0.89512	-0.88672	-0.87334	-0.86078	-0.84498	-0.81511	0.78
-0.76	-0.88495	-0.87579	-0.86122	-0.84756	-0.83040	-0.79803	0.76
-0.74	-0.87467	-0.86474	-0.84899	-0.83423	-0.81572	-0.78089	0.74
-0.72	-0.86427	-0.85359	-0.83665	-0.82080	-0.80096	-0.76371	0.72
-0.70	-0.85375	-0.84231	-0.82419	-0.80727	-0.78611	-0.74648	0.70
-0.68	-0.84311	-0.83092	-0.81163	-0.79363	-0.77117	-0.72919	0.68
-0.66	-0.83235	-0.81940	-0.79894	-0.77989	-0.75613	-0.71185	0.66
-0.64	-0.82146	-0.80777	-0.78615	-0.76603	-0.74100	-0.69446	0.64
-0.62	-0.81044	-0.79600	-0.77323	-0.75207	-0.72578	-0.67702	0.62
-0.60	-0.79930	-0.78411	-0.76019	-0.73800	-0.71047	-0.65953	0.60
-0.58	-0.78802	-0.77209	-0.74702	-0.72381	-0.69505	-0.64197	0.58
-0.56	-0.77661	-0.75993	-0.73373	-0.70951	-0.67954	-0.62437	0.56
-0.54	-0.76505	-0.74764	-0.72032	-0.69509	-0.66393	-0.60671	0.54
-0.52	-0.75336	-0.73521	-0.70677	-0.68055	-0.64822	-0.58899	0.52
-0.50	-0.74153	-0.72265	-0.69310	-0.66589	-0.63240	-0.57121	0.50
-0.48	-0.72954	-0.70994	-0.67929	-0.65111	-0.61649	-0.55338	0.48
-0.46	-0.71741	-0.69708	-0.66534	-0.63621	-0.60047	-0.53548	0.46
-0.44	-0.70512	-0.68407	-0.65126	-0.62118	-0.58434	-0.51753	0.44
-0.42	-0.69268	-0.67092	-0.63703	-0.60602	-0.56810	-0.49951	0.42
-0.40	-0.68008	-0.65761	-0.62266	-0.59073	-0.55176	-0.48144	0.40
-0.38	-0.66732	-0.64414	-0.60814	-0.57531	-0.53530	-0.46330	0.38
-0.36	-0.65439	-0.63051	-0.59348	-0.55976	-0.51873	-0.44510	0.36
-0.34	-0.64129	-0.61672	-0.57867	-0.54407	-0.50205	-0.42683	0.34
-0.32	-0.62802	-0.60276	-0.56370	-0.52824	-0.48525	-0.40850	0.32
-0.30	-0.61457	-0.58863	-0.54857	-0.51227	-0.46834	-0.39011	0.30
-0.28	-0.60094	-0.57433	-0.53329	-0.49616	-0.45131	-0.37165	0.28
-0.26	-0.58713	-0.55985	-0.51784	-0.47991	-0.43415	-0.35312	0.26
-0.24	-0.57313	-0.54518	-0.50223	-0.46350	-0.41688	-0.33452	0.24
-0.22	-0.55893	-0.53034	-0.48645	-0.44695	-0.39948	-0.31585	0.22
-0.20	-0.54454	-0.51530	-0.47050	-0.43024	-0.38195	-0.29712	0.20
-0.18	-0.52994	-0.50008	-0.45437	-0.41338	-0.36430	-0.27831	0.18
-0.16	-0.51514	-0.48465	-0.43807	-0.39636	-0.34651	-0.25943	0.16
-0.14	-0.50013	-0.46903	-0.42158	-0.37918	-0.32860	-0.24048	0.14
-0.12	-0.48490	-0.45320	-0.40491	-0.36184	-0.31055	-0.22145	0.12
-0.10	-0.46945	-0.43715	-0.38805	-0.34433	-0.29237	-0.20235	0.10
-0.08	-0.45377	-0.42090	-0.37100	-0.32665	-0.27404	-0.18317	0.08
-0.06	-0.43786	-0.40443	-0.35375	-0.30880	-0.25558	-0.16392	0.06
-0.04	-0.42172	-0.38773	-0.33630	-0.29078	-0.23698	-0.14459	0.04
-0.02	-0.40532	-0.37080	-0.31865	-0.27257	-0.21823	-0.12517	0.02

↑ r_o

α =	.995 ↑-	.99 ↑-	.975 ↑-	.95 ↑-	.90 ↑-	.75 ↑-

TABLE 40.2 SAMPLE SIZE = 42 (CONT.)

α =	.005 ↓+	.01 ↓+	.025 ↓+	.05 ↓+	.10 ↓+	.25 ↓+	
r_0 ↓							
0.00	-0.38868	-0.35364	-0.30079	-0.25419	-0.19934	-0.10568	0.00
0.02	-0.37179	-0.33623	-0.28272	-0.23562	-0.18030	-0.08611	-0.02
0.04	-0.35463	-0.31859	-0.26443	-0.21686	-0.16111	-0.06646	-0.04
0.06	-0.33720	-0.30069	-0.24593	-0.19792	-0.14176	-0.04672	-0.06
0.08	-0.31950	-0.28253	-0.22719	-0.17878	-0.12226	-0.02690	-0.08
0.10	-0.30151	-0.26411	-0.20823	-0.15944	-0.10260	-0.00699	-0.10
0.12	-0.28323	-0.24542	-0.18903	-0.13990	-0.08278	0.01300	-0.12
0.14	-0.26466	-0.22646	-0.16959	-0.12015	-0.06280	0.03308	-0.14
0.16	-0.24578	-0.20721	-0.14990	-0.10019	-0.04265	0.05325	-0.16
0.18	-0.22658	-0.18767	-0.12997	-0.08002	-0.02234	0.07351	-0.18
0.20	-0.20706	-0.16783	-0.10978	-0.05963	-0.00185	0.09386	-0.20
0.22	-0.18721	-0.14769	-0.08932	-0.03902	0.01881	0.11431	-0.22
0.24	-0.16702	-0.12724	-0.06860	-0.01819	0.03965	0.13485	-0.24
0.26	-0.14648	-0.10647	-0.04761	0.00288	0.06067	0.15548	-0.26
0.28	-0.12558	-0.08537	-0.02634	0.02418	0.08187	0.17621	-0.28
0.30	-0.10431	-0.06393	-0.00478	0.04572	0.10326	0.19704	-0.30
0.32	-0.08266	-0.04214	0.01707	0.06751	0.12484	0.21797	-0.32
0.34	-0.06061	-0.02000	0.03922	0.08955	0.14661	0.23899	-0.34
0.36	-0.03816	0.00251	0.06168	0.11184	0.16857	0.26012	-0.36
0.38	-0.01530	0.02539	0.08445	0.13439	0.19074	0.28136	-0.38
0.40	0.00800	0.04866	0.10754	0.15720	0.21310	0.30270	-0.40
0.42	0.03174	0.07232	0.13096	0.18029	0.23568	0.32414	-0.42
0.44	0.05593	0.09639	0.15472	0.20366	0.25846	0.34569	-0.44
0.46	0.08059	0.12088	0.17882	0.22730	0.28146	0.36736	-0.46
0.48	0.10574	0.14580	0.20328	0.25124	0.30467	0.38913	-0.48
0.50	0.13138	0.17117	0.22809	0.27546	0.32811	0.41101	-0.50
0.52	0.15754	0.19699	0.25328	0.29999	0.35177	0.43301	-0.52
0.54	0.18424	0.22328	0.27885	0.32483	0.37566	0.45513	-0.54
0.56	0.21149	0.25006	0.30481	0.34999	0.39979	0.47736	-0.56
0.58	0.23930	0.27733	0.33117	0.37546	0.42415	0.49971	-0.58
0.60	0.26771	0.30512	0.35794	0.40126	0.44876	0.52219	-0.60
0.62	0.29673	0.33344	0.38513	0.42740	0.47361	0.54478	-0.62
0.64	0.32637	0.36231	0.41275	0.45389	0.49871	0.56750	-0.64
0.66	0.35667	0.39174	0.44082	0.48072	0.52408	0.59035	-0.66
0.68	0.38765	0.42176	0.46935	0.50792	0.54970	0.61332	-0.68
0.70	0.41932	0.45237	0.49835	0.53548	0.57559	0.63643	-0.70
0.72	0.45172	0.48361	0.52783	0.56342	0.60176	0.65966	-0.72
0.74	0.48488	0.51549	0.55781	0.59175	0.62820	0.68303	-0.74
0.76	0.51882	0.54804	0.58829	0.62048	0.65493	0.70654	-0.76
0.78	0.55357	0.58127	0.61931	0.64961	0.68194	0.73018	-0.78
0.80	0.58917	0.61522	0.65086	0.67916	0.70925	0.75397	-0.80
0.82	0.62565	0.64990	0.68297	0.70913	0.73686	0.77789	-0.82
0.84	0.66304	0.68534	0.71565	0.73954	0.76478	0.80196	-0.84
0.86	0.70138	0.72158	0.74893	0.77040	0.79301	0.82618	-0.86
0.88	0.74071	0.75863	0.78281	0.80172	0.82156	0.85055	-0.88
0.90	0.78107	0.79653	0.81732	0.83351	0.85044	0.87506	-0.90
0.92	0.82250	0.83532	0.85247	0.86579	0.87966	0.89973	-0.92
0.94	0.86507	0.87502	0.88830	0.89856	0.90921	0.92456	-0.94
0.96	0.90880	0.91568	0.92481	0.93184	0.93911	0.94954	-0.96
0.98	0.95376	0.95732	0.96204	0.96565	0.96938	0.97469	-0.98

↑ r_0

α = .995 ↑-   .99 ↑-   .975 ↑-   .95 ↑-   .90 ↑-   .75 ↑-

TABLE 41.1   SAMPLE SIZE =   43

α =	.005 ↓+	.01 ↓+	.025 ↓+	.05 ↓+	.10 ↓+	.25 ↓+	
r_0 ↓							
-0.98	-0.99087	-0.99011	-0.98889	-0.98772	-0.98624	-0.98338	0.98
-0.96	-0.98165	-0.98013	-0.97768	-0.97536	-0.97241	-0.96672	0.96
-0.94	-0.97233	-0.97005	-0.96638	-0.96291	-0.95850	-0.95001	0.94
-0.92	-0.96292	-0.95987	-0.95499	-0.95037	-0.94451	-0.93326	0.92
-0.90	-0.95341	-0.94960	-0.94350	-0.93774	-0.93044	-0.91647	0.90
-0.88	-0.94379	-0.93922	-0.93192	-0.92502	-0.91630	-0.89963	0.88
-0.86	-0.93407	-0.92875	-0.92024	-0.91221	-0.90207	-0.88275	0.86
-0.84	-0.92425	-0.91816	-0.90845	-0.89930	-0.88776	-0.86582	0.84
-0.82	-0.91432	-0.90748	-0.89657	-0.88630	-0.87337	-0.84885	0.82
-0.80	-0.90429	-0.89668	-0.88458	-0.87321	-0.85890	-0.83182	0.80
-0.78	-0.89414	-0.88578	-0.87248	-0.86001	-0.84434	-0.81476	0.78
-0.76	-0.88388	-0.87476	-0.86028	-0.84672	-0.82970	-0.79764	0.76
-0.74	-0.87351	-0.86363	-0.84797	-0.83332	-0.81497	-0.78047	0.74
-0.72	-0.86302	-0.85239	-0.83556	-0.81982	-0.80015	-0.76326	0.72
-0.70	-0.85241	-0.84103	-0.82303	-0.80622	-0.78524	-0.74599	0.70
-0.68	-0.84168	-0.82955	-0.81038	-0.79252	-0.77024	-0.72868	0.68
-0.66	-0.83082	-0.81795	-0.79762	-0.77871	-0.75515	-0.71131	0.66
-0.64	-0.81985	-0.80623	-0.78475	-0.76479	-0.73997	-0.69389	0.64
-0.62	-0.80874	-0.79438	-0.77175	-0.75076	-0.72470	-0.67642	0.62
-0.60	-0.79751	-0.78240	-0.75864	-0.73662	-0.70933	-0.65890	0.60
-0.58	-0.78614	-0.77030	-0.74540	-0.72237	-0.69386	-0.64132	0.58
-0.56	-0.77463	-0.75806	-0.73204	-0.70800	-0.67830	-0.62368	0.56
-0.54	-0.76299	-0.74568	-0.71855	-0.69352	-0.66263	-0.60599	0.54
-0.52	-0.75121	-0.73317	-0.70493	-0.67892	-0.64687	-0.58824	0.52
-0.50	-0.73928	-0.72052	-0.69118	-0.66419	-0.63101	-0.57044	0.50
-0.48	-0.72721	-0.70772	-0.67730	-0.64935	-0.61504	-0.55258	0.48
-0.46	-0.71498	-0.69478	-0.66328	-0.63439	-0.59897	-0.53466	0.46
-0.44	-0.70261	-0.68170	-0.64912	-0.61930	-0.58279	-0.51668	0.44
-0.42	-0.69008	-0.66846	-0.63482	-0.60408	-0.56651	-0.49864	0.42
-0.40	-0.67739	-0.65506	-0.62038	-0.58873	-0.55012	-0.48054	0.40
-0.38	-0.66454	-0.64151	-0.60580	-0.57325	-0.53362	-0.46237	0.38
-0.36	-0.65152	-0.62781	-0.59107	-0.55764	-0.51701	-0.44415	0.36
-0.34	-0.63833	-0.61393	-0.57618	-0.54189	-0.50028	-0.42586	0.34
-0.32	-0.62498	-0.59989	-0.56115	-0.52601	-0.48344	-0.40751	0.32
-0.30	-0.61144	-0.58569	-0.54596	-0.50999	-0.46648	-0.38909	0.30
-0.28	-0.59772	-0.57130	-0.53061	-0.49382	-0.44941	-0.37061	0.28
-0.26	-0.58383	-0.55675	-0.51510	-0.47751	-0.43221	-0.35206	0.26
-0.24	-0.56974	-0.54201	-0.49942	-0.46105	-0.41490	-0.33344	0.24
-0.22	-0.55546	-0.52709	-0.48358	-0.44445	-0.39746	-0.31475	0.22
-0.20	-0.54098	-0.51198	-0.46756	-0.42769	-0.37989	-0.29600	0.20
-0.18	-0.52630	-0.49668	-0.45138	-0.41078	-0.36220	-0.27717	0.18
-0.16	-0.51142	-0.48118	-0.43501	-0.39371	-0.34438	-0.25827	0.16
-0.14	-0.49632	-0.46548	-0.41847	-0.37649	-0.32643	-0.23930	0.14
-0.12	-0.48102	-0.44958	-0.40174	-0.35910	-0.30835	-0.22026	0.12
-0.10	-0.46549	-0.43347	-0.38483	-0.34154	-0.29014	-0.20114	0.10
-0.08	-0.44973	-0.41715	-0.36772	-0.32382	-0.27178	-0.18195	0.08
-0.06	-0.43374	-0.40061	-0.35042	-0.30593	-0.25329	-0.16268	0.06
-0.04	-0.41752	-0.38384	-0.33292	-0.28787	-0.23466	-0.14334	0.04
-0.02	-0.40106	-0.36685	-0.31522	-0.26963	-0.21589	-0.12391	0.02

↑ r_0

α =   .995 ↑−   .99 ↑−   .975 ↑−   .95 ↑−   .90 ↑−   .75 ↑−

TABLE 41.2   SAMPLE SIZE =   43   (CONT.)

α =	.005 ↓+	.01 ↓+	.025 ↓+	.05 ↓+	.10 ↓+	.25 ↓+	
r_0 ↓							
0.00	-0.38434	-0.34963	-0.29732	-0.25121	-0.19697	-0.10441	0.00
0.02	-0.36738	-0.33216	-0.27920	-0.23261	-0.17791	-0.08483	-0.02
0.04	-0.35015	-0.31446	-0.26087	-0.21382	-0.15869	-0.06516	-0.04
0.06	-0.33265	-0.29650	-0.24232	-0.19484	-0.13933	-0.04542	-0.06
0.08	-0.31489	-0.27829	-0.22354	-0.17567	-0.11981	-0.02559	-0.08
0.10	-0.29684	-0.25982	-0.20454	-0.15630	-0.10013	-0.00568	-0.10
0.12	-0.27850	-0.24108	-0.18531	-0.13674	-0.08030	0.01432	-0.12
0.14	-0.25987	-0.22207	-0.16583	-0.11697	-0.06030	0.03441	-0.14
0.16	-0.24093	-0.20278	-0.14612	-0.09699	-0.04014	0.05458	-0.16
0.18	-0.22168	-0.18320	-0.12616	-0.07680	-0.01982	0.07485	-0.18
0.20	-0.20211	-0.16332	-0.10594	-0.05640	0.00068	0.09520	-0.20
0.22	-0.18222	-0.14315	-0.08546	-0.03577	0.02135	0.11565	-0.22
0.24	-0.16198	-0.12266	-0.06472	-0.01493	0.04219	0.13618	-0.24
0.26	-0.14140	-0.10186	-0.04372	0.00615	0.06321	0.15682	-0.26
0.28	-0.12047	-0.08073	-0.02243	0.02745	0.08441	0.17754	-0.28
0.30	-0.09916	-0.05927	-0.00086	0.04900	0.10579	0.19837	-0.30
0.32	-0.07748	-0.03747	0.02100	0.07078	0.12736	0.21929	-0.32
0.34	-0.05541	-0.01531	0.04315	0.09281	0.14912	0.24031	-0.34
0.36	-0.03294	0.00720	0.06560	0.11510	0.17108	0.26143	-0.36
0.38	-0.01006	0.03009	0.08837	0.13763	0.19323	0.28265	-0.38
0.40	0.01324	0.05336	0.11145	0.16044	0.21558	0.30398	-0.40
0.42	0.03698	0.07701	0.13485	0.18350	0.23813	0.32541	-0.42
0.44	0.06116	0.10107	0.15859	0.20684	0.26089	0.34695	-0.44
0.46	0.08582	0.12554	0.18266	0.23046	0.28386	0.36859	-0.46
0.48	0.11094	0.15044	0.20709	0.25436	0.30705	0.39035	-0.48
0.50	0.13657	0.17578	0.23187	0.27856	0.33045	0.41221	-0.50
0.52	0.16270	0.20156	0.25702	0.30305	0.35408	0.43419	-0.52
0.54	0.18935	0.22781	0.28254	0.32784	0.37793	0.45628	-0.54
0.56	0.21655	0.25453	0.30844	0.35294	0.40201	0.47849	-0.56
0.58	0.24431	0.28174	0.33474	0.37836	0.42633	0.50082	-0.58
0.60	0.27264	0.30946	0.36144	0.40410	0.45088	0.52326	-0.60
0.62	0.30158	0.33770	0.38856	0.43017	0.47568	0.54583	-0.62
0.64	0.33114	0.36648	0.41610	0.45658	0.50072	0.56851	-0.64
0.66	0.36133	0.39581	0.44409	0.48334	0.52602	0.59132	-0.66
0.68	0.39219	0.42572	0.47251	0.51045	0.55158	0.61426	-0.68
0.70	0.42374	0.45621	0.50140	0.53792	0.57740	0.63733	-0.70
0.72	0.45600	0.48732	0.53077	0.56577	0.60348	0.66052	-0.72
0.74	0.48899	0.51905	0.56062	0.59399	0.62984	0.68385	-0.74
0.76	0.52276	0.55143	0.59097	0.62260	0.65648	0.70731	-0.76
0.78	0.55732	0.58449	0.62183	0.65161	0.68340	0.73090	-0.78
0.80	0.59270	0.61824	0.65323	0.68102	0.71061	0.75463	-0.80
0.82	0.62894	0.65272	0.68516	0.71086	0.73812	0.77851	-0.82
0.84	0.66608	0.68793	0.71766	0.74112	0.76592	0.80252	-0.84
0.86	0.70414	0.72392	0.75074	0.77182	0.79404	0.82668	-0.86
0.88	0.74316	0.76071	0.78441	0.80297	0.82246	0.85098	-0.88
0.90	0.78319	0.79833	0.81870	0.83458	0.85121	0.87543	-0.90
0.92	0.82427	0.83681	0.85361	0.86667	0.88029	0.90004	-0.92
0.94	0.86644	0.87618	0.88918	0.89924	0.90970	0.92479	-0.94
0.96	0.90976	0.91648	0.92542	0.93231	0.93945	0.94970	-0.96
0.98	0.95426	0.95774	0.96235	0.96589	0.96954	0.97477	-0.98

↑ r_0

α =	.995 ↑−	.99 ↑−	.975 ↑−	.95 ↑−	.90 ↑−	.75 ↑−

TABLE 42.1   SAMPLE SIZE =   44

α =	.005 ↓+	.01 ↓+	.025 ↓+	.05 ↓+	.10 ↓+	.25 ↓+	
r_0 ↓							
-0.98	-0.99079	-0.99003	-0.98881	-0.98765	-0.98618	-0.98335	0.98
-0.96	-0.98148	-0.97996	-0.97753	-0.97522	-0.97229	-0.96665	0.96
-0.94	-0.97208	-0.96980	-0.96616	-0.96270	-0.95832	-0.94991	0.94
-0.92	-0.96258	-0.95954	-0.95469	-0.95010	-0.94428	-0.93313	0.92
-0.90	-0.95298	-0.94918	-0.94312	-0.93740	-0.93015	-0.91631	0.90
-0.88	-0.94327	-0.93873	-0.93146	-0.92461	-0.91595	-0.89944	0.88
-0.86	-0.93347	-0.92817	-0.91970	-0.91173	-0.90167	-0.88253	0.86
-0.84	-0.92356	-0.91750	-0.90784	-0.89876	-0.88730	-0.86557	0.84
-0.82	-0.91354	-0.90673	-0.89588	-0.88569	-0.87286	-0.84856	0.82
-0.80	-0.90342	-0.89585	-0.88382	-0.87252	-0.85833	-0.83151	0.80
-0.78	-0.89319	-0.88487	-0.87165	-0.85926	-0.84371	-0.81441	0.78
-0.76	-0.88284	-0.87377	-0.85937	-0.84590	-0.82902	-0.79726	0.76
-0.74	-0.87238	-0.86256	-0.84699	-0.83244	-0.81423	-0.78007	0.74
-0.72	-0.86180	-0.85123	-0.83450	-0.81888	-0.79936	-0.76282	0.72
-0.70	-0.85110	-0.83979	-0.82189	-0.80521	-0.78440	-0.74553	0.70
-0.68	-0.84029	-0.82823	-0.80917	-0.79144	-0.76935	-0.72818	0.68
-0.66	-0.82935	-0.81654	-0.79634	-0.77756	-0.75420	-0.71079	0.66
-0.64	-0.81828	-0.80474	-0.78339	-0.76358	-0.73897	-0.69334	0.64
-0.62	-0.80709	-0.79280	-0.77032	-0.74949	-0.72364	-0.67584	0.62
-0.60	-0.79576	-0.78075	-0.75714	-0.73528	-0.70822	-0.65829	0.60
-0.58	-0.78431	-0.76856	-0.74383	-0.72097	-0.69271	-0.64068	0.58
-0.56	-0.77271	-0.75623	-0.73039	-0.70654	-0.67709	-0.62302	0.56
-0.54	-0.76098	-0.74378	-0.71683	-0.69199	-0.66138	-0.60530	0.54
-0.52	-0.74911	-0.73119	-0.70314	-0.67733	-0.64557	-0.58753	0.52
-0.50	-0.73710	-0.71845	-0.68932	-0.66255	-0.62966	-0.56970	0.50
-0.48	-0.72494	-0.70558	-0.67537	-0.64765	-0.61364	-0.55181	0.48
-0.46	-0.71263	-0.69256	-0.66128	-0.63262	-0.59752	-0.53386	0.46
-0.44	-0.70017	-0.67939	-0.64705	-0.61747	-0.58130	-0.51586	0.44
-0.42	-0.68755	-0.66607	-0.63269	-0.60220	-0.56497	-0.49779	0.42
-0.40	-0.67478	-0.65260	-0.61818	-0.58679	-0.54854	-0.47967	0.40
-0.38	-0.66184	-0.63897	-0.60353	-0.57126	-0.53199	-0.46148	0.38
-0.36	-0.64874	-0.62518	-0.58873	-0.55559	-0.51533	-0.44323	0.36
-0.34	-0.63546	-0.61123	-0.57378	-0.53979	-0.49857	-0.42492	0.34
-0.32	-0.62202	-0.59712	-0.55868	-0.52385	-0.48168	-0.40655	0.32
-0.30	-0.60840	-0.58283	-0.54342	-0.50777	-0.46468	-0.38811	0.30
-0.28	-0.59460	-0.56838	-0.52801	-0.49155	-0.44757	-0.36960	0.28
-0.26	-0.58062	-0.55375	-0.51244	-0.47519	-0.43034	-0.35103	0.26
-0.24	-0.56645	-0.53893	-0.49670	-0.45869	-0.41298	-0.33239	0.24
-0.22	-0.55209	-0.52394	-0.48080	-0.44203	-0.39550	-0.31369	0.22
-0.20	-0.53753	-0.50876	-0.46472	-0.42523	-0.37790	-0.29491	0.20
-0.18	-0.52277	-0.49338	-0.44848	-0.40827	-0.36018	-0.27607	0.18
-0.16	-0.50781	-0.47782	-0.43206	-0.39115	-0.34233	-0.25715	0.16
-0.14	-0.49264	-0.46205	-0.41546	-0.37388	-0.32434	-0.23817	0.14
-0.12	-0.47725	-0.44608	-0.39868	-0.35645	-0.30623	-0.21911	0.12
-0.10	-0.46165	-0.42990	-0.38171	-0.33886	-0.28798	-0.19998	0.10
-0.08	-0.44582	-0.41352	-0.36455	-0.32109	-0.26960	-0.18077	0.08
-0.06	-0.42976	-0.39691	-0.34720	-0.30316	-0.25108	-0.16149	0.06
-0.04	-0.41346	-0.38009	-0.32965	-0.28506	-0.23243	-0.14213	0.04
-0.02	-0.39693	-0.36303	-0.31191	-0.26679	-0.21363	-0.12270	0.02

↑ r_0

α =	.995 ↑-	.99 ↑-	.975 ↑-	.95 ↑-	.90 ↑-	.75 ↑-

TABLE 42.2   SAMPLE SIZE =   44   (CONT.)

α =	.005 ↓+	.01 ↓+	.025 ↓+	.05 ↓+	.10 ↓+	.25 ↓+	
r_0 ↓							
0.00	−0.38014	−0.34575	−0.29396	−0.24833	−0.19469	−0.10318	0.00
0.02	−0.36311	−0.32823	−0.27579	−0.22969	−0.17560	−0.08359	−0.02
0.04	−0.34582	−0.31047	−0.25742	−0.21088	−0.15636	−0.06392	−0.04
0.06	−0.32826	−0.29246	−0.23883	−0.19187	−0.13698	−0.04416	−0.06
0.08	−0.31043	−0.27420	−0.22002	−0.17267	−0.11744	−0.02432	−0.08
0.10	−0.29232	−0.25568	−0.20098	−0.15328	−0.09775	−0.00440	−0.10
0.12	−0.27392	−0.23689	−0.18171	−0.13369	−0.07790	0.01560	−0.12
0.14	−0.25523	−0.21783	−0.16221	−0.11390	−0.05789	0.03569	−0.14
0.16	−0.23625	−0.19850	−0.14247	−0.09390	−0.03772	0.05587	−0.16
0.18	−0.21695	−0.17888	−0.12248	−0.07370	−0.01739	0.07613	−0.18
0.20	−0.19733	−0.15897	−0.10224	−0.05328	0.00311	0.09649	−0.20
0.22	−0.17739	−0.13876	−0.08174	−0.03264	0.02379	0.11693	−0.22
0.24	−0.15712	−0.11824	−0.06099	−0.01179	0.04463	0.13747	−0.24
0.26	−0.13650	−0.09741	−0.03996	0.00929	0.06565	0.15810	−0.26
0.28	−0.11553	−0.07626	−0.01866	0.03061	0.08685	0.17882	−0.28
0.30	−0.09420	−0.05478	0.00291	0.05215	0.10823	0.19964	−0.30
0.32	−0.07249	−0.03296	0.02477	0.07393	0.12980	0.22056	−0.32
0.34	−0.05040	−0.01080	0.04693	0.09596	0.15155	0.24157	−0.34
0.36	−0.02792	0.01173	0.06938	0.11823	0.17349	0.26269	−0.36
0.38	−0.00503	0.03462	0.09214	0.14076	0.19563	0.28390	−0.38
0.40	0.01828	0.05788	0.11521	0.16355	0.21796	0.30521	−0.40
0.42	0.04202	0.08153	0.13860	0.18659	0.24049	0.32663	−0.42
0.44	0.06621	0.10558	0.16231	0.20991	0.26323	0.34816	−0.44
0.46	0.09085	0.13003	0.18636	0.23350	0.28618	0.36979	−0.46
0.48	0.11596	0.15490	0.21076	0.25738	0.30933	0.39152	−0.48
0.50	0.14155	0.18021	0.23550	0.28153	0.33270	0.41337	−0.50
0.52	0.16765	0.20595	0.26061	0.30598	0.35629	0.43533	−0.52
0.54	0.19427	0.23215	0.28608	0.33073	0.38011	0.45740	−0.54
0.56	0.22141	0.25882	0.31194	0.35578	0.40415	0.47958	−0.56
0.58	0.24911	0.28598	0.33818	0.38115	0.42842	0.50188	−0.58
0.60	0.27738	0.31363	0.36481	0.40683	0.45292	0.52429	−0.60
0.62	0.30624	0.34179	0.39186	0.43284	0.47767	0.54683	−0.62
0.64	0.33571	0.37048	0.41932	0.45918	0.50266	0.56948	−0.64
0.66	0.36580	0.39972	0.44721	0.48586	0.52789	0.59226	−0.66
0.68	0.39655	0.42952	0.47555	0.51288	0.55338	0.61516	−0.68
0.70	0.42797	0.45989	0.50433	0.54027	0.57913	0.63819	−0.70
0.72	0.46009	0.49087	0.53358	0.56801	0.60514	0.66134	−0.72
0.74	0.49293	0.52246	0.56331	0.59613	0.63142	0.68463	−0.74
0.76	0.52653	0.55469	0.59353	0.62463	0.65797	0.70804	−0.76
0.78	0.56090	0.58757	0.62425	0.65352	0.68480	0.73159	−0.78
0.80	0.59608	0.62114	0.65549	0.68281	0.71192	0.75527	−0.80
0.82	0.63209	0.65541	0.68727	0.71251	0.73932	0.77909	−0.82
0.84	0.66898	0.69042	0.71959	0.74263	0.76702	0.80305	−0.84
0.86	0.70677	0.72617	0.75248	0.77318	0.79502	0.82716	−0.86
0.88	0.74551	0.76271	0.78595	0.80417	0.82333	0.85140	−0.88
0.90	0.78523	0.80005	0.82001	0.83561	0.85195	0.87579	−0.90
0.92	0.82596	0.83823	0.85470	0.86751	0.88089	0.90033	−0.92
0.94	0.86776	0.87729	0.89002	0.89989	0.91016	0.92501	−0.94
0.96	0.91067	0.91724	0.92600	0.93275	0.93976	0.94985	−0.96
0.98	0.95473	0.95813	0.96265	0.96612	0.96971	0.97485	−0.98

↑ r_0

α =	.995 ↑−	.99 ↑−	.975 ↑−	.95 ↑−	.90 ↑−	.75 ↑−

TABLE 43.1   SAMPLE SIZE =   45

α =	.005 ↓+	.01 ↓+	.025 ↓+	.05 ↓+	.10 ↓+	.25 ↓+	
r_o ↓							
−0.98	−0.99070	−0.98995	−0.98874	−0.98759	−0.98613	−0.98332	0.98
−0.96	−0.98131	−0.97980	−0.97738	−0.97509	−0.97218	−0.96659	0.96
−0.94	−0.97183	−0.96956	−0.96593	−0.96250	−0.95815	−0.94982	0.94
−0.92	−0.96224	−0.95922	−0.95439	−0.94983	−0.94405	−0.93301	0.92
−0.90	−0.95256	−0.94878	−0.94275	−0.93707	−0.92987	−0.91615	0.90
−0.88	−0.94277	−0.93824	−0.93102	−0.92421	−0.91562	−0.89925	0.88
−0.86	−0.93288	−0.92760	−0.91918	−0.91126	−0.90128	−0.88231	0.86
−0.84	−0.92289	−0.91686	−0.90725	−0.89822	−0.88686	−0.86532	0.84
−0.82	−0.91279	−0.90600	−0.89522	−0.88509	−0.87236	−0.84829	0.82
−0.80	−0.90258	−0.89505	−0.88308	−0.87186	−0.85778	−0.83120	0.80
−0.78	−0.89226	−0.88398	−0.87084	−0.85853	−0.84311	−0.81407	0.78
−0.76	−0.88183	−0.87280	−0.85849	−0.84511	−0.82836	−0.79690	0.76
−0.74	−0.87128	−0.86151	−0.84603	−0.83158	−0.81352	−0.77967	0.74
−0.72	−0.86062	−0.85010	−0.83347	−0.81796	−0.79859	−0.76240	0.72
−0.70	−0.84984	−0.83858	−0.82079	−0.80423	−0.78358	−0.74508	0.70
−0.68	−0.83893	−0.82694	−0.80800	−0.79039	−0.76848	−0.72770	0.68
−0.66	−0.82791	−0.81517	−0.79510	−0.77645	−0.75329	−0.71028	0.66
−0.64	−0.81676	−0.80329	−0.78208	−0.76241	−0.73800	−0.69280	0.64
−0.62	−0.80548	−0.79127	−0.76894	−0.74825	−0.72262	−0.67528	0.62
−0.60	−0.79407	−0.77914	−0.75568	−0.73399	−0.70715	−0.65770	0.60
−0.58	−0.78253	−0.76687	−0.74230	−0.71961	−0.69159	−0.64006	0.58
−0.56	−0.77085	−0.75447	−0.72879	−0.70512	−0.67593	−0.62237	0.56
−0.54	−0.75904	−0.74193	−0.71516	−0.69052	−0.66017	−0.60463	0.54
−0.52	−0.74708	−0.72926	−0.70141	−0.67580	−0.64431	−0.58683	0.52
−0.50	−0.73498	−0.71645	−0.68752	−0.66096	−0.62835	−0.56897	0.50
−0.48	−0.72274	−0.70349	−0.67350	−0.64600	−0.61229	−0.55106	0.48
−0.46	−0.71034	−0.69040	−0.65934	−0.63091	−0.59612	−0.53309	0.46
−0.44	−0.69780	−0.67715	−0.64505	−0.61571	−0.57986	−0.51506	0.44
−0.42	−0.68510	−0.66376	−0.63062	−0.60037	−0.56348	−0.49697	0.42
−0.40	−0.67224	−0.65021	−0.61604	−0.58491	−0.54700	−0.47883	0.40
−0.38	−0.65922	−0.63650	−0.60133	−0.56932	−0.53042	−0.46062	0.38
−0.36	−0.64603	−0.62264	−0.58646	−0.55360	−0.51372	−0.44235	0.36
−0.34	−0.63268	−0.60862	−0.57145	−0.53775	−0.49691	−0.42401	0.34
−0.32	−0.61916	−0.59443	−0.55629	−0.52176	−0.47998	−0.40562	0.32
−0.30	−0.60545	−0.58007	−0.54097	−0.50563	−0.46295	−0.38716	0.30
−0.28	−0.59158	−0.56554	−0.52550	−0.48936	−0.44579	−0.36863	0.28
−0.26	−0.57751	−0.55084	−0.50986	−0.47295	−0.42852	−0.35004	0.26
−0.24	−0.56326	−0.53595	−0.49407	−0.45639	−0.41113	−0.33138	0.24
−0.22	−0.54882	−0.52089	−0.47811	−0.43969	−0.39362	−0.31266	0.22
−0.20	−0.53419	−0.50564	−0.46198	−0.42284	−0.37598	−0.29387	0.20
−0.18	−0.51935	−0.49019	−0.44568	−0.40584	−0.35822	−0.27501	0.18
−0.16	−0.50431	−0.47456	−0.42920	−0.38868	−0.34034	−0.25607	0.16
−0.14	−0.48907	−0.45873	−0.41255	−0.37136	−0.32232	−0.23707	0.14
−0.12	−0.47361	−0.44269	−0.39571	−0.35389	−0.30418	−0.21800	0.12
−0.10	−0.45793	−0.42645	−0.37869	−0.33625	−0.28590	−0.19885	0.10
−0.08	−0.44203	−0.41000	−0.36149	−0.31845	−0.26749	−0.17963	0.08
−0.06	−0.42590	−0.39334	−0.34409	−0.30049	−0.24895	−0.16034	0.06
−0.04	−0.40953	−0.37645	−0.32649	−0.28235	−0.23027	−0.14097	0.04
−0.02	−0.39293	−0.35934	−0.30870	−0.26404	−0.21144	−0.12152	0.02

↑ r_o

α =   .995 ↑−   .99 ↑−   .975 ↑−   .95 ↑−   .90 ↑−   .75 ↑−

TABLE 43.2   SAMPLE SIZE =   45   (CONT.)

α =	.005 ↓+	.01 ↓+	.025 ↓+	.05 ↓+	.10 ↓+	.25 ↓+	
r_o ↓							
0.00	−0.37608	−0.34200	−0.29071	−0.24555	−0.19248	−0.10200	0.00
0.02	−0.35898	−0.32442	−0.27250	−0.22688	−0.17337	−0.08239	−0.02
0.04	−0.34162	−0.30661	−0.25409	−0.20803	−0.15412	−0.06271	−0.04
0.06	−0.32400	−0.28855	−0.23546	−0.18900	−0.13471	−0.04295	−0.06
0.08	−0.30612	−0.27024	−0.21662	−0.16977	−0.11516	−0.02310	−0.08
0.10	−0.28795	−0.25167	−0.19754	−0.15036	−0.09545	−0.00318	−0.10
0.12	−0.26950	−0.23284	−0.17824	−0.13074	−0.07559	0.01683	−0.12
0.14	−0.25076	−0.21374	−0.15871	−0.11093	−0.05557	0.03693	−0.14
0.16	−0.23172	−0.19436	−0.13894	−0.09092	−0.03539	0.05711	−0.16
0.18	−0.21237	−0.17471	−0.11893	−0.07070	−0.01504	0.07738	−0.18
0.20	−0.19272	−0.15476	−0.09866	−0.05027	0.00546	0.09773	−0.20
0.22	−0.17273	−0.13452	−0.07815	−0.02962	0.02614	0.11818	−0.22
0.24	−0.15242	−0.11398	−0.05738	−0.00876	0.04699	0.13871	−0.24
0.26	−0.13177	−0.09312	−0.03634	0.01233	0.06801	0.15934	−0.26
0.28	−0.11077	−0.07195	−0.01503	0.03364	0.08921	0.18006	−0.28
0.30	−0.08941	−0.05045	0.00655	0.05519	0.11058	0.20088	−0.30
0.32	−0.06768	−0.02862	0.02842	0.07697	0.13214	0.22178	−0.32
0.34	−0.04557	−0.00644	0.05057	0.09899	0.15388	0.24279	−0.34
0.36	−0.02307	0.01609	0.07302	0.12126	0.17581	0.26390	−0.36
0.38	−0.00017	0.03898	0.09577	0.14377	0.19794	0.28510	−0.38
0.40	0.02315	0.06224	0.11883	0.16654	0.22025	0.30640	−0.40
0.42	0.04689	0.08588	0.14220	0.18957	0.24277	0.32781	−0.42
0.44	0.07106	0.10991	0.16590	0.21287	0.26548	0.34932	−0.44
0.46	0.09569	0.13435	0.18993	0.23643	0.28840	0.37093	−0.46
0.48	0.12078	0.15920	0.21429	0.26027	0.31153	0.39265	−0.48
0.50	0.14635	0.18447	0.23900	0.28440	0.33487	0.41448	−0.50
0.52	0.17242	0.21018	0.26407	0.30881	0.35843	0.43642	−0.52
0.54	0.19899	0.23634	0.28950	0.33351	0.38221	0.45847	−0.54
0.56	0.22609	0.26296	0.31530	0.35852	0.40620	0.48063	−0.56
0.58	0.25374	0.29005	0.34148	0.38383	0.43043	0.50290	−0.58
0.60	0.28194	0.31763	0.36805	0.40945	0.45489	0.52529	−0.60
0.62	0.31072	0.34572	0.39503	0.43540	0.47958	0.54779	−0.62
0.64	0.34010	0.37433	0.42241	0.46167	0.50451	0.57042	−0.64
0.66	0.37010	0.40347	0.45022	0.48827	0.52969	0.59316	−0.66
0.68	0.40073	0.43316	0.47846	0.51522	0.55512	0.61603	−0.68
0.70	0.43203	0.46343	0.50715	0.54252	0.58080	0.63902	−0.70
0.72	0.46402	0.49427	0.53629	0.57017	0.60673	0.66214	−0.72
0.74	0.49671	0.52573	0.56590	0.59819	0.63293	0.68538	−0.74
0.76	0.53014	0.55780	0.59599	0.62658	0.65940	0.70875	−0.76
0.78	0.56433	0.59053	0.62657	0.65536	0.68615	0.73225	−0.78
0.80	0.59931	0.62392	0.65767	0.68453	0.71317	0.75589	−0.80
0.82	0.63511	0.65800	0.68928	0.71410	0.74047	0.77966	−0.82
0.84	0.67176	0.69279	0.72144	0.74408	0.76807	0.80357	−0.84
0.86	0.70930	0.72832	0.75414	0.77448	0.79596	0.82761	−0.86
0.88	0.74775	0.76461	0.78742	0.80531	0.82415	0.85180	−0.88
0.90	0.78717	0.80169	0.82128	0.83659	0.85265	0.87613	−0.90
0.92	0.82757	0.83959	0.85574	0.86832	0.88147	0.90061	−0.92
0.94	0.86901	0.87834	0.89082	0.90051	0.91060	0.92523	−0.94
0.96	0.91153	0.91797	0.92655	0.93318	0.94007	0.95000	−0.96
0.98	0.95518	0.95851	0.96293	0.96634	0.96986	0.97492	−0.98

↑ r_o

α =	.995 ↑−	.99 ↑−	.975 ↑−	.95 ↑−	.90 ↑−	.75 ↑−

TABLE 44.1   SAMPLE SIZE =   46

α =	.005 ↓+	.01 ↓+	.025 ↓+	.05 ↓+	.10 ↓+	.25 ↓+	
r_0 ↓							
-0.98	-0.99062	-0.98987	-0.98866	-0.98752	-0.98607	-0.98328	0.98
-0.96	-0.98115	-0.97965	-0.97724	-0.97496	-0.97207	-0.96653	0.96
-0.94	-0.97158	-0.96933	-0.96572	-0.96231	-0.95799	-0.94973	0.94
-0.92	-0.96191	-0.95891	-0.95410	-0.94957	-0.94384	-0.93289	0.92
-0.90	-0.95215	-0.94839	-0.94239	-0.93674	-0.92960	-0.91600	0.90
-0.88	-0.94228	-0.93777	-0.93059	-0.92382	-0.91529	-0.89907	0.88
-0.86	-0.93231	-0.92705	-0.91868	-0.91081	-0.90090	-0.88210	0.86
-0.84	-0.92223	-0.91623	-0.90668	-0.89771	-0.88643	-0.86508	0.84
-0.82	-0.91205	-0.90530	-0.89457	-0.88451	-0.87188	-0.84802	0.82
-0.80	-0.90176	-0.89426	-0.88236	-0.87122	-0.85724	-0.83091	0.80
-0.78	-0.89136	-0.88312	-0.87005	-0.85783	-0.84252	-0.81375	0.78
-0.76	-0.88084	-0.87186	-0.85763	-0.84434	-0.82772	-0.79654	0.76
-0.74	-0.87021	-0.86049	-0.84510	-0.83075	-0.81283	-0.77929	0.74
-0.72	-0.85947	-0.84900	-0.83247	-0.81706	-0.79785	-0.76199	0.72
-0.70	-0.84860	-0.83740	-0.81972	-0.80327	-0.78279	-0.74464	0.70
-0.68	-0.83762	-0.82568	-0.80686	-0.78938	-0.76764	-0.72724	0.68
-0.66	-0.82651	-0.81384	-0.79389	-0.77538	-0.75240	-0.70979	0.66
-0.64	-0.81528	-0.80188	-0.78080	-0.76127	-0.73706	-0.69229	0.64
-0.62	-0.80391	-0.78979	-0.76759	-0.74706	-0.72164	-0.67473	0.62
-0.60	-0.79242	-0.77757	-0.75426	-0.73273	-0.70612	-0.65712	0.60
-0.58	-0.78080	-0.76523	-0.74082	-0.71830	-0.69050	-0.63946	0.58
-0.56	-0.76904	-0.75275	-0.72724	-0.70375	-0.67480	-0.62175	0.56
-0.54	-0.75714	-0.74014	-0.71355	-0.68909	-0.65899	-0.60398	0.54
-0.52	-0.74510	-0.72739	-0.69972	-0.67431	-0.64308	-0.58616	0.52
-0.50	-0.73292	-0.71450	-0.68577	-0.65941	-0.62708	-0.56828	0.50
-0.48	-0.72060	-0.70147	-0.67168	-0.64439	-0.61097	-0.55034	0.48
-0.46	-0.70812	-0.68830	-0.65746	-0.62925	-0.59477	-0.53235	0.46
-0.44	-0.69550	-0.67498	-0.64310	-0.61399	-0.57846	-0.51429	0.44
-0.42	-0.68271	-0.66151	-0.62861	-0.59861	-0.56204	-0.49618	0.42
-0.40	-0.66977	-0.64789	-0.61397	-0.58309	-0.54552	-0.47801	0.40
-0.38	-0.65667	-0.63411	-0.59919	-0.56745	-0.52889	-0.45978	0.38
-0.36	-0.64341	-0.62017	-0.58427	-0.55168	-0.51215	-0.44149	0.36
-0.34	-0.62998	-0.60608	-0.56919	-0.53577	-0.49530	-0.42313	0.34
-0.32	-0.61637	-0.59182	-0.55397	-0.51973	-0.47834	-0.40472	0.32
-0.30	-0.60259	-0.57739	-0.53859	-0.50355	-0.46126	-0.38624	0.30
-0.28	-0.58864	-0.56279	-0.52306	-0.48723	-0.44407	-0.36769	0.28
-0.26	-0.57450	-0.54801	-0.50737	-0.47078	-0.42676	-0.34908	0.26
-0.24	-0.56017	-0.53306	-0.49152	-0.45417	-0.40934	-0.33041	0.24
-0.22	-0.54565	-0.51793	-0.47550	-0.43743	-0.39179	-0.31167	0.22
-0.20	-0.53094	-0.50261	-0.45931	-0.42053	-0.37412	-0.29286	0.20
-0.18	-0.51603	-0.48710	-0.44296	-0.40348	-0.35633	-0.27398	0.18
-0.16	-0.50092	-0.47140	-0.42643	-0.38628	-0.33841	-0.25503	0.16
-0.14	-0.48560	-0.45551	-0.40973	-0.36893	-0.32037	-0.23601	0.14
-0.12	-0.47007	-0.43941	-0.39284	-0.35141	-0.30220	-0.21692	0.12
-0.10	-0.45432	-0.42311	-0.37577	-0.33374	-0.28389	-0.19776	0.10
-0.08	-0.43835	-0.40660	-0.35852	-0.31590	-0.26546	-0.17853	0.08
-0.06	-0.42215	-0.38987	-0.34107	-0.29790	-0.24688	-0.15922	0.06
-0.04	-0.40572	-0.37293	-0.32344	-0.27972	-0.22818	-0.13984	0.04
-0.02	-0.38905	-0.35576	-0.30560	-0.26138	-0.20933	-0.12038	0.02

↑ r_0

α =	.995 ↑-	.99 ↑-	.975 ↑-	.95 ↑-	.90 ↑-	.75 ↑-

TABLE 44.2   SAMPLE SIZE = 46   (CONT.)

α =	.005 ↓+	.01 ↓+	.025 ↓+	.05 ↓+	.10 ↓+	.25 ↓+	
r_o ↓							
0.00	−0.37214	−0.33837	−0.28756	−0.24286	−0.19034	−0.10085	0.00
0.02	−0.35498	−0.32074	−0.26932	−0.22416	−0.17122	−0.08124	−0.02
0.04	−0.33756	−0.30287	−0.25087	−0.20528	−0.15194	−0.06155	−0.04
0.06	−0.31989	−0.28476	−0.23221	−0.18622	−0.13252	−0.04178	−0.06
0.08	−0.30194	−0.26641	−0.21332	−0.16697	−0.11295	−0.02193	−0.08
0.10	−0.28372	−0.24779	−0.19422	−0.14753	−0.09323	−0.00199	−0.10
0.12	−0.26522	−0.22892	−0.17489	−0.12790	−0.07335	0.01802	−0.12
0.14	−0.24643	−0.20978	−0.15533	−0.10807	−0.05332	0.03812	−0.14
0.16	−0.22734	−0.19037	−0.13553	−0.08804	−0.03313	0.05831	−0.16
0.18	−0.20795	−0.17068	−0.11550	−0.06780	−0.01278	0.07857	−0.18
0.20	−0.18825	−0.15070	−0.09521	−0.04736	0.00773	0.09893	−0.20
0.22	−0.16823	−0.13043	−0.07468	−0.02670	0.02841	0.11938	−0.22
0.24	−0.14788	−0.10986	−0.05389	−0.00583	0.04926	0.13991	−0.24
0.26	−0.12720	−0.08898	−0.03284	0.01526	0.07028	0.16054	−0.26
0.28	−0.10616	−0.06778	−0.01152	0.03658	0.09148	0.18125	−0.28
0.30	−0.08478	−0.04627	0.01007	0.05812	0.11285	0.20206	−0.30
0.32	−0.06303	−0.02442	0.03194	0.07990	0.13440	0.22297	−0.32
0.34	−0.04090	−0.00224	0.05409	0.10192	0.15614	0.24397	−0.34
0.36	−0.01839	0.02030	0.07653	0.12417	0.17806	0.26506	−0.36
0.38	0.00452	0.04319	0.09928	0.14668	0.20017	0.28626	−0.38
0.40	0.02784	0.06645	0.12232	0.16943	0.22247	0.30755	−0.40
0.42	0.05158	0.09008	0.14568	0.19244	0.24496	0.32894	−0.42
0.44	0.07575	0.11410	0.16936	0.21572	0.26766	0.35044	−0.44
0.46	0.10036	0.13851	0.19336	0.23925	0.29055	0.37204	−0.46
0.48	0.12543	0.16334	0.21769	0.26307	0.31365	0.39374	−0.48
0.50	0.15098	0.18858	0.24237	0.28716	0.33697	0.41555	−0.50
0.52	0.17701	0.21425	0.26740	0.31153	0.36049	0.43747	−0.52
0.54	0.20355	0.24036	0.29278	0.33619	0.38423	0.45950	−0.54
0.56	0.23060	0.26693	0.31853	0.36115	0.40819	0.48163	−0.56
0.58	0.25818	0.29397	0.34466	0.38641	0.43237	0.50388	−0.58
0.60	0.28632	0.32149	0.37117	0.41198	0.45678	0.52625	−0.60
0.62	0.31503	0.34950	0.39808	0.43786	0.48142	0.54872	−0.62
0.64	0.34432	0.37803	0.42539	0.46407	0.50630	0.57132	−0.64
0.66	0.37422	0.40708	0.45311	0.49060	0.53142	0.59403	−0.66
0.68	0.40475	0.43667	0.48126	0.51747	0.55678	0.61687	−0.68
0.70	0.43593	0.46682	0.50985	0.54468	0.58240	0.63982	−0.70
0.72	0.46779	0.49755	0.53889	0.57225	0.60827	0.66290	−0.72
0.74	0.50034	0.52887	0.56838	0.60017	0.63439	0.68610	−0.74
0.76	0.53360	0.56080	0.59835	0.62846	0.66078	0.70943	−0.76
0.78	0.56762	0.59336	0.62880	0.65713	0.68744	0.73289	−0.78
0.80	0.60241	0.62658	0.65975	0.68618	0.71437	0.75648	−0.80
0.82	0.63800	0.66048	0.69122	0.71562	0.74158	0.78020	−0.82
0.84	0.67443	0.69507	0.72321	0.74547	0.76908	0.80406	−0.84
0.86	0.71171	0.73038	0.75574	0.77573	0.79687	0.82806	−0.86
0.88	0.74990	0.76644	0.78883	0.80641	0.82495	0.85219	−0.88
0.90	0.78902	0.80327	0.82249	0.83753	0.85333	0.87646	−0.90
0.92	0.82911	0.84090	0.85674	0.86909	0.88202	0.90087	−0.92
0.94	0.87021	0.87935	0.89160	0.90111	0.91103	0.92543	−0.94
0.96	0.91237	0.91867	0.92708	0.93359	0.94036	0.95014	−0.96
0.98	0.95561	0.95887	0.96321	0.96655	0.97001	0.97499	−0.98

↑ r_o

α =	.995 ↑−	.99 ↑−	.975 ↑−	.95 ↑−	.90 ↑−	.75 ↑−

TABLE 45.1   SAMPLE SIZE =   47

α =	.005 ↓+	.01 ↓+	.025 ↓+	.05 ↓+	.10 ↓+	.25 ↓+	
r_o ↓							
−0.98	−0.99054	−0.98979	−0.98859	−0.98746	−0.98602	−0.98326	0.98
−0.96	−0.98099	−0.97949	−0.97710	−0.97483	−0.97196	−0.96647	0.96
−0.94	−0.97134	−0.96910	−0.96551	−0.96212	−0.95783	−0.94964	0.94
−0.92	−0.96160	−0.95860	−0.95382	−0.94932	−0.94362	−0.93277	0.92
−0.90	−0.95175	−0.94801	−0.94204	−0.93643	−0.92934	−0.91586	0.90
−0.88	−0.94180	−0.93731	−0.93017	−0.92344	−0.91498	−0.89890	0.88
−0.86	−0.93175	−0.92652	−0.91819	−0.91037	−0.90053	−0.88190	0.86
−0.84	−0.92160	−0.91562	−0.90612	−0.89721	−0.88601	−0.86485	0.84
−0.82	−0.91133	−0.90461	−0.89394	−0.88395	−0.87141	−0.84776	0.82
−0.80	−0.90096	−0.89350	−0.88166	−0.87059	−0.85672	−0.83062	0.80
−0.78	−0.89048	−0.88228	−0.86928	−0.85714	−0.84195	−0.81343	0.78
−0.76	−0.87988	−0.87094	−0.85679	−0.84359	−0.82710	−0.79620	0.76
−0.74	−0.86917	−0.85950	−0.84420	−0.82994	−0.81216	−0.77892	0.74
−0.72	−0.85835	−0.84794	−0.83150	−0.81620	−0.79713	−0.76159	0.72
−0.70	−0.84740	−0.83626	−0.81868	−0.80235	−0.78202	−0.74422	0.70
−0.68	−0.83634	−0.82446	−0.80575	−0.78839	−0.76682	−0.72679	0.68
−0.66	−0.82515	−0.81254	−0.79271	−0.77433	−0.75153	−0.70931	0.66
−0.64	−0.81383	−0.80051	−0.77956	−0.76017	−0.73615	−0.69178	0.64
−0.62	−0.80239	−0.78834	−0.76628	−0.74590	−0.72068	−0.67420	0.62
−0.60	−0.79082	−0.77605	−0.75289	−0.73151	−0.70511	−0.65657	0.60
−0.58	−0.77912	−0.76363	−0.73938	−0.71702	−0.68945	−0.63888	0.58
−0.56	−0.76728	−0.75108	−0.72574	−0.70242	−0.67370	−0.62114	0.56
−0.54	−0.75530	−0.73839	−0.71198	−0.68770	−0.65785	−0.60335	0.54
−0.52	−0.74318	−0.72557	−0.69809	−0.67286	−0.64190	−0.58550	0.52
−0.50	−0.73092	−0.71261	−0.68407	−0.65791	−0.62585	−0.56760	0.50
−0.48	−0.71852	−0.69951	−0.66992	−0.64284	−0.60970	−0.54964	0.48
−0.46	−0.70596	−0.68626	−0.65563	−0.62765	−0.59345	−0.53162	0.46
−0.44	−0.69326	−0.67287	−0.64121	−0.61233	−0.57710	−0.51355	0.44
−0.42	−0.68040	−0.65933	−0.62666	−0.59689	−0.56064	−0.49541	0.42
−0.40	−0.66738	−0.64563	−0.61196	−0.58133	−0.54408	−0.47722	0.40
−0.38	−0.65420	−0.63178	−0.59712	−0.56563	−0.52741	−0.45897	0.38
−0.36	−0.64086	−0.61778	−0.58213	−0.54981	−0.51063	−0.44066	0.36
−0.34	−0.62735	−0.60361	−0.56700	−0.53385	−0.49374	−0.42228	0.34
−0.32	−0.61367	−0.58928	−0.55172	−0.51776	−0.47674	−0.40385	0.32
−0.30	−0.59981	−0.57478	−0.53629	−0.50154	−0.45963	−0.38535	0.30
−0.28	−0.58578	−0.56011	−0.52070	−0.48517	−0.44240	−0.36678	0.28
−0.26	−0.57157	−0.54527	−0.50495	−0.46867	−0.42506	−0.34816	0.26
−0.24	−0.55717	−0.53026	−0.48904	−0.45202	−0.40760	−0.32946	0.24
−0.22	−0.54258	−0.51506	−0.47297	−0.43523	−0.39002	−0.31070	0.22
−0.20	−0.52779	−0.49967	−0.45673	−0.41829	−0.37232	−0.29188	0.20
−0.18	−0.51281	−0.48410	−0.44033	−0.40120	−0.35450	−0.27298	0.18
−0.16	−0.49763	−0.46834	−0.42375	−0.38396	−0.33655	−0.25402	0.16
−0.14	−0.48224	−0.45238	−0.40699	−0.36656	−0.31848	−0.23499	0.14
−0.12	−0.46664	−0.43623	−0.39006	−0.34901	−0.30028	−0.21588	0.12
−0.10	−0.45082	−0.41986	−0.37294	−0.33130	−0.28194	−0.19671	0.10
−0.08	−0.43479	−0.40330	−0.35564	−0.31343	−0.26348	−0.17746	0.08
−0.06	−0.41852	−0.38651	−0.33815	−0.29539	−0.24489	−0.15815	0.06
−0.04	−0.40203	−0.36952	−0.32047	−0.27718	−0.22616	−0.13875	0.04
−0.02	−0.38530	−0.35229	−0.30260	−0.25881	−0.20729	−0.11928	0.02

↑ r_o

α =	.995 ↑−	.99 ↑−	.975 ↑−	.95 ↑−	.90 ↑−	.75 ↑−

TABLE 45.2 SAMPLE SIZE = 47 (CONT.)

$\alpha =$	.005 ↓+	.01 ↓+	.025 ↓+	.05 ↓+	.10 ↓+	.25 ↓+	
r_0 ↓							
0.00	−0.36833	−0.33485	−0.28452	−0.24026	−0.18828	−0.09974	0.00
0.02	−0.35111	−0.31717	−0.26624	−0.22153	−0.16913	−0.08012	−0.02
0.04	−0.33363	−0.29926	−0.24775	−0.20262	−0.14984	−0.06042	−0.04
0.06	−0.31590	−0.28110	−0.22905	−0.18354	−0.13040	−0.04064	−0.06
0.08	−0.29790	−0.26270	−0.21014	−0.16426	−0.11082	−0.02079	−0.08
0.10	−0.27962	−0.24404	−0.19100	−0.14480	−0.09108	−0.00085	−0.10
0.12	−0.26107	−0.22513	−0.17164	−0.12515	−0.07119	0.01917	−0.12
0.14	−0.24223	−0.20595	−0.15206	−0.10530	−0.05115	0.03927	−0.14
0.16	−0.22310	−0.18650	−0.13224	−0.08525	−0.03095	0.05946	−0.16
0.18	−0.20367	−0.16677	−0.11218	−0.06500	−0.01059	0.07973	−0.18
0.20	−0.18393	−0.14677	−0.09187	−0.04455	0.00993	0.10009	−0.20
0.22	−0.16387	−0.12647	−0.07133	−0.02388	0.03061	0.12054	−0.22
0.24	−0.14349	−0.10587	−0.05052	−0.00300	0.05146	0.14107	−0.24
0.26	−0.12277	−0.08497	−0.02946	0.01809	0.07248	0.16169	−0.26
0.28	−0.10171	−0.06376	−0.00813	0.03941	0.09367	0.18241	−0.28
0.30	−0.08030	−0.04223	0.01346	0.06096	0.11504	0.20321	−0.30
0.32	−0.05853	−0.02037	0.03533	0.08273	0.13659	0.22411	−0.32
0.34	−0.03639	0.00182	0.05748	0.10474	0.15831	0.24510	−0.34
0.36	−0.01387	0.02436	0.07992	0.12699	0.18022	0.26619	−0.36
0.38	0.00905	0.04725	0.10266	0.14948	0.20232	0.28737	−0.38
0.40	0.03237	0.07051	0.12569	0.17222	0.22460	0.30866	−0.40
0.42	0.05611	0.09413	0.14904	0.19521	0.24708	0.33004	−0.42
0.44	0.08027	0.11814	0.17269	0.21846	0.26975	0.35152	−0.44
0.46	0.10487	0.14253	0.19667	0.24198	0.29263	0.37311	−0.46
0.48	0.12992	0.16733	0.22098	0.26576	0.31570	0.39479	−0.48
0.50	0.15544	0.19254	0.24562	0.28982	0.33898	0.41659	−0.50
0.52	0.18144	0.21818	0.27061	0.31415	0.36247	0.43849	−0.52
0.54	0.20793	0.24425	0.29595	0.33878	0.38618	0.46049	−0.54
0.56	0.23494	0.27077	0.32165	0.36369	0.41010	0.48261	−0.56
0.58	0.26247	0.29775	0.34772	0.38890	0.43424	0.50483	−0.58
0.60	0.29054	0.32520	0.37417	0.41441	0.45860	0.52717	−0.60
0.62	0.31917	0.35314	0.40101	0.44023	0.48320	0.54962	−0.62
0.64	0.34839	0.38159	0.42825	0.46638	0.50802	0.57219	−0.64
0.66	0.37820	0.41055	0.45590	0.49284	0.53309	0.59487	−0.66
0.68	0.40862	0.44004	0.48396	0.51963	0.55839	0.61767	−0.68
0.70	0.43969	0.47008	0.51245	0.54676	0.58394	0.64059	−0.70
0.72	0.47141	0.50069	0.54139	0.57424	0.60974	0.66363	−0.72
0.74	0.50382	0.53188	0.57077	0.60207	0.63579	0.68680	−0.74
0.76	0.53693	0.56367	0.60062	0.63026	0.66211	0.71009	−0.76
0.78	0.57078	0.59609	0.63095	0.65882	0.68868	0.73351	−0.78
0.80	0.60538	0.62914	0.66176	0.68776	0.71553	0.75705	−0.80
0.82	0.64078	0.66285	0.69308	0.71709	0.74265	0.78073	−0.82
0.84	0.67698	0.69725	0.72491	0.74681	0.77005	0.80454	−0.84
0.86	0.71403	0.73235	0.75727	0.77693	0.79774	0.82848	−0.86
0.88	0.75196	0.76819	0.79018	0.80747	0.82571	0.85256	−0.88
0.90	0.79080	0.80478	0.82365	0.83844	0.85398	0.87678	−0.90
0.92	0.83059	0.84214	0.85769	0.86984	0.88256	0.90113	−0.92
0.94	0.87136	0.88032	0.89233	0.90168	0.91144	0.92563	−0.94
0.96	0.91316	0.91934	0.92759	0.93398	0.94064	0.95027	−0.96
0.98	0.95602	0.95922	0.96347	0.96675	0.97016	0.97506	−0.98

↑ r_0

$\alpha =$	.995 ↑−	.99 ↑−	.975 ↑−	.95 ↑−	.90 ↑−	.75 ↑−

TABLE 46.1   SAMPLE SIZE =   48

α =	.005 ↓+	.01 ↓+	.025 ↓+	.05 ↓+	.10 ↓+	.25 ↓+	
r_o ↓							
-0.98	-0.99047	-0.98972	-0.98853	-0.98740	-0.98597	-0.98323	0.98
-0.96	-0.98084	-0.97934	-0.97696	-0.97471	-0.97186	-0.96641	0.96
-0.94	-0.97111	-0.96887	-0.96530	-0.96193	-0.95768	-0.94956	0.94
-0.92	-0.96129	-0.95830	-0.95355	-0.94907	-0.94342	-0.93266	0.92
-0.90	-0.95136	-0.94764	-0.94170	-0.93612	-0.92908	-0.91571	0.90
-0.88	-0.94134	-0.93687	-0.92976	-0.92308	-0.91467	-0.89873	0.88
-0.86	-0.93121	-0.92600	-0.91771	-0.90994	-0.90018	-0.88170	0.86
-0.84	-0.92097	-0.91502	-0.90557	-0.89672	-0.88560	-0.86462	0.84
-0.82	-0.91063	-0.90394	-0.89333	-0.88340	-0.87095	-0.84751	0.82
-0.80	-0.90018	-0.89276	-0.88099	-0.86999	-0.85621	-0.83034	0.80
-0.78	-0.88962	-0.88146	-0.86854	-0.85647	-0.84140	-0.81313	0.78
-0.76	-0.87895	-0.87005	-0.85598	-0.84287	-0.82649	-0.79587	0.76
-0.74	-0.86816	-0.85853	-0.84332	-0.82916	-0.81151	-0.77856	0.74
-0.72	-0.85726	-0.84690	-0.83055	-0.81535	-0.79643	-0.76121	0.72
-0.70	-0.84623	-0.83514	-0.81767	-0.80144	-0.78128	-0.74380	0.70
-0.68	-0.83509	-0.82327	-0.80468	-0.78743	-0.76603	-0.72635	0.68
-0.66	-0.82382	-0.81128	-0.79157	-0.77332	-0.75069	-0.70885	0.66
-0.64	-0.81243	-0.79917	-0.77835	-0.75910	-0.73526	-0.69129	0.64
-0.62	-0.80091	-0.78693	-0.76501	-0.74477	-0.71975	-0.67369	0.62
-0.60	-0.78926	-0.77457	-0.75155	-0.73033	-0.70414	-0.65603	0.60
-0.58	-0.77748	-0.76208	-0.73798	-0.71578	-0.68843	-0.63832	0.58
-0.56	-0.76556	-0.74945	-0.72428	-0.70112	-0.67263	-0.62056	0.56
-0.54	-0.75351	-0.73669	-0.71045	-0.68635	-0.65674	-0.60274	0.54
-0.52	-0.74131	-0.72380	-0.69650	-0.67146	-0.64075	-0.58487	0.52
-0.50	-0.72897	-0.71077	-0.68242	-0.65645	-0.62466	-0.56694	0.50
-0.48	-0.71649	-0.69759	-0.66820	-0.64133	-0.60847	-0.54896	0.48
-0.46	-0.70386	-0.68428	-0.65386	-0.62609	-0.59218	-0.53092	0.46
-0.44	-0.69108	-0.67081	-0.63938	-0.61072	-0.57578	-0.51282	0.44
-0.42	-0.67814	-0.65720	-0.62476	-0.59523	-0.55929	-0.49467	0.42
-0.40	-0.66505	-0.64344	-0.61001	-0.57961	-0.54268	-0.47646	0.40
-0.38	-0.65180	-0.62952	-0.59511	-0.56387	-0.52598	-0.45818	0.38
-0.36	-0.63838	-0.61545	-0.58006	-0.54800	-0.50916	-0.43985	0.36
-0.34	-0.62479	-0.60121	-0.56487	-0.53199	-0.49223	-0.42146	0.34
-0.32	-0.61104	-0.58682	-0.54954	-0.51586	-0.47520	-0.40300	0.32
-0.30	-0.59711	-0.57225	-0.53405	-0.49959	-0.45805	-0.38448	0.30
-0.28	-0.58301	-0.55752	-0.51840	-0.48318	-0.44079	-0.36590	0.28
-0.26	-0.56872	-0.54261	-0.50260	-0.46663	-0.42341	-0.34726	0.26
-0.24	-0.55425	-0.52753	-0.48664	-0.44994	-0.40592	-0.32855	0.24
-0.22	-0.53959	-0.51227	-0.47052	-0.43310	-0.38831	-0.30977	0.22
-0.20	-0.52474	-0.49682	-0.45423	-0.41612	-0.37058	-0.29093	0.20
-0.18	-0.50969	-0.48119	-0.43777	-0.39899	-0.35272	-0.27202	0.18
-0.16	-0.49443	-0.46537	-0.42114	-0.38171	-0.33475	-0.25304	0.16
-0.14	-0.47898	-0.44935	-0.40434	-0.36428	-0.31664	-0.23399	0.14
-0.12	-0.46331	-0.43314	-0.38736	-0.34669	-0.29841	-0.21488	0.12
-0.10	-0.44743	-0.41672	-0.37020	-0.32894	-0.28006	-0.19569	0.10
-0.08	-0.43133	-0.40009	-0.35286	-0.31103	-0.26157	-0.17643	0.08
-0.06	-0.41500	-0.38326	-0.33532	-0.29296	-0.24295	-0.15710	0.06
-0.04	-0.39845	-0.36621	-0.31760	-0.27472	-0.22420	-0.13770	0.04
-0.02	-0.38166	-0.34893	-0.29969	-0.25631	-0.20531	-0.11822	0.02

↑ r_o

α =   .995 ↑−   .99 ↑−   .975 ↑−   .95 ↑−   .90 ↑−   .75 ↑−

TABLE 46.2   SAMPLE SIZE =   48   (CONT.)

α =	.005 ↓+	.01 ↓+	.025 ↓+	.05 ↓+	.10 ↓+	.25 ↓+	
r_0 ↓							
0.00	-0.36462	-0.33144	-0.28157	-0.23773	-0.18628	-0.09867	0.00
0.02	-0.34735	-0.31371	-0.26325	-0.21898	-0.16711	-0.07904	-0.02
0.04	-0.32982	-0.29575	-0.24473	-0.20005	-0.14780	-0.05933	-0.04
0.06	-0.31203	-0.27755	-0.22600	-0.18094	-0.12835	-0.03955	-0.06
0.08	-0.29398	-0.25910	-0.20705	-0.16164	-0.10875	-0.01968	-0.08
0.10	-0.27566	-0.24041	-0.18789	-0.14216	-0.08900	0.00026	-0.10
0.12	-0.25706	-0.22145	-0.16850	-0.12249	-0.06910	0.02028	-0.12
0.14	-0.23817	-0.20224	-0.14889	-0.10262	-0.04905	0.04039	-0.14
0.16	-0.21900	-0.18276	-0.12905	-0.08256	-0.02884	0.06058	-0.16
0.18	-0.19953	-0.16300	-0.10897	-0.06230	-0.00848	0.08085	-0.18
0.20	-0.17975	-0.14296	-0.08865	-0.04183	0.01205	0.10121	-0.20
0.22	-0.15965	-0.12263	-0.06808	-0.02115	0.03273	0.12166	-0.22
0.24	-0.13924	-0.10201	-0.04726	-0.00027	0.05359	0.14219	-0.24
0.26	-0.11849	-0.08109	-0.02619	0.02038	0.07461	0.16281	-0.26
0.28	-0.09741	-0.05986	-0.00486	0.04215	0.09580	0.18352	-0.28
0.30	-0.07598	-0.03832	0.01674	0.06370	0.11716	0.20432	-0.30
0.32	-0.05419	-0.01645	0.03862	0.08547	0.13870	0.22521	-0.32
0.34	-0.03203	0.00575	0.06077	0.10747	0.16041	0.24620	-0.34
0.36	-0.00949	0.02829	0.08320	0.12971	0.18231	0.26728	-0.36
0.38	0.01343	0.05118	0.10593	0.15219	0.20440	0.28845	-0.38
0.40	0.03675	0.07443	0.12895	0.17491	0.22666	0.30973	-0.40
0.42	0.06048	0.09804	0.15228	0.19789	0.24912	0.33110	-0.42
0.44	0.08463	0.12203	0.17591	0.22112	0.27178	0.35257	-0.44
0.46	0.10922	0.14641	0.19987	0.24461	0.29463	0.37414	-0.46
0.48	0.13425	0.17118	0.22415	0.26836	0.31768	0.39581	-0.48
0.50	0.15974	0.19636	0.24876	0.29239	0.34093	0.41758	-0.50
0.52	0.18571	0.22196	0.27371	0.31669	0.36439	0.43947	-0.52
0.54	0.21217	0.24799	0.29901	0.34127	0.38806	0.46145	-0.54
0.56	0.23912	0.27446	0.32466	0.36614	0.41194	0.48355	-0.56
0.58	0.26660	0.30139	0.35068	0.39130	0.43604	0.50575	-0.58
0.60	0.29461	0.32878	0.37707	0.41676	0.46036	0.52806	-0.60
0.62	0.32317	0.35665	0.40384	0.44252	0.48491	0.55049	-0.62
0.64	0.35230	0.38502	0.43101	0.46860	0.50968	0.57302	-0.64
0.66	0.38202	0.41389	0.45858	0.49500	0.53469	0.59568	-0.66
0.68	0.41234	0.44329	0.48656	0.52172	0.55994	0.61845	-0.68
0.70	0.44330	0.47323	0.51496	0.54877	0.58543	0.64133	-0.70
0.72	0.47490	0.50372	0.54379	0.57617	0.61116	0.66434	-0.72
0.74	0.50717	0.53479	0.57307	0.60391	0.63714	0.68747	-0.74
0.76	0.54013	0.56644	0.60281	0.63200	0.66338	0.71072	-0.76
0.78	0.57382	0.59871	0.63301	0.66046	0.68988	0.73410	-0.78
0.80	0.60824	0.63160	0.66369	0.68929	0.71665	0.75760	-0.80
0.82	0.64344	0.66514	0.69486	0.71850	0.74368	0.78123	-0.82
0.84	0.67943	0.69935	0.72654	0.74809	0.77099	0.80500	-0.84
0.86	0.71625	0.73425	0.75874	0.77809	0.79857	0.82889	-0.86
0.88	0.75394	0.76987	0.79148	0.80849	0.82645	0.85292	-0.88
0.90	0.79251	0.80622	0.82476	0.83931	0.85461	0.87708	-0.90
0.92	0.83200	0.84334	0.85861	0.87055	0.88307	0.90138	-0.92
0.94	0.87246	0.88125	0.89304	0.90223	0.91184	0.92582	-0.94
0.96	0.91392	0.91998	0.92807	0.93436	0.94091	0.95040	-0.96
0.98	0.95642	0.95955	0.96372	0.96694	0.97029	0.97513	-0.98

↑ r_0

α =	.995 ↑−	.99 ↑−	.975 ↑−	.95 ↑−	.90 ↑−	.75 ↑−

TABLE 47.1   SAMPLE SIZE =   49

$a =$	.005 ↓+	.01 ↓+	.025 ↓+	.05 ↓+	.10 ↓+	.25 ↓+	
r_0 ↓							
-0.98	-0.99039	-0.98965	-0.98846	-0.98734	-0.98592	-0.98320	0.98
-0.96	-0.98069	-0.97920	-0.97683	-0.97459	-0.97176	-0.96636	0.96
-0.94	-0.97089	-0.96866	-0.96510	-0.96175	-0.95753	-0.94947	0.94
-0.92	-0.96099	-0.95801	-0.95328	-0.94883	-0.94322	-0.93255	0.92
-0.90	-0.95099	-0.94727	-0.94137	-0.93582	-0.92883	-0.91558	0.90
-0.88	-0.94088	-0.93643	-0.92936	-0.92272	-0.91437	-0.89856	0.88
-0.86	-0.93068	-0.92549	-0.91725	-0.90953	-0.89983	-0.88151	0.86
-0.84	-0.92037	-0.91444	-0.90504	-0.89624	-0.88521	-0.86441	0.84
-0.82	-0.90995	-0.90329	-0.89274	-0.88287	-0.87051	-0.84726	0.82
-0.80	-0.89943	-0.89203	-0.88032	-0.86939	-0.85572	-0.83007	0.80
-0.78	-0.88879	-0.88066	-0.86781	-0.85583	-0.84086	-0.81283	0.78
-0.76	-0.87804	-0.86918	-0.85519	-0.84216	-0.82591	-0.79554	0.76
-0.74	-0.86717	-0.85759	-0.84247	-0.82840	-0.81087	-0.77821	0.74
-0.72	-0.85619	-0.84588	-0.82963	-0.81453	-0.79576	-0.76083	0.72
-0.70	-0.84509	-0.83406	-0.81669	-0.80057	-0.78055	-0.74340	0.70
-0.68	-0.83387	-0.82212	-0.80363	-0.78650	-0.76526	-0.72592	0.68
-0.66	-0.82253	-0.81006	-0.79046	-0.77233	-0.74988	-0.70840	0.66
-0.64	-0.81106	-0.79787	-0.77718	-0.75805	-0.73440	-0.69082	0.64
-0.62	-0.79947	-0.78556	-0.76377	-0.74367	-0.71884	-0.67319	0.62
-0.60	-0.78774	-0.77313	-0.75025	-0.72918	-0.70319	-0.65551	0.60
-0.58	-0.77588	-0.76056	-0.73661	-0.71458	-0.68744	-0.63778	0.58
-0.56	-0.76389	-0.74787	-0.72285	-0.69986	-0.67160	-0.61999	0.56
-0.54	-0.75176	-0.73504	-0.70896	-0.68504	-0.65566	-0.60215	0.54
-0.52	-0.73949	-0.72208	-0.69495	-0.67010	-0.63963	-0.58426	0.52
-0.50	-0.72708	-0.70898	-0.68081	-0.65504	-0.62350	-0.56631	0.50
-0.48	-0.71452	-0.69573	-0.66654	-0.63986	-0.60727	-0.54830	0.48
-0.46	-0.70181	-0.68235	-0.65213	-0.62457	-0.59094	-0.53024	0.46
-0.44	-0.68896	-0.66882	-0.63760	-0.60915	-0.57450	-0.51212	0.44
-0.42	-0.67595	-0.65514	-0.62292	-0.59361	-0.55797	-0.49395	0.42
-0.40	-0.66278	-0.64131	-0.60811	-0.57795	-0.54133	-0.47571	0.40
-0.38	-0.64945	-0.62732	-0.59315	-0.56216	-0.52458	-0.45742	0.38
-0.36	-0.63597	-0.61318	-0.57805	-0.54624	-0.50773	-0.43907	0.36
-0.34	-0.62231	-0.59888	-0.56281	-0.53019	-0.49077	-0.42066	0.34
-0.32	-0.60848	-0.58442	-0.54741	-0.51401	-0.47370	-0.40218	0.32
-0.30	-0.59448	-0.56979	-0.53187	-0.49769	-0.45652	-0.38365	0.30
-0.28	-0.58031	-0.55500	-0.51617	-0.48124	-0.43922	-0.36505	0.28
-0.26	-0.56595	-0.54003	-0.50032	-0.46464	-0.42181	-0.34638	0.26
-0.24	-0.55141	-0.52488	-0.48431	-0.44791	-0.40429	-0.32766	0.24
-0.22	-0.53668	-0.50956	-0.46813	-0.43103	-0.38664	-0.30887	0.22
-0.20	-0.52176	-0.49405	-0.45180	-0.41401	-0.36888	-0.29001	0.20
-0.18	-0.50665	-0.47836	-0.43529	-0.39684	-0.35100	-0.27108	0.18
-0.16	-0.49133	-0.46248	-0.41862	-0.37953	-0.33299	-0.25209	0.16
-0.14	-0.47581	-0.44641	-0.40177	-0.36206	-0.31487	-0.23303	0.14
-0.12	-0.46008	-0.43014	-0.38474	-0.34443	-0.29661	-0.21390	0.12
-0.10	-0.44413	-0.41366	-0.36754	-0.32665	-0.27823	-0.19470	0.10
-0.08	-0.42797	-0.39698	-0.35015	-0.30871	-0.25972	-0.17543	0.08
-0.06	-0.41158	-0.38010	-0.33258	-0.29060	-0.24108	-0.15609	0.06
-0.04	-0.39497	-0.36299	-0.31482	-0.27233	-0.22230	-0.13668	0.04
-0.02	-0.37812	-0.34567	-0.29686	-0.25390	-0.20339	-0.11719	0.02

↑ r_0

$a =$	.995 ↑-	.99 ↑-	.975 ↑-	.95 ↑-	.90 ↑-	.75 ↑-

TABLE 47.2   SAMPLE SIZE = 49 (CONT.)

α =	.005 ↓+	.01 ↓+	.025 ↓+	.05 ↓+	.10 ↓+	.25 ↓+	
r_0 ↓							
0.00	-0.36103	-0.32813	-0.27871	-0.23529	-0.18434	-0.09763	0.00
0.02	-0.34370	-0.31036	-0.26036	-0.21651	-0.16516	-0.07799	-0.02
0.04	-0.32612	-0.29235	-0.24180	-0.19755	-0.14583	-0.05828	-0.04
0.06	-0.30828	-0.27411	-0.22304	-0.17842	-0.12636	-0.03849	-0.06
0.08	-0.29018	-0.25562	-0.20406	-0.15910	-0.10675	-0.01862	-0.08
0.10	-0.27181	-0.23688	-0.18487	-0.13960	-0.08699	0.00133	-0.10
0.12	-0.25316	-0.21789	-0.16546	-0.11991	-0.06708	0.02136	-0.12
0.14	-0.23423	-0.19864	-0.14582	-0.10003	-0.04702	0.04147	-0.14
0.16	-0.21502	-0.17913	-0.12596	-0.07995	-0.02680	0.06166	-0.16
0.18	-0.19551	-0.15934	-0.10586	-0.05967	-0.00643	0.08194	-0.18
0.20	-0.17569	-0.13927	-0.08552	-0.03920	0.01410	0.10230	-0.20
0.22	-0.15557	-0.11892	-0.06494	-0.01851	0.03479	0.12274	-0.22
0.24	-0.13512	-0.09828	-0.04411	0.00238	0.05564	0.14327	-0.24
0.26	-0.11435	-0.07734	-0.02303	0.02348	0.07666	0.16389	-0.26
0.28	-0.09324	-0.05610	-0.00169	0.04480	0.09785	0.18460	-0.28
0.30	-0.07179	-0.03454	0.01992	0.06635	0.11921	0.20539	-0.30
0.32	-0.04998	-0.01266	0.04179	0.08811	0.14074	0.22628	-0.32
0.34	-0.02781	0.00955	0.06394	0.11011	0.16245	0.24726	-0.34
0.36	-0.00527	0.03209	0.08637	0.13234	0.18434	0.26833	-0.36
0.38	0.01766	0.05498	0.10909	0.15481	0.20640	0.28950	-0.38
0.40	0.04098	0.07822	0.13210	0.17752	0.22866	0.31076	-0.40
0.42	0.06471	0.10182	0.15541	0.20047	0.25110	0.33212	-0.42
0.44	0.08885	0.12580	0.17903	0.22368	0.27373	0.35358	-0.44
0.46	0.11342	0.15016	0.20296	0.24715	0.29656	0.37513	-0.46
0.48	0.13844	0.17491	0.22721	0.27087	0.31958	0.39679	-0.48
0.50	0.16390	0.20006	0.25179	0.29487	0.34281	0.41855	-0.50
0.52	0.18984	0.22562	0.27670	0.31913	0.36624	0.44041	-0.52
0.54	0.21625	0.25161	0.30196	0.34367	0.38987	0.46238	-0.54
0.56	0.24316	0.27803	0.32756	0.36850	0.41372	0.48445	-0.56
0.58	0.27058	0.30490	0.35353	0.39361	0.43778	0.50663	-0.58
0.60	0.29853	0.33223	0.37986	0.41902	0.46206	0.52892	-0.60
0.62	0.32702	0.36003	0.40658	0.44473	0.48656	0.55132	-0.62
0.64	0.35608	0.38832	0.43367	0.47075	0.51129	0.57383	-0.64
0.66	0.38571	0.41711	0.46116	0.49708	0.53624	0.59646	-0.66
0.68	0.41593	0.44642	0.48906	0.52373	0.56143	0.61920	-0.68
0.70	0.44678	0.47625	0.51737	0.55071	0.58686	0.64205	-0.70
0.72	0.47826	0.50664	0.54611	0.57802	0.61253	0.66503	-0.72
0.74	0.51040	0.53758	0.57529	0.60567	0.63845	0.68812	-0.74
0.76	0.54321	0.56910	0.60491	0.63367	0.66462	0.71133	-0.76
0.78	0.57674	0.60123	0.63499	0.66203	0.69104	0.73467	-0.78
0.80	0.61099	0.63396	0.66554	0.69076	0.71772	0.75813	-0.80
0.82	0.64600	0.66734	0.69658	0.71985	0.74467	0.78172	-0.82
0.84	0.68179	0.70137	0.72812	0.74933	0.77189	0.80544	-0.84
0.86	0.71839	0.73607	0.76016	0.77920	0.79938	0.82928	-0.86
0.88	0.75583	0.77148	0.79273	0.80947	0.82716	0.85326	-0.88
0.90	0.79414	0.80761	0.82583	0.84014	0.85522	0.87737	-0.90
0.92	0.83336	0.84449	0.85950	0.87124	0.88357	0.90162	-0.92
0.94	0.87352	0.88214	0.89373	0.90276	0.91222	0.92600	-0.94
0.96	0.91465	0.92059	0.92854	0.93472	0.94117	0.95053	-0.96
0.98	0.95680	0.95987	0.96396	0.96713	0.97043	0.97519	-0.98

↑ r_0

α =   .995 ↑−   .99 ↑−   .975 ↑−   .95 ↑−   .90 ↑−   .75 ↑−

TABLE 48.1   SAMPLE SIZE =   50

$\alpha =$	.005 ↓+	.01 ↓+	.025 ↓+	.05 ↓+	.10 ↓+	.25 ↓+	
r_o ↓							
−0.98	−0.99032	−0.98958	−0.98839	−0.98728	−0.98587	−0.98317	0.98
−0.96	−0.98054	−0.97906	−0.97670	−0.97447	−0.97166	−0.96630	0.96
−0.94	−0.97066	−0.96844	−0.96491	−0.96158	−0.95738	−0.94939	0.94
−0.92	−0.96069	−0.95773	−0.95302	−0.94860	−0.94302	−0.93244	0.92
−0.90	−0.95062	−0.94692	−0.94105	−0.93553	−0.92859	−0.91544	0.90
−0.88	−0.94044	−0.93601	−0.92897	−0.92237	−0.91408	−0.89840	0.88
−0.86	−0.93016	−0.92500	−0.91680	−0.90912	−0.89949	−0.88132	0.86
−0.84	−0.91978	−0.91388	−0.90453	−0.89578	−0.88482	−0.86419	0.84
−0.82	−0.90929	−0.90266	−0.89216	−0.88235	−0.87007	−0.84702	0.82
−0.80	−0.89869	−0.89132	−0.87968	−0.86882	−0.85524	−0.82980	0.80
−0.78	−0.88797	−0.87989	−0.86710	−0.85519	−0.84033	−0.81254	0.78
−0.76	−0.87715	−0.86834	−0.85442	−0.84147	−0.82534	−0.79523	0.76
−0.74	−0.86621	−0.85667	−0.84163	−0.82765	−0.81026	−0.77787	0.74
−0.72	−0.85516	−0.84490	−0.82874	−0.81374	−0.79510	−0.76047	0.72
−0.70	−0.84398	−0.83300	−0.81573	−0.79972	−0.77985	−0.74301	0.70
−0.68	−0.83269	−0.82099	−0.80261	−0.78559	−0.76451	−0.72551	0.68
−0.66	−0.82127	−0.80886	−0.78938	−0.77137	−0.74908	−0.70796	0.66
−0.64	−0.80973	−0.79661	−0.77603	−0.75704	−0.73357	−0.69036	0.64
−0.62	−0.79806	−0.78423	−0.76257	−0.74260	−0.71796	−0.67270	0.62
−0.60	−0.78626	−0.77173	−0.74899	−0.72806	−0.70227	−0.65500	0.60
−0.58	−0.77433	−0.75909	−0.73529	−0.71340	−0.68648	−0.63724	0.58
−0.56	−0.76226	−0.74633	−0.72147	−0.69864	−0.67059	−0.61944	0.56
−0.54	−0.75006	−0.73343	−0.70752	−0.68376	−0.65462	−0.60157	0.54
−0.52	−0.73772	−0.72040	−0.69345	−0.66877	−0.63854	−0.58366	0.52
−0.50	−0.72523	−0.70723	−0.67925	−0.65366	−0.62237	−0.56569	0.50
−0.48	−0.71260	−0.69392	−0.66492	−0.63844	−0.60610	−0.54766	0.48
−0.46	−0.69982	−0.68047	−0.65046	−0.62309	−0.58973	−0.52958	0.46
−0.44	−0.68689	−0.66687	−0.63586	−0.60753	−0.57326	−0.51144	0.44
−0.42	−0.67381	−0.65313	−0.62113	−0.59204	−0.55669	−0.49325	0.42
−0.40	−0.66057	−0.63923	−0.60626	−0.57633	−0.54001	−0.47499	0.40
−0.38	−0.64718	−0.62519	−0.59125	−0.56050	−0.52323	−0.45668	0.38
−0.36	−0.63362	−0.61098	−0.57610	−0.54453	−0.50634	−0.43831	0.36
−0.34	−0.61989	−0.59662	−0.56080	−0.52844	−0.48935	−0.41988	0.34
−0.32	−0.60600	−0.58209	−0.54535	−0.51221	−0.47224	−0.40139	0.32
−0.30	−0.59193	−0.56740	−0.52976	−0.49585	−0.45503	−0.38283	0.30
−0.28	−0.57768	−0.55254	−0.51401	−0.47935	−0.43770	−0.36422	0.28
−0.26	−0.56326	−0.53751	−0.49810	−0.46272	−0.42026	−0.34554	0.26
−0.24	−0.54865	−0.52231	−0.48204	−0.44594	−0.40270	−0.32680	0.24
−0.22	−0.53386	−0.50693	−0.46582	−0.42903	−0.38503	−0.30799	0.22
−0.20	−0.51887	−0.49136	−0.44944	−0.41197	−0.36724	−0.28912	0.20
−0.18	−0.50369	−0.47562	−0.43288	−0.39476	−0.34933	−0.27018	0.18
−0.16	−0.48831	−0.45968	−0.41616	−0.37741	−0.33130	−0.25117	0.16
−0.14	−0.47273	−0.44355	−0.39927	−0.35990	−0.31314	−0.23210	0.14
−0.12	−0.45693	−0.42722	−0.38220	−0.34224	−0.29486	−0.21296	0.12
−0.10	−0.44093	−0.41070	−0.36496	−0.32443	−0.27646	−0.19375	0.10
−0.08	−0.42471	−0.39397	−0.34753	−0.30645	−0.25792	−0.17447	0.08
−0.06	−0.40826	−0.37703	−0.32992	−0.28832	−0.23926	−0.15511	0.06
−0.04	−0.39159	−0.35988	−0.31212	−0.27002	−0.22046	−0.13569	0.04
−0.02	−0.37468	−0.34251	−0.29412	−0.25155	−0.20153	−0.11619	0.02

↑ r_o

$\alpha =$	.995 ↑−	.99 ↑−	.975 ↑−	.95 ↑−	.90 ↑−	.75 ↑−

TABLE 48.2   SAMPLE SIZE =   50   (CONT.)

α =	.005 ↓+	.01 ↓+	.025 ↓+	.05 ↓+	.10 ↓+	.25 ↓+	
r_0 ↓							
0.00	−0.35754	−0.32492	−0.27594	−0.23292	−0.18247	−0.09662	0.00
0.02	−0.34016	−0.30710	−0.25755	−0.21411	−0.16326	−0.07698	−0.02
0.04	−0.32252	−0.28905	−0.23896	−0.19513	−0.14392	−0.05726	−0.04
0.06	−0.30464	−0.27076	−0.22017	−0.17598	−0.12444	−0.03746	−0.06
0.08	−0.28649	−0.25224	−0.20116	−0.15664	−0.10481	−0.01758	−0.08
0.10	−0.26807	−0.23346	−0.18194	−0.13712	−0.08504	0.00237	−0.10
0.12	−0.24938	−0.21444	−0.16251	−0.11741	−0.06512	0.02240	−0.12
0.14	−0.23042	−0.19516	−0.14285	−0.09751	−0.04505	0.04252	−0.14
0.16	−0.21116	−0.17561	−0.12296	−0.07742	−0.02482	0.06271	−0.16
0.18	−0.19161	−0.15579	−0.10284	−0.05713	−0.00445	0.08299	−0.18
0.20	−0.17176	−0.13570	−0.08249	−0.03665	0.01609	0.10335	−0.20
0.22	−0.15161	−0.11533	−0.06189	−0.01596	0.03678	0.12379	−0.22
0.24	−0.13113	−0.09466	−0.04105	0.00494	0.05763	0.14432	−0.24
0.26	−0.11033	−0.07371	−0.01996	0.02605	0.07865	0.16493	−0.26
0.28	−0.08920	−0.05244	0.00138	0.04737	0.09984	0.18564	−0.28
0.30	−0.06773	−0.03087	0.02299	0.06891	0.12119	0.20643	−0.30
0.32	−0.04591	−0.00899	0.04487	0.09068	0.14272	0.22731	−0.32
0.34	−0.02372	0.01322	0.06701	0.11267	0.16442	0.24828	−0.34
0.36	−0.00117	0.03577	0.08944	0.13489	0.18629	0.26935	−0.36
0.38	0.02176	0.05866	0.11215	0.15734	0.20835	0.29051	−0.38
0.40	0.04508	0.08189	0.13514	0.18004	0.23059	0.31176	−0.40
0.42	0.06880	0.10548	0.15844	0.20298	0.25301	0.33311	−0.42
0.44	0.09293	0.12944	0.18204	0.22616	0.27562	0.35455	−0.44
0.46	0.11749	0.15378	0.20594	0.24960	0.29843	0.37610	−0.46
0.48	0.14248	0.17851	0.23017	0.27330	0.32143	0.39774	−0.48
0.50	0.16792	0.20363	0.25471	0.29726	0.34463	0.41948	−0.50
0.52	0.19382	0.22915	0.27959	0.32149	0.36802	0.44133	−0.52
0.54	0.22020	0.25510	0.30481	0.34600	0.39163	0.46327	−0.54
0.56	0.24706	0.28148	0.33037	0.37078	0.41544	0.48533	−0.56
0.58	0.27443	0.30829	0.35628	0.39585	0.43946	0.50749	−0.58
0.60	0.30232	0.33556	0.38256	0.42121	0.46370	0.52975	−0.60
0.62	0.33074	0.36330	0.40921	0.44686	0.48816	0.55213	−0.62
0.64	0.35972	0.39151	0.43624	0.47282	0.51283	0.57461	−0.64
0.66	0.38926	0.42022	0.46366	0.49909	0.53774	0.59721	−0.66
0.68	0.41939	0.44944	0.49148	0.52567	0.56288	0.61992	−0.68
0.70	0.45013	0.47917	0.51970	0.55257	0.58825	0.64274	−0.70
0.72	0.48149	0.50945	0.54835	0.57981	0.61385	0.66569	−0.72
0.74	0.51350	0.54027	0.57742	0.60738	0.63970	0.68874	−0.74
0.76	0.54618	0.57167	0.60694	0.63529	0.66580	0.71192	−0.76
0.78	0.57955	0.60365	0.63690	0.66355	0.69215	0.73522	−0.78
0.80	0.61364	0.63624	0.66733	0.69217	0.71876	0.75864	−0.80
0.82	0.64846	0.66945	0.69824	0.72116	0.74563	0.78219	−0.82
0.84	0.68406	0.70331	0.72963	0.75053	0.77276	0.80586	−0.84
0.86	0.72044	0.73783	0.76152	0.78027	0.80016	0.82966	−0.86
0.88	0.75765	0.77303	0.79393	0.81041	0.82784	0.85359	−0.88
0.90	0.79572	0.80895	0.82687	0.84095	0.85580	0.87766	−0.90
0.92	0.83466	0.84559	0.86034	0.87190	0.88404	0.90185	−0.92
0.94	0.87453	0.88300	0.89438	0.90327	0.91258	0.92618	−0.94
0.96	0.91535	0.92118	0.92899	0.93507	0.94142	0.95065	−0.96
0.98	0.95716	0.96017	0.96419	0.96731	0.97055	0.97525	−0.98

↑ r_0

α =	.995 ↑−	.99 ↑−	.975 ↑−	.95 ↑−	.90 ↑−	.75 ↑−

TABLE 49.1   SAMPLE SIZE = 51

α =	.005 ↓+	.01 ↓+	.025 ↓+	.05 ↓+	.10 ↓+	.25 ↓+	
r_0 ↓							
-0.98	-0.99025	-0.98951	-0.98833	-0.98722	-0.98582	-0.98315	0.98
-0.96	-0.98040	-0.97892	-0.97657	-0.97436	-0.97157	-0.96625	0.96
-0.94	-0.97045	-0.96824	-0.96472	-0.96141	-0.95724	-0.94931	0.94
-0.92	-0.96040	-0.95746	-0.95277	-0.94837	-0.94283	-0.93233	0.92
-0.90	-0.95026	-0.94658	-0.94073	-0.93525	-0.92835	-0.91531	0.90
-0.88	-0.94001	-0.93560	-0.92859	-0.92203	-0.91380	-0.89825	0.88
-0.86	-0.92966	-0.92451	-0.91636	-0.90873	-0.89916	-0.88114	0.86
-0.84	-0.91920	-0.91333	-0.90402	-0.89533	-0.88445	-0.86399	0.84
-0.82	-0.90864	-0.90204	-0.89159	-0.88184	-0.86965	-0.84679	0.82
-0.80	-0.89796	-0.89064	-0.87905	-0.86826	-0.85478	-0.82955	0.80
-0.78	-0.88718	-0.87913	-0.86642	-0.85458	-0.83982	-0.81226	0.78
-0.76	-0.87628	-0.86751	-0.85367	-0.84080	-0.82478	-0.79492	0.76
-0.74	-0.86527	-0.85578	-0.84082	-0.82693	-0.80966	-0.77754	0.74
-0.72	-0.85415	-0.84393	-0.82786	-0.81296	-0.79445	-0.76011	0.72
-0.70	-0.84290	-0.83197	-0.81480	-0.79889	-0.77916	-0.74264	0.70
-0.68	-0.83153	-0.81989	-0.80162	-0.78471	-0.76378	-0.72511	0.68
-0.66	-0.82004	-0.80769	-0.78833	-0.77043	-0.74831	-0.70753	0.66
-0.64	-0.80843	-0.79537	-0.77492	-0.75605	-0.73275	-0.68991	0.64
-0.62	-0.79669	-0.78293	-0.76140	-0.74156	-0.71711	-0.67223	0.62
-0.60	-0.78482	-0.77036	-0.74776	-0.72697	-0.70137	-0.65451	0.60
-0.58	-0.77281	-0.75766	-0.73400	-0.71226	-0.68554	-0.63673	0.58
-0.56	-0.76067	-0.74483	-0.72012	-0.69745	-0.66962	-0.61890	0.56
-0.54	-0.74840	-0.73187	-0.70611	-0.68252	-0.65360	-0.60102	0.54
-0.52	-0.73599	-0.71877	-0.69199	-0.66748	-0.63749	-0.58308	0.52
-0.50	-0.72343	-0.70553	-0.67773	-0.65232	-0.62128	-0.56509	0.50
-0.48	-0.71073	-0.69216	-0.66334	-0.63705	-0.60497	-0.54704	0.48
-0.46	-0.69788	-0.67864	-0.64883	-0.62166	-0.58856	-0.52894	0.46
-0.44	-0.68488	-0.66498	-0.63418	-0.60615	-0.57206	-0.51078	0.44
-0.42	-0.67173	-0.65117	-0.61939	-0.59052	-0.55545	-0.49257	0.42
-0.40	-0.65842	-0.63721	-0.60447	-0.57476	-0.53873	-0.47429	0.40
-0.38	-0.64496	-0.62310	-0.58940	-0.55888	-0.52192	-0.45596	0.38
-0.36	-0.63133	-0.60884	-0.57420	-0.54287	-0.50500	-0.43757	0.36
-0.34	-0.61754	-0.59441	-0.55885	-0.52673	-0.48797	-0.41913	0.34
-0.32	-0.60357	-0.57983	-0.54335	-0.51046	-0.47083	-0.40062	0.32
-0.30	-0.58944	-0.56507	-0.52770	-0.49406	-0.45358	-0.38205	0.30
-0.28	-0.57513	-0.55016	-0.51190	-0.47752	-0.43622	-0.36341	0.28
-0.26	-0.56064	-0.53507	-0.49595	-0.46085	-0.41875	-0.34472	0.26
-0.24	-0.54597	-0.51981	-0.47984	-0.44403	-0.40116	-0.32596	0.24
-0.22	-0.53111	-0.50437	-0.46357	-0.42708	-0.38346	-0.30714	0.22
-0.20	-0.51606	-0.48875	-0.44714	-0.40998	-0.36564	-0.28825	0.20
-0.18	-0.50082	-0.47294	-0.43054	-0.39274	-0.34771	-0.26930	0.18
-0.16	-0.48537	-0.45695	-0.41378	-0.37535	-0.32965	-0.25028	0.16
-0.14	-0.46973	-0.44077	-0.39684	-0.35781	-0.31147	-0.23120	0.14
-0.12	-0.45388	-0.42439	-0.37973	-0.34012	-0.29316	-0.21204	0.12
-0.10	-0.43781	-0.40781	-0.36245	-0.32227	-0.27473	-0.19282	0.10
-0.08	-0.42153	-0.39103	-0.34498	-0.30426	-0.25618	-0.17353	0.08
-0.06	-0.40503	-0.37404	-0.32733	-0.28610	-0.23749	-0.15416	0.06
-0.04	-0.38830	-0.35685	-0.30949	-0.26777	-0.21868	-0.13473	0.04
-0.02	-0.37135	-0.33943	-0.29146	-0.24928	-0.19973	-0.11522	0.02

↑ r_0

α =	.995 ↑−	.99 ↑−	.975 ↑−	.95 ↑−	.90 ↑−	.75 ↑−

TABLE 49.2   SAMPLE SIZE =   51   (CONT.)

α =	.005 ↓+	.01 ↓+	.025 ↓+	.05 ↓+	.10 ↓+	.25 ↓+	
r_0 ↓							
0.00	-0.35415	-0.32180	-0.27324	-0.23062	-0.18064	-0.09565	0.00
0.02	-0.33672	-0.30394	-0.25483	-0.21179	-0.16142	-0.07599	-0.02
0.04	-0.31903	-0.28585	-0.23621	-0.19279	-0.14207	-0.05626	-0.04
0.06	-0.30110	-0.26752	-0.21738	-0.17361	-0.12257	-0.03646	-0.06
0.08	-0.28290	-0.24896	-0.19835	-0.15425	-0.10293	-0.01658	-0.08
0.10	-0.26445	-0.23015	-0.17911	-0.13471	-0.08315	0.00338	-0.10
0.12	-0.24572	-0.21109	-0.15965	-0.11499	-0.06321	0.02341	-0.12
0.14	-0.22671	-0.19177	-0.13996	-0.09508	-0.04314	0.04353	-0.14
0.16	-0.20742	-0.17220	-0.12006	-0.07497	-0.02291	0.06372	-0.16
0.18	-0.18783	-0.15235	-0.09992	-0.05467	-0.00252	0.08400	-0.18
0.20	-0.16795	-0.13224	-0.07955	-0.03418	0.01801	0.10436	-0.20
0.22	-0.14776	-0.11184	-0.05894	-0.01348	0.03871	0.12481	-0.22
0.24	-0.12726	-0.09116	-0.03809	0.00742	0.05956	0.14533	-0.24
0.26	-0.10644	-0.07018	-0.01699	0.02853	0.08058	0.16595	-0.26
0.28	-0.08529	-0.04891	0.00436	0.04986	0.10176	0.18665	-0.28
0.30	-0.06380	-0.02733	0.02597	0.07140	0.12311	0.20743	-0.30
0.32	-0.04196	-0.00543	0.04784	0.09316	0.14463	0.22831	-0.32
0.34	-0.01977	0.01679	0.06999	0.11514	0.16632	0.24928	-0.34
0.36	0.00279	0.03933	0.09241	0.13735	0.18819	0.27034	-0.36
0.38	0.02573	0.06221	0.11511	0.15979	0.21023	0.29148	-0.38
0.40	0.04905	0.08544	0.13809	0.18247	0.23245	0.31273	-0.40
0.42	0.07276	0.10902	0.16137	0.20540	0.25486	0.33406	-0.42
0.44	0.09688	0.13297	0.18495	0.22856	0.27746	0.35550	-0.44
0.46	0.12142	0.15729	0.20883	0.25198	0.30024	0.37703	-0.46
0.48	0.14640	0.18199	0.23303	0.27565	0.32321	0.39866	-0.48
0.50	0.17181	0.20708	0.25754	0.29958	0.34638	0.42038	-0.50
0.52	0.19768	0.23257	0.28239	0.32378	0.36975	0.44221	-0.52
0.54	0.22402	0.25848	0.30756	0.34825	0.39333	0.46414	-0.54
0.56	0.25084	0.28481	0.33308	0.37299	0.41710	0.48617	-0.56
0.58	0.27815	0.31157	0.35894	0.39801	0.44109	0.50831	-0.58
0.60	0.30598	0.33878	0.38517	0.42332	0.46528	0.53056	-0.60
0.62	0.33433	0.36645	0.41176	0.44892	0.48970	0.55291	-0.62
0.64	0.36323	0.39460	0.43872	0.47482	0.51433	0.57537	-0.64
0.66	0.39269	0.42322	0.46607	0.50103	0.53919	0.59794	-0.66
0.68	0.42273	0.45235	0.49381	0.52754	0.56427	0.62062	-0.68
0.70	0.45336	0.48199	0.52195	0.55438	0.58958	0.64341	-0.70
0.72	0.48461	0.51216	0.55051	0.58153	0.61513	0.66632	-0.72
0.74	0.51650	0.54287	0.57949	0.60902	0.64092	0.68935	-0.74
0.76	0.54904	0.57414	0.60890	0.63685	0.66695	0.71249	-0.76
0.78	0.58226	0.60599	0.63875	0.66502	0.69323	0.73576	-0.78
0.80	0.61619	0.63844	0.66906	0.69354	0.71976	0.75914	-0.80
0.82	0.65084	0.67149	0.69984	0.72242	0.74655	0.78265	-0.82
0.84	0.68624	0.70518	0.73109	0.75168	0.77360	0.80628	-0.84
0.86	0.72242	0.73952	0.76284	0.78130	0.80091	0.83003	-0.86
0.88	0.75941	0.77453	0.79509	0.81132	0.82850	0.85392	-0.88
0.90	0.79723	0.81023	0.82786	0.84173	0.85636	0.87793	-0.90
0.92	0.83592	0.84666	0.86116	0.87254	0.88450	0.90207	-0.92
0.94	0.87550	0.88382	0.89501	0.90376	0.91293	0.92635	-0.94
0.96	0.91602	0.92175	0.92942	0.93540	0.94166	0.95076	-0.96
0.98	0.95751	0.96046	0.96442	0.96748	0.97068	0.97531	-0.98

↑ r_0

α =	.995 ↑−	.99 ↑−	.975 ↑−	.95 ↑−	.90 ↑−	.75 ↑−

TABLE 50.1   SAMPLE SIZE =   52

α =	.005 ↓+	.01 ↓+	.025 ↓+	.05 ↓+	.10 ↓+	.25 ↓+	
r_o ↓							
-0.98	-0.99018	-0.98944	-0.98827	-0.98717	-0.98577	-0.98312	0.98
-0.96	-0.98025	-0.97879	-0.97645	-0.97425	-0.97147	-0.96620	0.96
-0.94	-0.97024	-0.96804	-0.96453	-0.96124	-0.95710	-0.94924	0.94
-0.92	-0.96012	-0.95719	-0.95253	-0.94815	-0.94265	-0.93223	0.92
-0.90	-0.94990	-0.94624	-0.94042	-0.93497	-0.92812	-0.91519	0.90
-0.88	-0.93959	-0.93519	-0.92822	-0.92170	-0.91352	-0.89810	0.88
-0.86	-0.92916	-0.92404	-0.91593	-0.90834	-0.89884	-0.88096	0.86
-0.84	-0.91864	-0.91279	-0.90353	-0.89489	-0.88408	-0.86379	0.84
-0.82	-0.90800	-0.90143	-0.89104	-0.88135	-0.86925	-0.84656	0.82
-0.80	-0.89726	-0.88996	-0.87844	-0.86771	-0.85433	-0.82930	0.80
-0.78	-0.88641	-0.87839	-0.86574	-0.85398	-0.83933	-0.81198	0.78
-0.76	-0.87544	-0.86670	-0.85294	-0.84015	-0.82424	-0.79463	0.76
-0.74	-0.86436	-0.85491	-0.84003	-0.82623	-0.80908	-0.77722	0.74
-0.72	-0.85316	-0.84300	-0.82701	-0.81220	-0.79383	-0.75977	0.72
-0.70	-0.84184	-0.83097	-0.81389	-0.79808	-0.77849	-0.74227	0.70
-0.68	-0.83041	-0.81882	-0.80065	-0.78385	-0.76307	-0.72472	0.68
-0.66	-0.81885	-0.80656	-0.78730	-0.76952	-0.74756	-0.70712	0.66
-0.64	-0.80716	-0.79417	-0.77384	-0.75509	-0.73196	-0.68947	0.64
-0.62	-0.79535	-0.78166	-0.76026	-0.74055	-0.71628	-0.67178	0.62
-0.60	-0.78341	-0.76902	-0.74656	-0.72591	-0.70050	-0.65403	0.60
-0.58	-0.77133	-0.75626	-0.73274	-0.71115	-0.68463	-0.63623	0.58
-0.56	-0.75913	-0.74337	-0.71881	-0.69629	-0.66867	-0.61838	0.56
-0.54	-0.74678	-0.73034	-0.70475	-0.68131	-0.65261	-0.60047	0.54
-0.52	-0.73430	-0.71718	-0.69056	-0.66622	-0.63646	-0.58251	0.52
-0.50	-0.72167	-0.70388	-0.67625	-0.65102	-0.62021	-0.56450	0.50
-0.48	-0.70891	-0.69044	-0.66181	-0.63570	-0.60387	-0.54644	0.48
-0.46	-0.69599	-0.67686	-0.64724	-0.62027	-0.58742	-0.52832	0.46
-0.44	-0.68292	-0.66314	-0.63253	-0.60471	-0.57088	-0.51014	0.44
-0.42	-0.66970	-0.64927	-0.61769	-0.58903	-0.55424	-0.49190	0.42
-0.40	-0.65633	-0.63525	-0.60272	-0.57323	-0.53749	-0.47361	0.40
-0.38	-0.64280	-0.62108	-0.58760	-0.55730	-0.52064	-0.45527	0.38
-0.36	-0.62910	-0.60675	-0.57235	-0.54125	-0.50368	-0.43686	0.36
-0.34	-0.61524	-0.59226	-0.55694	-0.52507	-0.48662	-0.41839	0.34
-0.32	-0.60121	-0.57762	-0.54140	-0.50876	-0.46945	-0.39987	0.32
-0.30	-0.58701	-0.56281	-0.52570	-0.49232	-0.45217	-0.38128	0.30
-0.28	-0.57264	-0.54783	-0.50986	-0.47574	-0.43479	-0.36263	0.28
-0.26	-0.55809	-0.53269	-0.49386	-0.45903	-0.41728	-0.34392	0.26
-0.24	-0.54335	-0.51737	-0.47770	-0.44218	-0.39967	-0.32515	0.24
-0.22	-0.52843	-0.50188	-0.46138	-0.42519	-0.38194	-0.30631	0.22
-0.20	-0.51332	-0.48620	-0.44491	-0.40805	-0.36409	-0.28741	0.20
-0.18	-0.49802	-0.47034	-0.42827	-0.39077	-0.34613	-0.26845	0.18
-0.16	-0.48252	-0.45430	-0.41146	-0.37335	-0.32805	-0.24941	0.16
-0.14	-0.46681	-0.43806	-0.39448	-0.35578	-0.30984	-0.23032	0.14
-0.12	-0.45090	-0.42164	-0.37733	-0.33805	-0.29151	-0.21115	0.12
-0.10	-0.43478	-0.40501	-0.36001	-0.32017	-0.27306	-0.19192	0.10
-0.08	-0.41845	-0.38818	-0.34250	-0.30214	-0.25448	-0.17262	0.08
-0.06	-0.40189	-0.37115	-0.32482	-0.28394	-0.23578	-0.15324	0.06
-0.04	-0.38511	-0.35390	-0.30694	-0.26559	-0.21694	-0.13380	0.04
-0.02	-0.36810	-0.33644	-0.28888	-0.24707	-0.19798	-0.11428	0.02

↑ r_o

α =	.995 ↑-	.99 ↑-	.975 ↑-	.95 ↑-	.90 ↑-	.75 ↑-

TABLE 50.2   SAMPLE SIZE =   52   (CONT.)

α =	.005 ↓+	.01 ↓+	.025 ↓+	.05 ↓+	.10 ↓+	.25 ↓+	
r_o ↓							
0.00	-0.35086	-0.31876	-0.27063	-0.22839	-0.17888	-0.09470	0.00
0.02	-0.33337	-0.30086	-0.25218	-0.20953	-0.15964	-0.07504	-0.02
0.04	-0.31564	-0.28273	-0.23353	-0.19051	-0.14027	-0.05530	-0.04
0.06	-0.29766	-0.26437	-0.21468	-0.17131	-0.12076	-0.03549	-0.06
0.08	-0.27942	-0.24577	-0.19562	-0.15193	-0.10111	-0.01561	-0.08
0.10	-0.26092	-0.22693	-0.17635	-0.13238	-0.08131	0.00435	-0.10
0.12	-0.24215	-0.20783	-0.15687	-0.11264	-0.06137	0.02439	-0.12
0.14	-0.22311	-0.18849	-0.13717	-0.09271	-0.04128	0.04451	-0.14
0.16	-0.20378	-0.16888	-0.11724	-0.07259	-0.02105	0.06471	-0.16
0.18	-0.18416	-0.14901	-0.09709	-0.05229	-0.00066	0.08499	-0.18
0.20	-0.16425	-0.12887	-0.07670	-0.03178	0.01988	0.10535	-0.20
0.22	-0.14403	-0.10846	-0.05608	-0.01108	0.04058	0.12579	-0.22
0.24	-0.12351	-0.08775	-0.03522	0.00983	0.06143	0.14632	-0.24
0.26	-0.10266	-0.06676	-0.01411	0.03094	0.08245	0.16693	-0.26
0.28	-0.08149	-0.04548	0.00724	0.05227	0.10363	0.18763	-0.28
0.30	-0.05998	-0.02388	0.02886	0.07381	0.12497	0.20841	-0.30
0.32	-0.03813	-0.00198	0.05073	0.09556	0.14648	0.22928	-0.32
0.34	-0.01593	0.02024	0.07287	0.11754	0.16817	0.25024	-0.34
0.36	0.00663	0.04278	0.09528	0.13974	0.19002	0.27129	-0.36
0.38	0.02957	0.06566	0.11798	0.16217	0.21205	0.29243	-0.38
0.40	0.05289	0.08888	0.14095	0.18484	0.23426	0.31366	-0.40
0.42	0.07660	0.11246	0.16421	0.20774	0.25665	0.33499	-0.42
0.44	0.10071	0.13639	0.18777	0.23089	0.27923	0.35641	-0.44
0.46	0.12523	0.16068	0.21163	0.25428	0.30199	0.37793	-0.46
0.48	0.15018	0.18536	0.23580	0.27793	0.32494	0.39954	-0.48
0.50	0.17557	0.21042	0.26028	0.30183	0.34809	0.42126	-0.50
0.52	0.20141	0.23588	0.28509	0.32599	0.37143	0.44307	-0.52
0.54	0.22771	0.26174	0.31023	0.35042	0.39497	0.46498	-0.54
0.56	0.25448	0.28803	0.33570	0.37512	0.41871	0.48699	-0.56
0.58	0.28175	0.31474	0.36152	0.40010	0.44266	0.50911	-0.58
0.60	0.30952	0.34190	0.38769	0.42537	0.46682	0.53133	-0.60
0.62	0.33781	0.36950	0.41422	0.45092	0.49119	0.55366	-0.62
0.64	0.36663	0.39757	0.44112	0.47676	0.51578	0.57610	-0.64
0.66	0.39601	0.42612	0.46840	0.50291	0.54059	0.59864	-0.66
0.68	0.42595	0.45517	0.49606	0.52936	0.56562	0.62130	-0.68
0.70	0.45649	0.48471	0.52413	0.55612	0.59088	0.64406	-0.70
0.72	0.48762	0.51478	0.55259	0.58320	0.61637	0.66694	-0.72
0.74	0.51939	0.54538	0.58148	0.61061	0.64209	0.68993	-0.74
0.76	0.55180	0.57653	0.61079	0.63835	0.66806	0.71304	-0.76
0.78	0.58488	0.60825	0.64053	0.66643	0.69427	0.73627	-0.78
0.80	0.61864	0.64055	0.67072	0.69486	0.72073	0.75962	-0.80
0.82	0.65312	0.67346	0.70138	0.72364	0.74744	0.78308	-0.82
0.84	0.68834	0.70698	0.73250	0.75279	0.77441	0.80667	-0.84
0.86	0.72432	0.74115	0.76411	0.78230	0.80164	0.83039	-0.86
0.88	0.76109	0.77597	0.79621	0.81219	0.82913	0.85423	-0.88
0.90	0.79868	0.81147	0.82882	0.84248	0.85690	0.87819	-0.90
0.92	0.83712	0.84768	0.86195	0.87315	0.88495	0.90229	-0.92
0.94	0.87644	0.88461	0.89562	0.90423	0.91327	0.92652	-0.94
0.96	0.91667	0.92229	0.92984	0.93573	0.94189	0.95088	-0.96
0.98	0.95784	0.96074	0.96463	0.96765	0.97080	0.97537	-0.98

↑ r_o

α =	.995 ↑-	.99 ↑-	.975 ↑-	.95 ↑-	.90 ↑-	.75 ↑-

TABLE 51.1   SAMPLE SIZE =   53

α =	.005 ↓+	.01 ↓+	.025 ↓+	.05 ↓+	.10 ↓+	.25 ↓+	
r_0 ↓							
-0.98	-0.99011	-0.98937	-0.98821	-0.98711	-0.98573	-0.98309	0.98
-0.96	-0.98012	-0.97865	-0.97633	-0.97414	-0.97138	-0.96615	0.96
-0.94	-0.97003	-0.96784	-0.96435	-0.96108	-0.95696	-0.94916	0.94
-0.92	-0.95985	-0.95692	-0.95228	-0.94794	-0.94247	-0.93213	0.92
-0.90	-0.94956	-0.94591	-0.94012	-0.93470	-0.92790	-0.91506	0.90
-0.88	-0.93917	-0.93480	-0.92786	-0.92138	-0.91325	-0.89795	0.88
-0.86	-0.92868	-0.92358	-0.91551	-0.90797	-0.89853	-0.88079	0.86
-0.84	-0.91809	-0.91226	-0.90306	-0.89447	-0.88373	-0.86359	0.84
-0.82	-0.90738	-0.90084	-0.89050	-0.88087	-0.86885	-0.84634	0.82
-0.80	-0.89657	-0.88931	-0.87785	-0.86718	-0.85389	-0.82905	0.80
-0.78	-0.88565	-0.87767	-0.86509	-0.85340	-0.83884	-0.81172	0.78
-0.76	-0.87461	-0.86592	-0.85223	-0.83952	-0.82372	-0.79434	0.76
-0.74	-0.86346	-0.85406	-0.83926	-0.82554	-0.80851	-0.77691	0.74
-0.72	-0.85220	-0.84208	-0.82618	-0.81147	-0.79322	-0.75943	0.72
-0.70	-0.84081	-0.82999	-0.81300	-0.79729	-0.77784	-0.74191	0.70
-0.68	-0.82930	-0.81778	-0.79971	-0.78302	-0.76238	-0.72434	0.68
-0.66	-0.81768	-0.80545	-0.78630	-0.76864	-0.74683	-0.70672	0.66
-0.64	-0.80592	-0.79300	-0.77278	-0.75415	-0.73119	-0.68905	0.64
-0.62	-0.79404	-0.78042	-0.75914	-0.73957	-0.71547	-0.67133	0.62
-0.60	-0.78203	-0.76772	-0.74539	-0.72487	-0.69965	-0.65356	0.60
-0.58	-0.76989	-0.75490	-0.73152	-0.71007	-0.68374	-0.63574	0.58
-0.56	-0.75762	-0.74194	-0.71753	-0.69516	-0.66774	-0.61787	0.56
-0.54	-0.74520	-0.72885	-0.70341	-0.68014	-0.65165	-0.59994	0.54
-0.52	-0.73265	-0.71563	-0.68917	-0.66500	-0.63546	-0.58197	0.52
-0.50	-0.71996	-0.70227	-0.67481	-0.64975	-0.61918	-0.56393	0.50
-0.48	-0.70713	-0.68877	-0.66031	-0.63439	-0.60280	-0.54585	0.48
-0.46	-0.69414	-0.67513	-0.64569	-0.61891	-0.58632	-0.52771	0.46
-0.44	-0.68101	-0.66134	-0.63093	-0.60331	-0.56974	-0.50951	0.44
-0.42	-0.66773	-0.64741	-0.61604	-0.58758	-0.55306	-0.49126	0.42
-0.40	-0.65429	-0.63333	-0.60102	-0.57174	-0.53628	-0.47295	0.40
-0.38	-0.64069	-0.61910	-0.58585	-0.55577	-0.51940	-0.45459	0.38
-0.36	-0.62693	-0.60471	-0.57054	-0.53968	-0.50241	-0.43616	0.36
-0.34	-0.61301	-0.59017	-0.55509	-0.52346	-0.48532	-0.41768	0.34
-0.32	-0.59891	-0.57547	-0.53950	-0.50711	-0.46812	-0.39914	0.32
-0.30	-0.58465	-0.56060	-0.52376	-0.49063	-0.45081	-0.38054	0.30
-0.28	-0.57022	-0.54557	-0.50786	-0.47401	-0.43339	-0.36187	0.28
-0.26	-0.55560	-0.53037	-0.49182	-0.45726	-0.41586	-0.34315	0.26
-0.24	-0.54081	-0.51500	-0.47562	-0.44037	-0.39822	-0.32436	0.24
-0.22	-0.52582	-0.49945	-0.45926	-0.42334	-0.38046	-0.30551	0.22
-0.20	-0.51066	-0.48372	-0.44274	-0.40618	-0.36259	-0.28660	0.20
-0.18	-0.49529	-0.46781	-0.42606	-0.38886	-0.34460	-0.26762	0.18
-0.16	-0.47973	-0.45172	-0.40921	-0.37140	-0.32649	-0.24857	0.16
-0.14	-0.46397	-0.43543	-0.39219	-0.35380	-0.30826	-0.22946	0.14
-0.12	-0.44801	-0.41896	-0.37500	-0.33604	-0.28991	-0.21029	0.12
-0.10	-0.43183	-0.40228	-0.35764	-0.31813	-0.27144	-0.19104	0.10
-0.08	-0.41544	-0.38541	-0.34009	-0.30007	-0.25284	-0.17173	0.08
-0.06	-0.39884	-0.36833	-0.32237	-0.28185	-0.23411	-0.15235	0.06
-0.04	-0.38200	-0.35104	-0.30446	-0.26347	-0.21526	-0.13290	0.04
-0.02	-0.36495	-0.33354	-0.28637	-0.24493	-0.19628	-0.11337	0.02

↑ r_0

α =	.995 ↑-	.99 ↑-	.975 ↑-	.95 ↑-	.90 ↑-	.75 ↑-

TABLE 51.2   SAMPLE SIZE =   53   (CONT.)

α =	.005 ↓+	.01 ↓+	.025 ↓+	.05 ↓+	.10 ↓+	.25 ↓+	
r_0 ↓							
0.00	−0.34765	−0.31582	−0.26809	−0.22622	−0.17716	−0.09378	0.00
0.02	−0.33012	−0.29787	−0.24961	−0.20734	−0.15791	−0.07411	−0.02
0.04	−0.31235	−0.27971	−0.23093	−0.18830	−0.13852	−0.05437	−0.04
0.06	−0.29432	−0.26131	−0.21205	−0.16908	−0.11900	−0.03456	−0.06
0.08	−0.27604	−0.24267	−0.19297	−0.14968	−0.09933	−0.01467	−0.08
0.10	−0.25750	−0.22380	−0.17368	−0.13011	−0.07953	0.00530	−0.10
0.12	−0.23869	−0.20467	−0.15417	−0.11035	−0.05958	0.02534	−0.12
0.14	−0.21961	−0.18530	−0.13445	−0.09041	−0.03949	0.04546	−0.14
0.16	−0.20025	−0.16567	−0.11450	−0.07029	−0.01924	0.06566	−0.16
0.18	−0.18060	−0.14577	−0.09434	−0.04997	0.00115	0.08594	−0.18
0.20	−0.16066	−0.12561	−0.07394	−0.02945	0.02169	0.10630	−0.20
0.22	−0.14041	−0.10517	−0.05331	−0.00874	0.04239	0.12675	−0.22
0.24	−0.11986	−0.08445	−0.03243	0.01217	0.06325	0.14727	−0.24
0.26	−0.09900	−0.06345	−0.01132	0.03328	0.08426	0.16788	−0.26
0.28	−0.07780	−0.04215	0.01004	0.05461	0.10544	0.18857	−0.28
0.30	−0.05628	−0.02055	0.03166	0.07614	0.12678	0.20935	−0.30
0.32	−0.03442	0.00136	0.05353	0.09789	0.14828	0.23022	−0.32
0.34	−0.01221	0.02359	0.07567	0.11986	0.16996	0.25117	−0.34
0.36	0.01036	0.04613	0.09807	0.14205	0.19180	0.27222	−0.36
0.38	0.03330	0.06901	0.12076	0.16447	0.21382	0.29335	−0.38
0.40	0.05661	0.09222	0.14372	0.18713	0.23602	0.31457	−0.40
0.42	0.08032	0.11578	0.16696	0.21001	0.25839	0.33589	−0.42
0.44	0.10442	0.13969	0.19050	0.23314	0.28095	0.35730	−0.44
0.46	0.12893	0.16397	0.21434	0.25651	0.30369	0.37880	−0.46
0.48	0.15386	0.18862	0.23848	0.28013	0.32662	0.40040	−0.48
0.50	0.17922	0.21366	0.26294	0.30400	0.34974	0.42210	−0.50
0.52	0.20502	0.23908	0.28771	0.32813	0.37305	0.44390	−0.52
0.54	0.23128	0.26491	0.31281	0.35253	0.39656	0.46579	−0.54
0.56	0.25801	0.29115	0.33824	0.37719	0.42027	0.48779	−0.56
0.58	0.28523	0.31781	0.36401	0.40213	0.44418	0.50989	−0.58
0.60	0.31294	0.34491	0.39013	0.42734	0.46830	0.53209	−0.60
0.62	0.34116	0.37245	0.41660	0.45284	0.49263	0.55439	−0.62
0.64	0.36992	0.40045	0.44344	0.47863	0.51718	0.57680	−0.64
0.66	0.39921	0.42893	0.47065	0.50472	0.54194	0.59932	−0.66
0.68	0.42907	0.45789	0.49824	0.53111	0.56692	0.62195	−0.68
0.70	0.45950	0.48734	0.52623	0.55781	0.59213	0.64469	−0.70
0.72	0.49053	0.51731	0.55461	0.58482	0.61756	0.66754	−0.72
0.74	0.52218	0.54780	0.58340	0.61215	0.64323	0.69050	−0.74
0.76	0.55446	0.57884	0.61261	0.63981	0.66913	0.71358	−0.76
0.78	0.58740	0.61043	0.64225	0.66780	0.69527	0.73677	−0.78
0.80	0.62101	0.64260	0.67233	0.69614	0.72166	0.76008	−0.80
0.82	0.65533	0.67535	0.70286	0.72482	0.74830	0.78351	−0.82
0.84	0.69037	0.70872	0.73386	0.75386	0.77519	0.80706	−0.84
0.86	0.72616	0.74271	0.76533	0.78326	0.80234	0.83073	−0.86
0.88	0.76272	0.77736	0.79728	0.81304	0.82975	0.85453	−0.88
0.90	0.80009	0.81267	0.82974	0.84320	0.85743	0.87845	−0.90
0.92	0.83828	0.84866	0.86271	0.87374	0.88538	0.90250	−0.92
0.94	0.87734	0.88537	0.89621	0.90469	0.91360	0.92668	−0.94
0.96	0.91729	0.92282	0.93024	0.93604	0.94211	0.95098	−0.96
0.98	0.95816	0.96102	0.96484	0.96781	0.97091	0.97542	−0.98

↑ r_0

α =	.995 ↑−	.99 ↑−	.975 ↑−	.95 ↑−	.90 ↑−	.75 ↑−

TABLE 52.1   SAMPLE SIZE =   54

α =	.005 ↓+	.01 ↓+	.025 ↓+	.05 ↓+	.10 ↓+	.25 ↓+	
r_0 ↓							
-0.98	-0.99004	-0.98931	-0.98815	-0.98706	-0.98568	-0.98307	0.98
-0.96	-0.97998	-0.97853	-0.97621	-0.97403	-0.97129	-0.96610	0.96
-0.94	-0.96983	-0.96765	-0.96418	-0.96092	-0.95683	-0.94909	0.94
-0.92	-0.95958	-0.95667	-0.95205	-0.94772	-0.94229	-0.93204	0.92
-0.90	-0.94923	-0.94559	-0.93983	-0.93444	-0.92768	-0.91494	0.90
-0.88	-0.93877	-0.93441	-0.92751	-0.92107	-0.91299	-0.89780	0.88
-0.86	-0.92821	-0.92313	-0.91510	-0.90760	-0.89823	-0.88062	0.86
-0.84	-0.91755	-0.91175	-0.90259	-0.89405	-0.88338	-0.86340	0.84
-0.82	-0.90678	-0.90026	-0.88998	-0.88040	-0.86846	-0.84613	0.82
-0.80	-0.89590	-0.88867	-0.87727	-0.86666	-0.85346	-0.82882	0.80
-0.78	-0.88491	-0.87696	-0.86445	-0.85283	-0.83837	-0.81146	0.78
-0.76	-0.87381	-0.86515	-0.85153	-0.83890	-0.82321	-0.79405	0.76
-0.74	-0.86259	-0.85322	-0.83851	-0.82487	-0.80796	-0.77660	0.74
-0.72	-0.85125	-0.84119	-0.82538	-0.81075	-0.79263	-0.75911	0.72
-0.70	-0.83980	-0.82903	-0.81214	-0.79652	-0.77721	-0.74156	0.70
-0.68	-0.82823	-0.81676	-0.79879	-0.78220	-0.76171	-0.72397	0.68
-0.66	-0.81653	-0.80437	-0.78532	-0.76777	-0.74612	-0.70633	0.66
-0.64	-0.80471	-0.79185	-0.77175	-0.75324	-0.73044	-0.68864	0.64
-0.62	-0.79277	-0.77922	-0.75806	-0.73861	-0.71468	-0.67090	0.62
-0.60	-0.78069	-0.76646	-0.74425	-0.72387	-0.69882	-0.65311	0.60
-0.58	-0.76848	-0.75357	-0.73032	-0.70902	-0.68288	-0.63527	0.58
-0.56	-0.75614	-0.74055	-0.71628	-0.69406	-0.66684	-0.61737	0.56
-0.54	-0.74367	-0.72740	-0.70211	-0.67899	-0.65071	-0.59943	0.54
-0.52	-0.73105	-0.71411	-0.68782	-0.66381	-0.63449	-0.58143	0.52
-0.50	-0.71829	-0.70069	-0.67340	-0.64852	-0.61817	-0.56338	0.50
-0.48	-0.70539	-0.68713	-0.65886	-0.63311	-0.60175	-0.54528	0.48
-0.46	-0.69234	-0.67343	-0.64418	-0.61758	-0.58524	-0.52712	0.46
-0.44	-0.67915	-0.65959	-0.62937	-0.60194	-0.56863	-0.50890	0.44
-0.42	-0.66580	-0.64560	-0.61443	-0.58618	-0.55192	-0.49064	0.42
-0.40	-0.65230	-0.63146	-0.59936	-0.57029	-0.53510	-0.47231	0.40
-0.38	-0.63863	-0.61717	-0.58414	-0.55428	-0.51819	-0.45393	0.38
-0.36	-0.62481	-0.60273	-0.56879	-0.53815	-0.50117	-0.43549	0.36
-0.34	-0.61083	-0.58813	-0.55329	-0.52189	-0.48405	-0.41699	0.34
-0.32	-0.59667	-0.57337	-0.53765	-0.50550	-0.46682	-0.39843	0.32
-0.30	-0.58235	-0.55845	-0.52186	-0.48898	-0.44948	-0.37981	0.30
-0.28	-0.56785	-0.54337	-0.50592	-0.47232	-0.43203	-0.36113	0.28
-0.26	-0.55318	-0.52811	-0.48983	-0.45554	-0.41447	-0.34239	0.26
-0.24	-0.53832	-0.51269	-0.47359	-0.43861	-0.39680	-0.32359	0.24
-0.22	-0.52328	-0.49709	-0.45719	-0.42155	-0.37902	-0.30473	0.22
-0.20	-0.50806	-0.48131	-0.44062	-0.40435	-0.36112	-0.28580	0.20
-0.18	-0.49264	-0.46535	-0.42390	-0.38700	-0.34311	-0.26681	0.18
-0.16	-0.47702	-0.44920	-0.40701	-0.36951	-0.32498	-0.24776	0.16
-0.14	-0.46121	-0.43287	-0.38996	-0.35188	-0.30673	-0.22863	0.14
-0.12	-0.44519	-0.41635	-0.37273	-0.33409	-0.28835	-0.20945	0.12
-0.10	-0.42896	-0.39963	-0.35533	-0.31615	-0.26986	-0.19019	0.10
-0.08	-0.41252	-0.38271	-0.33775	-0.29806	-0.25124	-0.17087	0.08
-0.06	-0.39586	-0.36558	-0.31999	-0.27981	-0.23250	-0.15148	0.06
-0.04	-0.37898	-0.34825	-0.30206	-0.26141	-0.21362	-0.13202	0.04
-0.02	-0.36187	-0.33071	-0.28393	-0.24284	-0.19462	-0.11249	0.02

↑ r_0

α =	.995 ↑−	.99 ↑−	.975 ↑−	.95 ↑−	.90 ↑−	.75 ↑−

TABLE 52.2   SAMPLE SIZE =   54   (CONT.)

α =	.005 ↓+	.01 ↓+	.025 ↓+	.05 ↓+	.10 ↓+	.25 ↓+	
r_0 ↓							
0.00	-0.34453	-0.31295	-0.26561	-0.22411	-0.17549	-0.09288	0.00
0.02	-0.32696	-0.29497	-0.24711	-0.20521	-0.15623	-0.07321	-0.02
0.04	-0.30914	-0.27676	-0.22840	-0.18614	-0.13683	-0.05346	-0.04
0.06	-0.29107	-0.25833	-0.20950	-0.16691	-0.11729	-0.03364	-0.06
0.08	-0.27275	-0.23966	-0.19039	-0.14749	-0.09761	-0.01375	-0.08
0.10	-0.25417	-0.22075	-0.17107	-0.12791	-0.07780	0.00622	-0.10
0.12	-0.23532	-0.20160	-0.15155	-0.10814	-0.05784	0.02627	-0.12
0.14	-0.21621	-0.18220	-0.13181	-0.08818	-0.03774	0.04639	-0.14
0.16	-0.19681	-0.16254	-0.11185	-0.06804	-0.01749	0.06659	-0.16
0.18	-0.17713	-0.14262	-0.09166	-0.04772	0.00291	0.08687	-0.18
0.20	-0.15716	-0.12244	-0.07125	-0.02720	0.02345	0.10723	-0.20
0.22	-0.13689	-0.10198	-0.05061	-0.00648	0.04415	0.12767	-0.22
0.24	-0.11632	-0.08124	-0.02973	0.01443	0.06501	0.14820	-0.24
0.26	-0.09543	-0.06022	-0.00861	0.03555	0.08602	0.16880	-0.26
0.28	-0.07423	-0.03891	0.01276	0.05688	0.10719	0.18949	-0.28
0.30	-0.05269	-0.01730	0.03438	0.07841	0.12853	0.21027	-0.30
0.32	-0.03081	0.00461	0.05625	0.10015	0.15003	0.23113	-0.32
0.34	-0.00859	0.02684	0.07838	0.12212	0.17169	0.25208	-0.34
0.36	0.01398	0.04938	0.10078	0.14430	0.19353	0.27311	-0.36
0.38	0.03692	0.07225	0.12345	0.16671	0.21553	0.29424	-0.38
0.40	0.06023	0.09546	0.14640	0.18935	0.23772	0.31545	-0.40
0.42	0.08392	0.11900	0.16963	0.21222	0.26007	0.33676	-0.42
0.44	0.10801	0.14290	0.19315	0.23532	0.28261	0.35816	-0.44
0.46	0.13251	0.16716	0.21697	0.25867	0.30533	0.37965	-0.46
0.48	0.15741	0.19179	0.24109	0.28227	0.32824	0.40124	-0.48
0.50	0.18275	0.21679	0.26551	0.30611	0.35133	0.42292	-0.50
0.52	0.20852	0.24218	0.29025	0.33021	0.37462	0.44470	-0.52
0.54	0.23474	0.26797	0.31531	0.35457	0.39810	0.46658	-0.54
0.56	0.26143	0.29417	0.34070	0.37920	0.42178	0.48856	-0.56
0.58	0.28860	0.32078	0.36642	0.40409	0.44566	0.51064	-0.58
0.60	0.31625	0.34782	0.39249	0.42926	0.46974	0.53282	-0.60
0.62	0.34441	0.37531	0.41891	0.45471	0.49403	0.55510	-0.62
0.64	0.37309	0.40324	0.44569	0.48045	0.51853	0.57749	-0.64
0.66	0.40231	0.43164	0.47283	0.50648	0.54325	0.59998	-0.66
0.68	0.43208	0.46052	0.50035	0.53281	0.56818	0.62259	-0.68
0.70	0.46242	0.48988	0.52826	0.55944	0.59334	0.64530	-0.70
0.72	0.49335	0.51975	0.55656	0.58638	0.61872	0.66812	-0.72
0.74	0.52488	0.55014	0.58526	0.61363	0.64433	0.69105	-0.74
0.76	0.55704	0.58106	0.61438	0.64121	0.67017	0.71409	-0.76
0.78	0.58984	0.61253	0.64391	0.66912	0.69625	0.73725	-0.78
0.80	0.62330	0.64457	0.67389	0.69737	0.72257	0.76053	-0.80
0.82	0.65746	0.67718	0.70430	0.72596	0.74913	0.78392	-0.82
0.84	0.69233	0.71040	0.73517	0.75490	0.77595	0.80743	-0.84
0.86	0.72793	0.74423	0.76651	0.78419	0.80302	0.83106	-0.86
0.88	0.76429	0.77870	0.79833	0.81386	0.83034	0.85482	-0.88
0.90	0.80144	0.81382	0.83063	0.84390	0.85793	0.87870	-0.90
0.92	0.83940	0.84961	0.86344	0.87432	0.88579	0.90270	-0.92
0.94	0.87821	0.88611	0.89677	0.90513	0.91392	0.92683	-0.94
0.96	0.91789	0.92332	0.93063	0.93634	0.94233	0.95109	-0.96
0.98	0.95847	0.96128	0.96504	0.96796	0.97102	0.97548	-0.98

↑ r_0

α =	.995 ↑-	.99 ↑-	.975 ↑-	.95 ↑-	.90 ↑-	.75 ↑-

TABLE 53.1   SAMPLE SIZE =   55

α =	.005 ↓+	.01 ↓+	.025 ↓+	.05 ↓+	.10 ↓+	.25 ↓+	
r_0 ↓							
-0.98	-0.98997	-0.98925	-0.98809	-0.98701	-0.98564	-0.98305	0.98
-0.96	-0.97985	-0.97840	-0.97609	-0.97393	-0.97121	-0.96605	0.96
-0.94	-0.96963	-0.96746	-0.96400	-0.96077	-0.95670	-0.94902	0.94
-0.92	-0.95932	-0.95642	-0.95182	-0.94752	-0.94212	-0.93194	0.92
-0.90	-0.94890	-0.94528	-0.93954	-0.93418	-0.92747	-0.91482	0.90
-0.88	-0.93838	-0.93404	-0.92717	-0.92076	-0.91274	-0.89766	0.88
-0.86	-0.92775	-0.92270	-0.91470	-0.90725	-0.89793	-0.88046	0.86
-0.84	-0.91702	-0.91125	-0.90213	-0.89364	-0.88304	-0.86321	0.84
-0.82	-0.90619	-0.89970	-0.88947	-0.87995	-0.86808	-0.84592	0.82
-0.80	-0.89524	-0.88804	-0.87670	-0.86616	-0.85304	-0.82858	0.80
-0.78	-0.88419	-0.87628	-0.86383	-0.85227	-0.83791	-0.81120	0.78
-0.76	-0.87302	-0.86440	-0.85085	-0.83829	-0.82271	-0.79378	0.76
-0.74	-0.86173	-0.85241	-0.83777	-0.82422	-0.80742	-0.77631	0.74
-0.72	-0.85033	-0.84031	-0.82459	-0.81005	-0.79205	-0.75879	0.72
-0.70	-0.83882	-0.82810	-0.81129	-0.79578	-0.77659	-0.74122	0.70
-0.68	-0.82718	-0.81576	-0.79789	-0.78140	-0.76105	-0.72361	0.68
-0.66	-0.81542	-0.80331	-0.78437	-0.76693	-0.74542	-0.70594	0.66
-0.64	-0.80353	-0.79074	-0.77074	-0.75235	-0.72971	-0.68823	0.64
-0.62	-0.79152	-0.77804	-0.75700	-0.73767	-0.71391	-0.67047	0.62
-0.60	-0.77938	-0.76522	-0.74314	-0.72288	-0.69802	-0.65266	0.60
-0.58	-0.76711	-0.75227	-0.72916	-0.70799	-0.68204	-0.63480	0.58
-0.56	-0.75470	-0.73919	-0.71506	-0.69299	-0.66596	-0.61689	0.56
-0.54	-0.74216	-0.72598	-0.70084	-0.67788	-0.64980	-0.59893	0.54
-0.52	-0.72948	-0.71264	-0.68650	-0.66265	-0.63354	-0.58091	0.52
-0.50	-0.71666	-0.69916	-0.67203	-0.64731	-0.61719	-0.56284	0.50
-0.48	-0.70370	-0.68554	-0.65744	-0.63186	-0.60074	-0.54472	0.48
-0.46	-0.69059	-0.67178	-0.64271	-0.61630	-0.58419	-0.52654	0.46
-0.44	-0.67733	-0.65788	-0.62785	-0.60061	-0.56755	-0.50831	0.44
-0.42	-0.66392	-0.64384	-0.61287	-0.58480	-0.55080	-0.49003	0.42
-0.40	-0.65035	-0.62964	-0.59774	-0.56888	-0.53396	-0.47168	0.40
-0.38	-0.63663	-0.61530	-0.58248	-0.55283	-0.51701	-0.45328	0.38
-0.36	-0.62275	-0.60080	-0.56708	-0.53666	-0.49996	-0.43483	0.36
-0.34	-0.60870	-0.58614	-0.55153	-0.52036	-0.48281	-0.41631	0.34
-0.32	-0.59449	-0.57133	-0.53585	-0.50393	-0.46555	-0.39774	0.32
-0.30	-0.58010	-0.55636	-0.52001	-0.48737	-0.44818	-0.37911	0.30
-0.28	-0.56555	-0.54122	-0.50403	-0.47068	-0.43071	-0.36041	0.28
-0.26	-0.55082	-0.52591	-0.48790	-0.45386	-0.41312	-0.34166	0.26
-0.24	-0.53590	-0.51044	-0.47161	-0.43690	-0.39543	-0.32285	0.24
-0.22	-0.52081	-0.49478	-0.45517	-0.41981	-0.37762	-0.30397	0.22
-0.20	-0.50552	-0.47896	-0.43857	-0.40257	-0.35970	-0.28503	0.20
-0.18	-0.49005	-0.46295	-0.42180	-0.38519	-0.34166	-0.26603	0.18
-0.16	-0.47438	-0.44676	-0.40488	-0.36767	-0.32351	-0.24696	0.16
-0.14	-0.45851	-0.43038	-0.38778	-0.35001	-0.30523	-0.22783	0.14
-0.12	-0.44244	-0.41381	-0.37052	-0.33219	-0.28684	-0.20863	0.12
-0.10	-0.42616	-0.39704	-0.35308	-0.31422	-0.26832	-0.18936	0.10
-0.08	-0.40967	-0.38008	-0.33547	-0.29611	-0.24969	-0.17003	0.08
-0.06	-0.39296	-0.36291	-0.31768	-0.27783	-0.23092	-0.15063	0.06
-0.04	-0.37603	-0.34554	-0.29971	-0.25940	-0.21203	-0.13116	0.04
-0.02	-0.35888	-0.32795	-0.28155	-0.24081	-0.19302	-0.11163	0.02

↑ r_0

α =	.995 ↑-	.99 ↑-	.975 ↑-	.95 ↑-	.90 ↑-	.75 ↑-

TABLE 53.2   SAMPLE SIZE = 55   (CONT.)

α =	.005 ↓+	.01 ↓+	.025 ↓+	.05 ↓+	.10 ↓+	.25 ↓+	
r_0 ↓							
0.00	−0.34150	−0.31016	−0.26321	−0.22206	−0.17387	−0.09202	0.00
0.02	−0.32388	−0.29214	−0.24467	−0.20314	−0.15459	−0.07234	−0.02
0.04	−0.30601	−0.27390	−0.22594	−0.18405	−0.13518	−0.05258	−0.04
0.06	−0.28791	−0.25543	−0.20701	−0.16480	−0.11563	−0.03276	−0.06
0.08	−0.26955	−0.23673	−0.18788	−0.14537	−0.09594	−0.01286	−0.08
0.10	−0.25093	−0.21779	−0.16855	−0.12576	−0.07612	0.00711	−0.10
0.12	−0.23205	−0.19861	−0.14900	−0.10598	−0.05615	0.02716	−0.12
0.14	−0.21290	−0.17918	−0.12924	−0.08602	−0.03604	0.04729	−0.14
0.16	−0.19347	−0.15950	−0.10926	−0.06587	−0.01579	0.06749	−0.16
0.18	−0.17377	−0.13956	−0.08907	−0.04553	0.00461	0.08777	−0.18
0.20	−0.15377	−0.11935	−0.06864	−0.02500	0.02516	0.10814	−0.20
0.22	−0.13347	−0.09888	−0.04799	−0.00428	0.04586	0.12858	−0.22
0.24	−0.11288	−0.07813	−0.02710	0.01664	0.06672	0.14910	−0.24
0.26	−0.09197	−0.05709	−0.00597	0.03776	0.08773	0.16970	−0.26
0.28	−0.07075	−0.03577	0.01540	0.05908	0.10890	0.19039	−0.28
0.30	−0.04920	−0.01416	0.03701	0.08061	0.13023	0.21116	−0.30
0.32	−0.02731	0.00776	0.05889	0.10235	0.15172	0.23202	−0.32
0.34	−0.00509	0.02999	0.08102	0.12431	0.17338	0.25296	−0.34
0.36	0.01749	0.05253	0.10341	0.14648	0.19520	0.27399	−0.36
0.38	0.04043	0.07540	0.12607	0.16888	0.21720	0.29510	−0.38
0.40	0.06374	0.09860	0.14901	0.19150	0.23937	0.31631	−0.40
0.42	0.08742	0.12213	0.17222	0.21435	0.26171	0.33761	−0.42
0.44	0.11150	0.14602	0.19572	0.23744	0.28423	0.35899	−0.44
0.46	0.13598	0.17025	0.21952	0.26077	0.30693	0.38047	−0.46
0.48	0.16086	0.19486	0.24361	0.28434	0.32981	0.40205	−0.48
0.50	0.18617	0.21983	0.26801	0.30815	0.35288	0.42372	−0.50
0.52	0.21191	0.24519	0.29271	0.33222	0.37614	0.44548	−0.52
0.54	0.23810	0.27094	0.31773	0.35655	0.39960	0.46734	−0.54
0.56	0.26475	0.29709	0.34308	0.38114	0.42324	0.48930	−0.56
0.58	0.29186	0.32366	0.36876	0.40599	0.44709	0.51136	−0.58
0.60	0.31946	0.35065	0.39478	0.43112	0.47114	0.53352	−0.60
0.62	0.34756	0.37807	0.42114	0.45652	0.49539	0.55579	−0.62
0.64	0.37617	0.40594	0.44786	0.48221	0.51985	0.57815	−0.64
0.66	0.40531	0.43427	0.47494	0.50818	0.54452	0.60062	−0.66
0.68	0.43500	0.46306	0.50240	0.53445	0.56941	0.62320	−0.68
0.70	0.46524	0.49234	0.53023	0.56102	0.59451	0.64588	−0.70
0.72	0.49607	0.52212	0.55845	0.58789	0.61984	0.66868	−0.72
0.74	0.52749	0.55241	0.58706	0.61507	0.64539	0.69158	−0.74
0.76	0.55952	0.58322	0.61608	0.64258	0.67117	0.71459	−0.76
0.78	0.59219	0.61457	0.64552	0.67040	0.69719	0.73772	−0.78
0.80	0.62552	0.64648	0.67539	0.69856	0.72344	0.76096	−0.80
0.82	0.65952	0.67895	0.70569	0.72706	0.74994	0.78432	−0.82
0.84	0.69422	0.71202	0.73644	0.75590	0.77668	0.80779	−0.84
0.86	0.72964	0.74569	0.76765	0.78509	0.80367	0.83139	−0.86
0.88	0.76580	0.77999	0.79933	0.81465	0.83092	0.85510	−0.88
0.90	0.80274	0.81493	0.83149	0.84457	0.85842	0.87893	−0.90
0.92	0.84048	0.85053	0.86415	0.87487	0.88619	0.90289	−0.92
0.94	0.87904	0.88682	0.89732	0.90556	0.91423	0.92698	−0.94
0.96	0.91846	0.92381	0.93101	0.93663	0.94254	0.95119	−0.96
0.98	0.95877	0.96153	0.96523	0.96811	0.97113	0.97553	−0.98

↑ r_0

α =	.995 ↑−	.99 ↑−	.975 ↑−	.95 ↑−	.90 ↑−	.75 ↑−

TABLE 54.1   SAMPLE SIZE =   56

α =	.005 ↓+	.01 ↓+	.025 ↓+	.05 ↓+	.10 ↓+	.25 ↓+	
r_o ↓							
-0.98	-0.98991	-0.98919	-0.98804	-0.98696	-0.98560	-0.98302	0.98
-0.96	-0.97972	-0.97828	-0.97598	-0.97383	-0.97112	-0.96601	0.96
-0.94	-0.96944	-0.96727	-0.96384	-0.96062	-0.95658	-0.94895	0.94
-0.92	-0.95906	-0.95617	-0.95160	-0.94732	-0.94196	-0.93185	0.92
-0.90	-0.94858	-0.94497	-0.93926	-0.93393	-0.92726	-0.91471	0.90
-0.88	-0.93799	-0.93367	-0.92684	-0.92046	-0.91249	-0.89753	0.88
-0.86	-0.92730	-0.92227	-0.91431	-0.90690	-0.89764	-0.88030	0.86
-0.84	-0.91651	-0.91076	-0.90169	-0.89324	-0.88271	-0.86303	0.84
-0.82	-0.90561	-0.89915	-0.88897	-0.87950	-0.86771	-0.84572	0.82
-0.80	-0.89460	-0.88743	-0.87614	-0.86566	-0.85263	-0.82836	0.80
-0.78	-0.88348	-0.87560	-0.86322	-0.85173	-0.83746	-0.81096	0.78
-0.76	-0.87225	-0.86367	-0.85019	-0.83771	-0.82222	-0.79351	0.76
-0.74	-0.86090	-0.85162	-0.83706	-0.82358	-0.80689	-0.77602	0.74
-0.72	-0.84944	-0.83946	-0.82382	-0.80936	-0.79148	-0.75848	0.72
-0.70	-0.83785	-0.82718	-0.81047	-0.79504	-0.77599	-0.74089	0.70
-0.68	-0.82615	-0.81479	-0.79701	-0.78063	-0.76041	-0.72326	0.68
-0.66	-0.81433	-0.80228	-0.78344	-0.76611	-0.74475	-0.70557	0.66
-0.64	-0.80238	-0.78964	-0.76976	-0.75148	-0.72900	-0.68784	0.64
-0.62	-0.79030	-0.77689	-0.75596	-0.73676	-0.71316	-0.67006	0.62
-0.60	-0.77810	-0.76401	-0.74205	-0.72193	-0.69723	-0.65223	0.60
-0.58	-0.76576	-0.75100	-0.72802	-0.70699	-0.68122	-0.63435	0.58
-0.56	-0.75330	-0.73786	-0.71388	-0.69194	-0.66511	-0.61642	0.56
-0.54	-0.74069	-0.72460	-0.69961	-0.67679	-0.64891	-0.59844	0.54
-0.52	-0.72795	-0.71119	-0.68521	-0.66152	-0.63262	-0.58040	0.52
-0.50	-0.71507	-0.69766	-0.67069	-0.64614	-0.61623	-0.56232	0.50
-0.48	-0.70204	-0.68398	-0.65605	-0.63065	-0.59975	-0.54418	0.48
-0.46	-0.68887	-0.67017	-0.64128	-0.61504	-0.58317	-0.52598	0.46
-0.44	-0.67555	-0.65621	-0.62637	-0.59931	-0.56649	-0.50774	0.44
-0.42	-0.66208	-0.64211	-0.61134	-0.58347	-0.54972	-0.48943	0.42
-0.40	-0.64845	-0.62786	-0.59617	-0.56750	-0.53284	-0.47107	0.40
-0.38	-0.63467	-0.61346	-0.58086	-0.55141	-0.51586	-0.45266	0.38
-0.36	-0.62073	-0.59891	-0.56541	-0.53520	-0.49879	-0.43419	0.36
-0.34	-0.60662	-0.58421	-0.54982	-0.51887	-0.48160	-0.41566	0.34
-0.32	-0.59235	-0.56934	-0.53409	-0.50240	-0.46432	-0.39707	0.32
-0.30	-0.57791	-0.55431	-0.51821	-0.48581	-0.44692	-0.37842	0.30
-0.28	-0.56330	-0.53912	-0.50219	-0.46908	-0.42942	-0.35971	0.28
-0.26	-0.54851	-0.52377	-0.48601	-0.45223	-0.41181	-0.34095	0.26
-0.24	-0.53354	-0.50824	-0.46969	-0.43524	-0.39409	-0.32212	0.24
-0.22	-0.51839	-0.49254	-0.45320	-0.41811	-0.37626	-0.30323	0.22
-0.20	-0.50306	-0.47666	-0.43656	-0.40084	-0.35831	-0.28428	0.20
-0.18	-0.48753	-0.46061	-0.41976	-0.38343	-0.34025	-0.26526	0.18
-0.16	-0.47181	-0.44437	-0.40280	-0.36588	-0.32207	-0.24619	0.16
-0.14	-0.45589	-0.42794	-0.38567	-0.34818	-0.30378	-0.22704	0.14
-0.12	-0.43976	-0.41133	-0.36837	-0.33034	-0.28536	-0.20783	0.12
-0.10	-0.42343	-0.39452	-0.35090	-0.31235	-0.26683	-0.18856	0.10
-0.08	-0.40689	-0.37751	-0.33325	-0.29420	-0.24817	-0.16922	0.08
-0.06	-0.39014	-0.36031	-0.31543	-0.27590	-0.22939	-0.14981	0.06
-0.04	-0.37317	-0.34290	-0.29743	-0.25745	-0.21049	-0.13033	0.04
-0.02	-0.35597	-0.32527	-0.27924	-0.23884	-0.19145	-0.11079	0.02

↑ r_o

α =	.995 ↑-	.99 ↑-	.975 ↑-	.95 ↑-	.90 ↑-	.75 ↑-

TABLE 54.2   SAMPLE SIZE = 56   (CONT.)

α =	.005 ↓+	.01 ↓+	.025 ↓+	.05 ↓+	.10 ↓+	.25 ↓+	
r_0 ↓							
0.00	−0.33854	−0.30744	−0.26087	−0.22006	−0.17229	−0.09117	0.00
0.02	−0.32088	−0.28939	−0.24231	−0.20112	−0.15300	−0.07149	−0.02
0.04	−0.30297	−0.27111	−0.22355	−0.18202	−0.13357	−0.05173	−0.04
0.06	−0.28483	−0.25261	−0.20460	−0.16275	−0.11401	−0.03190	−0.06
0.08	−0.26643	−0.23388	−0.18544	−0.14330	−0.09432	−0.01199	−0.08
0.10	−0.24778	−0.21491	−0.16609	−0.12368	−0.07448	0.00798	−0.10
0.12	−0.22886	−0.19570	−0.14652	−0.10388	−0.05451	0.02804	−0.12
0.14	−0.20968	−0.17625	−0.12674	−0.08391	−0.03439	0.04816	−0.14
0.16	−0.19022	−0.15654	−0.10675	−0.06375	−0.01413	0.06837	−0.16
0.18	−0.17049	−0.13658	−0.08654	−0.04340	0.00627	0.08865	−0.18
0.20	−0.15046	−0.11636	−0.06610	−0.02287	0.02682	0.10901	−0.20
0.22	−0.13015	−0.09586	−0.04544	−0.00214	0.04753	0.12945	−0.22
0.24	−0.10953	−0.07510	−0.02455	0.01878	0.06838	0.14997	−0.24
0.26	−0.08861	−0.05405	−0.00341	0.03990	0.08939	0.17057	−0.26
0.28	−0.06737	−0.03272	0.01796	0.06122	0.11056	0.19126	−0.28
0.30	−0.04580	−0.01110	0.03958	0.08275	0.13188	0.21202	−0.30
0.32	−0.02391	0.01083	0.06145	0.10448	0.15337	0.23287	−0.32
0.34	−0.00168	0.03306	0.08357	0.12643	0.17502	0.25381	−0.34
0.36	0.02090	0.05560	0.10596	0.14860	0.19683	0.27483	−0.36
0.38	0.04384	0.07846	0.12861	0.17098	0.21881	0.29594	−0.38
0.40	0.06715	0.10165	0.15154	0.19359	0.24097	0.31714	−0.40
0.42	0.09082	0.12517	0.17474	0.21643	0.26329	0.33843	−0.42
0.44	0.11489	0.14904	0.19822	0.23950	0.28580	0.35980	−0.44
0.46	0.13935	0.17326	0.22199	0.26280	0.30848	0.38127	−0.46
0.48	0.16421	0.19784	0.24606	0.28635	0.33134	0.40283	−0.48
0.50	0.18949	0.22278	0.27043	0.31014	0.35439	0.42449	−0.50
0.52	0.21520	0.24811	0.29510	0.33418	0.37762	0.44624	−0.52
0.54	0.24136	0.27382	0.32009	0.35847	0.40105	0.46809	−0.54
0.56	0.26796	0.29993	0.34539	0.38302	0.42466	0.49003	−0.56
0.58	0.29503	0.32645	0.37103	0.40784	0.44848	0.51207	−0.58
0.60	0.32257	0.35339	0.39700	0.43292	0.47249	0.53421	−0.60
0.62	0.35061	0.38075	0.42331	0.45828	0.49670	0.55645	−0.62
0.64	0.37915	0.40856	0.44997	0.48391	0.52112	0.57880	−0.64
0.66	0.40822	0.43681	0.47699	0.50983	0.54575	0.60125	−0.66
0.68	0.43782	0.46553	0.50437	0.53604	0.57059	0.62380	−0.68
0.70	0.46798	0.49473	0.53213	0.56255	0.59565	0.64646	−0.70
0.72	0.49870	0.52441	0.56027	0.58935	0.62092	0.66922	−0.72
0.74	0.53001	0.55460	0.58880	0.61647	0.64642	0.69209	−0.74
0.76	0.56193	0.58530	0.61774	0.64389	0.67215	0.71508	−0.76
0.78	0.59447	0.61654	0.64708	0.67164	0.69810	0.73817	−0.78
0.80	0.62765	0.64832	0.67684	0.69972	0.72429	0.76138	−0.80
0.82	0.66150	0.68067	0.70704	0.72812	0.75072	0.78470	−0.82
0.84	0.69604	0.71359	0.73767	0.75687	0.77739	0.80814	−0.84
0.86	0.73129	0.74711	0.76876	0.78597	0.80431	0.83170	−0.86
0.88	0.76726	0.78124	0.80030	0.81541	0.83147	0.85537	−0.88
0.90	0.80400	0.81600	0.83233	0.84523	0.85890	0.87917	−0.90
0.92	0.84152	0.85142	0.86484	0.87541	0.88658	0.90308	−0.92
0.94	0.87985	0.88750	0.89785	0.90597	0.91453	0.92712	−0.94
0.96	0.91902	0.92428	0.93137	0.93691	0.94274	0.95129	−0.96
0.98	0.95906	0.96177	0.96541	0.96826	0.97123	0.97558	−0.98

↑ r_0

α =	.995 ↑−	.99 ↑−	.975 ↑−	.95 ↑−	.90 ↑−	.75 ↑−

TABLE 55.1   SAMPLE SIZE =   57

α =	.005 ↓+	.01 ↓+	.025 ↓+	.05 ↓+	.10 ↓+	.25 ↓+	
r_0 ↓							
-0.98	-0.98985	-0.98913	-0.98798	-0.98691	-0.98556	-0.98300	0.98
-0.96	-0.97960	-0.97816	-0.97587	-0.97373	-0.97104	-0.96596	0.96
-0.94	-0.96925	-0.96709	-0.96367	-0.96047	-0.95646	-0.94888	0.94
-0.92	-0.95881	-0.95593	-0.95138	-0.94712	-0.94179	-0.93176	0.92
-0.90	-0.94826	-0.94467	-0.93899	-0.93369	-0.92706	-0.91460	0.90
-0.88	-0.93761	-0.93331	-0.92651	-0.92017	-0.91225	-0.89739	0.88
-0.86	-0.92686	-0.92185	-0.91393	-0.90656	-0.89736	-0.88014	0.86
-0.84	-0.91601	-0.91028	-0.90125	-0.89286	-0.88239	-0.86285	0.84
-0.82	-0.90504	-0.89861	-0.88848	-0.87906	-0.86735	-0.84552	0.82
-0.80	-0.89397	-0.88683	-0.87560	-0.86518	-0.85223	-0.82814	0.80
-0.78	-0.88279	-0.87495	-0.86262	-0.85120	-0.83703	-0.81072	0.78
-0.76	-0.87149	-0.86295	-0.84954	-0.83713	-0.82174	-0.79325	0.76
-0.74	-0.86008	-0.85085	-0.83636	-0.82296	-0.80638	-0.77573	0.74
-0.72	-0.84856	-0.83863	-0.82306	-0.80870	-0.79093	-0.75817	0.72
-0.70	-0.83691	-0.82629	-0.80966	-0.79433	-0.77540	-0.74057	0.70
-0.68	-0.82515	-0.81384	-0.79615	-0.77987	-0.75979	-0.72291	0.68
-0.66	-0.81326	-0.80127	-0.78253	-0.76530	-0.74409	-0.70521	0.66
-0.64	-0.80125	-0.78858	-0.76880	-0.75064	-0.72830	-0.68746	0.64
-0.62	-0.78911	-0.77576	-0.75496	-0.73587	-0.71243	-0.66966	0.62
-0.60	-0.77685	-0.76283	-0.74099	-0.72099	-0.69647	-0.65181	0.60
-0.58	-0.76445	-0.74976	-0.72691	-0.70601	-0.68042	-0.63391	0.58
-0.56	-0.75192	-0.73657	-0.71272	-0.69092	-0.66427	-0.61596	0.56
-0.54	-0.73926	-0.72324	-0.69840	-0.67572	-0.64804	-0.59796	0.54
-0.52	-0.72645	-0.70979	-0.68396	-0.66042	-0.63171	-0.57991	0.52
-0.50	-0.71351	-0.69619	-0.66939	-0.64500	-0.61529	-0.56181	0.50
-0.48	-0.70043	-0.68247	-0.65470	-0.62946	-0.59878	-0.54365	0.48
-0.46	-0.68719	-0.66860	-0.63988	-0.61381	-0.58217	-0.52544	0.46
-0.44	-0.67381	-0.65459	-0.62493	-0.59805	-0.56546	-0.50717	0.44
-0.42	-0.66028	-0.64043	-0.60984	-0.58216	-0.54866	-0.48885	0.42
-0.40	-0.64660	-0.62613	-0.59463	-0.56616	-0.53175	-0.47048	0.40
-0.38	-0.63276	-0.61168	-0.57927	-0.55003	-0.51475	-0.45205	0.38
-0.36	-0.61876	-0.59707	-0.56378	-0.53379	-0.49764	-0.43356	0.36
-0.34	-0.60460	-0.58231	-0.54815	-0.51741	-0.48043	-0.41502	0.34
-0.32	-0.59027	-0.56740	-0.53238	-0.50091	-0.46311	-0.39641	0.32
-0.30	-0.57577	-0.55232	-0.51646	-0.48429	-0.44569	-0.37775	0.30
-0.28	-0.56111	-0.53708	-0.50039	-0.46753	-0.42817	-0.35903	0.28
-0.26	-0.54626	-0.52167	-0.48418	-0.45064	-0.41053	-0.34025	0.26
-0.24	-0.53124	-0.50610	-0.46781	-0.43361	-0.39278	-0.32141	0.24
-0.22	-0.51604	-0.49035	-0.45129	-0.41645	-0.37493	-0.30251	0.22
-0.20	-0.50065	-0.47443	-0.43461	-0.39915	-0.35696	-0.28355	0.20
-0.18	-0.48507	-0.45833	-0.41777	-0.38171	-0.33887	-0.26452	0.18
-0.16	-0.46929	-0.44204	-0.40077	-0.36413	-0.32068	-0.24543	0.16
-0.14	-0.45332	-0.42557	-0.38360	-0.34641	-0.30236	-0.22628	0.14
-0.12	-0.43715	-0.40892	-0.36627	-0.32854	-0.28393	-0.20706	0.12
-0.10	-0.42077	-0.39206	-0.34877	-0.31052	-0.26537	-0.18778	0.10
-0.08	-0.40419	-0.37502	-0.33109	-0.29235	-0.24670	-0.16843	0.08
-0.06	-0.38739	-0.35777	-0.31324	-0.27403	-0.22790	-0.14901	0.06
-0.04	-0.37037	-0.34032	-0.29520	-0.25555	-0.20898	-0.12953	0.04
-0.02	-0.35313	-0.32266	-0.27699	-0.23691	-0.18993	-0.10997	0.02

↑ r_0

α =	.995 ↑-	.99 ↑-	.975 ↑-	.95 ↑-	.90 ↑-	.75 ↑-

TABLE 55.2   SAMPLE SIZE = 57   (CONT.)

α =	.005 ↓+	.01 ↓+	.025 ↓+	.05 ↓+	.10 ↓+	.25 ↓+	
r_o ↓							
0.00	−0.33566	−0.30479	−0.25859	−0.21812	−0.17075	−0.09035	0.00
0.02	−0.31795	−0.28671	−0.24000	−0.19916	−0.15145	−0.07066	−0.02
0.04	−0.30001	−0.26840	−0.22122	−0.18004	−0.13201	−0.05089	−0.04
0.06	−0.28183	−0.24987	−0.20224	−0.16075	−0.11244	−0.03106	−0.06
0.08	−0.26339	−0.23110	−0.18307	−0.14129	−0.09273	−0.01115	−0.08
0.10	−0.24471	−0.21211	−0.16369	−0.12165	−0.07289	0.00883	−0.10
0.12	−0.22576	−0.19287	−0.14411	−0.10185	−0.05291	0.02888	−0.12
0.14	−0.20655	−0.17339	−0.12431	−0.08186	−0.03279	0.04901	−0.14
0.16	−0.18706	−0.15366	−0.10431	−0.06169	−0.01252	0.06922	−0.16
0.18	−0.16730	−0.13368	−0.08408	−0.04134	0.00788	0.08950	−0.18
0.20	−0.14725	−0.11344	−0.06364	−0.02080	0.02844	0.10986	−0.20
0.22	−0.12691	−0.09293	−0.04296	−0.00006	0.04914	0.13030	−0.22
0.24	−0.10628	−0.07215	−0.02206	0.02086	0.07000	0.15082	−0.24
0.26	−0.08533	−0.05109	−0.00092	0.04198	0.09101	0.17142	−0.26
0.28	−0.06408	−0.02975	0.02045	0.06330	0.11217	0.19210	−0.28
0.30	−0.04250	−0.00812	0.04207	0.08483	0.13349	0.21286	−0.30
0.32	−0.02060	0.01381	0.06394	0.10656	0.15497	0.23371	−0.32
0.34	0.00164	0.03603	0.08606	0.12850	0.17661	0.25464	−0.34
0.36	0.02422	0.05857	0.10844	0.15066	0.19841	0.27566	−0.36
0.38	0.04716	0.08143	0.13108	0.17303	0.22038	0.29676	−0.38
0.40	0.07046	0.10461	0.15399	0.19562	0.24252	0.31795	−0.40
0.42	0.09413	0.12812	0.17718	0.21845	0.26484	0.33922	−0.42
0.44	0.11818	0.15197	0.20065	0.24149	0.28732	0.36059	−0.44
0.46	0.14262	0.17617	0.22440	0.26478	0.30998	0.38205	−0.46
0.48	0.16746	0.20073	0.24844	0.28830	0.33283	0.40360	−0.48
0.50	0.19272	0.22565	0.27278	0.31206	0.35585	0.42524	−0.50
0.52	0.21840	0.25094	0.29742	0.33607	0.37906	0.44698	−0.52
0.54	0.24451	0.27662	0.32237	0.36033	0.40246	0.46881	−0.54
0.56	0.27108	0.30269	0.34764	0.38485	0.42604	0.49073	−0.56
0.58	0.29810	0.32916	0.37323	0.40962	0.44983	0.51276	−0.58
0.60	0.32559	0.35604	0.39915	0.43467	0.47380	0.53488	−0.60
0.62	0.35357	0.38335	0.42541	0.45998	0.49798	0.55710	−0.62
0.64	0.38205	0.41109	0.45202	0.48556	0.52236	0.57942	−0.64
0.66	0.41104	0.43928	0.47897	0.51143	0.54695	0.60185	−0.66
0.68	0.44056	0.46792	0.50629	0.53759	0.57174	0.62438	−0.68
0.70	0.47062	0.49704	0.53398	0.56403	0.59675	0.64701	−0.70
0.72	0.50125	0.52663	0.56204	0.59077	0.62198	0.66975	−0.72
0.74	0.53246	0.55672	0.59049	0.61782	0.64742	0.69259	−0.74
0.76	0.56426	0.58732	0.61934	0.64517	0.67309	0.71555	−0.76
0.78	0.59667	0.61845	0.64859	0.67284	0.69899	0.73861	−0.78
0.80	0.62972	0.65011	0.67825	0.70084	0.72512	0.76179	−0.80
0.82	0.66343	0.68232	0.70834	0.72916	0.75148	0.78508	−0.82
0.84	0.69781	0.71511	0.73886	0.75781	0.77808	0.80848	−0.84
0.86	0.73288	0.74847	0.76983	0.78681	0.80492	0.83200	−0.86
0.88	0.76868	0.78245	0.80125	0.81616	0.83201	0.85564	−0.88
0.90	0.80522	0.81704	0.83313	0.84586	0.85936	0.87939	−0.90
0.92	0.84253	0.85227	0.86550	0.87593	0.88696	0.90327	−0.92
0.94	0.88063	0.88816	0.89836	0.90637	0.91481	0.92726	−0.94
0.96	0.91956	0.92473	0.93172	0.93719	0.94294	0.95138	−0.96
0.98	0.95934	0.96201	0.96559	0.96839	0.97133	0.97563	−0.98

↑ r_o

α =	.995 ↑−	.99 ↑−	.975 ↑−	.95 ↑−	.90 ↑−	.75 ↑−

TABLE 56.1   SAMPLE SIZE =   58

α =	.005 ↓+	.01 ↓+	.025 ↓+	.05 ↓+	.10 ↓+	.25 ↓+	
r₀ ↓							
-0.98	-0.98979	-0.98907	-0.98793	-0.98686	-0.98552	-0.98298	0.98
-0.96	-0.97948	-0.97804	-0.97576	-0.97364	-0.97096	-0.96592	0.96
-0.94	-0.96907	-0.96692	-0.96351	-0.96033	-0.95634	-0.94882	0.94
-0.92	-0.95856	-0.95570	-0.95116	-0.94693	-0.94163	-0.93167	0.92
-0.90	-0.94795	-0.94438	-0.93872	-0.93345	-0.92686	-0.91449	0.90
-0.88	-0.93725	-0.93296	-0.92619	-0.91988	-0.91201	-0.89726	0.88
-0.86	-0.92643	-0.92144	-0.91356	-0.90622	-0.89708	-0.87999	0.86
-0.84	-0.91551	-0.90981	-0.90083	-0.89248	-0.88208	-0.86268	0.84
-0.82	-0.90449	-0.89808	-0.88800	-0.87864	-0.86700	-0.84532	0.82
-0.80	-0.89336	-0.88625	-0.87507	-0.86471	-0.85184	-0.82792	0.80
-0.78	-0.88211	-0.87430	-0.86204	-0.85069	-0.83660	-0.81048	0.78
-0.76	-0.87075	-0.86225	-0.84891	-0.83657	-0.82128	-0.79299	0.76
-0.74	-0.85928	-0.85009	-0.83567	-0.82235	-0.80588	-0.77546	0.74
-0.72	-0.84770	-0.83781	-0.82233	-0.80804	-0.79040	-0.75788	0.72
-0.70	-0.83599	-0.82542	-0.80888	-0.79364	-0.77483	-0.74025	0.70
-0.68	-0.82417	-0.81291	-0.79532	-0.77913	-0.75918	-0.72258	0.68
-0.66	-0.81222	-0.80028	-0.78165	-0.76452	-0.74344	-0.70486	0.66
-0.64	-0.80015	-0.78753	-0.76787	-0.74981	-0.72762	-0.68709	0.64
-0.62	-0.78795	-0.77466	-0.75397	-0.73500	-0.71172	-0.66927	0.62
-0.60	-0.77562	-0.76167	-0.73996	-0.72008	-0.69572	-0.65140	0.60
-0.58	-0.76317	-0.74855	-0.72583	-0.70506	-0.67963	-0.63348	0.58
-0.56	-0.75058	-0.73530	-0.71158	-0.68993	-0.66346	-0.61552	0.56
-0.54	-0.73785	-0.72192	-0.69722	-0.67469	-0.64719	-0.59750	0.54
-0.52	-0.72499	-0.70841	-0.68273	-0.65934	-0.63084	-0.57943	0.52
-0.50	-0.71199	-0.69477	-0.66812	-0.64388	-0.61438	-0.56131	0.50
-0.48	-0.69885	-0.68098	-0.65338	-0.62830	-0.59784	-0.54313	0.48
-0.46	-0.68556	-0.66706	-0.63851	-0.61262	-0.58120	-0.52491	0.46
-0.44	-0.67212	-0.65300	-0.62352	-0.59681	-0.56446	-0.50663	0.44
-0.42	-0.65853	-0.63879	-0.60839	-0.58089	-0.54763	-0.48829	0.42
-0.40	-0.64479	-0.62443	-0.59313	-0.56485	-0.53069	-0.46990	0.40
-0.38	-0.63089	-0.60993	-0.57773	-0.54869	-0.51366	-0.45145	0.38
-0.36	-0.61684	-0.59527	-0.56220	-0.53240	-0.49652	-0.43295	0.36
-0.34	-0.60262	-0.58046	-0.54652	-0.51600	-0.47928	-0.41439	0.34
-0.32	-0.58824	-0.56550	-0.53071	-0.49946	-0.46194	-0.39578	0.32
-0.30	-0.57369	-0.55037	-0.51475	-0.48280	-0.44450	-0.37710	0.30
-0.28	-0.55896	-0.53508	-0.49864	-0.46601	-0.42694	-0.35837	0.28
-0.26	-0.54407	-0.51963	-0.48238	-0.44908	-0.40928	-0.33958	0.26
-0.24	-0.52899	-0.50401	-0.46598	-0.43203	-0.39151	-0.32072	0.24
-0.22	-0.51374	-0.48821	-0.44942	-0.41484	-0.37363	-0.30181	0.22
-0.20	-0.49829	-0.47225	-0.43270	-0.39751	-0.35564	-0.28283	0.20
-0.18	-0.48266	-0.45610	-0.41583	-0.38004	-0.33754	-0.26380	0.18
-0.16	-0.46684	-0.43977	-0.39879	-0.36243	-0.31932	-0.24470	0.16
-0.14	-0.45082	-0.42326	-0.38159	-0.34468	-0.30098	-0.22553	0.14
-0.12	-0.43460	-0.40656	-0.36422	-0.32678	-0.28253	-0.20631	0.12
-0.10	-0.41818	-0.38967	-0.34669	-0.30874	-0.26396	-0.18701	0.10
-0.08	-0.40155	-0.37258	-0.32898	-0.29054	-0.24526	-0.16766	0.08
-0.06	-0.38470	-0.35530	-0.31110	-0.27220	-0.22645	-0.14823	0.06
-0.04	-0.36764	-0.33781	-0.29304	-0.25370	-0.20751	-0.12874	0.04
-0.02	-0.35036	-0.32012	-0.27479	-0.23504	-0.18845	-0.10918	0.02

↑ r₀

α =	.995 ↑-	.99 ↑-	.975 ↑-	.95 ↑-	.90 ↑-	.75 ↑-

TABLE 56.2   SAMPLE SIZE = 58 (CONT.)

α =	.005 ↓+	.01 ↓+	.025 ↓+	.05 ↓+	.10 ↓+	.25 ↓+	
r_0 ↓							
0.00	−0.33284	−0.30221	−0.25637	−0.21623	−0.16926	−0.08955	0.00
0.02	−0.31510	−0.28409	−0.23776	−0.19725	−0.14994	−0.06985	−0.02
0.04	−0.29712	−0.26575	−0.21895	−0.17811	−0.13049	−0.05008	−0.04
0.06	−0.27890	−0.24719	−0.19995	−0.15880	−0.11091	−0.03024	−0.06
0.08	−0.26043	−0.22840	−0.18076	−0.13933	−0.09119	−0.01033	−0.08
0.10	−0.24171	−0.20938	−0.16136	−0.11968	−0.07134	0.00965	−0.10
0.12	−0.22273	−0.19012	−0.14176	−0.09986	−0.05135	0.02971	−0.12
0.14	−0.20349	−0.17061	−0.12195	−0.07986	−0.03123	0.04984	−0.14
0.16	−0.18398	−0.15086	−0.10193	−0.05968	−0.01096	0.07005	−0.16
0.18	−0.16419	−0.13086	−0.08169	−0.03932	0.00945	0.09033	−0.18
0.20	−0.14412	−0.11060	−0.06123	−0.01878	0.03001	0.11069	−0.20
0.22	−0.12376	−0.09007	−0.04055	0.00196	0.05072	0.13113	−0.22
0.24	−0.10311	−0.06928	−0.01964	0.02289	0.07157	0.15165	−0.24
0.26	−0.08215	−0.04821	0.00150	0.04401	0.09258	0.17224	−0.26
0.28	−0.06088	−0.02686	0.02288	0.06533	0.11373	0.19292	−0.28
0.30	−0.03929	−0.00523	0.04450	0.08685	0.13505	0.21368	−0.30
0.32	−0.01738	0.01670	0.06636	0.10857	0.15652	0.23452	−0.32
0.34	0.00486	0.03893	0.08848	0.13051	0.17815	0.25545	−0.34
0.36	0.02745	0.06147	0.11085	0.15266	0.19995	0.27646	−0.36
0.38	0.05038	0.08432	0.13349	0.17502	0.22191	0.29755	−0.38
0.40	0.07367	0.10749	0.15638	0.19760	0.24404	0.31873	−0.40
0.42	0.09734	0.13099	0.17955	0.22040	0.26633	0.34000	−0.42
0.44	0.12137	0.15482	0.20300	0.24344	0.28880	0.36136	−0.44
0.46	0.14580	0.17900	0.22673	0.26670	0.31145	0.38280	−0.46
0.48	0.17062	0.20354	0.25075	0.29020	0.33427	0.40434	−0.48
0.50	0.19585	0.22843	0.27506	0.31393	0.35727	0.42597	−0.50
0.52	0.22150	0.25369	0.29967	0.33791	0.38045	0.44769	−0.52
0.54	0.24758	0.27933	0.32459	0.36214	0.40382	0.46951	−0.54
0.56	0.27410	0.30536	0.34982	0.38662	0.42738	0.49142	−0.56
0.58	0.30108	0.33179	0.37537	0.41136	0.45113	0.51342	−0.58
0.60	0.32852	0.35862	0.40124	0.43636	0.47508	0.53553	−0.60
0.62	0.35644	0.38587	0.42745	0.46163	0.49922	0.55773	−0.62
0.64	0.38485	0.41355	0.45400	0.48717	0.52356	0.58003	−0.64
0.66	0.41377	0.44167	0.48090	0.51299	0.54811	0.60243	−0.66
0.68	0.44321	0.47024	0.50815	0.53908	0.57286	0.62494	−0.68
0.70	0.47319	0.49927	0.53577	0.56547	0.59782	0.64755	−0.70
0.72	0.50372	0.52878	0.56376	0.59215	0.62300	0.67026	−0.72
0.74	0.53483	0.55878	0.59213	0.61913	0.64839	0.69308	−0.74
0.76	0.56651	0.58928	0.62089	0.64641	0.67401	0.71601	−0.76
0.78	0.59881	0.62029	0.65005	0.67401	0.69985	0.73904	−0.78
0.80	0.63173	0.65184	0.67962	0.70192	0.72591	0.76219	−0.80
0.82	0.66529	0.68393	0.70960	0.73016	0.75221	0.78544	−0.82
0.84	0.69951	0.71657	0.74001	0.75872	0.77875	0.80881	−0.84
0.86	0.73443	0.74980	0.77086	0.78763	0.80552	0.83229	−0.86
0.88	0.77004	0.78362	0.80216	0.81687	0.83254	0.85589	−0.88
0.90	0.80639	0.81804	0.83391	0.84647	0.85980	0.87961	−0.90
0.92	0.84350	0.85310	0.86614	0.87643	0.88732	0.90344	−0.92
0.94	0.88138	0.88880	0.89885	0.90675	0.91509	0.92740	−0.94
0.96	0.92008	0.92517	0.93206	0.93745	0.94313	0.95148	−0.96
0.98	0.95961	0.96223	0.96577	0.96853	0.97143	0.97568	−0.98

↑ r_0

α =	.995 ↑−	.99 ↑−	.975 ↑−	.95 ↑−	.90 ↑−	.75 ↑−

TABLE 57.1   SAMPLE SIZE =   59

α =	.005 ↓+	.01 ↓+	.025 ↓+	.05 ↓+	.10 ↓+	.25 ↓+	
r_0 ↓							
-0.98	-0.98973	-0.98901	-0.98787	-0.98681	-0.98548	-0.98296	0.98
-0.96	-0.97935	-0.97793	-0.97566	-0.97354	-0.97088	-0.96587	0.96
-0.94	-0.96889	-0.96674	-0.96335	-0.96019	-0.95622	-0.94875	0.94
-0.92	-0.95832	-0.95547	-0.95095	-0.94674	-0.94148	-0.93159	0.92
-0.90	-0.94765	-0.94409	-0.93846	-0.93322	-0.92667	-0.91438	0.90
-0.88	-0.93688	-0.93261	-0.92588	-0.91960	-0.91178	-0.89713	0.88
-0.86	-0.92601	-0.92104	-0.91319	-0.90590	-0.89681	-0.87984	0.86
-0.84	-0.91503	-0.90935	-0.90041	-0.89211	-0.88177	-0.86251	0.84
-0.82	-0.90395	-0.89757	-0.88753	-0.87822	-0.86665	-0.84513	0.82
-0.80	-0.89275	-0.88568	-0.87455	-0.86425	-0.85146	-0.82772	0.80
-0.78	-0.88145	-0.87368	-0.86147	-0.85018	-0.83618	-0.81025	0.78
-0.76	-0.87003	-0.86157	-0.84829	-0.83602	-0.82083	-0.79274	0.76
-0.74	-0.85850	-0.84935	-0.83500	-0.82176	-0.80539	-0.77519	0.74
-0.72	-0.84685	-0.83701	-0.82161	-0.80741	-0.78987	-0.75759	0.72
-0.70	-0.83509	-0.82456	-0.80811	-0.79296	-0.77427	-0.73994	0.70
-0.68	-0.82321	-0.81200	-0.79450	-0.77841	-0.75859	-0.72225	0.68
-0.66	-0.81120	-0.79932	-0.78078	-0.76375	-0.74282	-0.70451	0.66
-0.64	-0.79907	-0.78652	-0.76695	-0.74900	-0.72696	-0.68672	0.64
-0.62	-0.78681	-0.77359	-0.75301	-0.73415	-0.71102	-0.66889	0.62
-0.60	-0.77443	-0.76054	-0.73895	-0.71919	-0.69499	-0.65100	0.60
-0.58	-0.76191	-0.74737	-0.72477	-0.70413	-0.67887	-0.63307	0.58
-0.56	-0.74927	-0.73406	-0.71048	-0.68895	-0.66267	-0.61508	0.56
-0.54	-0.73648	-0.72063	-0.69607	-0.67368	-0.64637	-0.59705	0.54
-0.52	-0.72356	-0.70707	-0.68153	-0.65829	-0.62998	-0.57896	0.52
-0.50	-0.71050	-0.69337	-0.66687	-0.64279	-0.61350	-0.56082	0.50
-0.48	-0.69730	-0.67953	-0.65209	-0.62718	-0.59692	-0.54263	0.48
-0.46	-0.68396	-0.66556	-0.63718	-0.61145	-0.58025	-0.52439	0.46
-0.44	-0.67046	-0.65144	-0.62214	-0.59561	-0.56348	-0.50609	0.44
-0.42	-0.65682	-0.63718	-0.60697	-0.57965	-0.54662	-0.48774	0.42
-0.40	-0.64302	-0.62278	-0.59166	-0.56357	-0.52966	-0.46934	0.40
-0.38	-0.62907	-0.60822	-0.57622	-0.54737	-0.51259	-0.45087	0.38
-0.36	-0.61496	-0.59352	-0.56065	-0.53106	-0.49543	-0.43236	0.36
-0.34	-0.60069	-0.57866	-0.54493	-0.51461	-0.47817	-0.41379	0.34
-0.32	-0.58625	-0.56365	-0.52907	-0.49804	-0.46080	-0.39516	0.32
-0.30	-0.57165	-0.54847	-0.51307	-0.48135	-0.44333	-0.37647	0.30
-0.28	-0.55687	-0.53314	-0.49693	-0.46452	-0.42575	-0.35772	0.28
-0.26	-0.54192	-0.51764	-0.48064	-0.44757	-0.40807	-0.33892	0.26
-0.24	-0.52680	-0.50197	-0.46419	-0.43048	-0.39027	-0.32005	0.24
-0.22	-0.51149	-0.48613	-0.44759	-0.41326	-0.37237	-0.30113	0.22
-0.20	-0.49600	-0.47012	-0.43084	-0.39590	-0.35436	-0.28214	0.20
-0.18	-0.48032	-0.45393	-0.41393	-0.37841	-0.33623	-0.26309	0.18
-0.16	-0.46445	-0.43756	-0.39686	-0.36077	-0.31799	-0.24398	0.16
-0.14	-0.44838	-0.42100	-0.37963	-0.34299	-0.29964	-0.22481	0.14
-0.12	-0.43212	-0.40426	-0.36223	-0.32507	-0.28117	-0.20557	0.12
-0.10	-0.41565	-0.38733	-0.34466	-0.30700	-0.26258	-0.18627	0.10
-0.08	-0.39897	-0.37021	-0.32693	-0.28878	-0.24387	-0.16690	0.08
-0.06	-0.38208	-0.35289	-0.30901	-0.27042	-0.22504	-0.14747	0.06
-0.04	-0.36498	-0.33536	-0.29092	-0.25189	-0.20608	-0.12797	0.04
-0.02	-0.34765	-0.31763	-0.27266	-0.23322	-0.18701	-0.10841	0.02

↑ r_0

α =	.995 ↑-	.99 ↑-	.975 ↑-	.95 ↑-	.90 ↑-	.75 ↑-

TABLE 57.2   SAMPLE SIZE =   59   (CONT.)

α =	.005 ↓+	.01 ↓+	.025 ↓+	.05 ↓+	.10 ↓+	.25 ↓+	
r_0 ↓							
0.00	−0.33010	−0.29970	−0.25420	−0.21438	−0.16780	−0.08877	0.00
0.02	−0.31232	−0.28154	−0.23557	−0.19539	−0.14847	−0.06907	−0.02
0.04	−0.29431	−0.26317	−0.21674	−0.17623	−0.12901	−0.04929	−0.04
0.06	−0.27605	−0.24458	−0.19772	−0.15691	−0.10942	−0.02945	−0.06
0.08	−0.25755	−0.22576	−0.17850	−0.13742	−0.08969	−0.00954	−0.08
0.10	−0.23880	−0.20671	−0.15909	−0.11776	−0.06984	0.01045	−0.10
0.12	−0.21979	−0.18743	−0.13947	−0.09793	−0.04984	0.03051	−0.12
0.14	−0.20052	−0.16790	−0.11965	−0.07792	−0.02971	0.05065	−0.14
0.16	−0.18098	−0.14813	−0.09961	−0.05773	−0.00943	0.07085	−0.16
0.18	−0.16116	−0.12811	−0.07936	−0.03736	0.01098	0.09114	−0.18
0.20	−0.14107	−0.10783	−0.05889	−0.01681	0.03154	0.11150	−0.20
0.22	−0.12069	−0.08729	−0.03820	0.00393	0.05225	0.13194	−0.22
0.24	−0.10002	−0.06649	−0.01729	0.02486	0.07310	0.15245	−0.24
0.26	−0.07904	−0.04541	0.00386	0.04598	0.09410	0.17305	−0.26
0.28	−0.05776	−0.02405	0.02524	0.06730	0.11526	0.19372	−0.28
0.30	−0.03617	−0.00241	0.04686	0.08882	0.13657	0.21447	−0.30
0.32	−0.01425	0.01952	0.06872	0.11054	0.15803	0.23531	−0.32
0.34	0.00800	0.04175	0.09083	0.13246	0.17966	0.25623	−0.34
0.36	0.03058	0.06428	0.11320	0.15460	0.20144	0.27723	−0.36
0.38	0.05352	0.08713	0.13582	0.17695	0.22339	0.29832	−0.38
0.40	0.07680	0.11029	0.15871	0.19952	0.24551	0.31949	−0.40
0.42	0.10046	0.13378	0.18186	0.22231	0.26779	0.34075	−0.42
0.44	0.12448	0.15760	0.20529	0.24532	0.29024	0.36210	−0.44
0.46	0.14889	0.18176	0.22900	0.26857	0.31287	0.38354	−0.46
0.48	0.17369	0.20627	0.25299	0.29204	0.33567	0.40506	−0.48
0.50	0.19890	0.23113	0.27728	0.31575	0.35865	0.42668	−0.50
0.52	0.22452	0.25636	0.30186	0.33970	0.38181	0.44838	−0.52
0.54	0.25056	0.28197	0.32674	0.36390	0.40515	0.47019	−0.54
0.56	0.27704	0.30796	0.35193	0.38835	0.42868	0.49208	−0.56
0.58	0.30397	0.33434	0.37744	0.41305	0.45240	0.51407	−0.58
0.60	0.33136	0.36112	0.40327	0.43801	0.47632	0.53616	−0.60
0.62	0.35922	0.38832	0.42943	0.46323	0.50042	0.55834	−0.62
0.64	0.38757	0.41594	0.45593	0.48873	0.52473	0.58062	−0.64
0.66	0.41642	0.44399	0.48277	0.51449	0.54923	0.60300	−0.66
0.68	0.44578	0.47249	0.50996	0.54054	0.57394	0.62548	−0.68
0.70	0.47568	0.50145	0.53751	0.56687	0.59886	0.64807	−0.70
0.72	0.50612	0.53087	0.56543	0.59349	0.62399	0.67076	−0.72
0.74	0.53712	0.56078	0.59372	0.62040	0.64934	0.69355	−0.74
0.76	0.56870	0.59118	0.62240	0.64762	0.67490	0.71645	−0.76
0.78	0.60088	0.62208	0.65147	0.67514	0.70068	0.73945	−0.78
0.80	0.63367	0.65351	0.68094	0.70297	0.72669	0.76257	−0.80
0.82	0.66709	0.68548	0.71082	0.73113	0.75292	0.78579	−0.82
0.84	0.70117	0.71800	0.74113	0.75961	0.77939	0.80913	−0.84
0.86	0.73592	0.75108	0.77186	0.78842	0.80610	0.83258	−0.86
0.88	0.77137	0.78475	0.80304	0.81757	0.83304	0.85614	−0.88
0.90	0.80753	0.81902	0.83467	0.84707	0.86023	0.87982	−0.90
0.92	0.84444	0.85390	0.86676	0.87692	0.88767	0.90362	−0.92
0.94	0.88211	0.88942	0.89933	0.90713	0.91536	0.92753	−0.94
0.96	0.92058	0.92560	0.93238	0.93771	0.94331	0.95156	−0.96
0.98	0.95987	0.96245	0.96594	0.96866	0.97152	0.97572	−0.98

↑ r_0

α =	.995 ↑−	.99 ↑−	.975 ↑−	.95 ↑−	.90 ↑−	.75 ↑−

TABLE 58.1   SAMPLE SIZE =   60

α =	.005 ↓+	.01 ↓+	.025 ↓+	.05 ↓+	.10 ↓+	.25 ↓+	
r₀ ↓							
-0.98	-0.98967	-0.98895	-0.98782	-0.98677	-0.98544	-0.98294	0.98
-0.96	-0.97924	-0.97781	-0.97556	-0.97345	-0.97081	-0.96583	0.96
-0.94	-0.96871	-0.96658	-0.96320	-0.96005	-0.95610	-0.94869	0.94
-0.92	-0.95808	-0.95524	-0.95075	-0.94656	-0.94133	-0.93150	0.92
-0.90	-0.94736	-0.94381	-0.93821	-0.93299	-0.92648	-0.91428	0.90
-0.88	-0.93653	-0.93228	-0.92557	-0.91933	-0.91155	-0.89701	0.88
-0.86	-0.92560	-0.92064	-0.91284	-0.90558	-0.89655	-0.87970	0.86
-0.84	-0.91456	-0.90891	-0.90001	-0.89175	-0.88147	-0.86235	0.84
-0.82	-0.90342	-0.89706	-0.88708	-0.87782	-0.86632	-0.84495	0.82
-0.80	-0.89216	-0.88512	-0.87405	-0.86380	-0.85109	-0.82751	0.80
-0.78	-0.88080	-0.87306	-0.86092	-0.84969	-0.83577	-0.81003	0.78
-0.76	-0.86933	-0.86090	-0.84768	-0.83548	-0.82038	-0.79250	0.76
-0.74	-0.85774	-0.84862	-0.83435	-0.82118	-0.80491	-0.77493	0.74
-0.72	-0.84603	-0.83623	-0.82091	-0.80678	-0.78936	-0.75731	0.72
-0.70	-0.83421	-0.82373	-0.80736	-0.79229	-0.77372	-0.73964	0.70
-0.68	-0.82226	-0.81111	-0.79370	-0.77770	-0.75800	-0.72193	0.68
-0.66	-0.81020	-0.79837	-0.77993	-0.76301	-0.74220	-0.70417	0.66
-0.64	-0.79801	-0.78552	-0.76606	-0.74821	-0.72631	-0.68637	0.64
-0.62	-0.78570	-0.77254	-0.75207	-0.73332	-0.71034	-0.66851	0.62
-0.60	-0.77326	-0.75944	-0.73796	-0.71832	-0.69428	-0.65061	0.60
-0.58	-0.76068	-0.74621	-0.72374	-0.70321	-0.67813	-0.63266	0.58
-0.56	-0.74798	-0.73285	-0.70940	-0.68800	-0.66189	-0.61466	0.56
-0.54	-0.73514	-0.71937	-0.69494	-0.67269	-0.64556	-0.59660	0.54
-0.52	-0.72216	-0.70575	-0.68036	-0.65726	-0.62914	-0.57850	0.52
-0.50	-0.70905	-0.69200	-0.66566	-0.64172	-0.61263	-0.56035	0.50
-0.48	-0.69579	-0.67811	-0.65083	-0.62607	-0.59602	-0.54214	0.48
-0.46	-0.68239	-0.66409	-0.63587	-0.61031	-0.57932	-0.52388	0.46
-0.44	-0.66884	-0.64992	-0.62079	-0.59443	-0.56253	-0.50557	0.44
-0.42	-0.65514	-0.63561	-0.60558	-0.57844	-0.54564	-0.48720	0.42
-0.40	-0.64129	-0.62116	-0.59023	-0.56232	-0.52865	-0.46878	0.40
-0.38	-0.62728	-0.60656	-0.57475	-0.54609	-0.51156	-0.45031	0.38
-0.36	-0.61312	-0.59180	-0.55913	-0.52974	-0.49437	-0.43178	0.36
-0.34	-0.59880	-0.57690	-0.54338	-0.51326	-0.47708	-0.41319	0.34
-0.32	-0.58431	-0.56184	-0.52748	-0.49666	-0.45968	-0.39455	0.32
-0.30	-0.56965	-0.54661	-0.51144	-0.47993	-0.44219	-0.37585	0.30
-0.28	-0.55483	-0.53123	-0.49526	-0.46308	-0.42459	-0.35709	0.28
-0.26	-0.53983	-0.51569	-0.47893	-0.44609	-0.40688	-0.33827	0.26
-0.24	-0.52465	-0.49998	-0.46245	-0.42898	-0.38906	-0.31940	0.24
-0.22	-0.50929	-0.48409	-0.44581	-0.41173	-0.37114	-0.30046	0.22
-0.20	-0.49376	-0.46804	-0.42903	-0.39434	-0.35310	-0.28146	0.20
-0.18	-0.47803	-0.45181	-0.41208	-0.37682	-0.33496	-0.26241	0.18
-0.16	-0.46211	-0.43540	-0.39498	-0.35915	-0.31670	-0.24329	0.16
-0.14	-0.44600	-0.41880	-0.37771	-0.34135	-0.29833	-0.22410	0.14
-0.12	-0.42969	-0.40202	-0.36029	-0.32340	-0.27984	-0.20486	0.12
-0.10	-0.41318	-0.38505	-0.34269	-0.30531	-0.26123	-0.18555	0.10
-0.08	-0.39646	-0.36789	-0.32492	-0.28707	-0.24250	-0.16617	0.08
-0.06	-0.37953	-0.35053	-0.30698	-0.26868	-0.22366	-0.14673	0.06
-0.04	-0.36238	-0.33297	-0.28886	-0.25013	-0.20469	-0.12723	0.04
-0.02	-0.34502	-0.31521	-0.27057	-0.23144	-0.18560	-0.10765	0.02

↑ r₀

α =	.995 ↑-	.99 ↑-	.975 ↑-	.95 ↑-	.90 ↑-	.75 ↑-

TABLE 58.2   SAMPLE SIZE = 60   (CONT.)

α =	.005 ↓+	.01 ↓+	.025 ↓+	.05 ↓+	.10 ↓+	.25 ↓+	
r_0 ↓							
0.00	−0.32743	−0.29724	−0.25209	−0.21258	−0.16638	−0.08801	0.00
0.02	−0.30961	−0.27906	−0.23343	−0.19357	−0.14704	−0.06830	−0.02
0.04	−0.29156	−0.26066	−0.21458	−0.17440	−0.12757	−0.04853	−0.04
0.06	−0.27327	−0.24204	−0.19554	−0.15506	−0.10797	−0.02868	−0.06
0.08	−0.25474	−0.22319	−0.17631	−0.13556	−0.08823	−0.00876	−0.08
0.10	−0.23595	−0.20412	−0.15688	−0.11589	−0.06837	0.01123	−0.10
0.12	−0.21691	−0.18481	−0.13724	−0.09604	−0.04836	0.03129	−0.12
0.14	−0.19762	−0.16526	−0.11740	−0.07602	−0.02823	0.05143	−0.14
0.16	−0.17805	−0.14547	−0.09735	−0.05583	−0.00795	0.07164	−0.16
0.18	−0.15822	−0.12543	−0.07709	−0.03545	0.01247	0.09192	−0.18
0.20	−0.13810	−0.10514	−0.05662	−0.01490	0.03303	0.11228	−0.20
0.22	−0.11770	−0.08458	−0.03592	0.00584	0.05374	0.13272	−0.22
0.24	−0.09701	−0.06377	−0.01500	0.02678	0.07459	0.15323	−0.24
0.26	−0.07602	−0.04268	0.00615	0.04790	0.09559	0.17383	−0.26
0.28	−0.05473	−0.02131	0.02754	0.06922	0.11674	0.19450	−0.28
0.30	−0.03312	0.00033	0.04915	0.09073	0.13805	0.21525	−0.30
0.32	−0.01120	0.02227	0.07102	0.11245	0.15951	0.23608	−0.32
0.34	0.01105	0.04450	0.09312	0.13437	0.18112	0.25699	−0.34
0.36	0.03364	0.06703	0.11548	0.15650	0.20290	0.27799	−0.36
0.38	0.05657	0.08986	0.13809	0.17884	0.22484	0.29907	−0.38
0.40	0.07985	0.11302	0.16097	0.20139	0.24694	0.32024	−0.40
0.42	0.10349	0.13649	0.18411	0.22416	0.26921	0.34149	−0.42
0.44	0.12751	0.16030	0.20752	0.24716	0.29164	0.36282	−0.44
0.46	0.15190	0.18444	0.23121	0.27038	0.31425	0.38425	−0.46
0.48	0.17668	0.20892	0.25518	0.29383	0.33703	0.40576	−0.48
0.50	0.20186	0.23376	0.27943	0.31752	0.35999	0.42737	−0.50
0.52	0.22745	0.25896	0.30398	0.34144	0.38313	0.44906	−0.52
0.54	0.25346	0.28453	0.32883	0.36561	0.40645	0.47085	−0.54
0.56	0.27990	0.31048	0.35399	0.39003	0.42995	0.49273	−0.56
0.58	0.30678	0.33682	0.37946	0.41469	0.45364	0.51470	−0.58
0.60	0.33412	0.36355	0.40524	0.43961	0.47752	0.53677	−0.60
0.62	0.36193	0.39070	0.43135	0.46479	0.50159	0.55893	−0.62
0.64	0.39021	0.41826	0.45780	0.49024	0.52586	0.58119	−0.64
0.66	0.41899	0.44625	0.48458	0.51596	0.55033	0.60355	−0.66
0.68	0.44828	0.47467	0.51171	0.54195	0.57500	0.62601	−0.68
0.70	0.47810	0.50355	0.53920	0.56823	0.59987	0.64858	−0.70
0.72	0.50845	0.53290	0.56705	0.59479	0.62496	0.67124	−0.72
0.74	0.53935	0.56271	0.59526	0.62164	0.65025	0.69401	−0.74
0.76	0.57082	0.59302	0.62386	0.64878	0.67576	0.71688	−0.76
0.78	0.60288	0.62382	0.65284	0.67624	0.70149	0.73986	−0.78
0.80	0.63555	0.65514	0.68222	0.70399	0.72744	0.76294	−0.80
0.82	0.66884	0.68699	0.71201	0.73207	0.75362	0.78614	−0.82
0.84	0.70277	0.71938	0.74221	0.76046	0.78002	0.80944	−0.84
0.86	0.73737	0.75232	0.77284	0.78919	0.80666	0.83286	−0.86
0.88	0.77265	0.78585	0.80390	0.81824	0.83354	0.85638	−0.88
0.90	0.80863	0.81996	0.83540	0.84764	0.86065	0.88003	−0.90
0.92	0.84535	0.85468	0.86736	0.87739	0.88802	0.90378	−0.92
0.94	0.88282	0.89002	0.89979	0.90749	0.91563	0.92766	−0.94
0.96	0.92106	0.92601	0.93270	0.93795	0.94349	0.95165	−0.96
0.98	0.96012	0.96266	0.96610	0.96879	0.97161	0.97577	−0.98

↑ r_0

α =	.995 ↑−	.99 ↑−	.975 ↑−	.95 ↑−	.90 ↑−	.75 ↑−

TABLE 59.1   SAMPLE SIZE =   62

r_0 ↓	α = .005 ↓+	.01 ↓+	.025 ↓+	.05 ↓+	.10 ↓+	.25 ↓+	
-0.98	-0.98955	-0.98884	-0.98772	-0.98668	-0.98537	-0.98289	0.98
-0.96	-0.97901	-0.97759	-0.97536	-0.97327	-0.97066	-0.96575	0.96
-0.94	-0.96837	-0.96625	-0.96290	-0.95978	-0.95588	-0.94857	0.94
-0.92	-0.95763	-0.95481	-0.95035	-0.94621	-0.94103	-0.93134	0.92
-0.90	-0.94679	-0.94326	-0.93771	-0.93255	-0.92611	-0.91407	0.90
-0.88	-0.93584	-0.93162	-0.92498	-0.91880	-0.91111	-0.89677	0.88
-0.86	-0.92480	-0.91988	-0.91215	-0.90497	-0.89604	-0.87942	0.86
-0.84	-0.91365	-0.90804	-0.89922	-0.89104	-0.88089	-0.86203	0.84
-0.82	-0.90239	-0.89609	-0.88619	-0.87703	-0.86567	-0.84459	0.82
-0.80	-0.89102	-0.88403	-0.87307	-0.86293	-0.85037	-0.82711	0.80
-0.78	-0.87954	-0.87187	-0.85984	-0.84873	-0.83499	-0.80959	0.78
-0.76	-0.86795	-0.85960	-0.84651	-0.83444	-0.81953	-0.79203	0.76
-0.74	-0.85625	-0.84722	-0.83308	-0.82006	-0.80399	-0.77442	0.74
-0.72	-0.84443	-0.83472	-0.81955	-0.80558	-0.78837	-0.75676	0.72
-0.70	-0.83250	-0.82211	-0.80590	-0.79101	-0.77267	-0.73906	0.70
-0.68	-0.82044	-0.80939	-0.79215	-0.77633	-0.75688	-0.72132	0.68
-0.66	-0.80827	-0.79655	-0.77830	-0.76156	-0.74102	-0.70352	0.66
-0.64	-0.79597	-0.78359	-0.76433	-0.74669	-0.72506	-0.68568	0.64
-0.62	-0.78354	-0.77050	-0.75024	-0.73171	-0.70903	-0.66779	0.62
-0.60	-0.77099	-0.75730	-0.73605	-0.71664	-0.69290	-0.64986	0.60
-0.58	-0.75831	-0.74397	-0.72174	-0.70146	-0.67669	-0.63187	0.58
-0.56	-0.74549	-0.73051	-0.70731	-0.68617	-0.66039	-0.61384	0.56
-0.54	-0.73254	-0.71693	-0.69276	-0.67078	-0.64400	-0.59575	0.54
-0.52	-0.71946	-0.70321	-0.67810	-0.65527	-0.62752	-0.57762	0.52
-0.50	-0.70623	-0.68936	-0.66331	-0.63966	-0.61095	-0.55943	0.50
-0.48	-0.69287	-0.67537	-0.64839	-0.62394	-0.59429	-0.54119	0.48
-0.46	-0.67936	-0.66125	-0.63335	-0.60811	-0.57753	-0.52291	0.46
-0.44	-0.66570	-0.64699	-0.61819	-0.59216	-0.56068	-0.50456	0.44
-0.42	-0.65190	-0.63258	-0.60289	-0.57610	-0.54374	-0.48617	0.42
-0.40	-0.63794	-0.61803	-0.58747	-0.55991	-0.52670	-0.46772	0.40
-0.38	-0.62383	-0.60333	-0.57191	-0.54362	-0.50955	-0.44922	0.38
-0.36	-0.60957	-0.58849	-0.55621	-0.52720	-0.49231	-0.43066	0.36
-0.34	-0.59514	-0.57349	-0.54038	-0.51066	-0.47497	-0.41205	0.34
-0.32	-0.58055	-0.55834	-0.52441	-0.49399	-0.45753	-0.39338	0.32
-0.30	-0.56580	-0.54303	-0.50829	-0.47720	-0.43999	-0.37466	0.30
-0.28	-0.55087	-0.52756	-0.49203	-0.46029	-0.42234	-0.35587	0.28
-0.26	-0.53578	-0.51192	-0.47563	-0.44324	-0.40459	-0.33703	0.26
-0.24	-0.52050	-0.49613	-0.45908	-0.42607	-0.38673	-0.31813	0.24
-0.22	-0.50505	-0.48016	-0.44238	-0.40876	-0.36876	-0.29918	0.22
-0.20	-0.48942	-0.46402	-0.42553	-0.39132	-0.35069	-0.28016	0.20
-0.18	-0.47360	-0.44771	-0.40852	-0.37374	-0.33250	-0.26108	0.18
-0.16	-0.45760	-0.43122	-0.39135	-0.35603	-0.31421	-0.24194	0.16
-0.14	-0.44140	-0.41455	-0.37402	-0.33818	-0.29580	-0.22274	0.14
-0.12	-0.42500	-0.39770	-0.35653	-0.32018	-0.27727	-0.20348	0.12
-0.10	-0.40841	-0.38065	-0.33888	-0.30204	-0.25863	-0.18415	0.10
-0.08	-0.39160	-0.36342	-0.32105	-0.28376	-0.23988	-0.16476	0.08
-0.06	-0.37459	-0.34599	-0.30306	-0.26533	-0.22100	-0.14531	0.06
-0.04	-0.35737	-0.32837	-0.28489	-0.24674	-0.20200	-0.12579	0.04
-0.02	-0.33993	-0.31054	-0.26655	-0.22801	-0.18289	-0.10620	0.02

↑ r_0

α =   .995 ↑-   .99 ↑-   .975 ↑-   .95 ↑-   .90 ↑-   .75 ↑-

TABLE 59.2   SAMPLE SIZE = 62   (CONT.)

α =	.005 ↓+	.01 ↓+	.025 ↓+	.05 ↓+	.10 ↓+	.25 ↓+	
r_o ↓							
0.00	−0.32227	−0.29251	−0.24803	−0.20912	−0.16365	−0.08655	0.00
0.02	−0.30438	−0.27427	−0.22932	−0.19008	−0.14428	−0.06683	−0.02
0.04	−0.28626	−0.25581	−0.21043	−0.17087	−0.12479	−0.04704	−0.04
0.06	−0.26791	−0.23714	−0.19135	−0.15151	−0.10517	−0.02719	−0.06
0.08	−0.24931	−0.21824	−0.17208	−0.13198	−0.08542	−0.00726	−0.08
0.10	−0.23047	−0.19912	−0.15261	−0.11228	−0.06554	0.01273	−0.10
0.12	−0.21138	−0.17976	−0.13295	−0.09242	−0.04552	0.03280	−0.12
0.14	−0.19203	−0.16018	−0.11308	−0.07238	−0.02538	0.05294	−0.14
0.16	−0.17241	−0.14035	−0.09301	−0.05217	−0.00509	0.07315	−0.16
0.18	−0.15253	−0.12027	−0.07272	−0.03178	0.01533	0.09344	−0.18
0.20	−0.13238	−0.09995	−0.05223	−0.01121	0.03590	0.11379	−0.20
0.22	−0.11194	−0.07937	−0.03152	0.00953	0.05660	0.13423	−0.22
0.24	−0.09122	−0.05853	−0.01059	0.03047	0.07746	0.15474	−0.24
0.26	−0.07021	−0.03742	0.01057	0.05159	0.09845	0.17533	−0.26
0.28	−0.04889	−0.01605	0.03195	0.07291	0.11960	0.19599	−0.28
0.30	−0.02727	0.00561	0.05357	0.09441	0.14089	0.21674	−0.30
0.32	−0.00533	0.02755	0.07543	0.11612	0.16234	0.23756	−0.32
0.34	0.01693	0.04977	0.09753	0.13803	0.18394	0.25846	−0.34
0.36	0.03951	0.07230	0.11987	0.16014	0.20570	0.27945	−0.36
0.38	0.06244	0.09512	0.14246	0.18245	0.22761	0.30051	−0.38
0.40	0.08571	0.11826	0.16531	0.20498	0.24969	0.32166	−0.40
0.42	0.10933	0.14170	0.18842	0.22773	0.27193	0.34290	−0.42
0.44	0.13332	0.16548	0.21180	0.25069	0.29434	0.36421	−0.44
0.46	0.15767	0.18958	0.23545	0.27387	0.31691	0.38562	−0.46
0.48	0.18241	0.21402	0.25937	0.29728	0.33965	0.40711	−0.48
0.50	0.20754	0.23881	0.28357	0.32091	0.36257	0.42869	−0.50
0.52	0.23307	0.26395	0.30807	0.34478	0.38566	0.45036	−0.52
0.54	0.25901	0.28945	0.33285	0.36889	0.40893	0.47212	−0.54
0.56	0.28538	0.31532	0.35793	0.39324	0.43238	0.49397	−0.56
0.58	0.31217	0.34157	0.38332	0.41784	0.45601	0.51591	−0.58
0.60	0.33941	0.36822	0.40902	0.44268	0.47983	0.53794	−0.60
0.62	0.36711	0.39525	0.43504	0.46778	0.50384	0.56007	−0.62
0.64	0.39528	0.42270	0.46139	0.49314	0.52804	0.58229	−0.64
0.66	0.42392	0.45056	0.48806	0.51877	0.55243	0.60462	−0.66
0.68	0.45307	0.47886	0.51507	0.54466	0.57702	0.62703	−0.68
0.70	0.48272	0.50759	0.54243	0.57083	0.60181	0.64955	−0.70
0.72	0.51290	0.53677	0.57014	0.59727	0.62680	0.67217	−0.72
0.74	0.54361	0.56642	0.59822	0.62400	0.65201	0.69489	−0.74
0.76	0.57488	0.59654	0.62666	0.65102	0.67742	0.71771	−0.76
0.78	0.60672	0.62714	0.65548	0.67834	0.70304	0.74063	−0.78
0.80	0.63915	0.65825	0.68468	0.70595	0.72888	0.76366	−0.80
0.82	0.67218	0.68986	0.71428	0.73387	0.75494	0.78679	−0.82
0.84	0.70583	0.72201	0.74428	0.76211	0.78123	0.81004	−0.84
0.86	0.74013	0.75470	0.77470	0.79066	0.80774	0.83339	−0.86
0.88	0.77509	0.78794	0.80553	0.81954	0.83448	0.85685	−0.88
0.90	0.81074	0.82175	0.83680	0.84874	0.86146	0.88042	−0.90
0.92	0.84708	0.85616	0.86851	0.87829	0.88867	0.90410	−0.92
0.94	0.88416	0.89117	0.90068	0.90818	0.91613	0.92790	−0.94
0.96	0.92199	0.92680	0.93331	0.93843	0.94383	0.95182	−0.96
0.98	0.96059	0.96307	0.96641	0.96903	0.97179	0.97585	−0.98

↑ r_o

α =	.995 ↑−	.99 ↑−	.975 ↑−	.95 ↑−	.90 ↑−	.75 ↑−

TABLE 60.1   SAMPLE SIZE =   64

α =	.005 ↓+	.01 ↓+	.025 ↓+	.05 ↓+	.10 ↓+	.25 ↓+	
r_o ↓							
-0.98	-0.98944	-0.98874	-0.98763	-0.98659	-0.98529	-0.98286	0.98
-0.96	-0.97879	-0.97738	-0.97517	-0.97310	-0.97052	-0.96567	0.96
-0.94	-0.96804	-0.96593	-0.96262	-0.95953	-0.95567	-0.94845	0.94
-0.92	-0.95719	-0.95439	-0.94997	-0.94587	-0.94075	-0.93119	0.92
-0.90	-0.94624	-0.94274	-0.93724	-0.93213	-0.92576	-0.91388	0.90
-0.88	-0.93518	-0.93100	-0.92441	-0.91830	-0.91069	-0.89654	0.88
-0.86	-0.92403	-0.91915	-0.91148	-0.90438	-0.89555	-0.87915	0.86
-0.84	-0.91277	-0.90720	-0.89846	-0.89037	-0.88034	-0.86172	0.84
-0.82	-0.90140	-0.89515	-0.88534	-0.87628	-0.86505	-0.84425	0.82
-0.80	-0.88992	-0.88299	-0.87213	-0.86209	-0.84968	-0.82674	0.80
-0.78	-0.87834	-0.87073	-0.85881	-0.84782	-0.83423	-0.80918	0.78
-0.76	-0.86664	-0.85835	-0.84539	-0.83345	-0.81871	-0.79158	0.76
-0.74	-0.85483	-0.84587	-0.83187	-0.81899	-0.80311	-0.77393	0.74
-0.72	-0.84290	-0.83327	-0.81824	-0.80443	-0.78742	-0.75624	0.72
-0.70	-0.83086	-0.82056	-0.80451	-0.78978	-0.77166	-0.73851	0.70
-0.68	-0.81869	-0.80774	-0.79067	-0.77502	-0.75581	-0.72073	0.68
-0.66	-0.80641	-0.79480	-0.77673	-0.76018	-0.73988	-0.70290	0.66
-0.64	-0.79400	-0.78174	-0.76267	-0.74523	-0.72387	-0.68503	0.64
-0.62	-0.78147	-0.76855	-0.74850	-0.73018	-0.70777	-0.66710	0.62
-0.60	-0.76881	-0.75525	-0.73422	-0.71503	-0.69159	-0.64913	0.60
-0.58	-0.75602	-0.74182	-0.71982	-0.69977	-0.67532	-0.63112	0.58
-0.56	-0.74310	-0.72827	-0.70531	-0.68441	-0.65896	-0.61305	0.56
-0.54	-0.73005	-0.71459	-0.69068	-0.66895	-0.64251	-0.59494	0.54
-0.52	-0.71686	-0.70077	-0.67593	-0.65338	-0.62598	-0.57677	0.52
-0.50	-0.70353	-0.68683	-0.66106	-0.63770	-0.60935	-0.55856	0.50
-0.48	-0.69007	-0.67275	-0.64606	-0.62190	-0.59264	-0.54029	0.48
-0.46	-0.67645	-0.65853	-0.63094	-0.60600	-0.57583	-0.52197	0.46
-0.44	-0.66270	-0.64417	-0.61570	-0.58999	-0.55892	-0.50360	0.44
-0.42	-0.64879	-0.62968	-0.60033	-0.57386	-0.54193	-0.48518	0.42
-0.40	-0.63474	-0.61504	-0.58482	-0.55761	-0.52483	-0.46671	0.40
-0.38	-0.62053	-0.60025	-0.56919	-0.54125	-0.50764	-0.44818	0.38
-0.36	-0.60617	-0.58531	-0.55342	-0.52477	-0.49036	-0.42960	0.36
-0.34	-0.59164	-0.57023	-0.53751	-0.50817	-0.47297	-0.41096	0.34
-0.32	-0.57696	-0.55499	-0.52147	-0.49144	-0.45548	-0.39227	0.32
-0.30	-0.56211	-0.53959	-0.50528	-0.47459	-0.43789	-0.37352	0.30
-0.28	-0.54709	-0.52404	-0.48895	-0.45762	-0.42020	-0.35471	0.28
-0.26	-0.53190	-0.50833	-0.47248	-0.44052	-0.40241	-0.33585	0.26
-0.24	-0.51654	-0.49245	-0.45587	-0.42329	-0.38451	-0.31693	0.24
-0.22	-0.50100	-0.47640	-0.43910	-0.40593	-0.36650	-0.29795	0.22
-0.20	-0.48528	-0.46019	-0.42218	-0.38844	-0.34839	-0.27892	0.20
-0.18	-0.46937	-0.44380	-0.40511	-0.37081	-0.33016	-0.25982	0.18
-0.16	-0.45328	-0.42723	-0.38788	-0.35305	-0.31183	-0.24066	0.16
-0.14	-0.43700	-0.41049	-0.37050	-0.33515	-0.29339	-0.22144	0.14
-0.12	-0.42052	-0.39356	-0.35295	-0.31711	-0.27483	-0.20217	0.12
-0.10	-0.40385	-0.37645	-0.33524	-0.29893	-0.25616	-0.18282	0.10
-0.08	-0.38697	-0.35915	-0.31737	-0.28060	-0.23737	-0.16342	0.08
-0.06	-0.36988	-0.34166	-0.29932	-0.26213	-0.21847	-0.14395	0.06
-0.04	-0.35259	-0.32397	-0.28111	-0.24351	-0.19945	-0.12442	0.04
-0.02	-0.33508	-0.30608	-0.26272	-0.22474	-0.18031	-0.10482	0.02

↑ r_o

α =	.995 ↑-	.99 ↑-	.975 ↑-	.95 ↑-	.90 ↑-	.75 ↑-

TABLE 60.2   SAMPLE SIZE =   64   (CONT.)

α =	.005 ↓+	.01 ↓+	.025 ↓+	.05 ↓+	.10 ↓+	.25 ↓+	
r_0 ↓							
0.00	−0.31735	−0.28799	−0.24415	−0.20582	−0.16104	−0.08516	0.00
0.02	−0.29939	−0.26970	−0.22540	−0.18675	−0.14166	−0.06543	−0.02
0.04	−0.28121	−0.25119	−0.20647	−0.16751	−0.12214	−0.04564	−0.04
0.06	−0.26279	−0.23246	−0.18735	−0.14812	−0.10251	−0.02577	−0.06
0.08	−0.24414	−0.21352	−0.16805	−0.12857	−0.08274	−0.00584	−0.08
0.10	−0.22524	−0.19435	−0.14855	−0.10885	−0.06285	0.01416	−0.10
0.12	−0.20610	−0.17496	−0.12886	−0.08896	−0.04282	0.03423	−0.12
0.14	−0.18670	−0.15533	−0.10896	−0.06891	−0.02266	0.05437	−0.14
0.16	−0.16704	−0.13547	−0.08887	−0.04869	−0.00237	0.07459	−0.16
0.18	−0.14712	−0.11536	−0.06857	−0.02829	0.01806	0.09487	−0.18
0.20	−0.12693	−0.09501	−0.04806	−0.00771	0.03862	0.11523	−0.20
0.22	−0.10646	−0.07441	−0.02733	0.01304	0.05933	0.13566	−0.22
0.24	−0.08571	−0.05355	−0.00639	0.03398	0.08018	0.15617	−0.24
0.26	−0.06467	−0.03243	0.01477	0.05510	0.10117	0.17675	−0.26
0.28	−0.04334	−0.01104	0.03616	0.07641	0.12231	0.19741	−0.28
0.30	−0.02170	0.01062	0.05777	0.09791	0.14359	0.21815	−0.30
0.32	0.00024	0.03256	0.07962	0.11961	0.16503	0.23896	−0.32
0.34	0.02251	0.05479	0.10171	0.14150	0.18661	0.25985	−0.34
0.36	0.04509	0.07730	0.12404	0.16359	0.20835	0.28083	−0.36
0.38	0.06801	0.10011	0.14661	0.18589	0.23025	0.30188	−0.38
0.40	0.09127	0.12323	0.16944	0.20839	0.25230	0.32302	−0.40
0.42	0.11487	0.14665	0.19252	0.23111	0.27451	0.34423	−0.42
0.44	0.13883	0.17039	0.21586	0.25403	0.29689	0.36553	−0.44
0.46	0.16315	0.19446	0.23947	0.27718	0.31943	0.38692	−0.46
0.48	0.18785	0.21886	0.26335	0.30054	0.34214	0.40839	−0.48
0.50	0.21293	0.24359	0.28750	0.32413	0.36501	0.42994	−0.50
0.52	0.23840	0.26867	0.31193	0.34795	0.38806	0.45159	−0.52
0.54	0.26428	0.29411	0.33666	0.37200	0.41128	0.47332	−0.54
0.56	0.29056	0.31991	0.36167	0.39629	0.43468	0.49514	−0.56
0.58	0.31728	0.34608	0.38699	0.42082	0.45826	0.51705	−0.58
0.60	0.34442	0.37263	0.41260	0.44559	0.48202	0.53905	−0.60
0.62	0.37201	0.39957	0.43853	0.47062	0.50596	0.56115	−0.62
0.64	0.40006	0.42690	0.46478	0.49589	0.53010	0.58334	−0.64
0.66	0.42858	0.45465	0.49135	0.52143	0.55442	0.60562	−0.66
0.68	0.45759	0.48281	0.51825	0.54722	0.57893	0.62800	−0.68
0.70	0.48709	0.51140	0.54549	0.57329	0.60364	0.65047	−0.70
0.72	0.51710	0.54043	0.57307	0.59963	0.62855	0.67305	−0.72
0.74	0.54763	0.56992	0.60101	0.62624	0.65367	0.69572	−0.74
0.76	0.57871	0.59986	0.62930	0.65314	0.67898	0.71849	−0.76
0.78	0.61034	0.63028	0.65796	0.68032	0.70451	0.74136	−0.78
0.80	0.64254	0.66118	0.68700	0.70780	0.73024	0.76434	−0.80
0.82	0.67532	0.69258	0.71642	0.73558	0.75619	0.78742	−0.82
0.84	0.70872	0.72449	0.74624	0.76366	0.78236	0.81060	−0.84
0.86	0.74273	0.75693	0.77645	0.79205	0.80875	0.83389	−0.86
0.88	0.77739	0.78991	0.80708	0.82075	0.83537	0.85729	−0.88
0.90	0.81271	0.82345	0.83812	0.84978	0.86221	0.88079	−0.90
0.92	0.84872	0.85755	0.86960	0.87914	0.88929	0.90441	−0.92
0.94	0.88542	0.89224	0.90151	0.90884	0.91660	0.92814	−0.94
0.96	0.92286	0.92754	0.93388	0.93887	0.94416	0.95198	−0.96
0.98	0.96104	0.96345	0.96670	0.96926	0.97195	0.97593	−0.98

↑ r_0

α =	.995 ↑−	.99 ↑−	.975 ↑−	.95 ↑−	.90 ↑−	.75 ↑−

TABLE 61.1   SAMPLE SIZE =   66

α =	.005 ↓+	.01 ↓+	.025 ↓+	.05 ↓+	.10 ↓+	.25 ↓+	
r_0 ↓							
-0.98	-0.98934	-0.98864	-0.98754	-0.98651	-0.98523	-0.98282	0.98
-0.96	-0.97858	-0.97718	-0.97498	-0.97294	-0.97038	-0.96560	0.96
-0.94	-0.96772	-0.96563	-0.96234	-0.95928	-0.95547	-0.94834	0.94
-0.92	-0.95676	-0.95398	-0.94961	-0.94554	-0.94048	-0.93104	0.92
-0.90	-0.94571	-0.94224	-0.93678	-0.93172	-0.92542	-0.91370	0.90
-0.88	-0.93455	-0.93039	-0.92386	-0.91781	-0.91029	-0.89632	0.88
-0.86	-0.92329	-0.91845	-0.91085	-0.90381	-0.89509	-0.87889	0.86
-0.84	-0.91192	-0.90640	-0.89774	-0.88973	-0.87981	-0.86143	0.84
-0.82	-0.90045	-0.89425	-0.88453	-0.87556	-0.86445	-0.84392	0.82
-0.80	-0.88887	-0.88199	-0.87123	-0.86129	-0.84902	-0.82637	0.80
-0.78	-0.87718	-0.86963	-0.85782	-0.84694	-0.83351	-0.80878	0.78
-0.76	-0.86537	-0.85715	-0.84431	-0.83250	-0.81792	-0.79115	0.76
-0.74	-0.85346	-0.84457	-0.83071	-0.81796	-0.80226	-0.77347	0.74
-0.72	-0.84143	-0.83188	-0.81699	-0.80332	-0.78651	-0.75574	0.72
-0.70	-0.82928	-0.81907	-0.80318	-0.78860	-0.77069	-0.73798	0.70
-0.68	-0.81701	-0.80615	-0.78925	-0.77377	-0.75478	-0.72016	0.68
-0.66	-0.80463	-0.79311	-0.77522	-0.75885	-0.73879	-0.70230	0.66
-0.64	-0.79212	-0.77996	-0.76108	-0.74383	-0.72272	-0.68440	0.64
-0.62	-0.77948	-0.76668	-0.74683	-0.72871	-0.70657	-0.66645	0.62
-0.60	-0.76672	-0.75328	-0.73246	-0.71349	-0.69033	-0.64845	0.60
-0.58	-0.75383	-0.73976	-0.71799	-0.69816	-0.67400	-0.63040	0.58
-0.56	-0.74081	-0.72611	-0.70339	-0.68273	-0.65759	-0.61230	0.56
-0.54	-0.72766	-0.71234	-0.68868	-0.66720	-0.64109	-0.59416	0.54
-0.52	-0.71437	-0.69843	-0.67385	-0.65156	-0.62450	-0.57596	0.52
-0.50	-0.70094	-0.68440	-0.65890	-0.63581	-0.60782	-0.55772	0.50
-0.48	-0.68738	-0.67023	-0.64383	-0.61996	-0.59105	-0.53943	0.48
-0.46	-0.67367	-0.65592	-0.62864	-0.60399	-0.57419	-0.52108	0.46
-0.44	-0.65981	-0.64148	-0.61332	-0.58791	-0.55724	-0.50269	0.44
-0.42	-0.64581	-0.62689	-0.59787	-0.57172	-0.54019	-0.48424	0.42
-0.40	-0.63166	-0.61217	-0.58229	-0.55541	-0.52305	-0.46574	0.40
-0.38	-0.61736	-0.59729	-0.56658	-0.53899	-0.50582	-0.44719	0.38
-0.36	-0.60290	-0.58227	-0.55074	-0.52245	-0.48848	-0.42858	0.36
-0.34	-0.58829	-0.56710	-0.53477	-0.50579	-0.47105	-0.40992	0.34
-0.32	-0.57351	-0.55178	-0.51865	-0.48900	-0.45352	-0.39120	0.32
-0.30	-0.55857	-0.53631	-0.50240	-0.47210	-0.43589	-0.37243	0.30
-0.28	-0.54347	-0.52067	-0.48601	-0.45507	-0.41816	-0.35361	0.28
-0.26	-0.52819	-0.50488	-0.46947	-0.43792	-0.40032	-0.33472	0.26
-0.24	-0.51274	-0.48893	-0.45279	-0.42064	-0.38238	-0.31578	0.24
-0.22	-0.49712	-0.47280	-0.43596	-0.40323	-0.36434	-0.29679	0.22
-0.20	-0.48131	-0.45651	-0.41899	-0.38569	-0.34619	-0.27773	0.20
-0.18	-0.46533	-0.44005	-0.40186	-0.36802	-0.32793	-0.25862	0.18
-0.16	-0.44915	-0.42342	-0.38457	-0.35021	-0.30956	-0.23944	0.16
-0.14	-0.43279	-0.40661	-0.36713	-0.33226	-0.29109	-0.22021	0.14
-0.12	-0.41624	-0.38961	-0.34953	-0.31418	-0.27250	-0.20091	0.12
-0.10	-0.39949	-0.37244	-0.33177	-0.29596	-0.25380	-0.18156	0.10
-0.08	-0.38253	-0.35507	-0.31384	-0.27759	-0.23499	-0.16214	0.08
-0.06	-0.36538	-0.33752	-0.29575	-0.25908	-0.21606	-0.14266	0.06
-0.04	-0.34801	-0.31977	-0.27749	-0.24043	-0.19701	-0.12311	0.04
-0.02	-0.33043	-0.30183	-0.25906	-0.22163	-0.17784	-0.10351	0.02

↑ r_0

α =	.995 ↑-	.99 ↑-	.975 ↑-	.95 ↑-	.90 ↑-	.75 ↑-

TABLE 61.2   SAMPLE SIZE =   66   (CONT.)

α =	.005 ↓+	.01 ↓+	.025 ↓+	.05 ↓+	.10 ↓+	.25 ↓+	
r_o ↓							
0.00	-0.31264	-0.28368	-0.24045	-0.20267	-0.15856	-0.08383	0.00
0.02	-0.29462	-0.26533	-0.22166	-0.18357	-0.13915	-0.06410	-0.02
0.04	-0.27638	-0.24677	-0.20269	-0.16431	-0.11962	-0.04429	-0.04
0.06	-0.25791	-0.22800	-0.18354	-0.14489	-0.09997	-0.02442	-0.06
0.08	-0.23920	-0.20901	-0.16421	-0.12531	-0.08019	-0.00448	-0.08
0.10	-0.22025	-0.18981	-0.14468	-0.10557	-0.06028	0.01552	-0.10
0.12	-0.20106	-0.17037	-0.12496	-0.08567	-0.04025	0.03560	-0.12
0.14	-0.18162	-0.15071	-0.10504	-0.06560	-0.02008	0.05574	-0.14
0.16	-0.16192	-0.13081	-0.08493	-0.04537	0.00022	0.07595	-0.16
0.18	-0.14196	-0.11068	-0.06461	-0.02496	0.02065	0.09624	-0.18
0.20	-0.12174	-0.09030	-0.04408	-0.00438	0.04122	0.11660	-0.20
0.22	-0.10124	-0.06968	-0.02335	0.01638	0.06193	0.13703	-0.22
0.24	-0.08046	-0.04880	-0.00240	0.03732	0.08277	0.15753	-0.24
0.26	-0.05940	-0.02767	0.01877	0.05844	0.10376	0.17811	-0.26
0.28	-0.03805	-0.00627	0.04016	0.07975	0.12489	0.19876	-0.28
0.30	-0.01640	0.01540	0.06177	0.10124	0.14617	0.21949	-0.30
0.32	0.00555	0.03734	0.08361	0.12293	0.16759	0.24030	-0.32
0.34	0.02782	0.05956	0.10569	0.14481	0.18916	0.26118	-0.34
0.36	0.05040	0.08206	0.12800	0.16688	0.21088	0.28214	-0.36
0.38	0.07331	0.10486	0.15056	0.18916	0.23276	0.30318	-0.38
0.40	0.09655	0.12796	0.17336	0.21163	0.25479	0.32430	-0.40
0.42	0.12014	0.15136	0.19641	0.23432	0.27697	0.34550	-0.42
0.44	0.14407	0.17507	0.21972	0.25721	0.29932	0.36679	-0.44
0.46	0.16836	0.19910	0.24329	0.28032	0.32183	0.38815	-0.46
0.48	0.19301	0.22345	0.26712	0.30365	0.34450	0.40960	-0.48
0.50	0.21805	0.24814	0.29123	0.32719	0.36734	0.43114	-0.50
0.52	0.24346	0.27316	0.31561	0.35096	0.39034	0.45276	-0.52
0.54	0.26927	0.29853	0.34027	0.37496	0.41352	0.47446	-0.54
0.56	0.29549	0.32426	0.36522	0.39919	0.43687	0.49626	-0.56
0.58	0.32212	0.35035	0.39046	0.42365	0.46040	0.51814	-0.58
0.60	0.34917	0.37681	0.41600	0.44836	0.48410	0.54011	-0.60
0.62	0.37666	0.40366	0.44185	0.47330	0.50798	0.56217	-0.62
0.64	0.40460	0.43089	0.46800	0.49850	0.53205	0.58433	-0.64
0.66	0.43300	0.45852	0.49447	0.52395	0.55631	0.60657	-0.66
0.68	0.46187	0.48656	0.52126	0.54965	0.58075	0.62891	-0.68
0.70	0.49122	0.51502	0.54839	0.57562	0.60538	0.65135	-0.70
0.72	0.52107	0.54390	0.57585	0.60186	0.63021	0.67388	-0.72
0.74	0.55144	0.57323	0.60365	0.62836	0.65524	0.69651	-0.74
0.76	0.58232	0.60300	0.63180	0.65514	0.68047	0.71923	-0.76
0.78	0.61375	0.63324	0.66032	0.68220	0.70590	0.74206	-0.78
0.80	0.64574	0.66395	0.68920	0.70955	0.73154	0.76498	-0.80
0.82	0.67829	0.69515	0.71845	0.73719	0.75738	0.78800	-0.82
0.84	0.71144	0.72684	0.74808	0.76512	0.78344	0.81113	-0.84
0.86	0.74519	0.75904	0.77811	0.79336	0.80972	0.83437	-0.86
0.88	0.77956	0.79177	0.80854	0.82191	0.83621	0.85770	-0.88
0.90	0.81458	0.82504	0.83937	0.85077	0.86293	0.88115	-0.90
0.92	0.85025	0.85887	0.87062	0.87995	0.88988	0.90470	-0.92
0.94	0.88661	0.89326	0.90230	0.90946	0.91705	0.92836	-0.94
0.96	0.92368	0.92823	0.93442	0.93930	0.94446	0.95213	-0.96
0.98	0.96146	0.96381	0.96698	0.96947	0.97211	0.97601	-0.98

↑ r_o

α =	.995 ↑-	.99 ↑-	.975 ↑-	.95 ↑-	.90 ↑-	.75 ↑-

TABLE 62.1   SAMPLE SIZE =   68

α =	.005 ↓+	.01 ↓+	.025 ↓+	.05 ↓+	.10 ↓+	.25 ↓+	
r_0 ↓							
−0.98	−0.98923	−0.98854	−0.98745	−0.98643	−0.98516	−0.98278	0.98
−0.96	−0.97837	−0.97699	−0.97481	−0.97278	−0.97025	−0.96553	0.96
−0.94	−0.96741	−0.96534	−0.96208	−0.95905	−0.95527	−0.94823	0.94
−0.92	−0.95635	−0.95360	−0.94926	−0.94523	−0.94022	−0.93090	0.92
−0.90	−0.94520	−0.94175	−0.93634	−0.93133	−0.92510	−0.91352	0.90
−0.88	−0.93394	−0.92981	−0.92334	−0.91734	−0.90991	−0.89610	0.88
−0.86	−0.92257	−0.91777	−0.91024	−0.90327	−0.89464	−0.87865	0.86
−0.84	−0.91111	−0.90563	−0.89704	−0.88911	−0.87930	−0.86115	0.84
−0.82	−0.89953	−0.89338	−0.88375	−0.87486	−0.86388	−0.84361	0.82
−0.80	−0.88785	−0.88103	−0.87036	−0.86053	−0.84839	−0.82603	0.80
−0.78	−0.87606	−0.86857	−0.85687	−0.84610	−0.83282	−0.80840	0.78
−0.76	−0.86416	−0.85600	−0.84328	−0.83158	−0.81717	−0.79073	0.76
−0.74	−0.85214	−0.84333	−0.82959	−0.81697	−0.80145	−0.77302	0.74
−0.72	−0.84001	−0.83054	−0.81579	−0.80226	−0.78565	−0.75527	0.72
−0.70	−0.82776	−0.81764	−0.80189	−0.78746	−0.76976	−0.73747	0.70
−0.68	−0.81540	−0.80463	−0.78789	−0.77257	−0.75380	−0.71962	0.68
−0.66	−0.80291	−0.79150	−0.77378	−0.75758	−0.73775	−0.70173	0.66
−0.64	−0.79030	−0.77825	−0.75955	−0.74249	−0.72163	−0.68380	0.64
−0.62	−0.77757	−0.76488	−0.74522	−0.72730	−0.70542	−0.66581	0.62
−0.60	−0.76471	−0.75140	−0.73078	−0.71201	−0.68912	−0.64779	0.60
−0.58	−0.75173	−0.73778	−0.71622	−0.69662	−0.67274	−0.62971	0.58
−0.56	−0.73861	−0.72405	−0.70155	−0.68112	−0.65628	−0.61159	0.56
−0.54	−0.72536	−0.71018	−0.68677	−0.66552	−0.63973	−0.59341	0.54
−0.52	−0.71198	−0.69619	−0.67186	−0.64982	−0.62309	−0.57519	0.52
−0.50	−0.69845	−0.68207	−0.65684	−0.63401	−0.60636	−0.55692	0.50
−0.48	−0.68479	−0.66781	−0.64169	−0.61809	−0.58954	−0.53860	0.48
−0.46	−0.67099	−0.65342	−0.62642	−0.60206	−0.57263	−0.52023	0.46
−0.44	−0.65705	−0.63889	−0.61103	−0.58592	−0.55563	−0.50181	0.44
−0.42	−0.64295	−0.62422	−0.59551	−0.56967	−0.53854	−0.48334	0.42
−0.40	−0.62871	−0.60941	−0.57986	−0.55330	−0.52135	−0.46481	0.40
−0.38	−0.61432	−0.59446	−0.56409	−0.53682	−0.50407	−0.44624	0.38
−0.36	−0.59977	−0.57936	−0.54818	−0.52022	−0.48669	−0.42761	0.36
−0.34	−0.58507	−0.56411	−0.53214	−0.50351	−0.46922	−0.40892	0.34
−0.32	−0.57021	−0.54871	−0.51596	−0.48667	−0.45164	−0.39019	0.32
−0.30	−0.55518	−0.53316	−0.49964	−0.46971	−0.43397	−0.37139	0.30
−0.28	−0.53999	−0.51745	−0.48319	−0.45263	−0.41620	−0.35255	0.28
−0.26	−0.52463	−0.50158	−0.46659	−0.43543	−0.39833	−0.33364	0.26
−0.24	−0.50910	−0.48555	−0.44985	−0.41810	−0.38035	−0.31468	0.24
−0.22	−0.49339	−0.46936	−0.43296	−0.40064	−0.36227	−0.29567	0.22
−0.20	−0.47751	−0.45300	−0.41592	−0.38306	−0.34409	−0.27660	0.20
−0.18	−0.46145	−0.43647	−0.39874	−0.36534	−0.32579	−0.25746	0.18
−0.16	−0.44520	−0.41977	−0.38140	−0.34749	−0.30740	−0.23827	0.16
−0.14	−0.42876	−0.40289	−0.36391	−0.32950	−0.28889	−0.21903	0.14
−0.12	−0.41213	−0.38583	−0.34626	−0.31138	−0.27027	−0.19972	0.12
−0.10	−0.39531	−0.36859	−0.32845	−0.29312	−0.25154	−0.18035	0.10
−0.08	−0.37829	−0.35117	−0.31047	−0.27471	−0.23270	−0.16092	0.08
−0.06	−0.36106	−0.33356	−0.29234	−0.25617	−0.21375	−0.14142	0.06
−0.04	−0.34363	−0.31575	−0.27403	−0.23748	−0.19468	−0.12187	0.04
−0.02	−0.32599	−0.29775	−0.25556	−0.21865	−0.17549	−0.10225	0.02

↑ r_0

α =	.995 ↑−	.99 ↑−	.975 ↑−	.95 ↑−	.90 ↑−	.75 ↑−

TABLE 62.2   SAMPLE SIZE = 68   (CONT.)

α =	.005 ↓+	.01 ↓+	.025 ↓+	.05 ↓+	.10 ↓+	.25 ↓+	
r_o ↓							
0.00	−0.30814	−0.27956	−0.23691	−0.19967	−0.15619	−0.08257	0.00
0.02	−0.29006	−0.26116	−0.21809	−0.18053	−0.13676	−0.06282	−0.02
0.04	−0.27177	−0.24255	−0.19908	−0.16125	−0.11721	−0.04301	−0.04
0.06	−0.25324	−0.22374	−0.17990	−0.14181	−0.09755	−0.02313	−0.06
0.08	−0.23448	−0.20471	−0.16053	−0.12221	−0.07775	−0.00319	−0.08
0.10	−0.21549	−0.18546	−0.14098	−0.10245	−0.05783	0.01682	−0.10
0.12	−0.19625	−0.16599	−0.12124	−0.08253	−0.03779	0.03690	−0.12
0.14	−0.17676	−0.14630	−0.10130	−0.06245	−0.01761	0.05704	−0.14
0.16	−0.15703	−0.12637	−0.08116	−0.04220	0.00269	0.07726	−0.16
0.18	−0.13703	−0.10621	−0.06083	−0.02178	0.02313	0.09755	−0.18
0.20	−0.11678	−0.08581	−0.04029	−0.00119	0.04370	0.11790	−0.20
0.22	−0.09625	−0.06517	−0.01954	0.01957	0.06440	0.13833	−0.22
0.24	−0.07545	−0.04427	0.00141	0.04051	0.08525	0.15883	−0.24
0.26	−0.05437	−0.02313	0.02258	0.06163	0.10623	0.17940	−0.26
0.28	−0.03300	−0.00172	0.04397	0.08293	0.12735	0.20005	−0.28
0.30	−0.01134	0.01995	0.06558	0.10442	0.14862	0.22077	−0.30
0.32	0.01062	0.04189	0.08741	0.12609	0.17003	0.24157	−0.32
0.34	0.03288	0.06411	0.10948	0.14796	0.19158	0.26244	−0.34
0.36	0.05546	0.08660	0.13178	0.17001	0.21329	0.28339	−0.36
0.38	0.07836	0.10939	0.15431	0.19227	0.23514	0.30442	−0.38
0.40	0.10159	0.13246	0.17709	0.21472	0.25715	0.32553	−0.40
0.42	0.12515	0.15584	0.20012	0.23738	0.27931	0.34672	−0.42
0.44	0.14906	0.17952	0.22339	0.26024	0.30163	0.36798	−0.44
0.46	0.17331	0.20351	0.24693	0.28331	0.32411	0.38933	−0.46
0.48	0.19793	0.22782	0.27072	0.30660	0.34675	0.41076	−0.48
0.50	0.22291	0.25246	0.29478	0.33010	0.36955	0.43227	−0.50
0.52	0.24827	0.27743	0.31910	0.35382	0.39251	0.45387	−0.52
0.54	0.27402	0.30274	0.34371	0.37777	0.41565	0.47555	−0.54
0.56	0.30017	0.32840	0.36859	0.40194	0.43895	0.49732	−0.56
0.58	0.32672	0.35441	0.39377	0.42634	0.46243	0.51918	−0.58
0.60	0.35368	0.38079	0.41923	0.45098	0.48608	0.54112	−0.60
0.62	0.38108	0.40754	0.44499	0.47586	0.50990	0.56315	−0.62
0.64	0.40891	0.43467	0.47106	0.50098	0.53391	0.58527	−0.64
0.66	0.43719	0.46219	0.49743	0.52635	0.55810	0.60748	−0.66
0.68	0.46593	0.49011	0.52412	0.55196	0.58247	0.62979	−0.68
0.70	0.49514	0.51844	0.55114	0.57784	0.60704	0.65218	−0.70
0.72	0.52484	0.54719	0.57848	0.60397	0.63179	0.67467	−0.72
0.74	0.55504	0.57636	0.60616	0.63037	0.65674	0.69726	−0.74
0.76	0.58575	0.60598	0.63418	0.65704	0.68188	0.71994	−0.76
0.78	0.61699	0.63605	0.66255	0.68399	0.70722	0.74271	−0.78
0.80	0.64877	0.66657	0.69128	0.71121	0.73276	0.76559	−0.80
0.82	0.68110	0.69757	0.72037	0.73872	0.75851	0.78857	−0.82
0.84	0.71401	0.72906	0.74983	0.76651	0.78446	0.81164	−0.84
0.86	0.74751	0.76104	0.77968	0.79461	0.81063	0.83482	−0.86
0.88	0.78161	0.79353	0.80992	0.82300	0.83701	0.85810	−0.88
0.90	0.81634	0.82655	0.84055	0.85170	0.86361	0.88148	−0.90
0.92	0.85171	0.86011	0.87159	0.88071	0.89043	0.90497	−0.92
0.94	0.88774	0.89421	0.90304	0.91004	0.91748	0.92856	−0.94
0.96	0.92445	0.92889	0.93493	0.93969	0.94475	0.95227	−0.96
0.98	0.96186	0.96414	0.96724	0.96968	0.97226	0.97608	−0.98

↑ r_o

α =	.995 ↑−	.99 ↑−	.975 ↑−	.95 ↑−	.90 ↑−	.75 ↑−

TABLE 63.1   SAMPLE SIZE =   70

α =	.005 ↓+	.01 ↓+	.025 ↓+	.05 ↓+	.10 ↓+	.25 ↓+	
r_0 ↓							
-0.98	-0.98914	-0.98845	-0.98736	-0.98636	-0.98510	-0.98275	0.98
-0.96	-0.97817	-0.97680	-0.97464	-0.97263	-0.97013	-0.96546	0.96
-0.94	-0.96712	-0.96506	-0.96182	-0.95882	-0.95509	-0.94813	0.94
-0.92	-0.95596	-0.95322	-0.94892	-0.94493	-0.93998	-0.93076	0.92
-0.90	-0.94470	-0.94129	-0.93592	-0.93096	-0.92479	-0.91335	0.90
-0.88	-0.93335	-0.92925	-0.92283	-0.91690	-0.90954	-0.89590	0.88
-0.86	-0.92189	-0.91712	-0.90965	-0.90275	-0.89421	-0.87841	0.86
-0.84	-0.91032	-0.90488	-0.89637	-0.88852	-0.87881	-0.86088	0.84
-0.82	-0.89865	-0.89254	-0.88300	-0.87420	-0.86333	-0.84331	0.82
-0.80	-0.88687	-0.88010	-0.86953	-0.85979	-0.84778	-0.82569	0.80
-0.78	-0.87498	-0.86755	-0.85595	-0.84529	-0.83215	-0.80804	0.78
-0.76	-0.86298	-0.85490	-0.84228	-0.83070	-0.81645	-0.79034	0.76
-0.74	-0.85087	-0.84213	-0.82851	-0.81602	-0.80067	-0.77260	0.74
-0.72	-0.83864	-0.82925	-0.81464	-0.80125	-0.78481	-0.75481	0.72
-0.70	-0.82630	-0.81626	-0.80066	-0.78638	-0.76887	-0.73698	0.70
-0.68	-0.81384	-0.80316	-0.78658	-0.77141	-0.75285	-0.71911	0.68
-0.66	-0.80126	-0.78994	-0.77239	-0.75636	-0.73676	-0.70119	0.66
-0.64	-0.78856	-0.77661	-0.75809	-0.74120	-0.72057	-0.68322	0.64
-0.62	-0.77573	-0.76315	-0.74368	-0.72594	-0.70431	-0.66521	0.62
-0.60	-0.76278	-0.74958	-0.72916	-0.71059	-0.68797	-0.64715	0.60
-0.58	-0.74970	-0.73588	-0.71453	-0.69513	-0.67153	-0.62905	0.58
-0.56	-0.73649	-0.72206	-0.69979	-0.67957	-0.65502	-0.61090	0.56
-0.54	-0.72315	-0.70811	-0.68493	-0.66391	-0.63842	-0.59270	0.54
-0.52	-0.70967	-0.69403	-0.66995	-0.64815	-0.62173	-0.57445	0.52
-0.50	-0.69606	-0.67983	-0.65485	-0.63227	-0.60495	-0.55615	0.50
-0.48	-0.68231	-0.66549	-0.63964	-0.61629	-0.58809	-0.53781	0.48
-0.46	-0.66842	-0.65101	-0.62430	-0.60021	-0.57113	-0.51941	0.46
-0.44	-0.65438	-0.63640	-0.60884	-0.58401	-0.55408	-0.50097	0.44
-0.42	-0.64020	-0.62166	-0.59325	-0.56770	-0.53695	-0.48247	0.42
-0.40	-0.62588	-0.60677	-0.57754	-0.55128	-0.51972	-0.46392	0.40
-0.38	-0.61140	-0.59174	-0.56169	-0.53474	-0.50239	-0.44533	0.38
-0.36	-0.59676	-0.57656	-0.54572	-0.51809	-0.48497	-0.42667	0.36
-0.34	-0.58198	-0.56123	-0.52961	-0.50132	-0.46746	-0.40797	0.34
-0.32	-0.56703	-0.54576	-0.51337	-0.48443	-0.44984	-0.38921	0.32
-0.30	-0.55192	-0.53013	-0.49699	-0.46743	-0.43213	-0.37040	0.30
-0.28	-0.53665	-0.51435	-0.48048	-0.45030	-0.41433	-0.35153	0.28
-0.26	-0.52121	-0.49841	-0.46382	-0.43304	-0.39642	-0.33261	0.26
-0.24	-0.50560	-0.48231	-0.44702	-0.41567	-0.37840	-0.31363	0.24
-0.22	-0.48982	-0.46605	-0.43008	-0.39816	-0.36029	-0.29460	0.22
-0.20	-0.47386	-0.44962	-0.41299	-0.38053	-0.34207	-0.27551	0.20
-0.18	-0.45772	-0.43303	-0.39575	-0.36277	-0.32375	-0.25636	0.18
-0.16	-0.44140	-0.41626	-0.37836	-0.34488	-0.30532	-0.23716	0.16
-0.14	-0.42489	-0.39932	-0.36082	-0.32685	-0.28679	-0.21789	0.14
-0.12	-0.40820	-0.38220	-0.34312	-0.30869	-0.26814	-0.19857	0.12
-0.10	-0.39131	-0.36491	-0.32526	-0.29040	-0.24939	-0.17919	0.10
-0.08	-0.37422	-0.34743	-0.30725	-0.27196	-0.23052	-0.15975	0.08
-0.06	-0.35693	-0.32976	-0.28907	-0.25338	-0.21154	-0.14024	0.06
-0.04	-0.33944	-0.31190	-0.27072	-0.23466	-0.19245	-0.12068	0.04
-0.02	-0.32174	-0.29385	-0.25221	-0.21580	-0.17324	-0.10105	0.02

↑ r_0

α =	.995 ↑-	.99 ↑-	.975 ↑-	.95 ↑-	.90 ↑-	.75 ↑-

TABLE 63.2   SAMPLE SIZE =   70   (CONT.)

α =	.005 ↓+	.01 ↓+	.025 ↓+	.05 ↓+	.10 ↓+	.25 ↓+	
r_o ↓							
0.00	−0.30382	−0.27561	−0.23352	−0.19679	−0.15392	−0.08136	0.00
0.02	−0.28569	−0.25716	−0.21467	−0.17763	−0.13447	−0.06160	−0.02
0.04	−0.26734	−0.23851	−0.19563	−0.15832	−0.11491	−0.04179	−0.04
0.06	−0.24877	−0.21966	−0.17642	−0.13886	−0.09523	−0.02190	−0.06
0.08	−0.22996	−0.20059	−0.15702	−0.11924	−0.07542	−0.00196	−0.08
0.10	−0.21092	−0.18130	−0.13744	−0.09946	−0.05549	0.01806	−0.10
0.12	−0.19164	−0.16180	−0.11768	−0.07953	−0.03544	0.03814	−0.12
0.14	−0.17212	−0.14208	−0.09772	−0.05943	−0.01526	0.05829	−0.14
0.16	−0.15235	−0.12212	−0.07756	−0.03917	0.00505	0.07851	−0.16
0.18	−0.13232	−0.10194	−0.05721	−0.01875	0.02549	0.09879	−0.18
0.20	−0.11203	−0.08152	−0.03666	0.00185	0.04606	0.11915	−0.20
0.22	−0.09148	−0.06085	−0.01591	0.02262	0.06677	0.13957	−0.22
0.24	−0.07066	−0.03995	0.00505	0.04356	0.08761	0.16007	−0.24
0.26	−0.04956	−0.01879	0.02622	0.06467	0.10859	0.18064	−0.26
0.28	−0.02818	0.00262	0.04761	0.08597	0.12970	0.20128	−0.28
0.30	−0.00651	0.02430	0.06922	0.10745	0.15096	0.22200	−0.30
0.32	0.01545	0.04624	0.09104	0.12911	0.17236	0.24279	−0.32
0.34	0.03772	0.06845	0.11310	0.15096	0.19390	0.26365	−0.34
0.36	0.06029	0.09093	0.13538	0.17300	0.21559	0.28459	−0.36
0.38	0.08318	0.11370	0.15790	0.19524	0.23742	0.30561	−0.38
0.40	0.10639	0.13676	0.18065	0.21767	0.25941	0.32670	−0.40
0.42	0.12993	0.16011	0.20365	0.24030	0.28155	0.34787	−0.42
0.44	0.15381	0.18376	0.22690	0.26313	0.30384	0.36912	−0.44
0.46	0.17803	0.20772	0.25039	0.28617	0.32628	0.39045	−0.46
0.48	0.20261	0.23199	0.27414	0.30942	0.34889	0.41186	−0.48
0.50	0.22755	0.25657	0.29816	0.33287	0.37165	0.43336	−0.50
0.52	0.25286	0.28149	0.32243	0.35655	0.39458	0.45493	−0.52
0.54	0.27854	0.30674	0.34698	0.38044	0.41767	0.47659	−0.54
0.56	0.30462	0.33233	0.37181	0.40456	0.44093	0.49833	−0.56
0.58	0.33109	0.35827	0.39691	0.42891	0.46436	0.52016	−0.58
0.60	0.35797	0.38457	0.42230	0.45348	0.48796	0.54208	−0.60
0.62	0.38527	0.41123	0.44799	0.47829	0.51173	0.56408	−0.62
0.64	0.41300	0.43827	0.47397	0.50334	0.53568	0.58617	−0.64
0.66	0.44117	0.46568	0.50025	0.52862	0.55981	0.60835	−0.66
0.68	0.46978	0.49349	0.52684	0.55416	0.58412	0.63062	−0.68
0.70	0.49887	0.52170	0.55375	0.57995	0.60861	0.65297	−0.70
0.72	0.52842	0.55031	0.58098	0.60599	0.63329	0.67543	−0.72
0.74	0.55846	0.57935	0.60854	0.63229	0.65816	0.69797	−0.74
0.76	0.58900	0.60881	0.63643	0.65885	0.68322	0.72061	−0.76
0.78	0.62006	0.63871	0.66467	0.68568	0.70847	0.74334	−0.78
0.80	0.65164	0.66906	0.69325	0.71279	0.73393	0.76617	−0.80
0.82	0.68377	0.69988	0.72219	0.74017	0.75958	0.78910	−0.82
0.84	0.71645	0.73116	0.75150	0.76783	0.78544	0.81212	−0.84
0.86	0.74971	0.76293	0.78117	0.79579	0.81150	0.83525	−0.86
0.88	0.78355	0.79520	0.81123	0.82404	0.83777	0.85847	−0.88
0.90	0.81800	0.82798	0.84167	0.85258	0.86426	0.88180	−0.90
0.92	0.85308	0.86128	0.87251	0.88144	0.89096	0.90523	−0.92
0.94	0.88880	0.89512	0.90375	0.91060	0.91788	0.92876	−0.94
0.96	0.92518	0.92951	0.93541	0.94007	0.94503	0.95240	−0.96
0.98	0.96224	0.96446	0.96749	0.96987	0.97240	0.97615	−0.98

↑ r_o

α =	.995 ↑−	.99 ↑−	.975 ↑−	.95 ↑−	.90 ↑−	.75 ↑−

TABLE 64.1   SAMPLE SIZE =   72

α =	.005 ↓+	.01 ↓+	.025 ↓+	.05 ↓+	.10 ↓+	.25 ↓+	
r_o ↓							
−0.98	−0.98904	−0.98836	−0.98728	−0.98628	−0.98504	−0.98271	0.98
−0.96	−0.97798	−0.97662	−0.97447	−0.97248	−0.97001	−0.96539	0.96
−0.94	−0.96683	−0.96479	−0.96158	−0.95860	−0.95491	−0.94803	0.94
−0.92	−0.95558	−0.95286	−0.94859	−0.94464	−0.93974	−0.93063	0.92
−0.90	−0.94423	−0.94084	−0.93552	−0.93059	−0.92450	−0.91319	0.90
−0.88	−0.93278	−0.92871	−0.92235	−0.91646	−0.90918	−0.89570	0.88
−0.86	−0.92122	−0.91649	−0.90908	−0.90225	−0.89380	−0.87818	0.86
−0.84	−0.90956	−0.90417	−0.89573	−0.88795	−0.87834	−0.86062	0.84
−0.82	−0.89780	−0.89174	−0.88227	−0.87356	−0.86281	−0.84302	0.82
−0.80	−0.88593	−0.87921	−0.86872	−0.85908	−0.84720	−0.82537	0.80
−0.78	−0.87394	−0.86657	−0.85507	−0.84451	−0.83152	−0.80769	0.78
−0.76	−0.86185	−0.85383	−0.84133	−0.82986	−0.81576	−0.78996	0.76
−0.74	−0.84965	−0.84097	−0.82748	−0.81511	−0.79992	−0.77219	0.74
−0.72	−0.83733	−0.82801	−0.81353	−0.80027	−0.78401	−0.75437	0.72
−0.70	−0.82489	−0.81494	−0.79947	−0.78533	−0.76802	−0.73651	0.70
−0.68	−0.81234	−0.80175	−0.78531	−0.77030	−0.75195	−0.71861	0.68
−0.66	−0.79967	−0.78845	−0.77105	−0.75518	−0.73580	−0.70066	0.66
−0.64	−0.78687	−0.77503	−0.75668	−0.73996	−0.71956	−0.68267	0.64
−0.62	−0.77396	−0.76149	−0.74220	−0.72464	−0.70325	−0.66463	0.62
−0.60	−0.76092	−0.74783	−0.72761	−0.70922	−0.68685	−0.64655	0.60
−0.58	−0.74775	−0.73405	−0.71290	−0.69371	−0.67037	−0.62841	0.58
−0.56	−0.73445	−0.72014	−0.69809	−0.67809	−0.65381	−0.61024	0.56
−0.54	−0.72102	−0.70611	−0.68316	−0.66236	−0.63716	−0.59201	0.54
−0.52	−0.70745	−0.69196	−0.66811	−0.64654	−0.62042	−0.57374	0.52
−0.50	−0.69376	−0.67767	−0.65294	−0.63061	−0.60360	−0.55542	0.50
−0.48	−0.67992	−0.66325	−0.63766	−0.61457	−0.58669	−0.53705	0.48
−0.46	−0.66594	−0.64870	−0.62226	−0.59843	−0.56969	−0.51863	0.46
−0.44	−0.65182	−0.63401	−0.60673	−0.58217	−0.55260	−0.50016	0.44
−0.42	−0.63756	−0.61919	−0.59108	−0.56581	−0.53542	−0.48164	0.42
−0.40	−0.62315	−0.60422	−0.57530	−0.54933	−0.51815	−0.46307	0.40
−0.38	−0.60858	−0.58912	−0.55939	−0.53274	−0.50078	−0.44445	0.38
−0.36	−0.59387	−0.57387	−0.54336	−0.51604	−0.48332	−0.42578	0.36
−0.34	−0.57900	−0.55847	−0.52719	−0.49922	−0.46577	−0.40705	0.34
−0.32	−0.56398	−0.54292	−0.51089	−0.48228	−0.44812	−0.38828	0.32
−0.30	−0.54879	−0.52723	−0.49445	−0.46523	−0.43037	−0.36945	0.30
−0.28	−0.53344	−0.51137	−0.47788	−0.44805	−0.41253	−0.35056	0.28
−0.26	−0.51793	−0.49537	−0.46117	−0.43076	−0.39458	−0.33162	0.26
−0.24	−0.50224	−0.47920	−0.44431	−0.41333	−0.37654	−0.31263	0.24
−0.22	−0.48639	−0.46288	−0.42732	−0.39579	−0.35839	−0.29358	0.22
−0.20	−0.47035	−0.44638	−0.41018	−0.37811	−0.34014	−0.27447	0.20
−0.18	−0.45415	−0.42973	−0.39289	−0.36031	−0.32179	−0.25531	0.18
−0.16	−0.43775	−0.41290	−0.37545	−0.34238	−0.30333	−0.23609	0.16
−0.14	−0.42118	−0.39590	−0.35786	−0.32432	−0.28477	−0.21681	0.14
−0.12	−0.40442	−0.37872	−0.34011	−0.30612	−0.26610	−0.19747	0.12
−0.10	−0.38746	−0.36137	−0.32221	−0.28779	−0.24732	−0.17808	0.10
−0.08	−0.37031	−0.34384	−0.30415	−0.26932	−0.22843	−0.15863	0.08
−0.06	−0.35296	−0.32612	−0.28593	−0.25071	−0.20943	−0.13911	0.06
−0.04	−0.33541	−0.30821	−0.26755	−0.23196	−0.19031	−0.11954	0.04
−0.02	−0.31765	−0.29011	−0.24900	−0.21307	−0.17109	−0.09990	0.02

↑ r_o

α =	.995 ↑−	.99 ↑−	.975 ↑−	.95 ↑−	.90 ↑−	.75 ↑−

TABLE 64.2   SAMPLE SIZE =   72   (CONT.)

α =	.005 ↓+	.01 ↓+	.025 ↓+	.05 ↓+	.10 ↓+	.25 ↓+	
r_o ↓							
0.00	-0.29969	-0.27182	-0.23028	-0.19403	-0.15174	-0.08020	0.00
0.02	-0.28150	-0.25333	-0.21139	-0.17485	-0.13228	-0.06044	-0.02
0.04	-0.26310	-0.23464	-0.19232	-0.15551	-0.11271	-0.04061	-0.04
0.06	-0.24448	-0.21574	-0.17308	-0.13603	-0.09301	-0.02073	-0.06
0.08	-0.22563	-0.19664	-0.15366	-0.11639	-0.07319	-0.00077	-0.08
0.10	-0.20655	-0.17732	-0.13406	-0.09660	-0.05326	0.01925	-0.10
0.12	-0.18723	-0.15779	-0.11427	-0.07665	-0.03319	0.03933	-0.12
0.14	-0.16767	-0.13803	-0.09429	-0.05655	-0.01300	0.05948	-0.14
0.16	-0.14786	-0.11805	-0.07412	-0.03628	0.00731	0.07970	-0.16
0.18	-0.12781	-0.09785	-0.05376	-0.01584	0.02775	0.09998	-0.18
0.20	-0.10749	-0.07741	-0.03320	0.00476	0.04833	0.12034	-0.20
0.22	-0.08692	-0.05673	-0.01243	0.02553	0.06903	0.14076	-0.22
0.24	-0.06608	-0.03581	0.00853	0.04647	0.08987	0.16126	-0.24
0.26	-0.04496	-0.01464	0.02971	0.06758	0.11084	0.18182	-0.26
0.28	-0.02357	0.00678	0.05109	0.08888	0.13195	0.20246	-0.28
0.30	-0.00189	0.02845	0.07270	0.11035	0.15319	0.22317	-0.30
0.32	0.02007	0.05039	0.09451	0.13200	0.17458	0.24395	-0.32
0.34	0.04234	0.07260	0.11656	0.15384	0.19611	0.26480	-0.34
0.36	0.06491	0.09507	0.13883	0.17586	0.21778	0.28573	-0.36
0.38	0.08779	0.11782	0.16132	0.19807	0.23960	0.30674	-0.38
0.40	0.11098	0.14086	0.18406	0.22048	0.26156	0.32782	-0.40
0.42	0.13450	0.16419	0.20703	0.24309	0.28368	0.34897	-0.42
0.44	0.15835	0.18781	0.23024	0.26589	0.30594	0.37021	-0.44
0.46	0.18254	0.21173	0.25370	0.28889	0.32836	0.39152	-0.46
0.48	0.20708	0.23596	0.27741	0.31210	0.35093	0.41292	-0.48
0.50	0.23197	0.26050	0.30138	0.33552	0.37367	0.43439	-0.50
0.52	0.25723	0.28537	0.32561	0.35915	0.39656	0.45595	-0.52
0.54	0.28286	0.31056	0.35011	0.38300	0.41961	0.47758	-0.54
0.56	0.30886	0.33609	0.37487	0.40706	0.44282	0.49930	-0.56
0.58	0.33526	0.36196	0.39991	0.43135	0.46620	0.52110	-0.58
0.60	0.36206	0.38817	0.42523	0.45586	0.48975	0.54299	-0.60
0.62	0.38927	0.41475	0.45084	0.48061	0.51347	0.56497	-0.62
0.64	0.41690	0.44169	0.47674	0.50558	0.53737	0.58702	-0.64
0.66	0.44496	0.46901	0.50293	0.53080	0.56144	0.60917	-0.66
0.68	0.47346	0.49670	0.52943	0.55625	0.58568	0.63141	-0.68
0.70	0.50241	0.52479	0.55624	0.58195	0.61011	0.65373	-0.70
0.72	0.53182	0.55328	0.58336	0.60790	0.63472	0.67615	-0.72
0.74	0.56171	0.58218	0.61081	0.63411	0.65951	0.69865	-0.74
0.76	0.59209	0.61150	0.63858	0.66057	0.68450	0.72125	-0.76
0.78	0.62297	0.64124	0.66668	0.68729	0.70967	0.74394	-0.78
0.80	0.65437	0.67143	0.69513	0.71429	0.73504	0.76673	-0.80
0.82	0.68630	0.70206	0.72392	0.74155	0.76060	0.78961	-0.82
0.84	0.71876	0.73316	0.75307	0.76909	0.78636	0.81258	-0.84
0.86	0.75179	0.76473	0.78259	0.79691	0.81233	0.83566	-0.86
0.88	0.78539	0.79679	0.81247	0.82502	0.83850	0.85883	-0.88
0.90	0.81958	0.82934	0.84273	0.85342	0.86488	0.88210	-0.90
0.92	0.85438	0.86240	0.87338	0.88212	0.89146	0.90548	-0.92
0.94	0.88980	0.89598	0.90442	0.91112	0.91827	0.92895	-0.94
0.96	0.92587	0.93010	0.93587	0.94043	0.94529	0.95253	-0.96
0.98	0.96259	0.96477	0.96772	0.97006	0.97253	0.97621	-0.98

† r_o

α =	.995 ↑-	.99 ↑-	.975 ↑-	.95 ↑-	.90 ↑-	.75 ↑-

TABLE 65.1   SAMPLE SIZE =   74

α =	.005 ↓+	.01 ↓+	.025 ↓+	.05 ↓+	.10 ↓+	.25 ↓+	
r_0 ↓							
-0.98	-0.98895	-0.98827	-0.98720	-0.98621	-0.98498	-0.98268	0.98
-0.96	-0.97780	-0.97645	-0.97432	-0.97234	-0.96989	-0.96533	0.96
-0.94	-0.96656	-0.96453	-0.96134	-0.95839	-0.95473	-0.94793	0.94
-0.92	-0.95521	-0.95251	-0.94828	-0.94436	-0.93951	-0.93050	0.92
-0.90	-0.94377	-0.94040	-0.93512	-0.93025	-0.92421	-0.91303	0.90
-0.88	-0.93223	-0.92819	-0.92188	-0.91605	-0.90884	-0.89552	0.88
-0.86	-0.92058	-0.91589	-0.90854	-0.90176	-0.89340	-0.87796	0.86
-0.84	-0.90883	-0.90348	-0.89510	-0.88740	-0.87788	-0.86037	0.84
-0.82	-0.89698	-0.89096	-0.88157	-0.87294	-0.86230	-0.84274	0.82
-0.80	-0.88501	-0.87835	-0.86795	-0.85840	-0.84664	-0.82507	0.80
-0.78	-0.87294	-0.86563	-0.85423	-0.84376	-0.83090	-0.80735	0.78
-0.76	-0.86076	-0.85280	-0.84040	-0.82904	-0.81509	-0.78959	0.76
-0.74	-0.84847	-0.83986	-0.82648	-0.81423	-0.79920	-0.77179	0.74
-0.72	-0.83606	-0.82681	-0.81246	-0.79932	-0.78324	-0.75395	0.72
-0.70	-0.82353	-0.81366	-0.79833	-0.78433	-0.76719	-0.73606	0.70
-0.68	-0.81089	-0.80039	-0.78410	-0.76924	-0.75107	-0.71813	0.68
-0.66	-0.79813	-0.78700	-0.76976	-0.75405	-0.73487	-0.70016	0.66
-0.64	-0.78525	-0.77350	-0.75532	-0.73877	-0.71859	-0.68214	0.64
-0.62	-0.77225	-0.75988	-0.74077	-0.72339	-0.70223	-0.66407	0.62
-0.60	-0.75912	-0.74614	-0.72611	-0.70791	-0.68578	-0.64596	0.60
-0.58	-0.74586	-0.73228	-0.71134	-0.69233	-0.66926	-0.62781	0.58
-0.56	-0.73248	-0.71830	-0.69645	-0.67666	-0.65265	-0.60960	0.56
-0.54	-0.71896	-0.70419	-0.68145	-0.66088	-0.63595	-0.59135	0.54
-0.52	-0.70532	-0.68996	-0.66634	-0.64499	-0.61917	-0.57306	0.52
-0.50	-0.69153	-0.67559	-0.65111	-0.62901	-0.60230	-0.55471	0.50
-0.48	-0.67761	-0.66110	-0.63576	-0.61291	-0.58535	-0.53632	0.48
-0.46	-0.66355	-0.64647	-0.62029	-0.59672	-0.56831	-0.51788	0.46
-0.44	-0.64935	-0.63171	-0.60470	-0.58041	-0.55118	-0.49939	0.44
-0.42	-0.63501	-0.61681	-0.58898	-0.56399	-0.53396	-0.48085	0.42
-0.40	-0.62052	-0.60177	-0.57314	-0.54746	-0.51664	-0.46226	0.40
-0.38	-0.60588	-0.58660	-0.55718	-0.53083	-0.49924	-0.44361	0.38
-0.36	-0.59109	-0.57128	-0.54108	-0.51407	-0.48174	-0.42492	0.36
-0.34	-0.57614	-0.55581	-0.52486	-0.49720	-0.46415	-0.40618	0.34
-0.32	-0.56104	-0.54019	-0.50850	-0.48022	-0.44646	-0.38738	0.32
-0.30	-0.54578	-0.52443	-0.49201	-0.46312	-0.42868	-0.36853	0.30
-0.28	-0.53035	-0.50851	-0.47538	-0.44590	-0.41080	-0.34963	0.28
-0.26	-0.51476	-0.49244	-0.45862	-0.42856	-0.39282	-0.33067	0.26
-0.24	-0.49901	-0.47621	-0.44171	-0.41109	-0.37475	-0.31166	0.24
-0.22	-0.48308	-0.45982	-0.42466	-0.39350	-0.35657	-0.29259	0.22
-0.20	-0.46698	-0.44327	-0.40747	-0.37579	-0.33829	-0.27347	0.20
-0.18	-0.45071	-0.42655	-0.39013	-0.35795	-0.31991	-0.25430	0.18
-0.16	-0.43425	-0.40967	-0.37265	-0.33998	-0.30142	-0.23506	0.16
-0.14	-0.41761	-0.39261	-0.35501	-0.32188	-0.28283	-0.21577	0.14
-0.12	-0.40078	-0.37538	-0.33723	-0.30365	-0.26414	-0.19642	0.12
-0.10	-0.38377	-0.35798	-0.31928	-0.28528	-0.24533	-0.17702	0.10
-0.08	-0.36656	-0.34039	-0.30118	-0.26678	-0.22642	-0.15755	0.08
-0.06	-0.34915	-0.32262	-0.28292	-0.24814	-0.20740	-0.13803	0.06
-0.04	-0.33154	-0.30467	-0.26450	-0.22937	-0.18826	-0.11844	0.04
-0.02	-0.31373	-0.28652	-0.24592	-0.21045	-0.16902	-0.09880	0.02

↑ r_0

α =	.995 ↑-	.99 ↑-	.975 ↑-	.95 ↑-	.90 ↑-	.75 ↑-

TABLE 65.2   SAMPLE SIZE = 74   (CONT.)

α =	.005 ↓+	.01 ↓+	.025 ↓+	.05 ↓+	.10 ↓+	.25 ↓+	
r_o ↓							
0.00	-0.29571	-0.26819	-0.22716	-0.19139	-0.14966	-0.07909	0.00
0.02	-0.27748	-0.24966	-0.20824	-0.17218	-0.13019	-0.05932	-0.02
0.04	-0.25903	-0.23092	-0.18915	-0.15282	-0.11059	-0.03949	-0.04
0.06	-0.24037	-0.21199	-0.16988	-0.13332	-0.09089	-0.01960	-0.06
0.08	-0.22147	-0.19285	-0.15043	-0.11367	-0.07106	0.00036	-0.08
0.10	-0.20235	-0.17350	-0.13081	-0.09386	-0.05111	0.02038	-0.10
0.12	-0.18300	-0.15394	-0.11100	-0.07390	-0.03104	0.04047	-0.12
0.14	-0.16340	-0.13416	-0.09100	-0.05378	-0.01085	0.06062	-0.14
0.16	-0.14357	-0.11415	-0.07082	-0.03350	0.00947	0.08084	-0.16
0.18	-0.12348	-0.09393	-0.05044	-0.01306	0.02992	0.10113	-0.18
0.20	-0.10314	-0.07347	-0.02987	0.00755	0.05049	0.12148	-0.20
0.22	-0.08255	-0.05277	-0.00910	0.02832	0.07120	0.14190	-0.22
0.24	-0.06169	-0.03184	0.01187	0.04926	0.09203	0.16239	-0.24
0.26	-0.04056	-0.01067	0.03304	0.07037	0.11300	0.18295	-0.26
0.28	-0.01915	0.01076	0.05443	0.09166	0.13410	0.20358	-0.28
0.30	0.00253	0.03243	0.07602	0.11312	0.15534	0.22428	-0.30
0.32	0.02450	0.05437	0.09784	0.13476	0.17671	0.24506	-0.32
0.34	0.04676	0.07657	0.11987	0.15658	0.19823	0.26590	-0.34
0.36	0.06932	0.09903	0.14212	0.17859	0.21988	0.28682	-0.36
0.38	0.09219	0.12177	0.16460	0.20079	0.24168	0.30782	-0.38
0.40	0.11537	0.14479	0.18731	0.22317	0.26363	0.32889	-0.40
0.42	0.13887	0.16809	0.21026	0.24575	0.28572	0.35003	-0.42
0.44	0.16269	0.19168	0.23344	0.26852	0.30796	0.37125	-0.44
0.46	0.18685	0.21557	0.25687	0.29150	0.33035	0.39255	-0.46
0.48	0.21135	0.23976	0.28054	0.31467	0.35289	0.41393	-0.48
0.50	0.23620	0.26426	0.30446	0.33805	0.37559	0.43538	-0.50
0.52	0.26140	0.28907	0.32865	0.36164	0.39844	0.45691	-0.52
0.54	0.28697	0.31421	0.35309	0.38544	0.42146	0.47853	-0.54
0.56	0.31292	0.33967	0.37780	0.40945	0.44463	0.50023	-0.56
0.58	0.33924	0.36547	0.40277	0.43369	0.46797	0.52200	-0.58
0.60	0.36596	0.39161	0.42803	0.45814	0.49147	0.54387	-0.60
0.62	0.39308	0.41811	0.45356	0.48282	0.51514	0.56581	-0.62
0.64	0.42062	0.44496	0.47938	0.50773	0.53898	0.58784	-0.64
0.66	0.44857	0.47218	0.50549	0.53287	0.56299	0.60996	-0.66
0.68	0.47695	0.49977	0.53190	0.55825	0.58718	0.63216	-0.68
0.70	0.50578	0.52774	0.55861	0.58387	0.61154	0.65445	-0.70
0.72	0.53506	0.55611	0.58563	0.60973	0.63608	0.67683	-0.72
0.74	0.56481	0.58488	0.61297	0.63584	0.66081	0.69930	-0.74
0.76	0.59503	0.61406	0.64062	0.66221	0.68571	0.72186	-0.76
0.78	0.62575	0.64365	0.66860	0.68883	0.71081	0.74451	-0.78
0.80	0.65697	0.67368	0.69692	0.71572	0.73610	0.76725	-0.80
0.82	0.68870	0.70415	0.72557	0.74287	0.76157	0.79009	-0.82
0.84	0.72096	0.73506	0.75458	0.77029	0.78724	0.81302	-0.84
0.86	0.75377	0.76644	0.78393	0.79798	0.81312	0.83605	-0.86
0.88	0.78714	0.79829	0.81365	0.82596	0.83919	0.85917	-0.88
0.90	0.82108	0.83063	0.84374	0.85422	0.86546	0.88239	-0.90
0.92	0.85561	0.86346	0.87421	0.88278	0.89194	0.90571	-0.92
0.94	0.89076	0.89680	0.90506	0.91163	0.91863	0.92913	-0.94
0.96	0.92652	0.93066	0.93630	0.94078	0.94554	0.95265	-0.96
0.98	0.96293	0.96505	0.96795	0.97023	0.97266	0.97627	-0.98

↑ r_o

α =	.995 ↑-	.99 ↑-	.975 ↑-	.95 ↑-	.90 ↑-	.75 ↑-

TABLE 66.1   SAMPLE SIZE =   76

α =	.005 ↓+	.01 ↓+	.025 ↓+	.05 ↓+	.10 ↓+	.25 ↓+	
r_o ↓							
-0.98	-0.98886	-0.98819	-0.98713	-0.98614	-0.98492	-0.98265	0.98
-0.96	-0.97762	-0.97628	-0.97416	-0.97221	-0.96978	-0.96527	0.96
-0.94	-0.96629	-0.96428	-0.96111	-0.95819	-0.95457	-0.94784	0.94
-0.92	-0.95486	-0.95218	-0.94798	-0.94409	-0.93928	-0.93038	0.92
-0.90	-0.94333	-0.93998	-0.93475	-0.92991	-0.92393	-0.91288	0.90
-0.88	-0.93170	-0.92769	-0.92142	-0.91565	-0.90851	-0.89534	0.88
-0.86	-0.91996	-0.91530	-0.90801	-0.90130	-0.89301	-0.87775	0.86
-0.84	-0.90813	-0.90281	-0.89450	-0.88686	-0.87745	-0.86013	0.84
-0.82	-0.89618	-0.89021	-0.88090	-0.87234	-0.86181	-0.84247	0.82
-0.80	-0.88413	-0.87751	-0.86720	-0.85774	-0.84610	-0.82477	0.80
-0.78	-0.87197	-0.86471	-0.85341	-0.84304	-0.83031	-0.80702	0.78
-0.76	-0.85971	-0.85180	-0.83951	-0.82826	-0.81445	-0.78924	0.76
-0.74	-0.84732	-0.83879	-0.82552	-0.81338	-0.79851	-0.77141	0.74
-0.72	-0.83483	-0.82566	-0.81142	-0.79842	-0.78249	-0.75354	0.72
-0.70	-0.82222	-0.81242	-0.79723	-0.78336	-0.76640	-0.73563	0.70
-0.68	-0.80949	-0.79907	-0.78293	-0.76821	-0.75023	-0.71767	0.68
-0.66	-0.79665	-0.78561	-0.76852	-0.75296	-0.73398	-0.69967	0.66
-0.64	-0.78368	-0.77203	-0.75401	-0.73762	-0.71765	-0.68162	0.64
-0.62	-0.77060	-0.75833	-0.73939	-0.72218	-0.70125	-0.66353	0.62
-0.60	-0.75738	-0.74452	-0.72466	-0.70665	-0.68476	-0.64540	0.60
-0.58	-0.74405	-0.73058	-0.70982	-0.69101	-0.66818	-0.62722	0.58
-0.56	-0.73058	-0.71652	-0.69487	-0.67528	-0.65153	-0.60899	0.56
-0.54	-0.71698	-0.70234	-0.67981	-0.65944	-0.63479	-0.59072	0.54
-0.52	-0.70325	-0.68803	-0.66463	-0.64351	-0.61797	-0.57240	0.52
-0.50	-0.68939	-0.67359	-0.64934	-0.62747	-0.60106	-0.55403	0.50
-0.48	-0.67539	-0.65902	-0.63393	-0.61132	-0.58406	-0.53562	0.48
-0.46	-0.66125	-0.64432	-0.61840	-0.59507	-0.56698	-0.51715	0.46
-0.44	-0.64697	-0.62949	-0.60275	-0.57871	-0.54981	-0.49864	0.44
-0.42	-0.63255	-0.61452	-0.58697	-0.56224	-0.53255	-0.48008	0.42
-0.40	-0.61798	-0.59942	-0.57107	-0.54567	-0.51519	-0.46147	0.40
-0.38	-0.60327	-0.58417	-0.55505	-0.52898	-0.49775	-0.44281	0.38
-0.36	-0.58840	-0.56878	-0.53890	-0.51218	-0.48022	-0.42410	0.36
-0.34	-0.57338	-0.55325	-0.52261	-0.49526	-0.46259	-0.40533	0.34
-0.32	-0.55820	-0.53757	-0.50620	-0.47824	-0.44487	-0.38652	0.32
-0.30	-0.54287	-0.52174	-0.48966	-0.46109	-0.42705	-0.36765	0.30
-0.28	-0.52738	-0.50576	-0.47298	-0.44383	-0.40914	-0.34873	0.28
-0.26	-0.51172	-0.48962	-0.45616	-0.42644	-0.39113	-0.32976	0.26
-0.24	-0.49589	-0.47333	-0.43921	-0.40894	-0.37302	-0.31073	0.24
-0.22	-0.47990	-0.45688	-0.42211	-0.39131	-0.35482	-0.29165	0.22
-0.20	-0.46374	-0.44027	-0.40487	-0.37356	-0.33651	-0.27252	0.20
-0.18	-0.44739	-0.42350	-0.38749	-0.35568	-0.31810	-0.25332	0.18
-0.16	-0.43087	-0.40656	-0.36996	-0.33767	-0.29959	-0.23408	0.16
-0.14	-0.41417	-0.38945	-0.35228	-0.31954	-0.28097	-0.21477	0.14
-0.12	-0.39729	-0.37217	-0.33445	-0.30128	-0.26225	-0.19541	0.12
-0.10	-0.38021	-0.35471	-0.31647	-0.28288	-0.24343	-0.17600	0.10
-0.08	-0.36295	-0.33707	-0.29833	-0.26435	-0.22449	-0.15652	0.08
-0.06	-0.34549	-0.31926	-0.28003	-0.24568	-0.20545	-0.13699	0.06
-0.04	-0.32783	-0.30126	-0.26158	-0.22688	-0.18630	-0.11739	0.04
-0.02	-0.30996	-0.28307	-0.24296	-0.20793	-0.16704	-0.09774	0.02

↑ r_o

α =	.995 ↑-	.99 ↑-	.975 ↑-	.95 ↑-	.90 ↑-	.75 ↑-

TABLE 66.2   SAMPLE SIZE =   76   (CONT.)

α =	.005 ↓+	.01 ↓+	.025 ↓+	.05 ↓+	.10 ↓+	.25 ↓+	
r_o ↓							
0.00	−0.29189	−0.26469	−0.22417	−0.18885	−0.14766	−0.07802	0.00
0.02	−0.27362	−0.24612	−0.20522	−0.16962	−0.12817	−0.05825	−0.02
0.04	−0.25512	−0.22736	−0.18610	−0.15024	−0.10857	−0.03841	−0.04
0.06	−0.23641	−0.20839	−0.16681	−0.13072	−0.08885	−0.01851	−0.06
0.08	−0.21748	−0.18921	−0.14734	−0.11105	−0.06901	0.00145	−0.08
0.10	−0.19832	−0.16983	−0.12769	−0.09123	−0.04905	0.02147	−0.10
0.12	−0.17893	−0.15024	−0.10786	−0.07126	−0.02897	0.04156	−0.12
0.14	−0.15930	−0.13044	−0.08785	−0.05113	−0.00877	0.06172	−0.14
0.16	−0.13944	−0.11041	−0.06765	−0.03084	0.01155	0.08194	−0.16
0.18	−0.11933	−0.09016	−0.04727	−0.01039	0.03200	0.10222	−0.18
0.20	−0.09897	−0.06969	−0.02669	0.01022	0.05257	0.12257	−0.20
0.22	−0.07835	−0.04898	−0.00591	0.03099	0.07327	0.14299	−0.22
0.24	−0.05747	−0.02804	0.01507	0.05193	0.09410	0.16348	−0.24
0.26	−0.03633	−0.00685	0.03624	0.07304	0.11506	0.18404	−0.26
0.28	−0.01492	0.01457	0.05762	0.09432	0.13616	0.20466	−0.28
0.30	0.00677	0.03625	0.07922	0.11578	0.15739	0.22536	−0.30
0.32	0.02874	0.05818	0.10102	0.13741	0.17875	0.24612	−0.32
0.34	0.05100	0.08037	0.12304	0.15922	0.20025	0.26696	−0.34
0.36	0.07355	0.10282	0.14528	0.18121	0.22189	0.28787	−0.36
0.38	0.09641	0.12555	0.16774	0.20339	0.24368	0.30885	−0.38
0.40	0.11957	0.14854	0.19043	0.22575	0.26560	0.32991	−0.40
0.42	0.14305	0.17182	0.21335	0.24830	0.28767	0.35104	−0.42
0.44	0.16685	0.19538	0.23650	0.27105	0.30988	0.37225	−0.44
0.46	0.19097	0.21924	0.25989	0.29399	0.33225	0.39353	−0.46
0.48	0.21543	0.24339	0.28353	0.31713	0.35476	0.41489	−0.48
0.50	0.24024	0.26785	0.30741	0.34047	0.37743	0.43633	−0.50
0.52	0.26539	0.29261	0.33155	0.36402	0.40025	0.45784	−0.52
0.54	0.29091	0.31769	0.35594	0.38777	0.42322	0.47944	−0.54
0.56	0.31679	0.34310	0.38059	0.41174	0.44636	0.50111	−0.56
0.58	0.34305	0.36883	0.40551	0.43592	0.46965	0.52287	−0.58
0.60	0.36969	0.39490	0.43070	0.46032	0.49311	0.54470	−0.60
0.62	0.39673	0.42131	0.45616	0.48493	0.51673	0.56662	−0.62
0.64	0.42416	0.44808	0.48191	0.50978	0.54052	0.58863	−0.64
0.66	0.45202	0.47520	0.50794	0.53485	0.56448	0.61071	−0.66
0.68	0.48029	0.50269	0.53426	0.56016	0.58861	0.63289	−0.68
0.70	0.50900	0.53056	0.56088	0.58570	0.61291	0.65514	−0.70
0.72	0.53815	0.55881	0.58780	0.61148	0.63739	0.67749	−0.72
0.74	0.56776	0.58745	0.61503	0.63750	0.66204	0.69992	−0.74
0.76	0.59784	0.61650	0.64257	0.66378	0.68688	0.72244	−0.76
0.78	0.62839	0.64595	0.67043	0.69030	0.71190	0.74506	−0.78
0.80	0.65944	0.67583	0.69862	0.71708	0.73711	0.76776	−0.80
0.82	0.69099	0.70613	0.72715	0.74412	0.76250	0.79055	−0.82
0.84	0.72306	0.73687	0.75601	0.77143	0.78809	0.81344	−0.84
0.86	0.75566	0.76807	0.78522	0.79901	0.81387	0.83642	−0.86
0.88	0.78880	0.79972	0.81478	0.82686	0.83984	0.85950	−0.88
0.90	0.82251	0.83185	0.84470	0.85499	0.86602	0.88267	−0.90
0.92	0.85679	0.86446	0.87500	0.88340	0.89240	0.90594	−0.92
0.94	0.89166	0.89757	0.90566	0.91210	0.91898	0.92930	−0.94
0.96	0.92714	0.93119	0.93671	0.94110	0.94578	0.95277	−0.96
0.98	0.96325	0.96533	0.96816	0.97040	0.97278	0.97633	−0.98

↑ r_o

α =	.995 ↑−	.99 ↑−	.975 ↑−	.95 ↑−	.90 ↑−	.75 ↑−

TABLE 67.1   SAMPLE SIZE =   78

α =	.005 ↓+	.01 ↓+	.025 ↓+	.05 ↓+	.10 ↓+	.25 ↓+	
r_0 ↓							
-0.98	-0.98877	-0.98810	-0.98705	-0.98608	-0.98487	-0.98262	0.98
-0.96	-0.97745	-0.97611	-0.97402	-0.97208	-0.96967	-0.96521	0.96
-0.94	-0.96603	-0.96403	-0.96089	-0.95800	-0.95441	-0.94775	0.94
-0.92	-0.95452	-0.95185	-0.94768	-0.94383	-0.93907	-0.93026	0.92
-0.90	-0.94290	-0.93958	-0.93438	-0.92959	-0.92367	-0.91273	0.90
-0.88	-0.93118	-0.92721	-0.92099	-0.91526	-0.90819	-0.89516	0.88
-0.86	-0.91937	-0.91473	-0.90750	-0.90085	-0.89264	-0.87755	0.86
-0.84	-0.90744	-0.90216	-0.89392	-0.88635	-0.87703	-0.85990	0.84
-0.82	-0.89541	-0.88949	-0.88025	-0.87177	-0.86134	-0.84221	0.82
-0.80	-0.88328	-0.87671	-0.86648	-0.85710	-0.84557	-0.82448	0.80
-0.78	-0.87104	-0.86383	-0.85262	-0.84234	-0.82974	-0.80671	0.78
-0.76	-0.85869	-0.85084	-0.83865	-0.82750	-0.81383	-0.78890	0.76
-0.74	-0.84622	-0.83775	-0.82459	-0.81256	-0.79784	-0.77104	0.74
-0.72	-0.83364	-0.82454	-0.81043	-0.79754	-0.78178	-0.75315	0.72
-0.70	-0.82095	-0.81123	-0.79616	-0.78242	-0.76564	-0.73521	0.70
-0.68	-0.80814	-0.79780	-0.78179	-0.76721	-0.74942	-0.71723	0.68
-0.66	-0.79522	-0.78426	-0.76732	-0.75191	-0.73313	-0.69920	0.66
-0.64	-0.78217	-0.77061	-0.75275	-0.73651	-0.71675	-0.68113	0.64
-0.62	-0.76900	-0.75684	-0.73806	-0.72102	-0.70030	-0.66302	0.62
-0.60	-0.75571	-0.74295	-0.72327	-0.70543	-0.68376	-0.64486	0.60
-0.58	-0.74229	-0.72894	-0.70837	-0.68974	-0.66715	-0.62665	0.58
-0.56	-0.72874	-0.71480	-0.69335	-0.67395	-0.65045	-0.60840	0.56
-0.54	-0.71507	-0.70055	-0.67823	-0.65806	-0.63367	-0.59011	0.54
-0.52	-0.70126	-0.68617	-0.66299	-0.64207	-0.61680	-0.57177	0.52
-0.50	-0.68732	-0.67166	-0.64763	-0.62598	-0.59985	-0.55338	0.50
-0.48	-0.67325	-0.65702	-0.63216	-0.60978	-0.58282	-0.53494	0.48
-0.46	-0.65903	-0.64225	-0.61657	-0.59348	-0.56570	-0.51646	0.46
-0.44	-0.64468	-0.62735	-0.60086	-0.57707	-0.54849	-0.49793	0.44
-0.42	-0.63018	-0.61231	-0.58503	-0.56056	-0.53119	-0.47934	0.42
-0.40	-0.61554	-0.59714	-0.56907	-0.54394	-0.51380	-0.46071	0.40
-0.38	-0.60075	-0.58183	-0.55299	-0.52720	-0.49632	-0.44203	0.38
-0.36	-0.58581	-0.56637	-0.53679	-0.51036	-0.47875	-0.42330	0.36
-0.34	-0.57072	-0.55078	-0.52045	-0.49340	-0.46109	-0.40452	0.34
-0.32	-0.55547	-0.53503	-0.50399	-0.47632	-0.44334	-0.38569	0.32
-0.30	-0.54007	-0.51914	-0.48739	-0.45914	-0.42549	-0.36681	0.30
-0.28	-0.52451	-0.50310	-0.47066	-0.44183	-0.40755	-0.34787	0.28
-0.26	-0.50878	-0.48691	-0.45380	-0.42441	-0.38951	-0.32888	0.26
-0.24	-0.49289	-0.47056	-0.43679	-0.40686	-0.37137	-0.30984	0.24
-0.22	-0.47684	-0.45405	-0.41965	-0.38920	-0.35313	-0.29074	0.22
-0.20	-0.46061	-0.43739	-0.40237	-0.37141	-0.33480	-0.27159	0.20
-0.18	-0.44420	-0.42056	-0.38494	-0.35350	-0.31636	-0.25239	0.18
-0.16	-0.42762	-0.40356	-0.36737	-0.33546	-0.29783	-0.23313	0.16
-0.14	-0.41086	-0.38640	-0.34965	-0.31729	-0.27919	-0.21381	0.14
-0.12	-0.39392	-0.36907	-0.33178	-0.29899	-0.26044	-0.19444	0.12
-0.10	-0.37679	-0.35156	-0.31376	-0.28057	-0.24159	-0.17501	0.10
-0.08	-0.35947	-0.33388	-0.29558	-0.26201	-0.22264	-0.15553	0.08
-0.06	-0.34196	-0.31602	-0.27725	-0.24331	-0.20358	-0.13599	0.06
-0.04	-0.32425	-0.29798	-0.25876	-0.22448	-0.18441	-0.11638	0.04
-0.02	-0.30633	-0.27975	-0.24011	-0.20551	-0.16513	-0.09672	0.02

↑ r_0

α =   .995 ↑−   .99 ↑−   .975 ↑−   .95 ↑−   .90 ↑−   .75 ↑−

TABLE 67.2   SAMPLE SIZE =   78   (CONT.)

α =	.005 ↓+	.01 ↓+	.025 ↓+	.05 ↓+	.10 ↓+	.25 ↓+	
r_o ↓							
0.00	−0.28822	−0.26133	−0.22130	−0.18641	−0.14574	−0.07700	0.00
0.02	−0.26990	−0.24273	−0.20232	−0.16716	−0.12623	−0.05722	−0.02
0.04	−0.25136	−0.22392	−0.18317	−0.14776	−0.10662	−0.03738	−0.04
0.06	−0.23261	−0.20492	−0.16385	−0.12822	−0.08689	−0.01747	−0.06
0.08	−0.21364	−0.18572	−0.14436	−0.10854	−0.06704	0.00249	−0.08
0.10	−0.19445	−0.16631	−0.12470	−0.08870	−0.04707	0.02252	−0.10
0.12	−0.17502	−0.14669	−0.10485	−0.06872	−0.02699	0.04261	−0.12
0.14	−0.15537	−0.12686	−0.08482	−0.04858	−0.00678	0.06277	−0.14
0.16	−0.13547	−0.10682	−0.06461	−0.02828	0.01354	0.08299	−0.16
0.18	−0.11534	−0.08655	−0.04421	−0.00783	0.03399	0.10327	−0.18
0.20	−0.09495	−0.06606	−0.02363	0.01278	0.05456	0.12362	−0.20
0.22	−0.07432	−0.04534	−0.00284	0.03356	0.07526	0.14404	−0.22
0.24	−0.05343	−0.02439	0.01813	0.05450	0.09609	0.16452	−0.24
0.26	−0.03228	−0.00320	0.03931	0.07561	0.11705	0.18507	−0.26
0.28	−0.01086	0.01823	0.06069	0.09688	0.13814	0.20570	−0.28
0.30	0.01084	0.03991	0.08228	0.11833	0.15936	0.22638	−0.30
0.32	0.03281	0.06184	0.10407	0.13995	0.18071	0.24714	−0.32
0.34	0.05507	0.08402	0.12608	0.16174	0.20220	0.26797	−0.34
0.36	0.07761	0.10646	0.14830	0.18372	0.22382	0.28887	−0.36
0.38	0.10046	0.12917	0.17075	0.20588	0.24559	0.30985	−0.38
0.40	0.12360	0.15215	0.19341	0.22822	0.26749	0.33089	−0.40
0.42	0.14706	0.17540	0.21631	0.25075	0.28954	0.35201	−0.42
0.44	0.17083	0.19894	0.23943	0.27347	0.31173	0.37320	−0.44
0.46	0.19492	0.22276	0.26279	0.29638	0.33407	0.39447	−0.46
0.48	0.21935	0.24687	0.28639	0.31948	0.35656	0.41581	−0.48
0.50	0.24411	0.27128	0.31024	0.34279	0.37919	0.43723	−0.50
0.52	0.26922	0.29600	0.33433	0.36630	0.40198	0.45873	−0.52
0.54	0.29468	0.32103	0.35867	0.39001	0.42492	0.48030	−0.54
0.56	0.32050	0.34638	0.38327	0.41393	0.44802	0.50196	−0.56
0.58	0.34669	0.37204	0.40813	0.43805	0.47127	0.52369	−0.58
0.60	0.37325	0.39804	0.43326	0.46240	0.49468	0.54550	−0.60
0.62	0.40021	0.42438	0.45865	0.48696	0.51826	0.56740	−0.62
0.64	0.42756	0.45106	0.48433	0.51174	0.54200	0.58937	−0.64
0.66	0.45531	0.47810	0.51028	0.53675	0.56590	0.61143	−0.66
0.68	0.48348	0.50549	0.53652	0.56198	0.58997	0.63358	−0.68
0.70	0.51208	0.53325	0.56305	0.58745	0.61422	0.65580	−0.70
0.72	0.54111	0.56139	0.58987	0.61315	0.63863	0.67812	−0.72
0.74	0.57058	0.58991	0.61700	0.63909	0.66322	0.70052	−0.74
0.76	0.60051	0.61883	0.64444	0.66527	0.68799	0.72300	−0.76
0.78	0.63092	0.64815	0.67218	0.69170	0.71294	0.74558	−0.78
0.80	0.66180	0.67787	0.70025	0.71839	0.73807	0.76824	−0.80
0.82	0.69317	0.70802	0.72865	0.74532	0.76339	0.79100	−0.82
0.84	0.72505	0.73860	0.75738	0.77252	0.78889	0.81384	−0.84
0.86	0.75745	0.76962	0.78644	0.79998	0.81459	0.83678	−0.86
0.88	0.79039	0.80109	0.81586	0.82771	0.84047	0.85981	−0.88
0.90	0.82387	0.83302	0.84562	0.85572	0.86656	0.88293	−0.90
0.92	0.85791	0.86542	0.87575	0.88400	0.89284	0.90615	−0.92
0.94	0.89252	0.89831	0.90624	0.91256	0.91932	0.92947	−0.94
0.96	0.92773	0.93169	0.93711	0.94141	0.94600	0.95288	−0.96
0.98	0.96355	0.96559	0.96836	0.97056	0.97290	0.97639	−0.98

↑ r_o

α =	.995 ↑−	.99 ↑−	.975 ↑−	.95 ↑−	.90 ↑−	.75 ↑−

TABLE 68.1   SAMPLE SIZE =   80

α =	.005 ↓+	.01 ↓+	.025 ↓+	.05 ↓+	.10 ↓+	.25 ↓+	
r_0 ↓							
-0.98	-0.98869	-0.98802	-0.98698	-0.98602	-0.98482	-0.98259	0.98
-0.96	-0.97729	-0.97596	-0.97387	-0.97195	-0.96957	-0.96515	0.96
-0.94	-0.96578	-0.96379	-0.96068	-0.95781	-0.95425	-0.94767	0.94
-0.92	-0.95418	-0.95154	-0.94740	-0.94358	-0.93886	-0.93015	0.92
-0.90	-0.94249	-0.93919	-0.93403	-0.92928	-0.92341	-0.91259	0.90
-0.88	-0.93069	-0.92674	-0.92056	-0.91489	-0.90788	-0.89499	0.88
-0.86	-0.91879	-0.91419	-0.90701	-0.90041	-0.89229	-0.87736	0.86
-0.84	-0.90678	-0.90154	-0.89336	-0.88586	-0.87662	-0.85968	0.84
-0.82	-0.89467	-0.88879	-0.87962	-0.87121	-0.86088	-0.84196	0.82
-0.80	-0.88246	-0.87593	-0.86579	-0.85649	-0.84507	-0.82421	0.80
-0.78	-0.87013	-0.86298	-0.85185	-0.84167	-0.82919	-0.80641	0.78
-0.76	-0.85770	-0.84991	-0.83782	-0.82677	-0.81323	-0.78857	0.76
-0.74	-0.84515	-0.83674	-0.82369	-0.81177	-0.79719	-0.77069	0.74
-0.72	-0.83250	-0.82346	-0.80946	-0.79669	-0.78109	-0.75277	0.72
-0.70	-0.81972	-0.81008	-0.79513	-0.78152	-0.76490	-0.73481	0.70
-0.68	-0.80684	-0.79658	-0.78070	-0.76625	-0.74864	-0.71680	0.68
-0.66	-0.79383	-0.78296	-0.76617	-0.75089	-0.73230	-0.69875	0.66
-0.64	-0.78071	-0.76923	-0.75152	-0.73544	-0.71588	-0.68066	0.64
-0.62	-0.76746	-0.75539	-0.73678	-0.71989	-0.69939	-0.66252	0.62
-0.60	-0.75409	-0.74143	-0.72192	-0.70425	-0.68281	-0.64434	0.60
-0.58	-0.74059	-0.72735	-0.70696	-0.68851	-0.66615	-0.62611	0.58
-0.56	-0.72697	-0.71314	-0.69189	-0.67267	-0.64941	-0.60784	0.56
-0.54	-0.71322	-0.69882	-0.67670	-0.65673	-0.63259	-0.58952	0.54
-0.52	-0.69934	-0.68437	-0.66140	-0.64069	-0.61568	-0.57116	0.52
-0.50	-0.68532	-0.66979	-0.64599	-0.62455	-0.59869	-0.55275	0.50
-0.48	-0.67117	-0.65509	-0.63046	-0.60830	-0.58162	-0.53429	0.48
-0.46	-0.65688	-0.64025	-0.61481	-0.59195	-0.56446	-0.51579	0.46
-0.44	-0.64246	-0.62528	-0.59904	-0.57550	-0.54721	-0.49723	0.44
-0.42	-0.62789	-0.61018	-0.58316	-0.55893	-0.52988	-0.47863	0.42
-0.40	-0.61317	-0.59494	-0.56715	-0.54227	-0.51246	-0.45998	0.40
-0.38	-0.59832	-0.57957	-0.55101	-0.52549	-0.49494	-0.44129	0.38
-0.36	-0.58331	-0.56405	-0.53476	-0.50860	-0.47734	-0.42254	0.36
-0.34	-0.56815	-0.54839	-0.51837	-0.49160	-0.45965	-0.40374	0.34
-0.32	-0.55284	-0.53259	-0.50185	-0.47448	-0.44186	-0.38489	0.32
-0.30	-0.53737	-0.51664	-0.48521	-0.45725	-0.42398	-0.36599	0.30
-0.28	-0.52174	-0.50054	-0.46843	-0.43991	-0.40601	-0.34704	0.28
-0.26	-0.50595	-0.48429	-0.45152	-0.42244	-0.38794	-0.32804	0.26
-0.24	-0.49000	-0.46788	-0.43447	-0.40486	-0.36977	-0.30898	0.24
-0.22	-0.47388	-0.45132	-0.41728	-0.38716	-0.35151	-0.28987	0.22
-0.20	-0.45759	-0.43460	-0.39995	-0.36934	-0.33315	-0.27071	0.20
-0.18	-0.44113	-0.41772	-0.38248	-0.35139	-0.31469	-0.25149	0.18
-0.16	-0.42449	-0.40068	-0.36487	-0.33332	-0.29613	-0.23222	0.16
-0.14	-0.40767	-0.38347	-0.34711	-0.31512	-0.27747	-0.21289	0.14
-0.12	-0.39067	-0.36609	-0.32921	-0.29680	-0.25870	-0.19351	0.12
-0.10	-0.37349	-0.34853	-0.31115	-0.27834	-0.23983	-0.17407	0.10
-0.08	-0.35612	-0.33081	-0.29294	-0.25975	-0.22086	-0.15458	0.08
-0.06	-0.33855	-0.31290	-0.27457	-0.24103	-0.20178	-0.13502	0.06
-0.04	-0.32080	-0.29482	-0.25605	-0.22218	-0.18259	-0.11541	0.04
-0.02	-0.30284	-0.27655	-0.23737	-0.20319	-0.16329	-0.09574	0.02

↑ r_0

α =	.995 ↑−	.99 ↑−	.975 ↑−	.95 ↑−	.90 ↑−	.75 ↑−

TABLE 68.2   SAMPLE SIZE =   80   (CONT.)

α =	.005 ↓+	.01 ↓+	.025 ↓+	.05 ↓+	.10 ↓+	.25 ↓+	
r_o ↓							
0.00	-0.28468	-0.25810	-0.21853	-0.18406	-0.14389	-0.07602	0.00
0.02	-0.26631	-0.23945	-0.19952	-0.16479	-0.12437	-0.05623	-0.02
0.04	-0.24774	-0.22062	-0.18035	-0.14538	-0.10474	-0.03638	-0.04
0.06	-0.22895	-0.20159	-0.16101	-0.12582	-0.08500	-0.01647	-0.06
0.08	-0.20994	-0.18235	-0.14150	-0.10612	-0.06514	0.00350	-0.08
0.10	-0.19072	-0.16292	-0.12182	-0.08627	-0.04517	0.02353	-0.10
0.12	-0.17126	-0.14328	-0.10195	-0.06628	-0.02508	0.04362	-0.12
0.14	-0.15158	-0.12342	-0.08191	-0.04613	-0.00487	0.06378	-0.14
0.16	-0.13166	-0.10336	-0.06169	-0.02582	0.01546	0.08400	-0.16
0.18	-0.11150	-0.08307	-0.04128	-0.00537	0.03591	0.10428	-0.18
0.20	-0.09110	-0.06257	-0.02068	0.01525	0.05648	0.12463	-0.20
0.22	-0.07045	-0.04184	0.00010	0.03603	0.07718	0.14504	-0.22
0.24	-0.04954	-0.02088	0.02108	0.05696	0.09800	0.16553	-0.24
0.26	-0.02838	0.00032	0.04226	0.07807	0.11895	0.18607	-0.26
0.28	-0.00695	0.02175	0.06364	0.09934	0.14003	0.20669	-0.28
0.30	0.01475	0.04343	0.08522	0.12078	0.16125	0.22737	-0.30
0.32	0.03672	0.06535	0.10700	0.14239	0.18259	0.24812	-0.32
0.34	0.05897	0.08752	0.12900	0.16417	0.20406	0.26895	-0.34
0.36	0.08151	0.10995	0.15121	0.18613	0.22568	0.28984	-0.36
0.38	0.10434	0.13265	0.17363	0.20827	0.24742	0.31080	-0.38
0.40	0.12747	0.15561	0.19628	0.23059	0.26931	0.33183	-0.40
0.42	0.15090	0.17884	0.21915	0.25310	0.29134	0.35294	-0.42
0.44	0.17465	0.20234	0.24225	0.27579	0.31351	0.37412	-0.44
0.46	0.19871	0.22613	0.26558	0.29867	0.33582	0.39537	-0.46
0.48	0.22310	0.25021	0.28914	0.32174	0.35828	0.41670	-0.48
0.50	0.24782	0.27458	0.31295	0.34501	0.38088	0.43810	-0.50
0.52	0.27288	0.29925	0.33699	0.36848	0.40364	0.45958	-0.52
0.54	0.29829	0.32423	0.36129	0.39215	0.42654	0.48114	-0.54
0.56	0.32405	0.34952	0.38584	0.41602	0.44960	0.50277	-0.56
0.58	0.35017	0.37512	0.41064	0.44010	0.47282	0.52448	-0.58
0.60	0.37667	0.40106	0.43571	0.46439	0.49619	0.54627	-0.60
0.62	0.40354	0.42732	0.46104	0.48890	0.51972	0.56814	-0.62
0.64	0.43081	0.45392	0.48664	0.51362	0.54341	0.59009	-0.64
0.66	0.45847	0.48087	0.51252	0.53856	0.56726	0.61213	-0.66
0.68	0.48653	0.50817	0.53868	0.56373	0.59128	0.63424	-0.68
0.70	0.51502	0.53583	0.56512	0.58912	0.61547	0.65644	-0.70
0.72	0.54393	0.56386	0.59185	0.61475	0.63983	0.67872	-0.72
0.74	0.57328	0.59227	0.61889	0.64061	0.66436	0.70109	-0.74
0.76	0.60307	0.62106	0.64622	0.66671	0.68906	0.72354	-0.76
0.78	0.63333	0.65024	0.67386	0.69305	0.71394	0.74608	-0.78
0.80	0.66405	0.67983	0.70181	0.71963	0.73900	0.76871	-0.80
0.82	0.69526	0.70983	0.73009	0.74647	0.76424	0.79142	-0.82
0.84	0.72696	0.74025	0.75869	0.77356	0.78966	0.81423	-0.84
0.86	0.75917	0.77110	0.78762	0.80092	0.81527	0.83712	-0.86
0.88	0.79190	0.80239	0.81689	0.82853	0.84108	0.86011	-0.88
0.90	0.82516	0.83413	0.84650	0.85641	0.86707	0.88319	-0.90
0.92	0.85897	0.86634	0.87647	0.88457	0.89325	0.90636	-0.92
0.94	0.89335	0.89902	0.90679	0.91300	0.91964	0.92963	-0.94
0.96	0.92830	0.93218	0.93749	0.94171	0.94622	0.95299	-0.96
0.98	0.96384	0.96583	0.96855	0.97071	0.97301	0.97644	-0.98

↑ r_o

α =	.995 ↑-	.99 ↑-	.975 ↑-	.95 ↑-	.90 ↑-	.75 ↑-

TABLE 69.1   SAMPLE SIZE =   85

α =	.005 ↓+	.01 ↓+	.025 ↓+	.05 ↓+	.10 ↓+	.25 ↓+	
r_0 ↓							
-0.98	-0.98849	-0.98784	-0.98681	-0.98587	-0.98469	-0.98253	0.98
-0.96	-0.97689	-0.97558	-0.97354	-0.97165	-0.96932	-0.96501	0.96
-0.94	-0.96519	-0.96324	-0.96018	-0.95736	-0.95388	-0.94747	0.94
-0.92	-0.95340	-0.95079	-0.94673	-0.94299	-0.93838	-0.92988	0.92
-0.90	-0.94150	-0.93826	-0.93319	-0.92854	-0.92280	-0.91226	0.90
-0.88	-0.92951	-0.92562	-0.91957	-0.91400	-0.90716	-0.89459	0.88
-0.86	-0.91741	-0.91289	-0.90585	-0.89939	-0.89145	-0.87689	0.86
-0.84	-0.90521	-0.90006	-0.89204	-0.88469	-0.87566	-0.85915	0.84
-0.82	-0.89291	-0.88713	-0.87814	-0.86991	-0.85981	-0.84138	0.82
-0.80	-0.88050	-0.87409	-0.86414	-0.85504	-0.84388	-0.82356	0.80
-0.78	-0.86799	-0.86096	-0.85005	-0.84008	-0.82789	-0.80570	0.78
-0.76	-0.85536	-0.84772	-0.83586	-0.82504	-0.81182	-0.78780	0.76
-0.74	-0.84263	-0.83437	-0.82158	-0.80991	-0.79568	-0.76986	0.74
-0.72	-0.82978	-0.82091	-0.80719	-0.79470	-0.77946	-0.75188	0.72
-0.70	-0.81682	-0.80735	-0.79271	-0.77939	-0.76317	-0.73386	0.70
-0.68	-0.80375	-0.79368	-0.77812	-0.76399	-0.74680	-0.71579	0.68
-0.66	-0.79056	-0.77989	-0.76344	-0.74850	-0.73035	-0.69769	0.66
-0.64	-0.77725	-0.76599	-0.74865	-0.73292	-0.71383	-0.67954	0.64
-0.62	-0.76381	-0.75198	-0.73375	-0.71725	-0.69724	-0.66135	0.62
-0.60	-0.75026	-0.73785	-0.71875	-0.70148	-0.68056	-0.64311	0.60
-0.58	-0.73659	-0.72360	-0.70364	-0.68561	-0.66380	-0.62483	0.58
-0.56	-0.72278	-0.70923	-0.68843	-0.66965	-0.64697	-0.60651	0.56
-0.54	-0.70885	-0.69474	-0.67310	-0.65359	-0.63005	-0.58814	0.54
-0.52	-0.69479	-0.68013	-0.65766	-0.63743	-0.61305	-0.56973	0.52
-0.50	-0.68060	-0.66539	-0.64211	-0.62117	-0.59597	-0.55127	0.50
-0.48	-0.66628	-0.65053	-0.62644	-0.60481	-0.57881	-0.53276	0.48
-0.46	-0.65182	-0.63554	-0.61066	-0.58835	-0.56156	-0.51421	0.46
-0.44	-0.63722	-0.62041	-0.59477	-0.57179	-0.54422	-0.49561	0.44
-0.42	-0.62248	-0.60516	-0.57875	-0.55512	-0.52681	-0.47697	0.42
-0.40	-0.60760	-0.58977	-0.56261	-0.53834	-0.50930	-0.45827	0.40
-0.38	-0.59258	-0.57424	-0.54635	-0.52146	-0.49171	-0.43953	0.38
-0.36	-0.57741	-0.55858	-0.52997	-0.50446	-0.47403	-0.42074	0.36
-0.34	-0.56209	-0.54278	-0.51347	-0.48736	-0.45625	-0.40191	0.34
-0.32	-0.54662	-0.52683	-0.49683	-0.47015	-0.43839	-0.38302	0.32
-0.30	-0.53100	-0.51074	-0.48007	-0.45283	-0.42044	-0.36408	0.30
-0.28	-0.51522	-0.49451	-0.46318	-0.43539	-0.40240	-0.34509	0.28
-0.26	-0.49928	-0.47812	-0.44616	-0.41784	-0.38426	-0.32605	0.26
-0.24	-0.48318	-0.46159	-0.42900	-0.40017	-0.36603	-0.30696	0.24
-0.22	-0.46692	-0.44490	-0.41171	-0.38238	-0.34770	-0.28782	0.22
-0.20	-0.45049	-0.42806	-0.39428	-0.36448	-0.32928	-0.26863	0.20
-0.18	-0.43389	-0.41105	-0.37672	-0.34645	-0.31076	-0.24938	0.18
-0.16	-0.41712	-0.39389	-0.35901	-0.32830	-0.29215	-0.23008	0.16
-0.14	-0.40017	-0.37657	-0.34116	-0.31003	-0.27343	-0.21073	0.14
-0.12	-0.38304	-0.35908	-0.32316	-0.29164	-0.25461	-0.19132	0.12
-0.10	-0.36574	-0.34142	-0.30502	-0.27312	-0.23570	-0.17186	0.10
-0.08	-0.34824	-0.32359	-0.28673	-0.25447	-0.21668	-0.15234	0.08
-0.06	-0.33057	-0.30559	-0.26829	-0.23569	-0.19756	-0.13277	0.06
-0.04	-0.31270	-0.28741	-0.24970	-0.21678	-0.17833	-0.11314	0.04
-0.02	-0.29463	-0.26905	-0.23095	-0.19773	-0.15900	-0.09345	0.02

↑ r_0

α =	.995 ↑-	.99 ↑-	.975 ↑-	.95 ↑-	.90 ↑-	.75 ↑-

TABLE 69.2   SAMPLE SIZE =   85   (CONT.)

α =	.005 ↓+	.01 ↓+	.025 ↓+	.05 ↓+	.10 ↓+	.25 ↓+	
r_0 ↓							
0.00	-0.27637	-0.25051	-0.21204	-0.17855	-0.13956	-0.07371	0.00
0.02	-0.25791	-0.23178	-0.19298	-0.15924	-0.12001	-0.05391	-0.02
0.04	-0.23924	-0.21287	-0.17375	-0.13979	-0.10036	-0.03405	-0.04
0.06	-0.22036	-0.19376	-0.15435	-0.12019	-0.08059	-0.01413	-0.06
0.08	-0.20128	-0.17447	-0.13480	-0.10046	-0.06071	0.00584	-0.08
0.10	-0.18197	-0.15497	-0.11507	-0.08058	-0.04072	0.02588	-0.10
0.12	-0.16245	-0.13527	-0.09517	-0.06056	-0.02062	0.04598	-0.12
0.14	-0.14270	-0.11537	-0.07509	-0.04039	-0.00040	0.06614	-0.14
0.16	-0.12272	-0.09526	-0.05484	-0.02007	0.01994	0.08636	-0.16
0.18	-0.10251	-0.07494	-0.03441	0.00040	0.04039	0.10664	-0.18
0.20	-0.08207	-0.05440	-0.01380	0.02102	0.06096	0.12699	-0.20
0.22	-0.06138	-0.03365	0.00700	0.04180	0.08166	0.14740	-0.22
0.24	-0.04044	-0.01267	0.02798	0.06273	0.10247	0.16787	-0.24
0.26	-0.01926	0.00854	0.04915	0.08382	0.12341	0.18841	-0.26
0.28	0.00218	0.02997	0.07052	0.10508	0.14447	0.20901	-0.28
0.30	0.02389	0.05165	0.09209	0.12650	0.16566	0.22968	-0.30
0.32	0.04585	0.07356	0.11385	0.14808	0.18698	0.25042	-0.32
0.34	0.06809	0.09571	0.13582	0.16984	0.20843	0.27122	-0.34
0.36	0.09061	0.11811	0.15799	0.19176	0.23000	0.29209	-0.36
0.38	0.11341	0.14076	0.18038	0.21385	0.25171	0.31303	-0.38
0.40	0.13650	0.16368	0.20297	0.23613	0.27355	0.33403	-0.40
0.42	0.15988	0.18685	0.22578	0.25857	0.29553	0.35511	-0.42
0.44	0.18356	0.21029	0.24881	0.28120	0.31765	0.37626	-0.44
0.46	0.20755	0.23400	0.27207	0.30402	0.33990	0.39748	-0.46
0.48	0.23185	0.25799	0.29555	0.32701	0.36229	0.41877	-0.48
0.50	0.25647	0.28226	0.31926	0.35020	0.38483	0.44013	-0.50
0.52	0.28141	0.30682	0.34320	0.37357	0.40751	0.46157	-0.52
0.54	0.30669	0.33168	0.36739	0.39714	0.43033	0.48308	-0.54
0.56	0.33231	0.35683	0.39181	0.42091	0.45330	0.50467	-0.56
0.58	0.35828	0.38229	0.41649	0.44487	0.47643	0.52633	-0.58
0.60	0.38461	0.40806	0.44141	0.46904	0.49970	0.54806	-0.60
0.62	0.41129	0.43415	0.46659	0.49341	0.52312	0.56988	-0.62
0.64	0.43835	0.46056	0.49202	0.51800	0.54670	0.59177	-0.64
0.66	0.46579	0.48730	0.51773	0.54279	0.57044	0.61374	-0.66
0.68	0.49362	0.51438	0.54370	0.56780	0.59433	0.63579	-0.68
0.70	0.52184	0.54181	0.56994	0.59302	0.61839	0.65791	-0.70
0.72	0.55048	0.56959	0.59646	0.61847	0.64261	0.68012	-0.72
0.74	0.57953	0.59772	0.62326	0.64414	0.66699	0.70241	-0.74
0.76	0.60900	0.62623	0.65036	0.67004	0.69154	0.72479	-0.76
0.78	0.63891	0.65510	0.67774	0.69617	0.71626	0.74724	-0.78
0.80	0.66927	0.68437	0.70543	0.72253	0.74115	0.76978	-0.80
0.82	0.70009	0.71402	0.73342	0.74914	0.76621	0.79241	-0.82
0.84	0.73137	0.74407	0.76172	0.77599	0.79145	0.81512	-0.84
0.86	0.76313	0.77453	0.79033	0.80308	0.81687	0.83792	-0.86
0.88	0.79539	0.80541	0.81927	0.83043	0.84247	0.86080	-0.88
0.90	0.82815	0.83671	0.84853	0.85803	0.86825	0.88378	-0.90
0.92	0.86143	0.86846	0.87813	0.88589	0.89422	0.90684	-0.92
0.94	0.89524	0.90065	0.90807	0.91401	0.92038	0.92999	-0.94
0.96	0.92960	0.93329	0.93836	0.94240	0.94672	0.95324	-0.96
0.98	0.96451	0.96641	0.96900	0.97106	0.97326	0.97657	-0.98

↑ r_0

α =	.995 ↑-	.99 ↑-	.975 ↑-	.95 ↑-	.90 ↑-	.75 ↑-

TABLE 70.1   SAMPLE SIZE =   90

α =	.005 ↓+	.01 ↓+	.025 ↓+	.05 ↓+	.10 ↓+	.25 ↓+	
r_0 ↓							
-0.98	-0.98831	-0.98766	-0.98666	-0.98573	-0.98458	-0.98246	0.98
-0.96	-0.97652	-0.97524	-0.97323	-0.97138	-0.96910	-0.96489	0.96
-0.94	-0.96464	-0.96272	-0.95971	-0.95695	-0.95354	-0.94728	0.94
-0.92	-0.95267	-0.95010	-0.94611	-0.94244	-0.93793	-0.92963	0.92
-0.90	-0.94059	-0.93740	-0.93242	-0.92786	-0.92224	-0.91195	0.90
-0.88	-0.92842	-0.92459	-0.91864	-0.91319	-0.90649	-0.89423	0.88
-0.86	-0.91614	-0.91169	-0.90477	-0.89844	-0.89067	-0.87647	0.86
-0.84	-0.90376	-0.89869	-0.89082	-0.88361	-0.87478	-0.85867	0.84
-0.82	-0.89128	-0.88559	-0.87676	-0.86870	-0.85882	-0.84083	0.82
-0.80	-0.87869	-0.87239	-0.86262	-0.85370	-0.84279	-0.82296	0.80
-0.78	-0.86600	-0.85909	-0.84838	-0.83862	-0.82669	-0.80504	0.78
-0.76	-0.85320	-0.84568	-0.83405	-0.82345	-0.81052	-0.78709	0.76
-0.74	-0.84029	-0.83217	-0.81962	-0.80820	-0.79428	-0.76910	0.74
-0.72	-0.82727	-0.81855	-0.80509	-0.79286	-0.77796	-0.75106	0.72
-0.70	-0.81413	-0.80483	-0.79047	-0.77743	-0.76157	-0.73299	0.70
-0.68	-0.80089	-0.79100	-0.77574	-0.76191	-0.74510	-0.71487	0.68
-0.66	-0.78752	-0.77705	-0.76092	-0.74630	-0.72856	-0.69671	0.66
-0.64	-0.77404	-0.76299	-0.74599	-0.73060	-0.71195	-0.67851	0.64
-0.62	-0.76044	-0.74882	-0.73096	-0.71481	-0.69526	-0.66027	0.62
-0.60	-0.74672	-0.73454	-0.71582	-0.69893	-0.67849	-0.64198	0.60
-0.58	-0.73288	-0.72014	-0.70058	-0.68295	-0.66164	-0.62366	0.58
-0.56	-0.71891	-0.70562	-0.68524	-0.66687	-0.64472	-0.60528	0.56
-0.54	-0.70482	-0.69098	-0.66978	-0.65070	-0.62771	-0.58687	0.54
-0.52	-0.69059	-0.67621	-0.65421	-0.63443	-0.61063	-0.56841	0.52
-0.50	-0.67624	-0.66133	-0.63854	-0.61807	-0.59346	-0.54991	0.50
-0.48	-0.66176	-0.64632	-0.62275	-0.60160	-0.57622	-0.53136	0.48
-0.46	-0.64714	-0.63118	-0.60684	-0.58504	-0.55889	-0.51276	0.46
-0.44	-0.63239	-0.61592	-0.59083	-0.56837	-0.54148	-0.49412	0.44
-0.42	-0.61750	-0.60053	-0.57469	-0.55160	-0.52398	-0.47544	0.42
-0.40	-0.60247	-0.58500	-0.55844	-0.53473	-0.50640	-0.45671	0.40
-0.38	-0.58729	-0.56934	-0.54207	-0.51775	-0.48873	-0.43793	0.38
-0.36	-0.57197	-0.55354	-0.52557	-0.50067	-0.47098	-0.41910	0.36
-0.34	-0.55651	-0.53761	-0.50896	-0.48347	-0.45314	-0.40022	0.34
-0.32	-0.54090	-0.52153	-0.49222	-0.46617	-0.43521	-0.38130	0.32
-0.30	-0.52513	-0.50532	-0.47535	-0.44876	-0.41719	-0.36233	0.30
-0.28	-0.50921	-0.48896	-0.45836	-0.43124	-0.39909	-0.34331	0.28
-0.26	-0.49314	-0.47245	-0.44124	-0.41361	-0.38089	-0.32424	0.26
-0.24	-0.47690	-0.45580	-0.42398	-0.39586	-0.36260	-0.30512	0.24
-0.22	-0.46051	-0.43900	-0.40660	-0.37800	-0.34421	-0.28595	0.22
-0.20	-0.44395	-0.42204	-0.38908	-0.36002	-0.32574	-0.26673	0.20
-0.18	-0.42723	-0.40493	-0.37142	-0.34192	-0.30716	-0.24745	0.18
-0.16	-0.41033	-0.38766	-0.35363	-0.32371	-0.28850	-0.22813	0.16
-0.14	-0.39327	-0.37023	-0.33570	-0.30537	-0.26973	-0.20875	0.14
-0.12	-0.37603	-0.35264	-0.31762	-0.28691	-0.25087	-0.18932	0.12
-0.10	-0.35861	-0.33488	-0.29941	-0.26833	-0.23191	-0.16984	0.10
-0.08	-0.34101	-0.31696	-0.28104	-0.24962	-0.21285	-0.15030	0.08
-0.06	-0.32323	-0.29887	-0.26253	-0.23079	-0.19369	-0.13071	0.06
-0.04	-0.30526	-0.28061	-0.24387	-0.21183	-0.17443	-0.11106	0.04
-0.02	-0.28710	-0.26217	-0.22506	-0.19274	-0.15506	-0.09136	0.02

↑ r_0

α =	.995 ↑-	.99 ↑-	.975 ↑-	.95 ↑-	.90 ↑-	.75 ↑-

TABLE 70.2   SAMPLE SIZE =   90   (CONT.)

α =	.005 ↓+	.01 ↓+	.025 ↓+	.05 ↓+	.10 ↓+	.25 ↓+	
r_0 ↓							
0.00	−0.26875	−0.24355	−0.20610	−0.17352	−0.13560	−0.07160	0.00
0.02	−0.25020	−0.22475	−0.18698	−0.15416	−0.11602	−0.05179	−0.02
0.04	−0.23145	−0.20577	−0.16770	−0.13467	−0.09634	−0.03192	−0.04
0.06	−0.21250	−0.18660	−0.14826	−0.11505	−0.07656	−0.01200	−0.06
0.08	−0.19334	−0.16724	−0.12866	−0.09528	−0.05666	0.00799	−0.08
0.10	−0.17397	−0.14769	−0.10890	−0.07538	−0.03666	0.02803	−0.10
0.12	−0.15438	−0.12795	−0.08896	−0.05534	−0.01654	0.04813	−0.12
0.14	−0.13458	−0.10800	−0.06886	−0.03515	0.00369	0.06829	−0.14
0.16	−0.11455	−0.08786	−0.04859	−0.01482	0.02403	0.08852	−0.16
0.18	−0.09430	−0.06750	−0.02814	0.00566	0.04448	0.10880	−0.18
0.20	−0.07381	−0.04694	−0.00752	0.02628	0.06506	0.12914	−0.20
0.22	−0.05309	−0.02616	0.01329	0.04706	0.08574	0.14954	−0.22
0.24	−0.03213	−0.00517	0.03427	0.06799	0.10655	0.17001	−0.24
0.26	−0.01093	0.01604	0.05544	0.08908	0.12747	0.19054	−0.26
0.28	0.01052	0.03748	0.07680	0.11032	0.14852	0.21113	−0.28
0.30	0.03222	0.05914	0.09835	0.13171	0.16969	0.23178	−0.30
0.32	0.05418	0.08104	0.12010	0.15327	0.19098	0.25250	−0.32
0.34	0.07641	0.10317	0.14204	0.17500	0.21240	0.27329	−0.34
0.36	0.09890	0.12554	0.16417	0.19689	0.23394	0.29414	−0.36
0.38	0.12167	0.14816	0.18652	0.21894	0.25562	0.31506	−0.38
0.40	0.14471	0.17102	0.20906	0.24116	0.27742	0.33604	−0.40
0.42	0.16804	0.19414	0.23182	0.26356	0.29935	0.35709	−0.42
0.44	0.19166	0.21752	0.25479	0.28613	0.32141	0.37821	−0.44
0.46	0.21558	0.24116	0.27797	0.30888	0.34361	0.39940	−0.46
0.48	0.23980	0.26506	0.30137	0.33180	0.36595	0.42065	−0.48
0.50	0.26432	0.28924	0.32499	0.35491	0.38842	0.44198	−0.50
0.52	0.28916	0.31370	0.34885	0.37820	0.41103	0.46338	−0.52
0.54	0.31432	0.33844	0.37293	0.40168	0.43378	0.48485	−0.54
0.56	0.33981	0.36347	0.39724	0.42535	0.45667	0.50639	−0.56
0.58	0.36563	0.38879	0.42179	0.44921	0.47970	0.52800	−0.58
0.60	0.39180	0.41441	0.44658	0.47326	0.50288	0.54969	−0.60
0.62	0.41831	0.44034	0.47162	0.49751	0.52621	0.57145	−0.62
0.64	0.44518	0.46657	0.49690	0.52196	0.54969	0.59329	−0.64
0.66	0.47242	0.49313	0.52244	0.54662	0.57332	0.61520	−0.66
0.68	0.50002	0.52001	0.54824	0.57148	0.59710	0.63719	−0.68
0.70	0.52801	0.54722	0.57430	0.59655	0.62104	0.65926	−0.70
0.72	0.55639	0.57476	0.60063	0.62183	0.64513	0.68140	−0.72
0.74	0.58516	0.60265	0.62722	0.64733	0.66938	0.70362	−0.74
0.76	0.61435	0.63089	0.65410	0.67305	0.69379	0.72592	−0.76
0.78	0.64394	0.65949	0.68125	0.69899	0.71836	0.74830	−0.78
0.80	0.67397	0.68845	0.70869	0.72515	0.74310	0.77076	−0.80
0.82	0.70443	0.71779	0.73642	0.75155	0.76800	0.79331	−0.82
0.84	0.73534	0.74751	0.76445	0.77817	0.79308	0.81593	−0.84
0.86	0.76670	0.77762	0.79278	0.80504	0.81832	0.83864	−0.86
0.88	0.79853	0.80812	0.82142	0.83214	0.84374	0.86143	−0.88
0.90	0.83084	0.83903	0.85036	0.85948	0.86933	0.88431	−0.90
0.92	0.86364	0.87036	0.87963	0.88708	0.89510	0.90727	−0.92
0.94	0.89695	0.90211	0.90922	0.91492	0.92105	0.93032	−0.94
0.96	0.93076	0.93429	0.93914	0.94302	0.94718	0.95346	−0.96
0.98	0.96511	0.96692	0.96940	0.97138	0.97350	0.97669	−0.98

↑ r_0

α =	.995 ↑−	.99 ↑−	.975 ↑−	.95 ↑−	.90 ↑−	.75 ↑−

TABLE 71.1   SAMPLE SIZE =   95

α =	.005 ↓+	.01 ↓+	.025 ↓+	.05 ↓+	.10 ↓+	.25 ↓+	
r_o ↓							
-0.98	-0.98814	-0.98750	-0.98651	-0.98560	-0.98447	-0.98241	0.98
-0.96	-0.97618	-0.97491	-0.97294	-0.97112	-0.96889	-0.96478	0.96
-0.94	-0.96413	-0.96223	-0.95928	-0.95657	-0.95323	-0.94711	0.94
-0.92	-0.95199	-0.94946	-0.94554	-0.94194	-0.93751	-0.92941	0.92
-0.90	-0.93974	-0.93660	-0.93170	-0.92722	-0.92172	-0.91167	0.90
-0.88	-0.92740	-0.92363	-0.91778	-0.91243	-0.90587	-0.89389	0.88
-0.86	-0.91495	-0.91057	-0.90378	-0.89756	-0.88995	-0.87608	0.86
-0.84	-0.90241	-0.89742	-0.88968	-0.88261	-0.87396	-0.85822	0.84
-0.82	-0.88976	-0.88416	-0.87549	-0.86758	-0.85791	-0.84033	0.82
-0.80	-0.87701	-0.87081	-0.86121	-0.85246	-0.84178	-0.82241	0.80
-0.78	-0.86415	-0.85735	-0.84684	-0.83726	-0.82558	-0.80444	0.78
-0.76	-0.85119	-0.84379	-0.83237	-0.82198	-0.80932	-0.78643	0.76
-0.74	-0.83811	-0.83013	-0.81781	-0.80661	-0.79298	-0.76839	0.74
-0.72	-0.82493	-0.81636	-0.80315	-0.79115	-0.77657	-0.75030	0.72
-0.70	-0.81164	-0.80249	-0.78839	-0.77561	-0.76009	-0.73218	0.70
-0.68	-0.79823	-0.78851	-0.77354	-0.75998	-0.74354	-0.71401	0.68
-0.66	-0.78471	-0.77441	-0.75858	-0.74426	-0.72691	-0.69581	0.66
-0.64	-0.77107	-0.76021	-0.74353	-0.72845	-0.71021	-0.67756	0.64
-0.62	-0.75731	-0.74590	-0.72837	-0.71255	-0.69343	-0.65927	0.62
-0.60	-0.74343	-0.73147	-0.71311	-0.69656	-0.67658	-0.64094	0.60
-0.58	-0.72944	-0.71692	-0.69775	-0.68048	-0.65965	-0.62257	0.58
-0.56	-0.71531	-0.70226	-0.68228	-0.66430	-0.64264	-0.60416	0.56
-0.54	-0.70107	-0.68748	-0.66671	-0.64803	-0.62556	-0.58570	0.54
-0.52	-0.68670	-0.67259	-0.65102	-0.63166	-0.60839	-0.56720	0.52
-0.50	-0.67220	-0.65756	-0.63523	-0.61520	-0.59115	-0.54865	0.50
-0.48	-0.65757	-0.64242	-0.61933	-0.59864	-0.57383	-0.53006	0.48
-0.46	-0.64280	-0.62715	-0.60331	-0.58198	-0.55643	-0.51143	0.46
-0.44	-0.62791	-0.61176	-0.58718	-0.56522	-0.53894	-0.49275	0.44
-0.42	-0.61287	-0.59624	-0.57094	-0.54836	-0.52137	-0.47403	0.42
-0.40	-0.59770	-0.58058	-0.55458	-0.53139	-0.50372	-0.45526	0.40
-0.38	-0.58239	-0.56480	-0.53810	-0.51433	-0.48599	-0.43644	0.38
-0.36	-0.56694	-0.54888	-0.52151	-0.49716	-0.46817	-0.41758	0.36
-0.34	-0.55134	-0.53282	-0.50479	-0.47988	-0.45027	-0.39867	0.34
-0.32	-0.53559	-0.51663	-0.48795	-0.46250	-0.43228	-0.37972	0.32
-0.30	-0.51970	-0.50030	-0.47099	-0.44501	-0.41420	-0.36072	0.30
-0.28	-0.50365	-0.48383	-0.45390	-0.42741	-0.39603	-0.34167	0.28
-0.26	-0.48746	-0.46721	-0.43669	-0.40971	-0.37778	-0.32257	0.26
-0.24	-0.47110	-0.45045	-0.41935	-0.39189	-0.35943	-0.30342	0.24
-0.22	-0.45459	-0.43354	-0.40188	-0.37395	-0.34100	-0.28422	0.22
-0.20	-0.43791	-0.41648	-0.38428	-0.35591	-0.32247	-0.26497	0.20
-0.18	-0.42107	-0.39927	-0.36654	-0.33775	-0.30385	-0.24568	0.18
-0.16	-0.40407	-0.38190	-0.34867	-0.31947	-0.28514	-0.22633	0.16
-0.14	-0.38690	-0.36438	-0.33066	-0.30107	-0.26633	-0.20693	0.14
-0.12	-0.36955	-0.34670	-0.31252	-0.28256	-0.24743	-0.18748	0.12
-0.10	-0.35203	-0.32886	-0.29423	-0.26392	-0.22843	-0.16798	0.10
-0.08	-0.33434	-0.31085	-0.27580	-0.24516	-0.20933	-0.14842	0.08
-0.06	-0.31646	-0.29268	-0.25723	-0.22628	-0.19014	-0.12881	0.06
-0.04	-0.29840	-0.27434	-0.23851	-0.20728	-0.17084	-0.10915	0.04
-0.02	-0.28016	-0.25583	-0.21964	-0.18814	-0.15145	-0.08944	0.02

↑ r_o

α =	.995 ↑-	.99 ↑-	.975 ↑-	.95 ↑-	.90 ↑-	.75 ↑-

TABLE 71.2   SAMPLE SIZE = 95   (CONT.)

α =	.005 ↓+	.01 ↓+	.025 ↓+	.05 ↓+	.10 ↓+	.25 ↓+	
r_o ↓							
0.00	-0.26172	-0.23714	-0.20063	-0.16888	-0.13195	-0.06967	0.00
0.02	-0.24310	-0.21828	-0.18146	-0.14949	-0.11236	-0.04985	-0.02
0.04	-0.22427	-0.19924	-0.16214	-0.12997	-0.09265	-0.02997	-0.04
0.06	-0.20525	-0.18001	-0.14266	-0.11032	-0.07285	-0.01003	-0.06
0.08	-0.18603	-0.16060	-0.12302	-0.09053	-0.05294	0.00996	-0.08
0.10	-0.16660	-0.14100	-0.10322	-0.07060	-0.03292	0.03001	-0.10
0.12	-0.14696	-0.12122	-0.08326	-0.05054	-0.01279	0.05011	-0.12
0.14	-0.12711	-0.10123	-0.06314	-0.03034	0.00744	0.07027	-0.14
0.16	-0.10704	-0.08105	-0.04285	-0.01000	0.02778	0.09049	-0.16
0.18	-0.08675	-0.06067	-0.02238	0.01049	0.04824	0.11077	-0.18
0.20	-0.06623	-0.04009	-0.00175	0.03112	0.06881	0.13111	-0.20
0.22	-0.04549	-0.01930	0.01906	0.05189	0.08949	0.15151	-0.22
0.24	-0.02451	0.00171	0.04005	0.07282	0.11029	0.17197	-0.24
0.26	-0.00329	0.02292	0.06121	0.09389	0.13120	0.19249	-0.26
0.28	0.01816	0.04436	0.08256	0.11512	0.15223	0.21307	-0.28
0.30	0.03987	0.06601	0.10410	0.13650	0.17338	0.23371	-0.30
0.32	0.06182	0.08789	0.12582	0.15803	0.19465	0.25442	-0.32
0.34	0.08403	0.11000	0.14773	0.17973	0.21604	0.27518	-0.34
0.36	0.10650	0.13234	0.16983	0.20158	0.23755	0.29602	-0.36
0.38	0.12923	0.15492	0.19214	0.22360	0.25919	0.31691	-0.38
0.40	0.15223	0.17774	0.21464	0.24578	0.28095	0.33787	-0.40
0.42	0.17551	0.20081	0.23734	0.26812	0.30284	0.35890	-0.42
0.44	0.19907	0.22413	0.26025	0.29064	0.32486	0.37999	-0.44
0.46	0.22292	0.24770	0.28336	0.31332	0.34701	0.40115	-0.46
0.48	0.24706	0.27152	0.30669	0.33618	0.36929	0.42238	-0.48
0.50	0.27149	0.29562	0.33024	0.35922	0.39170	0.44367	-0.50
0.52	0.29623	0.31998	0.35400	0.38243	0.41424	0.46503	-0.52
0.54	0.32128	0.34461	0.37798	0.40582	0.43692	0.48647	-0.54
0.56	0.34665	0.36952	0.40219	0.42940	0.45974	0.50797	-0.56
0.58	0.37233	0.39471	0.42663	0.45316	0.48270	0.52954	-0.58
0.60	0.39835	0.42020	0.45130	0.47711	0.50580	0.55118	-0.60
0.62	0.42470	0.44597	0.47620	0.50125	0.52903	0.57289	-0.62
0.64	0.45140	0.47205	0.50135	0.52558	0.55242	0.59468	-0.64
0.66	0.47844	0.49843	0.52674	0.55011	0.57595	0.61654	-0.66
0.68	0.50585	0.52512	0.55238	0.57484	0.59962	0.63847	-0.68
0.70	0.53362	0.55213	0.57827	0.59977	0.62345	0.66048	-0.70
0.72	0.56176	0.57947	0.60442	0.62490	0.64742	0.68256	-0.72
0.74	0.59028	0.60713	0.63083	0.65024	0.67155	0.70472	-0.74
0.76	0.61920	0.63513	0.65750	0.67579	0.69584	0.72695	-0.76
0.78	0.64851	0.66347	0.68444	0.70156	0.72028	0.74926	-0.78
0.80	0.67823	0.69217	0.71166	0.72754	0.74487	0.77165	-0.80
0.82	0.70837	0.72121	0.73915	0.75374	0.76963	0.79412	-0.82
0.84	0.73893	0.75063	0.76693	0.78016	0.79455	0.81667	-0.84
0.86	0.76993	0.78041	0.79500	0.80681	0.81964	0.83930	-0.86
0.88	0.80137	0.81058	0.82336	0.83369	0.84489	0.86201	-0.88
0.90	0.83327	0.84113	0.85202	0.86081	0.87031	0.88480	-0.90
0.92	0.86564	0.87208	0.88099	0.88816	0.89589	0.90767	-0.92
0.94	0.89848	0.90343	0.91026	0.91575	0.92166	0.93062	-0.94
0.96	0.93182	0.93520	0.93985	0.94358	0.94759	0.95366	-0.96
0.98	0.96565	0.96738	0.96976	0.97167	0.97371	0.97679	-0.98

↑ r_o

α =	.995 ↑-	.99 ↑-	.975 ↑-	.95 ↑-	.90 ↑-	.75 ↑-

TABLE 72.1   SAMPLE SIZE =   100

a =	.005 ↓+	.01 ↓+	.025 ↓+	.05 ↓+	.10 ↓+	.25 ↓+	
r_0 ↓							
-0.98	-0.98798	-0.98735	-0.98638	-0.98548	-0.98438	-0.98235	0.98
-0.96	-0.97586	-0.97461	-0.97267	-0.97088	-0.96869	-0.96467	0.96
-0.94	-0.96365	-0.96178	-0.95888	-0.95621	-0.95294	-0.94695	0.94
-0.92	-0.95135	-0.94886	-0.94500	-0.94146	-0.93712	-0.92919	0.92
-0.90	-0.93895	-0.93585	-0.93104	-0.92664	-0.92124	-0.91140	0.90
-0.88	-0.92645	-0.92274	-0.91699	-0.91173	-0.90530	-0.89357	0.88
-0.86	-0.91385	-0.90953	-0.90285	-0.89675	-0.88928	-0.87571	0.86
-0.84	-0.90115	-0.89623	-0.88862	-0.88168	-0.87320	-0.85781	0.84
-0.82	-0.88834	-0.88283	-0.87430	-0.86654	-0.85706	-0.83987	0.82
-0.80	-0.87544	-0.86933	-0.85990	-0.85131	-0.84084	-0.82189	0.80
-0.78	-0.86242	-0.85573	-0.84540	-0.83600	-0.82456	-0.80388	0.78
-0.76	-0.84931	-0.84203	-0.83080	-0.82061	-0.80820	-0.78582	0.76
-0.74	-0.83608	-0.82823	-0.81612	-0.80513	-0.79178	-0.76773	0.74
-0.72	-0.82275	-0.81432	-0.80133	-0.78957	-0.77528	-0.74960	0.72
-0.70	-0.80930	-0.80031	-0.78646	-0.77392	-0.75872	-0.73143	0.70
-0.68	-0.79575	-0.78618	-0.77148	-0.75819	-0.74208	-0.71322	0.68
-0.66	-0.78208	-0.77196	-0.75641	-0.74237	-0.72537	-0.69497	0.66
-0.64	-0.76829	-0.75762	-0.74124	-0.72646	-0.70859	-0.67668	0.64
-0.62	-0.75439	-0.74317	-0.72597	-0.71046	-0.69173	-0.65835	0.62
-0.60	-0.74037	-0.72861	-0.71059	-0.69437	-0.67480	-0.63998	0.60
-0.58	-0.72623	-0.71393	-0.69511	-0.67819	-0.65780	-0.62156	0.58
-0.56	-0.71196	-0.69914	-0.67953	-0.66191	-0.64071	-0.60311	0.56
-0.54	-0.69758	-0.68423	-0.66385	-0.64555	-0.62355	-0.58461	0.54
-0.52	-0.68307	-0.66921	-0.64806	-0.62909	-0.60632	-0.56607	0.52
-0.50	-0.66843	-0.65406	-0.63216	-0.61253	-0.58901	-0.54749	0.50
-0.48	-0.65366	-0.63879	-0.61615	-0.59588	-0.57161	-0.52886	0.48
-0.46	-0.63877	-0.62340	-0.60003	-0.57914	-0.55414	-0.51019	0.46
-0.44	-0.62374	-0.60789	-0.58380	-0.56229	-0.53659	-0.49148	0.44
-0.42	-0.60857	-0.59225	-0.56745	-0.54535	-0.51896	-0.47272	0.42
-0.40	-0.59327	-0.57648	-0.55100	-0.52830	-0.50125	-0.45392	0.40
-0.38	-0.57783	-0.56058	-0.53442	-0.51116	-0.48345	-0.43507	0.38
-0.36	-0.56225	-0.54455	-0.51773	-0.49391	-0.46557	-0.41618	0.36
-0.34	-0.54653	-0.52838	-0.50092	-0.47656	-0.44761	-0.39724	0.34
-0.32	-0.53067	-0.51208	-0.48400	-0.45910	-0.42956	-0.37825	0.32
-0.30	-0.51465	-0.49564	-0.46695	-0.44154	-0.41143	-0.35922	0.30
-0.28	-0.49849	-0.47907	-0.44977	-0.42387	-0.39321	-0.34014	0.28
-0.26	-0.48218	-0.46235	-0.43248	-0.40609	-0.37490	-0.32102	0.26
-0.24	-0.46571	-0.44549	-0.41505	-0.38821	-0.35651	-0.30185	0.24
-0.22	-0.44909	-0.42848	-0.39750	-0.37021	-0.33802	-0.28262	0.22
-0.20	-0.43231	-0.41133	-0.37983	-0.35210	-0.31945	-0.26335	0.20
-0.18	-0.41536	-0.39402	-0.36202	-0.33388	-0.30078	-0.24403	0.18
-0.16	-0.39826	-0.37657	-0.34407	-0.31555	-0.28203	-0.22466	0.16
-0.14	-0.38099	-0.35896	-0.32600	-0.29710	-0.26318	-0.20525	0.14
-0.12	-0.36355	-0.34120	-0.30779	-0.27853	-0.24424	-0.18578	0.12
-0.10	-0.34594	-0.32328	-0.28944	-0.25984	-0.22521	-0.16626	0.10
-0.08	-0.32815	-0.30519	-0.27095	-0.24104	-0.20608	-0.14669	0.08
-0.06	-0.31019	-0.28695	-0.25232	-0.22211	-0.18685	-0.12706	0.06
-0.04	-0.29205	-0.26854	-0.23355	-0.20307	-0.16753	-0.10739	0.04
-0.02	-0.27373	-0.24996	-0.21463	-0.18390	-0.14811	-0.08766	0.02

↑ r_0

a =	.995 ↑−	.99 ↑−	.975 ↑−	.95 ↑−	.90 ↑−	.75 ↑−

TABLE 72.2   SAMPLE SIZE = 100 (CONT.)

α =	.005 ↓+	.01 ↓+	.025 ↓+	.05 ↓+	.10 ↓+	.25 ↓+	
r_0 ↓							
0.00	-0.25522	-0.23121	-0.19557	-0.16460	-0.12859	-0.06788	0.00
0.02	-0.23653	-0.21229	-0.17636	-0.14518	-0.10897	-0.04805	-0.02
0.04	-0.21764	-0.19319	-0.15700	-0.12563	-0.08925	-0.02816	-0.04
0.06	-0.19855	-0.17392	-0.13748	-0.10595	-0.06943	-0.00822	-0.06
0.08	-0.17927	-0.15446	-0.11781	-0.08614	-0.04950	0.01178	-0.08
0.10	-0.15979	-0.13482	-0.09799	-0.06619	-0.02947	0.03183	-0.10
0.12	-0.14010	-0.11500	-0.07800	-0.04611	-0.00934	0.05193	-0.12
0.14	-0.12021	-0.09498	-0.05785	-0.02590	0.01090	0.07210	-0.14
0.16	-0.10010	-0.07477	-0.03755	-0.00555	0.03125	0.09232	-0.16
0.18	-0.07978	-0.05437	-0.01707	0.01494	0.05170	0.11260	-0.18
0.20	-0.05923	-0.03377	0.00357	0.03557	0.07227	0.13293	-0.20
0.22	-0.03847	-0.01296	0.02438	0.05635	0.09295	0.15332	-0.22
0.24	-0.01748	0.00805	0.04537	0.07726	0.11373	0.17377	-0.24
0.26	0.00375	0.02927	0.06653	0.09833	0.13463	0.19429	-0.26
0.28	0.02521	0.05070	0.08787	0.11954	0.15565	0.21486	-0.28
0.30	0.04691	0.07234	0.10938	0.14090	0.17678	0.23549	-0.30
0.32	0.06885	0.09420	0.13108	0.16241	0.19803	0.25618	-0.32
0.34	0.09104	0.11629	0.15297	0.18408	0.21939	0.27693	-0.34
0.36	0.11348	0.13860	0.17504	0.20590	0.24088	0.29775	-0.36
0.38	0.13618	0.16115	0.19731	0.22788	0.26248	0.31862	-0.38
0.40	0.15915	0.18393	0.21976	0.25002	0.28421	0.33956	-0.40
0.42	0.18238	0.20694	0.24241	0.27232	0.30606	0.36057	-0.42
0.44	0.20588	0.23020	0.26527	0.29478	0.32803	0.38163	-0.44
0.46	0.22966	0.25370	0.28832	0.31741	0.35013	0.40276	-0.46
0.48	0.25372	0.27746	0.31158	0.34021	0.37236	0.42396	-0.48
0.50	0.27807	0.30147	0.33505	0.36317	0.39471	0.44522	-0.50
0.52	0.30272	0.32573	0.35873	0.38631	0.41720	0.46655	-0.52
0.54	0.32766	0.35027	0.38262	0.40963	0.43981	0.48795	-0.54
0.56	0.35291	0.37507	0.40673	0.43312	0.46256	0.50941	-0.56
0.58	0.37847	0.40014	0.43106	0.45679	0.48545	0.53095	-0.58
0.60	0.40435	0.42550	0.45562	0.48064	0.50847	0.55255	-0.60
0.62	0.43055	0.45114	0.48041	0.50468	0.53162	0.57422	-0.62
0.64	0.45709	0.47706	0.50543	0.52890	0.55492	0.59596	-0.64
0.66	0.48396	0.50328	0.53068	0.55331	0.57836	0.61777	-0.66
0.68	0.51117	0.52980	0.55617	0.57792	0.60194	0.63965	-0.68
0.70	0.53874	0.55663	0.58191	0.60271	0.62566	0.66160	-0.70
0.72	0.56667	0.58377	0.60789	0.62771	0.64953	0.68363	-0.72
0.74	0.59496	0.61122	0.63412	0.65290	0.67355	0.70573	-0.74
0.76	0.62363	0.63900	0.66061	0.67830	0.69771	0.72790	-0.76
0.78	0.65267	0.66711	0.68736	0.70391	0.72203	0.75015	-0.78
0.80	0.68211	0.69555	0.71437	0.72972	0.74650	0.77247	-0.80
0.82	0.71195	0.72434	0.74165	0.75574	0.77112	0.79487	-0.82
0.84	0.74220	0.75347	0.76920	0.78198	0.79590	0.81735	-0.84
0.86	0.77286	0.78296	0.79703	0.80844	0.82084	0.83990	-0.86
0.88	0.80395	0.81282	0.82514	0.83511	0.84594	0.86253	-0.88
0.90	0.83548	0.84304	0.85354	0.86201	0.87120	0.88524	-0.90
0.92	0.86745	0.87365	0.88223	0.88914	0.89662	0.90803	-0.92
0.94	0.89988	0.90463	0.91121	0.91650	0.92221	0.93090	-0.94
0.96	0.93277	0.93602	0.94050	0.94410	0.94797	0.95385	-0.96
0.98	0.96614	0.96780	0.97009	0.97193	0.97390	0.97688	-0.98

↑ r_0

| α = | .995 ↑- | .99 ↑- | .975 ↑- | .95 ↑- | .90 ↑- | .75 ↑- |

TABLE 73.1   SAMPLE SIZE =   110

α =	.005 ↓+	.01 ↓+	.025 ↓+	.05 ↓+	.10 ↓+	.25 ↓+	
r_0 ↓							
-0.98	-0.98769	-0.98708	-0.98613	-0.98526	-0.98420	-0.98225	0.98
-0.96	-0.97528	-0.97407	-0.97218	-0.97045	-0.96834	-0.96447	0.96
-0.94	-0.96278	-0.96096	-0.95814	-0.95557	-0.95241	-0.94666	0.94
-0.92	-0.95019	-0.94777	-0.94403	-0.94061	-0.93642	-0.92881	0.92
-0.90	-0.93750	-0.93449	-0.92982	-0.92557	-0.92037	-0.91093	0.90
-0.88	-0.92471	-0.92111	-0.91554	-0.91046	-0.90426	-0.89301	0.88
-0.86	-0.91183	-0.90764	-0.90116	-0.89527	-0.88808	-0.87505	0.86
-0.84	-0.89885	-0.89407	-0.88670	-0.88000	-0.87183	-0.85706	0.84
-0.82	-0.88576	-0.88041	-0.87215	-0.86465	-0.85552	-0.83903	0.82
-0.80	-0.87258	-0.86665	-0.85751	-0.84922	-0.83914	-0.82097	0.80
-0.78	-0.85929	-0.85279	-0.84279	-0.83372	-0.82270	-0.80287	0.78
-0.76	-0.84589	-0.83883	-0.82797	-0.81813	-0.80619	-0.78473	0.76
-0.74	-0.83239	-0.82477	-0.81306	-0.80246	-0.78961	-0.76655	0.74
-0.72	-0.81878	-0.81061	-0.79806	-0.78670	-0.77296	-0.74834	0.72
-0.70	-0.80507	-0.79635	-0.78296	-0.77087	-0.75624	-0.73008	0.70
-0.68	-0.79124	-0.78198	-0.76777	-0.75495	-0.73946	-0.71179	0.68
-0.66	-0.77731	-0.76750	-0.75248	-0.73894	-0.72260	-0.69346	0.66
-0.64	-0.76326	-0.75292	-0.73710	-0.72285	-0.70567	-0.67510	0.64
-0.62	-0.74909	-0.73823	-0.72162	-0.70667	-0.68867	-0.65669	0.62
-0.60	-0.73481	-0.72343	-0.70603	-0.69041	-0.67160	-0.63824	0.60
-0.58	-0.72041	-0.70852	-0.69035	-0.67405	-0.65446	-0.61975	0.58
-0.56	-0.70589	-0.69349	-0.67457	-0.65761	-0.63724	-0.60123	0.56
-0.54	-0.69125	-0.67835	-0.65869	-0.64107	-0.61995	-0.58266	0.54
-0.52	-0.67649	-0.66310	-0.64270	-0.62445	-0.60258	-0.56405	0.52
-0.50	-0.66161	-0.64773	-0.62661	-0.60773	-0.58514	-0.54540	0.50
-0.48	-0.64660	-0.63224	-0.61041	-0.59092	-0.56763	-0.52671	0.48
-0.46	-0.63146	-0.61663	-0.59411	-0.57402	-0.55003	-0.50797	0.46
-0.44	-0.61619	-0.60090	-0.57769	-0.55702	-0.53236	-0.48919	0.44
-0.42	-0.60079	-0.58504	-0.56117	-0.53992	-0.51461	-0.47038	0.42
-0.40	-0.58526	-0.56907	-0.54454	-0.52273	-0.49679	-0.45151	0.40
-0.38	-0.56960	-0.55296	-0.52779	-0.50544	-0.47888	-0.43261	0.38
-0.36	-0.55379	-0.53673	-0.51093	-0.48806	-0.46089	-0.41366	0.36
-0.34	-0.53785	-0.52036	-0.49396	-0.47057	-0.44283	-0.39466	0.34
-0.32	-0.52177	-0.50387	-0.47687	-0.45298	-0.42468	-0.37563	0.32
-0.30	-0.50554	-0.48724	-0.45967	-0.43529	-0.40645	-0.35654	0.30
-0.28	-0.48918	-0.47048	-0.44234	-0.41749	-0.38813	-0.33742	0.28
-0.26	-0.47266	-0.45358	-0.42490	-0.39960	-0.36973	-0.31824	0.26
-0.24	-0.45599	-0.43655	-0.40733	-0.38159	-0.35125	-0.29903	0.24
-0.22	-0.43918	-0.41937	-0.38964	-0.36348	-0.33268	-0.27976	0.22
-0.20	-0.42221	-0.40205	-0.37183	-0.34527	-0.31403	-0.26045	0.20
-0.18	-0.40508	-0.38459	-0.35389	-0.32694	-0.29529	-0.24109	0.18
-0.16	-0.38780	-0.36698	-0.33582	-0.30851	-0.27646	-0.22168	0.16
-0.14	-0.37036	-0.34922	-0.31762	-0.28996	-0.25754	-0.20223	0.14
-0.12	-0.35275	-0.33131	-0.29930	-0.27130	-0.23853	-0.18273	0.12
-0.10	-0.33498	-0.31325	-0.28084	-0.25253	-0.21944	-0.16318	0.10
-0.08	-0.31704	-0.29503	-0.26225	-0.23365	-0.20025	-0.14358	0.08
-0.06	-0.29893	-0.27666	-0.24352	-0.21464	-0.18097	-0.12393	0.06
-0.04	-0.28065	-0.25813	-0.22466	-0.19552	-0.16160	-0.10423	0.04
-0.02	-0.26219	-0.23943	-0.20565	-0.17629	-0.14213	-0.08448	0.02

↑ r_0

α =	.995 ↑−	.99 ↑−	.975 ↑−	.95 ↑−	.90 ↑−	.75 ↑−

TABLE 73.2   SAMPLE SIZE =   110   (CONT.)

α =	.005 ↓+	.01 ↓+	.025 ↓+	.05 ↓+	.10 ↓+	.25 ↓+	
r₀ ↓							
0.00	−0.24355	−0.22058	−0.18651	−0.15693	−0.12257	−0.06468	0.00
0.02	−0.22473	−0.20155	−0.16722	−0.13745	−0.10291	−0.04484	−0.02
0.04	−0.20573	−0.18236	−0.14779	−0.11785	−0.08316	−0.02493	−0.04
0.06	−0.18654	−0.16300	−0.12821	−0.09813	−0.06331	−0.00498	−0.06
0.08	−0.16716	−0.14347	−0.10849	−0.07828	−0.04336	0.01502	−0.08
0.10	−0.14759	−0.12376	−0.08861	−0.05830	−0.02331	0.03508	−0.10
0.12	−0.12782	−0.10387	−0.06859	−0.03820	−0.00316	0.05519	−0.12
0.14	−0.10786	−0.08380	−0.04841	−0.01796	0.01708	0.07536	−0.14
0.16	−0.08769	−0.06354	−0.02807	0.00240	0.03744	0.09558	−0.16
0.18	−0.06731	−0.04310	−0.00758	0.02290	0.05789	0.11585	−0.18
0.20	−0.04673	−0.02247	0.01307	0.04353	0.07845	0.13618	−0.20
0.22	−0.02593	−0.00165	0.03389	0.06430	0.09911	0.15656	−0.22
0.24	−0.00491	0.01937	0.05487	0.08520	0.11988	0.17700	−0.24
0.26	0.01632	0.04059	0.07601	0.10624	0.14076	0.19749	−0.26
0.28	0.03778	0.06200	0.09733	0.12743	0.16175	0.21804	−0.28
0.30	0.05947	0.08363	0.11881	0.14875	0.18284	0.23865	−0.30
0.32	0.08138	0.10546	0.14047	0.17022	0.20405	0.25932	−0.32
0.34	0.10354	0.12750	0.16231	0.19183	0.22536	0.28004	−0.34
0.36	0.12593	0.14976	0.18432	0.21360	0.24680	0.30083	−0.36
0.38	0.14856	0.17223	0.20651	0.23551	0.26834	0.32167	−0.38
0.40	0.17145	0.19492	0.22888	0.25757	0.29000	0.34257	−0.40
0.42	0.19459	0.21785	0.25145	0.27978	0.31178	0.36353	−0.42
0.44	0.21798	0.24099	0.27419	0.30215	0.33367	0.38455	−0.44
0.46	0.24164	0.26438	0.29713	0.32468	0.35568	0.40564	−0.46
0.48	0.26556	0.28799	0.32027	0.34736	0.37782	0.42678	−0.48
0.50	0.28975	0.31185	0.34360	0.37021	0.40007	0.44799	−0.50
0.52	0.31422	0.33595	0.36712	0.39321	0.42245	0.46926	−0.52
0.54	0.33898	0.36030	0.39085	0.41638	0.44495	0.49059	−0.54
0.56	0.36401	0.38491	0.41479	0.43972	0.46758	0.51199	−0.56
0.58	0.38934	0.40977	0.43893	0.46323	0.49033	0.53345	−0.58
0.60	0.41497	0.43489	0.46329	0.48690	0.51321	0.55497	−0.60
0.62	0.44090	0.46027	0.48786	0.51076	0.53622	0.57657	−0.62
0.64	0.46714	0.48593	0.51264	0.53478	0.55936	0.59822	−0.64
0.66	0.49370	0.51186	0.53765	0.55898	0.58264	0.61995	−0.66
0.68	0.52058	0.53808	0.56288	0.58337	0.60604	0.64174	−0.68
0.70	0.54778	0.56457	0.58833	0.60793	0.62958	0.66360	−0.70
0.72	0.57532	0.59136	0.61402	0.63268	0.65326	0.68552	−0.72
0.74	0.60320	0.61844	0.63994	0.65762	0.67708	0.70752	−0.74
0.76	0.63142	0.64582	0.66610	0.68274	0.70104	0.72958	−0.76
0.78	0.66000	0.67351	0.69250	0.70806	0.72513	0.75172	−0.78
0.80	0.68895	0.70151	0.71915	0.73357	0.74937	0.77392	−0.80
0.82	0.71826	0.72983	0.74605	0.75928	0.77376	0.79620	−0.82
0.84	0.74795	0.75847	0.77320	0.78519	0.79829	0.81855	−0.84
0.86	0.77802	0.78744	0.80060	0.81130	0.82297	0.84097	−0.86
0.88	0.80848	0.81675	0.82827	0.83762	0.84780	0.86346	−0.88
0.90	0.83935	0.84640	0.85620	0.86415	0.87278	0.88603	−0.90
0.92	0.87062	0.87639	0.88440	0.89088	0.89791	0.90867	−0.92
0.94	0.90232	0.90674	0.91288	0.91783	0.92319	0.93139	−0.94
0.96	0.93444	0.93746	0.94164	0.94500	0.94864	0.95418	−0.96
0.98	0.96700	0.96854	0.97067	0.97239	0.97424	0.97705	−0.98

↑ r₀

| α = | .995 ↑− | .99 ↑− | .975 ↑− | .95 ↑− | .90 ↑− | .75 ↑− |

TABLE 74.1   SAMPLE SIZE =   120

$\alpha =$	.005 ↓+	.01 ↓+	.025 ↓+	.05 ↓+	.10 ↓+	.25 ↓+	
r_0 ↓							
-0.98	-0.98743	-0.98683	-0.98591	-0.98507	-0.98404	-0.98217	0.98
-0.96	-0.97476	-0.97358	-0.97174	-0.97007	-0.96802	-0.96430	0.96
-0.94	-0.96201	-0.96024	-0.95749	-0.95500	-0.95195	-0.94641	0.94
-0.92	-0.94916	-0.94680	-0.94316	-0.93985	-0.93581	-0.92847	0.92
-0.90	-0.93621	-0.93328	-0.92875	-0.92463	-0.91960	-0.91051	0.90
-0.88	-0.92317	-0.91967	-0.91425	-0.90933	-0.90334	-0.89251	0.88
-0.86	-0.91004	-0.90596	-0.89967	-0.89396	-0.88701	-0.87447	0.86
-0.84	-0.89680	-0.89216	-0.88500	-0.87851	-0.87062	-0.85640	0.84
-0.82	-0.88347	-0.87827	-0.87025	-0.86299	-0.85417	-0.83829	0.82
-0.80	-0.87004	-0.86427	-0.85541	-0.84738	-0.83765	-0.82015	0.80
-0.78	-0.85650	-0.85019	-0.84048	-0.83170	-0.82106	-0.80198	0.78
-0.76	-0.84286	-0.83600	-0.82546	-0.81594	-0.80441	-0.78376	0.76
-0.74	-0.82912	-0.82172	-0.81036	-0.80010	-0.78770	-0.76551	0.74
-0.72	-0.81527	-0.80733	-0.79516	-0.78418	-0.77091	-0.74722	0.72
-0.70	-0.80132	-0.79285	-0.77987	-0.76818	-0.75407	-0.72890	0.70
-0.68	-0.78725	-0.77826	-0.76449	-0.75209	-0.73715	-0.71054	0.68
-0.66	-0.77308	-0.76357	-0.74901	-0.73593	-0.72016	-0.69214	0.66
-0.64	-0.75880	-0.74877	-0.73344	-0.71968	-0.70311	-0.67370	0.64
-0.62	-0.74440	-0.73387	-0.71778	-0.70334	-0.68599	-0.65523	0.62
-0.60	-0.72989	-0.71886	-0.70202	-0.68692	-0.66879	-0.63672	0.60
-0.58	-0.71527	-0.70374	-0.68616	-0.67041	-0.65153	-0.61817	0.58
-0.56	-0.70052	-0.68851	-0.67020	-0.65382	-0.63419	-0.59957	0.56
-0.54	-0.68567	-0.67316	-0.65414	-0.63714	-0.61679	-0.58094	0.54
-0.52	-0.67069	-0.65771	-0.63798	-0.62037	-0.59931	-0.56227	0.52
-0.50	-0.65558	-0.64214	-0.62173	-0.60351	-0.58176	-0.54357	0.50
-0.48	-0.64036	-0.62646	-0.60536	-0.58656	-0.56413	-0.52482	0.48
-0.46	-0.62501	-0.61066	-0.58890	-0.56952	-0.54643	-0.50603	0.46
-0.44	-0.60954	-0.59474	-0.57232	-0.55239	-0.52866	-0.48719	0.44
-0.42	-0.59393	-0.57870	-0.55565	-0.53517	-0.51081	-0.46832	0.42
-0.40	-0.57820	-0.56254	-0.53886	-0.51785	-0.49288	-0.44941	0.40
-0.38	-0.56234	-0.54626	-0.52197	-0.50043	-0.47488	-0.43045	0.38
-0.36	-0.54634	-0.52985	-0.50496	-0.48292	-0.45680	-0.41145	0.36
-0.34	-0.53021	-0.51331	-0.48785	-0.46532	-0.43864	-0.39241	0.34
-0.32	-0.51394	-0.49665	-0.47062	-0.44761	-0.42040	-0.37333	0.32
-0.30	-0.49753	-0.47986	-0.45328	-0.42981	-0.40209	-0.35420	0.30
-0.28	-0.48098	-0.46294	-0.43582	-0.41191	-0.38369	-0.33503	0.28
-0.26	-0.46429	-0.44589	-0.41825	-0.39391	-0.36521	-0.31582	0.26
-0.24	-0.44745	-0.42870	-0.40056	-0.37580	-0.34665	-0.29656	0.24
-0.22	-0.43047	-0.41137	-0.38275	-0.35760	-0.32801	-0.27726	0.22
-0.20	-0.41334	-0.39391	-0.36482	-0.33929	-0.30929	-0.25791	0.20
-0.18	-0.39606	-0.37631	-0.34677	-0.32087	-0.29048	-0.23852	0.18
-0.16	-0.37862	-0.35856	-0.32860	-0.30235	-0.27159	-0.21908	0.16
-0.14	-0.36103	-0.34068	-0.31030	-0.28372	-0.25261	-0.19960	0.14
-0.12	-0.34328	-0.32265	-0.29187	-0.26499	-0.23355	-0.18007	0.12
-0.10	-0.32537	-0.30447	-0.27332	-0.24614	-0.21440	-0.16049	0.10
-0.08	-0.30730	-0.28614	-0.25464	-0.22719	-0.19516	-0.14087	0.08
-0.06	-0.28907	-0.26766	-0.23583	-0.20812	-0.17584	-0.12120	0.06
-0.04	-0.27066	-0.24902	-0.21689	-0.18894	-0.15642	-0.10148	0.04
-0.02	-0.25209	-0.23023	-0.19781	-0.16965	-0.13692	-0.08172	0.02

↑ r_0

$\alpha =$ .995 ↑−   .99 ↑−   .975 ↑−   .95 ↑−   .90 ↑−   .75 ↑−

TABLE 74.2   SAMPLE SIZE =   120   (CONT.)

α =	.005 ↓+	.01 ↓+	.025 ↓+	.05 ↓+	.10 ↓+	.25 ↓+	
r_0 ↓							
0.00	−0.23335	−0.21129	−0.17860	−0.15024	−0.11732	−0.06190	0.00
0.02	−0.21443	−0.19218	−0.15925	−0.13072	−0.09763	−0.04204	−0.02
0.04	−0.19533	−0.17291	−0.13976	−0.11108	−0.07785	−0.02212	−0.04
0.06	−0.17606	−0.15348	−0.12013	−0.09132	−0.05798	−0.00216	−0.06
0.08	−0.15660	−0.13388	−0.10036	−0.07143	−0.03801	0.01785	−0.08
0.10	−0.13695	−0.11411	−0.08045	−0.05143	−0.01795	0.03791	−0.10
0.12	−0.11712	−0.09417	−0.06039	−0.03131	0.00221	0.05803	−0.12
0.14	−0.09710	−0.07406	−0.04019	−0.01106	0.02246	0.07819	−0.14
0.16	−0.07688	−0.05377	−0.01983	0.00931	0.04282	0.09841	−0.16
0.18	−0.05646	−0.03330	0.00068	0.02981	0.06327	0.11868	−0.18
0.20	−0.03584	−0.01265	0.02133	0.05044	0.08382	0.13900	−0.20
0.22	−0.01502	0.00819	0.04215	0.07120	0.10447	0.15937	−0.22
0.24	0.00600	0.02921	0.06312	0.09209	0.12522	0.17980	−0.24
0.26	0.02724	0.05042	0.08425	0.11311	0.14608	0.20028	−0.26
0.28	0.04869	0.07182	0.10554	0.13427	0.16704	0.22081	−0.28
0.30	0.07036	0.09342	0.12699	0.15556	0.18810	0.24140	−0.30
0.32	0.09225	0.11522	0.14861	0.17699	0.20927	0.26204	−0.32
0.34	0.11437	0.13721	0.17040	0.19856	0.23054	0.28274	−0.34
0.36	0.13671	0.15941	0.19235	0.22026	0.25193	0.30350	−0.36
0.38	0.15928	0.18182	0.21448	0.24211	0.27342	0.32431	−0.38
0.40	0.18209	0.20444	0.23678	0.26411	0.29502	0.34517	−0.40
0.42	0.20514	0.22728	0.25926	0.28624	0.31673	0.36610	−0.42
0.44	0.22844	0.25033	0.28191	0.30853	0.33855	0.38708	−0.44
0.46	0.25198	0.27360	0.30475	0.33096	0.36049	0.40812	−0.46
0.48	0.27578	0.29709	0.32777	0.35354	0.38254	0.42922	−0.48
0.50	0.29983	0.32081	0.35098	0.37628	0.40471	0.45038	−0.50
0.52	0.32414	0.34477	0.37437	0.39917	0.42699	0.47160	−0.52
0.54	0.34872	0.36895	0.39796	0.42222	0.44939	0.49288	−0.54
0.56	0.37357	0.39338	0.42174	0.44542	0.47191	0.51421	−0.56
0.58	0.39870	0.41805	0.44571	0.46878	0.49454	0.53561	−0.58
0.60	0.42410	0.44297	0.46989	0.49230	0.51730	0.55707	−0.60
0.62	0.44980	0.46813	0.49427	0.51599	0.54019	0.57860	−0.62
0.64	0.47578	0.49355	0.51885	0.53985	0.56319	0.60018	−0.64
0.66	0.50206	0.51923	0.54364	0.56387	0.58632	0.62183	−0.66
0.68	0.52864	0.54517	0.56864	0.58806	0.60958	0.64354	−0.68
0.70	0.55553	0.57138	0.59386	0.61242	0.63296	0.66531	−0.70
0.72	0.58273	0.59787	0.61929	0.63695	0.65648	0.68715	−0.72
0.74	0.61025	0.62463	0.64494	0.66166	0.68012	0.70906	−0.74
0.76	0.63809	0.65167	0.67081	0.68655	0.70390	0.73103	−0.76
0.78	0.66627	0.67899	0.69692	0.71162	0.72780	0.75307	−0.78
0.80	0.69478	0.70661	0.72325	0.73688	0.75185	0.77517	−0.80
0.82	0.72364	0.73453	0.74981	0.76232	0.77603	0.79734	−0.82
0.84	0.75284	0.76274	0.77661	0.78794	0.80034	0.81958	−0.84
0.86	0.78241	0.79126	0.80366	0.81376	0.82480	0.84189	−0.86
0.88	0.81234	0.82010	0.83094	0.83977	0.84939	0.86426	−0.88
0.90	0.84264	0.84925	0.85848	0.86597	0.87413	0.88671	−0.90
0.92	0.87332	0.87873	0.88626	0.89237	0.89901	0.90922	−0.92
0.94	0.90439	0.90854	0.91430	0.91897	0.92404	0.93181	−0.94
0.96	0.93585	0.93868	0.94260	0.94577	0.94921	0.95447	−0.96
0.98	0.96772	0.96917	0.97117	0.97278	0.97453	0.97720	−0.98

↑ r_0

α =	.995 ↑−	.99 ↑−	.975 ↑−	.95 ↑−	.90 ↑−	.75 ↑−

TABLE 75.1   SAMPLE SIZE =   130

α =	.005 ↓+	.01 ↓+	.025 ↓+	.05 ↓+	.10 ↓+	.25 ↓+	
r_o ↓							
−0.98	−0.98719	−0.98661	−0.98572	−0.98490	−0.98390	−0.98209	0.98
−0.96	−0.97430	−0.97314	−0.97136	−0.96973	−0.96775	−0.96415	0.96
−0.94	−0.96131	−0.95958	−0.95691	−0.95449	−0.95153	−0.94618	0.94
−0.92	−0.94823	−0.94594	−0.94239	−0.93917	−0.93526	−0.92817	0.92
−0.90	−0.93506	−0.93220	−0.92779	−0.92379	−0.91892	−0.91014	0.90
−0.88	−0.92179	−0.91837	−0.91311	−0.90833	−0.90252	−0.89206	0.88
−0.86	−0.90843	−0.90446	−0.89834	−0.89279	−0.88607	−0.87396	0.86
−0.84	−0.89497	−0.89045	−0.88349	−0.87719	−0.86954	−0.85581	0.84
−0.82	−0.88142	−0.87635	−0.86855	−0.86150	−0.85296	−0.83764	0.82
−0.80	−0.86776	−0.86215	−0.85353	−0.84574	−0.83632	−0.81943	0.80
−0.78	−0.85401	−0.84786	−0.83842	−0.82991	−0.81961	−0.80118	0.78
−0.76	−0.84015	−0.83347	−0.82323	−0.81399	−0.80284	−0.78290	0.76
−0.74	−0.82619	−0.81899	−0.80795	−0.79800	−0.78600	−0.76459	0.74
−0.72	−0.81213	−0.80440	−0.79258	−0.78193	−0.76910	−0.74624	0.72
−0.70	−0.79796	−0.78972	−0.77712	−0.76578	−0.75213	−0.72785	0.70
−0.68	−0.78368	−0.77494	−0.76157	−0.74955	−0.73510	−0.70943	0.68
−0.66	−0.76930	−0.76005	−0.74592	−0.73324	−0.71800	−0.69097	0.66
−0.64	−0.75481	−0.74506	−0.73019	−0.71685	−0.70083	−0.67247	0.64
−0.62	−0.74021	−0.72997	−0.71436	−0.70038	−0.68360	−0.65394	0.62
−0.60	−0.72550	−0.71477	−0.69844	−0.68382	−0.66630	−0.63537	0.60
−0.58	−0.71067	−0.69947	−0.68242	−0.66718	−0.64893	−0.61676	0.58
−0.56	−0.69573	−0.68406	−0.66631	−0.65046	−0.63149	−0.59811	0.56
−0.54	−0.68068	−0.66854	−0.65010	−0.63364	−0.61398	−0.57943	0.54
−0.52	−0.66550	−0.65291	−0.63379	−0.61675	−0.59640	−0.56070	0.52
−0.50	−0.65021	−0.63717	−0.61738	−0.59976	−0.57875	−0.54194	0.50
−0.48	−0.63480	−0.62131	−0.60087	−0.58269	−0.56103	−0.52314	0.48
−0.46	−0.61926	−0.60534	−0.58427	−0.56553	−0.54324	−0.50430	0.46
−0.44	−0.60361	−0.58926	−0.56755	−0.54828	−0.52537	−0.48542	0.44
−0.42	−0.58782	−0.57306	−0.55074	−0.53094	−0.50743	−0.46650	0.42
−0.40	−0.57191	−0.55673	−0.53382	−0.51351	−0.48942	−0.44754	0.40
−0.38	−0.55587	−0.54029	−0.51679	−0.49599	−0.47133	−0.42854	0.38
−0.36	−0.53971	−0.52373	−0.49966	−0.47837	−0.45317	−0.40950	0.36
−0.34	−0.52341	−0.50705	−0.48242	−0.46066	−0.43493	−0.39042	0.34
−0.32	−0.50697	−0.49024	−0.46507	−0.44286	−0.41662	−0.37130	0.32
−0.30	−0.49040	−0.47331	−0.44761	−0.42496	−0.39823	−0.35213	0.30
−0.28	−0.47370	−0.45625	−0.43004	−0.40696	−0.37976	−0.33293	0.28
−0.26	−0.45685	−0.43906	−0.41236	−0.38887	−0.36121	−0.31368	0.26
−0.24	−0.43987	−0.42174	−0.39456	−0.37068	−0.34259	−0.29439	0.24
−0.22	−0.42274	−0.40428	−0.37664	−0.35239	−0.32389	−0.27505	0.22
−0.20	−0.40547	−0.38670	−0.35862	−0.33400	−0.30510	−0.25568	0.20
−0.18	−0.38805	−0.36897	−0.34047	−0.31550	−0.28624	−0.23625	0.18
−0.16	−0.37048	−0.35111	−0.32220	−0.29691	−0.26729	−0.21679	0.16
−0.14	−0.35276	−0.33311	−0.30381	−0.27821	−0.24826	−0.19728	0.14
−0.12	−0.33489	−0.31498	−0.28531	−0.25941	−0.22915	−0.17773	0.12
−0.10	−0.31686	−0.29669	−0.26667	−0.24050	−0.20995	−0.15813	0.10
−0.08	−0.29868	−0.27827	−0.24792	−0.22149	−0.19067	−0.13848	0.08
−0.06	−0.28034	−0.25969	−0.22903	−0.20236	−0.17131	−0.11879	0.06
−0.04	−0.26183	−0.24097	−0.21002	−0.18313	−0.15186	−0.09906	0.04
−0.02	−0.24316	−0.22210	−0.19088	−0.16379	−0.13232	−0.07928	0.02

↑ r_o

α =	.995 ↑−	.99 ↑−	.975 ↑−	.95 ↑−	.90 ↑−	.75 ↑−

TABLE 75.2   SAMPLE SIZE =   130   (CONT.)

α =	.005 ↓+	.01 ↓+	.025 ↓+	.05 ↓+	.10 ↓+	.25 ↓+	
r_o ↓							
0.00	−0.22433	−0.20308	−0.17161	−0.14434	−0.11269	−0.05945	0.00
0.02	−0.20532	−0.18390	−0.15221	−0.12478	−0.09298	−0.03957	−0.02
0.04	−0.18615	−0.16457	−0.13268	−0.10510	−0.07318	−0.01965	−0.04
0.06	−0.16680	−0.14507	−0.11301	−0.08531	−0.05329	0.00032	−0.06
0.08	−0.14727	−0.12542	−0.09320	−0.06541	−0.03330	0.02034	−0.08
0.10	−0.12757	−0.10560	−0.07326	−0.04538	−0.01323	0.04040	−0.10
0.12	−0.10768	−0.08562	−0.05317	−0.02524	0.00694	0.06052	−0.12
0.14	−0.08761	−0.06547	−0.03295	−0.00498	0.02720	0.08068	−0.14
0.16	−0.06735	−0.04516	−0.01258	0.01540	0.04755	0.10090	−0.16
0.18	−0.04691	−0.02467	0.00794	0.03590	0.06800	0.12116	−0.18
0.20	−0.02627	−0.00400	0.02860	0.05652	0.08854	0.14147	−0.20
0.22	−0.00543	0.01684	0.04941	0.07727	0.10918	0.16184	−0.22
0.24	0.01560	0.03786	0.07037	0.09815	0.12991	0.18225	−0.24
0.26	0.03684	0.05906	0.09148	0.11915	0.15075	0.20272	−0.26
0.28	0.05828	0.08044	0.11275	0.14028	0.17168	0.22324	−0.28
0.30	0.07993	0.10202	0.13417	0.16154	0.19272	0.24381	−0.30
0.32	0.10179	0.12378	0.15576	0.18293	0.21385	0.26443	−0.32
0.34	0.12387	0.14573	0.17750	0.20445	0.23509	0.28511	−0.34
0.36	0.14616	0.16788	0.19940	0.22611	0.25643	0.30584	−0.36
0.38	0.16868	0.19023	0.22147	0.24790	0.27787	0.32662	−0.38
0.40	0.19142	0.21278	0.24370	0.26983	0.29942	0.34746	−0.40
0.42	0.21439	0.23553	0.26610	0.29190	0.32107	0.36835	−0.42
0.44	0.23759	0.25849	0.28867	0.31411	0.34283	0.38930	−0.44
0.46	0.26103	0.28166	0.31142	0.33646	0.36470	0.41030	−0.46
0.48	0.28471	0.30505	0.33433	0.35896	0.38668	0.43136	−0.48
0.50	0.30863	0.32865	0.35743	0.38160	0.40876	0.45247	−0.50
0.52	0.33280	0.35246	0.38071	0.40438	0.43096	0.47365	−0.52
0.54	0.35723	0.37651	0.40416	0.42731	0.45327	0.49487	−0.54
0.56	0.38191	0.40078	0.42780	0.45040	0.47569	0.51616	−0.56
0.58	0.40685	0.42528	0.45163	0.47363	0.49823	0.53751	−0.58
0.60	0.43206	0.45001	0.47565	0.49702	0.52088	0.55891	−0.60
0.62	0.45754	0.47498	0.49986	0.52056	0.54365	0.58037	−0.62
0.64	0.48329	0.50019	0.52426	0.54427	0.56654	0.60189	−0.64
0.66	0.50933	0.52564	0.54886	0.56812	0.58954	0.62347	−0.66
0.68	0.53564	0.55135	0.57366	0.59215	0.61267	0.64511	−0.68
0.70	0.56225	0.57731	0.59866	0.61633	0.63591	0.66682	−0.70
0.72	0.58916	0.60352	0.62387	0.64068	0.65928	0.68858	−0.72
0.74	0.61636	0.62999	0.64928	0.66519	0.68277	0.71041	−0.74
0.76	0.64387	0.65674	0.67491	0.68987	0.70639	0.73230	−0.76
0.78	0.67169	0.68374	0.70075	0.71473	0.73013	0.75425	−0.78
0.80	0.69983	0.71103	0.72680	0.73975	0.75400	0.77626	−0.80
0.82	0.72829	0.73859	0.75308	0.76495	0.77800	0.79834	−0.82
0.84	0.75708	0.76644	0.77958	0.79033	0.80213	0.82048	−0.84
0.86	0.78620	0.79457	0.80631	0.81589	0.82639	0.84269	−0.86
0.88	0.81567	0.82300	0.83326	0.84163	0.85078	0.86496	−0.88
0.90	0.84548	0.85172	0.86045	0.86755	0.87531	0.88730	−0.90
0.92	0.87565	0.88075	0.88787	0.89366	0.89997	0.90970	−0.92
0.94	0.90618	0.91009	0.91554	0.91996	0.92477	0.93218	−0.94
0.96	0.93707	0.93974	0.94344	0.94644	0.94970	0.95472	−0.96
0.98	0.96834	0.96971	0.97160	0.97312	0.97478	0.97732	−0.98

↑ r_o

α =	.995 ↑−	.99 ↑−	.975 ↑−	.95 ↑−	.90 ↑−	.75 ↑−

TABLE 76.1   SAMPLE SIZE =   140

α =	.005 ↓+	.01 ↓+	.025 ↓+	.05 ↓+	.10 ↓+	.25 ↓+	
r_0 ↓							
-0.98	-0.98698	-0.98642	-0.98554	-0.98475	-0.98378	-0.98202	0.98
-0.96	-0.97388	-0.97275	-0.97101	-0.96942	-0.96750	-0.96402	0.96
-0.94	-0.96068	-0.95899	-0.95639	-0.95403	-0.95116	-0.94598	0.94
-0.92	-0.94740	-0.94515	-0.94170	-0.93857	-0.93476	-0.92791	0.92
-0.90	-0.93402	-0.93122	-0.92693	-0.92303	-0.91831	-0.90980	0.90
-0.88	-0.92055	-0.91721	-0.91207	-0.90743	-0.90179	-0.89166	0.88
-0.86	-0.90698	-0.90310	-0.89714	-0.89175	-0.88521	-0.87349	0.86
-0.84	-0.89332	-0.88890	-0.88212	-0.87600	-0.86858	-0.85529	0.84
-0.82	-0.87956	-0.87461	-0.86702	-0.86017	-0.85188	-0.83705	0.82
-0.80	-0.86571	-0.86023	-0.85184	-0.84427	-0.83512	-0.81878	0.80
-0.78	-0.85175	-0.84576	-0.83657	-0.82829	-0.81830	-0.80048	0.78
-0.76	-0.83770	-0.83119	-0.82122	-0.81224	-0.80142	-0.78214	0.76
-0.74	-0.82355	-0.81652	-0.80578	-0.79612	-0.78448	-0.76376	0.74
-0.72	-0.80929	-0.80176	-0.79025	-0.77991	-0.76747	-0.74535	0.72
-0.70	-0.79493	-0.78690	-0.77464	-0.76363	-0.75040	-0.72691	0.70
-0.68	-0.78047	-0.77194	-0.75894	-0.74727	-0.73326	-0.70843	0.68
-0.66	-0.76590	-0.75689	-0.74315	-0.73084	-0.71606	-0.68992	0.66
-0.64	-0.75122	-0.74173	-0.72726	-0.71432	-0.69879	-0.67137	0.64
-0.62	-0.73643	-0.72647	-0.71129	-0.69772	-0.68146	-0.65278	0.62
-0.60	-0.72154	-0.71110	-0.69523	-0.68104	-0.66406	-0.63416	0.60
-0.58	-0.70653	-0.69563	-0.67907	-0.66428	-0.64660	-0.61550	0.58
-0.56	-0.69142	-0.68006	-0.66282	-0.64744	-0.62907	-0.59680	0.56
-0.54	-0.67619	-0.66438	-0.64647	-0.63051	-0.61147	-0.57807	0.54
-0.52	-0.66084	-0.64859	-0.63003	-0.61350	-0.59380	-0.55930	0.52
-0.50	-0.64538	-0.63269	-0.61349	-0.59641	-0.57606	-0.54049	0.50
-0.48	-0.62980	-0.61669	-0.59685	-0.57922	-0.55825	-0.52165	0.48
-0.46	-0.61410	-0.60057	-0.58011	-0.56196	-0.54038	-0.50276	0.46
-0.44	-0.59827	-0.58433	-0.56328	-0.54460	-0.52243	-0.48384	0.44
-0.42	-0.58233	-0.56799	-0.54634	-0.52716	-0.50441	-0.46488	0.42
-0.40	-0.56626	-0.55153	-0.52930	-0.50963	-0.48632	-0.44588	0.40
-0.38	-0.55007	-0.53495	-0.51216	-0.49201	-0.46816	-0.42684	0.38
-0.36	-0.53375	-0.51825	-0.49491	-0.47430	-0.44993	-0.40776	0.36
-0.34	-0.51730	-0.50143	-0.47756	-0.45650	-0.43162	-0.38864	0.34
-0.32	-0.50072	-0.48449	-0.46011	-0.43861	-0.41324	-0.36949	0.32
-0.30	-0.48401	-0.46743	-0.44254	-0.42062	-0.39478	-0.35029	0.30
-0.28	-0.46717	-0.45025	-0.42487	-0.40254	-0.37625	-0.33105	0.28
-0.26	-0.45019	-0.43294	-0.40709	-0.38437	-0.35765	-0.31177	0.26
-0.24	-0.43307	-0.41550	-0.38919	-0.36610	-0.33896	-0.29245	0.24
-0.22	-0.41582	-0.39794	-0.37119	-0.34774	-0.32020	-0.27308	0.22
-0.20	-0.39842	-0.38024	-0.35307	-0.32927	-0.30136	-0.25368	0.20
-0.18	-0.38088	-0.36241	-0.33484	-0.31071	-0.28245	-0.23423	0.18
-0.16	-0.36320	-0.34445	-0.31649	-0.29205	-0.26345	-0.21474	0.16
-0.14	-0.34537	-0.32635	-0.29803	-0.27329	-0.24438	-0.19521	0.14
-0.12	-0.32739	-0.30812	-0.27944	-0.25443	-0.22522	-0.17564	0.12
-0.10	-0.30926	-0.28975	-0.26074	-0.23547	-0.20599	-0.15602	0.10
-0.08	-0.29097	-0.27124	-0.24192	-0.21640	-0.18667	-0.13636	0.08
-0.06	-0.27254	-0.25259	-0.22297	-0.19723	-0.16727	-0.11665	0.06
-0.04	-0.25394	-0.23379	-0.20390	-0.17796	-0.14779	-0.09690	0.04
-0.02	-0.23519	-0.21485	-0.18471	-0.15857	-0.12822	-0.07711	0.02

↑ r_0

α =	.995 ↑-	.99 ↑-	.975 ↑-	.95 ↑-	.90 ↑-	.75 ↑-

TABLE 76.2 SAMPLE SIZE = 140 (CONT.)

α =	.005 ↓+	.01 ↓+	.025 ↓+	.05 ↓+	.10 ↓+	.25 ↓+	
r_0 ↓							
0.00	−0.21628	−0.19576	−0.16539	−0.13909	−0.10857	−0.05727	0.00
0.02	−0.19720	−0.17652	−0.14595	−0.11949	−0.08884	−0.03738	−0.02
0.04	−0.17795	−0.15713	−0.12637	−0.09978	−0.06902	−0.01745	−0.04
0.06	−0.15854	−0.13758	−0.10666	−0.07997	−0.04911	0.00253	−0.06
0.08	−0.13896	−0.11789	−0.08683	−0.06004	−0.02911	0.02255	−0.08
0.10	−0.11921	−0.09803	−0.06686	−0.04000	−0.00903	0.04262	−0.10
0.12	−0.09928	−0.07802	−0.04675	−0.01985	0.01114	0.06273	−0.12
0.14	−0.07917	−0.05784	−0.02651	0.00042	0.03140	0.08290	−0.14
0.16	−0.05888	−0.03750	−0.00613	0.02081	0.05175	0.10311	−0.16
0.18	−0.03841	−0.01699	0.01440	0.04131	0.07220	0.12337	−0.18
0.20	−0.01775	0.00368	0.03506	0.06193	0.09273	0.14368	−0.20
0.22	0.00310	0.02452	0.05586	0.08267	0.11336	0.16403	−0.22
0.24	0.02413	0.04554	0.07681	0.10353	0.13408	0.18444	−0.24
0.26	0.04536	0.06673	0.09791	0.12451	0.15489	0.20489	−0.26
0.28	0.06679	0.08809	0.11915	0.14561	0.17581	0.22540	−0.28
0.30	0.08842	0.10964	0.14054	0.16684	0.19681	0.24595	−0.30
0.32	0.11025	0.13137	0.16209	0.18820	0.21791	0.26655	−0.32
0.34	0.13229	0.15329	0.18379	0.20968	0.23912	0.28721	−0.34
0.36	0.15454	0.17539	0.20564	0.23129	0.26042	0.30792	−0.36
0.38	0.17700	0.19768	0.22765	0.25304	0.28182	0.32867	−0.38
0.40	0.19967	0.22016	0.24982	0.27491	0.30331	0.34948	−0.40
0.42	0.22257	0.24284	0.27216	0.29692	0.32492	0.37035	−0.42
0.44	0.24568	0.26572	0.29465	0.31906	0.34662	0.39126	−0.44
0.46	0.26903	0.28879	0.31731	0.34133	0.36843	0.41223	−0.46
0.48	0.29260	0.31208	0.34014	0.36374	0.39034	0.43325	−0.48
0.50	0.31640	0.33556	0.36313	0.38629	0.41235	0.45433	−0.50
0.52	0.34045	0.35926	0.38630	0.40899	0.43447	0.47546	−0.52
0.54	0.36473	0.38317	0.40964	0.43182	0.45670	0.49664	−0.54
0.56	0.38926	0.40730	0.43316	0.45480	0.47904	0.51788	−0.56
0.58	0.41404	0.43165	0.45686	0.47792	0.50149	0.53918	−0.58
0.60	0.43907	0.45621	0.48073	0.50118	0.52404	0.56053	−0.60
0.62	0.46436	0.48101	0.50479	0.52460	0.54671	0.58194	−0.62
0.64	0.48990	0.50603	0.52903	0.54816	0.56949	0.60341	−0.64
0.66	0.51572	0.53129	0.55346	0.57188	0.59238	0.62493	−0.66
0.68	0.54180	0.55678	0.57808	0.59575	0.61539	0.64651	−0.68
0.70	0.56816	0.58251	0.60289	0.61978	0.63851	0.66815	−0.70
0.72	0.59480	0.60849	0.62790	0.64396	0.66175	0.68984	−0.72
0.74	0.62173	0.63471	0.65311	0.66830	0.68511	0.71160	−0.74
0.76	0.64894	0.66119	0.67851	0.69280	0.70859	0.73341	−0.76
0.78	0.67644	0.68792	0.70412	0.71746	0.73218	0.75529	−0.78
0.80	0.70425	0.71490	0.72993	0.74228	0.75590	0.77722	−0.80
0.82	0.73236	0.74216	0.75595	0.76728	0.77974	0.79922	−0.82
0.84	0.76078	0.76968	0.78218	0.79244	0.80370	0.82128	−0.84
0.86	0.78952	0.79747	0.80863	0.81776	0.82779	0.84340	−0.86
0.88	0.81858	0.82553	0.83529	0.84327	0.85200	0.86558	−0.88
0.90	0.84796	0.85388	0.86218	0.86894	0.87634	0.88782	−0.90
0.92	0.87768	0.88252	0.88928	0.89479	0.90081	0.91013	−0.92
0.94	0.90773	0.91144	0.91661	0.92082	0.92541	0.93250	−0.94
0.96	0.93813	0.94066	0.94418	0.94703	0.95014	0.95493	−0.96
0.98	0.96889	0.97018	0.97197	0.97342	0.97500	0.97743	−0.98

↑ r_0

α =	.995 ↑−	.99 ↑−	.975 ↑−	.95 ↑−	.90 ↑−	.75 ↑−

TABLE 77.1   SAMPLE SIZE =   150

α =	.005 ↓+	.01 ↓+	.025 ↓+	.05 ↓+	.10 ↓+	.25 ↓+	
r_0 ↓							
-0.98	-0.98679	-0.98624	-0.98538	-0.98461	-0.98366	-0.98196	0.98
-0.96	-0.97349	-0.97239	-0.97069	-0.96915	-0.96727	-0.96389	0.96
-0.94	-0.96011	-0.95846	-0.95592	-0.95362	-0.95082	-0.94579	0.94
-0.92	-0.94663	-0.94444	-0.94107	-0.93802	-0.93432	-0.92766	0.92
-0.90	-0.93307	-0.93034	-0.92614	-0.92235	-0.91775	-0.90950	0.90
-0.88	-0.91941	-0.91615	-0.91114	-0.90661	-0.90113	-0.89130	0.88
-0.86	-0.90566	-0.90187	-0.89605	-0.89080	-0.88445	-0.87307	0.86
-0.84	-0.89181	-0.88750	-0.88088	-0.87492	-0.86770	-0.85481	0.84
-0.82	-0.87787	-0.87304	-0.86563	-0.85896	-0.85090	-0.83652	0.82
-0.80	-0.86384	-0.85849	-0.85030	-0.84293	-0.83404	-0.81819	0.80
-0.78	-0.84970	-0.84385	-0.83489	-0.82683	-0.81712	-0.79983	0.78
-0.76	-0.83547	-0.82911	-0.81940	-0.81066	-0.80014	-0.78144	0.76
-0.74	-0.82114	-0.81429	-0.80381	-0.79441	-0.78310	-0.76302	0.74
-0.72	-0.80671	-0.79936	-0.78815	-0.77809	-0.76599	-0.74455	0.72
-0.70	-0.79218	-0.78434	-0.77240	-0.76169	-0.74883	-0.72606	0.70
-0.68	-0.77754	-0.76923	-0.75656	-0.74521	-0.73160	-0.70753	0.68
-0.66	-0.76280	-0.75401	-0.74063	-0.72866	-0.71431	-0.68897	0.66
-0.64	-0.74796	-0.73870	-0.72462	-0.71203	-0.69695	-0.67037	0.64
-0.62	-0.73301	-0.72329	-0.70851	-0.69532	-0.67953	-0.65174	0.62
-0.60	-0.71795	-0.70777	-0.69232	-0.67853	-0.66204	-0.63307	0.60
-0.58	-0.70278	-0.69216	-0.67603	-0.66166	-0.64449	-0.61436	0.58
-0.56	-0.68750	-0.67644	-0.65966	-0.64471	-0.62688	-0.59562	0.56
-0.54	-0.67211	-0.66061	-0.64319	-0.62768	-0.60920	-0.57685	0.54
-0.52	-0.65661	-0.64468	-0.62662	-0.61057	-0.59145	-0.55803	0.52
-0.50	-0.64100	-0.62865	-0.60996	-0.59337	-0.57363	-0.53918	0.50
-0.48	-0.62526	-0.61250	-0.59321	-0.57609	-0.55575	-0.52030	0.48
-0.46	-0.60941	-0.59625	-0.57636	-0.55873	-0.53780	-0.50138	0.46
-0.44	-0.59345	-0.57988	-0.55941	-0.54128	-0.51978	-0.48241	0.44
-0.42	-0.57736	-0.56341	-0.54237	-0.52375	-0.50169	-0.46342	0.42
-0.40	-0.56115	-0.54682	-0.52522	-0.50613	-0.48354	-0.44438	0.40
-0.38	-0.54482	-0.53011	-0.50798	-0.48843	-0.46531	-0.42531	0.38
-0.36	-0.52837	-0.51329	-0.49063	-0.47063	-0.44701	-0.40620	0.36
-0.34	-0.51179	-0.49636	-0.47318	-0.45275	-0.42864	-0.38704	0.34
-0.32	-0.49508	-0.47930	-0.45563	-0.43478	-0.41020	-0.36786	0.32
-0.30	-0.47824	-0.46213	-0.43797	-0.41671	-0.39168	-0.34863	0.30
-0.28	-0.46127	-0.44484	-0.42020	-0.39856	-0.37309	-0.32936	0.28
-0.26	-0.44417	-0.42742	-0.40233	-0.38032	-0.35443	-0.31005	0.26
-0.24	-0.42694	-0.40988	-0.38436	-0.36198	-0.33570	-0.29070	0.24
-0.22	-0.40957	-0.39221	-0.36627	-0.34355	-0.31689	-0.27131	0.22
-0.20	-0.39206	-0.37442	-0.34807	-0.32502	-0.29800	-0.25189	0.20
-0.18	-0.37441	-0.35650	-0.32977	-0.30640	-0.27904	-0.23242	0.18
-0.16	-0.35663	-0.33845	-0.31135	-0.28768	-0.26000	-0.21291	0.16
-0.14	-0.33870	-0.32026	-0.29281	-0.26887	-0.24089	-0.19335	0.14
-0.12	-0.32062	-0.30195	-0.27417	-0.24995	-0.22170	-0.17376	0.12
-0.10	-0.30240	-0.28350	-0.25540	-0.23094	-0.20243	-0.15412	0.10
-0.08	-0.28403	-0.26491	-0.23652	-0.21183	-0.18308	-0.13445	0.08
-0.06	-0.26551	-0.24619	-0.21752	-0.19262	-0.16365	-0.11473	0.06
-0.04	-0.24684	-0.22733	-0.19840	-0.17330	-0.14414	-0.09496	0.04
-0.02	-0.22802	-0.20832	-0.17916	-0.15389	-0.12455	-0.07516	0.02

↑ r_0

α =	.995 ↑-	.99 ↑-	.975 ↑-	.95 ↑-	.90 ↑-	.75 ↑-

TABLE 77.2 SAMPLE SIZE = 150 (CONT.)

α =	.005 ↓+	.01 ↓+	.025 ↓+	.05 ↓+	.10 ↓+	.25 ↓+	
r_o ↓							
0.00	-0.20903	-0.18917	-0.15980	-0.13437	-0.10487	-0.05531	0.00
0.02	-0.18989	-0.16988	-0.14032	-0.11474	-0.08512	-0.03541	-0.02
0.04	-0.17059	-0.15044	-0.12071	-0.09501	-0.06528	-0.01548	-0.04
0.06	-0.15113	-0.13086	-0.10097	-0.07517	-0.04536	0.00451	-0.06
0.08	-0.13150	-0.11112	-0.08110	-0.05523	-0.02536	0.02453	-0.08
0.10	-0.11170	-0.09123	-0.06111	-0.03517	-0.00527	0.04460	-0.10
0.12	-0.09173	-0.07119	-0.04099	-0.01501	0.01491	0.06472	-0.12
0.14	-0.07159	-0.05099	-0.02073	0.00527	0.03517	0.08488	-0.14
0.16	-0.05128	-0.03063	-0.00034	0.02565	0.05552	0.10509	-0.16
0.18	-0.03079	-0.01012	0.02018	0.04615	0.07596	0.12535	-0.18
0.20	-0.01012	0.01056	0.04084	0.06677	0.09649	0.14565	-0.20
0.22	0.01074	0.03141	0.06164	0.08750	0.11710	0.16600	-0.22
0.24	0.03177	0.05242	0.08258	0.10834	0.13781	0.18639	-0.24
0.26	0.05300	0.07359	0.10366	0.12931	0.15861	0.20683	-0.26
0.28	0.07441	0.09494	0.12488	0.15039	0.17950	0.22733	-0.28
0.30	0.09602	0.11647	0.14624	0.17159	0.20048	0.24786	-0.30
0.32	0.11782	0.13816	0.16776	0.19291	0.22155	0.26845	-0.32
0.34	0.13982	0.16004	0.18942	0.21436	0.24272	0.28909	-0.34
0.36	0.16202	0.18209	0.21122	0.23593	0.26398	0.30977	-0.36
0.38	0.18443	0.20433	0.23318	0.25762	0.28534	0.33051	-0.38
0.40	0.20704	0.22675	0.25530	0.27944	0.30680	0.35130	-0.40
0.42	0.22987	0.24936	0.27757	0.30139	0.32835	0.37213	-0.42
0.44	0.25290	0.27216	0.29999	0.32347	0.35000	0.39302	-0.44
0.46	0.27616	0.29516	0.32257	0.34568	0.37176	0.41396	-0.46
0.48	0.29963	0.31835	0.34532	0.36802	0.39361	0.43494	-0.48
0.50	0.32333	0.34173	0.36822	0.39049	0.41556	0.45599	-0.50
0.52	0.34726	0.36532	0.39129	0.41310	0.43761	0.47708	-0.52
0.54	0.37141	0.38911	0.41453	0.43584	0.45977	0.49822	-0.54
0.56	0.39580	0.41311	0.43793	0.45872	0.48203	0.51942	-0.56
0.58	0.42043	0.43732	0.46151	0.48174	0.50439	0.54068	-0.58
0.60	0.44530	0.46174	0.48526	0.50489	0.52686	0.56198	-0.60
0.62	0.47041	0.48637	0.50918	0.52819	0.54944	0.58334	-0.62
0.64	0.49578	0.51123	0.53328	0.55164	0.57212	0.60476	-0.64
0.66	0.52140	0.53630	0.55755	0.57523	0.59492	0.62622	-0.66
0.68	0.54727	0.56161	0.58201	0.59896	0.61782	0.64775	-0.68
0.70	0.57341	0.58714	0.60665	0.62284	0.64083	0.66933	-0.70
0.72	0.59981	0.61290	0.63148	0.64687	0.66396	0.69097	-0.72
0.74	0.62648	0.63890	0.65650	0.67106	0.68719	0.71266	-0.74
0.76	0.65343	0.66513	0.68171	0.69539	0.71054	0.73441	-0.76
0.78	0.68065	0.69161	0.70711	0.71988	0.73401	0.75621	-0.78
0.80	0.70817	0.71834	0.73270	0.74453	0.75759	0.77808	-0.80
0.82	0.73596	0.74531	0.75850	0.76934	0.78128	0.80000	-0.82
0.84	0.76406	0.77254	0.78449	0.79430	0.80510	0.82198	-0.84
0.86	0.79245	0.80003	0.81069	0.81943	0.82903	0.84402	-0.86
0.88	0.82114	0.82778	0.83709	0.84472	0.85309	0.86612	-0.88
0.90	0.85015	0.85579	0.86371	0.87017	0.87726	0.88828	-0.90
0.92	0.87947	0.88408	0.89053	0.89580	0.90156	0.91051	-0.92
0.94	0.90911	0.91263	0.91757	0.92159	0.92598	0.93279	-0.94
0.96	0.93907	0.94147	0.94482	0.94755	0.95053	0.95513	-0.96
0.98	0.96937	0.97059	0.97230	0.97369	0.97520	0.97753	-0.98

↑ r_o

α =	.995 ↑-	.99 ↑-	.975 ↑-	.95 ↑-	.90 ↑-	.75 ↑-

TABLE 78.1   SAMPLE SIZE =   160

α =	.005 ↓+	.01 ↓+	.025 ↓+	.05 ↓+	.10 ↓+	.25 ↓+	
r_0 ↓							
−0.98	−0.98662	−0.98607	−0.98524	−0.98448	−0.98356	−0.98191	0.98
−0.96	−0.97314	−0.97206	−0.97040	−0.96889	−0.96707	−0.96378	0.96
−0.94	−0.95958	−0.95797	−0.95548	−0.95324	−0.95051	−0.94563	0.94
−0.92	−0.94594	−0.94379	−0.94049	−0.93751	−0.93391	−0.92744	0.92
−0.90	−0.93220	−0.92953	−0.92542	−0.92172	−0.91724	−0.90922	0.90
−0.88	−0.91837	−0.91518	−0.91028	−0.90586	−0.90052	−0.89097	0.88
−0.86	−0.90445	−0.90074	−0.89505	−0.88993	−0.88375	−0.87269	0.86
−0.84	−0.89044	−0.88621	−0.87975	−0.87393	−0.86691	−0.85438	0.84
−0.82	−0.87633	−0.87160	−0.86437	−0.85786	−0.85001	−0.83604	0.82
−0.80	−0.86213	−0.85690	−0.84890	−0.84172	−0.83306	−0.81766	0.80
−0.78	−0.84783	−0.84211	−0.83336	−0.82550	−0.81605	−0.79925	0.78
−0.76	−0.83344	−0.82722	−0.81773	−0.80922	−0.79898	−0.78081	0.76
−0.74	−0.81895	−0.81224	−0.80202	−0.79286	−0.78185	−0.76234	0.74
−0.72	−0.80436	−0.79717	−0.78623	−0.77642	−0.76465	−0.74383	0.72
−0.70	−0.78966	−0.78201	−0.77035	−0.75992	−0.74740	−0.72529	0.70
−0.68	−0.77487	−0.76675	−0.75439	−0.74333	−0.73009	−0.70671	0.68
−0.66	−0.75998	−0.75139	−0.73834	−0.72668	−0.71271	−0.68811	0.66
−0.64	−0.74498	−0.73594	−0.72220	−0.70994	−0.69527	−0.66946	0.64
−0.62	−0.72988	−0.72039	−0.70598	−0.69313	−0.67777	−0.65079	0.62
−0.60	−0.71467	−0.70474	−0.68967	−0.67624	−0.66021	−0.63208	0.60
−0.58	−0.69936	−0.68899	−0.67327	−0.65928	−0.64258	−0.61333	0.58
−0.56	−0.68393	−0.67314	−0.65678	−0.64223	−0.62489	−0.59455	0.56
−0.54	−0.66840	−0.65718	−0.64020	−0.62511	−0.60714	−0.57574	0.54
−0.52	−0.65276	−0.64112	−0.62353	−0.60790	−0.58932	−0.55688	0.52
−0.50	−0.63700	−0.62496	−0.60676	−0.59062	−0.57143	−0.53800	0.50
−0.48	−0.62113	−0.60869	−0.58990	−0.57325	−0.55348	−0.51908	0.48
−0.46	−0.60515	−0.59231	−0.57295	−0.55580	−0.53546	−0.50012	0.46
−0.44	−0.58905	−0.57583	−0.55590	−0.53827	−0.51738	−0.48112	0.44
−0.42	−0.57283	−0.55924	−0.53875	−0.52065	−0.49922	−0.46209	0.42
−0.40	−0.55650	−0.54253	−0.52151	−0.50295	−0.48100	−0.44302	0.40
−0.38	−0.54004	−0.52572	−0.50417	−0.48517	−0.46272	−0.42392	0.38
−0.36	−0.52346	−0.50879	−0.48673	−0.46730	−0.44436	−0.40477	0.36
−0.34	−0.50676	−0.49174	−0.46920	−0.44934	−0.42593	−0.38559	0.34
−0.32	−0.48994	−0.47458	−0.45156	−0.43130	−0.40744	−0.36638	0.32
−0.30	−0.47299	−0.45731	−0.43382	−0.41317	−0.38887	−0.34712	0.30
−0.28	−0.45591	−0.43992	−0.41597	−0.39495	−0.37023	−0.32783	0.28
−0.26	−0.43870	−0.42240	−0.39802	−0.37664	−0.35152	−0.30849	0.26
−0.24	−0.42136	−0.40477	−0.37997	−0.35824	−0.33274	−0.28912	0.24
−0.22	−0.40389	−0.38701	−0.36181	−0.33975	−0.31389	−0.26971	0.22
−0.20	−0.38628	−0.36913	−0.34354	−0.32116	−0.29496	−0.25026	0.20
−0.18	−0.36854	−0.35113	−0.32517	−0.30249	−0.27596	−0.23077	0.18
−0.16	−0.35066	−0.33300	−0.30668	−0.28372	−0.25688	−0.21124	0.16
−0.14	−0.33265	−0.31474	−0.28809	−0.26486	−0.23773	−0.19167	0.14
−0.12	−0.31449	−0.29635	−0.26938	−0.24590	−0.21850	−0.17206	0.12
−0.10	−0.29619	−0.27783	−0.25057	−0.22684	−0.19920	−0.15241	0.10
−0.08	−0.27774	−0.25918	−0.23163	−0.20769	−0.17982	−0.13272	0.08
−0.06	−0.25915	−0.24039	−0.21259	−0.18844	−0.16037	−0.11299	0.06
−0.04	−0.24041	−0.22147	−0.19342	−0.16909	−0.14083	−0.09321	0.04
−0.02	−0.22152	−0.20241	−0.17414	−0.14964	−0.12122	−0.07339	0.02

↑ r_0

α =   .995 ↑−   .99 ↑−   .975 ↑−   .95 ↑−   .90 ↑−   .75 ↑−

TABLE 78.2   SAMPLE SIZE = 160   (CONT.)

α =	.005 ↓+	.01 ↓+	.025 ↓+	.05 ↓+	.10 ↓+	.25 ↓+	
r_0 ↓							
0.00	-0.20247	-0.18321	-0.15474	-0.13009	-0.10153	-0.05354	0.00
0.02	-0.18328	-0.16388	-0.13522	-0.11044	-0.08176	-0.03364	-0.02
0.04	-0.16392	-0.14440	-0.11558	-0.09069	-0.06191	-0.01369	-0.04
0.06	-0.14441	-0.12477	-0.09582	-0.07084	-0.04197	0.00629	-0.06
0.08	-0.12474	-0.10500	-0.07593	-0.05088	-0.02196	0.02632	-0.08
0.10	-0.10491	-0.08508	-0.05592	-0.03081	-0.00186	0.04640	-0.10
0.12	-0.08491	-0.06502	-0.03578	-0.01064	0.01832	0.06651	-0.12
0.14	-0.06474	-0.04480	-0.01551	0.00964	0.03858	0.08668	-0.14
0.16	-0.04441	-0.02443	0.00488	0.03003	0.05893	0.10688	-0.16
0.18	-0.02390	-0.00390	0.02541	0.05053	0.07936	0.12713	-0.18
0.20	-0.00322	0.01678	0.04606	0.07114	0.09988	0.14743	-0.20
0.22	0.01763	0.03762	0.06686	0.09185	0.12048	0.16777	-0.22
0.24	0.03867	0.05862	0.08778	0.11269	0.14118	0.18815	-0.24
0.26	0.05988	0.07979	0.10884	0.13363	0.16196	0.20859	-0.26
0.28	0.08128	0.10112	0.13004	0.15469	0.18282	0.22907	-0.28
0.30	0.10287	0.12262	0.15138	0.17587	0.20378	0.24959	-0.30
0.32	0.12464	0.14428	0.17286	0.19716	0.22483	0.27016	-0.32
0.34	0.14660	0.16612	0.19449	0.21857	0.24597	0.29078	-0.34
0.36	0.16876	0.18813	0.21625	0.24010	0.26720	0.31145	-0.36
0.38	0.19112	0.21032	0.23817	0.26176	0.28852	0.33216	-0.38
0.40	0.21368	0.23269	0.26023	0.28353	0.30994	0.35293	-0.40
0.42	0.23643	0.25523	0.28243	0.30542	0.33145	0.37374	-0.42
0.44	0.25940	0.27796	0.30479	0.32744	0.35305	0.39460	-0.44
0.46	0.28257	0.30088	0.32731	0.34959	0.37475	0.41551	-0.46
0.48	0.30596	0.32398	0.34997	0.37186	0.39655	0.43647	-0.48
0.50	0.32955	0.34727	0.37280	0.39426	0.41844	0.45748	-0.50
0.52	0.35337	0.37076	0.39578	0.41679	0.44043	0.47854	-0.52
0.54	0.37741	0.39444	0.41892	0.43945	0.46252	0.49965	-0.54
0.56	0.40167	0.41832	0.44222	0.46224	0.48471	0.52081	-0.56
0.58	0.42616	0.44240	0.46569	0.48517	0.50701	0.54202	-0.58
0.60	0.45089	0.46669	0.48932	0.50823	0.52940	0.56328	-0.60
0.62	0.47584	0.49118	0.51312	0.53142	0.55189	0.58460	-0.62
0.64	0.50104	0.51588	0.53708	0.55475	0.57449	0.60597	-0.64
0.66	0.52648	0.54080	0.56122	0.57823	0.59719	0.62739	-0.66
0.68	0.55216	0.56593	0.58554	0.60184	0.62000	0.64886	-0.68
0.70	0.57810	0.59127	0.61003	0.62559	0.64291	0.67039	-0.70
0.72	0.60429	0.61684	0.63469	0.64949	0.66593	0.69197	-0.72
0.74	0.63073	0.64264	0.65954	0.67353	0.68906	0.71361	-0.74
0.76	0.65744	0.66866	0.68457	0.69772	0.71230	0.73530	-0.76
0.78	0.68442	0.69492	0.70978	0.72206	0.73564	0.75705	-0.78
0.80	0.71166	0.72140	0.73518	0.74654	0.75910	0.77885	-0.80
0.82	0.73918	0.74813	0.76077	0.77118	0.78267	0.80071	-0.82
0.84	0.76698	0.77510	0.78655	0.79597	0.80635	0.82262	-0.84
0.86	0.79506	0.80231	0.81253	0.82092	0.83015	0.84459	-0.86
0.88	0.82343	0.82978	0.83870	0.84602	0.85406	0.86662	-0.88
0.90	0.85210	0.85749	0.86507	0.87128	0.87809	0.88870	-0.90
0.92	0.88106	0.88547	0.89164	0.89669	0.90223	0.91084	-0.92
0.94	0.91033	0.91370	0.91842	0.92227	0.92649	0.93304	-0.94
0.96	0.93990	0.94220	0.94540	0.94802	0.95087	0.95530	-0.96
0.98	0.96979	0.97096	0.97260	0.97393	0.97538	0.97762	-0.98

↑ r_0

α =	.995 ↑−	.99 ↑−	.975 ↑−	.95 ↑−	.90 ↑−	.75 ↑−

TABLE 79.1   SAMPLE SIZE =   170

α =	.005 ↓+	.01 ↓+	.025 ↓+	.05 ↓+	.10 ↓+	.25 ↓+	
r_0 ↓							
−0.98	−0.98645	−0.98592	−0.98510	−0.98436	−0.98347	−0.98185	0.98
−0.96	−0.97282	−0.97176	−0.97013	−0.96866	−0.96688	−0.96368	0.96
−0.94	−0.95910	−0.95752	−0.95508	−0.95289	−0.95023	−0.94547	0.94
−0.92	−0.94529	−0.94319	−0.93996	−0.93705	−0.93353	−0.92724	0.92
−0.90	−0.93140	−0.92878	−0.92477	−0.92115	−0.91678	−0.90897	0.90
−0.88	−0.91741	−0.91428	−0.90949	−0.90518	−0.89997	−0.89067	0.88
−0.86	−0.90334	−0.89970	−0.89414	−0.88914	−0.88310	−0.87235	0.86
−0.84	−0.88917	−0.88503	−0.87871	−0.87303	−0.86618	−0.85399	0.84
−0.82	−0.87491	−0.87028	−0.86321	−0.85685	−0.84920	−0.83560	0.82
−0.80	−0.86056	−0.85544	−0.84762	−0.84060	−0.83216	−0.81717	0.80
−0.78	−0.84611	−0.84050	−0.83195	−0.82429	−0.81507	−0.79872	0.78
−0.76	−0.83157	−0.82548	−0.81621	−0.80790	−0.79791	−0.78023	0.76
−0.74	−0.81693	−0.81037	−0.80038	−0.79143	−0.78070	−0.76172	0.74
−0.72	−0.80219	−0.79516	−0.78447	−0.77490	−0.76343	−0.74317	0.72
−0.70	−0.78735	−0.77987	−0.76848	−0.75829	−0.74610	−0.72458	0.70
−0.68	−0.77242	−0.76448	−0.75240	−0.74161	−0.72871	−0.70597	0.68
−0.66	−0.75738	−0.74899	−0.73624	−0.72486	−0.71125	−0.68732	0.66
−0.64	−0.74225	−0.73341	−0.71999	−0.70803	−0.69374	−0.66864	0.64
−0.62	−0.72701	−0.71773	−0.70366	−0.69113	−0.67617	−0.64992	0.62
−0.60	−0.71166	−0.70196	−0.68724	−0.67415	−0.65853	−0.63117	0.60
−0.58	−0.69621	−0.68608	−0.67074	−0.65709	−0.64084	−0.61239	0.58
−0.56	−0.68066	−0.67011	−0.65415	−0.63996	−0.62308	−0.59357	0.56
−0.54	−0.66499	−0.65403	−0.63746	−0.62275	−0.60525	−0.57472	0.54
−0.52	−0.64922	−0.63786	−0.62069	−0.60546	−0.58737	−0.55584	0.52
−0.50	−0.63334	−0.62158	−0.60383	−0.58809	−0.56942	−0.53691	0.50
−0.48	−0.61735	−0.60520	−0.58687	−0.57065	−0.55140	−0.51796	0.48
−0.46	−0.60124	−0.58871	−0.56983	−0.55312	−0.53333	−0.49897	0.46
−0.44	−0.58502	−0.57212	−0.55269	−0.53551	−0.51518	−0.47994	0.44
−0.42	−0.56868	−0.55542	−0.53545	−0.51782	−0.49697	−0.46088	0.42
−0.40	−0.55223	−0.53861	−0.51812	−0.50005	−0.47869	−0.44178	0.40
−0.38	−0.53566	−0.52169	−0.50070	−0.48219	−0.46035	−0.42265	0.38
−0.36	−0.51898	−0.50466	−0.48318	−0.46426	−0.44194	−0.40348	0.36
−0.34	−0.50217	−0.48752	−0.46556	−0.44623	−0.42346	−0.38427	0.34
−0.32	−0.48524	−0.47027	−0.44784	−0.42812	−0.40492	−0.36503	0.32
−0.30	−0.46818	−0.45290	−0.43002	−0.40993	−0.38630	−0.34575	0.30
−0.28	−0.45100	−0.43542	−0.41210	−0.39165	−0.36762	−0.32643	0.28
−0.26	−0.43370	−0.41782	−0.39408	−0.37328	−0.34887	−0.30708	0.26
−0.24	−0.41626	−0.40010	−0.37596	−0.35483	−0.33004	−0.28768	0.24
−0.22	−0.39870	−0.38226	−0.35774	−0.33628	−0.31115	−0.26825	0.22
−0.20	−0.38100	−0.36431	−0.33941	−0.31765	−0.29218	−0.24878	0.20
−0.18	−0.36318	−0.34623	−0.32097	−0.29892	−0.27315	−0.22927	0.18
−0.16	−0.34522	−0.32802	−0.30243	−0.28011	−0.25404	−0.20973	0.16
−0.14	−0.32712	−0.30970	−0.28378	−0.26120	−0.23485	−0.19014	0.14
−0.12	−0.30889	−0.29124	−0.26502	−0.24220	−0.21560	−0.17052	0.12
−0.10	−0.29051	−0.27266	−0.24616	−0.22311	−0.19627	−0.15085	0.10
−0.08	−0.27200	−0.25395	−0.22718	−0.20392	−0.17686	−0.13115	0.08
−0.06	−0.25334	−0.23511	−0.20809	−0.18464	−0.15738	−0.11140	0.06
−0.04	−0.23454	−0.21613	−0.18888	−0.16526	−0.13783	−0.09162	0.04
−0.02	−0.21559	−0.19703	−0.16957	−0.14578	−0.11820	−0.07179	0.02

↑ r_0

α =	.995 ↑−	.99 ↑−	.975 ↑−	.95 ↑−	.90 ↑−	.75 ↑−

TABLE 79.2   SAMPLE SIZE = 170   (CONT.)

α =	.005 ↓+	.01 ↓+	.025 ↓+	.05 ↓+	.10 ↓+	.25 ↓+	
r_o ↓							
0.00	-0.19650	-0.17778	-0.15013	-0.12621	-0.09849	-0.05193	0.00
0.02	-0.17725	-0.15841	-0.13058	-0.10653	-0.07870	-0.03202	-0.02
0.04	-0.15785	-0.13889	-0.11092	-0.08676	-0.05884	-0.01207	-0.04
0.06	-0.13830	-0.11923	-0.09113	-0.06689	-0.03889	0.00792	-0.06
0.08	-0.11859	-0.09943	-0.07122	-0.04692	-0.01887	0.02795	-0.08
0.10	-0.09873	-0.07949	-0.05120	-0.02684	0.00123	0.04803	-0.10
0.12	-0.07870	-0.05940	-0.03105	-0.00666	0.02141	0.06815	-0.12
0.14	-0.05851	-0.03917	-0.01077	0.01362	0.04168	0.08831	-0.14
0.16	-0.03816	-0.01879	0.00963	0.03401	0.06202	0.10851	-0.16
0.18	-0.01764	0.00174	0.03016	0.05450	0.08245	0.12875	-0.18
0.20	0.00304	0.02243	0.05081	0.07510	0.10296	0.14904	-0.20
0.22	0.02390	0.04327	0.07159	0.09581	0.12355	0.16938	-0.22
0.24	0.04493	0.06426	0.09251	0.11663	0.14423	0.18976	-0.24
0.26	0.06613	0.08541	0.11355	0.13756	0.16500	0.21018	-0.26
0.28	0.08752	0.10673	0.13473	0.15860	0.18585	0.23065	-0.28
0.30	0.10908	0.12820	0.15605	0.17975	0.20678	0.25116	-0.30
0.32	0.13083	0.14984	0.17750	0.20102	0.22781	0.27172	-0.32
0.34	0.15276	0.17164	0.19908	0.22240	0.24892	0.29232	-0.34
0.36	0.17488	0.19361	0.22081	0.24389	0.27012	0.31297	-0.36
0.38	0.19718	0.21575	0.24268	0.26550	0.29141	0.33366	-0.38
0.40	0.21969	0.23806	0.26469	0.28723	0.31278	0.35441	-0.40
0.42	0.24238	0.26055	0.28685	0.30908	0.33425	0.37520	-0.42
0.44	0.26528	0.28322	0.30915	0.33104	0.35582	0.39603	-0.44
0.46	0.28838	0.30606	0.33159	0.35313	0.37747	0.41692	-0.46
0.48	0.31168	0.32908	0.35419	0.37534	0.39922	0.43785	-0.48
0.50	0.33518	0.35229	0.37694	0.39768	0.42106	0.45883	-0.50
0.52	0.35890	0.37568	0.39984	0.42014	0.44299	0.47986	-0.52
0.54	0.38283	0.39926	0.42289	0.44272	0.46502	0.50094	-0.54
0.56	0.40698	0.42304	0.44610	0.46543	0.48715	0.52206	-0.56
0.58	0.43134	0.44700	0.46946	0.48827	0.50937	0.54324	-0.58
0.60	0.45593	0.47116	0.49299	0.51124	0.53169	0.56446	-0.60
0.62	0.48075	0.49553	0.51668	0.53434	0.55411	0.58574	-0.62
0.64	0.50579	0.52009	0.54052	0.55757	0.57663	0.60707	-0.64
0.66	0.53106	0.54485	0.56454	0.58094	0.59925	0.62845	-0.66
0.68	0.55658	0.56983	0.58872	0.60444	0.62197	0.64987	-0.68
0.70	0.58233	0.59501	0.61307	0.62808	0.64479	0.67136	-0.70
0.72	0.60832	0.62040	0.63759	0.65185	0.66772	0.69289	-0.72
0.74	0.63456	0.64601	0.66228	0.67577	0.69075	0.71447	-0.74
0.76	0.66105	0.67184	0.68715	0.69982	0.71388	0.73611	-0.76
0.78	0.68780	0.69789	0.71220	0.72402	0.73712	0.75780	-0.78
0.80	0.71480	0.72417	0.73742	0.74836	0.76047	0.77954	-0.80
0.82	0.74207	0.75067	0.76282	0.77285	0.78392	0.80134	-0.82
0.84	0.76960	0.77740	0.78841	0.79748	0.80748	0.82319	-0.84
0.86	0.79741	0.80437	0.81419	0.82226	0.83116	0.84510	-0.86
0.88	0.82549	0.83158	0.84015	0.84719	0.85494	0.86706	-0.88
0.90	0.85385	0.85902	0.86630	0.87227	0.87883	0.88908	-0.90
0.92	0.88249	0.88671	0.89265	0.89750	0.90284	0.91115	-0.92
0.94	0.91142	0.91465	0.91919	0.92289	0.92695	0.93328	-0.94
0.96	0.94065	0.94285	0.94592	0.94844	0.95119	0.95546	-0.96
0.98	0.97017	0.97129	0.97286	0.97414	0.97554	0.97770	-0.98

↑ r_o

α =	.995 ↑−	.99 ↑−	.975 ↑−	.95 ↑−	.90 ↑−	.75 ↑−

TABLE 80.1   SAMPLE SIZE =   180

α =	.005 ↓+	.01 ↓+	.025 ↓+	.05 ↓+	.10 ↓+	.25 ↓+	
r_o ↓							
−0.98	−0.98630	−0.98578	−0.98498	−0.98425	−0.98338	−0.98181	0.98
−0.96	−0.97252	−0.97148	−0.96988	−0.96844	−0.96670	−0.96358	0.96
−0.94	−0.95865	−0.95710	−0.95472	−0.95257	−0.94997	−0.94533	0.94
−0.92	−0.94470	−0.94264	−0.93948	−0.93663	−0.93319	−0.92705	0.92
−0.90	−0.93066	−0.92809	−0.92416	−0.92062	−0.91635	−0.90874	0.90
−0.88	−0.91653	−0.91346	−0.90877	−0.90455	−0.89946	−0.89040	0.88
−0.86	−0.90231	−0.89875	−0.89330	−0.88841	−0.88251	−0.87203	0.86
−0.84	−0.88800	−0.88395	−0.87776	−0.87220	−0.86551	−0.85362	0.84
−0.82	−0.87360	−0.86906	−0.86214	−0.85592	−0.84845	−0.83519	0.82
−0.80	−0.85910	−0.85409	−0.84644	−0.83958	−0.83133	−0.81673	0.80
−0.78	−0.84452	−0.83903	−0.83066	−0.82316	−0.81416	−0.79823	0.78
−0.76	−0.82984	−0.82388	−0.81480	−0.80668	−0.79693	−0.77970	0.76
−0.74	−0.81506	−0.80864	−0.79887	−0.79012	−0.77965	−0.76115	0.74
−0.72	−0.80019	−0.79331	−0.78285	−0.77350	−0.76230	−0.74256	0.72
−0.70	−0.78522	−0.77789	−0.76675	−0.75680	−0.74490	−0.72393	0.70
−0.68	−0.77016	−0.76238	−0.75057	−0.74003	−0.72744	−0.70528	0.68
−0.66	−0.75499	−0.74677	−0.73430	−0.72319	−0.70991	−0.68660	0.66
−0.64	−0.73972	−0.73107	−0.71796	−0.70628	−0.69233	−0.66788	0.64
−0.62	−0.72436	−0.71528	−0.70153	−0.68929	−0.67469	−0.64913	0.62
−0.60	−0.70889	−0.69939	−0.68501	−0.67222	−0.65699	−0.63034	0.60
−0.58	−0.69331	−0.68340	−0.66841	−0.65509	−0.63923	−0.61152	0.58
−0.56	−0.67764	−0.66732	−0.65172	−0.63787	−0.62141	−0.59267	0.56
−0.54	−0.66185	−0.65114	−0.63495	−0.62059	−0.60352	−0.57379	0.54
−0.52	−0.64596	−0.63485	−0.61808	−0.60322	−0.58558	−0.55487	0.52
−0.50	−0.62997	−0.61847	−0.60113	−0.58578	−0.56757	−0.53592	0.50
−0.48	−0.61386	−0.60198	−0.58409	−0.56826	−0.54950	−0.51694	0.48
−0.46	−0.59764	−0.58539	−0.56695	−0.55066	−0.53136	−0.49791	0.46
−0.44	−0.58131	−0.56870	−0.54973	−0.53298	−0.51316	−0.47886	0.44
−0.42	−0.56487	−0.55190	−0.53242	−0.51522	−0.49490	−0.45977	0.42
−0.40	−0.54831	−0.53500	−0.51501	−0.49738	−0.47657	−0.44065	0.40
−0.38	−0.53164	−0.51799	−0.49750	−0.47946	−0.45818	−0.42149	0.38
−0.36	−0.51485	−0.50087	−0.47991	−0.46146	−0.43972	−0.40229	0.36
−0.34	−0.49794	−0.48364	−0.46221	−0.44338	−0.42120	−0.38306	0.34
−0.32	−0.48091	−0.46630	−0.44443	−0.42521	−0.40261	−0.36379	0.32
−0.30	−0.46376	−0.44885	−0.42654	−0.40696	−0.38395	−0.34449	0.30
−0.28	−0.44649	−0.43129	−0.40855	−0.38862	−0.36522	−0.32515	0.28
−0.26	−0.42910	−0.41361	−0.39047	−0.37020	−0.34643	−0.30578	0.26
−0.24	−0.41158	−0.39581	−0.37229	−0.35170	−0.32757	−0.28636	0.24
−0.22	−0.39393	−0.37790	−0.35400	−0.33310	−0.30864	−0.26691	0.22
−0.20	−0.37616	−0.35988	−0.33561	−0.31442	−0.28964	−0.24743	0.20
−0.18	−0.35825	−0.34173	−0.31712	−0.29566	−0.27057	−0.22790	0.18
−0.16	−0.34022	−0.32346	−0.29853	−0.27680	−0.25143	−0.20834	0.16
−0.14	−0.32205	−0.30507	−0.27983	−0.25785	−0.23222	−0.18874	0.14
−0.12	−0.30375	−0.28656	−0.26103	−0.23881	−0.21293	−0.16910	0.12
−0.10	−0.28531	−0.26792	−0.24211	−0.21969	−0.19358	−0.14942	0.10
−0.08	−0.26673	−0.24915	−0.22309	−0.20047	−0.17415	−0.12971	0.08
−0.06	−0.24802	−0.23026	−0.20397	−0.18115	−0.15465	−0.10995	0.06
−0.04	−0.22916	−0.21124	−0.18473	−0.16175	−0.13507	−0.09016	0.04
−0.02	−0.21016	−0.19209	−0.16538	−0.14224	−0.11542	−0.07033	0.02

↑ r_o

α =	.995 ↑−	.99 ↑−	.975 ↑−	.95 ↑−	.90 ↑−	.75 ↑−

TABLE 80.2   SAMPLE SIZE = 180   (CONT.)

α =	.005 ↓+	.01 ↓+	.025 ↓+	.05 ↓+	.10 ↓+	.25 ↓+	
r_0 ↓							
0.00	−0.19102	−0.17281	−0.14591	−0.12265	−0.09570	−0.05045	0.00
0.02	−0.17173	−0.15340	−0.12634	−0.10296	−0.07590	−0.03054	−0.02
0.04	−0.15229	−0.13385	−0.10665	−0.08317	−0.05603	−0.01059	−0.04
0.06	−0.13270	−0.11416	−0.08684	−0.06328	−0.03608	0.00941	−0.06
0.08	−0.11296	−0.09434	−0.06692	−0.04330	−0.01605	0.02944	−0.08
0.10	−0.09307	−0.07437	−0.04688	−0.02321	0.00406	0.04952	−0.10
0.12	−0.07302	−0.05427	−0.02671	−0.00303	0.02424	0.06964	−0.12
0.14	−0.05282	−0.03402	−0.00643	0.01726	0.04451	0.08979	−0.14
0.16	−0.03245	−0.01363	0.01397	0.03765	0.06485	0.10999	−0.16
0.18	−0.01192	0.00691	0.03450	0.05814	0.08527	0.13024	−0.18
0.20	0.00877	0.02759	0.05514	0.07873	0.10577	0.15052	−0.20
0.22	0.02962	0.04843	0.07592	0.09943	0.12636	0.17085	−0.22
0.24	0.05065	0.06941	0.09682	0.12024	0.14702	0.19122	−0.24
0.26	0.07184	0.09055	0.11785	0.14115	0.16777	0.21163	−0.26
0.28	0.09321	0.11184	0.13901	0.16217	0.18861	0.23209	−0.28
0.30	0.11475	0.13330	0.16030	0.18330	0.20952	0.25259	−0.30
0.32	0.13647	0.15490	0.18172	0.20454	0.23052	0.27313	−0.32
0.34	0.15837	0.17667	0.20328	0.22588	0.25161	0.29372	−0.34
0.36	0.18045	0.19861	0.22497	0.24735	0.27278	0.31436	−0.36
0.38	0.20271	0.22070	0.24680	0.26892	0.29404	0.33503	−0.38
0.40	0.22516	0.24297	0.26876	0.29061	0.31538	0.35576	−0.40
0.42	0.24780	0.26540	0.29087	0.31241	0.33681	0.37653	−0.42
0.44	0.27063	0.28800	0.31311	0.33433	0.35834	0.39734	−0.44
0.46	0.29366	0.31078	0.33550	0.35636	0.37995	0.41820	−0.46
0.48	0.31688	0.33373	0.35803	0.37852	0.40165	0.43911	−0.48
0.50	0.34031	0.35685	0.38071	0.40079	0.42344	0.46006	−0.50
0.52	0.36393	0.38016	0.40353	0.42318	0.44532	0.48106	−0.52
0.54	0.38776	0.40365	0.42650	0.44570	0.46730	0.50211	−0.54
0.56	0.41180	0.42732	0.44963	0.46834	0.48936	0.52321	−0.56
0.58	0.43605	0.45118	0.47290	0.49110	0.51152	0.54435	−0.58
0.60	0.46052	0.47523	0.49633	0.51398	0.53378	0.56554	−0.60
0.62	0.48520	0.49947	0.51991	0.53700	0.55613	0.58678	−0.62
0.64	0.51010	0.52391	0.54365	0.56014	0.57858	0.60807	−0.64
0.66	0.53523	0.54854	0.56755	0.58341	0.60112	0.62941	−0.66
0.68	0.56058	0.57336	0.59161	0.60681	0.62377	0.65079	−0.68
0.70	0.58616	0.59840	0.61583	0.63034	0.64651	0.67223	−0.70
0.72	0.61198	0.62363	0.64022	0.65400	0.66934	0.69372	−0.72
0.74	0.63803	0.64907	0.66477	0.67780	0.69228	0.71526	−0.74
0.76	0.66433	0.67472	0.68949	0.70173	0.71532	0.73685	−0.76
0.78	0.69086	0.70059	0.71438	0.72580	0.73847	0.75849	−0.78
0.80	0.71765	0.72667	0.73945	0.75001	0.76171	0.78018	−0.80
0.82	0.74469	0.75297	0.76469	0.77436	0.78506	0.80192	−0.82
0.84	0.77198	0.77949	0.79010	0.79885	0.80851	0.82372	−0.84
0.86	0.79953	0.80623	0.81569	0.82348	0.83207	0.84556	−0.86
0.88	0.82735	0.83320	0.84146	0.84825	0.85573	0.86746	−0.88
0.90	0.85543	0.86041	0.86742	0.87317	0.87951	0.88942	−0.90
0.92	0.88378	0.88784	0.89355	0.89824	0.90338	0.91143	−0.92
0.94	0.91241	0.91552	0.91988	0.92345	0.92737	0.93349	−0.94
0.96	0.94132	0.94343	0.94639	0.94882	0.95147	0.95560	−0.96
0.98	0.97052	0.97159	0.97310	0.97433	0.97568	0.97777	−0.98

↑ r_0

α =	.995 ↑−	.99 ↑−	.975 ↑−	.95 ↑−	.90 ↑−	.75 ↑−

TABLE 81.1   SAMPLE SIZE =   190

α =	.005 ↓+	.01 ↓+	.025 ↓+	.05 ↓+	.10 ↓+	.25 ↓+	
r_0 ↓							
−0.98	−0.98616	−0.98565	−0.98486	−0.98415	−0.98330	−0.98176	0.98
−0.96	−0.97224	−0.97122	−0.96966	−0.96825	−0.96654	−0.96350	0.96
−0.94	−0.95824	−0.95671	−0.95438	−0.95227	−0.94973	−0.94520	0.94
−0.92	−0.94415	−0.94212	−0.93902	−0.93624	−0.93287	−0.92688	0.92
−0.90	−0.92997	−0.92745	−0.92360	−0.92013	−0.91596	−0.90853	0.90
−0.88	−0.91571	−0.91269	−0.90810	−0.90396	−0.89899	−0.89014	0.88
−0.86	−0.90135	−0.89786	−0.89252	−0.88773	−0.88197	−0.87173	0.86
−0.84	−0.88691	−0.88294	−0.87687	−0.87143	−0.86489	−0.85329	0.84
−0.82	−0.87238	−0.86793	−0.86114	−0.85506	−0.84776	−0.83481	0.82
−0.80	−0.85776	−0.85284	−0.84534	−0.83863	−0.83057	−0.81631	0.80
−0.78	−0.84304	−0.83766	−0.82946	−0.82212	−0.81333	−0.79778	0.78
−0.76	−0.82823	−0.82239	−0.81350	−0.80555	−0.79603	−0.77921	0.76
−0.74	−0.81333	−0.80704	−0.79746	−0.78891	−0.77867	−0.76062	0.74
−0.72	−0.79834	−0.79159	−0.78135	−0.77220	−0.76126	−0.74199	0.72
−0.70	−0.78325	−0.77606	−0.76515	−0.75542	−0.74379	−0.72334	0.70
−0.68	−0.76806	−0.76044	−0.74887	−0.73857	−0.72626	−0.70465	0.68
−0.66	−0.75277	−0.74472	−0.73252	−0.72165	−0.70868	−0.68593	0.66
−0.64	−0.73739	−0.72891	−0.71608	−0.70465	−0.69103	−0.66718	0.64
−0.62	−0.72190	−0.71301	−0.69955	−0.68759	−0.67333	−0.64839	0.62
−0.60	−0.70632	−0.69702	−0.68295	−0.67045	−0.65557	−0.62958	0.60
−0.58	−0.69063	−0.68093	−0.66626	−0.65323	−0.63775	−0.61073	0.58
−0.56	−0.67484	−0.66474	−0.64948	−0.63595	−0.61987	−0.59185	0.56
−0.54	−0.65895	−0.64845	−0.63262	−0.61859	−0.60193	−0.57293	0.54
−0.52	−0.64295	−0.63207	−0.61567	−0.60115	−0.58393	−0.55398	0.52
−0.50	−0.62684	−0.61559	−0.59864	−0.58364	−0.56586	−0.53500	0.50
−0.48	−0.61063	−0.59901	−0.58151	−0.56605	−0.54774	−0.51599	0.48
−0.46	−0.59431	−0.58233	−0.56430	−0.54838	−0.52955	−0.49694	0.46
−0.44	−0.57788	−0.56555	−0.54700	−0.53064	−0.51131	−0.47786	0.44
−0.42	−0.56133	−0.54866	−0.52961	−0.51282	−0.49299	−0.45875	0.42
−0.40	−0.54468	−0.53167	−0.51213	−0.49492	−0.47462	−0.43960	0.40
−0.38	−0.52791	−0.51457	−0.49456	−0.47694	−0.45618	−0.42042	0.38
−0.36	−0.51103	−0.49737	−0.47689	−0.45888	−0.43768	−0.40120	0.36
−0.34	−0.49403	−0.48006	−0.45913	−0.44074	−0.41911	−0.38194	0.34
−0.32	−0.47691	−0.46264	−0.44128	−0.42252	−0.40048	−0.36266	0.32
−0.30	−0.45968	−0.44511	−0.42333	−0.40422	−0.38178	−0.34333	0.30
−0.28	−0.44233	−0.42747	−0.40528	−0.38584	−0.36302	−0.32397	0.28
−0.26	−0.42485	−0.40972	−0.38714	−0.36737	−0.34419	−0.30458	0.26
−0.24	−0.40725	−0.39186	−0.36890	−0.34881	−0.32529	−0.28515	0.24
−0.22	−0.38953	−0.37388	−0.35056	−0.33018	−0.30633	−0.26568	0.22
−0.20	−0.37168	−0.35579	−0.33212	−0.31145	−0.28730	−0.24618	0.20
−0.18	−0.35371	−0.33758	−0.31358	−0.29265	−0.26820	−0.22664	0.18
−0.16	−0.33560	−0.31925	−0.29494	−0.27375	−0.24903	−0.20706	0.16
−0.14	−0.31737	−0.30080	−0.27619	−0.25477	−0.22979	−0.18745	0.14
−0.12	−0.29901	−0.28224	−0.25734	−0.23570	−0.21048	−0.16780	0.12
−0.10	−0.28051	−0.26355	−0.23839	−0.21654	−0.19110	−0.14811	0.10
−0.08	−0.26188	−0.24474	−0.21933	−0.19729	−0.17165	−0.12838	0.08
−0.06	−0.24311	−0.22580	−0.20017	−0.17795	−0.15213	−0.10862	0.06
−0.04	−0.22420	−0.20674	−0.18090	−0.15851	−0.13254	−0.08882	0.04
−0.02	−0.20516	−0.18755	−0.16152	−0.13899	−0.11288	−0.06898	0.02

↑ r_0

α =	.995 ↑−	.99 ↑−	.975 ↑−	.95 ↑−	.90 ↑−	.75 ↑−

TABLE 81.2   SAMPLE SIZE = 190   (CONT.)

α =	.005 ↓+	.01 ↓+	.025 ↓+	.05 ↓+	.10 ↓+	.25 ↓+	
r_o ↓							
0.00	-0.18597	-0.16823	-0.14203	-0.11938	-0.09314	-0.04910	0.00
0.02	-0.16664	-0.14878	-0.12243	-0.09967	-0.07333	-0.02918	-0.02
0.04	-0.14717	-0.12921	-0.10272	-0.07986	-0.05344	-0.00922	-0.04
0.06	-0.12755	-0.10949	-0.08289	-0.05996	-0.03349	0.01078	-0.06
0.08	-0.10778	-0.08965	-0.06296	-0.03997	-0.01345	0.03081	-0.08
0.10	-0.08787	-0.06967	-0.04290	-0.01988	0.00666	0.05089	-0.10
0.12	-0.06780	-0.04955	-0.02273	0.00031	0.02684	0.07100	-0.12
0.14	-0.04758	-0.02929	-0.00245	0.02060	0.04711	0.09116	-0.14
0.16	-0.02720	-0.00889	0.01796	0.04099	0.06745	0.11136	-0.16
0.18	-0.00667	0.01165	0.03848	0.06147	0.08786	0.13160	-0.18
0.20	0.01403	0.03233	0.05913	0.08206	0.10836	0.15188	-0.20
0.22	0.03488	0.05316	0.07989	0.10275	0.12893	0.17220	-0.22
0.24	0.05590	0.07414	0.10078	0.12355	0.14959	0.19256	-0.24
0.26	0.07709	0.09527	0.12180	0.14444	0.17032	0.21297	-0.26
0.28	0.09844	0.11654	0.14294	0.16544	0.19114	0.23341	-0.28
0.30	0.11996	0.13797	0.16421	0.18655	0.21203	0.25390	-0.30
0.32	0.14165	0.15955	0.18560	0.20776	0.23301	0.27443	-0.32
0.34	0.16352	0.18129	0.20713	0.22908	0.25408	0.29501	-0.34
0.36	0.18556	0.20319	0.22878	0.25051	0.27522	0.31563	-0.36
0.38	0.20778	0.22524	0.25057	0.27205	0.29645	0.33629	-0.38
0.40	0.23018	0.24746	0.27250	0.29370	0.31776	0.35700	-0.40
0.42	0.25277	0.26984	0.29455	0.31546	0.33916	0.37775	-0.42
0.44	0.27554	0.29238	0.31675	0.33734	0.36064	0.39854	-0.44
0.46	0.29850	0.31510	0.33908	0.35932	0.38222	0.41938	-0.46
0.48	0.32165	0.33798	0.36155	0.38142	0.40387	0.44026	-0.48
0.50	0.34499	0.36103	0.38416	0.40364	0.42562	0.46119	-0.50
0.52	0.36853	0.38426	0.40691	0.42597	0.44745	0.48217	-0.52
0.54	0.39227	0.40766	0.42981	0.44842	0.46938	0.50319	-0.54
0.56	0.41621	0.43124	0.45285	0.47099	0.49139	0.52425	-0.56
0.58	0.44035	0.45501	0.47604	0.49368	0.51350	0.54537	-0.58
0.60	0.46470	0.47895	0.49938	0.51649	0.53569	0.56653	-0.60
0.62	0.48926	0.50308	0.52287	0.53943	0.55798	0.58773	-0.62
0.64	0.51404	0.52739	0.54651	0.56248	0.58036	0.60899	-0.64
0.66	0.53902	0.55190	0.57030	0.58566	0.60284	0.63029	-0.66
0.68	0.56423	0.57660	0.59425	0.60897	0.62541	0.65164	-0.68
0.70	0.58966	0.60149	0.61836	0.63240	0.64807	0.67303	-0.70
0.72	0.61531	0.62658	0.64262	0.65596	0.67083	0.69448	-0.72
0.74	0.64120	0.65186	0.66705	0.67965	0.69369	0.71598	-0.74
0.76	0.66731	0.67735	0.69163	0.70348	0.71664	0.73752	-0.76
0.78	0.69366	0.70305	0.71638	0.72743	0.73970	0.75911	-0.78
0.80	0.72024	0.72895	0.74130	0.75152	0.76285	0.78076	-0.80
0.82	0.74707	0.75506	0.76638	0.77574	0.78610	0.80245	-0.82
0.84	0.77414	0.78139	0.79164	0.80009	0.80945	0.82419	-0.84
0.86	0.80146	0.80793	0.81706	0.82459	0.83291	0.84599	-0.86
0.88	0.82904	0.83469	0.84266	0.84922	0.85646	0.86783	-0.88
0.90	0.85687	0.86167	0.86843	0.87399	0.88012	0.88973	-0.90
0.92	0.88496	0.88887	0.89438	0.89891	0.90389	0.91168	-0.92
0.94	0.91331	0.91630	0.92051	0.92396	0.92775	0.93368	-0.94
0.96	0.94193	0.94397	0.94682	0.94916	0.95173	0.95573	-0.96
0.98	0.97083	0.97187	0.97332	0.97451	0.97581	0.97784	-0.98

↑ r_o

α =	.995 ↑−	.99 ↑−	.975 ↑−	.95 ↑−	.90 ↑−	.75 ↑−

TABLE 82.1   SAMPLE SIZE =   200

α =	.005 ↓+	.01 ↓+	.025 ↓+	.05 ↓+	.10 ↓+	.25 ↓+	
r₀ ↓							
-0.98	-0.98603	-0.98553	-0.98476	-0.98406	-0.98322	-0.98172	0.98
-0.96	-0.97198	-0.97098	-0.96944	-0.96806	-0.96639	-0.96342	0.96
-0.94	-0.95785	-0.95635	-0.95406	-0.95200	-0.94951	-0.94508	0.94
-0.92	-0.94363	-0.94164	-0.93860	-0.93587	-0.93258	-0.92672	0.92
-0.90	-0.92933	-0.92685	-0.92307	-0.91968	-0.91559	-0.90833	0.90
-0.88	-0.91494	-0.91198	-0.90747	-0.90342	-0.89855	-0.88991	0.88
-0.86	-0.90046	-0.89703	-0.89180	-0.88710	-0.88146	-0.87146	0.86
-0.84	-0.88590	-0.88199	-0.87605	-0.87072	-0.86431	-0.85298	0.84
-0.82	-0.87124	-0.86688	-0.86022	-0.85426	-0.84711	-0.83447	0.82
-0.80	-0.85650	-0.85167	-0.84432	-0.83774	-0.82986	-0.81593	0.80
-0.78	-0.84167	-0.83638	-0.82834	-0.82116	-0.81255	-0.79736	0.78
-0.76	-0.82674	-0.82101	-0.81229	-0.80451	-0.79519	-0.77876	0.76
-0.74	-0.81172	-0.80554	-0.79616	-0.78779	-0.77777	-0.76013	0.74
-0.72	-0.79661	-0.79000	-0.77995	-0.77100	-0.76029	-0.74147	0.72
-0.70	-0.78141	-0.77436	-0.76367	-0.75414	-0.74276	-0.72278	0.70
-0.68	-0.76610	-0.75863	-0.74730	-0.73721	-0.72517	-0.70406	0.68
-0.66	-0.75071	-0.74281	-0.73085	-0.72021	-0.70753	-0.68531	0.66
-0.64	-0.73521	-0.72690	-0.71433	-0.70315	-0.68983	-0.66653	0.64
-0.62	-0.71962	-0.71090	-0.69772	-0.68601	-0.67207	-0.64771	0.62
-0.60	-0.70393	-0.69481	-0.68103	-0.66880	-0.65425	-0.62887	0.60
-0.58	-0.68814	-0.67862	-0.66426	-0.65151	-0.63638	-0.60999	0.58
-0.56	-0.67224	-0.66234	-0.64740	-0.63416	-0.61844	-0.59108	0.56
-0.54	-0.65625	-0.64596	-0.63046	-0.61673	-0.60045	-0.57214	0.54
-0.52	-0.64015	-0.62949	-0.61343	-0.59923	-0.58240	-0.55316	0.52
-0.50	-0.62394	-0.61292	-0.59632	-0.58165	-0.56428	-0.53416	0.50
-0.48	-0.60763	-0.59625	-0.57913	-0.56400	-0.54611	-0.51512	0.48
-0.46	-0.59121	-0.57948	-0.56184	-0.54628	-0.52788	-0.49604	0.46
-0.44	-0.57469	-0.56262	-0.54447	-0.52848	-0.50958	-0.47694	0.44
-0.42	-0.55806	-0.54565	-0.52701	-0.51060	-0.49123	-0.45780	0.42
-0.40	-0.54131	-0.52857	-0.50946	-0.49264	-0.47281	-0.43863	0.40
-0.38	-0.52446	-0.51140	-0.49182	-0.47461	-0.45433	-0.41942	0.38
-0.36	-0.50749	-0.49412	-0.47409	-0.45650	-0.43579	-0.40018	0.36
-0.34	-0.49041	-0.47674	-0.45627	-0.43831	-0.41718	-0.38091	0.34
-0.32	-0.47321	-0.45924	-0.43836	-0.42003	-0.39851	-0.36160	0.32
-0.30	-0.45589	-0.44165	-0.42035	-0.40168	-0.37977	-0.34226	0.30
-0.28	-0.43846	-0.42394	-0.40225	-0.38325	-0.36098	-0.32288	0.28
-0.26	-0.42091	-0.40612	-0.38405	-0.36474	-0.34211	-0.30347	0.26
-0.24	-0.40324	-0.38819	-0.36576	-0.34614	-0.32318	-0.28402	0.24
-0.22	-0.38545	-0.37016	-0.34737	-0.32747	-0.30419	-0.26454	0.22
-0.20	-0.36754	-0.35200	-0.32888	-0.30871	-0.28513	-0.24502	0.20
-0.18	-0.34950	-0.33374	-0.31029	-0.28986	-0.26600	-0.22547	0.18
-0.16	-0.33133	-0.31536	-0.29161	-0.27093	-0.24681	-0.20588	0.16
-0.14	-0.31304	-0.29686	-0.27282	-0.25191	-0.22754	-0.18626	0.14
-0.12	-0.29462	-0.27824	-0.25394	-0.23281	-0.20821	-0.16659	0.12
-0.10	-0.27607	-0.25951	-0.23495	-0.21362	-0.18881	-0.14690	0.10
-0.08	-0.25738	-0.24065	-0.21586	-0.19435	-0.16935	-0.12716	0.08
-0.06	-0.23857	-0.22167	-0.19666	-0.17498	-0.14981	-0.10739	0.06
-0.04	-0.21962	-0.20257	-0.17736	-0.15553	-0.13020	-0.08758	0.04
-0.02	-0.20053	-0.18335	-0.15796	-0.13598	-0.11052	-0.06773	0.02

↑ r₀

α =	.995 ↑–	.99 ↑–	.975 ↑–	.95 ↑–	.90 ↑–	.75 ↑–

TABLE 82.2   SAMPLE SIZE =   200   (CONT.)

α =	.005 ↓+	.01 ↓+	.025 ↓+	.05 ↓+	.10 ↓+	.25 ↓+	
r_o ↓							
0.00	-0.18131	-0.16400	-0.13844	-0.11635	-0.09077	-0.04785	0.00
0.02	-0.16194	-0.14452	-0.11882	-0.09663	-0.07095	-0.02792	-0.02
0.04	-0.14244	-0.12492	-0.09909	-0.07681	-0.05106	-0.00796	-0.04
0.06	-0.12279	-0.10518	-0.07925	-0.05690	-0.03109	0.01204	-0.06
0.08	-0.10300	-0.08532	-0.05930	-0.03689	-0.01105	0.03208	-0.08
0.10	-0.08306	-0.06532	-0.03924	-0.01680	0.00906	0.05215	-0.10
0.12	-0.06298	-0.04519	-0.01906	0.00340	0.02924	0.07227	-0.12
0.14	-0.04274	-0.02492	0.00123	0.02369	0.04951	0.09242	-0.14
0.16	-0.02235	-0.00451	0.02164	0.04407	0.06984	0.11262	-0.16
0.18	-0.00181	0.01603	0.04216	0.06455	0.09026	0.13285	-0.18
0.20	0.01888	0.03671	0.06280	0.08514	0.11074	0.15313	-0.20
0.22	0.03973	0.05753	0.08356	0.10582	0.13131	0.17344	-0.22
0.24	0.06075	0.07850	0.10444	0.12660	0.15195	0.19380	-0.24
0.26	0.08192	0.09961	0.12544	0.14748	0.17267	0.21420	-0.26
0.28	0.10325	0.12087	0.14656	0.16846	0.19347	0.23463	-0.28
0.30	0.12475	0.14228	0.16781	0.18955	0.21435	0.25511	-0.30
0.32	0.14642	0.16384	0.18918	0.21074	0.23531	0.27563	-0.32
0.34	0.16826	0.18554	0.21067	0.23203	0.25635	0.29620	-0.34
0.36	0.19027	0.20740	0.23230	0.25343	0.27747	0.31680	-0.36
0.38	0.21245	0.22942	0.25405	0.27494	0.29867	0.33745	-0.38
0.40	0.23480	0.25159	0.27593	0.29655	0.31996	0.35814	-0.40
0.42	0.25734	0.27392	0.29795	0.31827	0.34132	0.37887	-0.42
0.44	0.28005	0.29642	0.32009	0.34011	0.36277	0.39964	-0.44
0.46	0.30295	0.31907	0.34237	0.36205	0.38431	0.42046	-0.46
0.48	0.32603	0.34189	0.36478	0.38410	0.40592	0.44133	-0.48
0.50	0.34930	0.36487	0.38734	0.40626	0.42763	0.46223	-0.50
0.52	0.37276	0.38803	0.41002	0.42854	0.44942	0.48318	-0.52
0.54	0.39641	0.41135	0.43285	0.45093	0.47130	0.50418	-0.54
0.56	0.42026	0.43485	0.45582	0.47344	0.49326	0.52522	-0.56
0.58	0.44431	0.45852	0.47893	0.49606	0.51531	0.54630	-0.58
0.60	0.46855	0.48236	0.50219	0.51880	0.53745	0.56743	-0.60
0.62	0.49300	0.50639	0.52559	0.54166	0.55968	0.58861	-0.62
0.64	0.51765	0.53060	0.54914	0.56464	0.58200	0.60983	-0.64
0.66	0.54251	0.55499	0.57283	0.58774	0.60441	0.63110	-0.66
0.68	0.56758	0.57956	0.59668	0.61096	0.62692	0.65241	-0.68
0.70	0.59287	0.60432	0.62068	0.63430	0.64951	0.67377	-0.70
0.72	0.61837	0.62928	0.64483	0.65777	0.67220	0.69518	-0.72
0.74	0.64409	0.65442	0.66913	0.68136	0.69498	0.71664	-0.74
0.76	0.67004	0.67976	0.69359	0.70508	0.71785	0.73814	-0.76
0.78	0.69621	0.70530	0.71822	0.72892	0.74082	0.75969	-0.78
0.80	0.72261	0.73104	0.74300	0.75290	0.76389	0.78129	-0.80
0.82	0.74925	0.75698	0.76794	0.77700	0.78705	0.80294	-0.82
0.84	0.77612	0.78312	0.79304	0.80124	0.81031	0.82463	-0.84
0.86	0.80323	0.80948	0.81831	0.82561	0.83367	0.84638	-0.86
0.88	0.83058	0.83604	0.84375	0.85011	0.85713	0.86817	-0.88
0.90	0.85818	0.86282	0.86936	0.87474	0.88069	0.89002	-0.90
0.92	0.88603	0.88981	0.89514	0.89952	0.90435	0.91191	-0.92
0.94	0.91413	0.91702	0.92109	0.92443	0.92810	0.93386	-0.94
0.96	0.94249	0.94446	0.94722	0.94948	0.95197	0.95585	-0.96
0.98	0.97111	0.97211	0.97352	0.97467	0.97593	0.97790	-0.98

↑ r_o

α =	.995 ↑–	.99 ↑–	.975 ↑–	.95 ↑–	.90 ↑–	.75 ↑–

TABLE 83.1   SAMPLE SIZE =   225

α =	.005 ↓+	.01 ↓+	.025 ↓+	.05 ↓+	.10 ↓+	.25 ↓+	
r_0 ↓							
-0.98	-0.98574	-0.98526	-0.98452	-0.98386	-0.98306	-0.98163	0.98
-0.96	-0.97141	-0.97044	-0.96897	-0.96765	-0.96606	-0.96324	0.96
-0.94	-0.95699	-0.95555	-0.95335	-0.95138	-0.94902	-0.94481	0.94
-0.92	-0.94248	-0.94058	-0.93767	-0.93506	-0.93192	-0.92636	0.92
-0.90	-0.92790	-0.92553	-0.92191	-0.91867	-0.91478	-0.90789	0.90
-0.88	-0.91323	-0.91040	-0.90608	-0.90222	-0.89758	-0.88938	0.88
-0.86	-0.89848	-0.89519	-0.89018	-0.88570	-0.88034	-0.87085	0.86
-0.84	-0.88364	-0.87990	-0.87421	-0.86913	-0.86304	-0.85229	0.84
-0.82	-0.86872	-0.86453	-0.85817	-0.85249	-0.84569	-0.83370	0.82
-0.80	-0.85370	-0.84908	-0.84206	-0.83579	-0.82829	-0.81508	0.80
-0.78	-0.83860	-0.83355	-0.82587	-0.81902	-0.81083	-0.79643	0.78
-0.76	-0.82342	-0.81793	-0.80960	-0.80219	-0.79332	-0.77776	0.76
-0.74	-0.80814	-0.80223	-0.79327	-0.78529	-0.77577	-0.75905	0.74
-0.72	-0.79277	-0.78644	-0.77686	-0.76833	-0.75815	-0.74032	0.72
-0.70	-0.77731	-0.77057	-0.76037	-0.75130	-0.74048	-0.72155	0.70
-0.68	-0.76176	-0.75461	-0.74381	-0.73420	-0.72276	-0.70276	0.68
-0.66	-0.74612	-0.73857	-0.72716	-0.71704	-0.70499	-0.68394	0.66
-0.64	-0.73038	-0.72244	-0.71045	-0.69981	-0.68716	-0.66509	0.64
-0.62	-0.71454	-0.70622	-0.69365	-0.68251	-0.66927	-0.64621	0.62
-0.60	-0.69862	-0.68991	-0.67678	-0.66514	-0.65133	-0.62730	0.60
-0.58	-0.68259	-0.67351	-0.65982	-0.64771	-0.63334	-0.60835	0.58
-0.56	-0.66647	-0.65702	-0.64279	-0.63020	-0.61529	-0.58938	0.56
-0.54	-0.65024	-0.64044	-0.62567	-0.61263	-0.59718	-0.57038	0.54
-0.52	-0.63392	-0.62376	-0.60848	-0.59498	-0.57901	-0.55135	0.52
-0.50	-0.61750	-0.60700	-0.59120	-0.57726	-0.56079	-0.53228	0.50
-0.48	-0.60098	-0.59013	-0.57384	-0.55948	-0.54251	-0.51319	0.48
-0.46	-0.58435	-0.57318	-0.55640	-0.54162	-0.52417	-0.49406	0.46
-0.44	-0.56762	-0.55612	-0.53887	-0.52368	-0.50578	-0.47490	0.44
-0.42	-0.55078	-0.53897	-0.52126	-0.50568	-0.48732	-0.45571	0.42
-0.40	-0.53384	-0.52172	-0.50356	-0.48760	-0.46881	-0.43649	0.40
-0.38	-0.51680	-0.50437	-0.48578	-0.46945	-0.45024	-0.41724	0.38
-0.36	-0.49964	-0.48693	-0.46791	-0.45122	-0.43160	-0.39795	0.36
-0.34	-0.48238	-0.46938	-0.44995	-0.43292	-0.41291	-0.37863	0.34
-0.32	-0.46500	-0.45173	-0.43190	-0.41454	-0.39416	-0.35928	0.32
-0.30	-0.44751	-0.43398	-0.41377	-0.39608	-0.37535	-0.33990	0.30
-0.28	-0.42992	-0.41612	-0.39554	-0.37755	-0.35647	-0.32048	0.28
-0.26	-0.41220	-0.39816	-0.37723	-0.35894	-0.33753	-0.30103	0.26
-0.24	-0.39438	-0.38009	-0.35882	-0.34025	-0.31854	-0.28155	0.24
-0.22	-0.37643	-0.36192	-0.34032	-0.32149	-0.29948	-0.26203	0.22
-0.20	-0.35837	-0.34364	-0.32173	-0.30264	-0.28035	-0.24248	0.20
-0.18	-0.34019	-0.32525	-0.30305	-0.28372	-0.26116	-0.22290	0.18
-0.16	-0.32189	-0.30675	-0.28427	-0.26471	-0.24191	-0.20328	0.16
-0.14	-0.30347	-0.28814	-0.26540	-0.24562	-0.22260	-0.18363	0.14
-0.12	-0.28493	-0.26942	-0.24643	-0.22645	-0.20322	-0.16394	0.12
-0.10	-0.26626	-0.25058	-0.22736	-0.20720	-0.18378	-0.14422	0.10
-0.08	-0.24747	-0.23163	-0.20819	-0.18787	-0.16427	-0.12447	0.08
-0.06	-0.22855	-0.21257	-0.18893	-0.16845	-0.14469	-0.10468	0.06
-0.04	-0.20951	-0.19339	-0.16957	-0.14895	-0.12505	-0.08485	0.04
-0.02	-0.19033	-0.17409	-0.15010	-0.12936	-0.10534	-0.06499	0.02

↑ r_0

α =	.995 ↑-	.99 ↑-	.975 ↑-	.95 ↑-	.90 ↑-	.75 ↑-

TABLE 83.2   SAMPLE SIZE =   225   (CONT.)

α =	.005 ↓+	.01 ↓+	.025 ↓+	.05 ↓+	.10 ↓+	.25 ↓+	
r_0 ↓							
0.00	−0.17103	−0.15467	−0.13054	−0.10969	−0.08557	−0.04509	0.00
0.02	−0.15159	−0.13513	−0.11087	−0.08993	−0.06572	−0.02516	−0.02
0.04	−0.13202	−0.11547	−0.09110	−0.07009	−0.04581	−0.00519	−0.04
0.06	−0.11231	−0.09569	−0.07123	−0.05016	−0.02583	0.01481	−0.06
0.08	−0.09247	−0.07579	−0.05125	−0.03013	−0.00578	0.03485	−0.08
0.10	−0.07249	−0.05576	−0.03117	−0.01002	0.01433	0.05493	−0.10
0.12	−0.05237	−0.03560	−0.01098	0.01018	0.03452	0.07505	−0.12
0.14	−0.03210	−0.01531	0.00932	0.03047	0.05478	0.09520	−0.14
0.16	−0.01170	0.00510	0.02973	0.05085	0.07511	0.11539	−0.16
0.18	0.00885	0.02565	0.05024	0.07132	0.09551	0.13561	−0.18
0.20	0.02955	0.04632	0.07087	0.09188	0.11598	0.15588	−0.20
0.22	0.05039	0.06713	0.09161	0.11254	0.13653	0.17618	−0.22
0.24	0.07138	0.08807	0.11246	0.13330	0.15714	0.19652	−0.24
0.26	0.09252	0.10915	0.13343	0.15414	0.17783	0.21690	−0.26
0.28	0.11382	0.13037	0.15451	0.17509	0.19860	0.23731	−0.28
0.30	0.13527	0.15173	0.17570	0.19613	0.21944	0.25777	−0.30
0.32	0.15688	0.17322	0.19702	0.21726	0.24035	0.27827	−0.32
0.34	0.17865	0.19486	0.21845	0.23850	0.26134	0.29880	−0.34
0.36	0.20057	0.21664	0.24000	0.25983	0.28240	0.31937	−0.36
0.38	0.22266	0.23857	0.26167	0.28126	0.30354	0.33999	−0.38
0.40	0.24492	0.26065	0.28346	0.30280	0.32476	0.36064	−0.40
0.42	0.26734	0.28287	0.30537	0.32443	0.34606	0.38133	−0.42
0.44	0.28992	0.30524	0.32741	0.34617	0.36743	0.40207	−0.44
0.46	0.31268	0.32776	0.34957	0.36801	0.38888	0.42284	−0.46
0.48	0.33561	0.35044	0.37186	0.38995	0.41042	0.44366	−0.48
0.50	0.35872	0.37327	0.39428	0.41200	0.43203	0.46451	−0.50
0.52	0.38200	0.39626	0.41683	0.43415	0.45372	0.48541	−0.52
0.54	0.40546	0.41940	0.43950	0.45641	0.47549	0.50635	−0.54
0.56	0.42910	0.44271	0.46231	0.47878	0.49734	0.52733	−0.56
0.58	0.45292	0.46618	0.48524	0.50126	0.51928	0.54835	−0.58
0.60	0.47693	0.48981	0.50832	0.52384	0.54130	0.56942	−0.60
0.62	0.50113	0.51361	0.53152	0.54654	0.56340	0.59052	−0.62
0.64	0.52552	0.53758	0.55487	0.56934	0.58559	0.61168	−0.64
0.66	0.55010	0.56171	0.57835	0.59226	0.60786	0.63287	−0.66
0.68	0.57487	0.58602	0.60197	0.61529	0.63021	0.65411	−0.68
0.70	0.59985	0.61050	0.62573	0.63844	0.65265	0.67539	−0.70
0.72	0.62502	0.63516	0.64963	0.66170	0.67518	0.69671	−0.72
0.74	0.65039	0.65999	0.67368	0.68507	0.69780	0.71808	−0.74
0.76	0.67598	0.68500	0.69787	0.70857	0.72050	0.73949	−0.76
0.78	0.70177	0.71020	0.72220	0.73218	0.74329	0.76095	−0.78
0.80	0.72777	0.73558	0.74669	0.75591	0.76617	0.78245	−0.80
0.82	0.75398	0.76114	0.77132	0.77976	0.78914	0.80400	−0.82
0.84	0.78041	0.78690	0.79611	0.80373	0.81219	0.82559	−0.84
0.86	0.80706	0.81284	0.82104	0.82782	0.83534	0.84723	−0.86
0.88	0.83393	0.83898	0.84613	0.85204	0.85859	0.86891	−0.88
0.90	0.86103	0.86531	0.87138	0.87638	0.88192	0.89064	−0.90
0.92	0.88835	0.89185	0.89678	0.90085	0.90535	0.91242	−0.92
0.94	0.91591	0.91858	0.92234	0.92545	0.92887	0.93425	−0.94
0.96	0.94370	0.94551	0.94807	0.95017	0.95248	0.95612	−0.96
0.98	0.97173	0.97265	0.97395	0.97502	0.97619	0.97803	−0.98

↑ r_0

α =	.995 ↑−	.99 ↑−	.975 ↑−	.95 ↑−	.90 ↑−	.75 ↑−

TABLE 84.1   SAMPLE SIZE =   250

α =	.005 ↓+	.01 ↓+	.025 ↓+	.05 ↓+	.10 ↓+	.25 ↓+	
r_0 ↓							
−0.98	−0.98549	−0.98503	−0.98432	−0.98368	−0.98291	−0.98155	0.98
−0.96	−0.97091	−0.96998	−0.96857	−0.96730	−0.96578	−0.96308	0.96
−0.94	−0.95624	−0.95486	−0.95275	−0.95086	−0.94859	−0.94459	0.94
−0.92	−0.94150	−0.93966	−0.93686	−0.93436	−0.93136	−0.92606	0.92
−0.90	−0.92667	−0.92439	−0.92091	−0.91781	−0.91408	−0.90751	0.90
−0.88	−0.91176	−0.90904	−0.90489	−0.90119	−0.89675	−0.88893	0.88
−0.86	−0.89677	−0.89361	−0.88880	−0.88451	−0.87938	−0.87033	0.86
−0.84	−0.88170	−0.87810	−0.87264	−0.86777	−0.86195	−0.85170	0.84
−0.82	−0.86654	−0.86252	−0.85641	−0.85097	−0.84447	−0.83304	0.82
−0.80	−0.85130	−0.84686	−0.84012	−0.83411	−0.82694	−0.81435	0.80
−0.78	−0.83597	−0.83111	−0.82375	−0.81719	−0.80937	−0.79564	0.78
−0.76	−0.82056	−0.81529	−0.80731	−0.80020	−0.79174	−0.77690	0.76
−0.74	−0.80506	−0.79938	−0.79079	−0.78316	−0.77406	−0.75813	0.74
−0.72	−0.78947	−0.78340	−0.77421	−0.76604	−0.75633	−0.73933	0.72
−0.70	−0.77380	−0.76733	−0.75755	−0.74887	−0.73854	−0.72051	0.70
−0.68	−0.75803	−0.75118	−0.74082	−0.73163	−0.72071	−0.70166	0.68
−0.66	−0.74218	−0.73494	−0.72401	−0.71433	−0.70282	−0.68278	0.66
−0.64	−0.72623	−0.71862	−0.70713	−0.69696	−0.68489	−0.66387	0.64
−0.62	−0.71020	−0.70221	−0.69018	−0.67953	−0.66689	−0.64493	0.62
−0.60	−0.69407	−0.68572	−0.67315	−0.66203	−0.64885	−0.62596	0.60
−0.58	−0.67784	−0.66914	−0.65604	−0.64446	−0.63075	−0.60696	0.58
−0.56	−0.66152	−0.65247	−0.63885	−0.62683	−0.61260	−0.58794	0.56
−0.54	−0.64511	−0.63572	−0.62159	−0.60913	−0.59439	−0.56889	0.54
−0.52	−0.62860	−0.61887	−0.60425	−0.59136	−0.57613	−0.54980	0.52
−0.50	−0.61199	−0.60193	−0.58683	−0.57353	−0.55782	−0.53069	0.50
−0.48	−0.59529	−0.58491	−0.56933	−0.55562	−0.53945	−0.51155	0.48
−0.46	−0.57848	−0.56779	−0.55176	−0.53765	−0.52102	−0.49237	0.46
−0.44	−0.56158	−0.55058	−0.53410	−0.51961	−0.50254	−0.47317	0.44
−0.42	−0.54458	−0.53328	−0.51636	−0.50149	−0.48400	−0.45394	0.42
−0.40	−0.52747	−0.51588	−0.49854	−0.48331	−0.46541	−0.43467	0.40
−0.38	−0.51026	−0.49839	−0.48063	−0.46506	−0.44676	−0.41538	0.38
−0.36	−0.49295	−0.48080	−0.46264	−0.44673	−0.42805	−0.39605	0.36
−0.34	−0.47553	−0.46311	−0.44457	−0.42834	−0.40929	−0.37670	0.34
−0.32	−0.45800	−0.44533	−0.42642	−0.40987	−0.39047	−0.35731	0.32
−0.30	−0.44037	−0.42745	−0.40817	−0.39132	−0.37159	−0.33789	0.30
−0.28	−0.42263	−0.40947	−0.38985	−0.37271	−0.35265	−0.31844	0.28
−0.26	−0.40479	−0.39139	−0.37143	−0.35402	−0.33365	−0.29896	0.26
−0.24	−0.38683	−0.37320	−0.35293	−0.33525	−0.31459	−0.27945	0.24
−0.22	−0.36876	−0.35492	−0.33434	−0.31641	−0.29548	−0.25991	0.22
−0.20	−0.35058	−0.33653	−0.31567	−0.29750	−0.27630	−0.24033	0.20
−0.18	−0.33228	−0.31804	−0.29690	−0.27851	−0.25707	−0.22072	0.18
−0.16	−0.31387	−0.29945	−0.27804	−0.25944	−0.23777	−0.20108	0.16
−0.14	−0.29534	−0.28074	−0.25910	−0.24029	−0.21841	−0.18141	0.14
−0.12	−0.27670	−0.26193	−0.24006	−0.22107	−0.19899	−0.16170	0.12
−0.10	−0.25794	−0.24302	−0.22092	−0.20177	−0.17951	−0.14196	0.10
−0.08	−0.23906	−0.22399	−0.20170	−0.18238	−0.15997	−0.12219	0.08
−0.06	−0.22006	−0.20485	−0.18238	−0.16292	−0.14036	−0.10238	0.06
−0.04	−0.20093	−0.18561	−0.16297	−0.14338	−0.12069	−0.08255	0.04
−0.02	−0.18169	−0.16625	−0.14346	−0.12376	−0.10096	−0.06267	0.02

↑ r_0

α =	.995 ↑−	.99 ↑−	.975 ↑−	.95 ↑−	.90 ↑−	.75 ↑−

TABLE 84.2   SAMPLE SIZE =   250   (CONT.)

α =	.005 ↓+	.01 ↓+	.025 ↓+	.05 ↓+	.10 ↓+	.25 ↓+	
r₀ ↓							
0.00	−0.16232	−0.14677	−0.12385	−0.10406	−0.08116	−0.04277	0.00
0.02	−0.14282	−0.12719	−0.10415	−0.08427	−0.06130	−0.02283	−0.02
0.04	−0.12319	−0.10748	−0.08435	−0.06441	−0.04138	−0.00285	−0.04
0.06	−0.10344	−0.08767	−0.06445	−0.04446	−0.02139	0.01716	−0.06
0.08	−0.08356	−0.06773	−0.04445	−0.02442	−0.00133	0.03720	−0.08
0.10	−0.06355	−0.04767	−0.02436	−0.00430	0.01879	0.05728	−0.10
0.12	−0.04340	−0.02750	−0.00415	0.01590	0.03898	0.07739	−0.12
0.14	−0.02312	−0.00720	0.01615	0.03619	0.05923	0.09754	−0.14
0.16	−0.00270	0.01322	0.03655	0.05656	0.07955	0.11772	−0.16
0.18	0.01785	0.03376	0.05706	0.07703	0.09994	0.13794	−0.18
0.20	0.03854	0.05443	0.07768	0.09758	0.12040	0.15819	−0.20
0.22	0.05937	0.07522	0.09839	0.11821	0.14092	0.17848	−0.22
0.24	0.08034	0.09614	0.11922	0.13894	0.16152	0.19881	−0.24
0.26	0.10146	0.11719	0.14015	0.15976	0.18218	0.21917	−0.26
0.28	0.12272	0.13837	0.16120	0.18067	0.20291	0.23957	−0.28
0.30	0.14413	0.15968	0.18235	0.20167	0.22372	0.26001	−0.30
0.32	0.16568	0.18112	0.20361	0.22276	0.24459	0.28048	−0.32
0.34	0.18738	0.20270	0.22498	0.24394	0.26554	0.30099	−0.34
0.36	0.20924	0.22441	0.24647	0.26521	0.28655	0.32154	−0.36
0.38	0.23125	0.24626	0.26807	0.28658	0.30764	0.34212	−0.38
0.40	0.25341	0.26825	0.28978	0.30805	0.32881	0.36274	−0.40
0.42	0.27573	0.29038	0.31161	0.32961	0.35004	0.38340	−0.42
0.44	0.29820	0.31264	0.33356	0.35126	0.37135	0.40410	−0.44
0.46	0.32084	0.33505	0.35562	0.37302	0.39273	0.42484	−0.46
0.48	0.34364	0.35761	0.37780	0.39487	0.41419	0.44561	−0.48
0.50	0.36660	0.38031	0.40010	0.41682	0.43572	0.46643	−0.50
0.52	0.38973	0.40315	0.42253	0.43887	0.45733	0.48728	−0.52
0.54	0.41303	0.42615	0.44507	0.46101	0.47901	0.50817	−0.54
0.56	0.43649	0.44929	0.46774	0.48326	0.50077	0.52910	−0.56
0.58	0.46013	0.47259	0.49053	0.50561	0.52261	0.55007	−0.58
0.60	0.48394	0.49604	0.51345	0.52807	0.54453	0.57108	−0.60
0.62	0.50792	0.51964	0.53649	0.55062	0.56652	0.59213	−0.62
0.64	0.53208	0.54341	0.55966	0.57328	0.58859	0.61322	−0.64
0.66	0.55643	0.56733	0.58296	0.59605	0.61074	0.63436	−0.66
0.68	0.58095	0.59141	0.60639	0.61892	0.63297	0.65553	−0.68
0.70	0.60566	0.61565	0.62995	0.64190	0.65529	0.67674	−0.70
0.72	0.63056	0.64006	0.65364	0.66499	0.67768	0.69799	−0.72
0.74	0.65564	0.66463	0.67747	0.68818	0.70015	0.71929	−0.74
0.76	0.68091	0.68937	0.70143	0.71148	0.72271	0.74063	−0.76
0.78	0.70638	0.71428	0.72553	0.73490	0.74535	0.76200	−0.78
0.80	0.73205	0.73935	0.74977	0.75842	0.76807	0.78343	−0.80
0.82	0.75791	0.76461	0.77414	0.78206	0.79088	0.80489	−0.82
0.84	0.78397	0.79003	0.79866	0.80581	0.81377	0.82639	−0.84
0.86	0.81024	0.81564	0.82331	0.82967	0.83674	0.84794	−0.86
0.88	0.83671	0.84142	0.84811	0.85365	0.85980	0.86953	−0.88
0.90	0.86339	0.86739	0.87306	0.87775	0.88295	0.89117	−0.90
0.92	0.89028	0.89353	0.89815	0.90196	0.90618	0.91285	−0.92
0.94	0.91738	0.91987	0.92339	0.92629	0.92950	0.93457	−0.94
0.96	0.94470	0.94639	0.94877	0.95074	0.95291	0.95633	−0.96
0.98	0.97224	0.97310	0.97431	0.97531	0.97641	0.97815	−0.98

↑ r₀

α =	.995 ↑−	.99 ↑−	.975 ↑−	.95 ↑−	.90 ↑−	.75 ↑−

TABLE 85.1   SAMPLE SIZE =   275

α =	.005 ↓+	.01 ↓+	.025 ↓+	.05 ↓+	.10 ↓+	.25 ↓+	
r₀ ↓							
-0.98	-0.98527	-0.98482	-0.98414	-0.98352	-0.98279	-0.98149	0.98
-0.96	-0.97047	-0.96958	-0.96821	-0.96699	-0.96553	-0.96295	0.96
-0.94	-0.95559	-0.95425	-0.95222	-0.95040	-0.94823	-0.94439	0.94
-0.92	-0.94063	-0.93886	-0.93616	-0.93376	-0.93088	-0.92580	0.92
-0.90	-0.92560	-0.92339	-0.92004	-0.91705	-0.91348	-0.90719	0.90
-0.88	-0.91048	-0.90785	-0.90385	-0.90029	-0.89604	-0.88855	0.88
-0.86	-0.89528	-0.89223	-0.88760	-0.88347	-0.87854	-0.86988	0.86
-0.84	-0.88001	-0.87654	-0.87128	-0.86659	-0.86100	-0.85119	0.84
-0.82	-0.86465	-0.86077	-0.85489	-0.84966	-0.84342	-0.83247	0.82
-0.80	-0.84921	-0.84492	-0.83843	-0.83266	-0.82578	-0.81373	0.80
-0.78	-0.83368	-0.82900	-0.82190	-0.81560	-0.80809	-0.79496	0.78
-0.76	-0.81807	-0.81299	-0.80531	-0.79848	-0.79036	-0.77616	0.76
-0.74	-0.80238	-0.79691	-0.78864	-0.78131	-0.77258	-0.75733	0.74
-0.72	-0.78660	-0.78075	-0.77191	-0.76407	-0.75475	-0.73848	0.72
-0.70	-0.77074	-0.76451	-0.75510	-0.74677	-0.73686	-0.71961	0.70
-0.68	-0.75479	-0.74819	-0.73823	-0.72940	-0.71893	-0.70070	0.68
-0.66	-0.73875	-0.73178	-0.72128	-0.71198	-0.70095	-0.68177	0.66
-0.64	-0.72263	-0.71530	-0.70426	-0.69449	-0.68292	-0.66281	0.64
-0.62	-0.70641	-0.69873	-0.68716	-0.67694	-0.66483	-0.64382	0.62
-0.60	-0.69011	-0.68208	-0.67000	-0.65933	-0.64670	-0.62481	0.60
-0.58	-0.67371	-0.66534	-0.65276	-0.64165	-0.62851	-0.60576	0.58
-0.56	-0.65723	-0.64852	-0.63544	-0.62391	-0.61028	-0.58669	0.56
-0.54	-0.64065	-0.63162	-0.61805	-0.60610	-0.59199	-0.56760	0.54
-0.52	-0.62398	-0.61463	-0.60059	-0.58823	-0.57364	-0.54847	0.52
-0.50	-0.60721	-0.59755	-0.58305	-0.57029	-0.55525	-0.52931	0.50
-0.48	-0.59035	-0.58038	-0.56543	-0.55229	-0.53680	-0.51013	0.48
-0.46	-0.57340	-0.56312	-0.54774	-0.53422	-0.51830	-0.49092	0.46
-0.44	-0.55634	-0.54578	-0.52997	-0.51608	-0.49975	-0.47168	0.44
-0.42	-0.53919	-0.52834	-0.51212	-0.49788	-0.48114	-0.45240	0.42
-0.40	-0.52194	-0.51082	-0.49419	-0.47961	-0.46248	-0.43311	0.40
-0.38	-0.50460	-0.49320	-0.47618	-0.46127	-0.44376	-0.41378	0.38
-0.36	-0.48715	-0.47549	-0.45809	-0.44286	-0.42499	-0.39442	0.36
-0.34	-0.46960	-0.45769	-0.43992	-0.42438	-0.40616	-0.37503	0.34
-0.32	-0.45195	-0.43979	-0.42167	-0.40583	-0.38728	-0.35561	0.32
-0.30	-0.43419	-0.42180	-0.40334	-0.38722	-0.36835	-0.33617	0.30
-0.28	-0.41633	-0.40371	-0.38493	-0.36853	-0.34935	-0.31669	0.28
-0.26	-0.39837	-0.38553	-0.36643	-0.34977	-0.33030	-0.29718	0.26
-0.24	-0.38030	-0.36725	-0.34785	-0.33094	-0.31120	-0.27764	0.24
-0.22	-0.36212	-0.34887	-0.32918	-0.31204	-0.29203	-0.25808	0.22
-0.20	-0.34384	-0.33039	-0.31043	-0.29306	-0.27281	-0.23848	0.20
-0.18	-0.32545	-0.31182	-0.29160	-0.27401	-0.25353	-0.21885	0.18
-0.16	-0.30694	-0.29314	-0.27267	-0.25489	-0.23420	-0.19919	0.16
-0.14	-0.28833	-0.27436	-0.25366	-0.23570	-0.21480	-0.17949	0.14
-0.12	-0.26960	-0.25548	-0.23457	-0.21643	-0.19535	-0.15977	0.12
-0.10	-0.25076	-0.23649	-0.21538	-0.19708	-0.17584	-0.14002	0.10
-0.08	-0.23180	-0.21740	-0.19610	-0.17766	-0.15627	-0.12023	0.08
-0.06	-0.21273	-0.19820	-0.17674	-0.15817	-0.13663	-0.10041	0.06
-0.04	-0.19354	-0.17890	-0.15728	-0.13859	-0.11694	-0.08056	0.04
-0.02	-0.17424	-0.15949	-0.13774	-0.11894	-0.09719	-0.06068	0.02

↑ r₀

α =	.995 ↑−	.99 ↑−	.975 ↑−	.95 ↑−	.90 ↑−	.75 ↑−

TABLE 85.2   SAMPLE SIZE =   275   (CONT.)

α =	.005 ↓+	.01 ↓+	.025 ↓+	.05 ↓+	.10 ↓+	.25 ↓+	
r_0 ↓							
0.00	−0.15481	−0.13997	−0.11810	−0.09921	−0.07738	−0.04077	0.00
0.02	−0.13527	−0.12035	−0.09837	−0.07941	−0.05750	−0.02082	−0.02
0.04	−0.11560	−0.10061	−0.07854	−0.05952	−0.03757	−0.00084	−0.04
0.06	−0.09581	−0.08076	−0.05862	−0.03956	−0.01757	0.01917	−0.06
0.08	−0.07590	−0.06080	−0.03861	−0.01952	0.00249	0.03921	−0.08
0.10	−0.05586	−0.04073	−0.01850	0.00061	0.02262	0.05929	−0.10
0.12	−0.03569	−0.02054	0.00171	0.02082	0.04280	0.07940	−0.12
0.14	−0.01540	−0.00023	0.02201	0.04110	0.06305	0.09955	−0.14
0.16	0.00502	0.02019	0.04241	0.06147	0.08337	0.11973	−0.16
0.18	0.02557	0.04073	0.06291	0.08192	0.10375	0.13994	−0.18
0.20	0.04626	0.06138	0.08351	0.10246	0.12419	0.16018	−0.20
0.22	0.06708	0.08216	0.10422	0.12308	0.14470	0.18046	−0.22
0.24	0.08803	0.10306	0.12502	0.14378	0.16527	0.20078	−0.24
0.26	0.10912	0.12408	0.14592	0.16457	0.18591	0.22113	−0.26
0.28	0.13034	0.14523	0.16693	0.18545	0.20662	0.24151	−0.28
0.30	0.15171	0.16649	0.18804	0.20641	0.22739	0.26193	−0.30
0.32	0.17321	0.18789	0.20926	0.22746	0.24823	0.28238	−0.32
0.34	0.19486	0.20941	0.23058	0.24860	0.26914	0.30287	−0.34
0.36	0.21665	0.23106	0.25201	0.26982	0.29011	0.32339	−0.36
0.38	0.23859	0.25284	0.27355	0.29114	0.31115	0.34395	−0.38
0.40	0.26067	0.27475	0.29519	0.31254	0.33227	0.36455	−0.40
0.42	0.28290	0.29679	0.31695	0.33404	0.35345	0.38518	−0.42
0.44	0.30528	0.31897	0.33881	0.35562	0.37470	0.40585	−0.44
0.46	0.32781	0.34128	0.36079	0.37730	0.39602	0.42655	−0.46
0.48	0.35049	0.36373	0.38288	0.39907	0.41742	0.44729	−0.48
0.50	0.37333	0.38631	0.40508	0.42093	0.43888	0.46807	−0.50
0.52	0.39632	0.40903	0.42739	0.44289	0.46041	0.48888	−0.52
0.54	0.41948	0.43190	0.44983	0.46494	0.48202	0.50973	−0.54
0.56	0.44279	0.45490	0.47237	0.48709	0.50370	0.53062	−0.56
0.58	0.46626	0.47805	0.49504	0.50933	0.52545	0.55154	−0.58
0.60	0.48990	0.50134	0.51782	0.53167	0.54728	0.57251	−0.60
0.62	0.51370	0.52478	0.54072	0.55411	0.56918	0.59351	−0.62
0.64	0.53767	0.54837	0.56374	0.57664	0.59116	0.61455	−0.64
0.66	0.56181	0.57210	0.58689	0.59928	0.61321	0.63562	−0.66
0.68	0.58612	0.59599	0.61015	0.62201	0.63533	0.65674	−0.68
0.70	0.61060	0.62003	0.63354	0.64485	0.65753	0.67790	−0.70
0.72	0.63526	0.64422	0.65706	0.66779	0.67981	0.69909	−0.72
0.74	0.66009	0.66857	0.68070	0.69082	0.70216	0.72032	−0.74
0.76	0.68510	0.69307	0.70446	0.71397	0.72459	0.74159	−0.76
0.78	0.71030	0.71774	0.72836	0.73721	0.74710	0.76290	−0.78
0.80	0.73567	0.74256	0.75238	0.76056	0.76969	0.78426	−0.80
0.82	0.76123	0.76754	0.77654	0.78402	0.79236	0.80565	−0.82
0.84	0.78698	0.79269	0.80082	0.80758	0.81510	0.82708	−0.84
0.86	0.81292	0.81801	0.82524	0.83125	0.83793	0.84855	−0.86
0.88	0.83905	0.84349	0.84979	0.85502	0.86084	0.87006	−0.88
0.90	0.86538	0.86914	0.87448	0.87891	0.88382	0.89162	−0.90
0.92	0.89190	0.89496	0.89931	0.90290	0.90689	0.91321	−0.92
0.94	0.91862	0.92096	0.92427	0.92701	0.93005	0.93484	−0.94
0.96	0.94554	0.94713	0.94937	0.95123	0.95328	0.95652	−0.96
0.98	0.97267	0.97347	0.97462	0.97556	0.97660	0.97824	−0.98

↑ r_0

α =	.995 ↑−	.99 ↑−	.975 ↑−	.95 ↑−	.90 ↑−	.75 ↑−

TABLE 86.1   SAMPLE SIZE =   300

α =	.005 ↓+	.01 ↓+	.025 ↓+	.05 ↓+	.10 ↓+	.25 ↓+	
r_0 ↓							
-0.98	-0.98508	-0.98464	-0.98398	-0.98339	-0.98268	-0.98143	0.98
-0.96	-0.97009	-0.96922	-0.96790	-0.96672	-0.96532	-0.96283	0.96
-0.94	-0.95502	-0.95372	-0.95176	-0.95000	-0.94791	-0.94421	0.94
-0.92	-0.93987	-0.93815	-0.93555	-0.93323	-0.93045	-0.92557	0.92
-0.90	-0.92465	-0.92251	-0.91928	-0.91639	-0.91295	-0.90690	0.90
-0.88	-0.90935	-0.90680	-0.90294	-0.89951	-0.89541	-0.88821	0.88
-0.86	-0.89397	-0.89102	-0.88654	-0.88256	-0.87781	-0.86949	0.86
-0.84	-0.87851	-0.87516	-0.87008	-0.86556	-0.86017	-0.85074	0.84
-0.82	-0.86298	-0.85922	-0.85354	-0.84850	-0.84249	-0.83197	0.82
-0.80	-0.84736	-0.84321	-0.83695	-0.83138	-0.82476	-0.81318	0.80
-0.78	-0.83166	-0.82713	-0.82028	-0.81421	-0.80698	-0.79436	0.78
-0.76	-0.81588	-0.81097	-0.80355	-0.79697	-0.78915	-0.77551	0.76
-0.74	-0.80002	-0.79473	-0.78675	-0.77968	-0.77128	-0.75664	0.74
-0.72	-0.78407	-0.77842	-0.76989	-0.76233	-0.75336	-0.73774	0.72
-0.70	-0.76805	-0.76203	-0.75295	-0.74492	-0.73539	-0.71881	0.70
-0.68	-0.75193	-0.74556	-0.73595	-0.72745	-0.71737	-0.69986	0.68
-0.66	-0.73574	-0.72901	-0.71887	-0.70992	-0.69931	-0.68089	0.66
-0.64	-0.71945	-0.71238	-0.70173	-0.69233	-0.68119	-0.66188	0.64
-0.62	-0.70308	-0.69567	-0.68452	-0.67467	-0.66303	-0.64285	0.62
-0.60	-0.68663	-0.67888	-0.66723	-0.65696	-0.64482	-0.62380	0.60
-0.58	-0.67008	-0.66201	-0.64988	-0.63918	-0.62655	-0.60471	0.58
-0.56	-0.65345	-0.64505	-0.63245	-0.62135	-0.60824	-0.58560	0.56
-0.54	-0.63673	-0.62802	-0.61495	-0.60345	-0.58988	-0.56647	0.54
-0.52	-0.61991	-0.61090	-0.59738	-0.58548	-0.57146	-0.54730	0.52
-0.50	-0.60301	-0.59369	-0.57973	-0.56746	-0.55300	-0.52811	0.50
-0.48	-0.58601	-0.57640	-0.56201	-0.54937	-0.53449	-0.50889	0.48
-0.46	-0.56892	-0.55903	-0.54421	-0.53121	-0.51592	-0.48964	0.46
-0.44	-0.55174	-0.54156	-0.52634	-0.51299	-0.49730	-0.47037	0.44
-0.42	-0.53446	-0.52402	-0.50840	-0.49471	-0.47863	-0.45107	0.42
-0.40	-0.51709	-0.50638	-0.49038	-0.47636	-0.45991	-0.43174	0.40
-0.38	-0.49962	-0.48865	-0.47228	-0.45795	-0.44114	-0.41238	0.38
-0.36	-0.48206	-0.47084	-0.45410	-0.43947	-0.42231	-0.39299	0.36
-0.34	-0.46440	-0.45294	-0.43585	-0.42092	-0.40343	-0.37357	0.34
-0.32	-0.44663	-0.43494	-0.41752	-0.40230	-0.38450	-0.35413	0.32
-0.30	-0.42877	-0.41685	-0.39911	-0.38362	-0.36551	-0.33466	0.30
-0.28	-0.41081	-0.39868	-0.38062	-0.36487	-0.34647	-0.31516	0.28
-0.26	-0.39275	-0.38040	-0.36205	-0.34605	-0.32738	-0.29563	0.26
-0.24	-0.37458	-0.36204	-0.34340	-0.32717	-0.30823	-0.27607	0.24
-0.22	-0.35631	-0.34358	-0.32467	-0.30821	-0.28902	-0.25648	0.22
-0.20	-0.33794	-0.32502	-0.30585	-0.28918	-0.26976	-0.23686	0.20
-0.18	-0.31946	-0.30637	-0.28696	-0.27009	-0.25045	-0.21721	0.18
-0.16	-0.30088	-0.28762	-0.26798	-0.25092	-0.23108	-0.19753	0.16
-0.14	-0.28219	-0.26878	-0.24892	-0.23168	-0.21165	-0.17783	0.14
-0.12	-0.26339	-0.24983	-0.22977	-0.21237	-0.19217	-0.15809	0.12
-0.10	-0.24448	-0.23079	-0.21054	-0.19299	-0.17263	-0.13832	0.10
-0.08	-0.22546	-0.21164	-0.19122	-0.17354	-0.15304	-0.11852	0.08
-0.06	-0.20633	-0.19240	-0.17181	-0.15401	-0.13338	-0.09869	0.06
-0.04	-0.18709	-0.17305	-0.15232	-0.13441	-0.11367	-0.07883	0.04
-0.02	-0.16773	-0.15360	-0.13275	-0.11474	-0.09390	-0.05895	0.02

↑ r_0

α =	.995 ↑-	.99 ↑-	.975 ↑-	.95 ↑-	.90 ↑-	.75 ↑-

TABLE 86.2   SAMPLE SIZE = 300 (CONT.)

α =	.005 ↓+	.01 ↓+	.025 ↓+	.05 ↓+	.10 ↓+	.25 ↓+	
r_0 ↓							
0.00	−0.14826	−0.13404	−0.11308	−0.09499	−0.07407	−0.03902	0.00
0.02	−0.12868	−0.11438	−0.09332	−0.07516	−0.05419	−0.01907	−0.02
0.04	−0.10898	−0.09461	−0.07348	−0.05527	−0.03424	0.00091	−0.04
0.06	−0.08916	−0.07474	−0.05354	−0.03529	−0.01424	0.02092	−0.06
0.08	−0.06922	−0.05476	−0.03351	−0.01524	0.00583	0.04097	−0.08
0.10	−0.04916	−0.03467	−0.01340	0.00489	0.02595	0.06105	−0.10
0.12	−0.02898	−0.01447	0.00681	0.02510	0.04614	0.08116	−0.12
0.14	−0.00868	0.00584	0.02712	0.04538	0.06638	0.10130	−0.14
0.16	0.01175	0.02626	0.04752	0.06575	0.08669	0.12147	−0.16
0.18	0.03230	0.04679	0.06801	0.08619	0.10706	0.14168	−0.18
0.20	0.05298	0.06744	0.08860	0.10671	0.12749	0.16191	−0.20
0.22	0.07378	0.08820	0.10928	0.12731	0.14798	0.18218	−0.22
0.24	0.09471	0.10908	0.13006	0.14800	0.16853	0.20249	−0.24
0.26	0.11578	0.13007	0.15094	0.16876	0.18915	0.22282	−0.26
0.28	0.13697	0.15118	0.17192	0.18961	0.20983	0.24319	−0.28
0.30	0.15830	0.17241	0.19299	0.21054	0.23058	0.26360	−0.30
0.32	0.17976	0.19377	0.21417	0.23155	0.25139	0.28403	−0.32
0.34	0.20135	0.21524	0.23545	0.25265	0.27226	0.30450	−0.34
0.36	0.22309	0.23683	0.25682	0.27383	0.29320	0.32501	−0.36
0.38	0.24496	0.25855	0.27831	0.29509	0.31420	0.34554	−0.38
0.40	0.26696	0.28039	0.29989	0.31644	0.33527	0.36611	−0.40
0.42	0.28911	0.30236	0.32158	0.33788	0.35641	0.38672	−0.42
0.44	0.31141	0.32446	0.34337	0.35940	0.37761	0.40736	−0.44
0.46	0.33384	0.34668	0.36527	0.38102	0.39888	0.42804	−0.46
0.48	0.35642	0.36903	0.38728	0.40271	0.42022	0.44874	−0.48
0.50	0.37915	0.39151	0.40939	0.42450	0.44162	0.46949	−0.50
0.52	0.40203	0.41413	0.43161	0.44638	0.46309	0.49027	−0.52
0.54	0.42506	0.43688	0.45394	0.46835	0.48463	0.51108	−0.54
0.56	0.44823	0.45976	0.47639	0.49040	0.50624	0.53193	−0.56
0.58	0.47156	0.48278	0.49894	0.51255	0.52792	0.55282	−0.58
0.60	0.49505	0.50593	0.52160	0.53479	0.54967	0.57374	−0.60
0.62	0.51869	0.52922	0.54438	0.55713	0.57149	0.59470	−0.62
0.64	0.54249	0.55266	0.56727	0.57955	0.59338	0.61570	−0.64
0.66	0.56645	0.57623	0.59028	0.60207	0.61534	0.63673	−0.66
0.68	0.59057	0.59995	0.61341	0.62469	0.63737	0.65779	−0.68
0.70	0.61486	0.62381	0.63665	0.64740	0.65948	0.67890	−0.70
0.72	0.63931	0.64781	0.66000	0.67021	0.68165	0.70004	−0.72
0.74	0.66392	0.67197	0.68348	0.69311	0.70390	0.72122	−0.74
0.76	0.68871	0.69627	0.70708	0.71611	0.72622	0.74243	−0.76
0.78	0.71367	0.72072	0.73080	0.73921	0.74862	0.76368	−0.78
0.80	0.73879	0.74532	0.75464	0.76241	0.77109	0.78498	−0.80
0.82	0.76410	0.77007	0.77860	0.78571	0.79364	0.80630	−0.82
0.84	0.78958	0.79498	0.80269	0.80910	0.81626	0.82767	−0.84
0.86	0.81523	0.82005	0.82690	0.83260	0.83896	0.84908	−0.86
0.88	0.84107	0.84527	0.85124	0.85620	0.86173	0.87052	−0.88
0.90	0.86709	0.87065	0.87571	0.87991	0.88458	0.89200	−0.90
0.92	0.89329	0.89619	0.90031	0.90372	0.90751	0.91352	−0.92
0.94	0.91969	0.92190	0.92503	0.92763	0.93051	0.93508	−0.94
0.96	0.94627	0.94776	0.94989	0.95165	0.95360	0.95668	−0.96
0.98	0.97304	0.97380	0.97488	0.97577	0.97676	0.97832	−0.98

↑ r_0

α =	.995 ↑−	.99 ↑−	.975 ↑−	.95 ↑−	.90 ↑−	.75 ↑−

TABLE 87.1   SAMPLE SIZE =   350

α =	.005 ↓+	.01 ↓+	.025 ↓+	.05 ↓+	.10 ↓+	.25 ↓+	
r_o ↓							
-0.98	-0.98475	-0.98434	-0.98372	-0.98316	-0.98250	-0.98133	0.98
-0.96	-0.96943	-0.96862	-0.96737	-0.96627	-0.96495	-0.96264	0.96
-0.94	-0.95404	-0.95282	-0.95097	-0.94933	-0.94736	-0.94392	0.94
-0.92	-0.93858	-0.93696	-0.93451	-0.93233	-0.92973	-0.92518	0.92
-0.90	-0.92304	-0.92103	-0.91799	-0.91528	-0.91206	-0.90642	0.90
-0.88	-0.90743	-0.90503	-0.90140	-0.89818	-0.89434	-0.88763	0.88
-0.86	-0.89174	-0.88896	-0.88476	-0.88102	-0.87658	-0.86882	0.86
-0.84	-0.87598	-0.87282	-0.86805	-0.86381	-0.85878	-0.84999	0.84
-0.82	-0.86014	-0.85661	-0.85128	-0.84655	-0.84093	-0.83113	0.82
-0.80	-0.84423	-0.84033	-0.83444	-0.82923	-0.82304	-0.81225	0.80
-0.78	-0.82824	-0.82398	-0.81755	-0.81186	-0.80510	-0.79335	0.78
-0.76	-0.81217	-0.80755	-0.80059	-0.79443	-0.78712	-0.77442	0.76
-0.74	-0.79602	-0.79105	-0.78357	-0.77695	-0.76910	-0.75546	0.74
-0.72	-0.77980	-0.77448	-0.76648	-0.75941	-0.75103	-0.73649	0.72
-0.70	-0.76349	-0.75784	-0.74933	-0.74181	-0.73291	-0.71748	0.70
-0.68	-0.74711	-0.74112	-0.73211	-0.72416	-0.71475	-0.69846	0.68
-0.66	-0.73064	-0.72432	-0.71483	-0.70645	-0.69655	-0.67941	0.66
-0.64	-0.71410	-0.70745	-0.69748	-0.68868	-0.67830	-0.66033	0.64
-0.62	-0.69747	-0.69051	-0.68006	-0.67086	-0.66000	-0.64123	0.62
-0.60	-0.68075	-0.67348	-0.66258	-0.65298	-0.64165	-0.62210	0.60
-0.58	-0.66396	-0.65638	-0.64503	-0.63504	-0.62326	-0.60295	0.58
-0.56	-0.64708	-0.63921	-0.62741	-0.61704	-0.60482	-0.58378	0.56
-0.54	-0.63011	-0.62195	-0.60973	-0.59899	-0.58634	-0.56457	0.54
-0.52	-0.61306	-0.60462	-0.59197	-0.58087	-0.56781	-0.54535	0.52
-0.50	-0.59593	-0.58720	-0.57415	-0.56270	-0.54923	-0.52609	0.50
-0.48	-0.57871	-0.56971	-0.55626	-0.54446	-0.53060	-0.50682	0.48
-0.46	-0.56140	-0.55213	-0.53830	-0.52617	-0.51193	-0.48751	0.46
-0.44	-0.54400	-0.53448	-0.52026	-0.50781	-0.49320	-0.46818	0.44
-0.42	-0.52651	-0.51674	-0.50216	-0.48940	-0.47443	-0.44882	0.42
-0.40	-0.50893	-0.49892	-0.48398	-0.47092	-0.45561	-0.42944	0.40
-0.38	-0.49126	-0.48101	-0.46574	-0.45238	-0.43674	-0.41003	0.38
-0.36	-0.47351	-0.46303	-0.44742	-0.43378	-0.41782	-0.39060	0.36
-0.34	-0.45565	-0.44495	-0.42902	-0.41512	-0.39885	-0.37114	0.34
-0.32	-0.43771	-0.42680	-0.41056	-0.39639	-0.37984	-0.35165	0.32
-0.30	-0.41967	-0.40855	-0.39202	-0.37760	-0.36077	-0.33213	0.30
-0.28	-0.40154	-0.39022	-0.37340	-0.35875	-0.34165	-0.31259	0.28
-0.26	-0.38332	-0.37181	-0.35471	-0.33984	-0.32248	-0.29302	0.26
-0.24	-0.36499	-0.35330	-0.33595	-0.32086	-0.30326	-0.27343	0.24
-0.22	-0.34657	-0.33471	-0.31711	-0.30181	-0.28399	-0.25381	0.22
-0.20	-0.32806	-0.31603	-0.29819	-0.28270	-0.26467	-0.23415	0.20
-0.18	-0.30944	-0.29725	-0.27920	-0.26353	-0.24530	-0.21448	0.18
-0.16	-0.29073	-0.27839	-0.26013	-0.24428	-0.22587	-0.19477	0.16
-0.14	-0.27191	-0.25944	-0.24098	-0.22498	-0.20639	-0.17504	0.14
-0.12	-0.25299	-0.24039	-0.22175	-0.20560	-0.18686	-0.15528	0.12
-0.10	-0.23398	-0.22125	-0.20244	-0.18616	-0.16728	-0.13549	0.10
-0.08	-0.21485	-0.20202	-0.18306	-0.16665	-0.14765	-0.11567	0.08
-0.06	-0.19563	-0.18269	-0.16359	-0.14708	-0.12796	-0.09583	0.06
-0.04	-0.17630	-0.16327	-0.14404	-0.12743	-0.10821	-0.07595	0.04
-0.02	-0.15687	-0.14375	-0.12441	-0.10772	-0.08842	-0.05605	0.02

↑ r_o

| α = | .995 ↑- | .99 ↑- | .975 ↑- | .95 ↑- | .90 ↑- | .75 ↑- |

TABLE 87.2   SAMPLE SIZE =   350   (CONT.)

α =	.005 ↓+	.01 ↓+	.025 ↓+	.05 ↓+	.10 ↓+	.25 ↓+	
r_0 ↓							
0.00	−0.13733	−0.12413	−0.10470	−0.08794	−0.06857	−0.03612	0.00
0.02	−0.11768	−0.10442	−0.08491	−0.06809	−0.04866	−0.01616	−0.02
0.04	−0.09792	−0.08461	−0.06503	−0.04817	−0.02870	0.00383	−0.04
0.06	−0.07805	−0.06470	−0.04507	−0.02817	−0.00869	0.02385	−0.06
0.08	−0.05808	−0.04469	−0.02502	−0.00811	0.01138	0.04389	−0.08
0.10	−0.03799	−0.02458	−0.00489	0.01202	0.03150	0.06397	−0.10
0.12	−0.01779	−0.00437	0.01532	0.03223	0.05169	0.08408	−0.12
0.14	0.00252	0.01594	0.03562	0.05251	0.07192	0.10421	−0.14
0.16	0.02295	0.03636	0.05601	0.07286	0.09222	0.12437	−0.16
0.18	0.04349	0.05688	0.07649	0.09328	0.11257	0.14457	−0.18
0.20	0.06415	0.07751	0.09705	0.11378	0.13298	0.16479	−0.20
0.22	0.08493	0.09824	0.11770	0.13435	0.15344	0.18505	−0.22
0.24	0.10583	0.11908	0.13844	0.15500	0.17396	0.20533	−0.24
0.26	0.12684	0.14003	0.15928	0.17572	0.19454	0.22565	−0.26
0.28	0.14798	0.16108	0.18020	0.19652	0.21518	0.24599	−0.28
0.30	0.16924	0.18224	0.20121	0.21739	0.23588	0.26637	−0.30
0.32	0.19062	0.20352	0.22232	0.23834	0.25664	0.28678	−0.32
0.34	0.21212	0.22491	0.24352	0.25937	0.27745	0.30721	−0.34
0.36	0.23375	0.24640	0.26481	0.28047	0.29833	0.32768	−0.36
0.38	0.25551	0.26801	0.28619	0.30165	0.31927	0.34818	−0.38
0.40	0.27740	0.28974	0.30768	0.32291	0.34026	0.36872	−0.40
0.42	0.29941	0.31158	0.32925	0.34425	0.36132	0.38928	−0.42
0.44	0.32155	0.33354	0.35092	0.36567	0.38244	0.40987	−0.44
0.46	0.34383	0.35561	0.37269	0.38717	0.40362	0.43050	−0.46
0.48	0.36624	0.37780	0.39456	0.40875	0.42486	0.45116	−0.48
0.50	0.38878	0.40012	0.41653	0.43041	0.44616	0.47185	−0.50
0.52	0.41146	0.42255	0.43859	0.45216	0.46753	0.49257	−0.52
0.54	0.43427	0.44510	0.46076	0.47398	0.48896	0.51333	−0.54
0.56	0.45723	0.46778	0.48302	0.49589	0.51045	0.53412	−0.56
0.58	0.48032	0.49058	0.50539	0.51788	0.53200	0.55494	−0.58
0.60	0.50355	0.51350	0.52786	0.53996	0.55362	0.57579	−0.60
0.62	0.52692	0.53655	0.55043	0.56212	0.57531	0.59668	−0.62
0.64	0.55044	0.55973	0.57311	0.58436	0.59705	0.61760	−0.64
0.66	0.57410	0.58303	0.59589	0.60669	0.61887	0.63855	−0.66
0.68	0.59791	0.60647	0.61877	0.62911	0.64075	0.65954	−0.68
0.70	0.62187	0.63003	0.64177	0.65161	0.66269	0.68056	−0.70
0.72	0.64597	0.65373	0.66487	0.67420	0.68470	0.70161	−0.72
0.74	0.67023	0.67756	0.68807	0.69688	0.70678	0.72270	−0.74
0.76	0.69464	0.70152	0.71139	0.71965	0.72892	0.74382	−0.76
0.78	0.71920	0.72562	0.73482	0.74251	0.75113	0.76497	−0.78
0.80	0.74392	0.74986	0.75835	0.76545	0.77341	0.78617	−0.80
0.82	0.76879	0.77423	0.78200	0.78849	0.79575	0.80739	−0.82
0.84	0.79383	0.79874	0.80576	0.81162	0.81817	0.82865	−0.84
0.86	0.81902	0.82339	0.82963	0.83484	0.84065	0.84995	−0.86
0.88	0.84437	0.84819	0.85362	0.85815	0.86321	0.87127	−0.88
0.90	0.86989	0.87312	0.87773	0.88156	0.88583	0.89264	−0.90
0.92	0.89558	0.89820	0.90194	0.90505	0.90852	0.91404	−0.92
0.94	0.92143	0.92343	0.92628	0.92865	0.93128	0.93548	−0.94
0.96	0.94745	0.94880	0.95073	0.95234	0.95412	0.95695	−0.96
0.98	0.97364	0.97433	0.97531	0.97612	0.97702	0.97846	−0.98

↑ r_0

α =	.995 ↑−	.99 ↑−	.975 ↑−	.95 ↑−	.90 ↑−	.75 ↑−

TABLE 88.1   SAMPLE SIZE =   400

α =	.005 ↓+	.01 ↓+	.025 ↓+	.05 ↓+	.10 ↓+	.25 ↓+	
r_0 ↓							
-0.98	-0.98448	-0.98409	-0.98350	-0.98297	-0.98235	-0.98125	0.98
-0.96	-0.96890	-0.96812	-0.96694	-0.96590	-0.96466	-0.96248	0.96
-0.94	-0.95324	-0.95208	-0.95033	-0.94877	-0.94692	-0.94368	0.94
-0.92	-0.93752	-0.93598	-0.93366	-0.93160	-0.92915	-0.92487	0.92
-0.90	-0.92172	-0.91981	-0.91693	-0.91437	-0.91133	-0.90603	0.90
-0.88	-0.90586	-0.90358	-0.90014	-0.89710	-0.89348	-0.88717	0.88
-0.86	-0.88992	-0.88728	-0.88330	-0.87977	-0.87558	-0.86829	0.86
-0.84	-0.87391	-0.87091	-0.86639	-0.86239	-0.85764	-0.84938	0.84
-0.82	-0.85782	-0.85447	-0.84943	-0.84496	-0.83966	-0.83045	0.82
-0.80	-0.84167	-0.83797	-0.83240	-0.82748	-0.82164	-0.81150	0.80
-0.78	-0.82544	-0.82140	-0.81532	-0.80994	-0.80358	-0.79253	0.78
-0.76	-0.80913	-0.80476	-0.79817	-0.79236	-0.78547	-0.77353	0.76
-0.74	-0.79276	-0.78805	-0.78097	-0.77472	-0.76733	-0.75451	0.74
-0.72	-0.77630	-0.77127	-0.76370	-0.75703	-0.74914	-0.73547	0.72
-0.70	-0.75977	-0.75441	-0.74637	-0.73928	-0.73090	-0.71641	0.70
-0.68	-0.74316	-0.73749	-0.72898	-0.72148	-0.71263	-0.69732	0.68
-0.66	-0.72648	-0.72050	-0.71153	-0.70363	-0.69431	-0.67821	0.66
-0.64	-0.70972	-0.70343	-0.69401	-0.68572	-0.67594	-0.65907	0.64
-0.62	-0.69288	-0.68630	-0.67643	-0.66776	-0.65754	-0.63991	0.62
-0.60	-0.67596	-0.66909	-0.65879	-0.64974	-0.63909	-0.62073	0.60
-0.58	-0.65896	-0.65180	-0.64109	-0.63167	-0.62059	-0.60152	0.58
-0.56	-0.64188	-0.63444	-0.62332	-0.61355	-0.60205	-0.58229	0.56
-0.54	-0.62472	-0.61701	-0.60548	-0.59536	-0.58347	-0.56304	0.54
-0.52	-0.60748	-0.59950	-0.58758	-0.57713	-0.56484	-0.54376	0.52
-0.50	-0.59016	-0.58192	-0.56962	-0.55883	-0.54617	-0.52446	0.50
-0.48	-0.57275	-0.56426	-0.55158	-0.54048	-0.52745	-0.50514	0.48
-0.46	-0.55527	-0.54653	-0.53349	-0.52208	-0.50869	-0.48579	0.46
-0.44	-0.53769	-0.52871	-0.51532	-0.50361	-0.48988	-0.46641	0.44
-0.42	-0.52004	-0.51082	-0.49709	-0.48509	-0.47103	-0.44701	0.42
-0.40	-0.50230	-0.49286	-0.47879	-0.46651	-0.45213	-0.42759	0.40
-0.38	-0.48447	-0.47481	-0.46043	-0.44787	-0.43318	-0.40814	0.38
-0.36	-0.46655	-0.45668	-0.44199	-0.42917	-0.41419	-0.38867	0.36
-0.34	-0.44855	-0.43848	-0.42349	-0.41042	-0.39515	-0.36917	0.34
-0.32	-0.43046	-0.42019	-0.40491	-0.39161	-0.37607	-0.34964	0.32
-0.30	-0.41229	-0.40182	-0.38627	-0.37273	-0.35693	-0.33010	0.30
-0.28	-0.39402	-0.38337	-0.36756	-0.35380	-0.33775	-0.31052	0.28
-0.26	-0.37566	-0.36484	-0.34877	-0.33480	-0.31853	-0.29092	0.26
-0.24	-0.35722	-0.34622	-0.32992	-0.31575	-0.29925	-0.27130	0.24
-0.22	-0.33868	-0.32752	-0.31099	-0.29664	-0.27993	-0.25165	0.22
-0.20	-0.32004	-0.30874	-0.29199	-0.27746	-0.26055	-0.23197	0.20
-0.18	-0.30132	-0.28987	-0.27292	-0.25822	-0.24113	-0.21227	0.18
-0.16	-0.28250	-0.27092	-0.25378	-0.23892	-0.22167	-0.19254	0.16
-0.14	-0.26359	-0.25188	-0.23456	-0.21956	-0.20215	-0.17279	0.14
-0.12	-0.24458	-0.23275	-0.21527	-0.20014	-0.18258	-0.15301	0.12
-0.10	-0.22548	-0.21354	-0.19591	-0.18065	-0.16296	-0.13320	0.10
-0.08	-0.20628	-0.19424	-0.17647	-0.16110	-0.14330	-0.11337	0.08
-0.06	-0.18698	-0.17485	-0.15695	-0.14149	-0.12358	-0.09351	0.06
-0.04	-0.16759	-0.15537	-0.13736	-0.12181	-0.10382	-0.07363	0.04
-0.02	-0.14809	-0.13580	-0.11769	-0.10206	-0.08400	-0.05372	0.02

↑ r_0

α =	.995 ↑-	.99 ↑-	.975 ↑-	.95 ↑-	.90 ↑-	.75 ↑-

TABLE 88.2   SAMPLE SIZE =   400   (CONT.)

α =	.005 ↓+	.01 ↓+	.025 ↓+	.05 ↓+	.10 ↓+	.25 ↓+	
r_0 ↓							
0.00	−0.12850	−0.11614	−0.09795	−0.08226	−0.06413	−0.03378	0.00
0.02	−0.10880	−0.09639	−0.07812	−0.06239	−0.04421	−0.01381	−0.02
0.04	−0.08901	−0.07655	−0.05822	−0.04245	−0.02424	0.00618	−0.04
0.06	−0.06911	−0.05661	−0.03825	−0.02244	−0.00422	0.02620	−0.06
0.08	−0.04910	−0.03658	−0.01819	−0.00237	0.01585	0.04625	−0.08
0.10	−0.02900	−0.01646	0.00195	0.01776	0.03597	0.06632	−0.10
0.12	−0.00878	0.00376	0.02216	0.03797	0.05615	0.08642	−0.12
0.14	0.01154	0.02408	0.04246	0.05824	0.07638	0.10655	−0.14
0.16	0.03196	0.04449	0.06284	0.07858	0.09666	0.12671	−0.16
0.18	0.05249	0.06499	0.08330	0.09899	0.11700	0.14689	−0.18
0.20	0.07313	0.08560	0.10384	0.11946	0.13739	0.16711	−0.20
0.22	0.09389	0.10631	0.12447	0.14001	0.15783	0.18735	−0.22
0.24	0.11475	0.12711	0.14517	0.16062	0.17832	0.20762	−0.24
0.26	0.13572	0.14801	0.16597	0.18131	0.19887	0.22792	−0.26
0.28	0.15681	0.16902	0.18684	0.20206	0.21948	0.24824	−0.28
0.30	0.17801	0.19013	0.20780	0.22289	0.24014	0.26860	−0.30
0.32	0.19932	0.21134	0.22885	0.24379	0.26085	0.28898	−0.32
0.34	0.22075	0.23265	0.24999	0.26476	0.28162	0.30939	−0.34
0.36	0.24229	0.25407	0.27121	0.28580	0.30244	0.32983	−0.36
0.38	0.26396	0.27559	0.29251	0.30691	0.32333	0.35030	−0.38
0.40	0.28574	0.29722	0.31391	0.32810	0.34426	0.37080	−0.40
0.42	0.30764	0.31896	0.33539	0.34936	0.36526	0.39133	−0.42
0.44	0.32966	0.34080	0.35697	0.37069	0.38631	0.41189	−0.44
0.46	0.35181	0.36275	0.37863	0.39210	0.40741	0.43248	−0.46
0.48	0.37407	0.38481	0.40038	0.41358	0.42858	0.45309	−0.48
0.50	0.39646	0.40698	0.42223	0.43514	0.44980	0.47374	−0.50
0.52	0.41898	0.42927	0.44417	0.45678	0.47108	0.49442	−0.52
0.54	0.44162	0.45166	0.46619	0.47849	0.49242	0.51512	−0.54
0.56	0.46439	0.47417	0.48832	0.50027	0.51381	0.53586	−0.56
0.58	0.48728	0.49679	0.51053	0.52214	0.53527	0.55663	−0.58
0.60	0.51031	0.51953	0.53284	0.54408	0.55678	0.57743	−0.60
0.62	0.53347	0.54239	0.55525	0.56610	0.57836	0.59826	−0.62
0.64	0.55676	0.56536	0.57775	0.58820	0.59999	0.61912	−0.64
0.66	0.58018	0.58844	0.60035	0.61037	0.62168	0.64001	−0.66
0.68	0.60374	0.61165	0.62305	0.63263	0.64344	0.66093	−0.68
0.70	0.62743	0.63498	0.64584	0.65497	0.66525	0.68188	−0.70
0.72	0.65126	0.65843	0.66873	0.67739	0.68713	0.70286	−0.72
0.74	0.67523	0.68200	0.69172	0.69989	0.70907	0.72388	−0.74
0.76	0.69934	0.70569	0.71482	0.72247	0.73107	0.74493	−0.76
0.78	0.72358	0.72951	0.73801	0.74513	0.75313	0.76600	−0.78
0.80	0.74797	0.75345	0.76130	0.76787	0.77525	0.78711	−0.80
0.82	0.77251	0.77752	0.78470	0.79070	0.79744	0.80826	−0.82
0.84	0.79719	0.80172	0.80820	0.81362	0.81969	0.82943	−0.84
0.86	0.82201	0.82604	0.83180	0.83661	0.84200	0.85064	−0.86
0.88	0.84698	0.85049	0.85551	0.85970	0.86438	0.87188	−0.88
0.90	0.87211	0.87508	0.87932	0.88286	0.88682	0.89315	−0.90
0.92	0.89738	0.89979	0.90324	0.90612	0.90933	0.91445	−0.92
0.94	0.92280	0.92464	0.92727	0.92946	0.93190	0.93579	−0.94
0.96	0.94838	0.94963	0.95140	0.95288	0.95453	0.95716	−0.96
0.98	0.97411	0.97474	0.97565	0.97640	0.97723	0.97856	−0.98

↑ r_0

α =	.995 ↑−	.99 ↑−	.975 ↑−	.95 ↑−	.90 ↑−	.75 ↑−

TABLE 89.1   SAMPLE SIZE =   450

α =	.005 ↓+	.01 ↓+	.025 ↓+	.05 ↓+	.10 ↓+	.25 ↓+	
r_0 ↓							
-0.98	-0.98426	-0.98388	-0.98332	-0.98282	-0.98222	-0.98118	0.98
-0.96	-0.96845	-0.96770	-0.96658	-0.96559	-0.96441	-0.96234	0.96
-0.94	-0.95257	-0.95146	-0.94979	-0.94831	-0.94655	-0.94349	0.94
-0.92	-0.93663	-0.93516	-0.93295	-0.93099	-0.92866	-0.92460	0.92
-0.90	-0.92061	-0.91879	-0.91604	-0.91361	-0.91073	-0.90570	0.90
-0.88	-0.90453	-0.90236	-0.89909	-0.89619	-0.89276	-0.88678	0.88
-0.86	-0.88838	-0.88587	-0.88207	-0.87872	-0.87475	-0.86784	0.86
-0.84	-0.87217	-0.86931	-0.86500	-0.86120	-0.85670	-0.84887	0.84
-0.82	-0.85588	-0.85268	-0.84788	-0.84363	-0.83861	-0.82988	0.82
-0.80	-0.83952	-0.83599	-0.83069	-0.82601	-0.82048	-0.81088	0.80
-0.78	-0.82309	-0.81924	-0.81345	-0.80834	-0.80231	-0.79185	0.78
-0.76	-0.80659	-0.80242	-0.79615	-0.79063	-0.78410	-0.77279	0.76
-0.74	-0.79001	-0.78553	-0.77879	-0.77286	-0.76585	-0.75372	0.74
-0.72	-0.77337	-0.76857	-0.76138	-0.75504	-0.74756	-0.73463	0.72
-0.70	-0.75665	-0.75155	-0.74390	-0.73717	-0.72922	-0.71551	0.70
-0.68	-0.73986	-0.73446	-0.72637	-0.71925	-0.71085	-0.69637	0.68
-0.66	-0.72299	-0.71730	-0.70877	-0.70127	-0.69244	-0.67721	0.66
-0.64	-0.70605	-0.70007	-0.69112	-0.68325	-0.67398	-0.65802	0.64
-0.62	-0.68904	-0.68278	-0.67340	-0.66517	-0.65549	-0.63882	0.62
-0.60	-0.67194	-0.66541	-0.65563	-0.64705	-0.63695	-0.61959	0.60
-0.58	-0.65478	-0.64797	-0.63779	-0.62887	-0.61837	-0.60034	0.58
-0.56	-0.63753	-0.63046	-0.61990	-0.61063	-0.59975	-0.58106	0.56
-0.54	-0.62021	-0.61288	-0.60194	-0.59235	-0.58108	-0.56177	0.54
-0.52	-0.60281	-0.59523	-0.58392	-0.57401	-0.56238	-0.54245	0.52
-0.50	-0.58534	-0.57751	-0.56583	-0.55561	-0.54363	-0.52310	0.50
-0.48	-0.56778	-0.55972	-0.54769	-0.53717	-0.52483	-0.50374	0.48
-0.46	-0.55015	-0.54185	-0.52948	-0.51867	-0.50600	-0.48435	0.46
-0.44	-0.53243	-0.52391	-0.51121	-0.50011	-0.48712	-0.46494	0.44
-0.42	-0.51463	-0.50589	-0.49287	-0.48150	-0.46820	-0.44551	0.42
-0.40	-0.49676	-0.48780	-0.47447	-0.46284	-0.44923	-0.42605	0.40
-0.38	-0.47880	-0.46964	-0.45601	-0.44412	-0.43023	-0.40657	0.38
-0.36	-0.46076	-0.45139	-0.43748	-0.42534	-0.41117	-0.38706	0.36
-0.34	-0.44263	-0.43308	-0.41888	-0.40651	-0.39207	-0.36753	0.34
-0.32	-0.42442	-0.41468	-0.40022	-0.38763	-0.37293	-0.34798	0.32
-0.30	-0.40613	-0.39621	-0.38149	-0.36868	-0.35375	-0.32840	0.30
-0.28	-0.38775	-0.37766	-0.36270	-0.34968	-0.33452	-0.30880	0.28
-0.26	-0.36929	-0.35904	-0.34384	-0.33062	-0.31524	-0.28918	0.26
-0.24	-0.35074	-0.34033	-0.32491	-0.31151	-0.29592	-0.26953	0.24
-0.22	-0.33211	-0.32155	-0.30591	-0.29234	-0.27655	-0.24986	0.22
-0.20	-0.31338	-0.30268	-0.28685	-0.27311	-0.25714	-0.23016	0.20
-0.18	-0.29457	-0.28374	-0.26771	-0.25382	-0.23768	-0.21044	0.18
-0.16	-0.27567	-0.26471	-0.24851	-0.23447	-0.21818	-0.19070	0.16
-0.14	-0.25668	-0.24560	-0.22924	-0.21507	-0.19863	-0.17093	0.14
-0.12	-0.23760	-0.22641	-0.20990	-0.19560	-0.17903	-0.15113	0.12
-0.10	-0.21843	-0.20714	-0.19048	-0.17608	-0.15939	-0.13131	0.10
-0.08	-0.19916	-0.18779	-0.17100	-0.15649	-0.13970	-0.11147	0.08
-0.06	-0.17981	-0.16835	-0.15145	-0.13685	-0.11996	-0.09160	0.06
-0.04	-0.16036	-0.14882	-0.13182	-0.11714	-0.10017	-0.07171	0.04
-0.02	-0.14082	-0.12921	-0.11212	-0.09738	-0.08034	-0.05179	0.02

↑ r_0

α =	.995 ↑-	.99 ↑-	.975 ↑-	.95 ↑-	.90 ↑-	.75 ↑-

TABLE 89.2   SAMPLE SIZE =   450   (CONT.)

α =	.005 ↓+	.01 ↓+	.025 ↓+	.05 ↓+	.10 ↓+	.25 ↓+	
r_0 ↓							
0.00	-0.12118	-0.10952	-0.09235	-0.07755	-0.06046	-0.03184	0.00
0.02	-0.10145	-0.08974	-0.07251	-0.05766	-0.04053	-0.01187	-0.02
0.04	-0.08162	-0.06987	-0.05259	-0.03771	-0.02055	0.00812	-0.04
0.06	-0.06170	-0.04991	-0.03260	-0.01770	-0.00053	0.02815	-0.06
0.08	-0.04167	-0.02987	-0.01253	0.00237	0.01955	0.04819	-0.08
0.10	-0.02155	-0.00974	0.00761	0.02251	0.03967	0.06827	-0.10
0.12	-0.00133	0.01049	0.02783	0.04271	0.05984	0.08836	-0.12
0.14	0.01899	0.03080	0.04812	0.06298	0.08006	0.10849	-0.14
0.16	0.03941	0.05120	0.06849	0.08331	0.10034	0.12864	-0.16
0.18	0.05993	0.07170	0.08893	0.10370	0.12066	0.14882	-0.18
0.20	0.08055	0.09228	0.10945	0.12416	0.14103	0.16902	-0.20
0.22	0.10128	0.11296	0.13005	0.14468	0.16145	0.18925	-0.22
0.24	0.12211	0.13374	0.15073	0.16527	0.18192	0.20951	-0.24
0.26	0.14305	0.15461	0.17149	0.18592	0.20245	0.22979	-0.26
0.28	0.16409	0.17557	0.19233	0.20664	0.22302	0.25010	-0.28
0.30	0.18524	0.19663	0.21324	0.22743	0.24365	0.27043	-0.30
0.32	0.20649	0.21778	0.23424	0.24828	0.26433	0.29080	-0.32
0.34	0.22786	0.23903	0.25532	0.26920	0.28506	0.31119	-0.34
0.36	0.24933	0.26038	0.27648	0.29019	0.30584	0.33160	-0.36
0.38	0.27091	0.28183	0.29772	0.31124	0.32667	0.35205	-0.38
0.40	0.29261	0.30338	0.31904	0.33237	0.34756	0.37252	-0.40
0.42	0.31441	0.32502	0.34045	0.35356	0.36850	0.39302	-0.42
0.44	0.33633	0.34677	0.36194	0.37482	0.38949	0.41355	-0.44
0.46	0.35836	0.36862	0.38351	0.39616	0.41054	0.43410	-0.46
0.48	0.38051	0.39057	0.40517	0.41756	0.43164	0.45469	-0.48
0.50	0.40277	0.41263	0.42692	0.43903	0.45279	0.47530	-0.50
0.52	0.42515	0.43479	0.44875	0.46057	0.47400	0.49594	-0.52
0.54	0.44765	0.45705	0.47066	0.48219	0.49526	0.51660	-0.54
0.56	0.47026	0.47942	0.49267	0.50388	0.51658	0.53730	-0.56
0.58	0.49300	0.50189	0.51476	0.52563	0.53795	0.55802	-0.58
0.60	0.51585	0.52448	0.53694	0.54746	0.55938	0.57878	-0.60
0.62	0.53883	0.54717	0.55921	0.56937	0.58087	0.59956	-0.62
0.64	0.56193	0.56997	0.58156	0.59135	0.60240	0.62037	-0.64
0.66	0.58516	0.59288	0.60401	0.61340	0.62400	0.64121	-0.66
0.68	0.60850	0.61590	0.62655	0.63552	0.64565	0.66207	-0.68
0.70	0.63198	0.63903	0.64918	0.65772	0.66736	0.68297	-0.70
0.72	0.65558	0.66227	0.67190	0.68000	0.68913	0.70390	-0.72
0.74	0.67931	0.68563	0.69471	0.70235	0.71095	0.72485	-0.74
0.76	0.70317	0.70910	0.71762	0.72478	0.73283	0.74584	-0.76
0.78	0.72716	0.73269	0.74062	0.74728	0.75477	0.76685	-0.78
0.80	0.75128	0.75639	0.76372	0.76986	0.77677	0.78789	-0.80
0.82	0.77554	0.78021	0.78691	0.79252	0.79882	0.80897	-0.82
0.84	0.79993	0.80414	0.81019	0.81525	0.82094	0.83007	-0.84
0.86	0.82445	0.82820	0.83357	0.83807	0.84311	0.85121	-0.86
0.88	0.84911	0.85238	0.85705	0.86096	0.86534	0.87237	-0.88
0.90	0.87391	0.87667	0.88063	0.88393	0.88763	0.89357	-0.90
0.92	0.89884	0.90109	0.90430	0.90698	0.90999	0.91479	-0.92
0.94	0.92392	0.92563	0.92808	0.93012	0.93240	0.93605	-0.94
0.96	0.94914	0.95030	0.95195	0.95333	0.95487	0.95733	-0.96
0.98	0.97450	0.97508	0.97592	0.97662	0.97740	0.97865	-0.98

↑ r_0

α =	.995 ↑−	.99 ↑−	.975 ↑−	.95 ↑−	.90 ↑−	.75 ↑−

TABLE 90.1   SAMPLE SIZE =   500

α =	.005 ↓+	.01 ↓+	.025 ↓+	.05 ↓+	.10 ↓+	.25 ↓+	
r_0 ↓							
−0.98	−0.98406	−0.98370	−0.98316	−0.98268	−0.98212	−0.98113	0.98
−0.96	−0.96806	−0.96735	−0.96627	−0.96532	−0.96420	−0.96223	0.96
−0.94	−0.95199	−0.95093	−0.94933	−0.94792	−0.94624	−0.94332	0.94
−0.92	−0.93586	−0.93446	−0.93234	−0.93046	−0.92824	−0.92438	0.92
−0.90	−0.91967	−0.91792	−0.91529	−0.91296	−0.91021	−0.90543	0.90
−0.88	−0.90340	−0.90132	−0.89819	−0.89542	−0.89214	−0.88645	0.88
−0.86	−0.88707	−0.88466	−0.88103	−0.87783	−0.87403	−0.86745	0.86
−0.84	−0.87068	−0.86794	−0.86382	−0.86019	−0.85589	−0.84844	0.84
−0.82	−0.85421	−0.85115	−0.84655	−0.84250	−0.83770	−0.82940	0.82
−0.80	−0.83768	−0.83430	−0.82923	−0.82476	−0.81948	−0.81034	0.80
−0.78	−0.82108	−0.81739	−0.81186	−0.80698	−0.80122	−0.79127	0.78
−0.76	−0.80441	−0.80042	−0.79443	−0.78915	−0.78292	−0.77217	0.76
−0.74	−0.78767	−0.78338	−0.77694	−0.77127	−0.76459	−0.75305	0.74
−0.72	−0.77086	−0.76627	−0.75940	−0.75334	−0.74621	−0.73391	0.72
−0.70	−0.75398	−0.74910	−0.74180	−0.73537	−0.72780	−0.71474	0.70
−0.68	−0.73703	−0.73187	−0.72414	−0.71734	−0.70934	−0.69556	0.68
−0.66	−0.72001	−0.71457	−0.70642	−0.69927	−0.69085	−0.67636	0.66
−0.64	−0.70292	−0.69720	−0.68865	−0.68115	−0.67232	−0.65713	0.64
−0.62	−0.68576	−0.67977	−0.67082	−0.66297	−0.65374	−0.63789	0.62
−0.60	−0.66852	−0.66227	−0.65294	−0.64475	−0.63513	−0.61862	0.60
−0.58	−0.65121	−0.64471	−0.63499	−0.62648	−0.61648	−0.59933	0.58
−0.56	−0.63383	−0.62707	−0.61699	−0.60815	−0.59779	−0.58002	0.56
−0.54	−0.61637	−0.60937	−0.59892	−0.58978	−0.57905	−0.56068	0.54
−0.52	−0.59883	−0.59160	−0.58080	−0.57135	−0.56028	−0.54133	0.52
−0.50	−0.58123	−0.57376	−0.56262	−0.55288	−0.54147	−0.52195	0.50
−0.48	−0.56354	−0.55585	−0.54438	−0.53435	−0.52261	−0.50255	0.48
−0.46	−0.54578	−0.53787	−0.52607	−0.51577	−0.50371	−0.48313	0.46
−0.44	−0.52795	−0.51982	−0.50771	−0.49714	−0.48478	−0.46369	0.44
−0.42	−0.51003	−0.50169	−0.48928	−0.47846	−0.46580	−0.44423	0.42
−0.40	−0.49204	−0.48350	−0.47080	−0.45972	−0.44678	−0.42474	0.40
−0.38	−0.47397	−0.46524	−0.45225	−0.44093	−0.42772	−0.40523	0.38
−0.36	−0.45582	−0.44690	−0.43364	−0.42209	−0.40861	−0.38570	0.36
−0.34	−0.43760	−0.42849	−0.41497	−0.40320	−0.38946	−0.36615	0.34
−0.32	−0.41929	−0.41001	−0.39623	−0.38425	−0.37028	−0.34657	0.32
−0.30	−0.40090	−0.39145	−0.37743	−0.36525	−0.35105	−0.32697	0.30
−0.28	−0.38243	−0.37282	−0.35857	−0.34619	−0.33177	−0.30735	0.28
−0.26	−0.36388	−0.35411	−0.33964	−0.32708	−0.31246	−0.28770	0.26
−0.24	−0.34524	−0.33533	−0.32065	−0.30791	−0.29310	−0.26804	0.24
−0.22	−0.32653	−0.31648	−0.30160	−0.28869	−0.27369	−0.24834	0.22
−0.20	−0.30772	−0.29754	−0.28248	−0.26942	−0.25425	−0.22863	0.20
−0.18	−0.28884	−0.27853	−0.26329	−0.25009	−0.23476	−0.20889	0.18
−0.16	−0.26987	−0.25945	−0.24404	−0.23070	−0.21522	−0.18913	0.16
−0.14	−0.25081	−0.24028	−0.22472	−0.21126	−0.19565	−0.16935	0.14
−0.12	−0.23167	−0.22104	−0.20534	−0.19176	−0.17602	−0.14954	0.12
−0.10	−0.21245	−0.20172	−0.18589	−0.17221	−0.15636	−0.12971	0.10
−0.08	−0.19313	−0.18232	−0.16637	−0.15259	−0.13665	−0.10986	0.08
−0.06	−0.17373	−0.16284	−0.14678	−0.13292	−0.11689	−0.08998	0.06
−0.04	−0.15424	−0.14328	−0.12713	−0.11320	−0.09709	−0.07008	0.04
−0.02	−0.13466	−0.12363	−0.10741	−0.09341	−0.07724	−0.05015	0.02

↑ r_0

α =	.995 ↑−	.99 ↑−	.975 ↑−	.95 ↑−	.90 ↑−	.75 ↑−

TABLE 90.2   SAMPLE SIZE =   500   (CONT.)

α =	.005 ↓+	.01 ↓+	.025 ↓+	.05 ↓+	.10 ↓+	.25 ↓+	
r_0 ↓							
0.00	-0.11498	-0.10391	-0.08762	-0.07357	-0.05735	-0.03020	0.00
0.02	-0.09522	-0.08411	-0.06775	-0.05367	-0.03741	-0.01023	-0.02
0.04	-0.07537	-0.06422	-0.04782	-0.03371	-0.01743	0.00977	-0.04
0.06	-0.05543	-0.04425	-0.02782	-0.01369	0.00260	0.02979	-0.06
0.08	-0.03539	-0.02419	-0.00775	0.00639	0.02267	0.04984	-0.08
0.10	-0.01526	-0.00405	0.01239	0.02652	0.04279	0.06991	-0.10
0.12	0.00496	0.01617	0.03261	0.04672	0.06296	0.09000	-0.12
0.14	0.02528	0.03648	0.05289	0.06698	0.08318	0.11012	-0.14
0.16	0.04570	0.05688	0.07325	0.08730	0.10344	0.13027	-0.16
0.18	0.06621	0.07736	0.09368	0.10768	0.12375	0.15044	-0.18
0.20	0.08681	0.09793	0.11419	0.12812	0.14411	0.17063	-0.20
0.22	0.10752	0.11858	0.13477	0.14862	0.16451	0.19085	-0.22
0.24	0.12832	0.13933	0.15542	0.16918	0.18496	0.21110	-0.24
0.26	0.14923	0.16016	0.17615	0.18981	0.20546	0.23137	-0.26
0.28	0.17023	0.18109	0.19695	0.21050	0.22601	0.25166	-0.28
0.30	0.19133	0.20210	0.21783	0.23125	0.24661	0.27198	-0.30
0.32	0.21253	0.22321	0.23878	0.25207	0.26726	0.29233	-0.32
0.34	0.23384	0.24441	0.25981	0.27294	0.28795	0.31270	-0.34
0.36	0.25525	0.26570	0.28092	0.29389	0.30870	0.33310	-0.36
0.38	0.27676	0.28708	0.30210	0.31489	0.32949	0.35352	-0.38
0.40	0.29838	0.30856	0.32336	0.33597	0.35034	0.37397	-0.40
0.42	0.32011	0.33013	0.34470	0.35710	0.37123	0.39445	-0.42
0.44	0.34193	0.35179	0.36612	0.37830	0.39218	0.41495	-0.44
0.46	0.36387	0.37356	0.38762	0.39957	0.41317	0.43548	-0.46
0.48	0.38592	0.39541	0.40920	0.42090	0.43422	0.45603	-0.48
0.50	0.40807	0.41737	0.43086	0.44230	0.45531	0.47661	-0.50
0.52	0.43033	0.43942	0.45260	0.46377	0.47646	0.49722	-0.52
0.54	0.45271	0.46157	0.47442	0.48530	0.49766	0.51785	-0.54
0.56	0.47519	0.48382	0.49632	0.50690	0.51891	0.53851	-0.56
0.58	0.49779	0.50618	0.51831	0.52857	0.54021	0.55920	-0.58
0.60	0.52050	0.52863	0.54038	0.55031	0.56157	0.57991	-0.60
0.62	0.54333	0.55118	0.56253	0.57212	0.58297	0.60065	-0.62
0.64	0.56627	0.57383	0.58476	0.59399	0.60443	0.62142	-0.64
0.66	0.58932	0.59659	0.60708	0.61594	0.62595	0.64221	-0.66
0.68	0.61250	0.61945	0.62949	0.63795	0.64751	0.66304	-0.68
0.70	0.63579	0.64242	0.65198	0.66003	0.66913	0.68389	-0.70
0.72	0.65920	0.66549	0.67456	0.68219	0.69080	0.70476	-0.72
0.74	0.68273	0.68867	0.69722	0.70442	0.71253	0.72567	-0.74
0.76	0.70638	0.71195	0.71997	0.72671	0.73431	0.74660	-0.76
0.78	0.73015	0.73534	0.74281	0.74908	0.75615	0.76756	-0.78
0.80	0.75405	0.75884	0.76574	0.77152	0.77804	0.78855	-0.80
0.82	0.77807	0.78245	0.78875	0.79404	0.79998	0.80957	-0.82
0.84	0.80221	0.80617	0.81186	0.81662	0.82198	0.83061	-0.84
0.86	0.82648	0.83000	0.83505	0.83928	0.84404	0.85169	-0.86
0.88	0.85088	0.85395	0.85834	0.86202	0.86615	0.87279	-0.88
0.90	0.87541	0.87800	0.88172	0.88483	0.88831	0.89392	-0.90
0.92	0.90006	0.90217	0.90519	0.90771	0.91054	0.91508	-0.92
0.94	0.92485	0.92645	0.92875	0.93067	0.93282	0.93626	-0.94
0.96	0.94977	0.95085	0.95241	0.95370	0.95515	0.95748	-0.96
0.98	0.97482	0.97537	0.97616	0.97681	0.97755	0.97873	-0.98

↑ r_0

α =	.995 ↑−	.99 ↑−	.975 ↑−	.95 ↑−	.90 ↑−	.75 ↑−

TABLE 91.1   SAMPLE SIZE =   550

α =	.005 ↓+	.01 ↓+	.025 ↓+	.05 ↓+	.10 ↓+	.25 ↓+	
r_o ↓							
-0.98	-0.98389	-0.98355	-0.98303	-0.98257	-0.98202	-0.98108	0.98
-0.96	-0.96773	-0.96704	-0.96601	-0.96509	-0.96401	-0.96213	0.96
-0.94	-0.95149	-0.95047	-0.94893	-0.94758	-0.94597	-0.94317	0.94
-0.92	-0.93520	-0.93385	-0.93181	-0.93001	-0.92788	-0.92419	0.92
-0.90	-0.91884	-0.91716	-0.91463	-0.91240	-0.90976	-0.90519	0.90
-0.88	-0.90242	-0.90042	-0.89741	-0.89475	-0.89161	-0.88617	0.88
-0.86	-0.88593	-0.88361	-0.88013	-0.87705	-0.87342	-0.86712	0.86
-0.84	-0.86938	-0.86675	-0.86279	-0.85931	-0.85519	-0.84806	0.84
-0.82	-0.85276	-0.84982	-0.84541	-0.84152	-0.83693	-0.82898	0.82
-0.80	-0.83608	-0.83284	-0.82797	-0.82368	-0.81862	-0.80988	0.80
-0.78	-0.81933	-0.81579	-0.81048	-0.80580	-0.80029	-0.79076	0.78
-0.76	-0.80252	-0.79868	-0.79293	-0.78787	-0.78191	-0.77162	0.76
-0.74	-0.78564	-0.78151	-0.77533	-0.76990	-0.76350	-0.75246	0.74
-0.72	-0.76869	-0.76428	-0.75768	-0.75188	-0.74505	-0.73328	0.72
-0.70	-0.75167	-0.74698	-0.73997	-0.73381	-0.72656	-0.71408	0.70
-0.68	-0.73458	-0.72963	-0.72221	-0.71570	-0.70804	-0.69486	0.68
-0.66	-0.71743	-0.71220	-0.70439	-0.69754	-0.68948	-0.67562	0.66
-0.64	-0.70021	-0.69472	-0.68652	-0.67933	-0.67088	-0.65636	0.64
-0.62	-0.68291	-0.67717	-0.66859	-0.66107	-0.65224	-0.63708	0.62
-0.60	-0.66555	-0.65956	-0.65061	-0.64277	-0.63356	-0.61778	0.60
-0.58	-0.64812	-0.64188	-0.63257	-0.62441	-0.61485	-0.59846	0.58
-0.56	-0.63061	-0.62413	-0.61447	-0.60601	-0.59609	-0.57911	0.56
-0.54	-0.61304	-0.60632	-0.59632	-0.58756	-0.57730	-0.55975	0.54
-0.52	-0.59539	-0.58845	-0.57810	-0.56906	-0.55847	-0.54036	0.52
-0.50	-0.57767	-0.57051	-0.55984	-0.55051	-0.53960	-0.52096	0.50
-0.48	-0.55988	-0.55250	-0.54151	-0.53192	-0.52069	-0.50153	0.48
-0.46	-0.54201	-0.53442	-0.52313	-0.51327	-0.50174	-0.48208	0.46
-0.44	-0.52407	-0.51628	-0.50469	-0.49457	-0.48275	-0.46262	0.44
-0.42	-0.50605	-0.49807	-0.48618	-0.47583	-0.46373	-0.44313	0.42
-0.40	-0.48797	-0.47978	-0.46762	-0.45703	-0.44466	-0.42361	0.40
-0.38	-0.46980	-0.46143	-0.44901	-0.43818	-0.42555	-0.40408	0.38
-0.36	-0.45156	-0.44302	-0.43033	-0.41928	-0.40640	-0.38453	0.36
-0.34	-0.43324	-0.42453	-0.41159	-0.40033	-0.38721	-0.36495	0.34
-0.32	-0.41485	-0.40597	-0.39279	-0.38133	-0.36799	-0.34535	0.32
-0.30	-0.39638	-0.38734	-0.37393	-0.36228	-0.34872	-0.32573	0.30
-0.28	-0.37783	-0.36864	-0.35501	-0.34318	-0.32941	-0.30609	0.28
-0.26	-0.35920	-0.34986	-0.33603	-0.32402	-0.31006	-0.28643	0.26
-0.24	-0.34050	-0.33102	-0.31699	-0.30481	-0.29066	-0.26675	0.24
-0.22	-0.32171	-0.31210	-0.29788	-0.28555	-0.27123	-0.24704	0.22
-0.20	-0.30284	-0.29311	-0.27871	-0.26624	-0.25175	-0.22731	0.20
-0.18	-0.28390	-0.27405	-0.25948	-0.24687	-0.23224	-0.20756	0.18
-0.16	-0.26487	-0.25491	-0.24019	-0.22745	-0.21268	-0.18779	0.16
-0.14	-0.24576	-0.23570	-0.22084	-0.20798	-0.19308	-0.16799	0.14
-0.12	-0.22657	-0.21641	-0.20142	-0.18845	-0.17343	-0.14817	0.12
-0.10	-0.20729	-0.19705	-0.18193	-0.16887	-0.15375	-0.12833	0.10
-0.08	-0.18793	-0.17761	-0.16238	-0.14924	-0.13402	-0.10847	0.08
-0.06	-0.16849	-0.15809	-0.14277	-0.12955	-0.11425	-0.08859	0.06
-0.04	-0.14896	-0.13850	-0.12309	-0.10980	-0.09444	-0.06868	0.04
-0.02	-0.12935	-0.11883	-0.10335	-0.09000	-0.07458	-0.04875	0.02

↑ r_o

α =	.995 ↑-	.99 ↑-	.975 ↑-	.95 ↑-	.90 ↑-	.75 ↑-

TABLE 91.2   SAMPLE SIZE =   550   (CONT.)

α =	.005 ↓+	.01 ↓+	.025 ↓+	.05 ↓+	.10 ↓+	.25 ↓+	
r_0 ↓							
0.00	-0.10965	-0.09909	-0.08354	-0.07015	-0.05468	-0.02879	0.00
0.02	-0.08987	-0.07926	-0.06367	-0.05023	-0.03474	-0.00882	-0.02
0.04	-0.07000	-0.05936	-0.04372	-0.03027	-0.01475	0.01118	-0.04
0.06	-0.05004	-0.03938	-0.02371	-0.01024	0.00528	0.03121	-0.06
0.08	-0.02999	-0.01931	-0.00364	0.00984	0.02536	0.05125	-0.08
0.10	-0.00985	0.00083	0.01651	0.02997	0.04548	0.07132	-0.10
0.12	0.01038	0.02105	0.03672	0.05017	0.06564	0.09141	-0.12
0.14	0.03069	0.04136	0.05700	0.07042	0.08585	0.11153	-0.14
0.16	0.05110	0.06175	0.07735	0.09073	0.10611	0.13167	-0.16
0.18	0.07160	0.08222	0.09777	0.11110	0.12640	0.15183	-0.18
0.20	0.09219	0.10277	0.11826	0.13152	0.14675	0.17202	-0.20
0.22	0.11288	0.12341	0.13882	0.15201	0.16714	0.19223	-0.22
0.24	0.13365	0.14413	0.15944	0.17255	0.18757	0.21247	-0.24
0.26	0.15453	0.16493	0.18014	0.19315	0.20805	0.23273	-0.26
0.28	0.17549	0.18583	0.20092	0.21381	0.22858	0.25301	-0.28
0.30	0.19656	0.20680	0.22176	0.23453	0.24915	0.27332	-0.30
0.32	0.21771	0.22787	0.24268	0.25532	0.26977	0.29365	-0.32
0.34	0.23897	0.24902	0.26366	0.27616	0.29044	0.31400	-0.34
0.36	0.26032	0.27025	0.28472	0.29706	0.31115	0.33438	-0.36
0.38	0.28178	0.29158	0.30586	0.31802	0.33191	0.35479	-0.38
0.40	0.30333	0.31300	0.32707	0.33905	0.35272	0.37522	-0.40
0.42	0.32498	0.33450	0.34835	0.36014	0.37357	0.39567	-0.42
0.44	0.34673	0.35610	0.36971	0.38128	0.39448	0.41615	-0.44
0.46	0.36859	0.37778	0.39114	0.40250	0.41543	0.43665	-0.46
0.48	0.39054	0.39956	0.41265	0.42377	0.43643	0.45718	-0.48
0.50	0.41260	0.42143	0.43424	0.44511	0.45747	0.47773	-0.50
0.52	0.43476	0.44339	0.45590	0.46651	0.47857	0.49831	-0.52
0.54	0.45703	0.46544	0.47763	0.48797	0.49971	0.51892	-0.54
0.56	0.47940	0.48759	0.49945	0.50950	0.52090	0.53955	-0.56
0.58	0.50188	0.50983	0.52134	0.53109	0.54215	0.56020	-0.58
0.60	0.52447	0.53217	0.54332	0.55275	0.56344	0.58088	-0.60
0.62	0.54716	0.55461	0.56537	0.57447	0.58478	0.60159	-0.62
0.64	0.56997	0.57714	0.58750	0.59625	0.60617	0.62232	-0.64
0.66	0.59288	0.59976	0.60971	0.61811	0.62761	0.64308	-0.66
0.68	0.61590	0.62249	0.63200	0.64003	0.64910	0.66386	-0.68
0.70	0.63904	0.64531	0.65437	0.66201	0.67065	0.68467	-0.70
0.72	0.66228	0.66824	0.67682	0.68406	0.69224	0.70551	-0.72
0.74	0.68564	0.69126	0.69936	0.70618	0.71388	0.72637	-0.74
0.76	0.70911	0.71438	0.72198	0.72837	0.73558	0.74726	-0.76
0.78	0.73270	0.73761	0.74468	0.75062	0.75732	0.76817	-0.78
0.80	0.75640	0.76094	0.76746	0.77294	0.77912	0.78911	-0.80
0.82	0.78022	0.78437	0.79033	0.79533	0.80097	0.81008	-0.82
0.84	0.80416	0.80790	0.81328	0.81779	0.82287	0.83107	-0.84
0.86	0.82821	0.83154	0.83632	0.84032	0.84483	0.85209	-0.86
0.88	0.85239	0.85529	0.85944	0.86292	0.86684	0.87314	-0.88
0.90	0.87669	0.87913	0.88265	0.88559	0.88890	0.89422	-0.90
0.92	0.90110	0.90309	0.90594	0.90833	0.91101	0.91532	-0.92
0.94	0.92564	0.92716	0.92933	0.93114	0.93318	0.93645	-0.94
0.96	0.95030	0.95133	0.95280	0.95402	0.95540	0.95760	-0.96
0.98	0.97509	0.97561	0.97635	0.97698	0.97767	0.97879	-0.98

↑ r_0

α =	.995 ↑-	.99 ↑-	.975 ↑-	.95 ↑-	.90 ↑-	.75 ↑-

TABLE 92.1   SAMPLE SIZE =   600

α =	.005 ↓+	.01 ↓+	.025 ↓+	.05 ↓+	.10 ↓+	.25 ↓+	
r_0 ↓							
-0.98	-0.98374	-0.98341	-0.98291	-0.98247	-0.98194	-0.98103	0.98
-0.96	-0.96743	-0.96677	-0.96577	-0.96489	-0.96385	-0.96205	0.96
-0.94	-0.95105	-0.95007	-0.94858	-0.94727	-0.94573	-0.94304	0.94
-0.92	-0.93461	-0.93331	-0.93135	-0.92961	-0.92757	-0.92402	0.92
-0.90	-0.91811	-0.91649	-0.91406	-0.91191	-0.90937	-0.90498	0.90
-0.88	-0.90155	-0.89962	-0.89672	-0.89416	-0.89114	-0.88592	0.88
-0.86	-0.88493	-0.88269	-0.87933	-0.87637	-0.87288	-0.86684	0.86
-0.84	-0.86824	-0.86570	-0.86189	-0.85854	-0.85458	-0.84774	0.84
-0.82	-0.85149	-0.84865	-0.84440	-0.84066	-0.83624	-0.82862	0.82
-0.80	-0.83468	-0.83155	-0.82686	-0.82273	-0.81787	-0.80948	0.80
-0.78	-0.81780	-0.81438	-0.80927	-0.80477	-0.79947	-0.79033	0.78
-0.76	-0.80085	-0.79716	-0.79162	-0.78675	-0.78102	-0.77115	0.76
-0.74	-0.78385	-0.77987	-0.77392	-0.76870	-0.76255	-0.75196	0.74
-0.72	-0.76677	-0.76253	-0.75617	-0.75060	-0.74403	-0.73274	0.72
-0.70	-0.74964	-0.74512	-0.73837	-0.73245	-0.72548	-0.71351	0.70
-0.68	-0.73243	-0.72766	-0.72052	-0.71426	-0.70690	-0.69425	0.68
-0.66	-0.71516	-0.71013	-0.70261	-0.69602	-0.68827	-0.67498	0.66
-0.64	-0.69782	-0.69254	-0.68465	-0.67773	-0.66961	-0.65569	0.64
-0.62	-0.68042	-0.67489	-0.66663	-0.65940	-0.65092	-0.63638	0.62
-0.60	-0.66294	-0.65717	-0.64856	-0.64103	-0.63219	-0.61705	0.60
-0.58	-0.64540	-0.63940	-0.63044	-0.62260	-0.61342	-0.59770	0.58
-0.56	-0.62779	-0.62156	-0.61226	-0.60414	-0.59461	-0.57832	0.56
-0.54	-0.61011	-0.60366	-0.59403	-0.58562	-0.57577	-0.55893	0.54
-0.52	-0.59237	-0.58569	-0.57574	-0.56706	-0.55689	-0.53952	0.52
-0.50	-0.57455	-0.56766	-0.55740	-0.54845	-0.53797	-0.52009	0.50
-0.48	-0.55666	-0.54956	-0.53900	-0.52979	-0.51901	-0.50064	0.48
-0.46	-0.53870	-0.53140	-0.52055	-0.51108	-0.50002	-0.48117	0.46
-0.44	-0.52067	-0.51318	-0.50204	-0.49233	-0.48098	-0.46167	0.44
-0.42	-0.50257	-0.49489	-0.48347	-0.47353	-0.46191	-0.44216	0.42
-0.40	-0.48439	-0.47653	-0.46485	-0.45468	-0.44281	-0.42263	0.40
-0.38	-0.46615	-0.45811	-0.44617	-0.43578	-0.42366	-0.40308	0.38
-0.36	-0.44783	-0.43961	-0.42743	-0.41683	-0.40447	-0.38350	0.36
-0.34	-0.42943	-0.42106	-0.40863	-0.39783	-0.38525	-0.36391	0.34
-0.32	-0.41096	-0.40243	-0.38978	-0.37879	-0.36598	-0.34429	0.32
-0.30	-0.39242	-0.38374	-0.37087	-0.35969	-0.34668	-0.32466	0.30
-0.28	-0.37380	-0.36498	-0.35190	-0.34055	-0.32734	-0.30500	0.28
-0.26	-0.35511	-0.34614	-0.33287	-0.32135	-0.30796	-0.28532	0.26
-0.24	-0.33634	-0.32724	-0.31378	-0.30211	-0.28854	-0.26562	0.24
-0.22	-0.31750	-0.30827	-0.29463	-0.28281	-0.26908	-0.24590	0.22
-0.20	-0.29857	-0.28923	-0.27543	-0.26346	-0.24958	-0.22616	0.20
-0.18	-0.27957	-0.27012	-0.25616	-0.24407	-0.23004	-0.20640	0.18
-0.16	-0.26050	-0.25094	-0.23683	-0.22462	-0.21046	-0.18661	0.16
-0.14	-0.24134	-0.23169	-0.21744	-0.20512	-0.19084	-0.16681	0.14
-0.12	-0.22210	-0.21236	-0.19799	-0.18557	-0.17118	-0.14698	0.12
-0.10	-0.20279	-0.19297	-0.17848	-0.16596	-0.15147	-0.12713	0.10
-0.08	-0.18339	-0.17350	-0.15890	-0.14631	-0.13173	-0.10726	0.08
-0.06	-0.16392	-0.15395	-0.13927	-0.12660	-0.11195	-0.08737	0.06
-0.04	-0.14436	-0.13433	-0.11957	-0.10684	-0.09212	-0.06746	0.04
-0.02	-0.12472	-0.11464	-0.09981	-0.08703	-0.07226	-0.04752	0.02

↑ r_0

α =	.995 ↑-	.99 ↑-	.975 ↑-	.95 ↑-	.90 ↑-	.75 ↑-

TABLE 92.2   SAMPLE SIZE =   600   (CONT.)

α =	.005 ↓+	.01 ↓+	.025 ↓+	.05 ↓+	.10 ↓+	.25 ↓+	
r_0 ↓							
0.00	-0.10500	-0.09488	-0.07999	-0.06716	-0.05235	-0.02757	0.00
0.02	-0.08519	-0.07504	-0.06010	-0.04724	-0.03240	-0.00759	-0.02
0.04	-0.06531	-0.05512	-0.04015	-0.02727	-0.01241	0.01242	-0.04
0.06	-0.04533	-0.03513	-0.02013	-0.00724	0.00763	0.03244	-0.06
0.08	-0.02528	-0.01506	-0.00005	0.01284	0.02770	0.05248	-0.08
0.10	-0.00514	0.00509	0.02009	0.03298	0.04782	0.07255	-0.10
0.12	0.01509	0.02531	0.04030	0.05317	0.06798	0.09264	-0.12
0.14	0.03541	0.04561	0.06058	0.07342	0.08818	0.11275	-0.14
0.16	0.05581	0.06599	0.08092	0.09372	0.10843	0.13289	-0.16
0.18	0.07630	0.08645	0.10132	0.11407	0.12872	0.15305	-0.18
0.20	0.09687	0.10699	0.12180	0.13448	0.14905	0.17323	-0.20
0.22	0.11754	0.12761	0.14234	0.15495	0.16942	0.19343	-0.22
0.24	0.13829	0.14831	0.16295	0.17548	0.18984	0.21366	-0.24
0.26	0.15914	0.16909	0.18362	0.19606	0.21031	0.23391	-0.26
0.28	0.18007	0.18995	0.20437	0.21669	0.23081	0.25418	-0.28
0.30	0.20110	0.21089	0.22518	0.23739	0.25137	0.27447	-0.30
0.32	0.22222	0.23191	0.24606	0.25814	0.27196	0.29479	-0.32
0.34	0.24343	0.25302	0.26701	0.27895	0.29260	0.31513	-0.34
0.36	0.26473	0.27422	0.28804	0.29982	0.31329	0.33550	-0.36
0.38	0.28613	0.29549	0.30913	0.32075	0.33402	0.35589	-0.38
0.40	0.30763	0.31685	0.33029	0.34173	0.35479	0.37630	-0.40
0.42	0.32921	0.33830	0.35152	0.36277	0.37561	0.39673	-0.42
0.44	0.35090	0.35983	0.37282	0.38388	0.39648	0.41719	-0.44
0.46	0.37268	0.38145	0.39420	0.40504	0.41739	0.43768	-0.46
0.48	0.39456	0.40315	0.41565	0.42626	0.43835	0.45818	-0.48
0.50	0.41653	0.42495	0.43717	0.44754	0.45935	0.47871	-0.50
0.52	0.43861	0.44683	0.45876	0.46888	0.48040	0.49927	-0.52
0.54	0.46078	0.46880	0.48042	0.49029	0.50149	0.51985	-0.54
0.56	0.48306	0.49086	0.50216	0.51175	0.52264	0.54045	-0.56
0.58	0.50543	0.51301	0.52398	0.53327	0.54383	0.56108	-0.58
0.60	0.52791	0.53525	0.54587	0.55486	0.56506	0.58173	-0.60
0.62	0.55049	0.55758	0.56783	0.57651	0.58635	0.60240	-0.62
0.64	0.57317	0.58000	0.58987	0.59822	0.60768	0.62310	-0.64
0.66	0.59596	0.60251	0.61198	0.61999	0.62906	0.64383	-0.66
0.68	0.61885	0.62512	0.63418	0.64183	0.65048	0.66458	-0.68
0.70	0.64185	0.64782	0.65644	0.66372	0.67196	0.68535	-0.70
0.72	0.66495	0.67061	0.67879	0.68569	0.69348	0.70615	-0.72
0.74	0.68816	0.69350	0.70121	0.70771	0.71505	0.72698	-0.74
0.76	0.71148	0.71649	0.72371	0.72980	0.73668	0.74782	-0.76
0.78	0.73490	0.73957	0.74629	0.75196	0.75834	0.76870	-0.78
0.80	0.75844	0.76275	0.76895	0.77417	0.78006	0.78960	-0.80
0.82	0.78208	0.78602	0.79169	0.79646	0.80183	0.81052	-0.82
0.84	0.80584	0.80940	0.81451	0.81881	0.82365	0.83147	-0.84
0.86	0.82971	0.83287	0.83741	0.84122	0.84552	0.85245	-0.86
0.88	0.85369	0.85644	0.86039	0.86370	0.86743	0.87345	-0.88
0.90	0.87779	0.88011	0.88345	0.88625	0.88940	0.89448	-0.90
0.92	0.90200	0.90389	0.90660	0.90887	0.91142	0.91553	-0.92
0.94	0.92632	0.92776	0.92982	0.93155	0.93349	0.93661	-0.94
0.96	0.95076	0.95174	0.95313	0.95430	0.95561	0.95771	-0.96
0.98	0.97532	0.97582	0.97652	0.97711	0.97778	0.97884	-0.98

↑ r_0

α =	.995 ↑-	.99 ↑-	.975 ↑-	.95 ↑-	.90 ↑-	.75 ↑-

TABLE 93.1   SAMPLE SIZE =   700

α =	.005 ↓+	.01 ↓+	.025 ↓+	.05 ↓+	.10 ↓+	.25 ↓+	
r_o ↓							
-0.98	-0.98349	-0.98318	-0.98271	-0.98230	-0.98181	-0.98096	0.98
-0.96	-0.96693	-0.96631	-0.96538	-0.96455	-0.96358	-0.96190	0.96
-0.94	-0.95031	-0.94938	-0.94799	-0.94677	-0.94533	-0.94283	0.94
-0.92	-0.93363	-0.93240	-0.93056	-0.92895	-0.92704	-0.92373	0.92
-0.90	-0.91689	-0.91537	-0.91309	-0.91108	-0.90871	-0.90462	0.90
-0.88	-0.90009	-0.89828	-0.89556	-0.89318	-0.89036	-0.88550	0.88
-0.86	-0.88324	-0.88114	-0.87799	-0.87523	-0.87197	-0.86635	0.86
-0.84	-0.86632	-0.86394	-0.86037	-0.85724	-0.85355	-0.84719	0.84
-0.82	-0.84934	-0.84669	-0.84271	-0.83921	-0.83509	-0.82801	0.82
-0.80	-0.83231	-0.82938	-0.82499	-0.82114	-0.81661	-0.80881	0.80
-0.78	-0.81521	-0.81201	-0.80723	-0.80303	-0.79809	-0.78959	0.78
-0.76	-0.79805	-0.79459	-0.78942	-0.78487	-0.77953	-0.77035	0.76
-0.74	-0.78083	-0.77711	-0.77155	-0.76668	-0.76095	-0.75110	0.74
-0.72	-0.76355	-0.75958	-0.75364	-0.74844	-0.74232	-0.73183	0.72
-0.70	-0.74621	-0.74199	-0.73568	-0.73016	-0.72367	-0.71254	0.70
-0.68	-0.72881	-0.72434	-0.71767	-0.71183	-0.70498	-0.69323	0.68
-0.66	-0.71134	-0.70664	-0.69961	-0.69347	-0.68626	-0.67391	0.66
-0.64	-0.69381	-0.68887	-0.68151	-0.67506	-0.66750	-0.65456	0.64
-0.62	-0.67622	-0.67105	-0.66335	-0.65661	-0.64871	-0.63520	0.62
-0.60	-0.65856	-0.65317	-0.64513	-0.63811	-0.62988	-0.61582	0.60
-0.58	-0.64084	-0.63523	-0.62687	-0.61957	-0.61102	-0.59642	0.58
-0.56	-0.62305	-0.61723	-0.60856	-0.60099	-0.59213	-0.57700	0.56
-0.54	-0.60520	-0.59917	-0.59020	-0.58236	-0.57320	-0.55757	0.54
-0.52	-0.58729	-0.58106	-0.57178	-0.56369	-0.55423	-0.53811	0.52
-0.50	-0.56931	-0.56288	-0.55332	-0.54498	-0.53523	-0.51864	0.50
-0.48	-0.55126	-0.54464	-0.53480	-0.52622	-0.51620	-0.49914	0.48
-0.46	-0.53314	-0.52634	-0.51623	-0.50742	-0.49713	-0.47963	0.46
-0.44	-0.51496	-0.50798	-0.49760	-0.48857	-0.47802	-0.46010	0.44
-0.42	-0.49672	-0.48956	-0.47893	-0.46967	-0.45888	-0.44055	0.42
-0.40	-0.47840	-0.47107	-0.46020	-0.45074	-0.43970	-0.42098	0.40
-0.38	-0.46002	-0.45252	-0.44141	-0.43175	-0.42049	-0.40140	0.38
-0.36	-0.44156	-0.43392	-0.42258	-0.41272	-0.40124	-0.38179	0.36
-0.34	-0.42304	-0.41524	-0.40369	-0.39365	-0.38196	-0.36216	0.34
-0.32	-0.40445	-0.39651	-0.38474	-0.37453	-0.36264	-0.34252	0.32
-0.30	-0.38579	-0.37771	-0.36574	-0.35536	-0.34328	-0.32285	0.30
-0.28	-0.36706	-0.35885	-0.34669	-0.33614	-0.32389	-0.30317	0.28
-0.26	-0.34826	-0.33992	-0.32758	-0.31688	-0.30446	-0.28346	0.26
-0.24	-0.32939	-0.32093	-0.30842	-0.29758	-0.28499	-0.26374	0.24
-0.22	-0.31045	-0.30187	-0.28920	-0.27822	-0.26548	-0.24400	0.22
-0.20	-0.29143	-0.28275	-0.26993	-0.25882	-0.24594	-0.22424	0.20
-0.18	-0.27234	-0.26356	-0.25060	-0.23937	-0.22636	-0.20445	0.18
-0.16	-0.25318	-0.24431	-0.23121	-0.21988	-0.20675	-0.18465	0.16
-0.14	-0.23395	-0.22499	-0.21177	-0.20034	-0.18710	-0.16483	0.14
-0.12	-0.21464	-0.20560	-0.19227	-0.18074	-0.16740	-0.14499	0.12
-0.10	-0.19526	-0.18615	-0.17271	-0.16110	-0.14768	-0.12513	0.10
-0.08	-0.17581	-0.16662	-0.15309	-0.14142	-0.12791	-0.10524	0.08
-0.06	-0.15628	-0.14703	-0.13342	-0.12168	-0.10810	-0.08534	0.06
-0.04	-0.13667	-0.12738	-0.11369	-0.10189	-0.08826	-0.06542	0.04
-0.02	-0.11699	-0.10765	-0.09390	-0.08206	-0.06838	-0.04548	0.02

↑ r_o

α =	.995 ↑-	.99 ↑-	.975 ↑-	.95 ↑-	.90 ↑-	.75 ↑-

TABLE 93.2   SAMPLE SIZE =   700   (CONT.)

α =	.005 ↓+	.01 ↓+	.025 ↓+	.05 ↓+	.10 ↓+	.25 ↓+	
r_o ↓							
0.00	−0.09723	−0.08785	−0.07406	−0.06218	−0.04846	−0.02552	0.00
0.02	−0.07740	−0.06799	−0.05415	−0.04224	−0.02850	−0.00553	−0.02
0.04	−0.05748	−0.04805	−0.03419	−0.02226	−0.00851	0.01447	−0.04
0.06	−0.03749	−0.02804	−0.01416	−0.00223	0.01153	0.03449	−0.06
0.08	−0.01742	−0.00796	0.00592	0.01786	0.03160	0.05454	−0.08
0.10	0.00272	0.01218	0.02606	0.03799	0.05172	0.07460	−0.10
0.12	0.02295	0.03240	0.04627	0.05817	0.07187	0.09469	−0.12
0.14	0.04326	0.05270	0.06653	0.07841	0.09206	0.11479	−0.14
0.16	0.06365	0.07306	0.08686	0.09869	0.11230	0.13492	−0.16
0.18	0.08411	0.09350	0.10725	0.11903	0.13257	0.15507	−0.18
0.20	0.10466	0.11401	0.12769	0.13942	0.15288	0.17524	−0.20
0.22	0.12530	0.13460	0.14820	0.15986	0.17323	0.19543	−0.22
0.24	0.14601	0.15526	0.16878	0.18035	0.19363	0.21564	−0.24
0.26	0.16681	0.17599	0.18941	0.20089	0.21406	0.23587	−0.26
0.28	0.18769	0.19680	0.21011	0.22149	0.23453	0.25613	−0.28
0.30	0.20865	0.21768	0.23087	0.24214	0.25505	0.27640	−0.30
0.32	0.22970	0.23864	0.25170	0.26284	0.27560	0.29670	−0.32
0.34	0.25084	0.25968	0.27258	0.28360	0.29620	0.31702	−0.34
0.36	0.27206	0.28080	0.29354	0.30441	0.31684	0.33736	−0.36
0.38	0.29336	0.30199	0.31455	0.32527	0.33752	0.35772	−0.38
0.40	0.31476	0.32326	0.33563	0.34618	0.35824	0.37810	−0.40
0.42	0.33624	0.34460	0.35678	0.36715	0.37900	0.39850	−0.42
0.44	0.35781	0.36603	0.37799	0.38818	0.39980	0.41893	−0.44
0.46	0.37947	0.38754	0.39927	0.40926	0.42065	0.43938	−0.46
0.48	0.40121	0.40912	0.42062	0.43039	0.44154	0.45985	−0.48
0.50	0.42305	0.43079	0.44203	0.45158	0.46247	0.48034	−0.50
0.52	0.44498	0.45253	0.46351	0.47283	0.48344	0.50085	−0.52
0.54	0.46699	0.47436	0.48505	0.49413	0.50446	0.52139	−0.54
0.56	0.48910	0.49627	0.50666	0.51548	0.52551	0.54195	−0.56
0.58	0.51131	0.51826	0.52834	0.53690	0.54662	0.56253	−0.58
0.60	0.53360	0.54033	0.55009	0.55836	0.56776	0.58313	−0.60
0.62	0.55599	0.56249	0.57191	0.57989	0.58895	0.60376	−0.62
0.64	0.57847	0.58473	0.59380	0.60147	0.61018	0.62440	−0.64
0.66	0.60105	0.60706	0.61575	0.62311	0.63146	0.64507	−0.66
0.68	0.62372	0.62947	0.63778	0.64481	0.65278	0.66577	−0.68
0.70	0.64649	0.65196	0.65988	0.66656	0.67414	0.68648	−0.70
0.72	0.66935	0.67454	0.68204	0.68838	0.69555	0.70722	−0.72
0.74	0.69232	0.69721	0.70428	0.71025	0.71700	0.72798	−0.74
0.76	0.71538	0.71997	0.72659	0.73218	0.73849	0.74877	−0.76
0.78	0.73854	0.74281	0.74897	0.75416	0.76004	0.76957	−0.78
0.80	0.76179	0.76574	0.77142	0.77621	0.78162	0.79041	−0.80
0.82	0.78515	0.78876	0.79395	0.79832	0.80325	0.81126	−0.82
0.84	0.80861	0.81186	0.81654	0.82048	0.82493	0.83214	−0.84
0.86	0.83217	0.83506	0.83922	0.84271	0.84665	0.85304	−0.86
0.88	0.85584	0.85835	0.86196	0.86500	0.86842	0.87396	−0.88
0.90	0.87960	0.88173	0.88478	0.88735	0.89024	0.89491	−0.90
0.92	0.90347	0.90520	0.90767	0.90975	0.91210	0.91588	−0.92
0.94	0.92745	0.92876	0.93064	0.93222	0.93400	0.93687	−0.94
0.96	0.95152	0.95241	0.95369	0.95475	0.95595	0.95789	−0.96
0.98	0.97571	0.97616	0.97680	0.97735	0.97795	0.97893	−0.98

↑ r_o

α =	.995 ↑−	.99 ↑−	.975 ↑−	.95 ↑−	.90 ↑−	.75 ↑−

TABLE 94.1   SAMPLE SIZE =   800

α =	.005 ↓+	.01 ↓+	.025 ↓+	.05 ↓+	.10 ↓+	.25 ↓+	
r_0 ↓							
-0.98	-0.98329	-0.98299	-0.98255	-0.98216	-0.98170	-0.98090	0.98
-0.96	-0.96652	-0.96593	-0.96505	-0.96428	-0.96336	-0.96178	0.96
-0.94	-0.94970	-0.94882	-0.94751	-0.94636	-0.94500	-0.94265	0.94
-0.92	-0.93282	-0.93166	-0.92993	-0.92840	-0.92660	-0.92350	0.92
-0.90	-0.91589	-0.91445	-0.91230	-0.91041	-0.90818	-0.90434	0.90
-0.88	-0.89890	-0.89719	-0.89462	-0.89237	-0.88972	-0.88516	0.88
-0.86	-0.88185	-0.87987	-0.87690	-0.87430	-0.87123	-0.86596	0.86
-0.84	-0.86475	-0.86250	-0.85914	-0.85619	-0.85271	-0.84674	0.84
-0.82	-0.84759	-0.84508	-0.84133	-0.83804	-0.83416	-0.82751	0.82
-0.80	-0.83038	-0.82761	-0.82347	-0.81985	-0.81558	-0.80826	0.80
-0.78	-0.81310	-0.81008	-0.80557	-0.80162	-0.79697	-0.78899	0.78
-0.76	-0.79577	-0.79250	-0.78762	-0.78335	-0.77832	-0.76971	0.76
-0.74	-0.77838	-0.77487	-0.76963	-0.76504	-0.75965	-0.75041	0.74
-0.72	-0.76093	-0.75718	-0.75159	-0.74669	-0.74094	-0.73109	0.72
-0.70	-0.74343	-0.73944	-0.73350	-0.72830	-0.72220	-0.71176	0.70
-0.68	-0.72586	-0.72165	-0.71536	-0.70987	-0.70343	-0.69241	0.68
-0.66	-0.70823	-0.70380	-0.69718	-0.69140	-0.68462	-0.67304	0.66
-0.64	-0.69055	-0.68589	-0.67895	-0.67289	-0.66579	-0.65365	0.64
-0.62	-0.67280	-0.66793	-0.66068	-0.65434	-0.64692	-0.63425	0.62
-0.60	-0.65500	-0.64992	-0.64235	-0.63575	-0.62802	-0.61483	0.60
-0.58	-0.63713	-0.63184	-0.62398	-0.61711	-0.60908	-0.59539	0.58
-0.56	-0.61920	-0.61372	-0.60556	-0.59844	-0.59012	-0.57593	0.56
-0.54	-0.60121	-0.59553	-0.58709	-0.57972	-0.57112	-0.55646	0.54
-0.52	-0.58316	-0.57729	-0.56857	-0.56097	-0.55209	-0.53697	0.52
-0.50	-0.56505	-0.55900	-0.55000	-0.54217	-0.53302	-0.51746	0.50
-0.48	-0.54687	-0.54064	-0.53139	-0.52333	-0.51392	-0.49793	0.48
-0.46	-0.52863	-0.52223	-0.51272	-0.50445	-0.49479	-0.47839	0.46
-0.44	-0.51033	-0.50376	-0.49401	-0.48552	-0.47563	-0.45883	0.44
-0.42	-0.49197	-0.48523	-0.47524	-0.46655	-0.45643	-0.43925	0.42
-0.40	-0.47354	-0.46665	-0.45643	-0.44755	-0.43720	-0.41965	0.40
-0.38	-0.45504	-0.44800	-0.43756	-0.42849	-0.41793	-0.40004	0.38
-0.36	-0.43649	-0.42930	-0.41865	-0.40940	-0.39863	-0.38040	0.36
-0.34	-0.41786	-0.41054	-0.39968	-0.39026	-0.37930	-0.36075	0.34
-0.32	-0.39918	-0.39171	-0.38066	-0.37108	-0.35993	-0.34108	0.32
-0.30	-0.38042	-0.37283	-0.36160	-0.35185	-0.34053	-0.32140	0.30
-0.28	-0.36160	-0.35389	-0.34248	-0.33259	-0.32110	-0.30169	0.28
-0.26	-0.34271	-0.33488	-0.32331	-0.31327	-0.30163	-0.28197	0.26
-0.24	-0.32376	-0.31582	-0.30408	-0.29392	-0.28212	-0.26222	0.24
-0.22	-0.30474	-0.29670	-0.28481	-0.27452	-0.26258	-0.24246	0.22
-0.20	-0.28565	-0.27751	-0.26548	-0.25507	-0.24301	-0.22268	0.20
-0.18	-0.26650	-0.25826	-0.24610	-0.23559	-0.22340	-0.20289	0.18
-0.16	-0.24727	-0.23895	-0.22667	-0.21605	-0.20375	-0.18307	0.16
-0.14	-0.22798	-0.21958	-0.20718	-0.19648	-0.18408	-0.16323	0.14
-0.12	-0.20862	-0.20014	-0.18764	-0.17685	-0.16436	-0.14338	0.12
-0.10	-0.18918	-0.18064	-0.16805	-0.15718	-0.14461	-0.12351	0.10
-0.08	-0.16968	-0.16108	-0.14841	-0.13747	-0.12483	-0.10362	0.08
-0.06	-0.15011	-0.14145	-0.12870	-0.11771	-0.10501	-0.08371	0.06
-0.04	-0.13047	-0.12176	-0.10895	-0.09791	-0.08515	-0.06378	0.04
-0.02	-0.11075	-0.10201	-0.08914	-0.07806	-0.06526	-0.04383	0.02

↑ r_0

α =	.995 ↑−	.99 ↑−	.975 ↑−	.95 ↑−	.90 ↑−	.75 ↑−

TABLE 94.2 SAMPLE SIZE = 800 (CONT.)

α =	.005 ↓+	.01 ↓+	.025 ↓+	.05 ↓+	.10 ↓+	.25 ↓+	
r_0 ↓							
0.00	−0.09097	−0.08219	−0.06928	−0.05816	−0.04533	−0.02387	0.00
0.02	−0.07111	−0.06230	−0.04936	−0.03822	−0.02536	−0.00388	−0.02
0.04	−0.05118	−0.04235	−0.02938	−0.01823	−0.00536	0.01612	−0.04
0.06	−0.03117	−0.02233	−0.00935	0.00181	0.01467	0.03615	−0.06
0.08	−0.01110	−0.00225	0.01073	0.02189	0.03475	0.05619	−0.08
0.10	0.00905	0.01790	0.03087	0.04202	0.05486	0.07625	−0.10
0.12	0.02928	0.03811	0.05107	0.06220	0.07500	0.09634	−0.12
0.14	0.04958	0.05840	0.07133	0.08242	0.09519	0.11644	−0.14
0.16	0.06995	0.07875	0.09164	0.10270	0.11541	0.13656	−0.16
0.18	0.09040	0.09917	0.11201	0.12302	0.13567	0.15670	−0.18
0.20	0.11093	0.11966	0.13244	0.14339	0.15596	0.17686	−0.20
0.22	0.13153	0.14021	0.15292	0.16380	0.17630	0.19704	−0.22
0.24	0.15221	0.16084	0.17346	0.18427	0.19667	0.21724	−0.24
0.26	0.17297	0.18154	0.19406	0.20478	0.21708	0.23745	−0.26
0.28	0.19380	0.20230	0.21472	0.22534	0.23752	0.25769	−0.28
0.30	0.21472	0.22314	0.23544	0.24596	0.25801	0.27795	−0.30
0.32	0.23571	0.24405	0.25622	0.26662	0.27853	0.29823	−0.32
0.34	0.25678	0.26502	0.27705	0.28733	0.29909	0.31853	−0.34
0.36	0.27793	0.28607	0.29795	0.30809	0.31969	0.33885	−0.36
0.38	0.29916	0.30720	0.31891	0.32890	0.34032	0.35919	−0.38
0.40	0.32047	0.32839	0.33992	0.34976	0.36100	0.37954	−0.40
0.42	0.34187	0.34966	0.36100	0.37067	0.38172	0.39992	−0.42
0.44	0.36334	0.37100	0.38214	0.39163	0.40247	0.42032	−0.44
0.46	0.38490	0.39241	0.40334	0.41264	0.42326	0.44074	−0.46
0.48	0.40654	0.41390	0.42460	0.43371	0.44409	0.46118	−0.48
0.50	0.42826	0.43546	0.44592	0.45482	0.46497	0.48164	−0.50
0.52	0.45007	0.45710	0.46731	0.47599	0.48588	0.50212	−0.52
0.54	0.47196	0.47881	0.48875	0.49721	0.50683	0.52263	−0.54
0.56	0.49394	0.50060	0.51026	0.51847	0.52782	0.54315	−0.56
0.58	0.51600	0.52246	0.53184	0.53980	0.54885	0.56369	−0.58
0.60	0.53815	0.54440	0.55347	0.56117	0.56992	0.58425	−0.60
0.62	0.56038	0.56642	0.57517	0.58260	0.59103	0.60484	−0.62
0.64	0.58270	0.58851	0.59694	0.60408	0.61218	0.62545	−0.64
0.66	0.60511	0.61069	0.61877	0.62561	0.63338	0.64607	−0.66
0.68	0.62761	0.63294	0.64066	0.64719	0.65461	0.66672	−0.68
0.70	0.65019	0.65527	0.66262	0.66883	0.67588	0.68739	−0.70
0.72	0.67287	0.67768	0.68464	0.69053	0.69720	0.70808	−0.72
0.74	0.69563	0.70017	0.70673	0.71227	0.71855	0.72879	−0.74
0.76	0.71848	0.72274	0.72888	0.73407	0.73995	0.74952	−0.76
0.78	0.74143	0.74539	0.75110	0.75593	0.76139	0.77028	−0.78
0.80	0.76447	0.76812	0.77339	0.77784	0.78287	0.79105	−0.80
0.82	0.78759	0.79093	0.79574	0.79980	0.80439	0.81185	−0.82
0.84	0.81082	0.81383	0.81817	0.82182	0.82596	0.83267	−0.84
0.86	0.83413	0.83680	0.84065	0.84390	0.84756	0.85351	−0.86
0.88	0.85754	0.85986	0.86321	0.86603	0.86921	0.87437	−0.88
0.90	0.88104	0.88301	0.88584	0.88822	0.89090	0.89525	−0.90
0.92	0.90464	0.90624	0.90853	0.91046	0.91264	0.91616	−0.92
0.94	0.92834	0.92955	0.93129	0.93276	0.93441	0.93709	−0.94
0.96	0.95213	0.95295	0.95413	0.95512	0.95623	0.95804	−0.96
0.98	0.97601	0.97643	0.97703	0.97753	0.97809	0.97901	−0.98

↑ r_0

α =	.995 ↑−	.99 ↑−	.975 ↑−	.95 ↑−	.90 ↑−	.75 ↑−

TABLE 95.1   SAMPLE SIZE =   900

α =	.005 ↓+	.01 ↓+	.025 ↓+	.05 ↓+	.10 ↓+	.25 ↓+	
r_0 ↓							
-0.98	-0.98312	-0.98283	-0.98241	-0.98204	-0.98160	-0.98085	0.98
-0.96	-0.96618	-0.96562	-0.96478	-0.96405	-0.96318	-0.96169	0.96
-0.94	-0.94919	-0.94836	-0.94711	-0.94602	-0.94473	-0.94251	0.94
-0.92	-0.93215	-0.93104	-0.92940	-0.92795	-0.92624	-0.92331	0.92
-0.90	-0.91505	-0.91368	-0.91164	-0.90984	-0.90773	-0.90410	0.90
-0.88	-0.89790	-0.89627	-0.89384	-0.89170	-0.88919	-0.88487	0.88
-0.86	-0.88070	-0.87881	-0.87600	-0.87352	-0.87062	-0.86563	0.86
-0.84	-0.86344	-0.86130	-0.85811	-0.85531	-0.85202	-0.84637	0.84
-0.82	-0.84613	-0.84374	-0.84018	-0.83706	-0.83339	-0.82709	0.82
-0.80	-0.82876	-0.82613	-0.82221	-0.81877	-0.81473	-0.80780	0.80
-0.78	-0.81134	-0.80847	-0.80419	-0.80044	-0.79604	-0.78850	0.78
-0.76	-0.79386	-0.79076	-0.78613	-0.78207	-0.77732	-0.76917	0.76
-0.74	-0.77633	-0.77300	-0.76802	-0.76367	-0.75857	-0.74983	0.74
-0.72	-0.75874	-0.75518	-0.74987	-0.74523	-0.73979	-0.73048	0.72
-0.70	-0.74110	-0.73731	-0.73168	-0.72675	-0.72098	-0.71111	0.70
-0.68	-0.72340	-0.71940	-0.71344	-0.70823	-0.70214	-0.69172	0.68
-0.66	-0.70564	-0.70143	-0.69516	-0.68968	-0.68326	-0.67231	0.66
-0.64	-0.68782	-0.68340	-0.67683	-0.67108	-0.66436	-0.65289	0.64
-0.62	-0.66995	-0.66533	-0.65845	-0.65245	-0.64543	-0.63345	0.62
-0.60	-0.65202	-0.64720	-0.64003	-0.63378	-0.62647	-0.61400	0.60
-0.58	-0.63403	-0.62902	-0.62157	-0.61507	-0.60747	-0.59453	0.58
-0.56	-0.61599	-0.61079	-0.60306	-0.59632	-0.58845	-0.57504	0.56
-0.54	-0.59788	-0.59250	-0.58450	-0.57753	-0.56939	-0.55554	0.54
-0.52	-0.57972	-0.57416	-0.56590	-0.55870	-0.55030	-0.53602	0.52
-0.50	-0.56150	-0.55576	-0.54725	-0.53983	-0.53118	-0.51648	0.50
-0.48	-0.54322	-0.53731	-0.52855	-0.52093	-0.51203	-0.49693	0.48
-0.46	-0.52488	-0.51881	-0.50981	-0.50198	-0.49285	-0.47736	0.46
-0.44	-0.50647	-0.50025	-0.49102	-0.48299	-0.47364	-0.45777	0.44
-0.42	-0.48801	-0.48163	-0.47218	-0.46396	-0.45439	-0.43817	0.42
-0.40	-0.46949	-0.46296	-0.45329	-0.44489	-0.43511	-0.41855	0.40
-0.38	-0.45091	-0.44424	-0.43436	-0.42579	-0.41581	-0.39891	0.38
-0.36	-0.43226	-0.42546	-0.41538	-0.40664	-0.39646	-0.37925	0.36
-0.34	-0.41355	-0.40662	-0.39635	-0.38745	-0.37709	-0.35958	0.34
-0.32	-0.39479	-0.38772	-0.37728	-0.36822	-0.35769	-0.33989	0.32
-0.30	-0.37595	-0.36877	-0.35815	-0.34894	-0.33825	-0.32019	0.30
-0.28	-0.35706	-0.34977	-0.33898	-0.32963	-0.31878	-0.30046	0.28
-0.26	-0.33810	-0.33070	-0.31976	-0.31028	-0.29928	-0.28072	0.26
-0.24	-0.31908	-0.31158	-0.30048	-0.29088	-0.27974	-0.26096	0.24
-0.22	-0.30000	-0.29240	-0.28116	-0.27144	-0.26017	-0.24119	0.22
-0.20	-0.28085	-0.27316	-0.26179	-0.25196	-0.24057	-0.22140	0.20
-0.18	-0.26164	-0.25386	-0.24237	-0.23244	-0.22094	-0.20159	0.18
-0.16	-0.24236	-0.23450	-0.22290	-0.21288	-0.20127	-0.18176	0.16
-0.14	-0.22302	-0.21508	-0.20338	-0.19327	-0.18157	-0.16191	0.14
-0.12	-0.20361	-0.19561	-0.18381	-0.17363	-0.16184	-0.14205	0.12
-0.10	-0.18414	-0.17607	-0.16419	-0.15393	-0.14207	-0.12217	0.10
-0.08	-0.16460	-0.15648	-0.14452	-0.13420	-0.12227	-0.10227	0.08
-0.06	-0.14500	-0.13682	-0.12479	-0.11442	-0.10244	-0.08235	0.06
-0.04	-0.12532	-0.11711	-0.10502	-0.09460	-0.08257	-0.06242	0.04
-0.02	-0.10558	-0.09733	-0.08519	-0.07474	-0.06267	-0.04247	0.02

↑ r_0

α =	.995 ↑-	.99 ↑-	.975 ↑-	.95 ↑-	.90 ↑-	.75 ↑-

TABLE 95.2 SAMPLE SIZE = 900 (CONT.)

α =	.005 ↓+	.01 ↓+	.025 ↓+	.05 ↓+	.10 ↓+	.25 ↓+	
r_0 ↓							
0.00	-0.08577	-0.07749	-0.06532	-0.05483	-0.04273	-0.02250	0.00
0.02	-0.06590	-0.05759	-0.04539	-0.03488	-0.02277	-0.00251	-0.02
0.04	-0.04595	-0.03763	-0.02541	-0.01489	-0.00276	0.01749	-0.04
0.06	-0.02594	-0.01761	-0.00537	0.00515	0.01728	0.03752	-0.06
0.08	-0.00586	0.00248	0.01472	0.02523	0.03735	0.05756	-0.08
0.10	0.01429	0.02263	0.03486	0.04536	0.05746	0.07762	-0.10
0.12	0.03451	0.04284	0.05505	0.06553	0.07760	0.09770	-0.12
0.14	0.05481	0.06311	0.07529	0.08575	0.09777	0.11780	-0.14
0.16	0.07517	0.08345	0.09559	0.10601	0.11798	0.13791	-0.16
0.18	0.09560	0.10386	0.11595	0.12632	0.13823	0.15804	-0.18
0.20	0.11611	0.12433	0.13636	0.14667	0.15851	0.17820	-0.20
0.22	0.13669	0.14486	0.15682	0.16706	0.17883	0.19837	-0.22
0.24	0.15734	0.16546	0.17733	0.18751	0.19918	0.21855	-0.24
0.26	0.17806	0.18612	0.19791	0.20799	0.21957	0.23876	-0.26
0.28	0.19885	0.20685	0.21853	0.22853	0.23999	0.25899	-0.28
0.30	0.21972	0.22764	0.23921	0.24911	0.26045	0.27923	-0.30
0.32	0.24067	0.24850	0.25995	0.26974	0.28094	0.29950	-0.32
0.34	0.26168	0.26943	0.28075	0.29041	0.30148	0.31978	-0.34
0.36	0.28278	0.29043	0.30159	0.31113	0.32204	0.34008	-0.36
0.38	0.30394	0.31149	0.32250	0.33190	0.34264	0.36040	-0.38
0.40	0.32519	0.33262	0.34346	0.35271	0.36328	0.38074	-0.40
0.42	0.34651	0.35382	0.36448	0.37357	0.38396	0.40110	-0.42
0.44	0.36790	0.37509	0.38556	0.39448	0.40467	0.42147	-0.44
0.46	0.38938	0.39643	0.40669	0.41544	0.42542	0.44187	-0.46
0.48	0.41093	0.41783	0.42788	0.43644	0.44621	0.46228	-0.48
0.50	0.43256	0.43931	0.44913	0.45749	0.46703	0.48272	-0.50
0.52	0.45426	0.46085	0.47044	0.47859	0.48789	0.50317	-0.52
0.54	0.47605	0.48247	0.49181	0.49974	0.50879	0.52365	-0.54
0.56	0.49791	0.50416	0.51323	0.52094	0.52972	0.54414	-0.56
0.58	0.51986	0.52592	0.53471	0.54219	0.55069	0.56465	-0.58
0.60	0.54188	0.54775	0.55626	0.56348	0.57170	0.58518	-0.60
0.62	0.56399	0.56965	0.57786	0.58483	0.59275	0.60573	-0.62
0.64	0.58618	0.59165	0.59952	0.60622	0.61383	0.62630	-0.64
0.66	0.60845	0.61367	0.62124	0.62766	0.63496	0.64689	-0.66
0.68	0.63080	0.63579	0.64303	0.64916	0.65612	0.66751	-0.68
0.70	0.65323	0.65798	0.66487	0.67070	0.67732	0.68814	-0.70
0.72	0.67575	0.68025	0.68677	0.69229	0.69856	0.70879	-0.72
0.74	0.69835	0.70259	0.70874	0.71394	0.71983	0.72945	-0.74
0.76	0.72103	0.72501	0.73077	0.73563	0.74115	0.75014	-0.76
0.78	0.74380	0.74750	0.75286	0.75738	0.76250	0.77085	-0.78
0.80	0.76666	0.77007	0.77501	0.77918	0.78390	0.79158	-0.80
0.82	0.78960	0.79272	0.79722	0.80102	0.80533	0.81233	-0.82
0.84	0.81262	0.81544	0.81950	0.82292	0.82680	0.83310	-0.84
0.86	0.83574	0.83823	0.84184	0.84487	0.84831	0.85389	-0.86
0.88	0.85894	0.86111	0.86424	0.86688	0.86986	0.87471	-0.88
0.90	0.88222	0.88406	0.88670	0.88893	0.89145	0.89554	-0.90
0.92	0.90560	0.90709	0.90923	0.91104	0.91308	0.91639	-0.92
0.94	0.92907	0.93020	0.93183	0.93320	0.93475	0.93726	-0.94
0.96	0.95262	0.95339	0.95449	0.95541	0.95646	0.95815	-0.96
0.98	0.97626	0.97665	0.97721	0.97768	0.97821	0.97907	-0.98

↑ r_0

| α = | .995 ↑- | .99 ↑- | .975 ↑- | .95 ↑- | .90 ↑- | .75 ↑- |

TABLE 96.1   SAMPLE SIZE =   1000

α =	.005 ↓+	.01 ↓+	.025 ↓+	.05 ↓+	.10 ↓+	.25 ↓+	
r_0 ↓							
-0.98	-0.98297	-0.98270	-0.98230	-0.98194	-0.98153	-0.98081	0.98
-0.96	-0.96589	-0.96535	-0.96455	-0.96385	-0.96303	-0.96160	0.96
-0.94	-0.94875	-0.94796	-0.94677	-0.94572	-0.94450	-0.94238	0.94
-0.92	-0.93157	-0.93052	-0.92894	-0.92756	-0.92594	-0.92315	0.92
-0.90	-0.91434	-0.91303	-0.91108	-0.90937	-0.90735	-0.90390	0.90
-0.88	-0.89705	-0.89549	-0.89317	-0.89113	-0.88874	-0.88463	0.88
-0.86	-0.87971	-0.87791	-0.87522	-0.87287	-0.87010	-0.86535	0.86
-0.84	-0.86232	-0.86028	-0.85723	-0.85456	-0.85143	-0.84605	0.84
-0.82	-0.84488	-0.84260	-0.83920	-0.83622	-0.83273	-0.82674	0.82
-0.80	-0.82738	-0.82487	-0.82113	-0.81785	-0.81400	-0.80742	0.80
-0.78	-0.80984	-0.80710	-0.80301	-0.79944	-0.79525	-0.78807	0.78
-0.76	-0.79224	-0.78927	-0.78485	-0.78099	-0.77646	-0.76872	0.76
-0.74	-0.77458	-0.77140	-0.76666	-0.76251	-0.75765	-0.74934	0.74
-0.72	-0.75687	-0.75348	-0.74841	-0.74399	-0.73881	-0.72996	0.72
-0.70	-0.73911	-0.73550	-0.73013	-0.72543	-0.71994	-0.71055	0.70
-0.68	-0.72130	-0.71748	-0.71180	-0.70684	-0.70104	-0.69113	0.68
-0.66	-0.70343	-0.69941	-0.69343	-0.68821	-0.68211	-0.67170	0.66
-0.64	-0.68550	-0.68129	-0.67502	-0.66955	-0.66315	-0.65225	0.64
-0.62	-0.66752	-0.66312	-0.65656	-0.65085	-0.64417	-0.63278	0.62
-0.60	-0.64949	-0.64489	-0.63806	-0.63211	-0.62515	-0.61330	0.60
-0.58	-0.63140	-0.62662	-0.61952	-0.61333	-0.60610	-0.59380	0.58
-0.56	-0.61325	-0.60830	-0.60093	-0.59452	-0.58703	-0.57429	0.56
-0.54	-0.59505	-0.58992	-0.58230	-0.57566	-0.56792	-0.55476	0.54
-0.52	-0.57679	-0.57149	-0.56362	-0.55678	-0.54879	-0.53522	0.52
-0.50	-0.55848	-0.55302	-0.54490	-0.53785	-0.52962	-0.51566	0.50
-0.48	-0.54011	-0.53448	-0.52614	-0.51888	-0.51043	-0.49608	0.48
-0.46	-0.52168	-0.51590	-0.50733	-0.49988	-0.49120	-0.47649	0.46
-0.44	-0.50320	-0.49727	-0.48848	-0.48084	-0.47195	-0.45688	0.44
-0.42	-0.48465	-0.47858	-0.46958	-0.46176	-0.45266	-0.43725	0.42
-0.40	-0.46605	-0.45984	-0.45064	-0.44265	-0.43335	-0.41761	0.40
-0.38	-0.44739	-0.44104	-0.43165	-0.42349	-0.41400	-0.39795	0.38
-0.36	-0.42867	-0.42220	-0.41261	-0.40430	-0.39463	-0.37828	0.36
-0.34	-0.40990	-0.40330	-0.39353	-0.38506	-0.37522	-0.35859	0.34
-0.32	-0.39106	-0.38434	-0.37440	-0.36579	-0.35578	-0.33889	0.32
-0.30	-0.37216	-0.36533	-0.35523	-0.34648	-0.33632	-0.31916	0.30
-0.28	-0.35321	-0.34627	-0.33601	-0.32713	-0.31682	-0.29943	0.28
-0.26	-0.33419	-0.32715	-0.31675	-0.30774	-0.29729	-0.27967	0.26
-0.24	-0.31512	-0.30798	-0.29743	-0.28831	-0.27773	-0.25990	0.24
-0.22	-0.29598	-0.28875	-0.27807	-0.26884	-0.25813	-0.24011	0.22
-0.20	-0.27678	-0.26947	-0.25867	-0.24933	-0.23851	-0.22031	0.20
-0.18	-0.25752	-0.25013	-0.23921	-0.22978	-0.21886	-0.20049	0.18
-0.16	-0.23820	-0.23073	-0.21971	-0.21019	-0.19917	-0.18065	0.16
-0.14	-0.21882	-0.21128	-0.20016	-0.19056	-0.17945	-0.16079	0.14
-0.12	-0.19938	-0.19177	-0.18057	-0.17089	-0.15970	-0.14092	0.12
-0.10	-0.17987	-0.17221	-0.16092	-0.15118	-0.13992	-0.12104	0.10
-0.08	-0.16030	-0.15258	-0.14123	-0.13143	-0.12011	-0.10113	0.08
-0.06	-0.14066	-0.13291	-0.12149	-0.11164	-0.10027	-0.08121	0.06
-0.04	-0.12097	-0.11317	-0.10169	-0.09181	-0.08039	-0.06127	0.04
-0.02	-0.10121	-0.09337	-0.08186	-0.07193	-0.06048	-0.04132	0.02

↑ r_0

α =	.995 ↑-	.99 ↑-	.975 ↑-	.95 ↑-	.90 ↑-	.75 ↑-

TABLE 96.2   SAMPLE SIZE =   1000   (CONT.)

α =	.005 ↓+	.01 ↓+	.025 ↓+	.05 ↓+	.10 ↓+	.25 ↓+	
r_o ↓							
0.00	−0.08138	−0.07352	−0.06197	−0.05202	−0.04054	−0.02134	0.00
0.02	−0.06149	−0.05361	−0.04203	−0.03206	−0.02057	−0.00135	−0.02
0.04	−0.04154	−0.03364	−0.02204	−0.01206	−0.00056	0.01865	−0.04
0.06	−0.02152	−0.01361	−0.00200	0.00798	0.01948	0.03868	−0.06
0.08	−0.00143	0.00648	0.01808	0.02806	0.03955	0.05872	−0.08
0.10	0.01872	0.02662	0.03822	0.04818	0.05965	0.07878	−0.10
0.12	0.03894	0.04683	0.05841	0.06835	0.07979	0.09885	−0.12
0.14	0.05922	0.06710	0.07865	0.08856	0.09996	0.11894	−0.14
0.16	0.07958	0.08743	0.09894	0.10881	0.12016	0.13905	−0.16
0.18	0.10000	0.10782	0.11928	0.12910	0.14040	0.15918	−0.18
0.20	0.12048	0.12827	0.13967	0.14944	0.16067	0.17933	−0.20
0.22	0.14104	0.14878	0.16011	0.16982	0.18097	0.19949	−0.22
0.24	0.16166	0.16935	0.18060	0.19024	0.20130	0.21967	−0.24
0.26	0.18235	0.18999	0.20115	0.21071	0.22167	0.23987	−0.26
0.28	0.20311	0.21068	0.22175	0.23122	0.24208	0.26008	−0.28
0.30	0.22395	0.23144	0.24240	0.25177	0.26251	0.28031	−0.30
0.32	0.24485	0.25227	0.26310	0.27237	0.28298	0.30056	−0.32
0.34	0.26582	0.27315	0.28386	0.29301	0.30349	0.32083	−0.34
0.36	0.28686	0.29410	0.30467	0.31369	0.32403	0.34112	−0.36
0.38	0.30797	0.31511	0.32553	0.33442	0.34460	0.36142	−0.38
0.40	0.32916	0.33619	0.34645	0.35520	0.36521	0.38175	−0.40
0.42	0.35042	0.35733	0.36742	0.37602	0.38585	0.40209	−0.42
0.44	0.37174	0.37854	0.38844	0.39688	0.40653	0.42244	−0.44
0.46	0.39315	0.39981	0.40952	0.41779	0.42724	0.44282	−0.46
0.48	0.41462	0.42115	0.43065	0.43874	0.44799	0.46321	−0.48
0.50	0.43617	0.44255	0.45184	0.45974	0.46877	0.48363	−0.50
0.52	0.45779	0.46402	0.47308	0.48079	0.48958	0.50406	−0.52
0.54	0.47949	0.48555	0.49438	0.50188	0.51044	0.52451	−0.54
0.56	0.50126	0.50716	0.51573	0.52302	0.53132	0.54497	−0.56
0.58	0.52310	0.52882	0.53714	0.54420	0.55225	0.56546	−0.58
0.60	0.54503	0.55056	0.55860	0.56543	0.57320	0.58596	−0.60
0.62	0.56702	0.57237	0.58012	0.58670	0.59420	0.60649	−0.62
0.64	0.58910	0.59424	0.60170	0.60803	0.61523	0.62703	−0.64
0.66	0.61125	0.61618	0.62333	0.62939	0.63629	0.64759	−0.66
0.68	0.63348	0.63819	0.64502	0.65081	0.65739	0.66817	−0.68
0.70	0.65578	0.66027	0.66677	0.67227	0.67853	0.68876	−0.70
0.72	0.67817	0.68242	0.68857	0.69378	0.69970	0.70938	−0.72
0.74	0.70063	0.70463	0.71043	0.71534	0.72091	0.73002	−0.74
0.76	0.72317	0.72692	0.73235	0.73695	0.74216	0.75067	−0.76
0.78	0.74579	0.74928	0.75433	0.75860	0.76344	0.77134	−0.78
0.80	0.76849	0.77171	0.77637	0.78030	0.78476	0.79203	−0.80
0.82	0.79127	0.79421	0.79846	0.80205	0.80612	0.81274	−0.82
0.84	0.81414	0.81679	0.82061	0.82385	0.82751	0.83347	−0.84
0.86	0.83708	0.83943	0.84283	0.84569	0.84894	0.85422	−0.86
0.88	0.86011	0.86215	0.86510	0.86759	0.87041	0.87499	−0.88
0.90	0.88321	0.88494	0.88743	0.88953	0.89191	0.89578	−0.90
0.92	0.90640	0.90780	0.90982	0.91153	0.91345	0.91658	−0.92
0.94	0.92968	0.93074	0.93228	0.93357	0.93503	0.93741	−0.94
0.96	0.95303	0.95375	0.95479	0.95566	0.95665	0.95825	−0.96
0.98	0.97647	0.97684	0.97736	0.97781	0.97831	0.97912	−0.98

↑ r_o

α =	.995 ↑−	.99 ↑−	.975 ↑−	.95 ↑−	.90 ↑−	.75 ↑−

ABCDEFGHIJ −89876